# Learning and Memory

A subject collection from *Cold Spring Harbor Perspectives in Biology*

OTHER SUBJECT COLLECTIONS FROM *COLD SPRING HARBOR PERSPECTIVES IN BIOLOGY*

*DNA Recombination*

*Neurogenesis*

*Size Control in Biology: From Organelles to Organisms*

*Mitosis*

*Glia*

*Innate Immunity and Inflammation*

*The Genetics and Biology of Sexual Conflict*

*The Origin and Evolution of Eukaryotes*

*Endocytosis*

*Mitochondria*

*Signaling by Receptor Tyrosine Kinases*

*DNA Repair, Mutagenesis, and Other Responses to DNA Damage*

*Cell Survival and Cell Death*

*Immune Tolerance*

*DNA Replication*

*Endoplasmic Reticulum*

*Wnt Signaling*

SUBJECT COLLECTIONS FROM *COLD SPRING HARBOR PERSPECTIVES IN MEDICINE*

*Epilepsy: The Biology of a Spectrum Disorder*

*Molecular Approaches to Reproductive and Newborn Medicine*

*The Hepatitis B and Delta Viruses*

*Intellectual Property in Molecular Medicine*

*Retinal Disorders: Genetic Approaches to Diagnosis and Treatment*

*The Biology of Heart Disease*

*Human Fungal Pathogens*

*Tuberculosis*

*The Skin and Its Diseases*

*MYC and the Pathway to Cancer*

*Bacterial Pathogenesis*

*Transplantation*

*Cystic Fibrosis: A Trilogy of Biochemistry, Physiology, and Therapy*

*Hemoglobin and Its Diseases*

*Addiction*

*Parkinson's Disease*

*Type 1 Diabetes*

*Angiogenesis: Biology and Pathology*

# Learning and Memory

A subject collection from *Cold Spring Harbor Perspectives in Biology*

EDITED BY

### Eric R. Kandel
*Columbia University*

### Yadin Dudai
*The Weizmann Institute of Science
and New York University*

### Mark R. Mayford
*The Scripps Research Institute*

COLD SPRING HARBOR LABORATORY PRESS
Cold Spring Harbor, New York • www.cshlpress.org

## Learning and Memory

A subject collection from *Cold Spring Harbor Perspectives in Biology*
Articles online at www.cshperspectives.org

| | |
|---|---|
| Executive Editor | Richard Sever |
| Managing Editor | Maria Smit |
| Senior Project Manager | Barbara Acosta |
| Permissions Administrator | Carol Brown |
| Production Editor | Diane Schubach |
| Production Manager/Cover Designer | Denise Weiss |
| | |
| Publisher | John Inglis |

*Front cover artwork:* Schematic representation of the cerebellar cortex, characterized by a highly structured organization. The large Purkinje neurons determine the main output of the cerebellum and have elaborate dendritic arbors, pervaded with vast amounts of input fibers that provide sensory information. As we experience ourselves and the world we live in, Purkinje neurons continuously learn what sensory information they ought to respond to, aided by one single axonal climbing fiber that provides teaching signals that enforce neuroplasticity. (Cover caption kindly provided by Chris De Zeeuw and Bas Koekkoek [both from Erasmus MC, Rotterdam] and Michiel Ten Brinke [Netherlands Institute for Neuroscience]. Cover image designed by Sarah Mack [Columbia University].)

*Library of Congress Cataloging-in-Publication Data*

Learning and memory : a subject collection from cold spring harbor perspectives in biology / edited by Eric R. Kandel, Columbia University, Yadin Dudai, The Weizmann Institute of Science, and Mark R. Mayford, The Scripps Research Institute.
      pages cm
   Includes bibliographical references and index.
   ISBN 978-1-62182-091-8 (hardcover : alk. paper); 978-1-62182-160-1 (paper)
1. Memory. 2. Learning. I. Kandel, Eric R., editor. II. Dudai, Yadin, editor. III. Mayford, Mark R., editor.
   QP406.L4327 2015
   612.8'23312--dc23
                                        2015027429
10 9 8 7 6 5 4 3 2

All World Wide Web addresses are accurate to the best of our knowledge at the time of printing.

For a complete catalog of all Cold Spring Harbor Laboratory Press publications, visit our website at www.cshlpress.org.

# Contents

# Contents

# Preface

OVER THE LAST DECADE, the neurobiological study of learning and memory has progressed dramatically, fueled in part by the availability of new technologies such as gene knockouts, multi-unit recordings, human and animal brain imaging, and optogenetics. These biological methods have been supplemented by behavioral insights about the reliability and stability of memory. For example, the reconsolidation model of memory suggests that recall itself may evoke or reinstate an active neuronal plasticity process that is required for memory maintenance. Human psychological studies have highlighted that false memories can develop quite easily—a finding that has important societal implications.

With these advances in mind, Cold Spring Harbor Laboratory Press has invited us to bring together an updated review of the field, which we have attempted to do in the following 19 chapters, written by many of the leaders in the field. We have asked our authors to cover the four major areas of the field outlined below:

- encoding of experience-dependent information

- consolidation and off-line transformation of information

- maintenance, reconsolidation, and retrieval

- novel approaches to the study of plasticity and memory

For readers outside the field it might be useful to place the study of learning and memory into a broader context.

Learning and memory are two of the most magical capabilities of our mind. Learning is the biological process of acquiring new knowledge about the world, and memory is the process of retaining and reconstructing that knowledge over time. Most of our knowledge of the world and most of our skills are not innate, but learned. Thus, we are who we are in large part because of what we have learned and what we remember and forget.

During the last four decades, neuroscience, the biological study of the brain, has established a common, broad conceptual framework that extends from cell and molecular biology, on the one hand, to brain systems biology, psychophysics, psychology, and neuronal modeling, on the other. Within this new, interdisciplinary structure, the scope of memory research ranges from genes to cognition, from molecules to mind.

Throughout this volume we will emphasize that memory storage is not the result of a linear sequence of events that culminates in an indelible, long-term memory. Rather, it is the dynamic outcome of several interactive processes: encoding or acquisition of new information, short-term memory, intermediate-term memory, consolidation of long-term memory, maintenance of long-term memory, and destabilization and restabilization of memory in the course of retrieving, updating, and integrating a given memory with other memories. We can see these dynamics at work in multiple levels of analysis and brain organization and in varying degrees, from simple to complex memory systems. These dynamics are initiated by molecular and cellular modifications at the level of individual synaptic connections and the neuronal cell bodies to which these synapses relate. These dynamics then extend to more distributed changes through multiple synaptic connections of many neurons embedded in larger neuronal networks whose interactions are expressed at the behavioral level.

We have organized the book around these basic temporal components of memory. In Section I, we focus on information encoding or learning; in Section II, we focus on postlearning consolidation where the molecular and in some cases the anatomical nature of the memories change. In Section III, we cover the maintenance, stability, and retrieval of memories. Finally, in Section IV, we discuss several emerging technologies that we think will have a significant impact on future studies of memory. In each section we discuss studies in species ranging from invertebrates to human beings. This is an evolutionary perspective that is based on the finding that memory is implemented in and subserved by cellular and molecular plasticity processes that are conserved and that insights can be gained from studying these basic processes across species.

Even simple forms of reflex learning in invertebrates show encoding, consolidation, and retrieval of memory that are reflected in synaptic and cell-wide plastic changes that control the behavior. In higher organisms there is an increase in circuit complexity of several orders of magnitude. With this increased complexity we see increased anatomical segregation of sensory processing streams and functional specialization of brain regions. Nevertheless, mammalian neurons show synaptic plasticity with some molecular similarities to invertebrates, as well as some differences. At the behavioral level mammals show the same basic, generic processes of encoding, consolidation, and retrieval. The added circuit complexity of the mammalian brain allows for greater diversity in behavior but also makes a full description of the information processing from sensory input to motor output much more difficult.

This can be seen throughout in the approaches applied and the nature of the questions asked in the various systems. In human beings, our knowledge necessarily comes primarily from noninvasive, functional imaging techniques and from study of patients who have damage to specific brain areas that affect memory. Our molecular understanding of memory in human beings is thus quite limited, but with the advent of functional magnetic resonance imaging (fMRI), our knowledge of the anatomy and of activity patterns is sometimes quite detailed. In mice and rats, the availability of genetic and pharmacological tools has allowed a detailed dissection of molecular signaling events important in learning and memory. However, the complexity of the mammalian brain has made mapping this knowledge onto the precise neurons and circuits that control behavior and its modification difficult, although new approaches offer some promise in this area. Finally, in simple systems like *Aplysia*, a deep understanding of the molecular mechanisms operating in neurons to modify behavior can be obtained and mapped directly onto the circuits that control the behavior. By integrating studies from multiple systems and varied approaches, we hope to convey some general principles and theoretical frameworks that operate in the field that can be useful in whatever level of analysis or species most captures the reader's interest.

## SECTION I: ENCODING

In this section we cover the principles of encoding, the initial modification of behavior with experience. We begin with an examination of circuit complexity, anatomical specialization, and memory subtypes initially identified in humans. Squire and Dede discuss the two major memory systems postulated for the mammalian brain: (1) explicit (also called declarative) memory, for facts and events, people, places, and objects; and (2) implicit (also called nondeclarative) memory, for perceptual and motor skills. The critical neurobiological insight for this distinction came from studies of patient H.M. (Scoville and Milner 1957), which revealed that the major aspects of explicit memory require the hippocampus and adjacent cortex—and in humans normally involve conscious awareness—whereas implicit memory does not require conscious awareness and relies mostly on other brain systems: namely, the cerebellum, the striatum, the amygdala, and simple reflex pathways themselves. We discuss these memory subsystems, and some more recent developments in their conceptualization, in subsequent chapters of this section.

Although it was accepted by the early 1970s that there are two major types of memory, little was known about how either type is formed. We could not distinguish experimentally, for example, between two leading—and conflicting—approaches to the mechanisms of memory storage: the aggregate field approach advocated by Lashley in the 1950s and by Adey in the 1960s, which assumed that information is stored in the bioelectric field generated by the aggregate activity of many neurons; and the cellular connectionist approach, which derived from Cajal's idea that memory is stored as an anatomical change in the strength of synaptic connections (Cajal 1894). In 1948, Konorski (1948) renamed Cajal's idea synaptic plasticity (the ability of neurons to modulate the strength of their synapses as a result of use).

To distinguish between these disparate neurobiological approaches to memory storage it soon became clear that one needed to develop tractable behavioral systems. Such systems would make it more likely to see how specific changes in the neuronal components of a behavior cause modifications of that behavior during learning and memory storage. From 1964 to 1979 several simple model systems of implicit memory emerged: the flexion reflex of cats, the eye-blink response of rabbits, and a variety of simple forms of reflex learning in invertebrates—namely, the defensive gill-withdrawal reflex of *Aplysia*, olfactory learning in *Drosophila*, the escape reflex of *Tritonia*, and various behavioral modifications in *Hermissenda*, *Pleurobranchaea*, *Limax*, crayfish, and honeybees (see Kandel 1976 for more detailed discussion).

Hawkins and Byrne, in the second and fifth chapters, discuss the insights that emerged from this reductionist approach in *Aplysia*. The first was purely behavioral and revealed that even animals with relatively few nerve cells, approximately 20,000 in the central nervous system of *Aplysia*, have remarkable learning capabilities. This simple nervous system can give rise to a variety of elementary forms of learning: habituation, dishabituation, sensitization, classical conditioning, and operant conditioning. Each form of learning, in turn, gives rise to short- or long-term memory.

Single-trial learning and the formation of short-term memory in the gill and siphon-withdrawal reflex of *Aplysia* result in part from changes in the strength of a sensorimotor synapse that mediates a major component of the reflex behavior. This synaptic plasticity is due to the modulation of the release of chemical transmitters from the presynaptic sensory neuron. A decrease in the amount of transmitter released was found to be associated with short-term habituation, whereas an increase was associated with short-term dishabituation and sensitization.

Further studies of the synaptic connections between the sensory and motor neurons that control the withdrawal reflex in *Aplysia* revealed that a single sensitizing stimulus to the tail increases the strength of the synaptic connections between the sensory and motor neurons by activating modulatory neurons that release serotonin onto the sensory neuron. Serotonin, in turn, increases the concentration of cyclic adenosine monophosphate (cAMP) in the sensory cell. The cAMP molecules activate the cAMP-dependent protein kinase (PKA), which signals the sensory neuron to release more of the transmitter glutamate into the synaptic cleft, thus temporarily strengthening the connection between the sensory and motor neuron. In fact, simply injecting cAMP or PKA directly into the sensory neuron produces temporary strengthening of the sensorimotor connection.

Next, Hawkins and his colleagues (1983) and Walters and Byrne (1983) succeeded in producing classical conditioning of *Aplysia*'s withdrawal reflex and began to analyze the mechanisms underlying this form of learning. Paired training, in which the conditioned stimulus (stimulation of the siphon) is applied just before the unconditioned stimulus (a shock to the tail), produces a greater increase in the gill-withdrawal reflex than either stimulus alone or unpaired stimuli. This is in part because the firing of an action potential by the sensory neuron just before the tail shock causes an influx of $Ca^{2+}$ into the sensory neuron that enhances the response of the adenylyl cyclase to serotonin. This, in turn, increases the amount of cAMP in the sensory neuron, which causes greater facilitation of the synaptic connection between sensory and motor neurons, an action also known as activity-dependent enhancement of synaptic facilitation. These studies support a synaptic plasticity model for memory encoding and describe some of the molecular underpinnings of this plasticity, thereby providing a conceptual and experimental framework for subsequent studies of encoding in mammalian systems, some of which have revealed similar mechanisms (Yagishita et al. 2014; Johansen et al. 2014).

In the third chapter, De Zeeuw and Ten Brinke examine eye-blink conditioning, a form of implicit memory in the mammalian brain. This is produced by pairing a tone (the conditioned stimulus [CS]) with an aversive air puff to the eye (the unconditioned stimulus [US]), resulting in a learned eye blink response to the CS alone (Thompson et al. 1983). The circuit mechanisms underlying this form of learning are one of the best characterized in a mammalian system. Theoretical and experimental studies suggest that before learning, activation of cerebellar Purkinje neurons in response to the CS leads to an inhibition of neurons in the interpositus nucleus (one of the deep nuclei of the cerebellum), thereby inhibiting motor output. With conditioning there is a decrease in the activity of the Purkinje cell in response to the CS, resulting in disinhibition of the neurons of the interpositus nucleus, leading to an eye blink. This model is consistent with findings that Purkinje cell activity can be reduced as a result of long-term depression (LTD), a form of synaptic plasticity, at the parallel fiber excitatory synaptic input onto the Purkinje neurons (Ito 2001). This decrease in the strength of the parallel fibers occurs when the climbing fiber inputs to the cerebellum are activated in appropriate temporal proximity to parallel fiber activity. The Purkinje cells become less responsive to input, as a result of a down-regulation of AMPA receptors at the parallel fiber to Purkinje cell synapse.

In the fourth chapter, Graybiel and Grafton examine the basal ganglia, a group of subcortical nuclei that includes the striatum, which are important in the implicit learning of skills and habits. The ventral striatum contains dopaminergic circuitry that is a critical component of reward and is a major target for drugs of abuse. The dorsal striatum, which is the primary focus of this chapter, is involved in the control of movement and skill and habit learning. These are primarily stereotyped motor programs that may begin as explicit memories but eventually progress to unconscious skills. For example, a child learns to tie her shoes in an explicit manner following specific instructions, but over many repetitions this common task becomes an implicit skill or habit involving the striatum. This progression can be seen in the activity pattern of striatal neurons that bracket the habitual motor behavior, firing before and after completion, suggesting a signal for the initiation and successful completion of the task. This principal can be extended to the individual steps of a complex motor task as well as to higher cognitive functions in the process of chunking. In chunking, the task is broken up into smaller motor programs with bracketing neural activity signaling the initiation of the appropriate behavioral chunk and feedback reward activity indicating the degree of successful completion of the behavior.

In the sixth chapter, Fanselow and Wassum discuss the mechanisms of Pavlovian conditioning in vertebrates, focusing primarily on fear conditioning. As in invertebrates, Pavlovian fear conditioning is also a behavioral model widely used in rodent experiments where it is considered to have both implicit and explicit forms (see also LeDoux 2015). When an animal is presented with a tone that is followed by a shock to the foot, the animal exhibits a learned fear response that can be gauged by freezing (a natural fear behavior) in response to the tone alone. This is an implicit form of learning that involves the amygdala but not the hippocampus. However, animals also develop a fear response to the test chamber in which they received the shocks (the context). This contextual learning is dependent on the hippocampus and is a model of explicit learning accessible in rodents.

This important paradigm in mammalian learning and memory research offers the richness in behavioral features of Pavlovian conditioning (e.g., blocking, interference, second-order conditioning). It has been well studied at the anatomical and circuit level and provides some of the strongest evidence linking synaptic plasticity and learning in a mammalian system. Finally, the contextual version of the task is one of the very few hippocampus-dependent tasks in rodents that, in some cases, shows the temporal gradient of retrograde amnesia that was reported in studies of human hippocampal patients.

The hippocampus was also the brain area where a form of synaptic plasticity thought to be important for memory, long-term potentiation (LTP), was first described by Lomo (1966) and more extensively by Bliss and Lomo (1973). They found that high-frequency electrical stimulation of the perforant path input to the hippocampus resulted in an increase in the strength of the stimulated synapses that lasted for many days. Subsequent studies found that some forms of LTP displayed the elementary properties of associability and specificity formulated by Hebb (1949), according to which (a) only synapses that are active when the postsynaptic cell is strongly depolarized are potentiated and (b) inactive synapses were not potentiated. Thus, groups of synapses that are coordinately active and contribute together to the firing of the target postsynaptic neuron will be strengthened, providing a plausible mechanism for linking ensembles of neurons encoding different environmental features that are presented together and thereby forming memory associations.

These properties of LTP have made it the focus of intense interest as a model synaptic mechanism for memory encoding in mammals. Angelakos and Able discuss the use of genetically modified mice in neurobiology with a focus on attempts to test the role of LTP in learning and memory. Then Basu and Siegelbaum explore the molecular signaling and cellular mechanisms that underlie LTP and the complementary form of plasticity, LTD, as well as the studies of their role in learning and memory.

The mechanism for initial induction of LTP varies in different regions of the hippocampus and in the same region with different patterns of stimulation. In the CA1 region, 100-Hz stimulation induces a form of LTP that is dependent on $N$-methyl-D-aspartate (NMDA) receptor activation. Moreover the properties of this receptor were proposed to explain the associative and activity-dependent properties of LTP. NMDA receptors are both voltage- and ligand-gated, and, to become active, they require both depolarization of the postsynaptic membrane, in which they reside, and concurrent release of glutamate from the presynaptic terminal. Activated NMDA receptors produce a strong $Ca^{2+}$ influx into the postsynaptic cell that is required to induce LTP. This $Ca^{2+}$ signal can activate a wide range of signaling pathways in the postsynaptic cell including calcium/calmodulin protein kinase II (CaMKII), protein kinase C (PKC), PKA, and mitogen-activated protein kinase (MAPK), each of which has been implicated in the induction of LTP as well as in its later stabilization.

Attempts to understand the role of LTP in memory storage have focused on genetic and pharmacological interventions designed to block LTP and consequent examination of the impact of this LTP blockade on memory formation. Many of these studies have used pharmacological or genetic blockade of the NMDA receptors required for LTP initiation or of other molecular signaling components in the LTP pathway. The results of these studies have been inconclusive. Some studies show learning deficiencies in animals lacking LTP and others do not. These results highlight some of the difficulties in mammalian studies that attempt to relate synaptic plasticity to behavior. First, LTP is not a unitary phenomenon and its natural induction is not well understood so that in some cases of apparent inhibition of LTP, other behaviorally relevant forms of LTP may still be induced. Also, many of the studies have focused on one brain region, usually the hippocampus, and it is possible that other brain circuits may compensate when LTP in the hippocampus is impaired or that animals may solve the behavioral tasks in a hippocampus-independent manner. These are difficulties inherent in the study of complex behaviors in the absence of a complete understanding of the underlying processing circuits.

## SECTION II: CONSOLIDATION

Over the years, the term "memory consolidation" has been used to define two different yet interrelated processes (Dudai and Morris 2000). Synaptic, cellular, or immediate consolidation refers to the gene expression–dependent transformation of information into a long-term form in the neural circuit that encodes the memory. Systems consolidation refers to a slower postencoding reorganization of long-term memory over distributed brain circuits into remote memory lasting months to years, and it is commonly studied within the context of the corticohippocampal system that subserves explicit memory.

In the first chapter in this section, Alberini and Kandel examine the mechanisms of cellular and synaptic consolidation. The idea of a cellular memory consolidation process derives from the observation that inhibitors of protein synthesis, when given during or shortly after initial learning, do not disrupt short-term memory at 1 hour, but impair 24-hour memory. These and subsequent studies suggest that a transcriptional/translational program, initiated with learning, is required for the stabilization or consolidation of memories at later time points. This chapter focuses on a substantial body of work characterizing the molecular switch from short-term to long-term memory. A cannonical pathway found in both *Aplysia* and mammals involves the activation of the transcription factor cAMP response element-binding protein 1 (CREB-1). The CREB-mediated response to external stimuli can be activated by a number of kinases (PKA, CaMKII, CaMKIV, RSK2, MAPK, and PKC) and phosphatases, which suggests that CREB integrates signals from these various pathways. The activation of CREB-dependent transcription in turn leads to the expression of a variety of downstream effector molecules that help to stabilize synaptic changes and memory.

More recent studies have uncovered a rich regulatory complexity in the control of this memory-related transcriptional program. In addition to positive factors such as CREB-1, there are inhibitory transcription factors such as CREB-2, which act as suppressors to oppose the formation of long-term memory. The transcription required for consolidation is also controlled by epigenetic factors. Regulated covalent modifications of histones and DNA exert a strong effect on this transcriptional cascade. Finally, small regulatory RNAs contribute the control of the transcription required for memory consolidation.

One of the consequences of the transcriptional signaling for long-term memory consolidation is the formation and stabilization of new synaptic connections, which is discussed by Bailey, Kandel, and Harris. The authors examine structural changes in the *Aplysia* sensorimotor synapse and in the mammalian hippocampus with synaptic plasticity and memory storage. Long-term sensitization training in *Aplysia* results in an increase in the number of anatomical synaptic connections between the sensory and motor synapses controlling the behavior. This form of structural plasticity requires the CREB-based transcriptional cascade for memory consolidation interacting with local changes in various cell surface and cytoskeletal molecules to stabilize the new synapses.

In rodents, similar types of structural plasticity have been observed in the hippocampus with learning and with the induction of LTP. The current inability to identify the precise neurons involved in a given learning task in the mammalian brain makes it difficult to study these effects with behavior; however, new in vivo imaging techniques that allow individual neurons to be followed longitudinally are facilitating behavioral studies. There are now a series of studies showing that activity-dependent synaptic plasticity (LTP) produces increases in synapse number and size. There are also important developmental components to this regulation. Taken together, these results suggest that physical changes in synaptic connections are a major contributor to the long lasting stabilization of memory.

Finally, Squire, Genzel, Wixted, and Morris deal with proposed mechanisms of systems consolidation. Systems consolidation is a concept originally used to explain the temporal gradient of amnesia that was reported following hippocampal damage in humans and some animal models. Although hippocampal damage in humans produces a retrograde amnesia for explicit memories that were acquired within a year or two before the injury, older memories are reported to stay intact. Thus, a new memory requires the hippocampus for retrieval but some process of consolidation allows older memories to be accessed independently of the hippocampus. The "standard model" for this process posits that shortly after learning the hippocampus is required to recruit distributed cortical elements of the memory into a coherent recall event, but over time persistent spontaneous activity, or replay, within the cortical elements strengthens their synaptic connectivity to allow the memory to be retrieved independently of the hippocampus. This model and other, competing models developed to explain this fascinating characteristic of hippocampal amnesia are discussed.

## SECTION III: MAINTENANCE, RECONSOLIDATION, AND RETRIEVAL

In this section we cover the mechanisms of memory retrieval and its use in guiding behavior. Retrieval is critical in memory research, and in the current state of memory research, there is in fact no evidence of a memory's existence until it is retrieved. Retrieval is generally thought to involve the reactivation or reconstruction of components of the neural ensembles activated during the initial learning. In some behavioral paradigms it has become clear that retrieval is not a passive process but engages molecular signaling to maintain the memory in an active process called reconsolidation. The overall picture is that of a dynamic process involving circuit activity as well as molecular signaling to maintain stable memories.

In the first chapter in this section, Si and Kandel explore molecular mechanisms for the persistent maintenance of memory. As we have seen, the consolidation and stabilization of memories require a transcriptional and translational program initiated in the nucleus, but the actual physical substrate of memory is at the synapse. Because individual neurons can have thousands of individual synaptic connections carrying different information, how do the synapses relevant to a specific memory selectively make use of the products of this transcriptional program? Prions are a class of self-aggregating proteins in which a conformational shift from the native, nonaggregated form to the prion form can be transferred from one protein molecule to the other. Si and Kandel have found that a naturally occurring protein, CPEB, has this self-aggregating prion-like property. CPEB is localized to synapses where it regulates the translation of mRNA for specific target proteins. Evidence is presented in this chapter that during memory maintenance CPEB is converted at the relevant synapses to an active prion-like state, which is self-maintaining, and that this conformational change allows the translation of proteins required specifically at those synapses. This mechanism, first identified in *Aplysia*, has now also been found in *Drosophila* and mice and provides a candidate mechanism for long-term synapse-specific gene expression changes.

The dynamic molecular changes evoked with memory retrieval are considered in the chapter on reconsolidation by Nader. A major development in research on consolidation in the past decade has been the revitalization of the idea (Misanin et al. 1968) that consolidation does not occur just once per item, but that under some circumstances it can be actively recruited during later retrieval of that same item. In many cases, when inhibitors of protein synthesis are given in a short time window after memory retrieval, they disrupt the subsequent storage of the memory, similar to what is seen with consolidation of initial learning—hence the term reconsolidation. This chapter covers the current state of this research and discusses the similarity and differences between consolidation and reconsolidation. There is a great deal of clinical interest in reconsolidation as a potential means of disrupting pathological memories that are a hallmark of posttraumatic stress disorder (PTSD).

Ben-Yakov, Dudai, and Mayford discuss the current state of research on the retrieval process itself in mice and men. Our brain can retrieve complex information and act on it within a fraction of a second, but we still do not know how it does so. Behavioral models (Tulving 1983; Roediger et al. 2007) lead us to expect that the brain does this through a combination of sequential and parallel distributed processes that involve multiple brain circuits. Human fMRI studies examining activity during retrieval provide snapshots of this distributed retrieval network, which recruits prefrontal and parietal cortices as well as medial temporal lobe structures such as the hippocampus. Electrophysiological observations in human patients undergoing seizure monitoring are also contributing to better understanding of the fast distributed spatiotemporal configurations that could allow fast retrieval of complex information. In mice, new techniques that allow the observation of activity in large populations of neurons with single-cell resolution has revealed that specific neural ensembles are recruited during retrieval. This is consistent with the idea that the brain represents specific memories in the spatiotemporal pattern of neurons that are activated. This idea has been tested in several recent studies showing that artificial stimulation of the precise pattern of neurons activated during learning can directly produce apparent memory recall.

Moser, Rowland, and Moser explore the nature of information coding and processing in medial temporal lobe structures. Because the hippocampus was identified as critical for explicit memory based on studies of human amnesic patients, animal studies of the hippocampus focused on the nature of the information that the hippocampus is concerned with. Electrophysiological recording of hippocampal activity in freely behaving rats demonstrated that the most striking feature of hippocampal neurons is their spatially specific firing (O'Keefe and Dostrovsky 1971). When animals are allowed to move freely in an open space or on more restrictive tracks, individual hippocampal pyramidal neurons are "place cells"; that is, they are active only when the animal passes through a limited region of the environment, their place field, suggesting that the hippocampal neurons encode a map of the animal's spatial location. In 2005, Edvard and May-Britt Moser extended this idea when they found a precursor of the spatial map in the entorhinal cortex that is formed by a new class of cells known as "grid cells." Each of these space-encoding cells has a grid-like, hexagonal receptive field and conveys information to the hippocampus about position, direction, and distance (Fyhn et al. 2004).

Explicit memory involves recall of people, places and events, and the properties of this entorhinal–hippocampal network serves as a model for studying the encoding and retrieval of place at the systems level. The chapter focuses on the properties of these neurons and how they form their unique firing properties and could contribute to spatial navigation and memory. This is one of the most well-studied and tractable models of higher-order sensory coding where the neurons respond not to specific features of the environment but in an integrated manner to a determination of where they are in the environment. How this occurs at a circuit level and how this information is conveyed to other brain areas is critical for understanding explicit forms of memory.

Working memory refers to the memory system that holds information in temporary storage (seconds) during the planning and execution of a task (Baddeley 2012). It is considered to engage multiple components that combine perceptual information, short- and long-term memory, intentions, and decision making (D'Esposito and Postle 2015). Nyberg and Eriksson, who cover the topic of working memory and its neurobiological underpinnings in this book, emphasize that working memory underlies many basic cognitive abilities that far exceed mnemonic performance per se. They focus on human imaging studies and conclude that working-memory maintenance involves frontal–parietal regions and distributed representational areas, which can be based on persistent activity in reentrant loops, synchronous oscillations, or changes in synaptic strength. Manipulation of the content of working memory depends on dorsal frontal cortex, and updating is apparently realized by a frontostriatal gating function, which involves dopaminergic modulation. Goals and intentions are considered to be represented as cognitive and motivational contexts in the rostral frontal cortex. Working memory has rather limited capacity, and it appears that variations in this capacity as well as working memory deficits can largely be accounted for by the effectiveness and integrity of the basal ganglia and dopaminergic neurotransmission.

## SECTION IV: NOVEL APPROACHES

The final three chapters discuss newly developed techniques that we think will have a significant impact on the study of memory in complex nervous systems. The most difficult problem in mammalian systems is identifying the precise neurons within a circuit that participate in the processing of information in a specific behavioral task, how these neural ensembles encode that information and convey it from one brain region to the next to ultimately control motor behavior, and where within this dispersed circuitry the physical changes with experience occur to produce memories. The focus of much of this research, reflected for example in the goals of the U.S. BRAIN Initiative, is on the recording and manipulation of large ensembles of neurons in behaving animals and people. The larger the data set of neural activity recorded during behavior, the more likely it is to reveal principles of information encoding, particularly if combined with modeling.

In combination with optogenetics, these principles can then be directly tested, as we will begin to see in the remaining chapters.

In the first chapter in this section, Jercog, Rogerson, and Schnitzer discuss new optical methods for the simultaneous recording of activity in large ensembles of neurons repeatedly over long time periods. Classical electrophysiological techniques suffer from difficulties in recording activity from large numbers of neurons over prolonged time periods and with certainty that the same neuron is identified in repeated recording trials. With the advent of chemical and genetic voltage and calcium sensors that reflect neural activity with an optical signal, these difficulties can be overcome to allow the recording of hundreds of neurons repeatedly over weeks to months in intact behaving animals. This chapter covers these approaches with a specific focus on a form of microscopy that allows the imaging of deep brain structures in these freely moving animals. This combination of techniques opens the possibility of recording neural ensemble activity in any brain region during any behavioral task to see changes during encoding and retrieval of memories.

Recording studies have identified neurons with exquisitely tuned activity responses to external stimuli. For example, in human studies, neurons in the hippocampus have been identified that seem to respond only to a specific individual. This specific activity strongly implies that these neurons somehow participate in the brain's encoding of the individual to which the neuron responds, but this has been difficult to test directly. Mayford and Reijmers discuss a technique that allows this question to be addressed. By using regulatory elements from genes whose expression is induced by neural activity, essentially any exogenous gene can be expressed in the ensemble of neurons that are active in response to any behaviorally relevant sensory stimulus. This approach has been used in mice to show that the ensemble of neurons naturally activated in the hippocampus and cortex during learning of a specific place could, when artificially stimulated, substitute for the actual sensory experience of the place. This approach allows cellular, molecular, and electrophysiological studies to focus specifically on the sparse and distributed group of neurons that are naturally active during learning and should be useful in identifying the circuit modifications that occur to produce memory.

Recent advances in human noninvasive functional brain imaging have enhanced the ability to interrelate three major facets of systems neurosciences: brain activity, behavior, and computational modeling. Brown, Staresina, and Wagner describe how novel acquisition and analyses methods in fMRI advance mechanistic accounts of learning and memory in the human brain. These include hardware and protocol developments that now permit spatial resolution approaching 1 mm and in some cases even at the submillimeter range. Combination of high-spatial-resolution MRI with multivariate analyses techniques has markedly improved the capacity to tap into the representational content of brain activity signatures and hence contribute to better understanding the role of distinct brain areas in representing discrete stimuli and responses and the transformation of representational content of perceived events from their encoding to their retrieval.

An illustrative approach is representational similarity analysis (RSA), which examines the correlation between stimulus-specific patterns of brain activation, which can be used to query, for example, how the similarity of multivoxel patterns measured during encoding relate to subsequent memory. Another, related approach is multivoxel pattern analysis (MVPA), in which multivoxel activation patterns for two or more classes of stimuli are used to train a classifier to identify characteristic activity patterns that maximally discriminate between these stimulus classes. Pattern classifiers can also be used to generate probabilistic predictions about the new event's likely class, which provides a trial-specific quantitative measure of the strength of neural signatures. For example, it has been reported that the strength of content-specific (i.e., face vs. scene) neural signatures in visual cortex correlates with the magnitude of hippocampal fMRI activity at encoding and predicts whether that information will later be remembered or forgotten. This suggests that the success of hippocampally mediated encoding of event details covaries with the strength or fidelity of the corresponding cortical representations.

Multivariate pattern analyses hence open new possibilities in the study of representational content of types of stimuli in specific brain regions. It would be of interest to observe how this class of methods is harnessed to discern activity signatures of tokens within each stimulus type (e.g., individual face vs. generic face) or whether this would require human imaging methods with improved spatiotemporal resolution. In parallel, functional connectivity measures using currently available fMRI also provide new information on how different brain regions and subregions interact in support of encoding and retrieval of memory.

## SUMMARY

As these chapters illustrate, a great deal of progress has been made in recent years in tapping into the biological mechanisms of learning and memory. In simple circuits that control behavior, the tools of cellular and molecular biology have revealed how individual neurons and molecular signaling pathways are modified by learning. Changes in synaptic strength produced by specific patterns of electrical activity or the action of modulatory transmitters can alter the processing of information to control behavioral performance. Both memory storage and synaptic plasticity have varying temporal phases, with the switch from short- to long-lasting synaptic and behavioral memory requiring new gene expression. The long-term phase seems to be using a number of synaptic and cell-wide mechanisms, such as synaptic tagging, changes in protein synthesis at the synapse, protein kinase–based cascades, and functional self-perpetuating prion-like mechanisms and epigenetic alterations for maintenance.

We are also beginning to uncover the structure of neural circuits in more complex forms of memory, among them those that involve the hippocampus and neocortex. Recent techniques for the genetic manipulation of neurons based on their natural activity during learning and recall are enabling investigators to investigate the function of distributed neural ensembles and their role in generating representations in complex memory. Finally, advances in functional brain imaging, combined with new electrophysiological and computational techniques for assessing neural activity in populations of neurons, are helping us to determine what regions of the human brain are involved in complex memory and to explore the coding properties and the transformation of the experience-dependent representation over time in consolidation, maintenance, and reactivation or reconstruction in retrieval. The application of these approaches over the next decade offer an exciting opportunity to uncover the computational rules for coding complex information in the mammalian brain and the nature of its modification with learning and memory.

ERIC R. KANDEL
YADIN DUDAI
MARK R. MAYFORD

## REFERENCES

Baddeley A. 2012. Working memory: Theories, models, and controversies. *Annu Rev Psychol* **63**: 1–29.

Bliss T and Lomo T. 1973. Long-lasting potentiation of synaptic transmission in dentate area of anesthetized rabbit following stimulation of performant path. *J Physiol* **232**: 331–356.

Cajal SR. 1894. La fine structure des centres nerveux. *Proc R Soc Lond* **55**: 444–468.

D'Esposito M and Postle BR. 2015. The cognitive neuroscience of working memory. *Annu Rev Psychol* **66**: 115–142.

Dudai Y and Morris RGM. 2000. To consolidate or not to consolidate: What are the questions? In *Brain, perception, memory. Advances in cognitive sciences* (ed. Bolhuis JJ), pp. 149–162. Oxford University Press, Oxford.

Fyhn M, Molden S, Witter MP, Moser EI, and Moser MB. 2004. Spatial representation in the entorhinal cortex. *Science* **305**: 1258–1264.

Hawkins RD, Abrams TW, Carew TJ, and Kandel ER. 1983. A cellular mechanism of classical conditioning in *Aplysia*: Activity-dependent amplification of presynaptic facilitation. *Science* **219**: 400–415.

Hebb DO. 1949. *The organization of behavior: A neuropsychological theory*. Wiley, New York.

Ito M. 2001. Cerebellar long-term depression: Characterization, signal transduction, and functional roles. *Physiol Rev* **81**: 1143–1195.

Johansen JP, Diaz-Mataix L, Hamanaka H, Ozawa T, Ycu E, Koivumaa J, Kumar A, Hou M, Deisseroth K, Boyden ES, et al. 2014. Hebbian and neuromodulatory mechanisms interact to trigger associative memory formation. *Proc Natl Acad Sci* **111:** E5584–E5592.

Kandel ER. 1976. *Cellular basis of behavior: An introduction to behavioral neurobiology.* W.H. Freeman, San Francisco.

Konorski J. 1948. *Conditioned reflexes and neuronal organization.* Cambridge University Press, New York.

LeDoux J. 2015. *Anxious: Using the brain to understand and treat fear and anxiety.* Viking, New York.

Lomo T. 1966. Frequency potentiation of excitatory synaptic activity in dentate area of hippocampal formation. *Acta Physiol Scand* **68:** 277.

Misanin JR, Miller RR, and Lewis DJ. 1968. Retrograde amnesia produced by electroconvulsive shock after reactivation of consolidated memory trace. *Science* **160:** 554–555.

O'Keefe J and Dostrovsky J. 1971. The hippocampus as a spatial map. Preliminary evidence from unit activity in the freely-moving rat. *Brain Res* **34:** 171–175.

Roediger HL, Dudai Y, and Fitzpatrick SM eds. 2007. *Science of memory: Concepts.* Oxford University Press, New York.

Scoville WB and Milner B. 1957. Loss of recent memory after bilateral hippocampal lesions. *J Neurol Neurosurg Psych* **20:** 11–21.

Thompson RF, McCormick DA, Lavond DG, Clark GA, Kettner RE, and Mauk MD. 1983. Initial localization of the memory trace for a basic form of associative learning. *Prog Psychobiol Physiol Psychol* **10:** 167–196.

Tulving E. 1983. *Elements of episodic memory.* Clarendon Press, Oxford.

Walters ET and Byrnes JH. 1983. Associative conditioning of single sensory neurons suggests a cellular mechanism for learning. *Science* **219:** 405–408.

Yagishita S, Hayashi-Takagi A, Ellis-Davies GCR, Urakubo H, Ishii S, and Kasai H. 2014. A critical time window for dopamine actions on the structural plasticity of dendritic spines. *Science* **345:** 1616–1620.

# Conscious and Unconscious Memory Systems

Larry R. Squire[1,2,3] and Adam J.O. Dede[3]

[1]Veterans Affairs, San Diego Healthcare System, San Diego, La Jolla, California 92161

[2]Departments of Psychiatry and Neurosciences, University of California, San Diego, La Jolla, California 92093

[3]Department of Psychology, University of California, San Diego, La Jolla, California 92093

*Correspondence:* lsquire@ucsd.edu

The idea that memory is not a single mental faculty has a long and interesting history but became a topic of experimental and biologic inquiry only in the mid-20th century. It is now clear that there are different kinds of memory, which are supported by different brain systems. One major distinction can be drawn between working memory and long-term memory. Long-term memory can be separated into declarative (explicit) memory and a collection of nondeclarative (implicit) forms of memory that include habits, skills, priming, and simple forms of conditioning. These memory systems depend variously on the hippocampus and related structures in the parahippocampal gyrus, as well as on the amygdala, the striatum, cerebellum, and the neocortex. This work recounts the discovery of declarative and nondeclarative memory and then describes the nature of declarative memory, working memory, nondeclarative memory, and the relationship between memory systems.

The idea that memory is not a single faculty has a long history. In his *Principles of Psychology*, William James (1890) wrote separate chapters on memory and habit. Bergson (1910) similarly distinguished between a kind of memory that represents our past and memory that is not representational but nevertheless allows the effect of the past to persist into the present. One finds other antecedents as well. McDougall (1923) wrote about explicit and implicit recognition memory, and Tolman (1948) proposed that there is more than one kind of learning. These early proposals were often expressed as a dichotomy involving two forms of memory. The terminologies differed, but the ideas were similar. Thus, Ryle (1949) distinguished between knowing how and knowing that, and Bruner (1969) identified memory without record and

memory with record. Later, the artificial intelligence literature introduced a distinction between declarative and procedural knowledge (Winograd 1975). Yet constructs founded in philosophy and psychology are often abstract and have an uncertain connection to biology, that is, to how the brain actually stores information. History shows that as biological information becomes available about structure and function, understanding becomes more concrete and less dependent on terminology.

## THE DISCOVERY OF DECLARATIVE AND NONDECLARATIVE MEMORY SYSTEMS

Biological and experimental inquiry into these matters began with studies of the noted patient H.M. (Scoville and Milner 1957; Squire 2009).

H.M. developed profound memory impairment following a bilateral resection of the medial temporal lobe, which had been performed to relieve severe epilepsy. The resection included much of the hippocampus and the adjacent parahippocampal gyrus (Annese et al. 2014; Augustinack et al. 2014). H.M.'s memory impairment was disabling and affected all manner of material (scenes, words, faces, etc.), so it was quite unexpected when he proved capable of learning a hand–eye coordination skill (mirror drawing) over a period of 3 days (Milner 1962). He learned rapidly and efficiently but on each test day had no memory of having practiced the task before. This finding showed that memory is not a single entity. Yet, at the time, discussion tended to set aside motor skills as an exception, a less cognitive form of memory. The view was that all of the rest of memory was impaired in H.M. and that the rest of memory is of one piece.

There were early suggestions in the animal literature that more than just motor skills were intact after lesions of hippocampus or related structures (Gaffan 1974; Hirsh 1974; O'Keefe and Nadel 1978). However, these proposals differed from each other, and they came at a time when the findings in experimental animals did not conform well to the findings for human memory and amnesia. In particular, animals with hippocampal lesions often succeeded at tasks that were failed by patients with similar lesions. It gradually became clear that animals and humans can approach the same task with different strategies (and using different brain systems), and also that patients with medial temporal lobe lesions, like experimental animals with similar lesions, can in fact succeed at a wide range of learning and memory abilities.

First came the finding that memory-impaired patients could acquire, at a normal rate, the perceptual skill of reading mirror-reversed words, despite poor memory for the task and for the words that were read (Cohen and Squire 1980). Thus, perceptual skills, not just motor skills, were intact. This finding was presented in the framework of a brain-based distinction between two major forms of memory that afford either declarative or procedural knowledge.

Declarative knowledge is knowledge available as conscious recollection, and it can be brought to mind as remembered verbal or nonverbal material, such as an idea, sound, image, sensation, odor, or word. Procedural knowledge refers to skill-based information. What is learned is embedded in acquired procedures and is expressed through performance.

Subsequently, other forms of experience-dependent behaviors were found to be distinct from declarative memory. One important phenomenon was priming—the improved ability to detect, produce, or classify an item based on a recent encounter with the same or related item (Tulving and Schacter 1990; Schacter and Buckner 1998). For example, individuals will name objects faster on their second presentation, and independent of whether they recognize the objects as having been presented before.

Another important discovery was that the neostriatum (not the medial temporal lobe) is important for the sort of gradual feedback-guided learning that results in habit memory (Mishkin et al. 1984; Packard et al. 1989; Knowlton et al. 1996). Tasks that assess habit learning are often structured so that explicit memorization is not useful (e.g., because the outcome of each trial is determined probabilistically), and individuals must depend more on a gut feeling. After learning, it is more accurate to say that individuals have acquired a disposition to perform in a particular way than to say that they have acquired a fact (i.e., declarative knowledge) about the world.

Given the wide variety of learning and memory phenomena that could eventually be shown in patients (e.g., priming and habit learning as well as simple forms of classical conditioning), the perspective eventually shifted to a framework involving multiple memory systems rather than just two kinds of memory (Fig. 1). Accordingly, the term "nondeclarative" was introduced with the idea that nondeclarative memory is an umbrella term referring to multiple forms of memory that are not declarative (Squire and Zola-Morgan 1988). Nondeclarative memory includes skills and habits, simple forms of conditioning, priming, and perceptual learning, as well as phylogenetically

Cite this article as *Cold Spring Harb Perspect Biol* doi: 10.1101/cshperspect.a021667

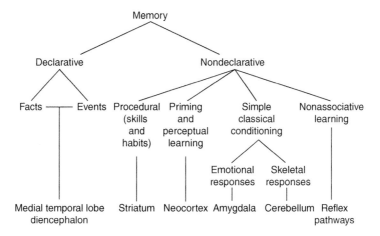

**Figure 1.** Organization of mammalian long-term memory systems. The figure lists the brain structures thought to be especially important for each form of memory. In addition to its central role in emotional learning, the amygdala is able to modulate the strength of both declarative and nondeclarative memory.

early forms of behavioral plasticity like habituation and sensitization that are well developed in invertebrates. The various memory systems can be distinguished in terms of the different kinds of information they process and the principles by which they operate.

Declarative memory provides a way to represent the external world. It is the kind of memory we typically have in mind when we use the term memory in everyday language. Declarative memory has two major components, semantic memory (facts about the world) and episodic memory (the ability to re-experience a time-and-place-specific event in its original context) (Tulving 1983). The acquisition of episodic memory requires the involvement of brain systems in addition to medial temporal lobe structures, especially the frontal lobes (Tulving 1989; Shimamura et al. 1991). There is some uncertainty around the issue of whether nonhuman animals have the capacity for episodic memory (i.e., the capacity for mental time travel that can return an animal to the scene of an earlier event), and the idea has been difficult to put to the test (Tulving 2005). For elegant demonstrations of episodic-like memory in nonhuman animals, see Clayton and Dickinson (1998).

Nondeclarative memory is dispositional and is expressed through performance rather than recollection. An important principle is the ability to gradually extract the common elements from a series of separate events. Nondeclarative memory provides for myriad unconscious ways of responding to the world. The unconscious status of nondeclarative memory creates some of the mystery of human experience. Here arise the habits and preferences that are inaccessible to conscious recollection, but they nevertheless are shaped by past events, they influence our current behavior and mental life, and they are a fundamental part of who we are.

Sherry and Schacter (1987) suggested that multiple memory systems evolved because they serve distinct and fundamentally different purposes. For example, the gradual changes that occur in birdsong learning are different from, and have a different function than, the rapid learning that occurs when a bird caches food for later recovery. These memory systems operate in parallel to support and guide behavior. For example, imagine an unpleasant event from early childhood, such as being knocked down by a large dog. Two independent consequences of the event could potentially persist into adulthood as declarative and nondeclarative memories. On the one hand, the individual might have a conscious, declarative memory of the event itself. On the other hand, the individual might have a fear of large dogs, quite independently of whether the event itself is remembered. Note that a fear of dogs would not be experienced as a memory but rather as a part of personality, a preference, or an attitude about the world.

## THE NATURE OF DECLARATIVE MEMORY

Studies of patients and experimental animals with medial temporal lobe damage have identified four task requirements that reliably reveal impaired memory: (1) tasks where learning occurs in a single trial or single study episode (Mishkin 1978); (2) tasks where associations between stimuli are learned across space and time (e.g., Higuchi and Miyashita 1996; Fortin et al. 2002); (3) tasks where the acquired information can be used flexibly (e.g., Bunsey and Eichenbaum 1996; Smith and Squire 2005); and (4) tasks where learning depends on awareness of what is being learned (Clark and Squire 1998; Smith and Squire 2008).

Declarative memory (sometimes termed explicit memory) is well adapted for the rapid learning of specific events. Declarative memory allows remembered material to be compared and contrasted. The stored representations are flexible, accessible to awareness, and can guide performance in multiple different contexts. The key structures that support declarative memory are the hippocampus and the adjacent entorhinal, perirhinal, and parahippocampal cortices, which make up much of the parahippocampal gyrus (Fig. 2) (Squire and Zola-Morgan 1991). These structures are organized hierarchically, and their anatomy suggests how the structures might contribute differently to the formation of declarative memory, for example, in the encoding of objects (perirhinal cortex) or scenes (parahippocampal cortex) and in the forming of associations between them (hippocampus) (Squire et al. 2004; Davachi 2006; Staresina et al. 2011).

Structures in the diencephalic midline (mammillary nuclei, medial dorsal nucleus, anterior thalamic nuclei, together with the internal medullary lamina and the mammillothalamic tract) are also important for declarative memo-

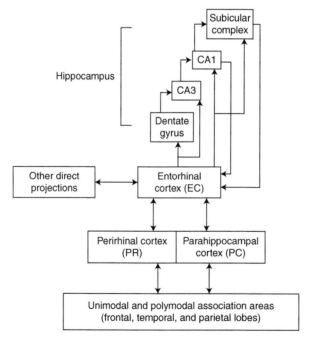

**Figure 2.** Schematic view of the medial temporal lobe memory system for declarative memory, which is composed of the hippocampus and the perirhinal, entorhinal, and parahippocampal cortices. In addition to the connections shown here, there are also weak projections from the perirhinal and parahippocampal cortices to the CA1-subiculum border.

ry. Damage to these structures causes the same core deficit as damage to the medial temporal lobe, probably because these nuclei and tracts are anatomically related to the medial temporal lobe (see Markowitsch 1988; Victor et al. 1989; Harding et al. 2000; Squire and Wixted 2011).

Discussion continues about the nature of declarative memory and about when exactly the hippocampus (and related structures) is involved in learning and memory. One proposal is that, whereas conscious recollection depends on the hippocampus as described above, the hippocampus is also important for unconscious memory under some circumstances (Henke 2010; Hannula and Greene 2012; Shohamy and Turk-Browne 2013). One way to explore this issue has been to record eye movements while volunteers are making behavioral memory judgments. In some situations (e.g., when a change has occurred in the layout of a scene), the eyes do not reveal signs of memory (by moving to the changed location) unless participants are aware of where the change occurred (Smith et al. 2008). Furthermore, these eye movement effects require the integrity of the hippocampus. Nevertheless, in other situations, eye movements can signal which item is correct, and correlate with hippocampal activity, even when behavioral memory judgments are incorrect and participants are therefore thought to be unaware (Hannula and Ranganath 2009). Such a finding could mean that eye movements (and hippocampal activity) can index unaware memory. Yet, it is also true that awareness is (presumably) continuous, and a low amount of awareness is not the same as a complete lack of awareness (Kumaran and Wagner 2009). Just as recognition memory can succeed when free recall fails, eye movements might reveal signs of aware memory when recognition fails. It would be instructive in this circumstance to obtain confidence ratings in association with memory judgments and ask whether there is any detectable awareness of which items are correct.

Other work has implicated medial temporal lobe structures in the unaware learning of sequences and other tasks with complex contingencies (Chun and Phelps 1999; Rose et al. 2002; Schendan et al. 2003). This idea is often based on functional magnetic resonance imaging (fMRI) evidence of medial temporal lobe activity during unaware learning. Strategic factors (e.g., explicit attempts to memorize) may explain some of these effects (Westerberg et al. 2011), together with the possibility that awareness may not be entirely absent (Poldrack and Rodriguez 2003). In addition, considering that fMRI data cannot establish a necessary role for a particular structure, it will be useful to supplement fMRI data with evidence that patients with hippocampal lesions are impaired at the same tasks that afford unaware learning. Interestingly, for some of these tasks, hippocampal patients were not impaired (Reber and Squire 1994; Manns and Squire 2001), but impairment has been reported in patients when the damage was undescribed or extended beyond the hippocampus (Chun and Phelps 1999). Studies that combine fMRI, patient data, and rigorous measures of awareness will be useful in pursuing this interesting issue.

## WORKING MEMORY

Working memory refers to the capacity to maintain a limited amount of information in mind, which can then support various cognitive abilities, including learning and reasoning (Baddeley 2003). Within cognitive neuroscience, the term "working memory" has largely replaced the less precise term "short-term memory." (Note that the term "short-term memory" remains useful in cellular neuroscience where it has a different and distinct meaning [Kandel et al. 2014].) Historically, working memory has been considered to be distinct from long-term memory and independent of the medial temporal lobe structures that support the formation of long-term memory (Drachman and Arbit 1966; Atkinson and Shiffrin 1968; Baddeley and Warrington 1970; Milner 1972). Long-term memory is needed when the capacity of working memory is exceeded or when working memory is disrupted by diverting attention to different material.

Uncertainty can arise when determining in any particular case, whether performance depends on working memory or whether so much information needs to be kept in mind that working memory capacity is exceeded and performance must rely on long-term memory. What is the situation when patients fail at tasks with short retention intervals, or no retention interval (Hannula et al. 2006; Olson et al. 2006; Warren et al. 2011, 2012; Yee et al. 2014)? Are these long-term memory tasks or, as has been suggested, do such findings show that working memory sometimes depends on the medial temporal lobe (Ranganath and Blumenfeld 2005; Graham et al. 2010). It is important to note that working memory cannot be defined in terms of any particular retention interval (Jeneson and Squire 2012). Even when the retention interval is measured in seconds, working memory capacity can be exceeded such that performance must depend, in part, on long-term memory (e.g., immediately after presentation of a list of 10 words).

Findings from a scene-location task illustrate the problem. The task involved scenes containing a number of different objects. On each trial, a scene was presented together with a question (e.g., is the plant on the table?). A few seconds later, the same scene was presented again but now the object queried about might or might not appear in a different location. In this condition, patients with hippocampal lesions were accurate at detecting whether or not the object had moved (Jeneson et al. 2011). However, when attention was drawn to four different objects, any one of which might move, patients were impaired (Yee et al. 2014). It is likely that the impairment in the second condition occurred because working memory capacity was exceeded. Visual working memory capacity is quite limited and, typically, even healthy young adults can maintain only three to four simple visual objects in working memory (Cowan 2001; Fukuda et al. 2010).

In any case, there are ways to distinguish working memory and long-term memory (Shrager et al. 2008; Jeneson et al. 2010). For example, one can vary the number of items or associations to be remembered and ask whether patients show a sharp discontinuity in performance as the number of items increases and working memory capacity is exceeded. In one study (Jeneson et al. 2010), patients with hip-

pocampal lesions or large medial temporal lobe lesions saw different numbers of objects (1 to 7) on a tabletop and then immediately tried to reproduce the array on an adjacent table. Performance was intact when only a few object locations needed to be remembered. However, there was an abrupt discontinuity in performance with larger numbers of object locations. One patient (who had large medial temporal lobe lesions similar to H.M.) learned one-, two-, or three-object locations as quickly as did controls, never needing more than one or two trials to succeed. Yet when four-object locations needed to be remembered, he did not succeed even after 10 trials with the same array. These findings indicate that the ability to maintain small numbers of object–place associations in memory is intact after medial temporal lobe lesions. An impairment was evident only when a capacity limit was reached, at which point performance needed to depend on long-term memory. A similar conclusion was reached in studies of a single patient with restricted hippocampal lesions. Performance was fully intact on an extensive working memory battery, including tasks of relational (associative) memory (Baddeley et al. 2010, 2011).

Spatial tasks like path integration can also be performed normally by patients with medial temporal lobe lesions, as long as the task can be managed within working memory (Shrager et al. 2008; Kim et al. 2013). The findings are different for rats with hippocampal lesions (Kim et al. 2013), either because for the rat the task relies on long-term memory or because the rat hippocampus is needed for some online spatial computations, as suggested previously (Whitlock et al. 2008).

fMRI findings are relevant to these issues, because medial temporal lobe activity is sometimes found in association with short-delay recognition memory tasks (Ranganath and D'Esposito 2001; Piekema et al. 2006; Toepper et al. 2010). Yet, it is noteworthy that the extent of medial temporal lobe activity in short-delay tasks can be modulated by memory demands. For example, in some studies, the medial temporal lobe activity that occurred while maintaining information in memory was correlated with subsequent retention of the material being learned (Schon et al. 2004; Ranganath et al. 2005; Nichols et al. 2006). In addition, in a study that required maintaining faces in memory, the connectivity between the hippocampus and the fusiform face area increased with higher mnemonic load (one face vs. four faces) (Rissman et al. 2008). Concurrently, with higher load, the connectivity decreased between frontal regions traditionally linked to working memory (Goldman-Rakic 1995; Postle 2006) and the fusiform face area. These findings suggest that fMRI activity in the medial temporal lobe reflects processes related to the formation of long-term memory rather than processes related to working memory itself (for review, see Jeneson and Squire 2012).

## NONDECLARATIVE MEMORY

Nondeclarative memory (sometimes termed implicit memory) refers to a collection of abilities that are expressed through performance without requiring conscious memory content. Study of nondeclarative memory began with motor skills and perceptual skills, as described above, but soon included additional abilities as well. The next of these to come under study was the phenomenon of priming. Priming is evident as improved access to items that have been recently presented or improved access to associates of those items. This improvement is unconscious and is experienced as part of perception, as perceptual fluency, not as an expression of memory. A key finding was that priming effects were intact in memory-impaired patients. For example, patients can perform normally on tests that use word stems as cues for recently presented words (e.g., study BRICK, CRATE; test with BRI___, CRA___). Importantly, performance was intact in patients only when they were instructed to complete each cue to form the first word that comes to mind. With conventional memory instructions (use the cue to help recall a recently presented word), healthy volunteers outperformed the patients (Graf et al. 1984). Priming can occur for material that has no preexisting memory representation (e.g., nonsense letter strings) (Hamann and

Squire 1997a) and for material that is related by meaning to recently studied items. Thus, when asked to free associate to a word (e.g., strap), volunteers produced a related word (e.g., belt) more than twice as often when that word (belt) was presented recently (Levy et al. 2004). Importantly, severely amnesic patients showed fully intact word priming, even while performing at chance levels in parallel memory tests (Hamann and Squire 1997b; Levy et al. 2004). Thus, priming occurs but it does not benefit conscious memory decisions. Indeed, direct measurements showed that priming provides only a weak and unreliable cue for conscious judgments of familiarity (Conroy et al. 2005).

Priming is presumably advantageous because animals evolved in a world where things that are encountered once are likely to be encountered again. Priming improves the speed and efficiency with which organisms interact with a familiar environment and may influence feature-based attentional processes (Hutchinson and Turk-Browne 2012; Theeuwes 2013). Evoked-potential studies indicate that the electrophysiological signature of priming occurs early and well before the activity that signals conscious recognition of a past event (Paller et al. 2003). In neuroimaging studies, priming is often associated with reduced activity in regions of neocortex relevant to the task (Squire et al. 1992; Schacter et al. 2007). A similar finding (repetition suppression) (Desimone 1996) has been described in nonhuman primates (a stimulus-specific attenuation in firing rate with repeated presentation of a stimulus), and may underlie the phenomenon of priming (Wiggs and Martin 1998). Models have been proposed to explain how a net reduction in cortical activity could allow for faster perceptual processing (i.e., priming) (Grill-Spector et al. 2006). Some studies have found a correlation between behavioral measures of priming and reduced activity in the prefrontal cortex (Maccotta and Buckner 2004). This result has not been found in ventral temporal cortex for either humans or nonhuman primates (Maccotta and Buckner 2004; McMahon and Olson 2007).

Changes in cortex also underlie the related phenomenon of perceptual learning (Gilbert et al. 2009). Perceptual learning refers to gradual improvement in the detection or discrimination of visual stimuli with repeated practice. Changes in cortical circuitry during perceptual learning are detectable as early as primary visual cortex (V1) and may depend in part on structural changes in the long-range horizontal connections formed by V1 pyramidal cells (Gilbert and Li 2012). This circuitry is under the control of bottom-up processes as well as top-down influences related to attention and behavioral context (Gilbert and Li 2013).

Another early example of nondeclarative memory was simple classical conditioning, best illustrated in the literature of delay eyeblink conditioning. In delay conditioning, a neutral conditioned stimulus (CS), such as a tone, is presented just before an unconditioned stimulus (US), such as an airpuff to the eye. The two stimuli then overlap and coterminate. Critically, delay eyeblink conditioning is intact in amnesia and is acquired independently of awareness (Gabrieli et al. 1995; Clark and Squire 1998). Participants who did not become aware of the relationship between the CS and US (i.e., that the CS predicts the US) learned just as well as volunteers who did become aware (Manns et al. 2001). Indeed, when CS–US association strength was varied (by changing the number of consecutive CS alone or CS–US presentations), the probability of a conditioned response increased with association strength but was inversely related to how much the US was expected (Clark et al. 2001). Largely on the basis of work with rabbits, delay eyeblink conditioning proved to depend on the cerebellum and associated brain stem circuitry (Thompson and Krupa 1994; Thompson and Steinmetz 2009). Forebrain structures are not necessary for acquisition or retention of classically conditioned eyeblink responses.

Evaluative information, that is, whether a stimulus has positive or negative value, is acquired largely as nondeclarative memory. Biological study of this kind of memory has focused especially on the associative learning of fear (Davis 2006; Adolphs 2013; LeDoux 2014). Its nondeclarative status is illustrated by the fact that, in humans, associative fear learning proceeded normally after hippocampal lesions, even though the CS–US pairings could not be reported (Bechara et al. 1995). The amygdala has a critical role in fear learning, and its function (as well as its connectivity) appears to be conserved widely across species. In human neuroimaging studies, the amygdala was activated not only by fear but by strongly positive emotions as well (Hamann et al. 2002). Thus, the amygdala appears to be critical for associating sensory stimuli with stimulus valence. Ordinarily, animals express fear learning by freezing behavior (immobility). However, in a task where learned fear must instead be expressed by executing an avoidance response (an escape), freezing is maladaptive. In this case, prefrontal cortex inhibits defense behaviors (such as freezing) that are mediated by the amygdala, thereby allowing the animal to escape (Moscarello and LeDoux 2013). Inhibitory action of the prefrontal cortex on the amygdala (from infralimbic prefrontal cortex in rat or from ventromedial prefrontal cortex in humans) has also been found to occur during the reversal of fear learning (i.e., extinction) (Milad and Quirk 2012). This work has relevance for clinical disorders, such as phobias and posttraumatic stress disorder (Davis 2011).

In addition to these functions, it is important to note that the amygdala also exerts a modulatory influence on both declarative and nondeclarative memory. This role of the amygdala is the basis for the fact that emotionally arousing events are typically remembered better than emotionally neutral events. The mechanism for this effect is understood and depends on the release of stress hormones from the adrenal gland, which affects the forebrain via the vagus nerve, the nucleus of the solitary tract, and the locus coeruleus. Ultimately, the effect is mediated by the amygdala through its basolateral nucleus (McGaugh and Roozendaal 2009).

The gradual trial-and-error learning that leads to the formation of habits was proposed in the 1980s to be supported by the striatum (Mishkin et al. 1984), and habit memory subsequently became an important focus of study (Yin and Knowlton 2006; Graybiel 2008; Liljeholm and O'Doherty 2012). Habit memory is characterized by automatized, repetitive behavior and, unlike declarative memory, is insensitive to changes in reward value (Dickinson 1985). An early demonstration of the distinction between declarative memory and habit memory came from rats with fornix lesions or caudate lesions tested on two ostensibly similar tasks. Rats with fornix lesions, which disrupt hippocampal function, failed when they needed to acquire a flexible behavior but succeeded when they needed to respond repetitively. Rats with caudate lesions showed the opposite pattern (Packard et al. 1989). A similar contrast between declarative memory and habit memory was shown for memory-impaired patients with hippocampal lesions and patients with nigrostriatal damage caused by Parkinson's disease (Knowlton et al. 1996). In the task, probabilistic classification, participants gradually learned which of two outcomes (sun or rain) would occur on each trial, given the particular combination of four cues that appeared. One, two, or three cues could appear on any trial, and each cue was independently and probabilistically related to the outcome. Patients with hippocampal lesions learned the task at a normal rate but could not report facts about the task. Parkinson patients remembered the facts but could not learn the task.

Tasks that can be learned quickly by memorization can also be learned by a trial-and-error, habit-based strategy, albeit much more slowly. In one study, healthy volunteers were able to learn eight separate pairs of "junk objects" within a single session of 40 trials (i.e., choose the correct object in each pair). Two severely amnesic patients with large medial temporal lobe lesions also learned but only gradually, requiring more than 25 test sessions and 1000 trials (Bayley et al. 2005). Unlike declarative memory, which is flexible and can guide behavior in different contexts, the acquired knowledge in this case was rigidly organized. Patient performance collapsed when the task format was altered by asking participants to sort the 16 objects into correct and incorrect groups (a trivial task for controls). In addition, although by the end of training the patients were consistently performing at a high level, at the start of each test day

they were never able to describe the task, the instructions, or the objects. Indeed, during testing they expressed surprise that they were performing so well. These findings provide particularly strong evidence for the distinction between declarative (conscious) and nondeclarative (unconscious) memory systems.

Reward-based learning of this kind depends on dopamine neurons in the midbrain (substantia nigra and ventral tegmental area), which project to the striatum and signal the information value of the reward (Schultz 2013). The dorsolateral striatum is crucial for the development of habits in coordination with other brain regions. Neurophysiological studies in mice during skill learning documented that the dorsolateral striatum was increasingly engaged as performance became more automatic and habit-like (Yin et al. 2009). In contrast, the dorsomedial striatum was engaged only early in training. Similarly, in rats learning a conditioned T-maze task, activity gradually increased in the dorsolateral striatum as training progressed, and this activity correlated with performance (Thorn et al. 2010). In the dorsomedial striatum, activity first increased but then decreased as training progressed. Increased activity in the dorsolateral striatum during the later stages of habit formation occurred together with late-developing activity in infralimbic cortex (Smith and Graybiel 2013). Moreover, disruption of infralimbic cortex during late training prevented habit formation. Thus, these two regions (dorsolateral striatum and infralimbic cortex) appear to work together to support a fully formed habit.

## THE RELATIONSHIP BETWEEN MEMORY SYSTEMS

The memory systems of the mammalian brain operate independently and in parallel to support behavior, and how one system or another gains control is a topic of considerable interest (Poldrack and Packard 2003; McDonald and Hong 2013; Packard and Goodman 2013). In some circumstances, memory systems are described as working cooperatively to optimize behavior and in other circumstances are described as working competitively. However, it is not easy to pin down what should count for or against cooperativity, competition, or independence in any particular case. For example, the fact that the manipulation of one memory system can affect the operation of another has been taken as evidence for competition between systems (Schwabe 2013). Yet, even for systems that are strictly independent, the loss of one system would be expected to affect the operation of another system by affording it more opportunity to control behavior.

Much of the experimental work on the relationship between memory systems has focused on hippocampus-dependent declarative memory and dorsolateral striatum-dependent habit memory. In an illustrative study (Packard and McGaugh 1996), rats were trained in a four-arm, plus-shaped maze to go left, always beginning in the south arm (and with the north arm blocked). In this situation, rats could learn either a place (the left arm) or a response (turn left). To discriminate between these two possibilities, rats were occasionally started in the north arm (with the south arm now blocked). When these north-arm starts were given early in training, rats tended to enter the same arm that had been rewarded. They had learned a place. However, with extended training, rats tended to repeat the learned left-turn response and enter the unrewarded arm. Place learning was abolished early in training by lidocaine infusions into the hippocampus. In this case, rats showed no preference for either arm. Correspondingly, later in training, response learning was abolished by lidocaine infusions into the caudate nucleus. In this case, however, rats showed place responding. In other words, even though behavior later in training was guided by caudate-dependent response learning, information remained available about place. When the caudate nucleus was inactive, the parallel memory system supported by the hippocampus was unmasked.

A similar circumstance has been described in humans performing a virtual navigation task that could be solved by either a spatial or nonspatial (habit-like) strategy (Iaria et al. 2003). At the outset, spatial and nonspatial strategies were adopted equally often, but as training prog-

ressed participants tended to shift to a nonspatial strategy. Participants who used the spatial strategy (navigating in relation to landmarks) showed increased activity in the right hippocampus early in training. Participants using the nonspatial strategy (counting maze arms) showed increased activity in the caudate nucleus, which emerged as training progressed.

Several factors increase the tendency to adopt a striatal strategy, including stress (Kim et al. 2001; Schwabe 2013), psychopathology (Wilkins et al. 2013), aging (Konishi et al. 2013), and a history of alcohol and drug use (Bohbot et al. 2013). Prefrontal cortex may also be important in determining which memory system gains control over behavior (McDonald and Hong 2013).

Although many tasks can be acquired by more than one memory system, other tasks strongly favor one system over another. In this circumstance, engaging the less optimal system can interfere with performance. Thus, fornix lesions in rats facilitated acquisition of a caudate-dependent maze habit that required repeated visits to designated arms (Packard et al. 1989). The fornix lesion presumably disrupted the tendency to use a nonoptimal declarative memory strategy. Similarly, a familiar feature of skill learning in humans is that trying to memorize, and use declarative memory, can disrupt performance.

Neuroimaging studies show that feedback-guided learning typically engages the striatum. A task described earlier, probabilistic classification, requires participants to make a guess on each trial based on cues that are only partially reliable. Participants typically begin by trying to memorize the task structure but then turn to the habit-like strategy of accumulating response strength in association with the cues. Correspondingly, fMRI revealed activity in the medial temporal lobe early during learning (Poldrack and Gabrieli 2001). As learning progressed, activity decreased in the medial temporal lobe, and activity increased in the striatum. Moreover, when the task was modified so as to encourage the use of declarative memory, less activity was observed in the striatum and more activity was observed in the medial temporal lobe.

## CONCLUSION

The memory system framework is fundamental to the contemporary study of learning and memory. Within this framework, the various memory systems have distinct purposes and distinct anatomy, and different species can solve the same task using different systems. Interestingly, efforts have been made to account for some findings (e.g., priming or classification learning) with models based on a single system (Zaki et al. 2003; Zaki 2004; Berry et al. 2012). Yet, these accounts have difficulty explaining double dissociations (e.g., Packard et al. 1989; Knowlton et al. 1996), chance performance on tests of declarative memory when priming is intact (Hamann and Squire 1997b), and successful habit learning in the face of expressed ignorance about the task (Bayley et al. 2005).

One implication of these facts is that the therapeutic targets for various kinds of memory disorders are quite different. For example, for extreme fear-based memories like phobias, one must target the amygdala, for strong habit-based memories like obsessive–compulsive disorders, one must target the striatum, and for severe forgetfulness, as in Alzheimer's disease, one must target the hippocampus and adjacent structures. The notion of multiple memory systems is now widely accepted and establishes an important organizing principle across species for investigations of the biology of memory.

## REFERENCES

Adolphs R. 2013. The biology of fear. *Curr Biol* **23:** R79–R93.

Annese J, Schenker-Ahmed NM, Bartsch H, Maechler P, Sheh C, Thomas N, Kayano J, Ghatan A, Bresler N, Frosch MP, et al. 2014. Postmortem examination of patient H.M.'s brain based on histological sectioning and digital 3D reconstruction. *Nat Commun* **5:** 3122.

Atkinson RC, Shiffrin RM. 1968. Human memory: A proposed system and its control processes. In *Psychology of learning and motivation: Advances in research and theory* (ed. Spence KW, Spence JT), pp. 89–195. Academic, New York.

Augustinack JC, van der Kouwe AJW, Salat DH, Benner T, Stevens AA, Annese J, Fischl B, Frosch MP, Corkin S. 2014. H.M.'s contributions to neuroscience: A review and autopsy studies. *Hippocampus* **24:** 1267–1286.

Baddeley A. 2003. Double dissociations: Not magic, but still useful. *Cortex* **39:** 129–131.

Baddeley AD, Warrington EK. 1970. Amnesia and the distinction between long- and short-term memory. *J Verb Learn Verb Behav* **9:** 176–189.

Baddeley A, Allen R, Vargha-Khadem F. 2010. Is the hippocampus necessary for visual and verbal binding in working memory? *Neuropsychologia* **48:** 1089–1095.

Baddeley A, Jarrold C, Vargha-Khadem F. 2011. Working memory and the hippocampus. *J Cog Neurosci* **23:** 3855–3861.

Bayley PJ, Frascino JC, Squire LR. 2005. Robust habit learning in the absence of awareness and independent of the medial temporal lobe. *Nature* **436:** 550–553.

Bechara A, Tranel D, Damasio H, Adolphs R, Rockland C, Damasio AR. 1995. Double dissociation of conditioning and declarative knowledge relative to the amygdala and hippocampus in humans. *Science* **269:** 1115–1118.

Bergson HL. 1910. *Matter and memory* (trans. Paul NM, Palmer WS). George Allen, London.

Berry CJ, Shanks DR, Speekenbrink M, Henson RNA. 2012. Models of recognition, repetition priming, and fluency: Exploring a new framework. *Psych Rev* **119:** 40–79.

Bohbot VD, Del Balso D, Conrad K, Konishi K, Leyton M. 2013. Caudate nucleus-dependent navigational strategies are associated with increased use of addictive drugs. *Hippocampus* **23:** 973–984.

Bruner JS. 1969. Modalities of memory. In *The pathology of memory* (ed. Talland GA, et al.). pp. 253–259. Academic, New York.

Bunsey M, Eichenbaum H. 1996. Conservation of hippocampal memory function in rats and humans. *Nature* **379:** 255–257.

Chun MM, Phelps EA. 1999. Memory deficits for implicit contextual information in amnesic subjects with hippocampal damage. *Nat Neurosci* **2:** 844–847.

Clark RE, Squire LR. 1998. Classical conditioning and brain systems: A key role for awareness. *Science* **280:** 77–81.

Clark RE, Manns JR, Squire LR. 2001. Trace and delay eyeblink conditioning: Contrasting phenomena of declarative and nondeclarative memory. *Psychol Sci* **12:** 304–308.

Clayton RS, Dickinson A. 1998. Episodic-like memory during cache recovery by scrub jays. *Nature* **395:** 272–274.

Cohen NJ, Squire LR. 1980. Preserved learning and retention of pattern analyzing skill in amnesia: Dissociation of knowing how and knowing that. *Science* **210:** 207–209.

Conroy MA, Hopkins RO, Squire LR. 2005. On the contribution of perceptual fluency and priming to recognition memory. *Cogn Affect Behav Neurosci* **5:** 14–20.

Cowan N. 2001. The magical number 4 in short-term memory: A reconsideration of mental storage capacity. *Behav Brain Sci* **24:** 87–185.

Davachi L. 2006. Item, context and relational episodic encoding in humans. *Curr Opin Neurobiol* **16:** 693–700.

Davis M. 2006. Neural systems involved in fear and anxiety measured with fear-potentiated startle. *Am Psychol* **61:** 741–756.

Davis M. 2011. NMDA receptors and fear extinction: Implications for cognitive behavioral therapy. *Dialogues Clin Neurosci* **13:** 463–474.

Desimone R. 1996. Neural mechanisms for visual memory and their role in attention. *Proc Natl Acad Sci* **93**: 13494–13499.

Dickinson A. 1985. Actions and habits: the development of behavioural autonomy. *Phil Trans R Soc Lond B Biol Sci* **308**: 67–78.

Drachman DA, Arbit J. 1966. Memory and the hippocampal complex. II. Is memory a multiple process? *Arch Neurol* **15**: 52–61.

Fortin NJ, Agster KL, Eichenbaum HB. 2002. Critical role of the hippocampus in memory for sequences of events. *Nat Neurosci* **5**: 458–462.

Fukuda K, Awh E, Vogel EK. 2010. Discrete capacity limits in visual working memory. *Curr Opin Neurobiol* **20**: 177–182.

Gabrieli JDE, McGlinchey-Berroth R, Carrillo MC, Gluck MA, Cermak LS, Disterhoft JF. 1995. Intact delay-eyeblink classical conditioning in amnesia. *Behav Neurosci* **109**: 819–827.

Gaffan D. 1974. Recognition impaired and association intact in the memory of monkeys after transaction of the fornix. *J Comp Physiol Psychol* **86**: 1100–1109.

Gilbert CD, Li W. 2012. Adult visual cortical plasticity. *Neuron* **75**: 250–264.

Gilbert CD, Li W. 2013. Top-down influences on visual processing. *Nat Rev Neurosci* **14**: 350–363.

Gilbert CD, Li W, Piech V. 2009. Perceptual learning and adult cortical plasticity. *J Physiol* **587**: 2743–2751.

Goldman-Rakic PS. 1995. Architecture of the prefrontal cortex and the central executive. *Ann NY Acad Sci* **769**: 71–83.

Graf P, Squire LR, Mandler G. 1984. The information that amnesic patients do not forget. *J Exp Psychol Learn Mem Cog* **10**: 164–178.

Graham KS, Barense MD, Lee AC. 2010. Going beyond LTM in the MTL: A synthesis of neuropsychological and neuroimaging findings on the role of the medial temporal lobe in memory and perception. *Neuropsychologia* **48**: 831–853.

Graybiel AM. 2008. Habits, rituals, and the evaluative brain. *Ann Rev Neurosci* **31**: 359–387.

Grill-Spector K, Henson R, Martin A. 2006. Repetition and the brain: Neural models of stimulus-specific effects. *Trends Cognit Sci* **10**: 14–23.

Hamann SB, Squire LR. 1997a. Intact priming for novel perceptual representations in amnesia. *J Cog Neurosci* **9**: 699–713.

Hamann SB, Squire LR. 1997b. Intact perceptual memory in the absence of conscious memory. *Behav Neurosci* **111**: 850–854.

Hamann SB, Ely TD, Hoffman JM, Kilts CD. 2002. Ecstasy and agony: Activation of the human amygdala in positive and negative emotion. *Psychol Sci* **13**: 135–141.

Hannula DE, Greene AJ. 2012. The hippocampus reevaluated in unconscious learning and memory: At a tipping point? *Front Human Neurosci* **6**: 80.

Hannula DE, Ranganath C. 2009. The eyes have it: Hippocampal activity predicts expression of memory in eye movements. *Neuron* **63**: 592–599.

Hannula DE, Tranel D, Cohen NJ. 2006. The long and the short of it: Relational memory impairments in amnesia, even at short lags. *J Neurosci* **26**: 8352–8359.

Harding A, Halliday G, Caine D, Krill J. 2000. Degeneraton of anterior thalamic nuclei differentiates alcoholics with amnesia. *Brain* **123**: 141–154.

Henke K. 2010. A model for memory systems based on processing modes rather than consciousness. *Nat Rev Neurosci* **11**: 523–532.

Higuchi S, Miyashita Y. 1996. Formation of mnemonic neuronal responses to visual paired associates in inferotemporal cortex is impaired by perirhinal and entorhinal lesions. *Proc Natl Acad Sci* **93**: 739–743.

Hirsh R. 1974. The hippocampus and contextual retrieval of information from memory: A theory. *Behav Biol* **12**: 421–444.

Hutchinson JB, Turk-Browne NB. 2012. Memory-guided attention: control from multiple memory systems. *Trends Cogn Sci* **16**: 576–579.

Iaria G, Petrides M, Dagher A, Pike B, Bohbot VD. 2003. Cognitive strategies dependent on the hippocampus and caudate nucleus in human navigation: Variability and change with practice. *J Neurosci* **13**: 5945–5952.

James W. 1890. *Principles of psychology.* Holt, New York.

Jeneson A, Squire LR. 2012. Working memory, long-term memory, and medial temporal lobe function. *Learn Mem* **19**: 15–25.

Jeneson A, Mauldin KN, Squire LR. 2010. Intact working memory for relational information after medial temporal lobe damage. *J Neurosci* **30**: 13624–13629.

Jeneson A, Mauldin KN, Hopkins RO. 2011. The role of the hippocampus in retaining relational information across short delays: The importance of memory load. *Learn Mem* **18**: 301–305.

Kandel ER, Dudai Y, Mayford MR. 2014. The molecular and systems biology of memory. *Cell* **157**: 163–186.

Kim JJ, Lee HJ, Han JS, Packard MG. 2001. Amygdala is critical for stress-induced modulation of hippocampal long-term potentiation and learning. *J Neurosci* **21**: 5222–5228.

Kim S, Sapiurka M, Clark R, Squire LR. 2013. Contrasting effects on path integration after hippocampal damage in humans and rats. *Proc Natl Acad Sci* **110**: 4732–4737.

Knowlton BJ, Mangels JA, Squire LR. 1996. A neostriatal habit learning system in humans. *Science* **273**: 1399–1402.

Konishi K, Etchamendy N, Roy S, Marighetto A, Rajah N, Bohbot VD. 2013. Decreased functional magnetic resonance imaging activity in the hippocampus in favor of the caudate nucleus in older adults tested in a virtual navigation task. *Hippocampus* **23**: 1005–1014.

Kumaran D, Wagner AD. 2009. It's in my eyes, but it doesn't look that way to me. *Neuron* **63**: 561–563.

LeDoux J. 2014. Coming to terms with fear. *Proc Natl Acad Sci* **111**: 2871–2878.

Levy DA, Stark CEL, Squire LR. 2004. Intact conceptual priming in the absence of declarative memory *Psychol Sci* **15**: 680–685.

Liljeholm M, O'Doherty JP. 2012. Contributions of the striatum to learning, motivation, and performance: An associative account. *Trends Cogn Sci* **16**: 467–475.

Cite this article as *Cold Spring Harb Perspect Biol* doi: 10.1101/cshperspect.a021667

Maccotta L, Buckner RL. 2004. Evidence for neural effects of repetition that directly correlate with behavioral priming. *J Cog Neurosci* **16:** 1625–1632.

Manns JR, Squire LR. 2001. Perceptual learning, awareness, and the hippocampus. *Hippocampus* **11:** 776–782.

Manns JM, Clark RE, Squire LR. 2001. Single-cue delay eyeblink classical conditioning is unrelated to awareness. *Cogn Affect Behav Neurosci* **1:** 192–198.

Markowitsch H. 1988. Diencephalic amnesia: A reorientation towards tracts? *Brain Res Rev* **13:** 351–370.

McDonald RJ, Hong NS. 2013. How does a specific learning and memory system in the mammalian brain gain control of behavior? *Hippocampus,* **23:** 1084–1102.

McDougall W. 1923. *Outline of psychology.* Scribners, New York.

McGaugh JL, Roozendaal B. 2009. Drug enhancement of memory consolidation: Historical perspective and neurobiological implications. *Psychopharmacology* **1:** 3–14.

McMahon DB, Olson CR. 2007. Repetition suppression in monkey inferotemporal cortex: Relation to behavioral priming. *J Neurophysiol* **97:** 3532–3543.

Milad MR, Quirk GJ. 2012. Fear extinction as a model for translational neuroscience: Ten years of progress. *Ann Rev Psychol* **63:** 129–151.

Milner B. 1962. Les troubles de la memoire accompagnant des lésions hippocampiques bilaterales. In *Physiologie de l'hippocampe,* pp. 257–272. Centre National de la Recherche Scientifique, Paris.

Milner B. 1972. Disorders of learning and memory after temporal lobe lesions in man. *Clin Neurosurg* **19:** 421–466.

Mishkin M. 1978. Memory in monkeys severely impaired by combined but not by separate removal of amygdala and hippocampus. *Nature* **273:** 297–298.

Mishkin M, Malamut B, Bachevalier J. 1984. Memories and habits: Two neural systems. In *Neurobiology of learning and memory* (ed. Lynch G, et al.), pp. 65–77. Guilford, New York.

Moscarello JM, LeDoux JE. 2013. Active avoidance learning requires prefrontal suppression of amygdala-mediated defensive reactions. *J Neurosci* **33:** 3815–3823.

Nichols EA, Kao YC, Verfaellie M, Gabrieli JD. 2006. Working memory and long-term memory for faces: Evidence from fMRI and global amnesia for involvement of the medial temporal lobes. *Hippocampus* **16:** 604–616.

O'Keefe J, Nadel L. 1978. *The hippocampus as a cognitive map.* Oxford University Press, London.

Olson IR, Page K, Moore KS, Chatterjee A, Verfaellie M. 2006. Working memory for conjunctions relies on the medial temporal lobe. *J Neurosci* **26:** 4596–4601.

Packard MG, Goodman J. 2013. Factors that influence the relative use of multiple memory systems. *Hippocampus* **23:** 1044–1052.

Packard MG, McGaugh JL. 1996. Inactivation of hippocampus or caudate nucleus with lidocaine differentially affects expression of place and response learning. *Neurobiol Learn Mem* **65:** 65–72.

Packard MG, Hirsh R, White NM. 1989. Differential effects of fornix and caudate nucleus lesions on two radial maze tasks: Evidence for multiple memory systems. *J Neurosci* **9:** 1465–1472.

Paller KA, Hutson CA, Miller BB, Boehm SG. 2003. Neural manifestations of memory with and without awareness. *Neuron* **38:** 507–516.

Piekema C, Kessels RP, Mars RB, et al. 2006. The right hippocampus participates in short-term memory maintenance of object–location associations. *NeuroImage* **33:** 374–382.

Poldrack RA, Gabrieli JD. 2001. Characterizing the neural mechanisms of skill learning and repetition priming: Evidence from mirror reading. *Brain* **124:** 67–82.

Poldrack RA, Packard MG. 2003. Competition among multiple memory systems: Converging evidence from animal and human brain studies. *Neuropsychologia* **41:** 245–251.

Poldrack RA, Rodriguez P. 2003. Sequence learning: What's the hippocampus to do? *Neuron* **37:** 891–893.

Poldrack RA, Clark J, Pare-Blagoev EJ, Shohamy D, Creso Moyano J, Myers C, Gluck MA. 2001. Interactive memory systems in the human brain. *Nature* **414:** 546–550.

Postle BR. 2006. Working memory as an emergent property of the mind and brain. *Neurosci* **139:** 23–38.

Ranganath C, Blumenfeld RS. 2005. Doubts about double dissociations between short- and long-term memory. *Trends Cogn Sci* **9:** 374–380.

Ranganath C, D'Esposito M. 2001. Medial temporal lobe activity associated with active maintenance of novel information. *Neuron* **31:** 865–873.

Ranganath C, Cohen MX, Brozinsky CJ. 2005. Working memory maintenance contributes to long-term memory formation: Neural and behavioral evidence. *J Cogn Neurosci* **17:** 994–1010.

Reber PJ, Squire LR. 1994. Parallel brain systems for learning with and without awareness. *Learn Mem* **1:** 217–229.

Rissman J, Gazzaley A, D'Esposito M. 2008. Dynamic adjustments in prefrontal, hippocampal, and inferior temporal interactions with increasing visual working memory load. *Cerebr Cortex* **18:** 1618–1629.

Rose M, Haider H, Weiller C, Buchel C. 2002. The role of medial temporal lobe structures in implicit learning: An event-related fMRI study. *Neuron* **36:** 1221–1231.

Ryle G. 1949. *The concept of mind.* Hutchinson, San Francisco.

Schacter DL, Buckner RL. 1998. Priming and the brain. *Neuron* **20:** 185–195.

Schacter DL, Wig GS, Stevens WD. 2007. Reductions in cortical activity during priming. *Curr Opin Neurobiol* **17:** 171–176.

Schendan HE, Searl MM, Melrose RJ, Stern CE. 2003. An fMRI study of the role of the medial temporal lobe in implicit and explicit sequence learning. *Neuron* **37:** 1013–1025.

Schon D, Lorber B, Spacal M, Semenza C. 2004. A selective deficit in the production of exact musical intervals following right-hemisphere damage. *Cogn Neuropsychol* **21:** 773–784.

Schultz W. 2013. Updating dopamine reward signals. *Curr Opin Neurobiol* **23:** 229–238.

Schwabe L. 2013. Stress and the engagement of multiple memory systems: Integration of animal and human studies. *Hippocampus* **23:** 1035–1043.

Scoville WB, Milner B. 1957. Loss of recent memory after bilateral hippocampal lesions. *J Neurol Neurosurg Psychiatry* **20:** 11–21.

Sherry DF, Schacter DL. 1987. The evolution of multiple memory systems. *Psychol Rev* **94:** 439–454.

Shimamura AP, Janowsky JS, Squire LR. 1991. What is the role of frontal lobe damage in memory disorders? In *Frontal lobe functioning and dysfunction* (ed. Levin HD, et al.), pp. 173–195. Oxford University Press, New York.

Shohamy D, Turk-Browne NB. 2013. Mechanisms for widespread hippocampal involvement in cognition. *J Exp Psychol Gen* **142:** 1159–1170.

Shrager Y, Levy DA, Hopkins RO, Squire LR. 2008. Working memory and the organization of brain systems. *J Neurosci* **28:** 4818–4822.

Smith KS, Graybiel AM. 2013. A dual operator view of habitual behavior reflecting cortical and striatal dynamics. *Neuron* **79:** 361–374.

Smith C, Squire LR. 2005. Declarative memory, awareness, and transitive inference. *J Neurosci* **25:** 10138–10146.

Smith CN, Squire LR. 2008. Experience-dependent eye movements reflect hippocampus-dependent (aware) memory. *J Neurosci* **28:** 12825–12833.

Squire LR. 2009. Memory and brain systems: 1969–2009. *J Neurosci* **29:** 12711–12716.

Squire LR, Wixted JT. 2011. The cognitive neuroscience of human memory since H.M. *Ann Rev Neurosci* **34:** 259–288.

Squire LR, Zola-Morgan S. 1988. Memory: Brain systems and behavior. *Trends Neurosci* **11:** 170–175.

Squire LR, Zola-Morgan S. 1991. The medial temporal lobe memory system. *Science* **253:** 1380–1386.

Squire LR, Ojemann JG, Miezin FM, Petersen SE, Videen TO, Raichle ME. 1992. Activation of the hippocampus in normal humans: A functional anatomical study of memory. *Proc Natl Acad Sci* **89:** 1837–1841.

Squire LR, Stark CEL, Clark RE. 2004. The medial temporal lobe. *Ann Rev Neurosci* **27:** 279–306.

Staresina BP, Duncan KD, Davachi L. 2011. Perirhinal and parahippocampal cortices differentially contribute to later recollection of object- and scene-related event details. *J Neurosci* **31:** 8739–8747.

Theeuwes J. 2013. Feature-based attention: It is all bottom-up priming. *Phil Trans R Soc B Biol Sci* **368:** 20130055.

Thompson RF, Krupa DJ. 1994. Organization of memory traces in the mammalian brain. *Ann Rev Neurosci* **17:** 519–550.

Thompson RF, Steinmetz JE. 2009. The role of the cerebellum in classical conditioning of discrete behavioral responses. *Neurosci* **162:** 732–755.

Thorn CA, Atallah H, Howe H, Graybiel AM. 2010. Differential dynamics of activity changes in dorsolateral and dorsomedial striatal loops during learning. *Neuron* **66:** 781–795.

Toepper M, Markowitsch HJ, Gebhardt H, Beblo T, Thomas C, Gallhofer B, Driessen M, Sammer G. 2010. Hippocampal involvement in working memory encoding of changing locations: An fMRI study. *Brain Res* **1354:** 91–99.

Tolman RC. 1948. Cognitive maps in rats and man. *Psychol Rev* **55:** 189–208.

Tulving E. 1983. *Elements of episodic memory.* Oxford University Press, Cambridge, MA.

Tulving E. 1989. Remembering and knowing the past. *American Scientist* **77:** 361–367.

Tulving E. 2005. Episodic memory and autonoesis: Uniquely human? In *The missing link in cognition: Evolution of self-knowing consciousness* (ed. Terrace H, et al.), pp. 3–56. Oxford University Press, New York.

Tulving E, Schacter DL. 1990. Priming and human memory systems. *Science* **247:** 301–306.

Victor M, Adams RD, Collins GH. 1989. *The Wernicke–Korsakoff syndrome and related neurological disorders due to alcoholism and malnutrition.* Davis, Philadelphia.

Warren DE, Duff MC, Tranel D, Cohen NJ. 2011. Observing degradation of visual representations over short intervals when medial temporal lobe is damaged. *J Cogn Neurosci* **23:** 3862–3873.

Warren DE, Duff MC, Jensen U, Tranel D, Cohen NJ. 2012. Hiding in plain view: Lesions of the medial temporal lobe impair online representation. *Hippocampus* **22:** 1577–1588.

Westerberg CE, Miller BB, Reber PJ, Cohen NJ, Paller KA. 2011. Neural correlates of contextual cueing are modulated by explicit learning. *Neuropsychologia* **49:** 3439–3447.

Whitlock JR, Sutherland RJ, Witter MP, Moser MB, Moser EI. 2008. Navigating from hippocampus to parietal cortex. *Proc Natl Acad Sci* **105:** 14755–14762.

Wiggs CL, Martin A. 1998. Properties and mechanisms of perceptual priming. *Curr Opin Neurobiol* **8:** 227–233.

Wilkins LK, Girard TA, Konishi K, King M, Herdman KA, King J, Christensen B, Bohbot VD. 2013. Selective deficit in spatial memory strategies contrast to intact response strategies in patients with schizophrenia spectrum disorders tested in a virtual navigation task. *Hippocampus* **23:** 1015–1024.

Winograd T. 1975. Frame representations and the declarative-procedural controversy. In *Representation and understanding: Studies in cognitive science* (ed. Bobrow D, et al.), pp. 185–210. Academic, New York.

Yee LT, Hannula DE, Tranel D, Cohen NJ. 2014. Short-term retention of relational memory in amnesia revisited: Accurate performance depends on hippocampal integrity. *Front Hum Neurosci* **8:** 16.

Yin HH, Knowlton BJ. 2006. The role of the basal ganglia in habit formation. *Nat Rev Neurosci* **7:** 464–476.

Yin HH, Mulcare SP, Hilário MRF, Clouse E, Holloway T, Davis MI, Hansson AC, Lovinger DM, Costa RM, et al. 2009. Dynamic reorganization of striatal circuits during the acquisition and consolidation of a skill. *Nat Neurosci* **12:** 333–341.

Zaki SR. 2004. Is categorization performance really intact in amnesia? A meta-analysis. *Pysch Bull Rev* **11:** 1048–1054.

Zaki SR, Nosofsky RM, Jessup NM, Unversagt FR. 2003. Categorization and recognition performance of a memory-impaired group: Evidence for single system models. *J Int Neuropsychol Soc* **9:** 94–406.

Cite this article as *Cold Spring Harb Perspect Biol* doi: 10.1101/cshperspect.a021667

# Nonassociative Learning in Invertebrates

John H. Byrne[1] and Robert D. Hawkins[2,3]

[1]Department of Neurobiology and Anatomy, The University of Texas Medical School at Houston, Houston, Texas 77030

[2]Department of Neuroscience, Columbia University, New York, New York 10032

[3]New York State Psychiatric Institute, New York, New York 10032

Correspondence: john.h.byrne@uth.tmc.edu

The simplicity and tractability of the neural circuits mediating behaviors in invertebrates have facilitated the cellular/molecular dissection of neural mechanisms underlying learning. The review has a particular focus on the general principles that have emerged from analyses of an example of nonassociative learning, sensitization in the marine mollusk *Aplysia*. Learning and memory rely on multiple mechanisms of plasticity at multiple sites of the neuronal circuits, with the relative contribution to memory of the different sites varying as a function of the extent of training and time after training. The same intracellular signaling cascades that induce short-term modifications in synaptic transmission can also be used to induce long-term changes. Although short-term memory relies on covalent modifications of preexisting proteins, long-term memory also requires regulated gene transcription and translation. Maintenance of long-term cellular memory involves both intracellular and extracellular feedback loops, which sustain the regulation of gene expression and the modification of targeted molecules.

Learning can be divided into two general categories: associative and nonassociative. Associative learning includes classical conditioning and operant conditioning, which are discussed by Hawkins and Byrne (2015). Nonassociative forms of learning include habituation and sensitization. Habituation, the simplest form of learning, is defined as the gradual waning of a behavioral response to a weak or moderate stimulus that is presented repeatedly. Following habituation, the response may be restored to its initial state either passively with time (i.e., spontaneous recovery), or with the presentation of a noxious stimulus. This latter phenomenon is called dishabituation, and its presence distinguishes habituation from simple fatigue (Thompson and Spencer 1966). Sensitization is defined as the enhancement of a behavioral response by strong or repeated stimulation. In one form of sensitization (also referred to as pseudo-conditioning), the behavioral response to a nonhabituated stimulus is enhanced by presentation of a noxious stimulus to another site, similar to dishabituation. In another form of sensitization, a behavioral response is enhanced by the repeated presentation of a moderate to strong intensity stimulus to the same site. This is operationally the opposite phenomenon of habituation, and is referred to as site-specific sensitization.

Invertebrates offer experimental advantages for analyzing the cellular and molecular mechanisms of learning. For example, many behaviors in invertebrates are mediated by relatively simple neural circuits, which can be analyzed with electrophysiological and optophysiological approaches. Once the circuit is specified, the neural locus for the particular example of learning can be found, and biophysical, biochemical, and molecular approaches can be used to identify mechanisms underlying the change. The relatively large size of some invertebrate neurons allows these analyses to take place at the level of individually identified neurons. In some cases, individual neurons can be surgically removed and assayed for changes in the levels of second messengers, protein phosphorylation, and RNA and protein synthesis. Moreover, peptides and nucleotides can be injected intracellularly or expressed in individual neurons via appropriate vectors.

This review will focus primarily on progress in understanding nonassociative learning in the marine mollusk *Aplysia*, but many other invertebrates have proven to be valuable model systems for the cellular and molecular analysis of learning and memory (for reviews, see Byrne 1987; Hawkins et al. 1987). Several of these model systems are described within. Each invertebrate model system has its own unique advantages. For example, *Aplysia* is excellent for applying cell biological approaches to the analysis of learning and memory mechanisms. Other invertebrate model systems, such as *Drosophila* and *Caenorhabditis elegans* offer tremendous advantages for obtaining insights into mechanisms of learning and memory through the application of molecular genetic approaches.

## NEURAL AND MOLECULAR MECHANISMS OF NONASSOCIATIVE LEARNING IN *Aplysia*

The mechanisms of several simple forms of learning have been studied extensively in *Aplysia*, which has a number of advantages for a reductionist approach (for additional references, see Hawkins et al. 2006). The nervous system of *Aplysia* consists of ~10,000 neurons, many of which are uniquely identifiable across individual animals. Studies of learning have focused primarily on defensive withdrawal reflexes, which have simple circuits consisting of only tens or perhaps hundreds of neurons. Many of those studies have examined the gill- and siphon-withdrawal reflex, in which a light touch to the siphon (an exhalant funnel for the gill) produces contraction of the gill and siphon, whereas others have examined the tail-withdrawal reflex or the tail-elicited siphon-withdrawal reflex. However, the results of all of these studies have generally been similar.

Despite their simplicity, the withdrawal reflexes undergo a variety of different forms of learning including habituation, dishabituation and sensitization. This review focuses on the mechanisms of sensitization. The memory for sensitization has multiple temporal domains that depend to a large extent on the training protocol. Typically, a single noxious stimulus, such as a shock produces short-term sensitization (STS) lasting minutes, whereas repeated shocks can produce long-term sensitization (LTS) lasting days. In addition, an intermediate-term stage has been identified that persists for hours. As we see in the literature, the different temporal phases of memory, the training contingencies that produce them, and even some of the underlying molecular mechanisms are conserved across species including humans.

## SHORT- AND INTERMEDIATE-TERM SENSITIZATION

### Mechanisms of Short-Term Sensitization

During STS, the withdrawal reflex elicited by a weak stimulus to one region of the animal's body is enhanced by a brief electric shock to another region (Carew et al. 1971; Walters et al. 1983b; Antonov et al. 1999; Philips et al. 2011). The neural circuits for many of the withdrawal reflexes consist in part of monosynaptic connections from sensory neurons (SN) to motor neurons (MN), as well as polysynaptic connections involving excitatory and inhibitory interneurons. It is possible to record the activity of these identified neurons and their synaptic connections during learning in semi-intact prepara-

Cite this article as *Cold Spring Harb Perspect Biol* doi: 10.1101/cshperspect.a021675

tions, and thus to examine the contribution of plasticity at different sites in the circuit to behavioral learning. Such experiments have shown, for example, that homosynaptic depression and heterosynaptic facilitation at the SN–MN synapses contribute to habituation, dishabituation, and sensitization of the siphon-withdrawal reflex, and that plasticity at other sites also contributes (Antonov et al. 1999). During sensitization by tail shock, siphon stimulation produces increased siphon withdrawal and increased activity in siphon motor neurons as a result of, in part, two mechanisms (Fig. 1). First, the same peripheral stimulus evokes a greater number of action potentials in the presynaptic SNs (i.e., enhanced excitability). Second, each action potential fired by a SN produces a greater excitatory postsynaptic potential (EPSP) in the MN (i.e., short-term facilitation [STF]). The changes in excitability and synaptic potentials are induced in part by the neuromodulator serotonin (5-HT). Both of these changes are mimicked by application of 5-HT, and 5-HT is released from modulatory interneurons during training (Brunelli et al. 1976; Walters et al. 1983b; Glanzman et al. 1989; Mackey et al. 1989; Levenson et al. 1999; Marinesco and Carew 2002; Philips et al. 2011).

The mechanisms of STF at the SN–MN synapses have been examined more extensively in neural analogs of learning in isolated ganglia or in cell culture, in which tail shock is replaced by either nerve shock or application of 5-HT. Early studies (reviewed in Byrne and Kandel 1996) found that STF produced by brief application of 5-HT to rested synapses (an analog of sensitization) involves cyclic adenosine-3-monophosphate (cAMP), protein kinase A (PKA), decreased $K^+$ current, and increases in spike width, $Ca^{2+}$ influx, and transmitter release from the SNs (Fig. 2, short term [ST]). In contrast, STF at depressed synapses (an analog of dishabituation) involves protein kinase C (PKC), which acts by a spike broadening-independent mechanism, perhaps vesicle mobilization (Fig. 2, DIS). These results suggested that although dishabituation and sensitization both involve facilitation at the SN–MN synapses, they may involve fundamentally different mechanisms at the molecular level.

## The Relationship between Short- and Long-Term Facilitation, and the Discovery of Intermediate-Term Facilitation

STF involves covalent modifications of proteins in existing synapses. In contrast, five tail shocks or five applications of 5-HT separated by 15 min produce long-term facilitation (LTF), which involves protein and RNA synthesis and

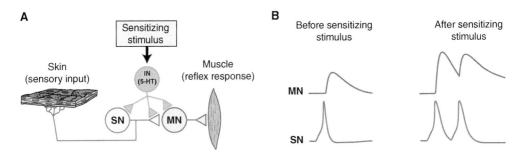

**Figure 1.** Heterosynaptic facilitation of the sensorimotor connection contributes to sensitization in *Aplysia*. (*A*) Sensitizing stimuli activate facilitatory interneurons (IN) that release modulatory transmitters, one of which is 5-HT. The modulator leads to an alteration of the properties of the sensory neuron (SN) and motor neuron (MN). (*B*) The enhanced synaptic input to the MN during sensitization results from enhanced sensory input, partly caused by two mechanisms. First, the same peripheral stimulus can evoke a greater number of action potentials in the presynaptic SN (i.e., enhanced excitability). Second, each action potential fired by an SN produces a stronger synaptic response in the MN (i.e., synaptic facilitation). A component of sensitization is also caused by the effects of 5-HT on the MN. (Based on data from Byrne and Kandel 1996.)

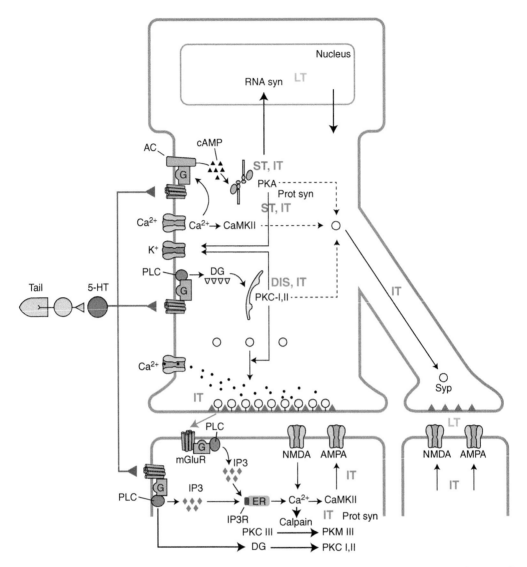

**Figure 2.** Cellular and molecular mechanisms of facilitation at sensory–motor neuron synapses that contribute to short- and intermediate-term learning in *Aplysia*. Dishabituation (DIS) involves presynaptic protein kinase C (PKC). Short-term (ST) sensitization involves presynaptic protein kinase A (PKA) and calmodulin-dependent protein kinase (CaMKII). Intermediate-term (IT) sensitization involves presynaptic PKA and CaMKII or PKC, protein synthesis (prot syn), and spontaneous transmitter release. In addition, it involves postsynaptic mGluRs, CaMKII or PKC, protein synthesis, and membrane insertion of α-amino-3-hydroxy-5-methyl-4-isoxazolepropionic acid (AMPA)-like receptors, as well as recruitment of pre- and postsynaptic proteins to new synaptic sites. In contrast, long-term (LT) sensitization involves gene regulation and growth of new synapses. AC, Adenyl cyclase; cAMP, cyclic adenosine-3-monophosphate; DG, diacylglycerol; NMDA, *N*-methyl-D-aspartate; PKM, protein kinase M; PLC, phospholipase C; RNA syn, RNA synthesis. (Based on data in Hawkins et al. 2013.)

Cite this article as *Cold Spring Harb Perspect Biol* doi: 10.1101/cshperspect.a021675

the growth of new synapses, and is thus fundamentally different from STF (Fig. 2, LT). However, it was not clear whether the different stages of plasticity are independent and induced in parallel, or one induces the other in series. Initial studies of facilitation in isolated ganglia suggested that STF and LTF are induced in parallel (Emptage and Carew 1993; Mauelshagen et al. 1996), but similar experiments in culture have suggested that they can be induced, at least partly, in series (Ghirardi et al. 1995).

During the course of these parametric studies of STF and LTF, Ghirardi et al. (1995) also obtained evidence for a third stage that they called intermediate-term facilitation (ITF), which is typically induced by an intermediate level of 5-HT (four to five pulses of a low concentration), persists for hours, and involves PKA and protein synthesis but not RNA synthesis (Fig. 2, intermediate term [IT]). They also found that a higher concentration of 5-HT induces both ITF and LTF, and that the mechanisms of the facilitation depend not only on the time after 5-HT but also on the concentration of 5-HT. For example, facilitation 30 min after the 5-HT, which is in the intermediate-term time range, can depend on PKA only (with 1 or 10 nM 5-HT), PKA and protein synthesis (with 50 nM 5-HT), or PKA, protein synthesis, and RNA synthesis (with 100 nM or 10 μM 5-HT). These results illustrate that ITF (like the other stages) is not a unitary entity but rather can involve a number of different mechanisms depending on the protocol and, therefore, suggest that it may be more meaningful to ask whether the mechanisms rather than stages are in parallel or series.

## Mechanisms of Induction, Maintenance, and Expression of Intermediate-Term Facilitation

The experiments of Ghirardi et al. (1995) did not distinguish between mechanisms of induction, maintenance, or expression of the facilitation, nor did they examine whether those mechanisms are pre- or postsynaptic. Several groups have addressed those questions using different experimental protocols for ITF. Carew and colleagues found that induction of intermediate-

term sensitization (ITS) and ITF with a repeated pulses protocol (five spaced tail shocks or five pulses of 5-HT) requires MAP kinase and protein but not RNA synthesis, and maintenance involves persistent activation of PKA but not PKC (Sutton and Carew 2000; Sutton et al. 2001). In contrast, induction of ITS or ITF with a site-specific protocol (e.g., spike activity in a SN and simultaneous 5-HT) requires PKC and MAP kinase but not protein synthesis, and maintenance involves persistent activation of PKC rather than PKA (Sutton and Carew 2000; Sutton et al. 2004; Zhao et al. 2006; Shobe et al. 2009). Like mammalian PKC, PKC in *Aplysia* has three isoforms: conventional (Apl I), novel (Apl II), and atypical (Apl III), which can be cleaved by calpain to form a persistently active kinase, protein kinase M (PKM) (Kruger et al. 1991; Bougie et al. 2009), which is critical for the maintenance of ITF (Bougie et al. 2012). Thus, induction of ITF with either protocol involves multiple mechanisms including activation of MAP kinase, whereas maintenance involves persistent kinase activity, but of different kinases (PKA or PKM) depending on the induction protocol used.

Glanzman and colleagues found that ITF induced by a single 10-min application of 5-HT involves postsynaptic $Ca^{2+}$, protein synthesis, and α-amino-3-hydroxy-5-methyl-4-isoxazolepropionic acid (AMPA) receptor insertion (Li et al. 2005; Villareal et al. 2007). To investigate those postsynaptic mechanisms independent of presynaptic mechanisms, they examined ITF of the response to focal application of glutamate (the glutamate excitatory potential or Glu-EP) in isolated MNs, and found that it also involves $Ca^{2+}$, protein synthesis, and AMPA receptor insertion (Chitwood et al. 2001; Villareal et al. 2007). Furthermore, induction of facilitation of the Glu-EP involves calpain-dependent proteolysis of PKC Apl III to form PKM, and maintenance involves persistent activation of PKM (Bougie et al. 2009; Villareal et al. 2009). Subsequent studies have shown that ITF of the SN–MN EPSP also involves postsynaptic PKC Apl III (Bougie et al. 2012).

These studies suggest that, although STF involves presynaptic mechanisms, ITF by 10-min

5-HT involves postsynaptic mechanisms. However, in each case, only one side of the synapse was examined, and it was not known whether the same protocol might involve mechanisms on both sides. Hawkins and colleagues addressed that question, and found that facilitation during STS in the semi-intact siphon-withdrawal preparation involves PKA, calmodulin-dependent protein kinase (CaMK)II, and transient spike broadening in the SN, but it does not involve $Ca^{2+}$ or CaMKII in the MN and, thus, appears to be entirely presynaptic. The facilitation during ITS also involves PKA, CaMKII, and transient spike broadening in the SN. However, it also involves $Ca^{2+}$ and CaMKII in the MN and protein synthesis in both neurons, and is thus both pre- and postsynaptic (Antonov et al. 2010).

Similarly, facilitation by 1-min application of 5-HT in cell culture (an analog of STS) involves PKA and CaMKII in the SN, but it does not involve $Ca^{2+}$ in the MN or PKC in either neuron (Jin et al. 2011). In contrast, facilitation by a 10-min application of 5-HT (an analog of intermediate-term sensitization) involves PKC (but not PKA or CaMKII) in the SN, in agreement with previous studies (Byrne and Kandel 1996). In addition, 10-min application of 5-HT involves $Ca^{2+}$ and CaMKII in the MN and protein synthesis in both neurons and is, thus, both pre- and postsynaptic. Collectively, these results suggest that ITF is the first stage to involve both pre- and postsynaptic molecular mechanisms.

ITF also involves recruitment of synaptic proteins. ITF induced by multiple applications of 5-HT is accompanied by filling of empty presynaptic varicosities with the vesicle-associated protein synaptophysin within 3 h, but not by the formation of new varicosities (Kim et al. 2003). Like facilitation of the postsynaptic potential (PSP) with this protocol (Ghirardi et al. 1995), the increase in clusters of synaptophysin does not persist for 24 h, and does not require protein or RNA synthesis. In contrast, LTF is accompanied by both filling of varicosities and the formation of new varicosities within 12–18 h. Again, like facilitation of the PSP and the increase in varicosities (Bailey et al. 1992), the increase in clusters of synaptophysin during LTF

persists for 24 h and requires protein synthesis. ITF and LTF are also accompanied by increases in clusters of the postsynaptic proteins ApGluR1 and ApNR1 within 12 h, whereas STF is not (Li et al. 2009). These results suggest that the intermediate-term stage is also the first to involve recruitment of both pre- and postsynaptic proteins, which could be initial steps in the formation of new synapses during long-term facilitation.

## Spontaneous Transmitter Release from the Presynaptic Neuron Recruits Postsynaptic Mechanisms of Intermediate- and Long-Term Facilitation

If STF is presynaptic but ITF and LTF involve both pre- and postsynaptic mechanisms, how are the postsynaptic mechanisms first recruited? There are at least two possibilities, which are not mutually exclusive: the pre- and postsynaptic mechanisms might be induced by activation of pre- and postsynaptic 5-HT receptors in parallel, or activation of presynaptic 5-HT receptors might increase spontaneous release of glutamate, which then activates postsynaptic glutamate receptors to induce the postsynaptic mechanisms in series (Fig. 2).

Partly because the postsynaptic mechanisms of ITF are similar to those induced by glutamate release during homosynaptic potentiation (Jin and Hawkins 2003), Jin et al. (2012a,b) investigated the possible role of spontaneous transmitter release from the presynaptic neuron as an anterograde signal for recruiting postsynaptic mechanisms of ITF and LTF. 5-HT produced a substantial increase in the frequency of spontaneous miniature excitatory postsynaptic currents (mEPSCs) and a more modest increase in their amplitude during, or shortly after, the induction of ITF or LTF in cell culture. Those increases correlated with subsequent facilitation of the evoked EPSP, consistent with the idea that spontaneous release contributes to the induction of the facilitation. In support of that idea, several manipulations that either reduced or enhanced spontaneous release (without affecting baseline evoked release) also reduced or enhanced ITF and LTF. These results suggested that spontaneous release is necessary for the induction of ITF and LTF, and acts synergistically with additional mechanisms (activated, e.g., by presynaptic cAMP) to produce the facilitation.

Further experiments showed that spontaneous release from the presynaptic neuron activates metabotropic glutamate receptors, which stimulate IP3 production and $Ca^{2+}$ release in the postsynaptic neuron (Fig. 2). In addition, expression of the latter part of facilitation may involve up-regulation of AMPA-like receptors (see also Li et al. 2005). To examine that mechanism more directly, Jin et al. (2012b) expressed the *Aplysia* homolog of the AMPA receptor subunit GluR1 in the MN. The application of 5-HT for 10 min produced an increase in membrane insertion of ApGluR1 into existing puncta as well as increases in the number of puncta of ApGluR1 and overlap with puncta of the presynaptic protein synaptophysin. Furthermore, all of those increases depended on spontaneous release and/or mGluRs. The increase in ApGluR1 puncta during ITF preceded an increase in synaptophysin puncta (Kim et al. 2003), and, therefore, may be a first step in a sequence that can lead to new synapse assembly during LTF.

## LONG-TERM SENSITIZATION

### Neuronal Correlates of Long-Term Sensitization

In addition to STS and ITS, withdrawal reflexes can also display LTS, lasting from several hours to weeks (Pinsker et al. 1973). Although STS (or its in vitro analogue STF) can be induced by brief treatments (lasting a few seconds to minutes), LTS- and 5-HT-induced LTF require more extensive training involving multiple trials spaced over hours or days.

One approach to investigating mechanisms of LTS is to examine its neural correlates—that is, to train an animal, remove the nervous system some time later, and test for differences in cellular properties compared with controls. Such studies have shown that mechanisms supporting LTS generally resemble the mechanisms supporting STS including changes in SN excitability (long-term enhanced excitability [LTEE]) and facilitation of sensorimotor synapses (i.e., LTF) (Frost et al. 1985; Scholz and Byrne 1987; Cleary et al. 1998). However, although STS is correlated with SN spike broadening, LTS is correlated with SN spike narrowing (Antzoulatos and Byrne 2007). The functional implications of spike narrowing are not obvious, but may involve an increase in the fidelity of the neuronal response to peripheral stimuli by decreasing the probability of spike failures. Another key difference between cellular correlates of STS and LTS is that LTS is associated with structural modifications of SNs, namely, outgrowth of neurites and remodeling of active zones, whereas STS is not. These data suggest that enhanced transmission is mediated by an increase in transmitter release and in number of synapses (Bailey and Chen 1983; Wainwright et al. 2002). LTS and LTF are also correlated with enhanced uptake of glutamate (Levenson et al. 2000), the endogenous transmitter of SNs (Dale and Kandel 1993; Antzoulatos and Byrne 2004). In addition, LTS is correlated with changes in the biophysical properties of MNs (Cleary et al. 1998), and an increased synthesis of postsynaptic receptors (Trudeau and Castellucci 1995; Zhu et al. 1997; Cai et al. 2008).

### Molecular Mechanisms Contributing to the Induction of LTF—cAMP-Response Element-Binding (CREB) and Gene Regulation

The molecular mechanisms of LTF have been examined more extensively in neural analogs in isolated ganglia or cell culture (Fig. 3). Such studies have shown that STF and LTF share some common cellular pathways during their induction. For example, both forms activate the cAMP/PKA cascade. However, although STF involves PKA-dependent covalent modifications of proteins involved in increasing spike width, excitability, and transmitter release, LTF involves the PKA-dependent regulation of gene transcription and new protein synthesis. Multiple training trials or repeated applications of 5-HT lead to a translocation of PKA to the nucleus, where it phosphorylates the transcriptional ac-

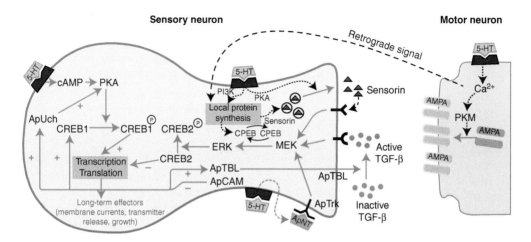

**Figure 3.** Simplified scheme of the mechanisms that contribute to long-term sensitization in *Aplysia*. Sensitization training leads to cyclic adenosine 3-monophosphate (cAMP)-dependent regulation of cAMP-response element binding (CREB)1. Serotonin also leads to activation of extracellular signal-regulated kinase (ERK), which regulates CREB2. Although CREB1 acts as an initiator of gene transcription, CREB2 acts as a repressor of gene transcription. The combined effects of activation of CREB1 and suppression of CREB2 lead to regulation of the synthesis of at least 10 proteins, only some of which are shown. *Aplysia* tolloid/BMP-like protein (ApTBL) is believed to activate latent forms of transforming growth factor (TGF)-β, which can then bind to receptors on the sensory neuron (SN). TGF-β activates ERK, which may act by initiating a second round of gene regulation by affecting CREB2-dependent pathways. Serotonin can also increase the local synthesis of the *Aplysia* homolog of cytoplasmic polyadenylation element-binding protein (ApCPEB) and the peptide sensorin through phosphoinositide-3-kinase (PI3K). ApCPEB can exist in two conformations, one of which dominates and allows ApCPEB to self-perpetuate. Sensorin release is dependent on type II protein kinase A (PKA). Sensorin binds to autoreceptors leading to further activation of ERK. Because increased synthesis of sensorin requires elevation of postsynaptic calcium, a retrograde signal is also postulated. In addition to the retrograde signal, 5-HT-induced postsynaptic signaling also leads to an increased number of glutamate receptors. AMPA, α-amino-3-hydroxy-5-methyl-4-isoxazolepropionic acid; ApCAM, *Aplysia* cell adhesion molecule; ApTrk, *Aplysia* tyrosine kinase autoreceptor; ApUch, *Aplysia* ubiquitin hydrolase; MEK, MAPK/ERK kinase; PKM, protein kinase M. (Based on data from Liu et al. 1997.)

tivator cAMP-responsive element-binding protein (CREB1). CREB1 binds to a regulatory region of genes known as CRE (cAMP-responsive element). A second transcription factor CREB2 also binds to the CRE, but unlike CREB1, CREB2 is a repressor of gene transcription. Although CREB1 is activated by PKA, CREB2 is inhibited in parallel by a 5-HT-induced increase in extracellular signal-regulated kinase (ERK) phosphorylation. In addition, CREB2 levels are reduced by a piRNA, which is a type of small regulatory RNA that can control gene expression through epigenetic mechanisms (Rajasethupathy et al. 2012). The role of transcription factors in long-term memory formation is not limited to the induction phase but may also extend to the consolidation phase, where consoli-

dation is defined as the time window during which RNA and protein synthesis are required for converting short- to long-term memory. For example, treatment of ganglia with five pulses of 5-HT over a 1.5-h period to mimic sensitization training leads to the binding of CREB1 to the promoter of its own gene and induces CREB1 synthesis, giving rise to a CREB1-positive feedback loop that supports memory consolidation (Liu et al. 2011).

### Feedback Loops also Contribute to LTF

As illustrated by the CREB1 feedback loop, the phosphorylation of CREB1 by PKA and CREB2 by ERK is not a simple serial cascade starting with binding of 5-HT to membrane receptors.

Indeed, the transduction process involves multiple feedback pathways. For example, levels of *Aplysia* ubiquitin hydrolase (ApUch) in SNs are increased, possibly via a CREB-based increase in ApUch transcription. The increased levels of ApUch increase the rate of degradation of proteins, via the ubiquitin–proteosome pathway, including the regulatory subunit of PKA (Chain et al. 1999). The catalytic subunit of PKA, when freed from the regulatory subunit, is highly active. Thus, increased ApUch triggered by the initial treatment of SNs with 5-HT will lead to an increase in PKA activity and a more protracted phosphorylation of CREB1 (Fig. 3). This phosphorylated CREB1 may act to further prolong ApUch expression, thus closing a positive feedback loop. Protein degradation, in general, and the role of ubiquitination in particular, is an emerging theme in recent studies on the neural basis of long-term memory (Fioravante and Byrne 2011).

Several extracellular feedback cascades also appear to operate to activate ERK and, thereby, regulate CREB2 (Martin et al. 1997; Guan et al. 2002). 5-HT stimulates secretion of a recently identified endogenous *Aplysia* neurotrophin (ApNT) from the SN (Kassabov et al. 2013), leading to the activation of an *Aplysia* tyrosine kinase autoreceptor (ApTrk) and subsequent activation of ERK in the SN (Fig. 3) (Purcell et al. 2003; Ormond et al. 2004; Sharma et al. 2006). A similar feedback loop occurs through the release of an *Aplysia* cysteine-rich neurotrophic factor (ApCRNF) (Pu et al. 2014). 5-HT also acts through feedback pathways involving the peptides sensorin and transforming growth factor (TGF)-β. A role for TGF-β was originally hypothesized based on the finding that LTF is associated with an increased expression of *Aplysia* tolloid/BMP-like (ApTBL)-1 protein. Tolloid and the related molecule BMP-1 appear to function as secreted $Zn^{2+}$ proteases, which activate members of the TGF-β family in some systems. In SNs, application of TGF-β mimics the effects of 5-HT in that it produces LTF (Fig. 3) (Zhang et al. 1997). Interestingly, TGF-β activates ERK in the SNs and induces its translocation to the nucleus. Thus, TGF-β could be part of an extracellular positive feedback loop,

possibly leading to another round of protein synthesis to further consolidate the memory (Zhang et al. 1997).

Another extracellular positive feedback loop that affects ERK involves the 5-HT-induced regulation of the release of the SN-specific neuropeptide sensorin (Fig. 3). Synthesis of sensorin is stimulated by 5-HT in a PI3 (phosphatidylinositol-3)-kinase–dependent manner, and sensorin binding to presynaptic autoreceptors activates ERK. Sensorin synthesis also requires elevation of postsynaptic calcium (Cai et al. 2008). The mechanism through which postsynaptic calcium regulates the local protein synthesis of sensorin in the presynaptic terminal is not fully understood, but presumably involves the release of a retrograde signal (Cai et al. 2008). Although the presence of a postsynaptic neuron seems to be required for long-term facilitation, it is not required for another correlate of long-term sensitization, increased SN excitability (Cleary et al. 1998; Liu et al. 2011). Thus, accumulating evidence suggests that expression of long-term memory in this simple system does not rely on a unitary mechanism, but on multiple mechanisms at multiple sites.

## Molecules Involved in the Maintenance and Expression of LTF

The combined effects of activation of CREB1 and removal of the repression of transcription by CREB2 lead to changes in the synthesis of specific proteins that allow for the maintenance and expression of LTF. The down-regulation of a homolog of a neuronal cell adhesion molecule (NCAM) shows that ApCAM plays a key role in the expression of long-term facilitation. This down-regulation has two components. First, the synthesis of ApCAM is reduced (Fig. 3). Second, preexisting ApCAM is internalized via increased endocytosis (not shown). The internalization and degradation of ApCAM allow for the restructuring of the axon arbor (Bailey and Kandel 2008). This restructuring allows the SN to form additional connections with the MN or with other cells. Structural changes associated with LTF also involve the presynaptic cell-adhesion protein neurexin, along with its postsynap-

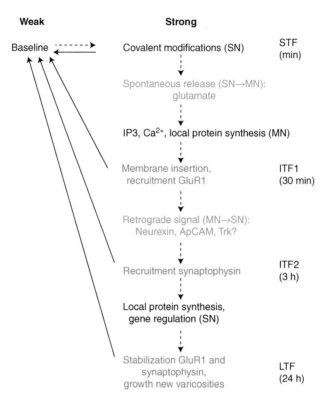

**Figure 4.** Cascade model of mechanisms contributing to the different stages of synaptic plasticity in *Aplysia*. In cascade models (Fusi et al. 2005), synapses have two levels of strength (weak and strong) and several increasingly long-lasting states. In *Aplysia,* relatively weak stimulation produces short-term facilitation (STF) that lasts minutes, stronger stimulation produces intermediate-term facilitation (ITF) that lasts minutes to hours, and even stronger stimulation produces long-term facilitation (LTF) that lasts days. The different stages of facilitation may involve a series or cascade of pre- and postsynaptic mechanisms that is initiated by spontaneous transmitter release during STF, progresses through two stages of ITF, and can culminate in synaptic growth during LTF. The mechanisms in this growth cascade are a subset of all mechanisms involved in facilitation, and some other mechanisms (not shown) may act in parallel and contribute only to specific stages. Thus, the idea of a cascade applies to the mechanisms and not the stages per se. In addition to this linear cascade, facilitation also involves feedforward and feedback loops. Dashed lines, transitions that are initiated by different durations or patterns of 5-HT; solid lines, spontaneous transitions; red, extracellular signaling molecules; blue, structural modifications. MN, motor neuron; SN, sensory neuron.

tic counterpart neuroligin and their transsynaptic interaction (Choi et al. 2011).

In addition, the persistence of LTF depends on two molecules that can maintain a change in functional state for long periods of time. First, the stabilization of new structures depends on the *Aplysia* homolog of cytoplasmic polyadenylation element-binding protein (ApCPEB) (Miniaci et al. 2008), which regulates local translation in the SNs (Fig. 3). ApCPEB has prion-like properties, meaning that it can exist in two conformations, one of which dominates and allows ApCPEB to self-perpetuate. Conversion of ApCPEB to the self-perpetuating state is enhanced by 5-HT and is required for the persistence of LTF (Si et al. 2010). Second, as discussed earlier for ITF, the persistence of LTF also depends on the persistent activation of PKC by cleavage to the PKM form (PKM Apl III) (Cai et al. 2011), which is thought to regulate the membrane insertion of AMPA receptors in the MN (Fig. 3).

## Possible Relationships between Mechanisms Contributing to STF, ITF, and LTF

Collectively, the results on STF, ITF, and LTF in *Aplysia* suggest that, like synapse formation (McAllister 2007), the different stages of synaptic plasticity may involve different pre- and postsynaptic mechanisms coordinated by back-and-forth signaling in a chain or cascade that can culminate in growth (Fig. 4). Consistent with that idea, spontaneous transmitter release from the presynaptic neuron during STF recruits postsynaptic molecular mechanisms of ITF including IP3, $Ca^{2+}$, and formation of clusters of AMPA-like glutamate receptors. Postsynaptic $Ca^{2+}$ is, in turn, necessary for LTF, perhaps through retrograde signaling to presynaptic ApCAM, neurexin, or Trk receptors (Purcell et al. 2003; Ormond et al. 2004; Sharma et al. 2006; Cai et al. 2008; Hu et al. 2010; Choi et al. 2011; Kassabov et al. 2013). The new postsynaptic clusters of AMPA-like receptors may also participate in retrograde signaling, and recruit presynaptic clusters of synaptophysin during a later stage of ITF and growth of presynaptic varicosities during LTF (Kim et al. 2003; Ripley et al. 2011; Lee et al. 2012). These ideas are similar to theoretical "cascade" models of memory storage that can show plasticity as well as long-term stability (Fusi et al. 2005), which would seem to be mutually exclusive but are both essential features of memory. In addition to this linear cascade, the mechanisms of plasticity also form feedforward loops involved in synaptic "tagging," which allows transcription-dependent LTF to be synapse specific (Casadio et al. 1999), as well as feedback loops involved in the long-term persistence of the functional and structural changes (Fig. 3).

## NEURAL AND MOLECULAR MECHANISMS OF NONASSOCIATIVE LEARNING IN OTHER INVERTEBRATE MODEL SYSTEMS

Analyses of nonassociative learning in other invertebrates have confirmed and extended the work on *Aplysia*. Thus, as in *Aplysia* (Castellucci et al. 1970; Castellucci and Kandel 1974), habituation in crayfish (Zucker 1972) and several other species (see below) appears to be caused by a progressive decrease in the amount of transmitter released by primary sensory neurons. In addition, there is apparent conservation of molecular mechanisms for sensitization, which often involve the engagement of 5-HT and the cAMP cascade. For example, the opisthobranch *Tritonia diomedea* initiates stereotypical rhythmic swimming to escape a noxious stimulus. The behavior shows both habituation and sensitization (Frost et al. 1996). Habituation appears to involve plasticity at multiple loci, including decrement at the first afferent synapse. Sensitization appears to involve enhanced excitability and synaptic strength in one of the central pattern–generating (CPG) interneurons. Modulation of interneurons can be mediated by 5-HT, which has diverse effects on multiple loci of the circuit (Sakurai et al. 2007). Sensitization of withdrawal reflexes of the land snail *Helix* also appears to be mediated by serotonergic modulatory cells whose spiking frequency increases following noxious stimulation (Balaban 2002). These serotonergic cells are electrically coupled so that they are recruited and fire synchronously in response to strong excitatory input.

The shortening reflex of the leech *Hirudo medicinalis* shows habituation and sensitization, and the neuronal changes underlying both occur in the pathway from mechanosensory neurons to the S cells (Sahley et al. 1994). Habituation correlates with decreased S-cell excitability (Burrell et al. 2001) and the reflex can be restored (dishabituation) following application of a single noxious stimulus (Boulis and Sahley 1988). The potentiation of the shortening reflex observed during sensitization is mediated by 5-HT through an increase of the levels of cAMP (Belardetti et al. 1982; Burrell and Sahley 2005), which also increases the excitability of S cells and spike after hyperpolarization (AHP) (Burrell et al. 2001; Burrell and Crisp 2008).

Habituation of the shortening reflex also involves depression of the synapses of touch (T) sensory neurons onto their follower target neurons. This synaptic depression is associated with an increase in the amplitude of the T-cell AHP that follows their discharge (Brunelli et al. 1997;

Scuri et al. 2002). The persistent increase in AHP amplitude, following low-frequency stimulation of T cells, has been attributed to increased activity of the electrogenic $Na^+$ pump, and requires activation of phospholipase A2 (Scuri et al. 2005; Zaccardi et al. 2012).

Nonassociative learning has been studied extensively in *C. elegans* by Catharine Rankin and colleagues. *C. elegans* is a valuable model system for cellular and molecular studies of learning because it has an extremely simple nervous system that consists of a total of 302 neurons, the anatomical connectivity of which has been described at the electron microscopy level. *C. elegans* responds to a vibratory stimulus applied to the medium in which they locomote by swimming backward. This reaction, known as the tap withdrawal reflex, shows habituation, dishabituation, sensitization, and long-term (24 h) retention of habituation training. Laser ablation studies have been used to elucidate the neural circuitry supporting the tap withdrawal reflex and to identify likely sites of plasticity within the network. Plastic changes during habituation appear to occur at the chemical synapses between presynaptic sensory neurons and postsynaptic command interneurons. Analysis of several *C. elegans* mutants has revealed that synapses at the locus of plasticity in the network may be glutamatergic (Ardiel and Rankin 2010; Bozorghmehr et al. 2013). CREB is required for long-term but not short-term habituation (Timbers and Rankin 2011).

## SUMMARY AND CONCLUSIONS

The simplicity and tractability of the neural circuits mediating behaviors in invertebrates have facilitated the cellular/molecular dissection of the underlying neural mechanisms of nonassociative learning and has illuminated several basic principles:

- Learning and memory rely on multiple mechanisms of plasticity at multiple sites of the neuronal circuits.

- The relative contribution to memory of the different sites varies as a function of the extent of training and time after training.

- Although the target proteins are different, the same intracellular signaling cascades that induce short-term modifications in synaptic transmission can also be used to induce long-term changes.

- Short-term memory relies on covalent modifications of preexisting proteins, but long-term memory also requires regulated gene transcription and translation.

- The induction of long-term memory requires both the activation of inducing signals, and the inhibition of inhibitory constraints imposed by other molecular pathways.

- Maintenance of long-term cellular memory involves both intracellular and extracellular feedback loops, which sustain the regulation of gene expression and the modification of targeted molecules.

- As described by Hawkins and Byrne (2015), associative forms of learning and memory can arise from the neural mechanisms that are used for nonassociative learning.

Since the 1960s, research on nonassociative learning in invertebrates has provided a wealth of information on the mechanisms of simple forms of learning. Although the learning studied is of the simplest type, the mechanisms have proven to be extremely rich and complex. The studies have provided important general principles that have proven applicable to all animals. Despite the great progress, the knowledge of memory mechanisms is still in its infancy. A more mature understanding will come from continued analyses of these model systems.

## ACKNOWLEDGMENTS

Preparation of this manuscript was supported by National Institutes of Health (NIH) Grants GM097502, NS019895, and NS083690.

## REFERENCES

*Reference is also in this collection.*

Antonov I, Kandel ER, Hawkins RD. 1999. The contribution of facilitation of monosynaptic PSPs to dishabituation and sensitization of the *Aplysia* siphon withdrawal reflex. *J Neurosci* 19: 10438–10450.

Antonov I, Kandel ER, Hawkins RD. 2010. Presynaptic and postsynaptic mechanisms of synaptic plasticity and metaplasticity during intermediate-term memory formation in *Aplysia. J Neurosci* 30: 5781–5791.

Antzoulatos EG, Byrne JH. 2004. Learning insights transmitted by glutamate. *Trends Neurosci* 27: 555–560.

Antzoulatos EG, Byrne JH. 2007. Long-term sensitization training produces spike narrowing in *Aplysia* sensory neurons. *J Neurosci* 27: 676–683.

Ardiel EL, Rankin CH. 2010. An elegant mind: Learning and memory in *Caenorhabditis elegans. Learn Mem* 17: 191–201.

Bailey CH, Chen M. 1983. Morphological basis of long-term habituation and sensitization in *Aplysia. Science* 220: 91–93.

Bailey CH, Kandel ER. 2008. Synaptic remodeling, synaptic growth and the storage of long-term memory in Aplysia. *Prog Brain Res.* 169: 179–198.

Bailey CH, Montarolo P, Chen M, Kandel ER, Schacher S. 1992. Inhibitors of protein and RNA synthesis block structural changes that accompany long-term heterosynaptic plasticity in *Aplysia. Neuron* 9: 749–758.

Balaban PM. 2002. Cellular mechanisms of behavioral plasticity in terrestrial snail. *Neurosci Biobehav Rev* 26: 597–630.

Belardetti F, Biondi C, Colombaioni L, Brunelli M, Trevisani A. 1982. Role of serotonin and cyclic AMP on facilitation of the fast conducting system activity in the leech *Hirudo medicinalis. Brain Res* 246: 89–103.

Bougie JK, Lim T, Farah CA, Manjunath V, Nagakura I, Ferraro GB, Sossin WS. 2009. The atypical protein kinase C in *Aplysia* can form a protein kinase M by cleavage. *J Neurochem* 109: 1129–1143.

Bougie JK, Cai D, Hastings M, Farah CA, Chen S, Fan X, McCamphill PK, Glanzman DL, Sossin WS. 2012. Serotonin-induced cleavage of the atypical protein kinase C Apl III in *Aplysia. J Neurosci* 32: 14630–14640.

Boulis NM, Sahley CL. 1988. A behavioral analysis of habituation and sensitization of shortening in the semi-intact leech. *J Neurosci* 8: 4621–4627.

Bozorgmehr T, Ardiel EL, McEwan AH, Rankin CH. 2013. Mechanisms of plasticity in a *Caenorhabditis elegans* mechanosensory circuit. *Front Physiol* 4: 88

Brunelli M, Castellucci V, Kandel ER. 1976. Synaptic facilitation and behavioral sensitization in *Aplysia*: Possible role of serotonin and cyclic AMP. *Science* 194: 1178–1181.

Brunelli M, Garcia-Gil M, Mozzachiodi R, Scuri R, Zaccardi ML. 1997. Neurobiological principles of learning and memory. *Arch Ital Biol* 135: 15–36.

Burrell BD, Crisp KM. 2008. Serotonergic modulation of afterhyperpolarization in a neuron that contributes to learning in the leech. *J Neurophysiol* 99: 605–616.

Burrell BD, Sahley CL. 2005. Serotonin mediates learning-induced potentiation of excitability. *J Neurophysiol* 94: 4002–4012.

Burrell BD, Sahley CL, Muller KJ. 2001. Non-associative learning and serotonin induce similar bi-directional changes in excitability of a neuron critical for learning in the medicinal leech. *J Neurosci* 21: 1401–1412.

Byrne JH. 1987. Cellular analysis of associative learning. *Physiol Rev* 67: 329–439.

Byrne JH, Kandel ER. 1996. Presynaptic facilitation revisited: State and time dependence. *J Neurosci* 16: 435–435.

Cai D, Chen S, Glanzman DL. 2008. Postsynaptic regulation of long-term facilitation in *Aplysia. Curr Biol* 18: 920–925.

Cai D, Pearce K, Chen S, Glanzman DL. 2011. Protein kinase M maintains long-term sensitization and long-term facilitation in *Aplysia. J Neurosci* 31: 6421–6431.

Carew TJ, Castellucci VF, Kandel ER. 1971. An analysis of dishabituation and sensitization of the gill-withdrawal reflex in *Aplysia. Int J Neurosci* 2: 79–98.

Casadio A, Martin KC, Giustetto M, Zhu H, Chen M, Bartsch D, Bailey CH, Kandel ER. 1999. A transient, neuron-wide form of CREB-mediated long-term facilitation can be stabilized at specific synapses by local protein synthesis. *Cell* 99: 221–237.

Castellucci VF, Kandel ER. 1974. A quantal analysis of the synaptic depression underlying habituation of the gill-withdrawal reflex in *Aplysia. Proc Natl Acad Sci* 71: 5004–5008.

Castellucci V, Pinsker H, Kupfermann I, Kandel ER. 1970. Neuronal mechanisms of habituation and dishabituation of the gill-withdrawal reflex in *Aplysia. Science* 167: 1745–1748.

Chain DG, Casadio A, Schacher S, Hegde AN, Valbrun M, Yamamoto N, Goldberg AL, Bartsch D, Kandel ER, Schwartz JH. 1999. Mechanisms for generating the autonomous cAMP-dependent protein kinase required for long-term facilitation in *Aplysia. Neuron* 22: 147–156.

Chitwood RA, Li Q, Glanzman DL. 2001. Serotonin facilitates AMPA-type responses in isolated siphon motor neurons of *Aplysia* in culture. *J Physiol* 534: 501–510.

Choi YB, Li HL, Kassabov SR, Jin I, Puthanveettil SV, Karl KA, Lu Y, Kim JH, Bailey CH, Kandel ER. 2011. Neurexin-neuroligin trans-synaptic interaction mediates learning-related synaptic remodeling and long-term facilitation in *Aplysia. Neuron* 70: 468–481.

Cleary LJ, Lee WL, Byrne JH. 1998. Cellular correlates of long-term sensitization in *Aplysia. J Neurosci* 18: 5988–5998.

Dale N, Kandel ER. 1993. L-glutamate may be the fast excitatory transmitter of *Aplysia* sensory neurons. *Proc Natl Acad Sci* 90: 7163–7167.

Emptage NJ, Carew TJ. 1993. Long-term synaptic facilitation in the absence of short-term facilitation in *Aplysia* neurons. *Science* 262: 253–256.

Fioravante D, Byrne JH. 2011. Protein degradation and memory formation. *Brain Res Bull* 85: 14–20.

Frost WN, Castellucci VF, Hawkins RD, Kandel ER. 1985. Monosynaptic connections made by the sensory neurons of the gill- and siphon-withdrawal reflex in *Aplysia* participate in the storage of long-term memory for sensitization. *Prod Natl Acad Sci* 82: 8266–8269.

Frost WN, Brown GD, Getting PA. 1996. Parametric features of habituation of swim cycle in the marine mollusc *Tritonia diomedea. Neurobiol Learn Mem* 65: 125–135.

Fusi S, Drew PJ, Abbott LF. 2005. Cascade models of synaptically stored memories. *Neuron* **45:** 599–611.

Ghirardi M, Montarolo PG, Kandel ER. 1995. A novel intermediate stage in the transition between short- and long-term facilitation in the sensory to motor neuron synapses of *Aplysia*. *Neuron* **14:** 413–420.

Glanzman DL, Mackey SL, Hawkins RD, Dyke AM, Lloyd PE, Kandel ER. 1989. Depletion of serotonin in the nervous system of *Aplysia* reduces the behavioral enhancement of gill withdrawal as well as the heterosynaptic facilitation produced by tail shock. *J Neurosci* **9:** 4200–4213.

Guan Z, Giustetto M, Lomvardas S, Kim JH, Miniaci MC, Schwartz JH, Thanos D, Kandel ER. 2002. Integration of long-term-memory-related synaptic plasticity involves bidirectional regulation of gene expression and chromatin structure. *Cell* **111:** 483–493.

* Hawkins RD, Byrne JH. 2015. Associative learning in invertebrates. *Cold Spring Harb Perspect Biol* **7:** a021709.

Hawkins RD, Clark GA, Kandel ER. 1987. Cell biological studies of learning in simple vertebrate and invertebrate systems. In *Handbook of physiology, section 1: The nervous system, Vol. V, Higher functions of the brain* (ed. Mountcastle VB, Plum F, Geiger SR), pp. 25–83. American Physiological Society, Bethesda, MD.

Hawkins RD, Kandel ER, Bailey CH. 2006. Molecular mechanisms of memory storage in *Aplysia*. *Biol Bull* **210:** 174–191.

Hawkins RD, Antonov I, Jin I. 2013. Mechanisms of short-term and intermediate-term memory in *Aplysia*. In *Invertebrate learning and memory* (ed. Menzel R, Benjamin P), pp. 194–205. Academic, London.

Hu JY, Chen Y, Bougie JK, Sossin WS, Schacher S. 2010. *Aplysia* cell adhesion molecule and a novel protein kinase C activity in the postsynaptic neuron are required for presynaptic growth and initial formation of specific synapses. *J Neurosci* **30:** 8353–8366.

Jin I, Hawkins RD. 2003. Presynaptic and postsynaptic mechanisms of a novel form of homosynaptic potentiation at *Aplysia* sensory-motor neuron synapses. *J Neurosci* **23:** 7288–7297.

Jin I, Kandel ER, Hawkins RD. 2011. Whereas short-term facilitation is presynaptic, intermediate-term facilitation involves both presynaptic and postsynaptic protein kinases and protein synthesis. *Learn Mem* **18:** 96–102.

Jin I, Puthenveettil S, Udo H, Karl K, Kandel ER, Hawkins RD. 2012a. Spontaneous transmitter release is critical for the induction of long-term and intermediate-term facilitation in *Aplysia*. *Proc Natl Acad Sci* **109:** 9131–9136.

Jin I, Udo H, Rayman JB, Puthenveettil S, Vishwasrao HD, Kandel ER, Hawkins RD. 2012b. Postsynaptic mechanisms recruited by spontaneous transmitter release during long-term and intermediate-term facilitation in *Aplysia*. *Proc Natl Acad Sci* **109:** 9137–9142.

Kassabov SR, Choi YB, Karl KA, Vishwasrao HD, Bailey CH, Kandel ER. 2013. A single *Aplysia* neurotrophin mediates synaptic facilitation via differentially processed isoforms. *Cell Rep* **3:** 1213–1227.

Kim J-H, Udo H, Li H-L, Toun TY, Chen M, Kandel ER, Bailey CH. 2003. Presynaptic activation of silent synapses and growth of new synapses contribute to intermediate and long-term facilitation in *Aplysia*. *Neuron* **40:** 151–165.

Kruger KE, Sossin WS, Sacktor TC, Bergold PJ, Beushausen S, Schwartz JH. 1991. Cloning and characterization of $Ca^{2+}$-dependent and $Ca^{2+}$-independent PKCs expressed in *Aplysia* sensory cells. *J Neurosci* **11:** 2302–2313.

Lee SJ, Uemura T, Yoshida T, Mishina M. 2012. GluRδ2 assembles four neurexins into *trans*-synaptic triad to trigger synapse formation. *J Neurosci* **32:** 4688–4701.

Levenson J, Endo S, Kategaya LS, Fernandez RI, Brabham DG, Chin L, Byrne JH, Eskin A. 2000. Long-term regulation of neuronal high-affinity glutamate and glutamine uptake in Aplysia. *Proc Natl Acad Sci* **97:** 12858–12863.

Li Q, Roberts AC, Glanzman DL. 2005. Synaptic facilitation and behavioral dishabituation in *Aplysia*: Dependence on release of $Ca^{2+}$ from postsynaptic intracellular stores, postsynaptic exocytosis, and modulation of postsynaptic AMPA receptor efficacy. *J Neurosci* **25:** 5623–5637.

Li HL, Huang BS, Vishwasrao H, Sutedja N, Chen W, Jin I, Hawkins RD, Bailey CH, Kandel ER. 2009. Dscam mediates remodeling of glutamate receptors in *Aplysia* during de novo and learning-related synapse formation. *Neuron* **61:** 527–540.

Liu Q-R, Hattar S, Endo S, MacPhee K, Zhang H, Cleary LJ, Byrne JH, Eskin A. 1997. A developmental gene (*Tolloid*/BMP-1) is regulated in *Aplysia* neurons by treatments that induce long-term sensitization. *J Neurosci* **17:** 755–764.

Liu RY, Cleary LJ, Byrne JH. 2011. The requirement for enhanced CREB1 expression in consolidation of long-term synaptic facilitation and long-term excitability in sensory neurons of *Aplysia*. *J Neurosci* **31:** 6871–6879.

Mackey SL, Kandel ER, Hawkins RD. 1989. Identified serotonergic neurons LCB1 and RCB1 in the cerebral ganglia of *Aplysia* produce presynaptic facilitation of siphon sensory neurons. *J Neurosci* **9:** 4227–4235.

Marinesco S, Carew TJ. 2002. Serotonin release evoked by tail nerve stimulation in the CNS of *Aplysia*: Characterization and relationship to heterosynaptic plasticity. *J Neurosci* **22:** 2299–2312.

Martin KC, Michael D, Rose JC, Barad M, Casadio A, Zhu H, Kandel ER. 1997. MAP kinase translocates into the nucleus of the presynaptic cell and is required for long-term facilitation in *Aplysia*. *Neuron* **18:** 899–912.

Mauelshagen J, Parker GR, Carew TJ. 1996. Dynamics of induction and expression of long-term synaptic facilitation in *Aplysia*. *J Neurosci* **16:** 7099–7108.

McAllister AK. 2007. Dynamic aspects of CNS synapse formation. *Annu Rev Neurosci* **30:** 425–450.

Miniaci MC, Kim JH, Puthanveettil SV, Si K, Zhu H, Kandel ER, Bailey CH. 2008. Sustained CPEB-dependent local protein synthesis is required to stabilize synaptic growth for persistence of long-term facilitation in *Aplysia*. *Neuron* **59:** 1024–1036.

Ormond J, Hislop J, Zhao Y, Webb N, Vaillaincourt F, Dyer JR, Ferraro G, Barker P, Martin KC, Sossin WS. 2004. ApTrkl, a Trk-like receptor, mediates serotonin-dependent ERK activation and long-term facilitation in *Aplysia* sensory neurons. *Neuron* **44:** 715–728.

Philips GT, Sherff CM, Menges SA, Carew TJ. 2011. The tail-elicited tail withdrawal reflex of *Aplysia* is mediated centrally at tail sensory-motor synapses and exhibits sensitization across multiple temporal domains. *Learn Mem* **18**: 272–282.

Pinsker HM, Hening WA, Carew TJ, Kandel ER. 1973. Long-term sensitization of a defensive withdrawal reflex in *Aplysia*. *Science* **182**: 1039–1042.

Pu L, Kopec AM, Boyle HD, Carew TJ. 2014. A novel cystine-rich neurotrophic factor in Aplysia facilitates growth, MAPK activation, and long-term synaptic facilitation. *Learn Mem* **21**: 215–222.

Purcell AL, Sharma SK, Bagnall MW, Sutton MA, Carew TJ. 2003. Activation of a tyrosine kinase-MAPK cascade enhances the induction of long-term synaptic facilitation and long-term memory in *Aplysia*. *Neuron* **37**: 473–484.

Rajasethupathy P, Antonov I, Sheridan R, Frey S, Sander C, Tuschl T, Kandel ER. 2012. A role for neuronal piRNAs in the epigenetic control of memory-related synaptic plasticity. *Cell* **149**: 693–707.

Ripley B, Otto S, Tiglio K, Williams ME, Ghosh A. 2011. Regulation of synaptic stability by AMPA receptor reverse signaling. *Proc Natl Acad Sci* **108**: 367–372.

Sahley CL, Modney BK, Boulis NM, Muller KJ. 1994. The S cell: An interneuron essential for sensitization and full dishabituation of leech shortening. *J Neurosci* **14**: 6715–6721.

Sakurai A, Calin-Jageman RJ, Katz PS. 2007. Potentiation phase of spike timing-dependent neuromodulation by a serotonergic interneuron involves an increase in the fraction of transmitter release. *J Neurophysiol* **98**: 1975–1987.

Scholz KP, Byrne JH. 1987. Long-term sensitization in *Aplysia*: Biophysical correlates in tail sensory neurons. *Science* **235**: 685–687.

Scuri R, Mozzachiodi R, Brunelli M. 2002. Activity-dependent increase of the AHP amplitude in T sensory neurons of the leech. *J Neurophysiol* **88**: 2490–2500.

Scuri R, Mozzachiodi R, Brunelli M. 2005. Role for calcium signaling and arachidonic acid metabolites in the activity-dependent increase of AHP amplitude in leech T sensory neurons. *J Neurophysiol* **94**: 1066–1073.

Sharma SK, Sherff CM, Stough S, Hsuan V, Carew TJ. 2006. A tropomyosin-related kinase B ligand is required for ERK activation, long-term synaptic facilitation, and long-term memory in *Aplysia*. *Proc Natl Acad Sci* **103**: 14206–14210.

Shobe JL, Zhao Y, Stough S, Ye X, Hsuan V, Martin KC, Carew TJ. 2009. Temporal phases of activity-dependent plasticity and memory are mediated by compartmentalized routing of MAPK signaling in *Aplysia* sensory neurons. *Neuron* **61**: 113–125.

Si K, Choi YB, White-Grindley E, Majumdar A, Kandel ER. 2010. *Aplysia* CPEB can form prion-like multimers in sensory neurons that contribute to long-term facilitation. *Cell* **140**: 421–435.

Sutton MA, Carew TJ. 2000. Parallel molecular pathways mediate expression of distinct forms of intermediate-term facilitation at tail sensory-motor synapses in *Aplysia*. *Neuron* **26**: 219–231.

Sutton MA, Masters SE, Bagnall MW, Carew TJ. 2001. Molecular mechanisms underlying a unique intermediate phase of memory in *Aplysia*. *Neuron* **31**: 143–154.

Sutton MA, Bagnall MW, Sharma SK, Shobe J, Carew TJ. 2004. Intermediate-term memory for site-specific sensitization in *Aplysia* is maintained by persistent activation of protein kinase C. *J Neurosci* **24**: 3600–3609.

Thompson RF, Spencer WA. 1966. Habituation: A model phenomenon for the study of neuronal substrates of behavior. *Psychol Rev* **73**: 16–43.

Timbers TA, Rankin CH. 2011. Tap withdrawal circuit interneurons require CREB for long-term habituation in *Caenorhabditis elegans*. *Behav Neurosci* **125**: 560–566.

Trudeau LE, Castellucci VF. 1995. Postsynaptic modifications in long-term facilitation in *Aplysia*: Upregulation of excitatory amino acid receptors. *J Neurosci* **15**: 1275–1284.

Villareal G, Li Q, Cai D, Glanzman DL. 2007. The role of rapid, local, postsynaptic protein synthesis in learning-related synaptic facilitation in *Aplysia*. *Curr Biol* **17**: 2073–2080.

Villareal G, Li Q, Cai D, Fink AE, Lim T, Bougie JK, Sossin WS, Glanzman DL. 2009. Role of protein kinase C in the induction and maintenance of serotonin-dependent enhancement of the glutamate response in isolated siphon motor neurons of *Aplysia californica*. *J Neurosci* **29**: 5100–5107.

Wainwright ML, Zhang H, Byrne JH, Cleary LJ. 2002. Localized neuronal outgrowth induced by long-term sensitization training in *Aplysia*. *J Neurosci* **22**: 4132–4141.

Walters ET, Byrne JH, Carew TJ, Kandel ER. 1983a. Mechanoafferent neurons innervating tail of *Aplysia*: I. Response properties and synaptic connections. *J Neurophysiol* **50**: 1522–1542.

Walters ET, Byrne JH, Carew TJ, Kandel ER. 1983b. Mechanoafferent neurons innervating tail of *Aplysia*: II. Modulation by sensitizing stimulation. *J Neurophysiol* **50**: 1543–1559.

Zaccardi ML, Mozzachiodi R, Traina G, Brunelli M, Scuri R. 2012. Molecular mechanisms of short-term habituation in the leech *Hirudo medicinalis*. *Behav Brain Res* **229**: 235–243.

Zhang F, Endo S, Cleary LJ, Eskin A, Byrne JH. 1997. Role of transforming growth factor-β in long-term synaptic facilitation in *Aplysia*. *Science* **275**: 1318–1320.

Zhao Y, Leal K, Abi-Farah C, Martin KC, Sossin WS, Klein M. 2006. Isoform specificity of PKC translocation in living *Aplysia* sensory neurons and a role for $Ca^{2+}$-dependent PKC APL I in the induction of intermediate-term facilitation. *J Neurosci* **26**: 8847–8856.

Zhu H, Wu F, Schacher S. 1997. Site-specific and sensory neuron-dependent increases in postsynaptic glutamate sensitivity accompany serotonin-induced long-term facilitation at *Aplysia* sensorimotor synapses. *J Neurosci* **17**: 4976–4986.

Zucker RS. 1972. Crayfish escape behavior and central synapses: II. Physiological mechanisms underlying behavioral habituation. *J Neurophysiol* **35**: 621–637.

# Motor Learning and the Cerebellum

## Chris I. De Zeeuw[1,2] and Michiel M. Ten Brinke[1]

[1]Department of Neuroscience, Erasmus Medical Center, 3015 GE Rotterdam, The Netherlands

[2]Netherlands Institute for Neuroscience, 1105 BA Amsterdam, The Netherlands

*Correspondence:* c.dezeeuw@erasmusmc.nl

Although our ability to store semantic declarative information can nowadays be readily surpassed by that of simple personal computers, our ability to learn and express procedural memories still outperforms that of supercomputers controlling the most advanced robots. To a large extent, our procedural memories are formed in the cerebellum, which embodies more than two-thirds of all neurons in our brain. In this review, we will focus on the emerging view that different modules of the cerebellum use different encoding schemes to form and express their respective memories. More specifically, zebrin-positive zones in the cerebellum, such as those controlling adaptation of the vestibulo-ocular reflex, appear to predominantly form their memories by potentiation mechanisms and express their memories via rate coding, whereas zebrin-negative zones, such as those controlling eyeblink conditioning, appear to predominantly form their memories by suppression mechanisms and express their memories in part by temporal coding using rebound bursting. Together, the different types of modules offer a rich repertoire to acquire and control sensorimotor processes with specific challenges in the spatiotemporal domain.

In the formation of procedural memories, the cerebellum shows at least two types of information coding within its massive neuronal networks (De Zeeuw et al. 2011; Person and Raman 2012; Heck et al. 2013; Yang and Lisberger 2013). Modulation of the average firing rate of neuronal spikes or "rate coding" is most often proposed as the predominant mechanism of information coding used for motor learning (Boyden et al. 2004; Lisberger 2009; Walter and Khodakhah 2009). However, spikes occur at millisecond precision, and their actual timing or "temporal coding" can increase the information content of spike trains and facilitate the entrainment of postsynaptic activity (Markram et al. 1997; De Zeeuw et al. 2011). To a large

extent, the coding mechanisms used in the cerebellum for learning and expressing a particular form of motor learning depend on the specific cerebellar module that is controlling the type of behavior involved. The existence of cerebellar modules was discovered half a century ago by Jan Voogd (1964). Simply by studying the thickness of myelinated Purkinje cell axons in the white matter of the cerebellar cortex, Voogd observed distinct differences that were consistently organized in sagittal zones (Fig. 1). Subsequent tracing and immunocytochemical experiments showed that each of these Purkinje cell zones provides an inhibitory projection to a distinct part of the cerebellar nuclei, which in turn inhibits a specific olivary subnucleus (Fig. 2)

Cite this article as *Cold Spring Harb Perspect Biol* doi: 10.1101/cshperspect.a021683

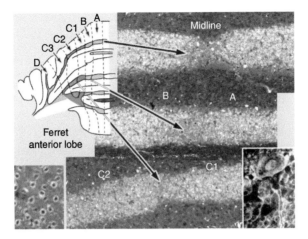

**Figure 1.** Original identification of Purkinje cell zones in the cerebellar cortex of ferrets using Haggquist staining (based on data from Voogd 1964).

(Groenewegen and Voogd 1977; De Zeeuw et al. 1994, 2011; Ruigrok and Voogd 2000; Ito 2002; Schonewille et al. 2006a). Because the climbing fibers originating from each olivary subnucleus project back to the Purkinje cells of the corresponding cerebellar cortical zone, these circuitries form precisely topographically organized three-element loops (Fig. 3A,B). They are referred to as the olivocerebellar modules and constitute the fundamental building blocks of the cerebellar system. The mossy fiber–parallel fiber system is superimposed in a largely orthogonal fashion on top of the sagittally oriented Purkinje cell zones (Fig. 3C); individual mossy fibers innervate multiple granule cells usually situated in multiple zones and the parallel fibers originating from these granule cells consistently traverse multiple zones in the molecular layer.

Over the past decades, it has gradually become clear that each module is concerned with control of specific tasks, such as execution of limb and finger movements, of trunk movements for balance, of compensatory eye movements about particular axes in space, reflexes of facial musculature, homeostasis of particular autonomic processes, and probably even specific cognitive tasks, such as time-sensitive decision making (De Zeeuw et al. 1994; Ito 2008; Jörntell et al. 2000; Apps and Hawkes 2009; Rahmati et al. 2014). However, it was not until recently that the specific intrinsic properties of different

categories of modules emerged (Zhou et al. 2014). Here, we review the intrinsic differences of cerebellar modules and the implications for the coding mechanisms involved in cerebellar motor learning.

## INTRINSIC DIFFERENCES AMONG CEREBELLAR MODULES

The sagittal zones of Purkinje cells in the cerebellar cortex can be identified based on the alternating presence and absence of expression of proteins, such as $5'$-nucleotidase, zebrin I (i.e., mabQ113 antigen) and zebrin II (i.e., aldolase C), phospholipase C$\beta$3 and $\beta$4, excitatory amino acid transporter 4 (EAAT4), GABA$_{B2}$ receptors, and splice variant b of the metabotropic glutamate receptor 1 (mGluR1b) (Brochu et al. 1990; Leclerc et al. 1990; Dehnes et al. 1998; Mateos et al. 2001; Wadiche and Jahr 2005; Apps and Hawkes 2009). These zebra-like patterns of protein distribution appear to be present in the cerebellum of all birds and mammals (Brochu et al. 1990; Sillitoe et al. 2003; Chung et al. 2007; Apps and Hawkes 2009; Graham and Wylie 2012), and in many cases they largely correspond to the organization of the olivocerebellar modules (Figs. 2,3A–C) (Sugihara and Shinoda 2004, 2007; Voogd and Ruigrok 2004; Pijpers et al. 2006; Sugihara et al. 2009; Sugihara 2011). For example, zebrin II, EAAT4, and

    Cite this article as *Cold Spring Harb Perspect Biol* doi: 10.1101/cshperspect.a021683

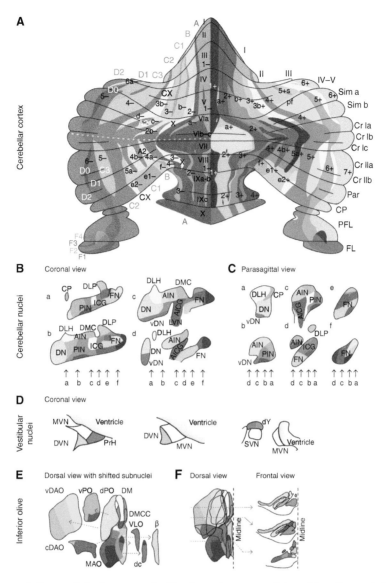

**Figure 2.** Olivocerebellar modules in mammals. The three-element modules of the olivocerebellar system are formed by a sagittal strip of Purkinje cells in the cerebellar cortex (*A*), which converge onto a particular set of cerebellar and/or vestibular nuclei (*B–D*), which, in turn, innervate the subnucleus in the inferior olive (*E,F*) that provides the climbing fibers to the corresponding strip of Purkinje cells, forming a closed triangular loop. (*A*) The left part of the cerebellar cortex indicates the original zones described by Voogd (1964), whereas the right part indicates the zebrin related groups described by Sugihara (2011). The color coding used in panels *B–F* is the same as that used for describing Sugihara's groups in *A* (i.e., *right* half), and together they reflect which parts within a particular olivocerebellar module are connected. For reference, we indicated the zebrin-positive strips with dark shading in Voogd's zones on the *left*. (*A*) 1–6(a/b/−/+) (see Sugihara and Shinoda 2004, 2007); I–X, lobules I–X; CP, copula pyramidis; Cr I/II, crus I/II of ansiform lobule; FL, flocculus; Par, paramedian lobule; PFL, paraflocculus; Sim, lobulus simplex. (*B,C*) AICG, anterior interstitial cell group; AIN, anterior interposed nucleus; CP, copula pyramidis; DLH, dorsolateral hump; DLP, dorsolateral protuberance; DMC, dorsomedial crest; (v)DN, (ventral) dentate nucleus; FN, fastigial nucleus; ICG, interstitial cell group; PIN, posterior interposed nucleus. (*D*) DVN, descending vestibular nucleus; dY, dorsal group Y; MVN, medial vestibular nucleus; PrH, prepositus hypoglossal nucleus; SVN, superior vestibular nucleus. (*E,F*) β, subnucleus β; (c/v)DAO, (central/ventral) dorsal accessory olive; dc, dorsal cap; DM, dorsomedial group; DMCC, dorsomedial cell column; MAO, medial accessory olive; (d/v)PO, (dorsal/ventral) principal olive; VLO, ventrolateral outgrowth. Note that the X/CX-zones have only been found at the electrophysiological level (Ekerot and Larson 1982).

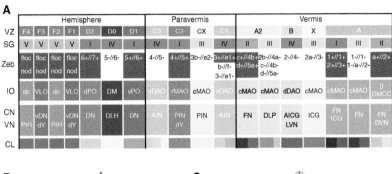

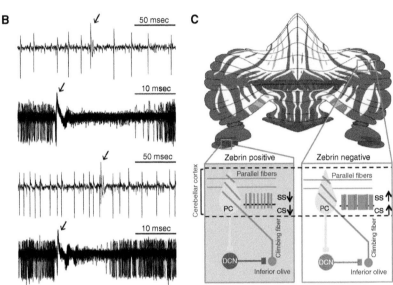

**Figure 3.** Olivocerebellar modules and Purkinje cell activity in relation to zebrin (II) distribution. (*A*) The VZ and SG rows refer to the zones and groups of Purkinje cells described by Voogd (1964; VZ) and Sugihara (2011; SG), respectively. The Zeb row indicates which zones and groups are zebrin positive (grey) and zebrin negative (white). The inferior olive (IO) row indicates which subnucleus of the IO is providing climbing fibers to a particular zone/group of Purkinje cells in the cerebellar cortex and collaterals to a particular part of the cerebellar nucleus, depicted in the same column. The CN and VN row indicates the parts of the cerebellar nuclei (CN) and vestibular nuclei (VN) that are innervated by the strip of Purkinje cells, depicted in the same column. It should be noted that only those vestibular nuclei are indicated that both receive a Purkinje cell input and provide a feedback projection to the inferior olive; because, for example, medial and superior vestibular nuclei do not project to the IO, they are not incorporated in this scheme. In addition, it should be noted that this overview is also incomplete in that some nuclei, such as the dorsomedial cell column (DMCC), may receive inhibitory feedback from multiple hindbrain regions. CL indicates the color legends used for Figure 2. AICG, anterior interstitial cell group; AIN, anterior interposed nucleus; β, subnucleus β; c/rMAO, caudal/rostral medial accessory olive; dc, dorsal cap; DLH, dorsolateral hump; DLP, dorsolateral protuberance; DM, dorsomedial group; DMC, dorsomedial crest; d/vDAO, dorsal/ventral dorsal accessory olive; DVN, descending vestibular nucleus; dY, dorsal group Y; floc, flocculus; FN, fastigial nucleus; ICG, interstitial cell group; LVN, lateral vestibular nucleus; nod, nodulus; PIN, posterior interposed nucleus; PrH, prepositus hypoglossal nucleus; vDN, ventral dentate nucleus; and VLO, ventrolateral outgrowth. (*B*) Examples of raw traces of Purkinje cell activity from zebrin-positive (*top* panels) and zebrin-negative (*bottom* panels) zones. Arrows indicate complex spike. (From Zhou et al. 2014; reprinted, with permission, from the authors.) (*C*) Intramodular connections via deep cerebellar nuclei (DCN) explaining why the complex spike (CS) activity within a module follows the intrinsic differences in simple spike (SS) activity of Purkinje cells (PC). PC and DCN are inhibitory, whereas climbing fibers are excitatory. (From Albergaria and Carey 2014; reprinted under the terms of the Creative Commons Attribution License, which permits unrestricted use and redistribution provided that the original author and source are credited.)

Cite this article as *Cold Spring Harb Perspect Biol* doi: 10.1101/cshperspect.a021683

GABA$_{B2}$ receptors are distributed in Purkinje cells of zones C2, D1, and D2, whereas mGluR1b is prominently expressed in zones B, C1, C3, and D0, providing a complementary pattern (Mateos et al. 2001; Chung et al. 2007; Apps and Hawkes 2009). Importantly, Zhou and colleagues (2014) recently showed that these distribution patterns determine the intrinsic simple spike activity of Purkinje cells (Fig. 3B,C).

During sensorimotor stimulation and natural behavior, the simple spikes can modulate as a consequence of excitation via the mossy fiber–parallel fiber pathway and inhibition via the molecular layer interneurons, but at-rest Purkinje cells show a relatively high level of intrinsic activity, which can reach levels up to 120 Hz. In Purkinje cell zones positive for zebrin II and EAAT4 (referred to as zebrin-positive zones), simple spike firing approximates 60 Hz, whereas in those zones positive for mGluR1b (referred to as zebrin-negative zones), the average firing rate reaches 90 Hz (Zhou et al. 2014). The intrinsic nature of this difference in simple spike activity at rest cannot be inferred only from the fact that it can be correlated with differential protein expression inside Purkinje cells, but also from the fact that this difference holds when excitatory or inhibitory inputs to Purkinje cells are blocked (Wulff et al. 2009; Galliano et al. 2013a; Zhou et al. 2014). The molecular mechanisms that determine the differences in firing frequencies in the zebrin-positive and zebrin-negative zones have been only partly resolved. Blocking transient receptor potential cation channel type C3 (TRPC3), which can be associated with zebrin-negative Purkinje cells and is required for the mGluR1-mediated slow excitatory postsynaptic currents (EPSCs) (Mateos et al. 2001; Hartmann et al. 2008; Chanda and Xu-Friedman 2011; Kim et al. 2012a,b; Nelson and Glitsch 2012), reduces simple spike activity of Purkinje cells in zebrin-negative, but not zebrin-positive Purkinje cells (Zhou et al. 2014). Thus, tonic activation of mGluR1b by ambient glutamate in zebrin-negative Purkinje cells might lead to opening of their TRPC3-channels and thereby to a relatively high level of simple spike activity (Yamakawa and Hirano 1999; Coesmans et al. 2003; Chanda and Xu-Friedman 2011). In contrast, similar glutamate-dependent increases may be prevented in zebrin-positive Purkinje cells, in which EAAT4 might help to keep glutamate concentrations relatively low (Dehnes et al. 1998; Auger and Attwell 2000; Wadiche and Jahr 2005; cf. Zhou et al. 2014). Downstream from mGluR1, proteins, such as the IP3-receptor (TRPC3 modulator), PLC-β3/4 (TRPC3 activator), protein kinase C (PKC)-δ, and NCS-1, play key roles in calcium release from intracellular calcium stores and consequently have electrophysiological impact in line with that of TRPC3. Because several of these proteins are expressed in zebrin-like bands (Barmack et al. 2000; Jinno et al. 2003; Sarna et al. 2006; Hartmann et al. 2008; Becker et al. 2009; Furutama et al. 2010; Wang et al. 2011; Kim et al. 2012a,b), it is possible that this entire pathway contributes to the high simple spike activity of zebrin-negative Purkinje cells. To what extent zebrin itself (i.e., zebrin II or aldolase C) contributes to this pathway is yet unknown, as the impact of its reaction products on simple spike firing is unclear (Zhou et al. 2014).

Whereas the simple spikes are to a large extent determined by the intrinsic activity of Purkinje cells, the all-or-none complex spike activity directly reflects activity in the afferent climbing fibers derived from neurons in the inferior olive (De Zeeuw et al. 2011; Albergaria and Carey 2014; Zhou et al. 2014). Interestingly, at rest, the firing frequency of the complex spikes is aligned with that of the simple spikes in that their firing frequency is also significantly higher in zebrin-negative zones compared with that in zebrin-positive zones (Fig. 3B,C). Thus, even though the complex spike activity copies perfectly the activity of olivary neurons at rest, it still follows the trend of simple spike activity, which is determined by the intrinsic activity of Purkinje cells. How can this come about? This alignment presumably results from a network effect within the olivocerebellar modules engaging the GABAergic neurons in the cerebellar nuclei (Chen et al. 2010; De Zeeuw et al. 2011). Enhanced simple spike activity, as observed in the zebrin-negative modules, will lead to reduced firing of these cerebellar nuclei neurons that inhibit the inferior olivary neurons leading to an increase in complex spike activity (De Zeeuw et al. 1988). Because an increase in complex spike activity suppresses simple spike frequency through cerebellar cortical interneurons in the molecular layer (Mathews et al. 2012; Coddington et al. 2013), this network effect ultimately provides an excellent way to mediate homeostasis of activity within the olivocerebellar modules (Fig. 3B,C).

Together, the intrinsically determined simple spike and complex spike activity at rest provide the baseline values around which the Purkinje cells are modulated during natural sensory stimulation, such as that used to induce motor learning. This raises the question as to whether motor learning in the different olivocerebellar modules is also dominated by different plasticity rules mechanisms. Given the baseline firing frequencies, one might expect that zebrin-positive modules with relatively low firing frequencies have ample room for mechanisms of potentiation, whereas zebrin-negative modules showing high simple spike activity could be more prone to suppression. Indeed, Wang and colleagues (2011) found in vivo that the activity of zebrin-positive, but not zebrin-negative, Purkinje cells can be readily enhanced, whereas Wadiche and Jahr (2005) found that, in vitro, the induction of long-term depression (LTD) at the parallel fiber to Purkinje cell synapse can be readily induced in zebrin-negative Purkinje cells in lobule III, but not in zebrin-positive cells in lobule X. Below we will review the dominant learning rules for both a zebrin-positive region, that is, the flocculus of the vestibulocerebellum controlling adaptation of the vestibulo-ocular reflex (Lisberger 1988; Ito 2002; De Zeeuw and Yeo 2005), and a zebrin-negative region, that is, hemispheral lobule VI controlling classical eyeblink conditioning (Hesslow 1994a,b; Thompson and Steinmetz 2009; Boele et al. 2010; Mostofi et al. 2010).

## MOTOR LEARNING IN A ZEBRIN-POSITIVE MODULE: ADAPTATION OF THE VESTIBULO-OCULAR REFLEX

The flocculus, like the nodulus of the vestibulocerebellum, is virtually completely zebrin posi-

tive, and indeed its Purkinje cells fire at an average of ∼60 Hz at rest (Fig. 3B). It contains five zones, one for controlling compensatory head movements (extension of the C2 zone) and four for controlling compensatory eye movements about different axes in space (extension of D1–D2 zones, but referred to as F zones) (Fig. 4A) (De Zeeuw et al. 1994; De Zeeuw and Koekkoek 1997; Schonewille et al. 2006a; Voogd et al. 2012). The vestibulo-ocular reflex translates head movement into compensatory eye movement so as to keep the observed image in the center of the visual field. By experimentally moving a subject's head while also moving the visual environment in the same or opposite direction (i.e., in or out of phase), this reflex will prove insufficient or exaggerated, until the new rules are integrated in the compensatory eye movements following a process of adaptation learning. Mechanical or genetic lesions of floccular Purkinje cells severely hamper adaptation of compensatory eye movements (Endo et al. 2009; Gao et al. 2012). Recordings of Purkinje cells in the flocculus of awake behaving mammals during a vestibulo-ocular reflex paradigm in the dark or light show simple spike modulation that correlates well with both maximum head velocity and maximum eye velocity (De Zeeuw et al. 1995). Adapting the reflex using gain-increase or phase-reversal training leads to an increment in the modulation amplitude of simple spikes (Clopath et al. 2014; K Voges and CI De Zeeuw, pers. comm.), whereas impairing the modulation amplitude of simple spikes by genetically attenuating the parallel fiber to Purkinje cell synapse leads to a reduction in the peak of simple spike modulation as well as in the adaptation and consolidation of the reflexive compensatory eye movements (Fig. 4B–E) (Galliano et al. 2013a).

Moreover, stimulating simple spike activity of Purkinje cells either pharmacologically or optogenetically leads to an increase in the excitatory phase of the modulation amplitude of the simple spikes as well as an increase in the gain of compensatory eye movements (van der Steen and Tan 1997; De Zeeuw et al. 2004; Nguyen-Vu et al. 2013). Thus, in line with the data obtained by Wang and colleagues (2011)

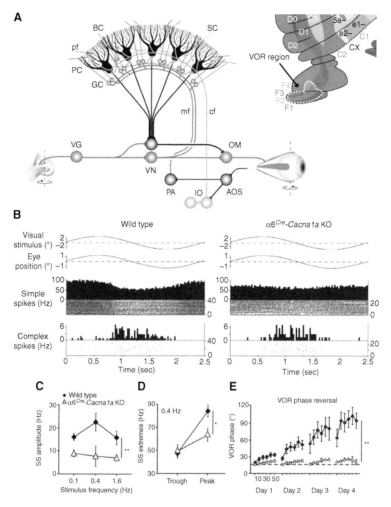

**Figure 4.** Circuit of the vestibulo-ocular reflex and prominent role of modulation amplitude. (*A*) The vestibulo-ocular reflex (VOR) is mediated by the three-neuron arc of Lorente de No in the brainstem. When the head rotates, the vestibular signals from the semicircular canals are transferred by the vestibular ganglion cells (VG) to the second-order vestibular neurons in the vestibular nuclei (VN), which in turn innervate the oculomotor neurons (OMs) driving the eyes to the opposite side. The vestibulocerebellum, which is superimposed on this three-neuron arc, is required to compensate for the delays introduced during input–output processing. To minimize retinal slip during head movements, the accessory optic system (AOS) relays slip signals through the climbing fiber (cf) system to the zebrin-positive Purkinje cells (PC) in floccular zones F1–F4 in the vestibu-locerebellar cortex (see *inset*), where the presence and absence of the climbing fiber activity is integrated with vestibular, optokinetic, and eye movement signals mediated by the mossy fiber (mf)–granule cell (GC)–parallel fiber (pf) pathway. The Purkinje cells in turn can inject well-calibrated, accelerating signals into the vestibular brainstem so as to precisely compensate for the delays. (*B*) $\alpha6^{Cre}$-*Cacna1a* knockout (KO) mice, in which transmission in the vast majority, but not all, of parallel fibers is blocked, show a normal amplitude of their optokinetic reflex when a visual stimulus is given, despite a reduced modulation of their simple spike activity. (*C,D*) The modulation of the simple spike, but not the complex spike, activity in $\alpha6^{Cre}$-*Cacna1a* KO is reduced over a wide range of frequencies, and these deficiencies are caused by a reduction in the peak of the modulation. (*E*) VOR phase reversal is a form of VOR adaptation, during which the phase of the VOR is reversed by providing an in-phase optokinetic stimulus that is greater in amplitude than the vestibular stimulus; $\alpha6^{Cre}$-*Cacna1a* KO mice have severe problems reversing the phase of their eye movements indicating that mammals have their abundance of GCs and pfs to control motor learning rather than basic motor performance. PA, pontine area; BC, basket cell; SC, stellate cell. (From Galliano et al. 2013a; modified, with permission, from the authors.)

and Wadiche and Jahr (2005) in other zebrin-positive areas of the cerebellum, these data suggest that strengthening the parallel fiber to Purkinje cell synapse (i.e., through long-term potentiation or LTP) or enhancing the intrinsic excitability of Purkinje cells form the dominating forms of plasticity in the zebrin-positive floccular zones controlling vestibulo-ocular reflex adaptation. Indeed, affecting both forms of potentiation simultaneously by deleting PP2B specifically in Purkinje cells results in deficits in various forms of adaptation of the reflex, such as gain increase, gain decrease, and phase-reversal adaptation (Schonewille et al. 2010). Along the same lines, enhancing Purkinje cell potentiation through an artificial or natural increase of estradiol also improves vestibulo-ocular reflex learning (Andreescu et al. 2007).

In contrast, blocking expression of LTD at the parallel fiber to Purkinje cell synapse by targeting proteins involved in late events of its signaling cascade at the level of GluRs (GluRd7 knockin and GluR2K882A knockin) or related proteins that control their trafficking (PICK1 knockout) does not lead to any obvious deficit in compensatory eye movement learning (Schonewille et al. 2011). These latter experiments indicate that LTD is not essential for vestibulo-ocular reflex adaptation, but they do not exclude the possibility that LTD contributes to this form of motor learning under physiological conditions. Possibly, the blockage of LTD expression at the parallel fiber to Purkinje cell synapse in the GluRd7 knockin, GluR2K882A knockin, and PICK1 knockout is compensated for by LTP at the parallel fiber to molecular layer interneuron synapse (Jörntell and Ekerot 2002; Gao et al. 2012; Tanaka et al. 2013). Even though motor learning in the zebrin-positive floccular zones may be dominated by postsynaptic and intrinsic potentiation of Purkinje cell activity, the olivocerebellar system is endowed with various distributed forms of plasticity that operate in a synergistic fashion and allow for ample compensation (Gao et al. 2012). This synergy results from the fact that virtually all major forms of plasticity in the cerebellar cortex are controlled by the climbing fibers, and climbing fiber activity is phase-dependent. For example, when an optokinetic pattern moves into temporonasal direction, the subsequent activation of complex spikes in the Purkinje cells of the floccular vertical-axis zones (Fig. 4A) enhances LTD at the parallel fiber to Purkinje cell synapse as well as (on the ipsilateral side) LTP at the parallel-fiber to molecular-layer interneuron synapse and potentiation at the molecular layer interneuron to Purkinje cell synapse (Gao et al. 2014). Yet, when the optokinetic stimulus moves in the opposite direction and the climbing fibers are virtually silent (while being active on the contralateral side), it will induce LTP at the parallel fiber to Purkinje cell synapse and LTD at the parallel-fiber to molecular-layer interneuron synapse (Gao et al. 2012). Together, these climbing-fiber-driven forms of plasticity are so prominent that selectively rerouting the climbing fibers from a contralateral to an ipsilateral projection, while maintaining the laterality of the mossy fiber system, completely reverses modulation of both Purkinje cells' simple spikes and molecular layer interneuron activity (Fig. 5) and induces dramatically ataxic motor behavior, which actually benefits from a cerebellectomy (Badura et al. 2013).

Downstream, it is probably the changes in simple spikes rather than the complex spikes that largely contribute to the changes in eye-movement behavior during adaptation of the vestibulo-ocular reflex (De Zeeuw et al. 2004). Comparison between recordings from floccular target neurons in the vestibular nuclei and floccular Purkinje cells indicates that it is the simple spikes that can relay the prediction signals required for this type of adaptation learning (De Zeeuw et al. 1995; Stahl and Simpson 1995). Indeed, through pure rate coding and plasticity mechanisms in both the flocculus and vestibular nucleus neurons (Nelson et al. 2005), one can explain normal vestibulo-ocular reflex learning and consolidation in regular wild-type animals as well as the specific behavioral phenotypes and simple spike firing characteristics observed in various mutant mice in which either the excitatory or inhibitory inputs to the Purkinje cells are affected (Clopath et al. 2014).

Cite this article as *Cold Spring Harb Perspect Biol* doi: 10.1101/cshperspect.a021683

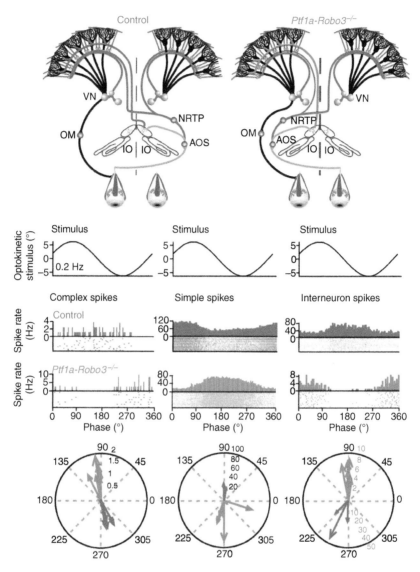

**Figure 5.** Climbing fibers dominate timing of simple spike firing. Although the complex spikes of Purkinje cells are modulated by activity in the climbing fiber system, the simple spikes are supposed to be largely driven by the mossy fiber system. The frequencies of these two types of spikes are often modulated reciprocally. An increase in complex spikes is associated with a decrease in simple spikes, and vice versa. This reciprocal firing is thought to be essential for motor behavior. Rerouting the climbing fiber system in *Ptf1a-Robo3* mice from a contralateral (dark blue line in *top* panel) to a predominantly ipsilateral projection (red line in *top* panel) does not only reverse the modulation of complex spike activity during natural optokinetic stimulation (see peristimulus–time histograms (PSTHs), raster, and polar plots in *left* panel), but also that of the simple spike activity (*middle* panel). Because the laterality of the mossy fiber projection is unaffected (green lines in *top* panels), these data show that the proper timing of the climbing fiber input is essential for well-coordinated motor performance by controlling the timing of simple spike firing. The phase of molecular layer interneurons is also reversed in the mutants (*right* panel), which suggests that climbing fibers evoke their effects on simple spike activity via molecular layer interneurons. VN, Vestibular nuclei; NRTP, nucleus reticularis tegmenti pontis; OM, oculomotor neuron; IO, inferior olive; AOS, accessory optic system. (From Badura et al. 2013; modified, with permission, from the authors.)

## MOTOR LEARNING IN A ZEBRIN-NEGATIVE MODULE: EYEBLINK CONDITIONING

The extensions of the zebrin-negative bands are more prominent in the rostral direction of the cerebellum compared with their caudal counterparts (Sugihara and Shinoda 2004), endowing hemispheric lobule VI, or simplex, with a substantial amount of zebrin-negative Purkinje cells that typically fire at ~90 Hz, subdivided across zones C1, C3, and D0 in particular (Sugihara and Shinoda 2004; Ten Brinke et al. 2014; Zhou et al. 2014). Together with zebrin-positive zone C2, zones C3 and D0 have been shown to respond to periocular stimulation (Hesslow 1994a,b; Mostofi et al. 2010). Through tracer, lesion, and stimulation studies, it has become apparent that cells in C2 are more generally receptive to different kinds of stimulation, whereas C3 and D0 are specifically engaged with eyelid behavior, with their Purkinje cell output ultimately tying in to the eyelid muscle circuitry (Yeo et al. 1985a,b,c, 1986; Hesslow 1994a,b; Attwell et al. 2001; Boele et al. 2010, 2013; Mostofi et al. 2010).

In the eyeblink-conditioning paradigm, a neutral stimulus leads to an eyeblink response on repeated pairing with a subsequent blink-inducing stimulus (McCormick and Thompson 1984; Yeo et al. 1986; Thompson and Steinmetz 2009; Boele et al. 2010). Eyeblink conditioning has been found to coincide with the development of a marked decrease in Purkinje cell simple spike firing with temporal characteristics similar to those of eyelid conditioned responses (CRs) (Fig. 6) (Albus 1971; Hesslow and Ivarsson 1994; Jirenhed et al. 2007; Ten Brinke et al. 2014). The conditioned stimulus ([CS], e.g., a light or tone) and unconditioned stimulus ([US], e.g., a corneal airpuff), in between which the CR occurs, find their respective physiological correlates in the activity of a myriad of parallel fibers and a single climbing fiber synapsing on the Purkinje cells. The repeated pairing of CS-related parallel fiber input with a subsequent climbing fiber signal, an efferent copy of the eyeblink reflex loop (Fig. 6A), sensitizes the Purkinje cells to the CS in that its simple spike activity gradually diminishes

as the conditioning proceeds (Ten Brinke et al. 2014). Importantly, this process is reversible; when the well-timed CS–US pairing is replaced with randomly paired conditioned and unconditioned stimuli, the conditioned eyeblink response and reduction in simple spike response are gradually and concomitantly extinguished (Fig. 6B, middle panel). Following this extinction, simple spike suppression reappears with a reoccurrence of the CRs in the reacquisition process (Fig. 6B, bottom panel). This suppression of simple spike activity in a zebrin-negative module, which necessitates plasticity reducing Purkinje cell activity, juxtaposes starkly with the predominantly simple spike-enhancing forms of plasticity implicated in vestibulo-ocular reflex learning that takes place in zebrin-positive areas.

Historically, the main plasticity mechanism thought to underlie the simple spike suppression during eyeblink conditioning was LTD at the parallel fiber to Purkinje cell synapse (Ito and Kano 1982; Hauge et al. 1998; Koekkoek et al. 2003). Indeed, LTD at this synapse occurs when parallel fibers and climbing fibers are activated conjunctively (Gao et al. 2012), which corresponds well to the situation created by the paired CS–US trials of the eyeblink-conditioning paradigm. However, when parallel fiber to Purkinje cell LTD is blocked following manipulation of the GluR2-AMPA receptors described above (i.e., GluRd7 and GluR2K882A knockin), acquisition of normal CRs is not significantly impaired (Schonewille et al. 2011; see also Welsh et al. 2005). This finding is in line with the fact that most parallel fibers are probably silent to begin with (Brunel et al. 2004; van Beugen et al. 2013) and that at-rest Purkinje cells fire intrinsically at virtually the same rate with intact parallel fiber input as they do without (Cerminara and Rawson 2004; Galliano et al. 2013a; Hesslow 2013). In terms of rate coding, this reduces the direct impact of the few depressed CS-conveying parallel fibers on the overall simple spike suppression to negligible proportions. Along the same line, Hesslow and colleagues found that the duration of the parallel fiber activation, through which a Purkinje cell is trained, does not deter-

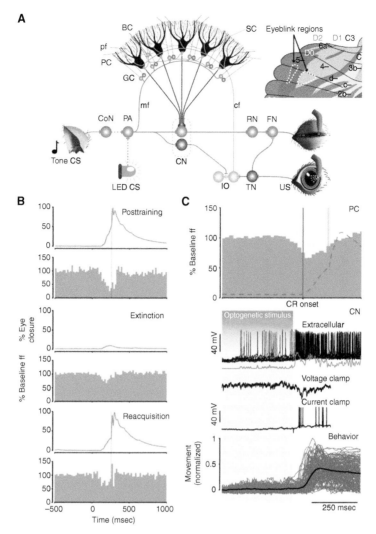

**Figure 6.** Eyeblink circuit and role of simple spike suppression. (*A*) During eyeblink conditioning, a conditioned stimulus (CS), such as a tone or LED light, is repetitively paired with an unconditioned stimuli (US), such as an airpuff, to learn a well-timed conditioned response (CR). CS and US sensory information converges at zebrin-negative Purkinje cells (PCs) in D0 and C3 (see *inset*) through the mossy fiber (mf)–granule cell (GC)–parallel fiber (pf) pathway and the climbing fibers (cfs) derived from the inferior olive (IO), respectively. Although the mossy fibers relay information on the CS from the pontine area (PA), the climbing fibers mediate efferent copies of signals evoked in the direct eyeblink reflex loop, which is formed by the orbital branch of the trigeminal nerve, trigeminal nucleus (TN), facial nucleus (FN), and eyelid muscle. When a fixed temporal relationship between parallel fiber and climbing fiber activation emerges, the same parallel fiber input starts to evoke a simple spike suppression that disinhibits the cerebellar nuclear (CN) cells, and consequently causes the eyelid to close before the US is about to occur. (*B*) Example of mean eyelid behavioral traces and simple spike frequency histograms of a mouse Purkinje cell from lobule HVI (zebrin-negative D0 zone) after training, extinction, and reacquisition (*top* to *bottom*). Note the concomitant changes in simple spike suppression and amplitude of the CRs. The green and red bands in the background depict CS and US duration, respectively. (*C*) (*Top*) Simple spike suppression precedes CR onset (blue line) and covaries with its course (dashed line). (*Bottom*) A rapid drop in Purkinje cell activity after stopping optogenetic stimulation elicits rebound burst activity in CN neurons (extracellular recording in vivo). The excitatory events in voltage and current clamp recordings of the same cell in vivo following optogenetic Purkinje cell stimulation show that this bursting may be facilitated by climbing and/or mossy fiber collaterals. Ultimately, CN rebound activity can reliably evoke a behavioral response. The data described under *C* are obtained from different experiments, but aligned at the same time scale to facilitate understanding of the course of events with respect to each other. CoN, cochlear nucleus; RN, red nucleus. (From Witter et al. 2013; modified, with permission, from the authors.)

mine the extent and duration of the simple spike suppression (Jirenhed and Hesslow 2011a,b).

Together, these findings suggest that there must be one or more mechanism(s) other than parallel fiber LTD that can actively suppress the simple spike activity when the conditioning signals have started to traverse across the parallel fibers. Two of these potential mechanisms include LTP at the parallel-fiber to molecular-layer interneuron synapse and potentiation at the molecular layer interneuron to Purkinje cell synapse, thereby facilitating inhibitory effects of these interneurons onto the Purkinje cells (Jörntell and Ekerot 2002; Gao et al. 2012). Indeed, blocking both mechanisms in effect by ablating the GABA-γ2 receptor specifically in Purkinje cells (Wulff et al. 2009) significantly reduces the percentage and amplitude of CRs (Boele 2014; Ten Brinke et al. 2014). Yet, this inhibitory effect on conditioning behavior is not complete (Boele 2014; Ten brinke et al. 2014), and the impact of gabazine on simple spike suppression in decerebrate ferrets is limited (Johansson et al. 2014), possibly because of extensive ephaptic inhibition at the pinceau-forming terminals of the basket cells (Blot and Barbour 2014). If the interneurons are indeed relevant for the simple spike suppression, LTD might still contribute to this process by reducing the excitation in Purkinje cells during the period in which their parallel fiber input also excites the adjacent molecular layer interneurons, evoking the active suppression. Other possibilities for active suppression include LTP at the parallel fiber to Purkinje cell synapse facilitating transmission of the CS signals and driving inhibitory intrinsic Purkinje cell mechanisms through, for example, metabotropic glutamate receptors and downstream PKC-mediated cascades and/or indirectly eliciting CS-related complex spike activity through the nuclei, which, in turn, adds to direct activation of the molecular layer interneurons (Berthier and Moore 1986; Angaut et al. 1996; Koekkoek et al. 2003; Schonewille et al. 2010; Johansson et al. 2014). In line with the large variety of potential mechanisms, current thoughts about cerebellar learning extend beyond the modification of

mere synaptic input of CS signals and are referred to as distributed synergistic plasticity (Gao et al. 2012).

By adopting a suppressive simple spike response, Purkinje cells disinhibit cerebellar interposed nuclear cells, and the subsequent rebound activity in these cells eventually feeds via the red nucleus into the eyelid muscles, effectively closing the eye in well-timed preparation just before the US occurs (Fig. 6C) (Gauck and Jaeger 2000; Boele et al. 2010; Witter et al. 2013). The simple spike suppression starts well before the onset of the conditioned eyeblink response (Ten Brinke et al. 2014), but the extent to which the subsequent increase in cerebellar nuclei firing determines the onset of the eyelid closure or dynamically controls the closure in an online fashion is not known (Sánchez-Campusano et al. 2011). Presumably, the rebound in the activity of cerebellar nuclei neurons following simple spike suppression is facilitated by excitatory inputs from mossy fiber and/or climbing fiber collaterals (Fig. 6C). Indeed, their excitatory inputs can be detected during whole-cell recordings in vivo at precisely the right moment showing coincidence with the internal rebound (Witter et al. 2013).

In contrast to the zebrin-positive modules controlling vestibulo-ocular reflex adaptation, which seem to entail predominantly rate coding, the zebrin-negative module-controlling eyeblink conditioning appears more prone to temporal coding. This is also supported by the fact that synapses of mossy fiber collaterals onto cerebellar nuclei neurons can show LTP following specific sequential activation (Pugh and Raman 2008) and that cerebellar nuclei neurons can be entrained by periods of synchronized simple spike activity at 50–80 Hz (De Zeeuw et al. 2008, 2011; Person and Raman 2012), that is, the firing rate level that zebrin-negative Purkinje cells acquire during simple spike suppression controlling the conditioned eyeblink response (Fig. 6B,C). Moreover, the mossy fiber collaterals on cerebellar nuclei neurons also show structural preterminal sprouting during conditioning, the amount of which correlates well with the amplitude of the conditioned responses (Boele et al. 2013).

The notion that the eyeblink-conditioning paradigm involves memory formation at both the cerebellar nuclear and cortical level may explain the considerable savings observed during reacquisition of the learned behavior after extinction (Fig. 6B) (Kehoe 1988; Ohyama et al. 2006). Thus, even though the molecular and cellular machinery behind the active suppression central to the activity in zebrin-negative cerebellar zones still poses questions, the evidence for its functional relation to both plasticity in the cerebellar nuclei and accurate behavioral output as well as for the crucial role of climbing fiber activity at all potential plasticity sites is compelling.

## CONCLUDING REMARKS

The data reviewed here establish the differential intrinsic activity of the different sagittal Purkinje cell zones in the cerebellum and the potential consequences for motor learning. In hindsight, the recent finding by Zhou and colleagues (2014) that the intrinsic simple spike-firing frequencies of zebrin-positive and zebrin-negative Purkinje cells differ dramatically (60 Hz vs. 90 Hz) could have been predicted by the original study performed 50 years ago by Voogd (1964). Voogd used Haggquist stainings and found that the cerebellar cortex can be divided in zones of Purkinje cells with thin myelinated axons (e.g., zones C2, D1, and D2, which later turned out to be zebrin positive) and Purkinje cells with thick myelinated axons (e.g., zones B, C1, C3, and D0, which turned out to be zebrin negative) (Fig. 1). Indeed, oligodendrogenesis and the thickness of a myelination sheath appear to depend on neuronal activity and firing rate (Gibson et al. 2014). Yet, we are still only just beginning to answer the 50-year-old question as to what the functional meaning of the cerebellar zones may be. The configuration of these different zones raise the possibility that different encoding schemes are used for motor learning. Indeed the zebrin-positive zones, such as the F1−F4 zones used for vestibulo-ocular reflex adaptation, appear well designed to use mainly potentiation to enhance simple spike firing rate and mediate motor learning through

rate-coding mechanisms (Fig. 4). Instead, the zebrin-negative zones, such as the C3 and D0 zones used for eyeblink conditioning, appear optimally designed to use mainly suppression to decrease simple spike firing rate and mediate motor learning, in part, through temporal coding mechanisms downstream in the cerebellar nuclei (Fig. 6), which is supported by input from collaterals of not only mossy fibers but also climbing fibers (Van der Want et al. 1989).

Given the enormous energy consumption of high levels of neuronal activity (Sengupta et al. 2014), one may wonder why a neurobiological system like the cerebellum uses such high levels of intrinsic activity to begin with. There must be obvious benefits preserved throughout evolution (Darwin 1859). Clearly, control of motor learning is one of the prime functions of the cerebellum (Ito 2002; De Zeeuw et al 2011; Gao et al. 2012) and exploiting diversity in intrinsic activity of its main output neurons like Purkinje cells and cerebellar nuclei neurons may benefit the execution of this function, but other regions like the cerebral cortex also have prime roles in learning, including both declarative and procedural memory formation, whereas their main output neurons, pyramidal cells, usually fire at a relatively low firing frequency at rest, preserving energy (Heck et al. 2013; Pouille et al. 2013).

What then is special about the cerebellar system in this regard? The most characteristic feature of all functions of the cerebellum is its ability to control timing at a high resolution. Across periods of hundreds of milliseconds, the cerebellum can regulate and fine-tune signal processing with a precision of ∼5 msec (D'Angelo and De Zeeuw 2009; De Zeeuw et al. 2011). This function appears critical for controlling not only relatively simple forms of learning-dependent timing, such as for vestibulo-ocular reflex phase-reversal learning (Fig. 4) and eyeblink conditioning (Fig. 6), but probably also for more complex, timing-sensitive processes involved in cognition and episodic memory formation (Ben-Yakov and Dudai 2011; Rahmati et al. 2014). For example, cerebellar cell-type-specific mouse mutants do not show deficits in general cognitive tasks like Morris water maze, fear conditioning, or open field (Galliano et al. 2013b), but the very same mutants start to show phenotypes in decision making when tight temporal response windows are inserted in go/no-go tasks (Rahmati et al. 2014). Likewise, the timing function of the olivocerebellar system is essential when acute reflexes need to be engaged following perturbations (Van Der Giessen et al. 2008). Therefore, it is parsimonious to assess the potential role(s) of high and varying levels of intrinsic Purkinje cell activity in the light of the overall function of the cerebellum in timing.

If one considers the motor domains and functions controlled by the various olivocerebellar modules in mammals (Fig. 2), the picture emerges that slower movements, such as compensatory eye and head movements, are controlled by zebrin-positive modules (e.g., vestibulocerebellum) operating at lower firing frequencies and using rate coding downstream (Clopath et al. 2014), whereas faster movements, such as eyeblink responses or limb activity during locomotion, may depend on zebrin-negative modules (e.g., 5-region in D0 and vermal lobule V, respectively) and fast rebound activity in the cerebellar nuclei (De Zeeuw et al. 2011; Witter et al. 2013). In general, one could state that the presence of intrinsic activity as in zebrin-positive modules allows for on-line modulation in both the excitatory and inhibitory direction with ample opportunity to expand in the excitatory domain during learning (i.e., potentiation) and that an excessive amount of intrinsic activity as in zebrin-negative modules also allows for on-line modulation in both the excitatory and inhibitory direction, but with ample opportunity to expand in the inhibitory domain (i.e., suppression) engaging fast rebound mechanisms in the nuclei downstream. The latter condition can be considered as a pulled string maintained at a high energy level that can be released on command and evoke very fast effects when needed, such as to protect one's eye with an eyeblink when a dangerous event is approaching.

The design of a system that allows for highly dynamic and precise control of temporal signal processing comes at the cost of continuous high levels of energy consumption, but apparently renders the system with a sufficiently improved survival rate. In this respect, it will be interesting to investigate the intrinsic activity of Purkinje cells during sleep. If the hypothesis described above is correct, one might expect to not waste energy during sleep and bring the cells into a down state (Loewenstein et al. 2005). These down states in Purkinje cells can indeed be induced by anesthetics, whereas they hardly occur in animals operating under physiological circumstances in the awake state (Schonewille et al. 2006b). In addition, it will be interesting to find out to what extent neighboring zebrin-positive and zebrin-negative zones can interact (see also Fig. 3). Interestingly, in the vestibulocerebellum of birds, neighboring zebrin-positive and zebrin-negative modules have been found to respond best to the same pattern of optic flow in 3D space (Graham and Wylie 2012). One could imagine that antagonistic movements characteristic of flying may have different temporal and thereby modulational demands depending on their relation with gravity, engaging the zebrin-positive and -negative modules under different circumstances.

The functional concept of cerebellar modules operating in different firing frequency domains outlined above is based on a dichotomy found in expression of zebrin and related proteins that may control intrinsic simple spike activity. However, even though the differences in firing frequency between zebrin-positive and zebrin-negative modules are highly significant, there is considerable overlap in the ranges of frequencies found (Zhou et al. 2014). This raises the possibility that the encoding schemes used in the various modules are more diverse than depicted here. Indeed, in some microzones, the level of zebrin expression is ambiguous (Mostofi et al. 2010), and the complete proteomics in Purkinje cells is probably sufficiently diverse to even make sagittal strips of single Purkinje cells unique (Voogd et al. 1996). Likewise, it should be noted that one cannot exclude the possibility that suppression and potentiation mechanisms may also take place in zebrin-positive and zebrin-negative modules, respectively (Yang and Lisberger 2013, 2014) and that the concomitant encoding schemes during learning may shift accordingly, if one switches from a chronic tonic form of learning to a more acute trial-by-trial form of conditioning. In addition, it should be noted that conversion mechanisms may take place at the input stage of both types of modules in that the enhanced input from the semicircular canals to the flocculus may both have to be suppressed during ipsiversive head movements (zebrin-positive zones) and that the active input signaling the conditioned stimulus has to be turned into simple spike suppression during expression of the conditioned response (zebrin-negative zones) (De Zeeuw et al. 2004; Johansson et al. 2014). Therefore, in future studies, it will be important to determine how refined the variety of encoding schemes for cerebellar motor learning really is and to what extent these schemes are related to the dynamics of the paradigm involved. By explaining the encoding schemes for the most widely studied form of motor learning in zebrin-positive modules, that is, adaptation of the vestibulo-ocular reflex in F1–F4, as well as of that in zebrin-negative modules, that is, eyeblink conditioning in C3–D0, this article provides a first step toward unraveling the various encoding schemes that can be used for cerebellar motor learning. Because the zebrin patterns and differences in baseline firing frequencies are consistently present throughout the cerebellar cortex, it is possible that these encoding schemes are applicable to all cerebellar learning functions.

## ACKNOWLEDGMENTS

We thank Bas Koekkoek, Martijn Schonewille, and Elise Buitenhuis Linssen for technical assistance. We thank Izumi Sugihara and Megan Carey for providing the zebrin template and related scheme and allowing us to use it for graphics. Support is provided by the Netherlands Organization for Scientific Research (NWO-ALW, MAGW, ZON-MW), Neuro-Basic, and European Union (ERC-advanced and ERC-POC).

## REFERENCES

Albergaria C, Carey MR. 2014. All Purkinje cells are not created equal. *eLife* **3**: e03285.

Albus JS. 1971. A theory of cerebellar function. *Math Biosci* **10**: 25–61.

Andreescu CE, Milojkovic BA, Haasdijk ED, Kramer P, De Jong FH, Krust A, De Zeeuw CI, De Jeu MT. 2007. Estradiol improves cerebellar memory formation by activating estrogen receptor β. *J Neurosci* **27**: 10832–10839.

Angaut P, Compoint C, Buisseret-Delmas C, Batini C. 1996. Synaptic connections of Purkinje cell axons with nucleocortical neurones in the cerebellar medial nucleus of the rat. *Neurosci Res* **26**: 345–348.

Apps R, Hawkes R. 2009. Cerebellar cortical organization: A one-map hypothesis. *Nat Rev Neurosci* **10**: 670–681.

Attwell PJ, Rahman S, Yeo CH. 2001. Acquisition of eyeblink conditioning is critically dependent on normal function in cerebellar cortical lobule HVI. *J Neurosci* **21**: 5715–5722.

Auger C, Attwell D. 2000. Fast removal of synaptic glutamate by postsynaptic transporters. *Neuron* **28**: 547–558.

Badura A, Schonewille M, Voges K, Galliano E, Renier N, Gao Z, Witter L, Hoebeek FE, Chedotal A, De Zeeuw CI. 2013. Climbing fiber input shapes reciprocity of Purkinje cell firing. *Neuron* **78**: 700–713.

Barmack NH, Qian Z, Yoshimura J. 2000. Regional and cellular distribution of protein kinase C in rat cerebellar Purkinje cells. *J Comp Neurol* **427**: 235–254.

Becker EB, Oliver PL, Glitsch MD, Banks GT, Achilli F, Hardy A, Nolan PM, Fisher EM, Davies KE. 2009. A point mutation in TRPC3 causes abnormal Purkinje cell development and cerebellar ataxia in moonwalker mice. *Proc Natl Acad Sci* **106**: 6706–6711.

Ben-Yakov A, Dudai Y. 2011. Constructing realistic engrams: Poststimulus activity of hippocampus and dorsal striatum predicts subsequent episodic memory. *J Neurosci* **31**: 9032–9042.

Berthier NE, Moore JW. 1986. Cerebellar Purkinje cell activity related to the classically conditioned nictitating membrane response. *Exp Brain Res* **63**: 341–350.

Blot A, Barbour B. 2014. Ultra-rapid axon–axon ephaptic inhibition of cerebellar Purkinje cells by the pinceau. *Nat Neurosci* **17**: 289–295.

Boele HJ. 2014. "Learning mechanisms of classical conditioning." PhD thesis, Erasmus University Medical Center, Rotterdam, The Netherlands.

Boele HJ, Koekkoek SK, De Zeeuw CI. 2010. Cerebellar and extracerebellar involvement in mouse eyeblink conditioning: The ACDC model. *Front Cell Neurosci* **3**: 19.

Boele HJ, Koekkoek SK, De Zeeuw CI, Ruigrok TJ. 2013. Axonal sprouting and formation of terminals in the adult cerebellum during associative motor learning. *J Neurosci* **33**: 17897–17907.

Boyden ES, Katoh A, Raymond JL. 2004. Cerebellum-dependent learning: The role of multiple plasticity mechanisms. *Annu Rev Neurosci* **27**: 581–609.

Brochu G, Maler L, Hawkes R. 1990. Zebrin II: A polypeptide antigen expressed selectively by Purkinje cells reveals compartments in rat and fish cerebellum. *J Comp Neurol* **291**: 538–552.

Brunel N, Hakim V, Isope P, Nadal JP, Barbour B. 2004. Optimal information storage and the distribution of synaptic weights: Perceptron versus Purkinje cell. *Neuron* **43**: 745–757.

Cerminara NL, Rawson JA. 2004. Evidence that climbing fibers control an intrinsic spike generator in cerebellar Purkinje cells. *J Neurosci* **24:** 4510–4517.

Chanda S, Xu-Friedman MA. 2011. Excitatory modulation in the cochlear nucleus through group I metabotropic glutamate receptor activation. *J Neurosci* **31:** 7450–7455.

Chen X, Kovalchuk Y, Adelsberger H, Henning HA, Sausbier M, Wietzorrek G, Ruth P, Yarom Y, Konnerth A. 2010. Disruption of the olivo-cerebellar circuit by Purkinje neuron-specific ablation of BK channels. *Proc Natl Acad Sci* **107:** 12323–12328.

Chung S, Zhang Y, Van Der Hoorn F, Hawkes R. 2007. The anatomy of the cerebellar nuclei in the normal and scrambler mouse as revealed by the expression of the microtubule-associated protein kinesin light chain 3. *Brain Res* **1140:** 120–131.

Clopath C, Badura A, De Zeeuw CI, Brunel N. 2014. A cerebellar learning model of vestibulo-ocular reflex adaptation in wild-type and mutant mice. *J Neurosci* **34:** 7203–7215.

Coddington LT, Rudolph S, Vande Lune P, Overstreet-Wadiche L, Wadiche JI. 2013. Spillover-mediated feedforward inhibition functionally segregates interneuron activity. *Neuron* **78:** 1050–1062.

Coesmans M, Smitt PA, Linden DJ, Shigemoto R, Hirano T, Yamakawa Y, van Alphen AM, Luo C, van der Geest JN, Kros JM, et al. 2003. Mechanisms underlying cerebellar motor deficits due to mGluR1-autoantibodies. *Ann Neurol* **53:** 325–336.

D'Angelo E, De Zeeuw CI. 2009. Timing and plasticity in the cerebellum: Focus on the granular layer. *Trends Neurosci* **32:** 30–40.

Darwin C. 1859. *On the origin of species*, pp. 502. John Murray, London.

Dehnes Y, Chaudhry FA, Ullensvang K, Lehre KP, Storm-Mathisen J, Danbolt NC. 1998. The glutamate transporter EAAT4 in rat cerebellar Purkinje cells: A glutamate-gated chloride channel concentrated near the synapse in parts of the dendritic membrane facing astroglia. *J Neurosci* **18:** 3606–3619.

De Zeeuw CI, Koekkoek SK. 1997. Signal processing in the C2 module of the flocculus and its role in head movement control. *Prog Brain Res* **114:** 299–320.

De Zeeuw CI, Yeo CH. 2005. Time and tide in cerebellar memory formation. *Curr Opin Neurobiol* **15:** 667–674.

De Zeeuw CI, Holstege JC, Calkoen F, Ruigrok TJ, Voogd J. 1988. A new combination of WGA-HRP anterograde tracing and GABA immunocytochemistry applied to afferents of the cat inferior olive at the ultrastructural level. *Brain Res* **447:** 369–375.

De Zeeuw CI, Wylie DR, DiGiorgi PL, Simpson JI. 1994. Projections of individual Purkinje cells of identified zones in the flocculus to the vestibular and cerebellar nuclei in the rabbit. *J Comp Neurol* **349:** 428–447.

De Zeeuw CI, Wylie DR, Stahl JS, Simpson JI. 1995. Phase relations of Purkinje cells in the rabbit flocculus during compensatory eye movements. *J Neurophysiol* **74:** 2051–2064.

De Zeeuw CI, Koekkoek SKE, van Alphen AM, Luo C, Hoebeek F, van der Steen J, Frens MA, Sun J, Goossens HHLM, Jaarsma D, et al. 2004. Gain and phase control of compensatory eye movements by the flocculus of the vestibulocerebellum, In *The vestibular system* (ed. Highstein SM, Fay RR, Popper AN), pp. 375–423. Springer, New York.

De Zeeuw CI, Hoebeek FE, Schonewille M. 2008. Causes and consequences of oscillations in the cerebellar cortex. *Neuron* **58:** 655–658.

De Zeeuw CI, Hoebeek FE, Bosman LW, Schonewille M, Witter L, Koekkoek SK. 2011. Spatiotemporal firing patterns in the cerebellum. *Nat Rev Neurosci* **12:** 327–344.

Ekerot CF, Larson B. 1982. Branching of olivary axons to innervate pairs of sagittal zones in the cerebellar anterior lobe of the cat. *Exp Brain Res* **48:** 185–198.

Endo S, Shutoh F, Dinh TL, Okamoto T, Ikeda T, Suzuki M, Kawahara S, Yanagihara D, Sato Y, Yamada K, et al. 2009. Dual involvement of G-substrate in motor learning revealed by gene deletion. *Proc Natl Acad Sci* **106:** 3525–3530.

Furutama D, Morita N, Takano R, Sekine Y, Sadakata T, Shinoda Y, Hayashi K, Mishima Y, Mikoshiba K, Hawkes R, et al. 2010. Expression of the *IP3R1* promoter-driven *nls-lacZ* transgene in Purkinje cell parasagittal arrays of developing mouse cerebellum. *J Neurosci Res* **88:** 2810–2825.

Galliano E, Gao Z, Schonewille M, Todorov B, Simons E, Pop AS, D'Angelo E, van den Maagdenberg AM, Hoebeek FE, De Zeeuw CI. 2013a. Silencing the majority of cerebellar granule cells uncovers their essential role in motor learning and consolidation. *Cell Rep* **3:** 1239–1251.

Galliano E, Potters JW, Elgersma Y, Wisden W, Kushner SA, De Zeeuw CI, Hoebeek FE. 2013b. Synaptic transmission and plasticity at inputs to murine cerebellar Purkinje cells are largely dispensable for standard nonmotor tasks. *J Neurosci* **33:** 12599–12618.

Gao Z, van Beugen BJ, De Zeeuw CI. 2012. Distributed synergistic plasticity and cerebellar learning. *Nat Rev Neurosci* **13:** 619–635.

Gao Z, vanWoerden GM, Elgersma Y, DeZeeuw CI, Hoebeek FE. 2014. Distinct roles of α- and βCaMKII in controlling long-term potentiation of GABA_A-receptor mediated transmission in murine Purkinje cells. *Front Cell Neurosci* doi: 10.3389/fncel.2014.00016.

Gauck V, Jaeger D. 2000. The control of rate and timing of spikes in the deep cerebellar nuclei by inhibition. *J Neurosci* **20:** 3006–3016.

Gibson EM, Purger D, Mount CW, Goldstein AK, Lin GL, Wood LS, Inema I, Miller SE, Bieri G, Zuchero JB, et al. 2014. Neuronal activity promotes oligodendrogenesis and adaptive myelination in the mammalian brain. *Science* **344:** 1252304.

Graham DJ, Wylie DR. 2012. Zebrin-immunopositive and -immunonegative stripe pairs represent functional units in the pigeon vestibulocerebellum. *J Neurosci* **32:** 12769–12779.

Groenewegen HJ, Voogd J. 1977. The parasagittal zonation within the olivocerebellar projection. I: Climbing fiber distribution in the vermis of cat cerebellum. *J Comp Neurol* **174:** 417–488.

Hartmann J, Dragicevic E, Adelsberger H, Henning HA, Sumser M, Abramowitz J, Blum R, Dietrich A, Freichel M, Flockerzi V, et al. 2008. TRPC3 channels are required

Cite this article as *Cold Spring Harb Perspect Biol* doi: 10.1101/cshperspect.a021683

for synaptic transmission and motor coordination. *Neuron* **59:** 392–398.

Hauge SA, Tracy JA, Baudry M, Thompson RF. 1998. Selective changes in AMPA receptors in rabbit cerebellum following classical conditioning of the eyelid-nictitating membrane response. *Brain Res* **803:** 9–18.

Heck DH, De Zeeuw CI, Jaeger D, Khodakhah K, Person AL. 2013. The neuronal code(s) of the cerebellum. *J Neurosci* **33:** 17603–17609.

Hesslow G. 1994a. Correspondence between climbing fibre input and motor output in eyeblink-related areas in cat cerebellar cortex. *J Physiol* **476:** 229–244.

Hesslow G. 1994b. Inhibition of classically conditioned eyeblink responses by stimulation of the cerebellar cortex in the decerebrate cat. *J Physiol* **476:** 245–256.

Hesslow G, Ivarsson M. 1994. Suppression of cerebellar Purkinje cells during conditioned responses in ferrets. *Neuroreport* **5:** 649–652.

Hesslow G, Jirenhed DA, Rasmussen A, Johansson F. 2013. Classical conditioning of motor responses: What is the learning mechanism? *Neural Netw* **47:** 81–87.

Ito M. 2002. Historical review of the significance of the cerebellum and the role of Purkinje cells in motor learning. *Ann NY Acad Sci* **978:** 273–288.

Ito M. 2008. Control of mental activities by internal models in the cerebellum. *Nat Rev Neurosci* **9:** 304–313.

Ito M, Kano M. 1982. Long-lasting depression of parallel fiber–Purkinje cell transmission induced by conjunctive stimulation of parallel fibers and climbing fibers in the cerebellar cortex. *Neurosci Lett* **33:** 253–258.

Jinno S, Jeromin A, Roder J, Kosaka T. 2003. Compartmentation of the mouse cerebellar cortex by neuronal calcium sensor-1. *J Comp Neurol* **458:** 412–424.

Jirenhed DA, Hesslow G. 2011a. Time course of classically conditioned Purkinje cell response is determined by initial part of conditioned stimulus. *J Neurosci* **31:** 9070–9074.

Jirenhed DA, Hesslow G. 2011b. Learning stimulus intervals—Adaptive timing of conditioned Purkinje cell responses. *Cerebellum* **10:** 523–535.

Jirenhed DA, Bengtsson F, Hesslow G. 2007. Acquisition, extinction, and reacquisition of a cerebellar cortical memory trace. *J Neurosci* **27:** 2493–2502.

Johansson F, Jirenhed DA, Rasmussen A, Zucca R, Hesslow G. 2014. Memory trace and timing mechanism localized to cerebellar Purkinje cells. *Proc Natl Acad Sci* **111:** 14930–14934.

Jörntell H, Ekerot CF. 2002. Reciprocal bidirectional plasticity of parallel fiber receptive fields in cerebellar Purkinje cells and their afferent interneurons. *Neuron* **34:** 797–806.

Jörntell H, Ekerot C, Garwicz M, Luo XL. 2000. Functional organization of climbing fibre projection to the cerebellar anterior lobe of the rat. *J Physiol* **522:** 297–309.

Kehoe EJ. 1988. A layered network model of associative learning: Learning to learn and configuration. *Psychol Rev* **95:** 411–433.

Kim CH, Oh SH, Lee JH, Chang SO, Kim J, Kim SJ. 2012a. Lobule-specific membrane excitability of cerebellar Purkinje cells. *J Physiol* **590:** 273–288.

Kim Y, Wong AC, Power JM, Tadros SF, Klugmann M, Moorhouse AJ, Bertrand PP, Housley GD. 2012b. Alternative splicing of the TRPC3 ion channel calmodulin/IP3 receptor-binding domain in the hindbrain enhances cation flux. *J Neurosci* **32:** 11414–11423.

Koekkoek SK, Hulscher HC, Dortland BR, Hensbroek RA, Elgersma Y, Ruigrok TJ, De Zeeuw CI. 2003. Cerebellar LTD and learning-dependent timing of conditioned eyelid responses. *Science* **301:** 1736–1739.

Leclerc N, Dore L, Parent A, Hawkes R. 1990. The compartmentalization of the monkey and rat cerebellar cortex: Zebrin I and cytochrome oxidase. *Brain Res* **506:** 70–78.

Lisberger SG. 1988. The neural basis for motor learning in the vestibulo-ocular reflex in monkeys. *Trends Neurosci* **11:** 147–152.

Lisberger SG. 2009. Internal models of eye movement in the floccular complex of the monkey cerebellum. *Neuroscience* **162:** 763–776.

Loewenstein Y, Mahon S, Chadderton P, Kitamura K, Sompolinsky H, Yarom Y, Hausser M. 2005. Bistability of cerebellar Purkinje cells modulated by sensory stimulation. *Nat Neurosci* **8:** 202–211.

Markram H, Lubke J, Frotscher M, Sakmann B. 1997. Regulation of synaptic efficacy by coincidence of postsynaptic APs and EPSPs. *Science* **275:** 213–215.

Mateos JM, Osorio A, Azkue JJ, Benitez R, Elezgarai I, Bilbao A, Diez J, Puente N, Kuhn R, Knopfel T, et al. 2001. Parasagittal compartmentalization of the metabotropic glutamate receptor mGluR1b in the cerebellar cortex. *Eur J Anat* **5:** 15–21.

Mathews PJ, Lee KH, Peng Z, Houser CR, Otis TS. 2012. Effects of climbing fiber driven inhibition on Purkinje neuron spiking. *J Neurosci* **32:** 17988–17997.

McCormick DA, Thompson RF. 1984. Cerebellum: Essential involvement in the classically conditioned eyelid response. *Science* **223:** 296–299.

Mostofi A, Holtzman T, Grout AS, Yeo CH, Edgley SA. 2010. Electrophysiological localization of eyeblink-related microzones in rabbit cerebellar cortex. *J Neurosci* **30:** 8920–8934.

Nelson C, Glitsch MD. 2012. Lack of kinase regulation of canonical transient receptor potential 3 (TRPC3) channel-dependent currents in cerebellar Purkinje cells. *J Biol Chem* **287:** 6326–6335.

Nelson AB, Gittis AH, du Lac S. 2005. Decreases in CaMKII activity trigger persistent potentiation of intrinsic excitability in spontaneously firing vestibular nucleus neurons. *Neuron* **46:** 623–631.

Nguyen-Vu TD, Kimpo RR, Rinaldi JM, Kohli A, Zeng H, Deisseroth K, Raymond JL. 2013. Cerebellar Purkinje cell activity drives motor learning. *Nat Neurosci* **16:** 1734–1736.

Ohyama T, Nores WL, Medina JF, Riusech FA, Mauk MD. 2006. Learning-induced plasticity in deep cerebellar nucleus. *J Neurosci* **26:** 12656–12663.

Person AL, Raman IM. 2012. Purkinje neuron synchrony elicits time-locked spiking in the cerebellar nuclei. *Nature* **481:** 502–505.

Pijpers A, Apps R, Pardoe J, Voogd J, Ruigrok TJ. 2006. Precise spatial relationships between mossy fibers and climbing fibers in rat cerebellar cortical zones. *J Neurosci* **26:** 12067–12080.

Pouille F, Watkinson O, Scanziani M, Trevelyan AJ. 2013. The contribution of synaptic location to inhibitory gain control in pyramidal cells. *Physiol Rep* **1:** e00067.

Pugh JR, Raman IM. 2008. Mechanisms of potentiation of mossy fiber EPSCs in the cerebellar nuclei by coincident synaptic excitation and inhibition. *J Neurosci* **28:** 10549–10560.

Rahmati N, Owens CB, Bosman LW, Spanke JK, Lindeman S, Gong W, Potters JW, Romano V, Voges K, Moscato L, et al. 2014. Cerebellar potentiation and learning a whisker-based object localization task with a time response window. *J Neurosci* **34:** 1949–1962.

Ruigrok TJ, Voogd J. 2000. Organization of projections from the inferior olive to the cerebellar nuclei in the rat. *J Comp Neurol* **426:** 209–228.

Sánchez-Campusano R, Gruart A, Delgado-Garcia JM. 2011. Dynamic changes in the cerebellar-interpositus/red-nucleus-motoneuron pathway during motor learning. *Cerebellum* **10:** 702–710.

Sarna JR, Marzban H, Watanabe M, Hawkes R. 2006. Complementary stripes of phospholipase Cβ3 and Cβ4 expression by Purkinje cell subsets in the mouse cerebellum. *J Comp Neurol* **496:** 303–313.

Schonewille M, Luo C, Ruigrok TJ, Voogd J, Schmolesky MT, Rutteman M, Hoebeek FE, De Jeu MT, De Zeeuw CI. 2006a. Zonal organization of the mouse flocculus: Physiology, input, and output. *J Comp Neurol* **497:** 670–682.

Schonewille M, Khosrovani S, Winkelman BH, Hoebeek FE, De Jeu MT, Larsen IM, Van der Burg J, Schmolesky MT, Frens MA, De Zeeuw CI. 2006b. Purkinje cells in awake behaving animals operate at the upstate membrane potential. *Nat Neurosci* **9:** 459–461; author reply 461.

Schonewille M, Belmeguenai A, Koekkoek SK, Houtman SH, Boele HJ, van Beugen BJ, Gao Z, Badura A, Ohtsuki G, Amerika WE, et al. 2010. Purkinje cell-specific knockout of the protein phosphatase PP2B impairs potentiation and cerebellar motor learning. *Neuron* **67:** 618–628.

Schonewille M, Gao Z, Boele HJ, Veloz MF, Amerika WE, Simek AA, De Jeu MT, Steinberg JP, Takamiya K, Hoebeek FE, et al. 2011. Reevaluating the role of LTD in cerebellar motor learning. *Neuron* **70:** 43–50.

Sengupta B, Laughlin SB, Niven JE. 2014. Consequences of converting graded to action potentials upon neural information coding and energy efficiency. *PLoS Comput Biol* **10:** e1003439.

Sillitoe RV, Kunzle H, Hawkes R. 2003. Zebrin II compartmentation of the cerebellum in a basal insectivore, the Madagascan hedgehog tenrec *Echinops telfairi*. *J Anat* **203:** 283–296.

Stahl JS, Simpson JI. 1995. Dynamics of rabbit vestibular nucleus neurons and the influence of the flocculus. *J Neurophysiol* **73:** 1396–1413.

Sugihara I. 2011. Compartmentalization of the deep cerebellar nuclei based on afferent projections and aldolase C expression. *Cerebellum* **10:** 449–463.

Sugihara I, Shinoda Y. 2004. Molecular, topographic, and functional organization of the cerebellar cortex: A study with combined aldolase C and olivocerebellar labeling. *J Neurosci* **24:** 8771–8785.

Sugihara I, Shinoda Y. 2007. Molecular, topographic, and functional organization of the cerebellar nuclei: Analysis by three-dimensional mapping of the olivonuclear projection and aldolase C labeling. *J Neurosci* **27:** 9696–9710.

Sugihara I, Fujita H, Na J, Quy PN, Li BY, Ikeda D. 2009. Projection of reconstructed single Purkinje cell axons in relation to the cortical and nuclear aldolase C compartments of the rat cerebellum. *J Comp Neurol* **512:** 282–304.

Tanaka S, Kawaguchi SY, Shioi G, Hirano T. 2013. Long-term potentiation of inhibitory synaptic transmission onto cerebellar Purkinje neurons contributes to adaptation of vestibulo-ocular reflex. *J Neurosci* **33:** 17209–17220.

Ten Brinke MM, Potters JW, Boele HJ, De Zeeuw CI. 2014. Linking cells to behavior: Purkinje cell electrophysiology during eyeblink conditioning in (transgenic) mice. In *9th Federation of European Neuroscience Societies (FENS), Abstract FENS-3629*. Milan, Italy, July 5–9.

Thompson RF, Steinmetz JE. 2009. The role of the cerebellum in classical conditioning of discrete behavioral responses. *Neuroscience* **162:** 732–755.

van Beugen BJ, Gao Z, Boele HJ, Hoebeek F, De Zeeuw CI. 2013. High frequency burst firing of granule cells ensures transmission at the parallel fiber to Purkinje cell synapse at the cost of temporal coding. *Front Neural Circuits* **7:** 95.

Van Der Giessen RS, Koekkoek SK, van Dorp S, De Gruijl JR, Cupido A, Khosrovani S, Dortland B, Wellershaus K, Degen J, Deuchars J, et al. 2008. Role of olivary electrical coupling in cerebellar motor learning. *Neuron* **58:** 599–612.

van der Steen J, Tan HS. 1997. Cholinergic control in the floccular cerebellum of the rabbit. *Prog Brain Res* **114:** 335–345.

Van der Want JJ, Wiklund L, Guegan M, Ruigrok T, Voogd J. 1989. Anterograde tracing of the rat olivocerebellar system with Phaseolus vulgaris leucoagglutinin (PHA-L). Demonstration of climbing fiber collateral innervation of the cerebellar nuclei. *J Comp Neurol* **288:** 1–18.

Voogd J. 1964. "The cerebellum of the cat. Structure and fiber connections." PhD thesis, Van Gorcum, Assen, The Netherlands.

Voogd J, Jaarsma DEM. 1996. The cerebellum, chemoarchitecture and anatomy. In *Handbook of chemical neuroanatomy* (ed. Swanson IW, Björklund A, Hökfelt T), pp. 1–369. Elsevier, Amsterdam.

Voogd J, Ruigrok TJ. 2004. The organization of the corticonuclear and olivocerebellar climbing fiber projections to the rat cerebellar vermis: The congruence of projection zones and the zebrin pattern. *J Neurocytol* **33:** 5–21.

Voogd J, Schraa-Tam CK, van der Geest JN, De Zeeuw CI. 2012. Visuomotor cerebellum in human and nonhuman primates. *Cerebellum* **11:** 392–410.

Wadiche JI, Jahr CE. 2005. Patterned expression of Purkinje cell glutamate transporters controls synaptic plasticity. *Nat Neurosci* **8:** 1329–1334.

Walter JT, Khodakhah K. 2009. The advantages of linear information processing for cerebellar computation. *Proc Natl Acad Sci* **106:** 4471–4476.

Wang X, Chen G, Gao W, Ebner TJ. 2011. Parasagittally aligned, mGluR1-dependent patches are evoked at long latencies by parallel fiber stimulation in the mouse cerebellar cortex in vivo. *J Neurophysiol* **105:** 1732–1746.

Welsh JP, Yamaguchi H, Zeng XH, Kojo M, Nakada Y, Takagi A, Sugimori M, Llinas RR. 2005. Normal motor learning during pharmacological prevention of Purkinje cell long-term depression. *Proc Natl Acad Sci* **102:** 17166–17171.

Witter L, Canto CB, Hoogland TM, de Gruijl JR, De Zeeuw CI. 2013. Strength and timing of motor responses mediated by rebound firing in the cerebellar nuclei after Purkinje cell activation. *Front Neural Circuits* **7:** 133.

Wulff P, Schonewille M, Renzi M, Viltono L, Sassoe-Pognetto M, Badura A, Gao Z, Hoebeek FE, van Dorp S, Wisden W, et al. 2009. Synaptic inhibition of Purkinje cells mediates consolidation of vestibulo-cerebellar motor learning. *Nat Neurosci* **12:** 1042–1049.

Yamakawa Y, Hirano T. 1999. Contribution of mGluR1 to the basal activity of a mouse cerebellar Purkinje neuron. *Neurosci Lett* **277:** 103–106.

Yang Y, Lisberger SG. 2013. Interaction of plasticity and circuit organization during the acquisition of cerebellum-dependent motor learning. *eLife* **2:** e01574.

Yang Y, Lisberger SG. 2014. Role of plasticity at different sites across the time course of cerebellar motor learning. *J Neurosci* **34:** 7077–7090.

Yeo CH, Hardiman MJ, Glickstein M. 1985a. Classical conditioning of the nictitating membrane response of the rabbit. I: Lesions of the cerebellar nuclei. *Exp Brain Res* **60:** 87–98.

Yeo CH, Hardiman MJ, Glickstein M. 1985b. Classical conditioning of the nictitating membrane response of the rabbit. II: Lesions of the cerebellar cortex. *Exp Brain Res* **60:** 99–113.

Yeo CH, Hardiman MJ, Glickstein M. 1985c. Classical conditioning of the nictitating membrane response of the rabbit. III: Connections of cerebellar lobule HVI. *Exp Brain Res* **60:** 114–126.

Yeo CH, Hardiman MJ, Glickstein M. 1986. Classical conditioning of the nictitating membrane response of the rabbit. IV: Lesions of the inferior olive. *Exp Brain Res* **63:** 81–92.

Zhou H, Lin Z, Voges K, Gao Z, Ju C, Bosman LWJ, Ruigrok TJ, Hoebeek FE, De Zeeuw CI, Schonewille M. 2014. Cerebellar modules operate at different frequencies. In *9th Federation of European Neuroscinece Societies (FENS), Abstract FENS-3573*, Milan, Italy, July 5–9.

# The Striatum: Where Skills and Habits Meet

Ann M. Graybiel[1,2] and Scott T. Grafton[3,4]

[1]McGovern Institute for Brain Research, Massachusetts Institute of Technology, Cambridge, Massachusetts 20139

[2]Department of Brain and Cognitive Sciences, Massachusetts Institute of Technology, Cambridge, Massachusetts 20139

[3]Institute for Collaborative Biotechnologies, University of California, Santa Barbara, California 93106-9660

[4]Department of Psychological and Brain Sciences, University of California, Santa Barbara, California 93106-9660

*Correspondence:* graybiel@mit.edu

After more than a century of work concentrating on the motor functions of the basal ganglia, new ideas have emerged, suggesting that the basal ganglia also have major functions in relation to learning habits and acquiring motor skills. We review the evidence supporting the role of the striatum in optimizing behavior by refining action selection and in shaping habits and skills as a modulator of motor repertoires. These findings challenge the notion that striatal learning processes are limited to the motor domain. The learning mechanisms supported by striatal circuitry generalize to other domains, including cognitive skills and emotion-related patterns of action.

The nuclei and interconnections of the basal ganglia are widely recognized for modulating motor behavior. Whether measured at the neuronal or regional level, the activities of neurons in the basal ganglia correlate with many movement parameters, particularly those that influence the vigor of an action, such as force and velocity. Pathology within different basal ganglia circuits predictably leads to either hypokinetic or hyperkinetic movement disorders. In parallel, however, the basal ganglia, and especially the striatum, are now widely recognized as being engaged in activity related to learning. Interactions between the dopamine-containing neurons of the midbrain and their targets in the striatum are critical to this function. A fundamental question is how these two capacities—(motor behavior and reinforcement-based learning)—relate to each other and what role the striatum and other basal ganglia nuclei have in forming new behavioral repertoires. Here, we consider relevant physiological properties of the striatum by contrasting two common forms of adaptation found in all mammals: the acquisition of behavioral habits and physical skills.

Without resorting to technical definitions, we all have an intuition of what habits and skills are. Tying one's shoes after putting them on is something we consider a habit—part of a behavioral routine. The capacity to tie the laces properly is a skill. Habits and skills have many common features. Habits are consistent behaviors triggered by appropriate events (typically, but not always, external stimuli) occurring within particular contexts. Physical skills are changes in a physical repertoire: new combinations of movements that lead to new capacities for goal-directed action. Both habits and skills can leverage reward-based learning, particularly during their initial acquisition. In either instance, after sufficient experience, the need for reward becomes lower and lower. With sufficient practice, both lead to "automaticity" and a resilience against competing actions that might lead to unlearning.

## THE DEGREES OF FREEDOM PROBLEM AND OPTIMAL PERFORMANCE

When acquiring a new habit or skill, an organism is faced with an enormous space of possibilities from which to choose. For habits, how does the organisms select from the many potential behaviors that it could perform? In the sections to follow, we review some of the evidence, indicating that the striatum has a principal function in learning-related plasticity associated with selecting one set of actions from many, resulting in the acquisition of habitual behavior. Similarly for skills, the motor system is also faced with an enormous set of possible solutions. There is an analogous problem of understanding how the organism narrows a search to find an effective solution when acquiring new motor skills. Of the many possible challenges in skill learning, two are renowned: the degrees of freedom problem (Bernstein 1967) and the problem of optimal control (Todorov and Jordan 2002).

How are behaviors or movements chosen so that the result is optimal? Optimality could be defined in a variety of ways, but there are two particularly relevant for habits and skills. First, optimality can be driven by the outcome of achieving a specific goal and receiving the reward. Monkeys, for example, will work with increased urgency to maximize the number of rewarded trials per hour. Second, optimality can also be determined by the particular ways in which motor behaviors are combined so that a cost is minimized. For example, an animal can

learn to find the shortest path to a reward, or find the most efficient combination of movements leading to a reward. This is at the heart of solving the degrees of freedom problem: a process that optimizes movement to a goal in terms of some metric, such as the energetics of the movement.

Much evidence points to the cerebellum and its reliance of online feedback to shape ongoing activity from multiple cortical motor regions, along with spinobulbar pattern generators so that the dynamics of movements are smoother, faster, and more efficient (Takemura et al. 2001; Gao et al. 2012). Notably, this error-based cerebellar shaping of behavior occurs independently of any reward signal. Here, we propose that the basal ganglia, of which the striatum is a main input station, also have a profound effect on optimizing behavior by implementing reinforcement-based feedback to allow effective combination of sequential motor elements. Thus, whether we speak of habits or skills, we see the striatum as a sort of learning machine dedicated to achieving success in behavior. We view this learning capacity as not only adhering to the main challenges of motor control, but also as extending beyond these to influence cognitive and emotional control.

## REINFORCEMENT-BASED LEARNING REPRESENTS A CORE MECHANISM THOUGHT TO UNDERLIE BEHAVIORAL OPTIMIZATION BY STRIATUM-BASED CIRCUITS

The behavioral literature on reinforcement learning shows that it is not the reward (or punishment) per se that reinforces (extinguishes) behaviors. Rather, it is the difference between the predicted value of future rewards or punishments and their ultimate reward or punishment. Learning theory has formalized the process by which these reinforcement contingencies, referred to as reward prediction errors (RPEs), drive behavioral change (Sutton and Barto 1998). These lead to the notion that, as an agent (actor) interacts with the environment, it develops state-specific behavioral policies. An influential idea is that these are instantiated in the

Cite this article as *Cold Spring Harb Perspect Biol* doi: 10.1101/cshperspect.a021691

brain according to algorithms, such as those in temporal difference models. Through experience, eligibility traces are built up and models of behavioral tasks can be optimized.

The reinforcement-related learning functions of the striatum are driven by evaluative circuits interconnecting the striatum both with the brainstem and with the neocortex and noncortical regions of the forebrain, especially the thalamus. The gradual selection of particular behavioral repertoires can lead toward optimal behavioral control by means of reducing the degrees of freedom normally used in navigating through daily behavior. As we note below, this evidence is mainly derived by recording from multiple neurons in the striatum and interconnected circuits, and in studies in which optogenetic methods are used to manipulate corticostriatal circuits.

Electrophysiological recordings in humans are rare, but it is possible to use functional magnetic resonance imaging (fMRI) to relate the activity of dopaminergic cell groups to metabolic signals recorded from the striatum. Results from a large number of fMRI studies suggest that the human ventral striatum changes its activity in relation to many different kinds of rewards, ranging from juice rewards to abstract social or esthetic qualities. In effect, evidence from these fMRI studies suggests that the ventral striatum is involved in learning by trial-and-error irrespective of the specific nature of the rewards (Daniel and Pollmann 2014). In tasks involving decision-making and economic games, there is overwhelming evidence that the ventral striatum and the ventral tegmental area (VTA) (Diuk et al. 2013) form a key circuit for encoding for RPE (for meta-analysis of 779 fMRI articles, see Garrison et al. 2013). Studies on experimental animals concur, but point to striking changes over the course of learning in signals related to correct or incorrect behavior (Atallah et al. 2014; K Smith and A Graybiel, unpubl.).

Following classic work suggesting that reward was the main driver the nigrostriatal system (Schultz 2002), studies began to show that nonrewarding, aversive drive also could be applied through this system. Experiments in rodents have suggested that nondopamine neurons expressing γ-aminobutyric acid (GABA) in the nigroventral tegmental region are sensitive to aversive stimuli (Bevan et al. 1996; Brown et al. 2012). In macaque monkeys, divisions of the substantia nigra pars compacta region have been identified as having differential positive or negative reinforcement sensitivities (Matsumoto and Hikosaka 2009). Thus, neurons in the midbrain dopamine-containing cell groups can show responses corresponding to the positive and negative RPEs of computational models. Within the striatum, as well, many neurons have spike responses that are related more to rewarding versus risky or aversive contexts (Yamada et al. 2013; Yanike and Ferrera 2014). New data suggest that some striatal neurons perform an integration of cost and benefit, which predicts natural behavioral learning (T Desrochers, K Amemori, and AM Graybiel, unpubl.).

Hence, the striatum is poised to be a hub for neuroplasticity, as it receives major input from aminergic fiber systems, including the dopamine-containing nigral innervation, and input from nearly every region of the neocortex. Not only nigrostriatal synapses, but also corticostriatal synapses are thought to be sites of neuroplasticity. Many neurons in the striatum fire in a given context as though encoding expectancy signals and priors—signals crucial for smooth behavioral performance with advanced planning (Hikosaka et al. 1989). These signals are likely generated as a result of experience-dependent plasticity in the circuits that form the input–output networks of this large region of the basal ganglia. This includes cholinergic interneurons (Doig et al. 2014). It is likely that through these and other network activities, including processing through the thalamus and neocortex and their projections to the striatum, neurons in the striatum build up selective responses to particular environmental events and certain behavioral actions and contexts.

## HABIT LEARNING: A MODEL FOR STUDYING BEHAVIORAL PLASTICITY INFLUENCED BY STRIATAL CIRCUITS

Two main lines of evidence have linked the striatum and its associated neural circuits with the development of habitual behaviors. First, a long line of lesion studies in rodents has shown that the striatum is necessary for habit formation (Balleine and Dickinson 1998; Yin and Knowlton 2006; Belin et al. 2009). Further, once acquired, habitual behaviors can be blocked or blunted by lesions of the striatum made after the habits are learned. These studies have shown that different districts within the striatum operate during habit formation. The ventral striatum is necessary for initial learning of motivated behaviors that could become habitual (Atallah et al. 2014). The dorsal striatum then becomes critical. First, behaviors are driven largely by the anticipated outcome of the behavior itself; this process, according to rodent lesion studies, requires the dorsomedial striatum ([DMS] in rodents). But then, according to these lesion studies, as the behaviors are repeated and bring about a positive outcome, the DMS is no longer required, but the dorsolateral parts of the striatum ([DLS] in rodents) is required for habitual performance.

A striking parallel to these behavioral findings on transitions occurring during habit learning has come from studies in which multiple simultaneous recordings have been made within the striatum on a daily basis as the acquisition of the habits occurs (Jog et al. 1999; Barnes et al. 2005; Thorn et al. 2010; Smith and Graybiel 2013; Atallah et al. 2014). Remarkable plasticity is seen in these recordings. For example, in T-maze learning studies, in which rodents learn to navigate a maze according to cues given midrun that instruct them about which side food reward will be given, recordings have been made in the DLS, the part of the striatum thought to be essential for postlearning habit performance, the DMS, the part of the striatum thought to be essential for initial goal-directed behavior during early acquisition, and in the ventromedial striatum, the region thought, along with the VTA, to be critical for initial acquisition. Striatal projection neuron ensembles in each of these regions develop different response patterns but, in all of the regions, the ensembles reflect the entire behavioral time.

These behavioral and physiological findings are important in supporting the view that there

is a critical transition period during habit formation. Before this transition, a given behavior being learned remains sensitive to outcome (usually tested as sensitivity to reward value). But, after this period, the same behavior becomes independent of the reward value. This distinction, introduced formally by Dickinson and his colleagues (Dickinson 1985) with reward devaluation paradigms, has been influential in models of habit formation and the shifts between goal-dependent and semiautomatic performance characteristics of habits. As we note below, this transition is marked by changes in the activity patterns of striatal neurons, and also of neurons in the prefrontal cortex.

Such transitions have not been tested directly in humans. However, brain imaging has suggested that, after conditioning, activity in the dorsal striatum is sensitive to the relative value of an action choice compared with other actions rather than to the relative value of rewards per se (Li and Daw 2011). Evidence suggests that the capacity to undergo this transition is influenced by the human *FOXP2* gene, a gene implicated in speech and language function in humans (Schreiweis et al. 2014). This influence is particularly intriguing because mutants of *FOXP2* do not alter motor performance or skill acquisition as tested, for example, by rotorod.

A second major line of work implicating the striatum in habitual behaviors comes from work on the neural origins of addictive behaviors. Much evidence suggests that the midbrain dopamine system, particularly the VTA, is influential in the initial stages of generation of these behaviors, and that the striatal target of the VTA system, which largely lies in the ventral striatum, is essential for the neural changes, leading to addiction. Remarkably, here too, with the progression of the addictive behavior, the dorsal striatum becomes more and more involved with time (Everitt and Robbins 2013). Thus, across very different domains of what in common parlance we call habits, there appear to be progressive stages for the ingraining of stereotyped action patterns into an individual's repertoire of behaviors, and a corresponding change in emphasis of striatal regions predominantly implicated (Graybiel 2008).

Cite this article as *Cold Spring Harb Perspect Biol* doi: 10.1101/cshperspect.a021691

Although not focusing in depth on addictive behavior, we emphasize that an important unresolved question is the degree to which addictions are extreme forms of habits, developed by trial-and-error learning and leading to distorted RPEs. In this context, addictions reveal fragility in this otherwise robust habit-learning mechanism, exposed by abnormal activation of the reward circuitry by exogenous chemicals, including cocaine and other psychomotor stimulants, ethanol, and nicotine. Positron emission tomography (PET) studies in humans have shown that drugs can induce rapid increases of striatal dopamine but that, once a person is addicted, these drug-induced dopamine increases (as well as their subjective effects on behavior) are blunted. By contrast, addicts experiencing craving to any one of a number of drugs can show a significant dopamine increase in striatum in response to drug-conditioned cues (such as thoughts leading to craving) with response levels that can be greater than those to the drug themselves (Volkow et al. 2011, 2014a,b). These cue-induced responses are particularly prominent in the dorsal striatum, including in the putamen, consistent with evidence that the dorsal striatum is heavily involved in "normal" habit learning. The cue responses likely reflect a profoundly distorted RPE.

Taken together, the findings from these two lines of work have sometimes been interpreted as suggesting that the striatum is "the seat of habitual behavior"—that the "habit," or its neural representation, is stored within the striatum itself. Classic methods, however, did not allow adequate testing of this notion.

An important detail of the anatomy of the striatum makes a definitive answer difficult to achieve. The projection neurons of the striatum, that is, the neurons that project to the pallidonigral output nuclei of the basal ganglia, also are the main striatal neurons receiving inputs to the striatum. Thus, any procedure, genetic or otherwise, affects circuits, and not just the striatum. We suggest that this is a critical distinction. It becomes impossible to say that habit representations are stored in the striatum because the striatum is only a node in larger networks. Neurobiological experimental techniques, such as optogenetics, now open the possibility of testing these assumptions directly.

## TIME SCALES OF LEARNING AND THE FORMATION OF MOTOR–MOTOR ASSOCIATIONS

Just as there are remarkable changes within striatal circuits occurring on multiple time scales during habit learning, of which we point out only some examples, during physical skill acquisition, there are analogous temporal shifts across circuits. For example, as nonhuman primates learn a novel behavior, such as a unique sequence of arm or finger movements, neural recordings typically show shifts in major activity from associative to sensorimotor districts of the striatum (Miyachi et al. 1997, 2002; Hikosaka et al. 1999), regions to which the DMS and DLS of rodents are thought to correspond (Graybiel 2008). Detailed modeling of reaction time behavior as the animal makes sequential reaches shows that they switch between two modes of control, consistent with the use of multiple motor control or prefrontal circuits to generate actions.

In rodents, profound changes in spike activity patterns of neurons occur simultaneously in the associative and sensorimotor striatum, and, as activity in the associative striatum declines, activity in the sensorimotor striatum becomes strong. These dynamics suggest that potentially competing circuits are organized to favor habit formation (Thorn et al. 2010). There is much evidence from classic rodent studies that different corticostriatal loops can compete with one another during the learning process and subsequent performance. This now has been found in humans. fMRI evidence shows that individuals who reduce activity in prefrontal regions sooner are those who acquire sequential skills more rapidly (Bassett et al. 2015).

In human brain-imaging experiments (Lehéricy et al. 2005; Grol et al. 2006; Doyon et al. 2009), a shift from anterior associative to sensorimotor striatum is also observed as people practice sequential finger movements. Initially, there is a broad recruitment of prefrontal, premotor, and sensorimotor cortex (along with the underlying corticostriatal target regions). With time, there is a progressive reduction of activity in prefrontal regions and associative striatum. These different cortical regions could potentially acquire, represent, or forget sequential information differently. For example, prefrontal regions that support working memory are invaluable for explicitly remembering a sequence of external cues that could guide movement. There are not yet sufficient studies of the striatum to relate all of these findings for the neocortex to selective changes in spike activities of different striatal regions, but we emphasize, again, that these dynamic changes occur across corticostriatal and other circuits.

A key advance in recent human brain-imaging methods is the ability to map activity that corresponds to a specific sequence or skill using either machine learning or repetition suppression methods (Wiestler and Diedrichsen 2013; Wymbs and Grafton 2014). These show that the degree to which different cortical areas represent a specific skill depends in large part on the depth of training experience, and not simply on time (Wymbs and Grafton 2014). Over a longer training horizon, there is less reliance on premotor areas to represent a sequence. Ultimately, a central feature of motor skill is the ability to guide actions without explicit memory or external stimuli through the creation of direct motor–motor associations. Not surprisingly, skill-specific changes also emerge within motor cortex (Karni et al. 1995; Wymbs and Grafton 2014). Thus, across habits and skills, different sorts of automaticity are gained. There are compelling parallels between these recording studies in animals during habit-learning and human-imaging experiments of skill learning.

Although the acquisition of physical skills is commonly attributed to online feedback-based error learning mediated by the cerebellum, allowing for powerful tuning of complex musculoskeletal dynamics, it is important to note that all of the associative and sensorimotor cortical areas that have been implicated in motor-skill acquisition and performance project directly to the striatum as parts of corticostriatal circuits. As shown by the recording work in rodents,

multiple corticostriatal loops, as judged by striatal projection neuron activity, are simultaneously active as reward-based learning occurs. Dopamine can modulate each of these loops. Dopamine-receptor blockade in nonhuman primates impairs skill acquisition (Tremblay et al. 2009). Deficits of dopamine signaling stemming from Parkinson's disease or dopamine-receptor blockade are both determinants of the rate of skill acquisition (Weickert et al. 2013). Genetically determined reductions of striatal dopamine function can also influence response to rewards and impact skill learning as well (Frank and Fossella 2011; Stice et al. 2012).

There is evidence from experiments in monkeys that well-practiced behaviors do not require the pallidal output nuclei of the basal ganglia for their expression (Desmurget and Turner 2010). This finding does not necessarily hold for habits, but it accords well with the proposal that the basal ganglia, and here we emphasize the striatum, is critical in the acquisition of action repertoires. Reinforcement-related signals reaching nonstriatal regions are very likely also important for influencing the formation of cortical or other connections that mediate motor–motor associations (in which the commands for one movement directly trigger the next command). It is important to note that there are significant striatal output connections that do not include the classical pallidal output nuclei of the basal ganglia. These connections could be part of the habits–skill performance circuitry even when the classic pallidal pathways are not required.

## BRACKETING: A READOUT THAT FRAMES AN ACTION

In the part of the rodent striatum thought to be necessary for habitual performance (i.e., DLS), a striking pattern of neuronal activity emerges. As animals learn to run in maze tasks, the striatal activity, at first, marks the full run time, but later begins to bracket the entire run. Activity becomes more and more prominent at the beginning and end of the runs, or beginning and end of the action through the turns. Surprisingly, at the same time, activity during the rest of the run time declines and may even be below prerun baseline levels. The kinematics of the runs are changing as the rats become more and more repetitive in their navigational routes, but the "end" activity can occur even after the rats are no longer running and, thus, cannot simply be attributed to velocity or acceleration signals. Similarly, the "beginning" or "start" activity can occur before the runs have begun.

Such beginning and end activity has been seen repeatedly in different experiments in rodents (Jog et al. 1999; Barnes et al. 2005; Thorn et al. 2010; Smith and Graybiel 2013), is long lasting, and has also been found in lever-pressing tasks (Jin and Costa 2010). Task-bracketing activities have also been recorded in the striatum (and prefrontal cortex) of macaque monkeys performing well-learned motor skills, including oculomotor sequential saccade tasks (Fujii and Graybiel 2003) and sequential arm-reaching tasks (J Feingold, D Gibson, AM Graybiel, et al., unpubl.; R Turner, pers. comm.). Concurrent phasic episodes of oscillatory local field potential (LFP) activity also occur in relationship to action boundaries (Howe et al. 2013). In patients with Parkinson's disease, LFP recordings within the subthalamic nucleus show anticipatory suppression of β-band activity at sequence boundaries that is linked to better performance (Herrojo Ruiz et al. 2014). This feature is also seen in monkeys performing arm-reaching tasks (J Feingold, D Gibson, AM Graybiel, et al., unpubl.).

Remarkably, a nearly inverse pattern of spike activity has been shown to develop in the associative, dorsomedial part of the rodent striatum, a region critical for goal-directed behavior. The projection neuron ensembles gradually develop increased firing during the runs, especially around the decision period of the task. There is much less activity at the beginning and end of the runs. Moreover, this decision-period activity then subsides during late learning—the very time that the "beginning-and-end" activity in the DLS is especially strong (Thorn et al. 2010).

Finally, in the ventromedial striatum, ensembles in the aggregate also show beginning-and-end responses. Yet, others fire throughout the runs, ramping up to the time of reward receipt (Atallah et al. 2014). These findings provide unequivocal evidence that striatal projection neurons have highly dynamic ensemble response patterns during habit learning. Especially notable is the fact that these patterns reflect entire behavioral sequences from beginning to end, which initially are goal directed, but after long training can become nearly autonomous except for being triggered by a start cue.

Protocols have now been used in rodents to determine how fixed the striatal task-bracketing patterns are. Extinction protocols, in which rewards were either removed or rarely given after acquisition and prolonged overtraining, nearly abolish the beginning-and-end pattern, but if the rewards are returned, the beginning-and-end pattern reappears almost immediately. Thus, the form of action boundary representation in the sensorimotor striatum somehow can be suppressed but cannot be erased by removal of rewards. After prolonged training, application of the classic reward devaluation procedure of Dickinson (1985), in which the reward is maintained, but is made unpalatable, hardly changes the striatal beginning-and-end pattern (Smith and Graybiel 2013). Thus, the task-bracketing pattern in the DLS is extremely resistant to degradation—it takes wholesale removal of rewards to block it fully, and even then, the pattern is latent, but not gone, and is rapidly retrievable.

Although the function of this bracketing activity remains unknown, a strong case can be made from the studies on habit learning that it is tied to feedback about how successful a sequence of actions within the bracket has been in gaining a desired outcome. This function is at the heart of the trial-and-error learning that leads to the forming of habits composed of multiple sequential actions. The end patterns found in these sequences could provide such outcome signals. These phasic end responses are themselves dynamic, changing or coming and going during the course of training (Smith and Graybiel 2013; Atallah et al. 2014). The mechanisms underlying these habit-related activity patterns are not understood, but evidence from the rodent experiments suggests that the activity of striatal interneurons is strongly modulated during habit learning. This result is notable because it indicates that intrastriatal networks undergo profound reorganizational changes. Thus, the ensemble patterns, such as task bracketing, cannot solely be attributed to the direct input connections of the projection neurons themselves; changes in local intrastriatal microcircuit occur as well. Below, we suggest that the bracketing could be a neural sign of the chunking of behaviors that have proven successful enough to the organism to merit prolonged expression.

## FROM BRACKETING TO CHUNKS: SHAPING THE ELEMENTS OF ACTION IN RELATION TO COSTS AND BENEFITS

The bracketing activity surrounding a habit formed from multiple behaviors is defined by both start- and end-related changes of neuronal activity. Although the end activity could clearly be used to predict a subsequent reward, the purpose of the start activity is less obvious. One possibility is that it serves as the opening of a bracketed behavioral unit. Interestingly, the beginning-and-end activities often are built and changed together, and sometimes appear in the responses of single striatal neurons. In monkeys, it has been shown that experimentally triggered changes in the end activity can induce changes in the accompanying start activity.

A key idea in control theory is that optimal behavior is determined not only by the reward obtained but also by the minimization of some cost function related to the set of actions needed to accomplish the action. If a behavior needs to be optimized over a particular time interval, then there needs to be an indication of when an action actually begins and ends. This estimate of a distinct time interval becomes increasingly important during transitions from habits to motor skills for which there is a further refinement of behavior at the level of kinematics and limb dynamics based on optimal control principles. Ultimately, optimal control requires an estimate of the physical or neural cost of performing a particular action. Together, the start-and-end activities could contribute to this estimation by providing a reading frame for labeling a given action.

A direct investigation of optimization within a habitual sequence of eye movements was examined in oculomotor scanning patterns generated by naïve, untrained monkeys (Desrochers et al. 2010). In this study, monkeys who were never trained experimentally were placed in a booth with a computer screen in front of them, on which colored discs appeared. The monkeys naturally looked around the display and, without experimental training, tended to acquire particular favored scan patterns that gradually changed over months of experience. The bit-by-bit changes in the monkeys' scanning resulted in a succession of favored spatiotemporal chunking patterns of the untrained habitual saccade sequences, and these habitual scanning patterns eventually became optimal or nearly optimal, as judged by models of the task. Remarkably, nearly all of these adjustments in the scanning patterns took place long after maximum reward had been obtained. Analysis showed that the behavior was driven by small trial-by-trial differences in the cost of the scans: the distance required. This result has led to the conclusion that extremely fine-grain, trial-by-trial monitoring of least cost can be a driver of habit learning, as well as a driver of skill learning. These results are in line with the notion that the brain has a natural tendency to reduce cost across many cognitive domains and that this tendency, in addition to sensitivity to reward, can drive the character of habits (Desrochers et al. 2010; Kool et al. 2010; Gepshtein et al. 2014). Recent electrophysiological recordings suggest that striatal projection neurons encode these outcome and cost signals in their end activity (T Desrochers, K Amemori, and AM Graybiel, unpubl.).

A set of behaviors that is reliably combined and expressed as a habit can ultimately be viewed as a "chunk," framed by the neuronal bracketing activity in the striatum. This notion of chunking, introduced by George Miller (1956) in reference to helping deal with memory load, invokes, for the motor system, the binding together of multiple behaviors into a single behavioral unit. There are different aspects to such packaging up of behaviors. First, there is a form of chunking (concatenation), bundling the in-

Cite this article as *Cold Spring Harb Perspect Biol* doi: 10.1101/cshperspect.a021691

dividual elements into a whole. As we note below, there is reason to think that the striatum and its circuits could be critical to this function (Graybiel 2008). Alongside this is another well-known phenomenon in cognitive science wherein adjacent elements of a long sequential stimulus or behavior are temporally divided up ( parsing), leading to detectable pauses between groups of adjacent elements. Chunking is often used strategically by humans to parse long strings of stimuli into smaller sets to facilitate memory, as proposed by Miller, by analogy to remembering sequences of numbers. For example, a U.S. phone number is strategically divided into a 3-3-4 pattern.

Pauses within a sequence of movements provide a useful way to identify chunks embedded within long sequences. New computational tools are emerging to identify chunks for habits or motor skills based on concatenation rather than parsing. Concatenation can be spotted by examining the covariation or timing of movements within a suspected chunk or by assessing the frequency with which errors are made at the boundary of adjacent chunks rather than within chunks (Acuna et al. 2014). Thus, it is becoming possible to assess the strength by which successive elements of a complex action are combined. Whether it is habits or skills, the animal is seeking to minimize some cost function over the time interval of the complete action. This development of smooth kinematics is universally observed as animals combine fragmented movements together. By concatenating muscle synergies or movements together in specific groups, it might be possible to improve efficiency. Interestingly, the learning of the kinematics can progress more rapidly than the progression of the behavior becoming a habit (Jog et al. 1999; Barnes et al. 2005; Smith and Graybiel 2013).

Based on the bracketing patterns that form in the striatum and elsewhere, and the behavioral changes that occur alongside them, it has been suggested that one function of the striatum (and, hence, of the basal ganglia) could be to facilitate such chunking as habits and routines form (Graybiel 1998, 2008). The key property of this process is the selection of behaviors that are successful, either through optimizing reward or cost or their integral. This notion fits well with our suggestion that the striatum and associated circuits could be important for achieving optimality in the performance of both habits and skills. Marking action-sequence boundaries allows the sequences to be represented as units that could then be released more readily than unconcatenated chains of elements. Importantly, this function should serve cognitive as well as motor packaging up of beneficial (or, in pathology, nonbeneficial) behaviors (Graybiel 2008).

Evidence for a role of the striatum in chunking movements stems, in part, from pharmaceutical blockade of dopamine receptors in monkeys with the dopamine D2-type receptor antagonist, raclopride. This manipulation does not interfere with well-learned sequences, but disrupts the formation of new chunks (Levesque et al. 2007). Chronic dopamine denervation in patients with Parkinson's disease can also lead to an impairment of chunking for new sequences of movement (Tremblay et al. 2010). There is also emerging evidence from human neuroimaging that the strength of activity in the associative striatum varies on a trial-by-trial basis with the degree to which subjects put together elements of motor sequences into chunks, the concatenation aspect of chunking (Wymbs et al. 2012).

Ultimately, if a string of behavioral elements is represented as a single unit, then there should be concomitant neuronal activity reflecting this. Not only is the representation of an action boundary present in the spiking of ensembles of striatal neurons, it is also observed in striatal neurons with maintained activity (Kubota et al. 2009; Barnes et al. 2011; Hernandez et al. 2013; Howe et al. 2013). Here, the action-related activity can be identified throughout the duration of the set of actions forming a chunk. This could represent a prolonged form of dopamine signaling, which, in the ventromedial striatum, can span the entire habitual behavior (Howe et al. 2013). Thus, the neuronal populations within rodent striatum associated with the bracketing and task-on activity could serve as neural signatures of the concatenation and parsing functions identified in human studies of chunking.

Collectively, these studies across multiple species and tasks suggest that striatal circuits can help to bring together advantageous behavioral segments into sequences that help to achieve behavioral goals. The result, in this view, is that the behaviors could be released readily as a complete "set" when the appropriate context calls for this release. We now know that these beginning-and-end patterns can develop in cortical regions and elsewhere, suggesting that this process is a network property, strongly evidenced in parts of the striatum and corresponding corticostriatal loops. This idea finds strong resonance in clinical observations, for example, in the problems that Parkinson's patients have in starting a sequential set of actions, such as walking, and then in ending the sequence once underway.

## WHERE ARE HABITS "STORED"? CIRCUIT DYNAMICS ARE CRITICAL TO HABIT LEARNING AND PERFORMANCE

Causal evidence for circuit-level control of habits is just emerging. Lesion studies show that the medial prefrontal cortical region, called infralimbic (IL) cortex, in rodents, like the DLS, is necessary for habits to be performed. New optogenetic studies have shown that the IL exerts online control of the performance of well-ingrained habits (Smith et al. 2012) and is necessary for their formation (Smith and Graybiel 2013). This work is critical to any account of the role of the striatum in habit formation, as it suggests a form of cortical control that can, on a moment-by-moment basis, determine whether a behavior is performed habitually or not.

Strikingly, IL develops, in its upper layers, a strong task-bracketing pattern during habit learning, one similar to that in the DLS. But, the cortical bracketing pattern, unlike the DLS bracketing pattern, is sensitive to reward devaluation; it is nearly lost. As IL does not project directly to the DLS, the results suggest that the online control is a circuit-level effect. This online control suggests that we need to rethink our ideas about the control of habits—both the

learning of these behaviors and their expression. At the very least, there are dual operators, cortical and subcortical, acting as habits become crystallized, and these act simultaneously with the cortical control being online (Smith and Graybiel 2013).

There is not yet comparable optogenetic evidence for the effects of perturbations within the striatum itself. This is key missing information. What can be said is that multiple circuits are simultaneously active as habits form, and these circuits have differential sensitivities and patterns of connectivity. There is no evidence for habits being "stored" in one site, such as the striatum, although local networks within the striatum acquire new activity patterns during habit learning and could be local controllers. These considerations deflate controversies pitting different individual regions as being most important for habit learning.

## CHUNKS, THE DEGREES-OF-FREEDOM PROBLEM, AND OPTIMAL CONTROL

We return to the key issues of the degrees-of-freedom problem and how to achieve optimal control. We suggest that a major advantage of habit formation is that this process allows many possible degrees of freedom to be essentially dropped from the animal's normal, habitual repertoire so long as the conditions surrounding the habit are not at odds with its performance. Change in these conditions could lead to a return to behavior typical of early acquisition, a kind of trial-and-error behavior akin to early language learning in children and to song learning in passerine birds. The degradation of the task-bracketing patterns with removal of positive outcomes fits with this reverse plasticity.

How does chunking relate to the problem of having unmanageable numbers of degrees of freedom and the need for optimality in motor control? In this instance, finding optimal solutions becomes increasingly difficult as the sequence of movements is lengthened. It is possible that by grouping fine-grained movement elements into chunks, the solutions for optimality become easier to compute. An alternative and intriguing possibility is that the pauses

Cite this article as *Cold Spring Harb Perspect Biol* doi: 10.1101/cshperspect.a021691

observed in complex movements are a result of optimal control. Preliminary evidence has been gathered in both humans and nonhuman primates as they perform five element sequential reaching tasks. For a given sequence, different monkeys will converge on the same pattern of chunking. Kinematic analysis suggests that the chunks lead to a more global pattern of movement efficiency than what is obtained otherwise (Ramkumar et al. 2014). However, these observations need to be tempered in light of the fact that, in many situations, optimality is not essential. Instead, muscle synergies seem to be built on habits that are "good enough" rather than optimal (de Rugy et al. 2012).

## BEYOND ACTIONS: THE UTILITY OF STRIATAL CIRCUITS FOR EMOTIONAL AND COGNITIVE HABITS AND SKILLS

As habitual actions become ingrained, the kinematics of the habitual actions, that is, the physical skills enabling these habits, become standardized. But there is a further connotation of the term "habit" to consider, one that invokes motivational processes that shape the expression of cognitive processes in particular contexts, irrespective of kinematics and physical skill. Although these kinds of habits are not restricted to motor acts or sequences of motor actions, they are likely to also rely on corticostriatal circuits that use contextual information to shape behavior. Further, these habits of thought can be powerfully shaped by complex social cues (Graybiel 2008). For example, striatal neurons are able to distinguish reward predictions intended for the monkey undergoing neuronal recordings from those destined for another animal (Báez-Mendoza et al. 2013). Habits of thought, at least in humans, are probably as common as motor habits and, like motor habits, are vulnerable to pathologic distortion. Our view is that such habits of mind can be created by cognitive pattern generators much as habits of action are generated (Graybiel 1997). Understanding these wider implications of learning repertoires of thought and action is an important goal for future work.

## ACKNOWLEDGMENTS

Supported in part by National Institutes of Health (NIH) Grants R01 EY012848, R01 NS025529, and R01 MH060379 (A.M.G.), Office of Naval Research Grant N00014-07-1-0903 (A.M.G.), Public Health Service (PHS) Grants NS44393 (S.T.G.), and Contract No. W911NF-09-D-0001 from the U.S. Army Research Office (S.T.G.).

## REFERENCES

Acuna DE, Wymbs NF, Reynolds CA, Picard N, Turner RS, Strick PL, Grafton ST, Körding K. 2014. Multi-faceted aspects of chunking enable robust algorithms. *J Neurophysiol* **112:** 1849–1856.

Atallah HE, McCool AD, Howe MW, Graybiel AM. 2014. Neurons in the ventral striatum exhibit cell-type-specific representations of outcome during learning. *Neuron* **82:** 1145–1156.

Báez-Mendoza R, Harris CJ, Schultz W. 2013. Activity of striatal neurons reflects social action and own reward. *Proc Natl Acad Sci* **110:** 16634–16639.

Balleine BW, Dickinson A. 1998. Goal-directed instrumental action: Contingency and incentive learning and their cortical substrates. *Neuropharmacology* **37:** 407–419.

Barnes TD, Kubota Y, Hu D, Jin DZ, Graybiel AM. 2005. Activity of striatal neurons reflects dynamic encoding and recoding of procedural memories. *Nature* **437:** 1158–1161.

Barnes TD, Mao J-B, Hu D, Kubota Y, Dreyer AA, Stamoulis C, Brown EN, Graybiel AM. 2011. Advance cueing produces enhanced action-boundary patterns of spike activity in the sensorimotor striatum. *J Neurophysiol* **105:** 1861–1878.

Bassett DS, Yang M, Wymbs NF, Grafton ST. 2015. Learning-induced autonomy of sensorimotor systems. *Nat Neurosci* **18:** 744–751.

Belin D, Jonkman S, Dickinson A, Robbins TW, Everitt BJ. 2009. Parallel and interactive learning processes within the basal ganglia: Relevance for the understanding of addiction. *Behav Brain Res* **199:** 89–102.

Bernstein NA. 1967. *The coordination and regulation of movements.* Pergamon, New York.

Bevan MD, Smith AD, Bolam JP. 1996. The substantia nigra as a site of synaptic integration of functionally diverse information arising from the ventral pallidum and the globus pallidus in the rat. *Neuroscience* **75:** 5–12.

Brown MT, Tan KR, O'Connor EC, Nikonenko I, Muller D, Luscher C. 2012. Ventral tegmental area GABA projections pause accumbal cholinergic interneurons to enhance associative learning. *Nature* **492:** 452–456.

Daniel R, Pollmann S. 2014. A universal role of the ventral striatum in reward-based learning: Evidence from human studies. *Neurobiol Learn Mem* **114:** 90–100.

de Rugy A, Loeb GE, Carroll TJ. 2012. Muscle coordination is habitual rather than optimal. *J Neurosci* **32:** 7384–7391.

Desmurget M, Turner RS. 2010. Motor sequences and the basal ganglia: Kinematics, not habits. *J Neurosci* **30:** 7685–7690.

Desrochers TM, Jin DZ, Goodman ND, Graybiel AM. 2010. Optimal habits can develop spontaneously through sensitivity to local cost. *Proc Natl Acad Sci* **107:** 20512–20517.

Dickinson A. 1985. Actions and habits: The development of behavioural autonomy. *Philos Trans R Soc Lond B Biol Sci* **308:** 67–78.

Diuk C, Tsai K, Wallis J, Botvinick M, Niv Y. 2013. Hierarchical learning induces two simultaneous, but separable, prediction errors in human basal ganglia. *J Neurosci* **33:** 5797–5805.

Doig NM, Magill PJ, Apicella P, Bolam JP, Sharott A. 2014. Cortical and thalamic excitation mediate the multiphasic responses of striatal cholinergic interneurons to motivationally salient stimuli. *J Neurosci* **34:** 3101–3117.

Doyon J, Bellec, Amsel, Penhune, Monchi, Carrier, Lehéricy S, Benali H. 2009. Contributions of the basal ganglia and functionally related brain structures to motor learning. *Behav Brain Res* **199:** 61–75.

Everitt BJ, Robbins TW. 2013. From the ventral to the dorsal striatum: Devolving views of their roles in drug addiction. *Neurosci Biobehav Rev* **37:** 1946–1954.

Frank MJ, Fossella JA. 2011. Neurogenetics and pharmacology of learning, motivation, and cognition. *Neuropsychopharmacology* **36:** 133–152.

Fujii N, Graybiel AM. 2003. Representation of action sequence boundaries by macaque prefrontal cortical neurons. *Science* **301:** 1246–1249.

Gao Z, van Beugen BJ, De Zeeuw CI. 2012. Distributed synergistic plasticity and cerebellar learning. *Nat Rev Neurosci* **13:** 619–635.

Garrison J, Erdeniz B, Done J. 2013. Prediction error in reinforcement learning: A meta-analysis of neuroimaging studies. *Neurosci Biobehav Rev* **37:** 1297–1310.

Gepshtein S, Li X, Snider J, Plank M, Lee D, Poizner H. 2014. Dopamine function and the efficiency of human movement. *J Cogn Neurosci* **26:** 645–657.

Graybiel AM. 1997. The basal ganglia and cognitive pattern generators. *Schizophr Bull* **23:** 459–469.

Graybiel A. 1998. The basal ganglia and chunking of action repertoires. *Neurobiol Learn Mem* **70:** 119–136.

Graybiel AM. 2008. Habits, rituals, and the evaluative brain. *Annu Rev Neurosci* **31:** 359–387.

Grol MJ, de Lange FP, Verstraten FA, Passingham RE, Toni I. 2006. Cerebral changes during performance of overlearned arbitrary visuomotor associations. *J Neurosci* **26:** 117–125.

Hernandez LF, Kubota Y, Hu D, Howe MW, Lemaire N, Graybiel AM. 2013. Selective effects of dopamine depletion and L-DOPA therapy on learning-related firing dynamics of striatal neurons. *J Neurosci* **33:** 4782–4795.

Herrojo Ruiz M, Rusconi M, Brücke C, Haynes J-D, Schönecker T, Kühn AA. 2014. Encoding of sequence boundaries in the subthalamic nucleus of patients with Parkinson's disease. *Brain* **137:** 2715–2730.

Hikosaka O, Sakamoto M, Usui S. 1989. Functional properties of monkey caudate neurons. III: Activities related to expectation of target and reward. *J Neurophysiol* **61:** 814–832.

Hikosaka O, Nakahara H, Rand MK, Sakai K, Lu X, Nakamura K, Miyachi S, Doya K. 1999. Parallel neural networks for learning sequential procedures. *Trends Neurosci* **22:** 464–471.

Howe MW, Tierney PL, Sandberg SG, Phillips PE, Graybiel AM. 2013. Prolonged dopamine signalling in striatum signals proximity and value of distant rewards. *Nature* **500:** 575–579.

Jin X, Costa RM. 2010. Start/stop signals emerge in nigrostriatal circuits during sequence learning. *Nature* **466:** 457–462.

Jog MS, Kubota Y, Connolly CI, Hillegaart V, Graybiel AM. 1999. Building neural representations of habits. *Science* **286:** 1745–1749.

Karni A, Meyer G, Jezzard P, Adams MM, Turner R, Ungerleider LG. 1995. Functional MRI evidence for adult motor cortex plasticity during motor skill learning. *Nature* **377:** 155–157.

Kool W, McGuire JT, Rosen ZB, Botvinick MM. 2010. Decision making and the avoidance of cognitive demand. *J Exp Psychol Gen* **139:** 665–682.

Kubota Y, Liu J, Hu D, DeCoteau WE, Eden UT, Smith AC, Graybiel AM. 2009. Stable encoding of task structure coexists with flexible coding of task events in sensorimotor striatum. *J Neurophysiol* **102:** 2142–2160.

Lehéricy S, Benali H, Van de Moortele PF, Pelegrini-Issac M, Waechter T, Ugurbil K, Doyon J. 2005. Distinct basal ganglia territories are engaged in early and advanced motor sequence learning. *Proc Natl Acad Sci* **102:** 12566–12571.

Levesque M, Bedard MA, Courtemanche R, Tremblay PL, Scherzer P, Blanchet PJ. 2007. Raclopride-induced motor consolidation impairment in primates: Role of the dopamine type-2 receptor in movement chunking into integrated sequences. *Exp Brain Res* **182:** 499–508.

Li J, Daw ND. 2011. Signals in human striatum are appropriate for policy update rather than value prediction. *J Neurosci* **31:** 5504–5511.

Matsumoto M, Hikosaka O. 2009. Two types of dopamine neuron distinctly convey positive and negative motivational signals. *Nature* **459:** 837–841.

Miller GA. 1956. The magic number seven plus or minus two: Some limits on our automatization of cognitive skills. *Psychol Rev* **63:** 81–97.

Miyachi S, Hikosaka O, Miyashita K, Karadi Z, Rand MK. 1997. Differential roles of monkey striatum in learning of sequential hand movement. *Exp Brain Res* **115:** 1–5.

Miyachi S, Hikosaka O, Lu X. 2002. Differential activation of monkey striatal neurons in the early and late stages of procedural learning. *Exp Brain Res* **146:** 122–126.

Ramkumar P, Acuna DE, Berniker M, Grafton S, Turner RS, Körding KP. 2014. Movement chunking as locally optimal control. In *Translational and Computational Motor Control (TCMC) 2014.* Washington, DC, November 14.

Schreiweis C, Bornschein U, Burguiere E, Kerimoglu C, Schreiter S, Dannemann M, Goyal S, Rea E, French CA, Puliyadi R, et al. 2014. Humanized Foxp2 accelerates learning by enhancing transitions from declarative to

 Cite this article as *Cold Spring Harb Perspect Biol* doi: 10.1101/cshperspect.a021691

procedural performance. *Proc Natl Acad Sci* **111:** 14253–14258.

Schultz W. 2002. Getting formal with dopamine and reward. *Neuron* **36:** 241–263.

Smith KS, Graybiel AM. 2013. A dual operator view of habitual behavior reflecting cortical and striatal dynamics. *Neuron* **79:** 361–374.

Smith KS, Virkud A, Deisseroth K, Graybiel AM. 2012. Reversible online control of habitual behavior by optogenetic perturbation of medial prefrontal cortex. *Proc Natl Acad Sci* **109:** 18932–18937.

Stice E, Yokum S, Burger K, Epstein L, Smolen A. 2012. Multilocus genetic composite reflecting dopamine signaling capacity predicts reward circuitry responsivity. *J Neurosci* **32:** 10093–10100.

Sutton RS, Barto AG. 1998. *Reinforcement learning*. MIT Press, Cambridge, MA.

Takemura A, Inoue Y, Gomi H, Kawato M, Kawano K. 2001. Change in neuronal firing patterns in the process of motor command generation for the ocular following response. *J Neurophysiol* **86:** 1750–1763.

Thorn CA, Atallah H, Howe M, Graybiel AM. 2010. Differential dynamics of activity changes in dorsolateral and dorsomedial striatal loops during learning. *Neuron* **66:** 781–795.

Todorov E, Jordan MI. 2002. Optimal feedback control as a theory of motor coordination. *Nat Neurosci* **5:** 1226–1235.

Tremblay P-L, Bedard M-A, Levesque M, Chebli M, Parent M, Courtemanche R, Blanchet PJ. 2009. Motor sequence learning in primate: Role of the D2 receptor in movement chunking during consolidation. *Behav Brain Res* **198:** 231–239.

Tremblay PL, Bedard MA, Langlois D, Blanchet PJ, Lemay M, Parent M. 2010. Movement chunking during sequence learning is a dopamine-dependent process: A study conducted in Parkinson's disease. *Exp Brain Res* **205:** 375–385.

Volkow ND, Wang G-J, Fowler JS, Tomasi D, Telang F. 2011. Addiction: Beyond dopamine reward circuitry. *Proc Natl Acad Sci* **108:** 15037–15042.

Volkow ND, Tomasi D, Wang GJ, Logan J, Alexoff DL, Jayne M, Fowler JS, Wong C, Yin P, Du C. 2014a. Stimulant-induced dopamine increases are markedly blunted in active cocaine abusers. *Mol Psychiatry* **19:** 1037–1043.

Volkow ND, Wang GJ, Telang F, Fowler JS, Alexoff D, Logan J, Jayne M, Wong C, Tomasi D. 2014b. Decreased dopamine brain reactivity in marijuana abusers is associated with negative emotionality and addiction severity. *Proc Natl Acad Sci* **111:** E3149–E3156.

Weickert TW, Mattay VS, Das S, Bigelow LB, Apud JA, Egan MF, Weinberger DR, Goldberg TE. 2013. Dopaminergic therapy removal differentially effects learning in schizophrenia and Parkinson's disease. *Schizophr Res* **149:** 162–166.

Wiestler T, Diedrichsen J. 2013. Skill learning strengthens cortical representations of motor sequences. *eLife* **2:** e00801.

Wymbs NF, Grafton ST. 2014. The human motor system supports sequence-specific representations over multiple training dependent time scales. *Cereb Cortex* doi: 10.1093/cercor/bhu144.

Wymbs NF, Bassett DS, Mucha PJ, Porter MA, Grafton ST. 2012. Differential recruitment of the sensorimotor putamen and frontoparietal cortex during motor chunking in humans. *Neuron* **74:** 936–946.

Yamada H, Inokawa H, Matsumoto N, Ueda Y, Enomoto K, Kimura M. 2013. Coding of the long-term value of multiple future rewards in the primate striatum. *J Neurophysiol* **109:** 1140–1151.

Yanike M, Ferrera VP. 2014. Representation of outcome risk and action in the anterior caudate nucleus. *J Neurosci* **34:** 3279–3290.

Yin HH, Knowlton BJ. 2006. The role of the basal ganglia in habit formation. *Nat Rev Neurosci* **7:** 464–476.

# Associative Learning in Invertebrates

Robert D. Hawkins[1,2] and John H. Byrne[3]

[1]Department of Neuroscience, Columbia University, New York, New York 10032

[2]New York State Psychiatric Institute, New York, New York 10032

[3]Department of Neurobiology and Anatomy, The University of Texas Medical School at Houston, Houston, Texas 77030

*Correspondence:* rdh1@columbia.edu

This work reviews research on neural mechanisms of two types of associative learning in the marine mollusk *Aplysia*, classical conditioning of the gill- and siphon-withdrawal reflex and operant conditioning of feeding behavior. Basic classical conditioning is caused in part by activity-dependent facilitation at sensory neuron–motor neuron (SN–MN) synapses and involves a hybrid combination of activity-dependent presynaptic facilitation and Hebbian potentiation, which are coordinated by *trans*-synaptic signaling. Classical conditioning also shows several higher-order features, which might be explained by the known circuit connections in *Aplysia*. Operant conditioning is caused in part by a different type of mechanism, an intrinsic increase in excitability of an identified neuron in the central pattern generator (CPG) for feeding. However, for both classical and operant conditioning, adenylyl cyclase is a molecular site of convergence of the two signals that are associated. Learning in other invertebrate preparations also involves many of the same mechanisms, which may contribute to learning in vertebrates as well.

Learning can be divided into two general categories: nonassociative learning, in which an animal learns about the properties or occurrence of a single stimulus (see Byrne and Hawkins 2015), and associative learning, in which an animal learns about the relationship between two stimuli or events. Associative learning includes classical and operant conditioning. During classical or Pavlovian conditioning, one stimulus (the unconditioned stimulus [US], e.g., meat powder) is repeatedly given shortly after another stimulus (the conditioned stimulus [CS], e.g., a bell). Following conditioning, the animal makes a response to the CS (the conditioned response or CR, e.g., salivation) that resembles the response to the US, and it is generally thought to have learned that the CS predicts the occurrence of the US. During operant conditioning, the animal is repeatedly given a reinforcement (e.g., a food pellet) shortly after it shows a particular behavior (e.g., lever pressing). Following appetitive operant conditioning, the animal increases the frequency and vigor of the reinforced behavior, and it is generally thought to have learned that its behavior predicts the occurrence of the reinforcement. Thus, these two types of conditioning can be thought of as prototypes of learning about the regularity and predictability of events in the world.

Cite this article as *Cold Spring Harb Perspect Biol* doi: 10.1101/cshperspect.a021709

The properties of classical and operant conditioning have been studied extensively in a wide range of species, and have been found to be basically similar throughout the animal kingdom. Thus, it is reasonable to suppose that at least some of the underlying mechanisms may be similar as well. That idea has encouraged the study of neural mechanisms of learning in relatively simple invertebrate species, which have a number of experimental advantages for such studies (Byrne and Hawkins 2015). In particular, many invertebrate nervous systems have comparatively few neurons, making it easier to elucidate the neural circuits for specific behaviors. Furthermore, some of the neurons are individually identifiable and many are very large, making it much easier to perform experimental manipulations, such as intracellular recording and injections, and to perform assays at the level if individual cells.

This review will focus on two types of associative learning in the marine mollusk *Aplysia*, classical conditioning of the gill- and siphon-withdrawal reflex, and operant conditioning of feeding behavior. The work will also briefly summarize progress on neural mechanisms of learning in several other invertebrate preparations that have made great contributions to the field as well.

## NEURAL AND MOLECULAR MECHANISMS OF CLASSICAL CONDITIONING OF THE GILL- AND SIPHON-WITHDRAWAL REFLEX IN *Aplysia*

### Behavioral Sensitization and Basic Classical Conditioning of the Withdrawal Reflex

The mechanisms of several simple forms of learning have been studied extensively in the marine mollusk *Aplysia*, which has a number of advantages for a reductionist approach (see Byrne and Hawkins 2015). Although conditioning has been shown for some more complex behaviors (e.g., Walters et al. 1981; Cook and Carew 1986), many of those studies have examined the gill- and siphon-withdrawal reflex, in which a light touch to the siphon (an exhalant funnel for the gill) produces contraction of the gill and siphon. As described in Byrne and Hawkins (2015), a noxious stimulus, such as a shock to the tail, produces an enhancement of subsequent responses to siphon stimulation or sensitization. In addition to this nonassociative form of learning, the reflex undergoes two associative forms of learning, operant conditioning (Hawkins et al. 2006) and classical conditioning, which has been studied much more extensively for this behavior. Classical conditioning resembles sensitization in that the response to stimulation of one pathway is enhanced by activity in another. Typically, in classical conditioning, an initially weak or ineffective CS becomes more effective in producing a behavioral response after it has been paired temporally with a strong US. What distinguishes classical conditioning from sensitization is the requirement for temporal pairing and contingency of the two stimuli during training (Rescorla 1967, 1968; Kamin 1969).

In most experiments on conditioning of the gill- and siphon-withdrawal reflex, the CS is a weak tactile stimulus to the siphon (which initially produces a weak withdrawal response), and the US is an electric shock to the tail (Fig. 1). If these two stimuli are paired for 20–30 trials, the siphon stimulation comes to elicit significantly larger gill and siphon withdrawals than if the two stimuli are presented in an unpaired or random fashion (Carew et al. 1981). This effect builds up during the training session and is retained for several days. The reflex also undergoes differential conditioning with stimulation of the siphon and mantle shelf (a region anterior to the gill), or two different sites on the siphon as the discriminative stimuli (Carew et al. 1983). Significant differential conditioning occurs either 15 min or 24 h after a single training trial, and there is stronger conditioning with five or 15 training trials. In addition, there is reliable conditioning when the CS precedes the US by 0.5 sec (the standard interstimulus interval) but no conditioning when the interval is 2 sec or longer or when the US precedes the CS (Hawkins et al. 1986). These results show stimulus and temporal specificities in conditioning of the reflex.

Conditioning of the reflex also shows response specificity. That is, *Aplysia* learn not

 Cite this article as *Cold Spring Harb Perspect Biol* doi: 10.1101/cshperspect.a021709

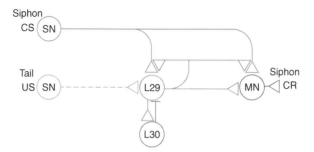

**Figure 1.** Partial circuit diagram for the *Aplysia* withdrawal reflex. The circuit can account for basic classical conditioning and a number of higher-order features of conditioning. Sensory neurons (SN) that are activated by the siphon conditioned stimulus (CS) make monosynaptic connections onto motor neurons (MN) that produce the gill- and siphon-withdrawal conditioned response (CR) and also onto several interneurons (not shown). The tail shock unconditioned stimulus (US) excites modulatory interneurons including the L29s. In addition, the L29s are excited by the siphon CS and excite the siphon motor neurons, so that they are also excitatory interneurons. Firing of the L29 neurons produces facilitation and activity-dependent facilitation at the SN–MN synapses, and also at the synapses from the SNs onto the L29 neurons themselves. Furthermore, the L29 neurons undergo spike accommodation during prolonged stimulation, caused in part by inhibitory feedback (horizontal bar) from L30 interneurons.

only to strengthen the magnitude of a previously existing reflex response, but they also learn to develop a new type of response to the CS that resembles the response to the US (Hawkins et al. 1989; Walters 1989). Thus, siphon stimulation initially produces straight contraction of the siphon, whereas tail shock produces backward bending of the siphon. When a siphon touch CS is paired with a tail shock US, the CS comes to produce backward bending as well. In all of these respects, conditioning of the siphon-withdrawal reflex is similar to many instances of vertebrate conditioning, such as conditioning of the rabbit eye-blink response (Gormezano 1972).

## Cellular and Molecular Mechanisms of Sensitization and Basic Classical Conditioning of the Withdrawal Reflex

The neural circuit for the reflex consists in part of monosynaptic connections from siphon sensory neurons (SNs) to gill and siphon motor neurons (MNs), as well as polysynaptic connections involving excitatory and inhibitory interneurons (Fig. 1). The circuit also includes several identified modulatory neurons. A pair of neurons in the cerebral ganglia (the CB1 neurons), and a group of about five neurons in the abdo-

minal ganglion (the L29 neurons) are excited by noxious stimulation, and produce facilitation of siphon SN–MN excitatory postsynaptic potentials (EPSPs) and broadening of action potentials in the SNs (Hawkins 1981; Hawkins et al. 1981b; Mackey et al. 1989). The CB1 neurons contain serotonin (5-HT), two to three of the five L29 neurons express the synthetic enzyme for nitric oxide (NO), and the other L29s may express an endogenous peptide (SCP), all of which contribute to facilitation and behavioral enhancement of the reflex (Abrams et al. 1984; Glanzman et al. 1989; Pieroni and Byrne 1992; Antonov et al. 2007).

It is possible to record the activity of these identified neurons and their synaptic connections during learning in a semi-intact preparation of the siphon withdrawal reflex, and thus to examine the contributions of plasticity at different sites in the circuit to behavioral learning. Such experiments have shown that heterosynaptic facilitation and activity-dependent facilitation at the SN–MN synapses contribute to sensitization and classical conditioning of the reflex, and that plasticity at other sites also contributes (Antonov et al. 1999, 2001). The mechanisms of facilitation at the SN–MN synapses have been examined more extensively in neural analogs of learning in isolated ganglia or in cell culture, in which tail shock is replaced by either nerve shock or application of an endogenous facilitatory transmitter, such as 5-HT, that is released following tail shock. As described in Byrne and Hawkins (2015), short-term facilitation by brief application of 5-HT to rested syn-apses (an analog of sensitization) involves pre-synaptic cyclic adenosine monophosphate (cAMP) and protein kinase A (PKA), which pro-duce decreased K⁺ current and increased spike width, Ca²⁺ influx, and transmitter release from the SNs (Fig. 2).

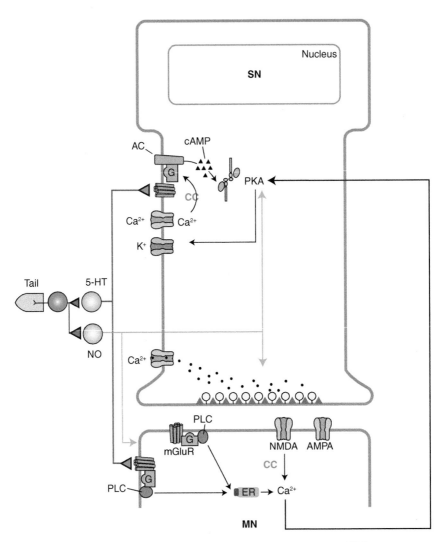

Figure 2. Cellular and molecular mechanisms of plasticity at sensory neuron (SN)–motor neuron (MN) synapses that contribute to basic classical conditioning. Classical conditioning (CC) involves a presynaptic associative mechanism, activity-dependent enhancement of presynaptic facilitation, which is due in part to $Ca^{2+}$ priming of adenylyl cyclase leading to increased production of cyclic adenosine monophosphate (cAMP) and increased activation of protein kinase A (PKA). Conditioning also involves a postsynaptic associative mechanism, Hebbian potentiation, which is caused by $Ca^{2+}$ influx through $N$-methyl-d-aspartate (NMDA) receptor channels. These two mechanisms interact through retrograde signaling. In addition, nitric oxide (NO) acts directly in both the sensory and motor neurons to affect different mechanisms of facilitation at the synapses between them. AC, Adenyl cyclase; ER, endoplasmic reticulum; PLC, phospholipase C.

Cite this article as *Cold Spring Harb Perspect Biol* doi: 10.1101/cshperspect.a021709

Because of the similarity of sensitization and classical conditioning, it was attractive to believe that conditioning might also involve heterosynaptic facilitation as a mechanism for strengthening the CS pathway. Specifically, the CS and US might converge at the level of individual neurons in the CS pathway, with the US producing greater facilitation of those neurons if they fire action potentials just before the US is delivered (as occurs during conditioning). Consistent with that idea, tail shock produces significantly greater facilitation of the monosynaptic EPSP from an SN to an MN if the shock is preceded by intracellularly produced spike activity in the SN than if it is either unpaired with spike activity or is presented alone (Hawkins et al. 1983; Walters and Byrne 1983). Like behavioral conditioning of siphon withdrawal, this effect builds up during the training session and is retained for at least 24 h (Buonomano and Byrne 1990). Also like behavioral conditioning, 0.5-sec forward pairing of the spike activity and tail shock is more effective than pairing with a longer interval or backward pairing (Clark et al. 1994). Activity-dependent facilitation can thus account for aspects of both the stimulus and temporal specificities of conditioning.

## Presynaptic Mechanisms

Activity-dependent facilitation could result from either a presynaptic or a postsynaptic mechanism. Because the facilitation underlying sensitization involves presynaptic mechanisms including broadening of action potentials in the SN, the initial studies of conditioning focused on that mechanism. In neural analogs of conditioning with SN spike activity as the CS and either tail shock or brief application of 5-HT as the US, paired training produced significantly greater broadening of action potentials in the SN than unpaired training (Hawkins et al. 1983; Eliot et al. 1994). Furthermore, forward pairing with 5-HT produced greater broadening than backward pairing (Clark et al. 1994). In experiments in isolated cell culture, paired training with 5-HT as the US also produced greater facilitation of the SN–MN EPSP than unpaired

training, but it did not enhance either the frequency or amplitude of spontaneous miniature EPSPs (Eliot et al. 1994; Bao et al. 1998). These results support the idea that pairing selectively affects some aspect of evoked, synchronized release of transmitter, such as presynaptic spike broadening. They also show that activity-dependent facilitation can occur at the level of individual neurons, and does not require additional neuronal circuitry.

Further studies suggested that the influx of $Ca^{2+}$ with each action potential is the critical aspect of spike activity, and that it "primes" the serotonin-sensitive adenylyl cyclase in the SNs so that the cyclase subsequently produces more cAMP in response to serotonin. Consistent with that idea, injection of either the slow $Ca^{2+}$ chelator ethylene glycol tetraacetic acid (EGTA) or a specific inhibitor of PKA into the SN blocks activity-dependent facilitation in culture (Bao et al. 1998). Furthermore, serotonin produces a greater increase in cAMP levels in sensory cells if it is preceded by spike activity in those cells than if it is not (Kandel et al. 1983; Ocorr et al. 1985). Likewise, forward pairing of $Ca^{2+}$ and serotonin produces a greater increase in cyclase activity than backward pairing in a cell-free membrane homogenate preparation (Abrams et al. 1991, 1998; Yovell et al. 1992). These results suggest that one site of convergence of the CS and US during conditioning is the cyclase molecule in the SNs.

## Postsynaptic Mechanisms and *Trans*-Synaptic Signaling

Another mechanism that contributes to classical conditioning is Hebbian potentiation, which is induced by near coincident firing of a SN (excited by the CS) and a MN (excited by the US). The requirement for MN firing could account for the response specificity of conditioning. The conjunction of presynaptic and postsynaptic firing (or depolarization) is necessary for $Ca^{2+}$ influx through postsynaptic N-methyl-D-aspartate (NMDA) receptor channels, which thus serve as another site of convergence of the CS and US during conditioning. The SN–MN EPSPs are glutamatergic and have

AMPA- and NMDA-like components (Dale and Kandel 1993; Trudeau and Castellucci 1993; Antonov et al. 2003; Antzoulatos and Byrne 2004), and tetanic stimulation of the SN produces Hebbian potentiation that is blocked by the NMDA antagonist APV, postsynaptic hyperpolarization, or injection of the $Ca^{2+}$ chelator BAPTA into the postsynaptic neuron (Lin and Glanzman 1994a,b). Activity-dependent facilitation in culture (Bao et al. 1998) or the ganglion (Murphy and Glanzman 1996,1997,1999) and pairing-specific facilitation during behavioral conditioning in the semi-intact preparation (Antonov et al. 2003) are also blocked by APV, postsynaptic hyperpolarization, or postsynaptic BAPTA, supporting a role for Hebbian potentiation in conditioning.

However, the facilitation in culture or the semi-intact preparation is also blocked by injection of a specific inhibitor of PKA into the presynaptic neuron (Bao et al. 1998; Antonov et al. 2003). Furthermore, postsynaptic hyperpolarization does not affect either the frequency or amplitude of spontaneous miniature EPSPs in culture, suggesting that it selectively blocks some aspect of evoked, synchronized release of transmitter from the presynaptic neuron, presumably through retrograde signaling (Bao et al. 1998). In support of that idea, postsynaptic BAPTA also blocks PKA-dependent, pairing-specific increases in evoked firing and membrane resistance of the presynaptic neuron during conditioning in the semi-intact preparation (Antonov et al. 2003). These results suggest that the facilitation during conditioning involves a hybrid combination of activity-dependent presynaptic facilitation and Hebbian potentiation, which are coordinated by *trans*-synaptic signaling (Fig. 2). Thus, facilitation of transmitter release from the SN depends on the combined actions and interactions of 5-HT, NO, SCP, activity, and a retrograde signal from the MN.

## Higher-Order Features of Conditioning

In addition to these basic features, classical conditioning in *Aplysia* and other animals shows higher-order features that have a cognitive flavor

and, therefore, may form a bridge to more advanced forms of learning. For example, studies of conditioning in vertebrates have shown that animals learn not only about the temporal pairing or contiguity of events but also about their correlation or contingency, that is, how well one event predicts another. Thus, presentation of extra, unpaired, or unpredicted USs during training, which decreases the degree to which the US is contingent on the CS, decreases conditioning (Rescorla 1968). A similar effect occurs in conditioning of the garden slug, *Limax maximus* (Sahley et al. 1981) and the siphon withdrawal reflex in *Aplysia* (Hawkins et al. 1986).

The gill-withdrawal reflex also shows second-order conditioning with two siphon CSs and a mantle shock US (Hawkins et al. 1998). Unlike first-order conditioning, which requires forward pairing of CS1 and the US, second-order conditioning occurs with either forward or simultaneous pairing of CS2 and CS1 in the second stage. Furthermore, following simultaneous second-order conditioning, extinction of CS1 produces a decrease in responding to CS2. That result is formally similar to a posttraining US exposure effect, and suggests that simultaneous second-order conditioning involves formation of a stimulus–stimulus (CS2–CS1) association.

These and other higher-order features of conditioning can be explained by a model based on known molecular mechanisms and circuit connections in *Aplysia*, in particular those of the L29 interneurons (Fig. 1) (Hawkins and Kandel 1984). The L29 neurons are excited by the CS as well as the US used in behavioral conditioning (Hawkins and Schacher 1989) and excite the siphon MNs, so that in addition to being facilitatory interneurons, the L29 neurons are also excitatory interneurons (SN–L29–MN) in the circuit for the siphon-withdrawal reflex (Fig. 1) (Hawkins et al. 1981a). A key feature of the model is that the L29 neurons produce facilitation and activity-dependent facilitation at all of the synapses of the SNs, including those onto the L29 neurons themselves (Hawkins 1981). In addition, the L29 neurons undergo spike accommodation during prolonged stimulation, caused in part by inhibitory feedback from L30 interneurons (Hawkins et al. 1981a). As a result, the L29 neurons should increase their firing to the CS and decrease their firing to the US during conditioning.

A computational model incorporating these circuit, cellular, and molecular mechanisms is able to simulate most of the known behavioral properties of habituation, dishabituation, sensitization, and basic classical conditioning, including both the stimulus and temporal specificity of conditioning (Hawkins 1989; see also Buonomano et al. 1990). The model provides a rudimentary cellular embodiment of the learning rule of Rescorla and Wagner (1972), and, thus, is also able to simulate the two higher-order features, contingency and second-order conditioning, which have been shown in *Aplysia*. In addition, the computational model can simulate several other higher-order features that have not yet been shown in *Aplysia* but have been shown in other invertebrates, including blocking, overshadowing, and CS and US pre-exposure effects.

Additional higher-order features of conditioning might be explained by two other properties of the circuit and synapses. First, different L29s respond to somewhat different USs (Hawkins and Schacher 1989). That result suggests that firing of an L29 can be considered the internal representation of a US, so that facilitation of SN-L29 synapses could be the basis for learning an association between a CS and the internal representation of the US. That idea is the basis for second-order conditioning in the model, and also suggests a possible mechanism for posttraining US exposure effects, in which exposure to the US after training alters the response to the CS (as happens following simultaneous second-order conditioning in *Aplysia*). Second, because Hebbian plasticity generally requires the near simultaneous pairing typical of stimulus–stimulus (S–S) learning, the Hebbian component of facilitation might contribute under conditions that are thought to involve S–S learning, such as simultaneous second-order conditioning.

Similar ideas might explain aspects of normal reward function and dysfunction in mammals as well. Dopamine (DA) neurons in the ventral tegmental area (VTA) are thought to mediate reward (Schultz et al. 1997; Tsai et al. 2009), and are thus analogous to the L29 neurons in *Aplysia* (except that they are generally activated by appetitive USs). The circuit properties of the VTA DA neurons are similar to those of the L29 neurons in *Aplysia*, and therefore might account for many of the same behavioral features of learning and reward (Hawkins 2013). In particular, they might explain why the VTA DA neurons increase their firing to the CS and decrease their firing to the US during conditioning, so that they come to fire in expectation of reward (Schultz et al. 1997).

## NEURAL AND MOLECULAR MECHANISMS OF CLASSICAL CONDITIONING IN OTHER INVERTEBRATE MODEL SYSTEMS

Studies of other invertebrates have shown that classical conditioning involves cellular and molecular mechanisms similar to those in *Aplysia* as well as additional mechanisms.

### *Drosophila*

The ease with which genetic studies are performed on *Drosophila* has made it an important system for studying associative learning. A commonly used protocol employs a differential odor–shock avoidance procedure in which animals learn to avoid odors paired (CS$^+$) with shock but not odors explicitly unpaired (CS$^-$). This learning is typically retained for 4–6 h, but retention for 24 h to 1 wk can be produced by a spaced training procedure. Analysis of several *Drosophila* mutants deficient in learning has revealed that elements of the cAMP signaling pathway are key in learning and memory, similar to classical conditioning in *Aplysia*. The formation of long-term memory in *Drosophila* involves stimulation of D1 dopamine receptors, increased cAMP levels, and activation of PKA and cAMP response element-binding protein (CREB) in mushroom bodies and other brain structures that are crucial sites for olfactory learning. The adenylyl cyclase *rutabaga* acts as a site of convergence for associative learning, and the phosphodiesterase *dunce* limits the spatial spread of cAMP (Tomchik and Davis 2009; Gervasi et al. 2010; Shuai et al. 2011; Berry et al. 2012; Chen et al. 2012; for reviews of this model system, see Dudai and Tully 2003; Davis 2005; Keene and Waddel 2007; Tomchik and Davis 2013).

### Honeybee (*Apis mellifera*)

Honeybees show classical conditioning of feeding behavior when a visual or olfactory CS is paired with application of sugar solution (US) to the antennae. Several regions of the brain necessary for this associative learning have been identified, including the antennal lobes and mushroom bodies. The different regions are thought to contribute to different aspects of a distributed engram. In addition, intracellular recordings have revealed that one identified cell that is thought to be octopaminergic, the ventral unpaired median (VUM) neuron, mediates reinforcement during olfactory conditioning and represents the neural correlate of the US. The learning has been dissected into several phases of memory, including short term, midterm, and long term. Numerous studies have revealed that, as in other species, the molecular mechanisms underlying memory formation in the honeybee involve up-regulation of the cAMP pathway and activation of PKA. Short-term memory involves brief activation of PKA, whereas long-term memory involves NO-cGMP signaling leading to longer activation of PKA and CREB-mediated transcription of downstream genes. In addition, midterm memory involves calpain-dependent cleavage of PKC to form PKM (for comprehensive reviews of this model system, see Menzel 2001, 2013; Menzel et al. 2006; Menzel and Benjamin 2013; Muller 2013).

### Hermissenda

The gastropod mollusk *Hermissenda* shows associative learning of light-elicited locomotion and changes in foot length (CRs). The conditioning procedure consists of pairing visual stimuli (light, the CS) with vestibular stimuli (high-speed rotation, the US). After condition-

ing, the CS suppresses normal light-elicited locomotion and elicits foot shortening (Crow and Alkon 1978; Lederhendler et al. 1986). The associative memory can be retained from days to weeks depending on the number of conditioning trials administered during initial acquisition. The type A and B photoreceptors and interneurons in the CS pathway have been identified as critical sites of plasticity for associative learning. The initial studies focused on the B photoreceptors, which show an increase in excitability caused by a decrease in $K^+$ current following training (Alkon et al. 1982). That change has been attributed to increased $Ca^{2+}$ activating CMKII (Alkon 1984), and also 5-HT activating PKC (Farley and Auerbach 1986) and possibly PKA (Alkon et al. 1983) or MAPK (Crow et al. 1998). Subsequent studies have shown that long-term retention depends on both protein and RNA synthesis (Crow and Forrester 1990; Crow et al. 1997). In addition, training produces changes in the membrane properties of the interneurons and changes in the strength of the synaptic connections between the SNs and the interneurons (reviewed in Crow and Jin 2013).

### Pond Snail (*Lymnaea stagnalis*)

The pulmonate *Lymnaea stagnalis* shows appetitive conditioning of feeding behavior when a neutral chemical or mechanical stimulus (CS) applied to the lips is paired with a strong stimulant of feeding, such as sucrose (US). Greater levels of rasping, a component of the feeding behavior, can be produced by a single trial, and this response can persist for at least 19 d. The early stages of learning involve PKA, MAPK, and CMKII, whereas long-term maintenance involves protein and RNA synthesis, CREB-dependent gene regulation, and up-regulation of NO. The circuit consists of command-like interneurons, a network of three types of central pattern generator (CPG) neurons, 10 types of MNs, and a variety of modulatory interneurons. An analog of the behavioral response occurs in the isolated central nervous system. The enhancement of the feeding motor program appears to be caused by facilitation of input to the interneurons, MNs, and presumably the CPG neurons, as well as persistent depolarization of identified modulatory neurons, resulting in increased activation of the CPG cells by mechanosensory inputs from the lips (reviewed in Kemenes 2013).

## Limax

The pulmonate *Limax* shows food avoidance learning when a preferred food odor (CS) is paired with a bitter taste (US). In addition to this example of basic classical conditioning, food avoidance in *Limax* shows higher-order features of classical conditioning including blocking and second-order conditioning. An analog of the learning occurs in the isolated central nervous system, facilitating subsequent cellular analyses of learning in *Limax*. The procerebral lobe in the cerebral ganglion processes olfactory information, and is a likely site for plasticity. Both neurogenesis and oscillations of local field potentials in the procerebral lobe are thought to be involved in learning and memory. The oscillatory activity of the procerebral lobe is modulated by several endogenous substances including the gaseous transmitter NO and the neuropeptide FMRFamide. FMRFamide and another neuropeptide found in the procerebral lobe (SCPB) have also been shown to affect activity of the rhythmic motor network underlying feeding behavior, making those neuropeptides attractive candidates for plasticity within that network as well (see Gelperin 2013 for a comprehensive review of this model system).

## NEURAL AND MOLECULAR MECHANISMS OF CLASSICAL AND OPERANT CONDITIONING OF FEEDING BEHAVIOR IN *Aplysia*

The study of feeding behavior of *Aplysia* has provided insights into the mechanisms underlying classical conditioning and also operant conditioning, which has been studied more extensively for this behavior. Feeding behavior in *Aplysia* shows several features that make it amenable to the study of learning. For example, the behavior occurs in an all-or-nothing manner and is there-fore easily quantified, and the CPG underlying the generation of the behavior is well characterized to the extent that many of the key individual neurons responsible for the generation of feeding movements have been identified. These technical advantages have been exploited to identify loci of plasticity and changes in membrane properties in the key neurons of the CPG that occur during associative learning.

In the first example of conditioning of feeding behavior in *Aplysia,* Susswein and colleagues (Susswein and Schwartz 1983; Susswein et al. 1986) developed a training procedure in which animals were presented with food (i.e., seaweed) that was made inedible by wrapping it in a plastic net. Wrapped food still elicits bites and is initially brought into the mouth and buccal cavity. However, because netted food cannot be swallowed, it triggers repetitive failed swallowing responses and is eventually rejected. Additional behavioral features of this conditioning have been elucidated (Schwarz et al. 1991; Botzer et al. 1998; Katzoff et al. 2002, 2010), as well as some of the underlying biochemical and molecular mechanisms (Cohen-Armon et al. 2004; Levitan et al. 2008; Michel et al. 2011, 2012).

Feeding behavior can also be modified by appetitive associative paradigms (i.e., classical and operant conditioning), which induce an increase of its expression (Lechner et al. 2000a; Brembs et al. 2002). During classical conditioning, tactile stimulation of the lips with a soft paintbrush serves as the CS and seaweed presentation serves as the US. Paired training produces a significant increase in the number of CS-evoked bites, compared with unpaired training, both 60 min and 24 h after training (Colwill et al. 1997; Lechner et al. 2000a,b; Lorenzetti et al. 2006). Further analyses found that afferent information related to the US is mediated by an anterior branch of the esophageal nerve En2 (Brembs et al. 2002), which projects to the foregut (Lechner et al. 2000a). This nerve is rich in dopamine-containing processes. Neural correlates of classical conditioning were identified by removing the buccal ganglia, where the CPG underlying feeding behavior is located, from recently trained animals and examining the change in cellular properties of key cells that mediate the behavior. Classical conditioning produced an increase in the CS-evoked excitatory synaptic drive to pattern-initiating neuron B31/32. In addition, training led to a change in the burst threshold of decision-making neuron B51 (Nargeot et al. 1999).

## Appetitive Operant Conditioning of Feeding

The finding that En2 was a reinforcement pathway facilitated the development of an appetitive operant conditioning behavioral protocol in which biting served as the operant, and electrical stimulation of En2 served as reinforcement (Brembs et al. 2002). During contingent training, bites were immediately followed by reinforcements (Fig. 3A), which were delivered via stimulating electrodes implanted on En2. During a 10-min training period, the En was stimulated each time the animal performed a spontaneous biting movement (contingent reinforcement) (Fig. 3B) (Brembs et al. 2002). Control animals received the same number of stimulations over the 10-min period; however, they were explicitly unpaired (i.e., yoked). The

group of animals that had received paired training showed a significantly larger number of spontaneous bites during test periods both 1 and 24 h after training compared with control animals.

An analysis of neuronal correlates of the conditioning revealed that the contingent-dependent increase in bites (i.e., ingestive buccal motor programs [iBMPs]) was associated with the regularization of the bursting activity of a cluster of pattern-initiating neurons, consisting of B30, B63, and B65 (Fig. 4A) (Nargeot et al. 2009). This synchronization of the pattern-initiating neurons appears to be produced by two distinct contingent-dependent mechanisms: (1) decreased burst threshold in B63, B30, and B65; and (2) enhanced electrical coupling between pairs of pattern-initiating neurons (Nargeot et al. 2009). In addition to the changes in the pattern-initiating neurons, operant conditioning is associated with modifications of the input resistance and burst threshold of B51 (Brembs et al. 2002), a neuron that is also modified by classical conditioning, but in opposite ways (Lorenzetti et al. 2006). The input resistance was significantly greater, and the burst

**A** *Aplysia* feeding behavior

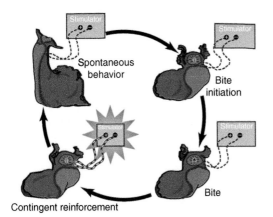

**B** Operant conditioning protocol

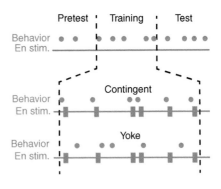

**Figure 3.** Operant conditioning of feeding behavior. (*A*) Throughout the experiment, the animal was observed and all bites were recorded. In the contingent reinforcement group, a bite was immediately followed by a brief electric stimulation of the esophageal nerve (En stim.). A control group received the same sequence of stimulations as the contingent group, but the stimulation was uncorrelated with the animal's behavior. (*B*) Experimental sessions consisted of a 5-min pretest, 10 min of training, and a final test period. In each period, the number of bites was recorded. The final test period was either 1 h or 24 h after training. The group of animals that had received contingent reinforcement showed a significantly larger number of spontaneous bites both 1 h and 24 h after training compared with control animals. (Modified from Brembs et al. 2002.)

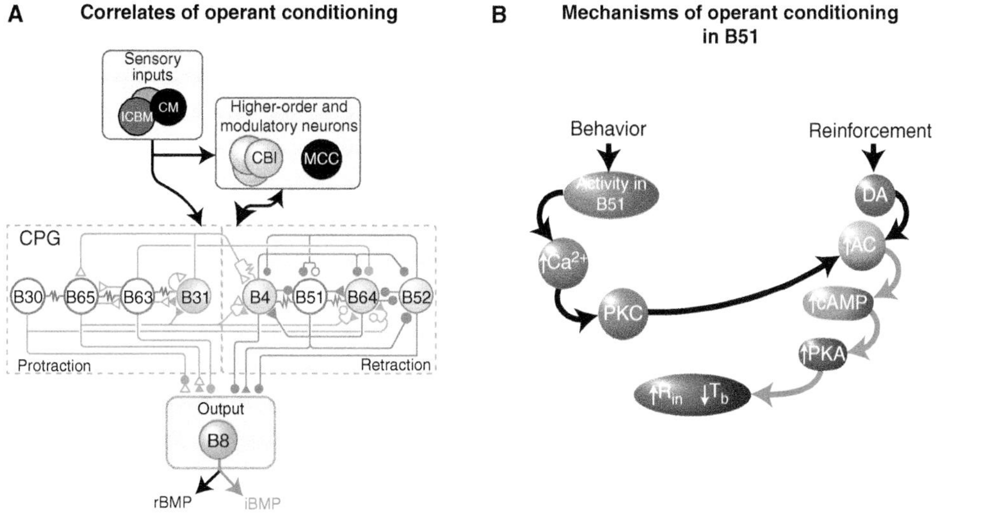

**A**     Correlates of operant conditioning       **B**     Mechanisms of operant conditioning in B51

Figure 4. Neuronal correlates of operant conditioning and mechanisms in neuron B51. (*A*) Simplified schematic of the feeding neural circuit. Sensory neurons (SNs) in the cerebral ganglia (interganglionic cerebral-buccal mechanoafferent [ICBM] and cerebral mechanoafferent [CM]) convey information to higher-order cells in the feeding circuit, such as the command-like cerebral-buccal interneurons (CBIs), and modulatory cells, such as the metacerebral cell (MCC), also located in the cerebral ganglia. Sensory information is also conveyed directly to the central pattern generator (CPG) in the buccal ganglia. In addition, the CPG receives inputs from the command-like cells and modulatory cells. Cells in the CPG can be classified, in part, by their activity during a buccal motor program (BMP). Some cells are active during the protraction phase (yellow shading), whereas others are active during the retraction phase (blue shading). Activity in radula closure motor neuron (MN) B8 occurs during the protraction phase in rejection BMPs (rBMPs) and during the retraction phase in ingestive BMPs (iBMPs). Identified loci of plasticity are indicated by red and white shading. Following operant conditioning, the electrical coupling among B30, B63, and B65 is strengthened, and the excitability of B51, B30, B63, and B65 is increased. (*B*) Model of the molecular mechanisms in B51 underlying operant conditioning. See text for details. (Modified from Lorenzetti et al. 2008.)

threshold was significantly lower in neurons from trained animals compared with untrained animals. These types of changes would serve to increase the probability that B51 would become active and would, therefore, facilitate the generation of the neural activity underlying biting movements in the trained animals.

To examine whether the changes in the properties of B51 are intrinsic to that cell, Brembs et al. (2002) used an analogue of operant conditioning in which an individual B51 cell was removed from a naïve ganglion and maintained in culture. Reinforcement was mimicked by the application of a brief "puff" of DA onto the cell, which was made contingent on a plateau potential that was elicited in B51 by injection of a brief depolarizing current pulse. Controls consisted of cells that received the DA puff 40 sec after the

plateau potential. Training produced a significant increase in input resistance and a significant decrease in burst threshold, similar to the changes observed in the neural correlates (Brembs et al. 2002) and in vitro analogues (Nargeot et al. 1999) of operant conditioning. These data suggest that B51 is an important locus of plasticity in operant conditioning of feeding behavior, and that intrinsic cell-wide plasticity may be one important mechanism underlying this type of learning.

A model of the molecular mechanisms underlying appetitive operant conditioning in neuron B51 is shown in Figure 4B. The CPG that mediates feeding behavior produces synaptic input to neuron B51. When this input is suprathreshold, it triggers an all-or-nothing sustained several-second burst of spikes (plateau potential) in B51, which is critical for the expression of ingestive behavior (i.e., an iBMP). A secondary consequence of the plateau potential is to produce an accumulation of $Ca^{2+}$ in B51, which leads to the activation of PKC. The activated PKC then weakly activates and primes a type II adenylyl cyclase. Reinforcement (reward) activates the dopaminergic modulatory system. DA binds to a D1-like receptor, but the DA-induced activation of the cAMP cascade is weak and insufficient to modulate downstream effectors (e.g., membrane channels regulating input resistance and burst threshold). However, if the ingestive behavior just precedes the delivery of the reward, as occurs during operant conditioning, the adenylyl cyclase will have been primed by PKC, and because of the synergistic interaction of the two pathways, the level of cAMP will be significantly greater than that produced by either behavior (activity in B51) alone or reinforcement (DA) alone. After a sufficient number of contingent reinforcements, the increased level of cAMP activates PKA sufficiently to produce the increase in input resistance and excitability of B51. Consequently, subsequent CPG-driven synaptic input to B51 is more likely to fire the cell and lead to the increase in ingestive behavior associated with operant conditioning.

Interestingly, the mechanisms of activity-dependent neuromodulation for this appetitive form of operant conditioning appear to be very similar to those observed in SNs of *Aplysia* during aversive classical conditioning of withdrawal reflexes (Fig. 2). In the SNs, the coincidence detection involves, at least in part, a synergistic interaction between a $Ca^{2+}$/calmodulin-sensitive adenylyl cyclase (type I) and a serotonin-activated cAMP cascade (Ocorr et al. 1985; Abrams et al. 1991, 1998; Yovell et al. 1992). Similarly, in *Drosophila*, a type I adenylyl cyclase is necessary for classical conditioning, but does not appear to be necessary for operant conditioning (Brembs and Plendl 2008). Although the specific isoform of adenylyl cyclase appears to differ (type I for classical conditioning and type II for operant conditioning), adenylyl cyclase appears to serve as a molecular site of convergence in both forms of learning. However, adenylyl cyclase is not the only coincidence detector for all

examples of classical and operant conditioning. For example, following classical conditioning of feeding behavior, the burst threshold of B51 increases rather than decreases as it does following operant conditioning and the input resistance does not change (Lorenzetti et al. 2006), indicating that the mechanisms underlying the modulation of B51 by classical conditioning involve a different as-yet-unidentified coincidence detector from that for operant conditioning. In addition, as described above for conditioning of the withdrawal reflex in *Aplysia* and in Basu and Siegelbaum (2015), NMDA receptors also serve as coincidence detectors in many CNS circuits.

Operant conditioning has also been examined in other invertebrates including *Drosophila* and *Lymnea*. PKC appears to play a key role in operant conditioning of turning behavior in *Drosophila* (Brembs and Plendl 2008) as it does in conditioning of feeding behavior in *Aplysia* (Fig. 4). Operant conditioning of aerial respiratory behavior in *Lymnea* (Lukowiak et al. 1996) also involves PKC as well as NMDA receptors and MAPK (Rosenegger and Lukowiak 2010), and long-term memory involves protein and RNA synthesis as well as changes in the activity of an identified neuron in the CPG, as in *Aplysia* (Spencer et al. 2002; Lukowiak et al. 2003).

Growing evidence indicates that many of the same pathways that mediate operant conditioning of feeding in *Aplysia* are also involved in vertebrate reward learning in the striatum (see also Graybiel 2015). For example, in vivo operant conditioning involves D1 dopamine receptors and cAMP/PKA in the nucleus accumbens (Smith-Roe and Kelley 2000; Baldwin et al. 2002). D1 dopamine receptors are also necessary for potentiation of corticostriatal synapses in an analogue of reward learning (Reynolds et al. 2001). In addition to synaptic plasticity, striatal neurons display an increased level of intrinsic excitability known as the "up state" that depends on expression of the CREB transcription factor (Dong et al. 2006).

## SUMMARY AND CONCLUSIONS

In this work, we have reviewed research on neural mechanisms of classical conditioning of the gill- and siphon-withdrawal reflex and operant conditioning of feeding behavior in *Aplysia*, and have briefly summarized progress in a few other important invertebrate preparations. It is instructive to compare and contrast mechanisms that have been described in these different preparations to try to determine to what extent they may be universal or unique:

- Classical conditioning of the withdrawal reflex and operant conditioning of feeding in *Aplysia* both involve changes in the biophysical properties of identified neurons in the circuits for the behaviors.

- In both cases, these changes include an increase in excitability that could be caused by a decrease in a $K^+$ current. Classical conditioning in *Hermissenda* also involves an increase in excitability that is thought to be caused by a decrease in $K^+$ current in identified cells.

- In addition, classical conditioning in *Aplysia* involves changes in the strength of excitatory synapses in the circuit. Synaptic plasticity is also thought to be involved in conditioning in *Hermissenda* and *Lymnea*.

- The changes in excitability and synaptic strength in *Aplysia* can both be reproduced in analogs of learning in single cells (or pairs of cells) in culture and, therefore, do not require additional circuitry.

- Modulatory transmitters carry the reinforcing signals—5-HT for classical conditioning and DA for operant conditioning in *Aplysia*. 5-HT, DA, and octopamine are also important for conditioning in *Drosophila*, *Apis*, and *Hermissenda*. In addition, NO is important for classical conditioning in *Aplysia* as well as conditioning in *Apis*, *Lymnea*, and *Limax*.

- For both classical and operant conditioning in *Aplysia*, adenylyl cyclase (AC) is a molecular site of convergence of the two signals that are associated. For classical conditioning, $Ca^{2+}$ acts synergistically with 5-HT to activate type I AC, whereas for operant conditioning, $Ca^{2+}$ stimulates PKC, which acts synergistically with DA to activate type II AC. Type I AC is also a molecular site of

convergence for classical conditioning in *Drosophila*.

- For both types of learning in *Aplysia*, the AC-cAMP-PKA pathway plays an important role in the initial stages of memory. This is true for conditioning in *Drosophila*, *Apis*, and *Lymnea* as well. Other kinases including PKC, MAPK, and CMKII are also important for conditioning in *Drosophila*, *Apis*, *Hermissenda*, and *Lymnea*.

- Long-term maintenance of conditioning involves PKA, CREB, and protein and RNA synthesis in *Drosophila*, *Apis*, *Hermissenda*, and *Lymnea*. This has not yet been tested for the two forms of conditioning in *Aplysia* reviewed here, but it is true for nonassociative learning of the same behaviors in *Aplysia* (Byrne and Hawkins 2015).

- Classical conditioning in *Aplysia* also involves NMDA receptors as another molecular site of convergence of the CS and US pathways. Furthermore, the two sites of convergence (presynaptic AC and postsynaptic NMDA receptors) act cooperatively through retrograde signaling. Operant conditioning in *Lymnea* also involves NMDA receptors.

- In addition, classical conditioning in *Aplysia* has higher-order features that might be explained at the neural circuit level. Classical conditioning in *Limax* shows similar features.

Thus, the neural mechanisms of different types of associative learning in these different invertebrate preparations have many shared features, but they also have some unique features that may be adapted to the particular type of learning or species. As we have noted, initial studies suggest that associative learning in vertebrates can share some of the same behavioral, circuit, cellular, and molecular mechanisms as well. Therefore, important future goals will be to decipher the logic of the shared and unique features in invertebrates, and to use the results of these studies to suggest new directions for research on mechanisms of learning in vertebrates.

## ACKNOWLEDGMENTS

Preparation of this manuscript is supported by National Institutes of Health (NIH)
Grants GM097502, MH058321, NS019895,
and NS083690.

## REFERENCES

*Reference is also in this collection.

Abrams TW, Castellucci VF, Camardo JS, Kandel ER, Lloyd PE. 1984. Two endogenous neuropeptides modulate the gill and siphon withdrawal reflex in Aplysia by presynaptic facilitation involving cAMP-dependent closure of a serotonin-sensitive potassium channel. Proc Natl Acad Sci 81: 7956–7960.

Abrams TW, Karl KA, Kandel ER. 1991. Biochemical studies of stimulus convergence during classical conditioning in Aplysia: Dual regulation of adenylate cyclase by Ca$^{2+}$/calmodulin and transmitter. J Neurosci 11: 2655–2665.

Abrams TW, Yovell Y, Onyike CU, Cohen JE, Jarrard HE. 1998. Analysis of sequence-dependent interactions between calcium and transmitter stimuli in activating adenylyl cyclase in Aplysia: Possible contributions to CS–US sequence requirement during conditioning. Learn Mem 4: 496–509.

Alkon DL. 1984. Calcium-mediated reduction of ionic currents: A biophysical memory trace. Science 226: 1037–1045.

Alkon DL, Lederhendler I, Shoukimas JJ. 1982. Primary changes of membrane currents during retention of associative learning. Science 215: 693–695.

Alkon DL, Acosta-Urquidi J, Olds J, Kuzma G, Neary JT. 1983. Protein kinase injection reduces voltage-dependent potassium currents. Science 219: 303–306.

Antonov I, Kandel ER, Hawkins RD. 1999. The contribution of facilitation of monosynaptic PSPs to dishabituation and sensitization of the Aplysia siphon withdrawal reflex. J Neurosci 19: 10438–10450.

Antonov I, Antonova I, Kandel ER, Hawkins RD. 2001. The contribution of activity-dependent synaptic plasticity to classical conditioning in Aplysia. J Neurosci 21: 6413–6422.

Antonov I, Antonova I, Kandel ER, Hawkins RD. 2003. Activity-dependent presynaptic facilitation and Hebbian LTP are both required and interact during classical conditioning in Aplysia. Neuron 37: 135–147.

Antonov I, Ha T, Antonova I, Moroz LL, Hawkins RD. 2007. Role of nitric oxide in classical conditioning of siphon withdrawal in Aplysia. J Neurosci 27: 10993–11002.

Antzoulatos EG, Byrne JH. 2004. Learning insights transmitted by glutamate. More than synaptic plasticity: Role of nonsynaptic plasticity in learning and memory. Trends Neurosci 27: 555–560.

Baldwin AE, Sadeghian K, Holahan MR, Kelley AE. 2002. Appetitive instrumental learning is impaired by inhibition of cAMP-dependent protein kinase within the nucleus accumbens. Neurobiol Learn Mem 77: 44–62.

Bao J-X, Kandel ER, Hawkins RD. 1998. Involvement of presynaptic and postsynaptic mechanisms in a cellular analog of classical conditioning at Aplysia sensory-motor neuron synapses in isolated cell culture. J Neurosci 18: 458–466.

*Basu J, Siegelbaum SA. 2015. The corticohippocampal circuit, synaptic plasticity, and memory. Cold Spring Harb Perspect Biol 7: a021733.

Berry JA, Cervantes-Sandoval I, Nicholas EP, Davis RL. 2012. Dopamine is required for learning and forgetting in Drosophila. Neuron 74: 530–542.

Botzer D, Markovich S, Susswein A. 1998. Multiple memory processes following training that a food is inedible in Aplysia. Learn Mem 5: 204–219.

Brembs B, Plendl W. 2008. Double dissociation of protein-kinase C and adenylyl cyclase manipulations on operant and classical learning in Drosophila. Curr Biol 18: 1168–1171.

Brembs B, Lorenzetti F, Reyes F, Baxter DA, Byrne JH. 2002. Operant reward learning in Aplysia: Neuronal correlates and mechanisms. Science 296: 1706–1709.

Buonomano DV, Byrne JH. 1990. Long-term synaptic changes produced by a cellular analogue of classical conditioning in Aplysia. Science 249: 420–423.

Buonomano DV, Baxter DA, Byrne JH. 1990. Small networks of empirically derived adaptive elements simulate some higher-order features of classical conditioning. Neural Networks 3: 507–523.

* Byrne JH, Hawkins RD. 2015. Nonassociative learning in invertebrates. Cold Spring Harb Perspect Biol 7: a021675.

Carew TJ, Walters ET, Kandel ER. 1981. Classical conditioning in a simple withdrawal reflex in Aplysia californica. J Neurosci 1: 1426–1437.

Carew TJ, Hawkins RD, Kandel ER. 1983. Differential classical conditioning of a defensive withdrawal reflex in Aplysia californica. Science 219: 397–400.

Chen CC, Wu JK, Lin HW, Pai TP, Fu TF, Wu CL, Tully T, Chiang AS. 2012. Visualizing long-term memory formation in two neurons of the Drosophila brain. Science 335: 678–685.

Clark GA, Hawkins RD, Kandel ER. 1994. Activity-dependent enhancement of presynaptic facilitation provides a cellular mechanism for the temporal specificity of classical conditioning in Aplysia. Learn Mem 1: 243–257.

Cohen-Armon M, Visochek L, Katzoff A, Levitan D, Susswein AJ, Klein R, Valbrun M, Schwartz JH. 2004. Long-term memory requires polyADP-ribosylation. Science 304: 1820–1822.

Colwill R, Goodrum K, Martin A. 1997. Pavlovian appetitive discriminative conditioning in Aplysia californica. Anim Learn Behav 25: 268–276.

Cook DG, Carew TJ. 1986. Operant conditioning of head waving in Aplysia. Proc Natl Acad Sci 83: 1120–1124.

Crow TJ, Alkon DL. 1978. Retention of an associative behavioral change in Hermissenda. Science 201: 1239–1241.

Crow T, Forrester J. 1990. Inhibition of protein synthesis blocks long-term enhancement of generator potentials produced by one-trial in vivo conditioning in Hermissenda. Proc Natl Acad Sci 87: 4490–4494.

Crow T, Jin NG. 2013. Multisite cellular and synaptic mechanisms in *Hermissenda* Pavlovian conditioning. In *Invertebrate learning and memory* (ed. Menzel R, Benjamin PR), pp. 236–250. Academic, San Diego.

Crow T, Siddiqi V, Dash PK. 1997. Long-term enhancement but not short-term in *Hermissenda* is dependent upon mRNA synthesis. *Neurobiol Learn Mem* **68:** 343–350.

Crow T, Xue-Bian JJ, Siddiqi V, Kang Y, Neary JT. 1998. Phosphorylation of mitogen-activated protein kinase by one-trial and multi-trial classical conditioning. *J Neurosci* **18:** 3480–3487.

Dale N, Kandel ER. 1993. L-glutamate may be the fast excitatory transmitter of *Aplysia* sensory neurons. *Proc Natl Acad Sci* **90:** 7163–7167.

Davis RL. 2005. Olfactory memory formation in *Drosophila*: From molecular to systems neuroscience. *Annu Rev Neurosci* **28:** 275–302.

Dong Y, Green T, Saal D, Marie H, Neve R, Nestler EJ, Malenka RC. 2006. CREB modulates excitability of nucleus accumbens neurons. *Nat Neurosci* **9:** 475–477.

Dudai Y, Tully T. 2003. Invertebrate learning. In *Learning and memory* (ed. Byrne JH), pp. 292–296. Macmillan, New York.

Eliot LS, Hawkins RD, Kandel ER, Schacher S. 1994. Pairing-specific, activity-dependent presynaptic facilitation at *Aplysia* sensory-motor neuron synapses in isolated cell cultures. *J Neurosci* **14:** 368–383.

Farley J, Auerbach S. 1986. Protein kinase C activation induces conductance changes in *Hermissenda* photoreceptors like those seen in associative learning. *Nature* **319:** 220–223.

Gelperin A. 2013. Associative memory mechanisms in terrestrial slugs and snails. In *Invertebrate learning and memory* (ed. Menzel R, Benjamin PR), pp. 280–292. Academic, San Diego.

Gervasi N, Tchénio P, Preat T. 2010. PKA dynamics in a *Drosophila* learning center: Coincidence detection by rutabaga adenylyl cyclase and spatial regulation by dunce phosphodiesterase. *Neuron* **65:** 516–529.

Glanzman DL, Mackey SL, Hawkins RD, Dyke AM, Lloyd PE, Kandel ER. 1989. Depletion of serotonin in the nervous system of *Aplysia* reduces the behavioral enhancement of gill withdrawal as well as the heterosynaptic facilitation produced by tail shock. *J Neurosci* **9:** 4200–4213.

Gormezano I. 1972. Investigations of defense and reward conditioning in the rabbit. In *Classical conditioning: II. Current research and theory* (ed. Black AH, Prokasy WF), pp. 151–181. Appleton-Century-Crofts, New York.

* Graybiel AM, Grafton ST. 2015. The striatum: Where skills and habits meet. *Cold Spring Harb Perspect Biol* **7:** a021691.

Hawkins RD. 1981. Interneurons involved in mediation and modulation of gill-withdrawal reflex in *Aplysia*: III. Identified facilitating neurons increase Ca²⁺ current in sensory neurons. *J Neurophysiol* **45:** 327–339.

Hawkins RD. 1989. A biologically based computational model for several simple forms of learning. *Psychol Learn Motiv* **23:** 65–108.

Hawkins RD. 2013. Possible contributions of a novel form of synaptic plasticity in *Aplysia* to reward, memory, and their dysfunctions in mammalian brain. *Learn Mem* **20:** 580–591.

Hawkins RD, Kandel ER. 1984. Is there a cell biological alphabet for simple forms of learning? *Psychol Rev* **91:** 375–391.

Hawkins RD, Schacher S. 1989. Identified facilitator neurons L29 and L28 are excited by cutaneous stimuli used in dishabituation, sensitization, and classical conditioning in *Aplysia*. *J Neurosci* **9:** 4236–4245.

Hawkins RD, Castellucci VF, Kandel ER. 1981a. Interneurons involved in mediation and modulation of gill-withdrawal reflex in *Aplysia*: I. Identification and characterization. *J Neurophysiol* **45:** 304–314.

Hawkins RD, Castellucci VF, Kandel ER. 1981b. Interneurons involved in mediation and modulation of gill-withdrawal reflex in *Aplysia*: II. Identified neurons produce heterosynaptic facilitation contributing to behavioral sensitization. *J Neurophysiol* **45:** 315–326.

Hawkins RD, Abrams TW, Carew TJ, Kandel ER. 1983. A cellular mechanism of classical conditioning in *Aplysia*: Activity-dependent amplification of presynaptic facilitation. *Science* **219:** 400–415.

Hawkins RD, Carew TJ, Kandel ER. 1986. Effects of interstimulus interval and contingency on classical conditioning of the *Aplysia* siphon withdrawal reflex. *J Neurosci* **6:** 1695–1701.

Hawkins RD, Lalevic N, Clark GA, Kandel ER. 1989. Classical conditioning of the *Aplysia* siphon-withdrawal reflex exhibits response specificity. *Proc Natl Acad Sci* **86:** 7620–7624.

Hawkins RD, Greene W, Kandel ER. 1998. Classical conditioning, differential conditioning, and second-order conditioning of the *Aplysia* gill-withdrawal reflex in a simplified mantle organ preparation. *Behav Neurosci* **112:** 636–645.

Hawkins RD, Clark GA, Kandel ER. 2006. Operant conditioning of gill withdrawal in *Aplysia*. *J Neurosci* **26:** 2443–2448.

Kamin LJ. 1969. Predictability, surprise, attention, and conditioning. In *Punishment and aversive behavior* (ed. Campbell BA, Church RM), pp. 279–296. Appleton-Century-Crofts, New York.

Kandel ER, Abrams T, Bernier L, Carew TJ, Hawkins RD, Schwartz JH. 1983. Classical conditioning and sensitization share aspects of the same molecular cascade in *Aplysia*. *Cold Spring Harbor Symp Quant Biol* **48:** 821–830.

Katzoff A, Ben-Gedalya T, Susswein AJ. 2002. Nitric oxide is necessary for multiple memory processes after learning that food is inedible in *Aplysia*. *J Neurosci* **22:** 9581–9594.

Katzoff A, Miller N, Susswein AJ. 2010. Nitric oxide and histamine signal attempts to swallow: A component of learning that food is inedible in *Aplysia*. *Learn Mem* **17:** 50–62.

Keene AC, Waddell S. 2007. *Drosophila* olfactory memory: Single genes to complex neural circuits. *Nat Rev Neurosci* **8:** 341–354.

Kemenes G. 2013. Molecular and cellular mechanisms of classical conditioning in the feeding system of *Lymnea*. In *Invertebrate learning and memory* (ed. Menzel R, Benjamin PR), pp. 251–264. Academic, San Diego.

Lechner HA, Baxter DA, Byrne JH. 2000a. Classical conditioning of feeding in *Aplysia*: I. Behavioral analysis. *J Neurosci* **20:** 3369–3376.

Lechner HA, Baxter DA, Byrne JH. 2000b. Classical conditioning of feeding in *Aplysia*: II. Neurophysiological correlates. *J Neurosci* **20:** 3377–3386.

Lederhendler II, Gart S, Alkon DL. 1986. Classical conditioning of Hermissenda: Origin of a new response. *J Neurosci* **6:** 1325–1331.

Levitan D, Lyons LC, Perelman A, Green CL, Motro B, Eskin A, Susswein AJ. 2008. Training with inedible food in *Aplysia* causes expression of C/EBP in the buccal but not cerebral ganglion. *Learn Mem* **15:** 412–416.

Lin XY, Glanzman DL. 1994a. Hebbian induction of long-term potentiation of *Aplysia* sensorimotor synapses: Partial requirement for activation of an NMDA-related receptor. *Proc Biol Sci* **255:** 215–221.

Lin XY, Glanzman DL. 1994b. Long-term potentiation of *Aplysia* sensorimotor synapses in cell culture: Regulation by postsynaptic voltage. *Proc Biol Sci* **255:** 113–118.

Lorenzetti FD, Mozzachiodi R, Baxter DA, Byrne JH. 2006. Classical and operant conditioning differentially modify the intrinsic properties of an identified neuron. *Nat Neurosci* **9:** 17–19.

Lorenzetti FD, Baxter DA, Byrne JH. 2008. Molecular mechanisms underlying a cellular analog of operant reward learning. *Neuron* **59:** 815–828.

Lukowiak K, Ringseis E, Spencer G, Wildering W, Syed N. 1996. Operant conditioning of aerial respiratory behaviour in *Lymnaea stagnalis*. *J Exp Biol* **199:** 683–691.

Lukowiak K, Sangha S, Scheibenstock A, Parvez K, McComb C, Rosenegger D, Varshney N, Sadamoto H. 2003. A molluscan model system in the search for the engram. *J Physiol Paris* **97:** 69–76.

Mackey SL, Kandel ER, Hawkins RD. 1989. Identified serotonergic neurons LCB1 and RCB1 in the cerebral ganglia of *Aplysia* produce presynaptic facilitation of siphon sensory neurons. *J Neurosci* **9:** 4227–4235.

Menzel R. 2001. Searching for the memory trace in a mini-brain, the honeybee. *Learn Mem* **8:** 53–62.

Menzel R. 2013. In search of the engram in the honeybee brain. In *Invertebrate learning and memory* (ed. Menzel R, Benjamin PR), pp. 397–415. Academic, San Diego.

Menzel R, Benjamin PR. 2013. Beyond the cellular alphabet of learning and memory in invertebrates. In *Invertebrate learning and memory* (ed. Menzel R, Benjamin PR), pp. 3–8. Academic, San Diego.

Menzel R, Leboulle G, Eisenhardt D. 2006. Small brains, bright minds. *Cell* **124:** 237–239.

Michel M, Green CL, Lyons LC. 2011. PKA and PKC are required for long-term but not short-term in vivo operant memory in *Aplysia*. *Learn Mem* **18:** 19–23.

Michel M, Green CL, Gardner JS, Organ CL, Lyons LC. 2012. Massed training-induced intermediate-term operant memory in *Aplysia* requires protein synthesis and multiple persistent kinase cascades. *J Neurosci* **32:** 4581–4591.

Muller U. 2013. Memory phases and signaling cascades in honeybees. In *Invertebrate learning and memory* (ed. Menzel R, Benjamin PR), pp. 433–441. Academic, San Diego.

Murphy GG, Glanzman DL. 1996. Enhancement of sensorimotor connections by conditioning-related stimulation in *Aplysia* depends on postsynaptic $Ca^{2+}$. *Proc Natl Acad Sci* **93:** 9931–9936.

Murphy GG, Glanzman DL. 1997. Mediation of classical conditioning in *Aplysia californica* by long-term potentiation of sensorimotor synapses. *Science* **278:** 467–471.

Murphy GG, Glanzman DL. 1999. Cellular analog of differential classical conditioning in *Aplysia*: Disruption by the NMDA receptor antagonist DL-2-amino-5-phosphonovalerate. *J Neurosci* **19:** 10595–10602.

Nargeot R, Baxter DA, Byrne JH. 1999. In vitro analog of operant conditioning in *Aplysia*: I. Contingent reinforcement modifies the functional dynamics of an identified neuron. *J Neurosci* **15:** 2247–2260.

Nargeot R, Le Bon-Jego M, Simmers J. 2009. Cellular and network mechanisms of operant learning-induced compulsive behavior in *Aplysia*. *Curr Biol* **19:** 975–998.

Ocorr KA, Walters ET, Byrne JH. 1985. Associative conditioning analog selectively increases cAMP levels of tail sensory neurons in *Aplysia*. *Proc Natl Acad Sci* **82:** 2548–2552.

Pieroni JP, Byrne JH. 1992. Differential effects of serotonin, FMRFamide and small cardioactive peptide on multiple, distributed processes modulating sensorimotor synaptic transmission in *Aplysia*. *J Neurosci* **12:** 2633–2647.

Rescorla RA. 1967. Pavlovian conditioning and its proper control procedures. *Psychol Rev* **74:** 71–80.

Rescorla RA. 1968. Probability of shock in the presence and absence of CS in fear conditioning. *J Comp Physiol Psychol* **66:** 1–5.

Rescorla RA, Wagner AR. 1972. A theory of Pavlovian conditioning: variations in the effectiveness of reinforcement and non-reinforcement. In *Classical conditioning II: Current research and theory* (ed. Black AH, Prokasy WF). Appleton-Century-Crofts, New York.

Reynolds JN, Hyland BI, Wickens JR. 2001. A cellular mechanism of reward-related learning. *Nature* **413:** 67–70.

Rosenegger D, Lukowiak K. 2010. The participation of NMDA receptors, PKC, and MAPK in the formation of memory following operant conditioning in *Lymnaea*. *Mol Brain* **3:** 24.

Sahley C, Rudy JW, Gelperin A. 1981. Analysis of associative learning in a terrestrial mollusk: I. Higher-order conditioning, blocking, and a transient US pre-exposure effect. *J Comp Physiol* **144:** 1–8.

Schultz W, Dayan P, Montague PR. 1997. A neural substrate of prediction and reward. *Science* **275:** 1593–1599.

Schwarz M, Feldman E, Susswein AJ. 1991. Variables affecting long-term memory of learning that a food is inedible in *Aplysia*. *Behav Neurosci* **105:** 193–201.

Shuai Y, Hu Y, Qin H, Campbell RA, Zhong Y. 2011. Distinct molecular underpinnings of *Drosophila* olfactory trace conditioning. *Proc Natl Acad Sci* **108:** 20201–20206.

Smith-Roe SL, Kelley AE. 2000. Coincident activation of NMDA and dopamine D1 receptors within the nucleus accumbens core is required for appetitive instrumental learning. *J Neurosci* **20:** 7737–7742.

Spencer GE, Kazmi MH, Syed NI, Lukowiak K. 2002. Changes in the activity of a CpG neuron after the reinforcement of an operantly conditioned behavior in *Lymnaea*. *J Neurophysiol* **88:** 1915–1923.

Susswein A, Schwartz M. 1983. A learned change of response to inedible food in *Aplysia*. *Behav Neural Biol* **39:** 1–6.

Susswein A, Schwartz M, Feldman E. 1986. Learned changes of feeding behavior in *Aplysia* in response to edible and inedible foods. *J Neurosci* **16:** 1513–1527.

Tomchik SM, Davis RL. 2009. Dynamics of learning-related cAMP signaling and stimulus integration in the Drosophila olfactory pathway. *Neuron* **64:** 510–521.

Tomchik SM, Davis RL. 2013. *Drosophila* memory research through four eras. In *Invertebrate learning and memory* (ed. Menzel R, Benjamin PR), pp. 359–377. Academic, San Diego.

Trudeau LE, Castellucci VF. 1993. Excitatory amino acid neurotransmission at sensory-motor and interneuronal synapses of *Aplysia californica*. *J Neurophysiol* **70:** 1221–1230.

Tsai HC, Zhang F, Adamantidis A, Stuber GD, Bonci A, de Lecea L, Deisseroth K. 2009. Phasic firing in dopaminergic neurons is sufficient for behavioral conditioning. *Science* **324:** 1080–1084.

Walters ET. 1989. Transformation of siphon responses during conditioning of *Aplysia* suggests a model of primitive stimulus-response association. *Proc Natl Acad Sci* **86:** 7616–7619.

Walters ET, Byrne JH. 1983. Associative conditioning of single sensory neurons suggests a cellular mechanism for learning. *Science* **219:** 405–408.

Walters ET, Carew TJ, Kandel ER. 1981. Associative learning in *Aplysia*: Evidence for conditioned fear in an invertebrate. *Science* **211:** 504–506.

Yovell Y, Kandel ER, Dudai Y, Abrams TW. 1992. A quantitative study of the $Ca^{2+}$/calmodulin sensitivity of adenylyl cyclase in *Aplysia*, *Drosophila*, and rat. *J Neurochem* **59:** 1736–1744.

# The Origins and Organization of Vertebrate Pavlovian Conditioning

Michael S. Fanselow and Kate M. Wassum

Department of Psychology, University of California Los Angeles, Los Angeles, California 90095-1563

*Correspondence:* fanselow@psych.ucla.edu

Pavlovian conditioning is the process by which we learn relationships between stimuli and thus constitutes a basic building block for how the brain constructs representations of the world. We first review the major concepts of Pavlovian conditioning and point out many of the pervasive misunderstandings about just what conditioning is. This brings us to a modern redefinition of conditioning as the process whereby experience with a conditional relationship between stimuli bestows these stimuli with the ability to promote adaptive behavior patterns that did not occur before the experience. Working from this framework, we provide an in-depth analysis of two examples, fear conditioning and food-based appetitive conditioning, which include a description of the only partially overlapping neural circuitry of each. We also describe how these circuits promote the basic characteristics that define Pavlovian conditioning, such as error-correction-driven regulation of learning.

## BASIC CONCEPTS OF ASSOCIATIVE LEARNING

Well before the birth of modern psychology and neuroscience, philosophers suggested that the way the mind creates ideas is by forming associations between events. Matters experienced would be joined because of their temporal proximity, common spatial locations, or perceived similarity. More complex thoughts would, in turn, be built from these basic associations. Although less discussed, the resulting associations would have to be stored in memory to impact cognition and action. Thus, there is a long history that acquired associations are at the core of the way the mind represents the world and that such associations provide the structure of memory itself.

## Environmental Relationships

Early in its history, psychology also emphasized the importance of acquired associations in shaping behavior. Associations arose from experiencing events in close temporal proximity. Experience with two types of environmental relationships fostered association formation. One relationship was when two stimuli were experienced close in time (Pavlov 1927); the other was when a behavior was followed closely by a stimulus (Thorndike 1898). Thus, we recognize two classes of associations, one caused by stimulus relationships the other caused by relationships between actions and the environment. This work focuses on the former class, stimulus-based associations. The modern neuroscientific study of associations began with the work of

Ivan Pavlov, who was concerned with stimulus associations and, therefore, the conditions that fostered such associations are appropriately called "Pavlovian conditioning." The latter class is called instrumental conditioning because, in such situations, behavioral action was instrumental in obtaining an outcome.

## The Procedure—Process-Mechanism Distinction

This instrumental versus Pavlovian distinction is based on the events that are experienced, the things that happen in the environment that cause associations to form. In his laboratory, to cause association formation, Pavlov paired two stimuli together; for example, a tone might be immediately followed by food. Hence, we are defining Pavlovian conditioning by a procedure. Evidence for the formation of an association was provided by a change in behavior to the first stimulus. The tone never caused salivation until it was paired with food. Note that this procedural definition is neutral with respect to what happens inside the organism to link experience with behavior. The theoretical construct used to explain this is what has been termed the "process." Some early proponents of associative learning suggested that a common internal process underlies both instrumental and Pavlovian associative learning (Watson 1916; Hull 1942). Others suggested that each procedure produced its behavioral effects through different psychological processes (Konorski and Miller 1937; Spence 1956). And there were those that suggested that we should focus only on procedures and not delve into processes, as only procedures and behaviors were observable (Skinner 1938). Psychological process models emphasize how certain components of the procedure are isomorphic with the mediating events. For example, Hull (1942) suggested that, regardless of whether the procedure was Pavlovian or instrumental, associations formed when a rapid temporal sequence of neutral stimulus→response→biologically significant stimulus was experienced. This experience caused a connection (association) between the mental representation of the stimulus and re-

sponse. We can distinguish such process models from mechanistic models, which describe how synapses within specific neurocircuits change with experience (i.e., the "mechanism" of learning). Of course, psychological process models can provide a framework for the discovery and understanding of the brain mechanisms of learning and the observed brain changes following learning can inform process theory. We take such an approach.

## The Learning—Performance Distinction

The only way to know that an association has formed is to observe a change in behavior following experience. Although behavior is a reflection of learning, it is, however, not learning itself. Learning resides in the process and/or mechanism that mediate the formation of associations between environment and behavior. But behavior will be affected by factors other than learning. For example, Pavlov's dog salivated more or less depending on its hunger status. Similarly, learning may occur but it may not alter behavior. A classic example of the learning–performance distinction is Tolman's latent learning experiment (Tolman 1951). In this experiment, a rat was allowed to explore an empty maze. The rat's behavior was aimless wandering about the maze trial after trial—there was no obvious change in behavior as a function of experience. But when food was suddenly introduced in one location, the rat immediately went to that location on the next trial. This did not happen if the rat never explored the maze without reward. Thus, the rat learned the stimulus configuration of the maze (i.e., formed a cognitive map) during its apparently aimless wandering, but never expressed this learning in performance until motivated to do so.

## The Misdefinition of Pavlovian Conditioning

Most definitions of Pavlovian conditioning are similar to this one taken from the Oxford Dictionary: "A learning process that occurs when two stimuli are repeatedly paired; a response that is at first elicited by the second stimulus is eventually elicited by the first stimulus alone."

Cite this article as *Cold Spring Harb Perspect Biol* doi: 10.1101/cshperspect.a021717

Virtually every aspect of this definition is incorrect. Here we point out the fallacies, later we will provide a more accurate, modern, definition.

1. The conditional response (CR) and the unconditional response (UR) are not the same; the learned response is often different from the one elicited by the unconditional stimulus (US) (see section on topography).

2. Repeated: Some of the most robust forms of conditioning occur with a single trial. The two most notable examples of conditioning with a single trial are fear conditioning (see section on fear conditioning) and taste aversion. In taste aversion, a novel taste is followed by an illness-producing stimulus. Following just a single experience there is a hedonic shift such that the novel taste, even if it was initially pleasing, becomes distasteful.

3. Pairing: This goes to the heart of why Pavlov used the term conditioning. In our example with a tone and food, Pavlov called the tone a conditional stimulus (CS) to differentiate it from the US, food. The response to the food, the UR, was "inborn" because the pathway that leads from stimulus to response "is already complete at birth." The tone, which initially does not produce the response of interest, is the CS and with experience comes to elicit a new response, which is labeled the CR. There is considerable controversy over the choice of the term conditioned versus conditional. In Anrep's translation, which we use here, the term conditioned is most frequently used. However, the literature has Pavlov saying the use of the term conditioned is "fully justified" because "these new reflexes actually depend on very many conditions." This accords better with the term conditional. Indeed, the stimulus relationship that produces association is well described as a situation in which the occurrence of the US depends, or is conditional, on the US.

Rescorla (1968) put this question to an empirical test. He asked whether co-occurrence of the CS and US was sufficient for learning or whether it was the conditional (dependent) relationship between the CS and US. He trained rats with a tone CS and a shock US. Several groups of rats received exactly the same number of CS–US pairings (e.g., 40% of the CSs were paired with shock). What differed between groups was the likelihood of the US in the absence of the CS; some groups had additional shocks delivered during the intertrial interval ([ITI] the time between CSs). When no shocks occurred during the ITI conditioning, the CS was at the maximum detectable level for the measure used; 40% CS–US pairing was sufficient to produce strong conditioning. However, if the probability of shock during the ITI was 0.10, conditioning was reduced by about half. And when the probability of shock during the ITI was the same 0.40 as during the CS, no conditioning occurred. Thus, the level of Pavlovian conditioning was determined by the "conditional" relationship of the CS and US. Therefore, the term "conditional" not only squares better with Pavlov's intention, it is also empirically supported. Converging evidence points to the fact that pairings are not the critical variable that determines conditioning.

## WHAT CAUSES CONDITIONING

### Pairing versus Contingency

During its early history, it was thought that contiguity of CS and US was the necessary and sufficient condition to cause association formation. This is the idea of pairing discussed above—all that matters is whether or not a sufficient number of pairings has occurred. If so, conditioning should occur; if not, there was no conditioning. Rescorla's (1968) experiments, showing that conditioning can be degraded by additional unpaired US presentations, severely challenged this view. There were several additional findings that also suggested pairing alone was not sufficient. One originally described by Pavlov (1927) was overshadowing. In overshadowing, a CS that conditions well (i.e., comes to produce a strong CR) on its own shows less conditioning when it is accompanied by another CS. The overshadowing effect is greater the more intense the added stimulus (Mackintosh 1976). Mackintosh suggested that this occurred because of competition for attentional resources (Sutherland and Mackintosh 1971). Intense CSs grab limited attentional resources away from an overshadowed CS that would alone command those resources. This account also fit well with a finding by Wagner and colleagues showing that a CS that was paired only 50% of the time with a US conditioned well if it was the best available predictor of the US, but failed to condition well when a better predictor was available (Wagner et al. 1968). Wagner and colleagues suggested that the subject allocated attentional resources to the most valid predictor in the situation and as a result learning about the less valid predictors suffered. In both cases, pairing alone was insufficient to explain the strength of conditioning. This shifted the interpretation of conditioning away from an emphasis on the US's ability to automatically reinforce association formation (contiguity theory) toward an attentional view that emphasized CS processing.

The idea of pairing was also challenged by a seminal series of experiments by Kamin (1969). Kamin followed a compound CS of noise and a light with a US and found that conditioning to the light varied with the prior history of training with the noise. If the noise had been paired with the US before compound training, no conditioning to the light would occur, a sort of exaggerated overshadowing effect (Table 1). The finding that prior conditioning to one element of a compound "blocks" conditioning to the other element also, at first blush, fits with the limited attention view. Because the subject first learns that the noise is a good predictor of the US, it grabs the attentional resources that could have been split between the two elements of the compound. However, a second experiment by Kamin dispelled that account. If US intensity was increased in the compound phase (relative to that during pretraining), the added element conditioned well (so called "magnitude unblocking"; Table 1). If attention was directed to the pretrained CS, then conditioning to the added element should still suffer. The fact that conditioning occurred to this element indicated that it must have garnered sufficient attention to support learning. Kamin offered an interpretation that tilted theory back toward an emphasis on the US. He suggested that a US only reinforces learning to the extent it is surprising. In the blocking design, the pretrained stimulus already predicts the US so the added element is never paired with a surprising US. In the unblocking design, increasing the US intensity means that the pretrained CS does not fully predict the new US that is paired with the compound; hence, US surprise is restored and the novel element receives the reinforcement that causes learning.

## The Rescorla–Wagner Model

Rescorla and Wagner formalized Kamin's notion of surprise and showed that this approach could account for both Rescorla's contingency effects and the attention-like phenomena described in the previous section (Rescorla and Wagner 1972). The model dictated that a given US could only support a limited amount of

**Table 1.** Conditioning phenomena: Several conditioning arrangements and their effects are schematized

| Phenomenon | Pretraining | Training | Test: Conditioning to light |
|---|---|---|---|
| Control | None | L+ | Strong |
| Overshadowing | None | TL+ | Weak |
| Blocking | T+ | TL+ | None |
| Unblocking | T+ | TL**+** | Strong |
| Overexpectation | T+, L+ | TL+ | Weak |
| Supernormal | V+, VT− | TL+ | Very strong |
| Latent inhibition | L− | L+ | Weak |

Conditional responses (CSs) are indicated by letters (T, tone; L, light; V, vibration). Two letters together (e.g., TL) indicate stimuli presented simultaneously. Reinforcement by pairing with an unconditioned stimulus (US) is indicated by a +; the bold + in unblocking indicates a larger US (e.g., more food or a more intense shock). Testing is always to the L stimulus and results are all relative to the control.

Cite this article as *Cold Spring Harb Perspect Biol* doi: 10.1101/cshperspect.a021717

associative strength and that this limit was determined by the intensity of the US. Surprise was the difference between the limit ($\lambda$) and the amount of conditioning that had already occurred to the stimuli present to support expectation of the US ($V$ = the amount of associative strength). Earlier contiguity models made similar assumptions (Hull 1942; Bush and Mosteller 1955), but Rescorla and Wagner put a slight twist on the idea that made a world of difference. Rather than assuming the change in conditioning to a specific stimulus on a trial was the difference between $\lambda$ and the associative strength of the stimulus in question, as previous contiguity models had, they postulated that the change in associative strength was the difference between $\lambda$ and the sum of the associative strength of all CSs present on that trial. The resultant model was that the change in associative strength to stimulus A on a trial ($\Delta V_A$) is a proportion ($\alpha$) of the difference between the intensity of the US on that trial ($\lambda$) of associative strength already conditioned to the CSs present on the trial ($V_\Sigma$).

Thus,

$$\Delta V_A = \alpha * (\lambda - V_\Sigma).$$

Kamin's blocking effect is easily predicted by this model. Because of the initial training the associative strength of the noise CS would be at or near $\lambda$. The novel light would enter the second phase with no associative strength, but because the associative strength of the noise contributes to $V_\Sigma$, the quantity ($\lambda - V_\Sigma$) is near 0, so the light receives no increment in its associative strength. Unblocking occurs because the increase in US intensity causes an increase in $\lambda$. The model obtained additional power from the assumption that the learning rate parameter $\alpha$ was determined by CS intensity or salience. This allowed the model to explain why overshadowing increases with CS intensity. In essence, the model put CSs in competition with each other for the reinforcing value of the US.

Another important aspect of the Rescorla–Wagner model was that it focused attention on the context as a significant contributor to conditioning. In any conditioning situation in addition to the explicit CS, there are also the situational cues within which conditioning takes place (i.e., the context). These cues would also naturally compete with CSs. Thus, Rescorla's contingency effects could be explained by the models prediction that giving the subject unpaired USs would drive the associative strength of the context and make it a significant competitor with the CS.

The Rescorla–Wagner model made predictions not only about increases in associative strength, but also decreases. Whenever the value of ($\lambda - V_\Sigma$) is negative, a decrement in associative strength occurs. This most frequently happens when an expected US is omitted, as in extinction, because the value of $\lambda$ drops to 0. This means that in some situations a stimulus can have a negative associative strength. Such CSs are Pavlovian inhibitors that have the ability to suppress a CR.

Further impact of the Rescorla–Wagner model came from its ability to predict new phenomena. Two examples are overexpectation and superconditioning (Table 1). One example of overexpectation occurs when two CSs are trained independently and then are put together as a compound that is reinforced with the same US that each previously independently predicted. Because each CS is near $\lambda$ alone, ($\lambda - V_\Sigma$) will be negative and result in a decrement in associative strength. Empirically, the prediction that the CR to the elements of the compound decrease in a manner proportional to their salience has been confirmed (Kamin and Gaioni 1974). Additionally, as predicted, if a novel stimulus is added to the compound, that novel stimulus becomes an inhibitor despite the fact that it was consistently reinforced (Kremer 1978).

## Error Correction and the Neural Instantiation of the Rescorla–Wagner Model

Kamin, Wagner, and Rescorla reframed the idea of a US and provided a modified version of contiguity theory. A US is defined as a surprising event and any stimulus contiguous with that surprise will be learned about. Indeed, in more recent versions of his theory, Wagner postulated that any surprising stimulus, even one that is hedonically neutral, will reinforce its associations with contiguous events (Wagner and Brandon 1989). This has the benefit of explaining a phenomenon such as sensory preconditioning in which two neutral events become associated. However, if a neutral stimulus is predicted it will be less able to promote association formation. For example, if a neutral stimulus is presented alone in a context, the context will come to predict the stimulus. If that previously preexposed stimulus is subsequently paired with a US, because the CS is not surprising, it will not enter into association with the US—a phenomenon called latent inhibition. Importantly, if the preexposed stimulus is rendered again surprising by presenting it in a novel context it will form associations, and latent inhibition is lost.

Bolles and Fanselow (1980) looked at surprise in a somewhat different way. They emphasized that the CS should produce accurate expectancies of the US and that inaccurate expectancies must be corrected. Their model of conditioning stated that "any error in the expectation is fed back so as to reduce future errors. If the amount of correction is directly proportional to the size of the error, then one has a learning system that will sooner or later correct its errors and generate accurate predictions." Thus, they described Pavlovian learning as an error-correction system driven by negative feedback. In this model, a comparison is made between the CS generated expectancy of the US and the actual US received. To match this to the Rescorla–Wagner equation, $V$ is the expectancy, which is also the strength of the CR, and this value is subtracted from the actual US received ($\lambda$). The negative feedback could be any CR that has the ability to oppose the reinforcing power of the US. Bolles and Fanselow (1980) also suggested a neural mechanism for this error correcting negative feedback. One CR to a fear CS is an analgesic response (Fanselow and Baackes 1982). Because shock conditions fear proportional to its painfulness, an analgesic CR would undermine the reinforcing effectiveness of the US. Such a finding is also consistent with the reduction in the UR that frequently

accompanies conditioning (Fanselow 1984; Canli and Donegan 1995). Additionally, pharmacological antagonism of endogenous opioids prevents error correction (Fanselow 1986a). We elaborate on this specific circuit in our detailed analysis of fear conditioning. Subsequently, a negative feedback circuit for eyeblink conditioning has been identified in the form of a γ-aminobutyric acid (GABA)ergic projection from the deep nuclei of the cerebellum, in which the CS–US association is formed, to the inferior olive, which is where the ascending reinforcement signal from the US is processed (Kim et al. 1998). Although the negative feedback circuit has not been fully identified in appetitive conditioning, groups of dopaminergic cells that behave as if they carry the error-correction signal (i.e., $\lambda - V_{\Sigma}$) have been described (Waelti et al. 2001) (see section on dopamine and reward prediction error).

### Beyond the Rescorla–Wagner model

As a heuristic device, the Rescorla–Wagner model captures a tremendous proportion of the variance found in conditioning phenomena. However, there are certain findings that the model, as initially proposed, has trouble with.

### Time

To perform the iterative calculations, the model breaks a conditioning session into a series of CS-length chunks. This makes predictions easy, but CS length is a rather arbitrary variable so there is no representation of exact time in that model. Sutton and Barto have developed a model that incorporates time within a Rescorla–Wagner-like calculation in their temporal difference reinforcement-learning model (Sutton 1988; Sutton and Barto 1990). Absolute time, in terms of weighted stimulus degradation parameters, features prominently in a model developed by Wagner and Brandon (1989).

### Attention

As mentioned above, preexposing a CS before conditioning reduces the rate of learning once that CS is reinforced (Lubow and Moore 1959;

Cite this article as *Cold Spring Harb Perspect Biol* doi: 10.1101/cshperspect.a021717

Rescorla 1971). The basic Rescorla–Wagner model does not predict this, because the absence of the US means that λ is 0, leaving associative strength ($V$) at the same 0 value it started with. Because associative strength is the only value carried over from preexposure to reinforcement, there is no reason for preexposure to have an effect. Rescorla (1971) recognized this early on and suggested that the learning rate parameter (α) must change because of non-reinforcement—whereas α should start at a value defined by the CSs salience; this value will decrease with nonreinforcement. The idea that this learning rate parameter changes over the course of learning is central to attentional models proposed by Mackintosh (1975) and Pearce and Hall (1980). For Pearce and Hall, associability (α) decreases for stimuli that predict no change in reinforcement and increases when there is a surprising change in reinforcement. Another approach, in Wagner's subsequent models described above, states that preexposure allows the context to predict the CS and because the CS is not surprising during the conditioning phase it receives less processing.

*Extinction*

Perhaps the major problem with the Rescorla–Wagner model is extinction. Although the model accurately predicts decrements in performance following extinction, it does so by causing a reduction in associative strength (unlearning). However, as first shown by Pavlov (1927) and substantially developed by Bouton (1993), associative strength is still intact after extinction. This is shown by recovery phenomena in which after extinction has weakened responding to the CS, the CR returns following a change in context (renewal), time elapsing between extinction and test (spontaneous recovery), and the administration of unpaired USs (reinstatement). Bouton (1993) suggests that rather than unlearning the original CS–US association, extinction causes the acquisition of an inhibitory CS–no US association that is context specific. This idea is formally incorporated into the Pearce–Hall model (Pearce and Hall 1980), although there is no specific account of why this

CS–no US association is context specific. We will address these issues when we consider the mechanisms underlying fear conditioning.

## THE CONTENT OF LEARNING

Shortly after the discovery of the Pavlovian conditioning phenomenon, a large group of scientists became intensely interested in the study of the procedures and behavioral output of conditioning (Skinner 1950). Many others, however, including Pavlov himself (Pavlov 1932), and quite prominently Jerzy Konorski (1948, 1967), argued that Pavlovian and other conditioned behaviors were a window into the brain mechanisms of behavior.

### Associative Structure and Its Diagnosis

Historically, associative conditioning has been thought to involve the formation of nodes (presumably in the brain) between the conditioned components. One primary theory was that Pavlovian (and instrumental) conditioning involves the formation of a stimulus–response (S–R) bond. The CS serves as the S node and with learning becomes capable of directly activating the motor program (the R node) innately generated by the US itself (Hull 1943; Spence 1956). This was encouraged by findings that the CR is often identical to the original UR. Interestingly, however, the CR can differ from the UR and, in some cases, can be entirely opposite. For example, morphine elicits an analgesic UR, but a morphine-predictive CS can actually elicit hyperalgesia (Siegel 1975a), a finding not easily reconciled by an S–R Pavlovian association because the hyperalgesia response "node" is never present during conditioning. At further odds with S–R theory, a Pavlovian CR can still develop if access to the US is blocked, preventing UR execution (Zentall and Hogan 1975). Also unfavorable to S–R theory is sensory preconditioning. In these experiments, two neutral stimuli are paired together (S1–S2) in the absence of any US. Later, S2 is paired with a US, and then in a third phase presentation of S1 alone elicits a CR (Brogden 1939; Rizley and Rescorla 1972; Rescorla 1980). S–R theory cannot explain this result because there was never an opportunity for S1 to become linked to the CR.

To reconcile these findings, Bolles (1972) argued for a more cognitive conditioned association (he was not the first to suggest this, c.f., Koffka 1935; Lewin 1936; Kohler 1940; Tolman 1949). Bolles suggested that, rather than an S–R relationship, during Pavlovian conditioning subjects form a stimulus–outcome (S–O) association (also termed stimulus–stimulus; S–S*) in which a link between the mental nodes representing the CS (S) and the specific US (i.e., outcome [O]) with which it is paired (Bolles 1972). This account was supported by many other learning theorists of the time (e.g., Rescorla 1973a) and suggested that Pavlovian CRs are elicited by a cognitive expectation of the predicted US. As a result, CRs can be more flexible. Indeed, this account allows the Pavlovian CR to take a form different than that directly elicited by the US itself, because it does not rely on a conditioned association to this original response. Sensory preconditioning is also well explained by this account by presuming that the neutral stimulus (S1) elicits a representation of the stimulus with which it was initially paired (S2), which in turn generates the response via a mental connection between the S2 and the US.

### Inflation and Devaluation

The essential difference between the theorized S–R and S–O associative structures is that details of the US's identity are encoded in the latter, but not the former. Therefore, the critical test between these theories is to evaluate CRs after making a specific change to the US, for example, altering its value (Rozeboom 1958;

Pickens and Holland 2004). Postconditioning devaluation of the US will modify CRs if such responding is guided by an S–O association, but not if it is guided by an S–R association (Pickens and Holland 2004). Rescorla put this to the test (Rescorla 1973b, 1974). In one experiment, a light CS was paired with a loud noise US. The US was then repeatedly presented alone to lower its value by habituation. Although the CS was never paired with the devalued US, its ability to generate a fear CR was reduced. Such behavior could only be generated if the CS aroused a memory of the US. In the converse experiment, Rescorla (1974) paired an auditory CS with a mild shock US. The rats were then exposed to a series of stronger shocks. At test, the tone alone elicited a stronger fear CR, as if it had been paired with the stronger US even though it had not. Again, the result is most congruent with an S–O association. Although aversive Pavlovian conditioning appears to be dominated by S–O associations, learning is not exclusively S–O. Within the same experiment, Rescorla showed that first- but not second-order associations were altered by an inflation procedure (Table 2). It is unclear why two different types of associations are formed. Potentially, the CS tends to become associated with the most salient aspect of the US. In first-order conditioning, the shock US is likely to be the most salient feature. In second-order conditioning, the stimulus serving as the "US" may be less salient than the emotional reaction it generates.

An even richer associative network underlies appetitive conditioning. In a typical devaluation experiment, a CS is paired with an appetitive US, often a food substance. Learning is shown when the subjects approach the location

Table 2. Inflation design and results in first- and second-order conditioning

| Training/test | Inflation group | Control group |
|---|---|---|
| First-order conditioning | Tone→mild shock | Tone→mild shock |
| Second-order conditioning | Light→tone | Light→tone |
| Inflation | Strong shocks alone | Context exposure |
| First-order test | Tone→strong CR | Tone→weak CR |
| Second-order test | Light→weak CR | Light→weak CR |

For simplicity, standard counterbalancing was omitted from the table (data based on Rescorla 1974). CR, Conditional response.

Cite this article as *Cold Spring Harb Perspect Biol* doi: 10.1101/cshperspect.a021717

of the US (the food port) when the CS is presented (but before food delivery). In some instances, if the CS is visual and localizable, the subject will approach the CS itself (i.e., sign tracking). Next, the value of the US is reduced either by selective satiation or by pairing its consumption with nausea (induced by lithium chloride [LiCl] injection) to form a taste aversion. Both of these treatments will result in complete rejection of the food. Postconditioning devaluation of a food US, by selective satiation (Holland and Rescorla 1975) or by taste aversion (Holland and Straub 1979), has been shown to reduce food port or CS (sign tracking) approach CRs in a probe test. This finding, replicated many times over in both humans (Gottfried et al. 2003; Bray et al. 2008) and rodents (Holland 1981; Colwill and Motzkin 1994), suggests that subjects mentally recall the devalued US when presented with the CS (Table 3). US revaluation can also turn aversive CRs appetitive. Normally, a CS predicting intraoral infusion of unpleasant high-sodium chloride solution will elicit escape-type CRs, but if the animal is put into a salt-appetite state the aversive CR will turn appetitive, that is, the animal will sign track to the CS (Robinson and Berridge 2013). In further support of the S–O account, presentation of an appetitive Pavlovian CS will bias instrumental action selection toward those actions earning the exact same outcome as predicted by the CS (Kruse et al. 1983; Colwill and Motzkin 1994). Because in these experiments the CS has never been directly paired with the instrumental action (no opportunity for S–R association), it is the cognitive expectation of the outcome elicited by the CS

that explains this selective Pavlovian instrumental transfer (PIT) effect. In all of these cases, the evidence suggests that US identity controls the CR and is, therefore, encoded in the associative structure guiding Pavlovian conditioned response.

The above evidence argues that a simple reflexive S–R associative structure cannot fully explain Pavlovian conditioning responding and that a cognitive, S–O associative structure can control such behavior. This, however, does not suggest that an S–O association is the exclusive association formed during Pavlovian conditioning. Indeed, it has been suggested that both associations develop and that they compete, or perhaps interact, to control conditioned responding (Holland and Rescorla 1975). Although satiety and taste-aversion devaluation produces complete rejection of the food, it does not often completely attenuate Pavlovian CRs, suggesting that some aspect of this responding may be driven by an S–R associative structure (or a less detailed S–O structure; see Dayan and Berridge 2014). Moreover, there are instances in which CRs are not sensitive to US devaluation. For example, if the basolateral amygdala (BLA) has been lesioned, rats will acquire a food-port approach CR, but this response will be insensitive to US devaluation (Hatfield et al. 1996), suggesting it is controlled by an S–R associative structure (more on this later). Which associative structure dominates behavioral control depends on a variety of factors including the CS form, type of pairing, CR form, and the requirement for detailed outcome discrimination.

### The Representation of Outcomes

It has been long recognized that stimuli, conditioned or otherwise, consist of many elements. Within an S–O framework, the CS may, therefore, become linked to one or more elements of US. This concept was formalized by Jerzy Konorski (1948, 1967) and later adapted by Anthony Dickinson (Dickinson and Balleine 2002). These and other investigators (Wagner and Brandon 2001; Delamater 2012; Dayan and Berridge 2014) have suggested that the ex-

Table 3. Devaluation design and results

|  | Devaluation group | Control group |
| --- | --- | --- |
| Phase 1 | Tone→food | Tone→food |
| Phase 2 | Food→LiCl | Food/LiCl unpaired |
|  | Food is rejected | Food not rejected |
| Test | Tone→Ø | Tone→Ø |
|  | Weak CR | Strong CR |

First-order appetitive Pavlovian associations are sensitive to devaluation of the US. LiCl, Lithium chloride; CR, conditional response.

tent to which a CS becomes associated with some or all properties of the US determines its influence over behavior. Take, for example, a grape-flavored sucrose solution US. This stimulus has very specific identifying features, for example, its grape taste, as well as features that may be more general (i.e., overlapping with other USs) including its fluidic and caloric properties and its general appetitive nature. Evidence suggests that the sensitivity of CRs to devaluation is mediated by CS retrieval of the identity-specific (e.g., specific taste) features of the US. This was deduced in a clever experiment in which rats were trained that one of two stimuli predicted one of two food pellets identical in all ways except for their specific flavor (Zener and McCurdy 1939; Holland and Rescorla 1975). After training, one of the food pellets was devalued. When the CS predicting the devalued food pellet was presented, rats showed attenuated food-port approach CRs, but when the other CS was presented rats continued to respond as normal, even though this CS predicted a reward

that was very similar to the one that had been devalued. The aforementioned selective PIT effect also provides evidence of encoding of more specific features of the US in the S−O association (Table 4). Moreover, that an animal given eyeblink conditioning displays the blink CR in only the eye on which the US was applied (Betts et al. 1996) supports the encoding of specific location information by the CS.

There is ample evidence that more than just the identity-specific information can be encoded during Pavlovian conditioning. In PIT, a Pavlovian relationship is first trained and then, subsequently, an instrumental action is trained. In the critical transfer test, the Pavlovian stimulus is presented while the subject performs the previously trained instrumental action and the effect of the CS on the instrumental action is observed. The first example of PIT was by Estes and Skinner (1941), who found that a tone previously paired with shock could suppress a rat's lever pressing for food. When the instrumental action and the Pavlovian relationship are

**Table 4.** Pavlovian instrumental transfer (PIT) design and results

|  | Training | Result |
| --- | --- | --- |
| **Selective PIT** | | |
| Pavlovian conditioning | Tone→food 1 | Tone→strong CR |
|  | Noise→food 2 | Noise→strong CR |
| Instrumental conditioning | Response 1→food 1 | Acquire both independent actions |
|  | Response 2→food 2 | |
| PIT test | ITI, tone, noise | ITI: response 1 ≅ response 2 |
|  | Response 1→∅ | Tone: response 1 > response 2/response 1 > ITI press |
|  | Response 2→∅ | Noise: response 1 < response 2/response 2 > ITI press |
| **General PIT** | | |
| Pavlovian conditioning | Tone→food 1 | Tone→strong CR |
|  | Noise→∅ | Noise→no CR |
| Instrumental conditioning | Response→food 2 | Acquire instrumental action |
| PIT test | ITI, tone, noise | Tone press > ITI press |
|  | Response→∅ | Noise press ≅ ITI press |
| Devaluation | Food→LiCl | Food is rejected |
| PIT test | ITI, tone, noise | Tone press > ITI press |
|  | Response→∅ | Noise press ≅ ITI press |
|  |  | Tone approach CR ≅ ITI approach CR |
|  |  | Noise approach CR ≅ ITI approach CR |

An appetitive conditioned stimulus (CS) can both bias the selection of instrumental action (outcome-specific PIT) by way of generating a detailed representation of the paired unconditioned stimulus (US), and can invigorate the performance of a nonselective range of instrumental actions by way of the CS acquiring general motivational properties. Counterbalancing is not represented. CR, Conditional response; LiCl, lithium chloride; ITI, intertrial interval.

 Cite this article as *Cold Spring Harb Perspect Biol* doi: 10.1101/cshperspect.a021717

trained with different rewarding outcomes, the task is referred to as general PIT because it assumes that it is the general motivational properties of the CS that are transferred allowing the CS to invigorate a nonselective range of reward-seeking behaviors (Balleine 1994; Corbit et al. 2007). These data support formation of an associative link between a CS and the general appetitive properties of the US. The idea that a Pavlovian CS provides a motivational influence over instrumental action is referred to as incentive motivation (Bolles and Zeigler 1967; Rescorla and Solomon 1967).

A similar conclusion is reached by exploring conditioned reinforcement effects in which an appetitive CS can serve to reinforce a new instrumental association in the absence of any US (Rescorla and Solomon 1967; Williams 1994). Both the general PIT and conditioned reinforcement phenomena suggest that the CS has taken on (or has access to) the general motivational value of the US. Interestingly, neither general PIT (Rescorla 1994; Holland 2004) nor conditioned reinforcement (Parkinson et al. 2005) are sensitive to US devaluation, suggesting that these behavioral responses are not mediated by a representation of the identity-specific details of the US. That general PIT (Balleine 1994) is sensitive to changes in motivational (e.g., hunger/thirst) state, suggests that some general features of the US important for determining its current biological significance (e.g., fluidic or caloric properties) are encoded in the Pavlovian associative structure that guides this form of CR (Balleine 1994).

These experiments provide evidence that many different features of the US can be encoded in the S–O associative structure guiding Pavlovian conditioning responding. These features may each have a different node that can be activated by external presentation of the US itself or by a CS (Delamater and Oakeshott 2007), or may exist in a single hierarchical representation of the US (Dayan and Berridge 2014). In either case, the level of detail accessed by the CS is determined by a variety of conditioning factors. In one interesting example of this (Vandercar and Schneiderman 1967), rabbits were conditioned that a tone predicted an eye shock and both heart rate and eye blink CRs were measured. If rabbits encoded the details of the shock US, they would be expected to show the very specific eye blink CR. If they encoded the general aversive nature of the US, the tone should elicit an increase in heart rate (fear CR). Results showed that the tone elevated both heart rate and the eye blink CRs, but only if it predicted the shock with a short latency. If the tone predicted the shock with a longer latency, only the heart rate CR was elevated, suggesting that the tone only had access to a fairly undetailed US representation. As mentioned above, general PIT also relies on a relatively undetailed US representation (i.e., insensitive to devaluation) (Balleine 1994), but this very same CS will also elicit food-port approach CRs that do require a detailed US representation (i.e., sensitive to devaluation).

This multifaceted conditioned responding suggests that multiple forms of learning may occur during Pavlovian conditioning. These different learning forms may be differentially recruited based on the nature of the CS–US predictive relationship. Wagner has proposed a computational model of conditioning that is based on the idea that both the specific information and the general motivational or emotional aspects of the US can develop their own associative links with elements of the CS and that these two different types of associative links form with different temporal dynamics (Wagner and Brandon 1989). Perhaps more importantly, this suggests that Pavlovian conditioning may use multifaceted neural mechanisms with different types of Pavlovian associations requiring different neural circuitry. We discuss this below in more detail (see section on appetitive conditioning).

## CR Deliberateness

Although Pavlovian conditioned responding can involve a cognitive expectation of a predicted rewarding or aversive stimulus, these are not deliberative actions intended to facilitate consumption or avoidance of the US. Holland discovered this by manipulating the CS–US contingency. He paired light with food delivery, a preparation in which rats quickly learn to approach the food-delivery port on presentation of the light. In this experiment, however, CR performance during the light (but before food delivery) would omit the US. Rats acquired and maintained an approach CR during the light to a similar degree as a yoked-control group for which there was no response contingency, even though this resulted in a considerable loss of available food (Holland 1979a). This and related results provide the critical distinction between Pavlovian and instrumental responding, which is sensitive to such contingency changes.

## THE TOPOGRAPHY OF THE CONDITIONED RESPONSE

In discussing S–O versus S–R associations above, we pointed out that contrary to the prediction of S–R theory that the CR and the UR should be similar to each other, they are often different. This is strikingly the case when tolerance-producing drugs are used as the US. The development of tolerance can largely be accounted for by the development of a CR that is antagonistic to the UR (Siegel 1991). With drugs that produce sensitization rather than tolerance (Robinson and Berridge 1993), the CR does resemble the UR and the summation of two similar responses results causes the sensitization. The finding that drug CRs can be similar or different is scientifically unsatisfying because an a priori prediction is difficult. Eikelboom and Stewart (1982) proposed a resolution to the issue by saying that the CR always resembles the UR, but that drugs have both a direct effect and also activate compensatory responses. They asserted that the UR was really the compensatory response and was, therefore, the same as the CR. However, one is still left in the position in which a priori it is unknown whether the UR to the drug is compensatory or not until one determines the direction of the CR. This is only part of the complexity. A perplexing example of additional complexity occurs when insulin is the US and changes in blood sugar are measured as a CR. Both hyper- (Siegel 1975b) and hypoglycemia (Woods et al. 1969) have been reported by different investigators, with the crit-

ical determinant of the CR's direction being the shape of the context used as a CS (Flaherty and Becker 1984).

The finding that CS form can sometimes dictate the form of the CR also occurs with straightforward appetitive conditioning. Holland (1977) discovered that a tone paired with food caused a "head-jerk" reaction, whereas a light paired with the same US caused a rearing response. This happens even when the two stimuli are conditioned in compound and then tested as elements. Further, if the tone and light are paired in a second-order conditioning procedure, the second-order CS still causes the CS-specific response even though it was paired with the response to the other CS. This topography of the CS-related CR seems to be related to the initial orienting response produced by these stimuli. Light presentation produces rearing, whereas tone produces a head jerk. These responses rapidly habituate but return if the CS is consistently paired with food. Interestingly, these CS-determined CRs do not occur if the same CSs are paired with a shock US. In that case, the CR is always freezing (Holland 1979b).

Fear conditioning provides another striking example of how CRs and URs are often unrelated. Electric footshock, the most common US in fear conditioning, produces an activity burst, but the CR to stimuli paired with the footshock is a freezing response (Fanselow 1980a, 1982). Freezing is never produced by the shock itself. For example, if a rat is placed into a chamber and given a shock immediately on placement in the chamber no conditioning occurs and no freezing occurs (Fanselow 1986b). Clearly, shock does not produce a freezing UR. Importantly, this lack of conditioning with immediate shock has been shown with fear-potentiated startle, fear-induced analgesia, and inhibitory avoidance (Fanselow et al. 1994; Kiernan et al. 1995). The reason no conditioning occurs to the context is because the animal requires some minimal time to process the context before it can serve as a CS. These unique aspects of context conditioning are reviewed elsewhere (see Fanselow 2010).

Given the absence of a single set of rules that can effectively specify the relationship between the behavioral topographies elicited by CS and US, we need to look elsewhere to understand just what a CR will look like. That understanding comes from putting Pavlovian conditioning in a broader functional perspective.

## EVOLUTIONARY FUNCTION OF PAVLOVIAN CONDITIONING

### Conditioning as Adaptation

Obviously, to the extent we can anticipate future events, we will have the opportunity to behave more adaptively. If Pavlovian learning is how we learn relationships between stimuli then this type of learning should allow us to anticipate and alter our behavior to help us either exploit or defend against significant future events. Salivating to stimuli that predict food allows the process of digestion to begin coincident with consumption. Indeed, Pavlov showed that the salivatory CR to a CS paired with food depended on the type of food. The type of saliva was one that aided in digestion of a meat US but helped dilute an acid US. Freezing in response to danger protects against visually guided predators. A conditioned compensatory CR, coming in advance of a drug, helps mitigate the deviation from homeostasis caused by the substance. Consistent with this, Pavlovian learning proceeds best if the CS occurs shortly before the US (Fanselow 2010). The exact temporal scale varies with the type of conditioning. For eyeblink conditioning, the ideal interval between CS onset and US onset is measured in tenths of seconds; in taste aversion, it is in the tens of minutes and sometimes hours. But both are better learned when CS precedes US. The differences in time scale also makes sense from a functional perspective; dust in a wind blast assaults our eyes much more rapidly than a toxin in food assaults our gut.

Explanations in terms of adaptive function often take the form of logical but unsubstantiatable "just-so stories." At their best, metrics of adaptability are rules of thumb—obtaining more calories with less effort must be adaptive. However, Pavlovian conditioning is one of the few areas in biology in which there is direct

experimental evidence of biological fitness. In an experiment with male blue gouramis, Hollis and colleagues (1989) paired a blue light CS with the opportunity to see, but not interact with a female US. Over the course of training, males acquired courting responses to the blue light. The critical test was when all fish were presented with the light and then the barrier separating the males and females was removed. Several days later, the number of offspring was counted and the paired males produced several orders of magnitude more fry than those for which the CS and US had been unpaired. This is a direct confirmation that Pavlovian conditioning enhances reproductive success! Matthews and colleagues (2007) found similar results in male quail, in which conditioning leads to increased sperm production and an increased number of fertilized eggs. It should also be noted that these experiments build on a literature that shows that virtually every aspect of reproduction from hormonal responses (Graham and Desjardins 1980) to sexual performance (Zamble et al. 1985) and to attraction (Domjan et al. 1988) are significantly influenced by Pavlovian conditioning.

### Functional Behavior Systems Approach to CR Topography

When we recognize that CRs have biological utility, then the topography of the CR must be one that is functional in that context. Bolles's recognition that Pavlovian fear conditioning activated defensive behavioral systems predicts that defensive behaviors such as freezing should be CRs (Bolles 1970). Functional systems are typically organized in a temporal sequence. To obtain food, we must decide to forage and then search for food. Once found, the food must be procured and only then can consumption ensue (Collier et al. 1972). Each of these phases of feeding requires completely different behaviors. Timberlake suggested that a CS for food should produce CRs that are appropriate for a particular phase depending on the temporal relationship between CS and US. When the CS–US interval is short the CR will be a consummatory response (e.g., salivation); when it is long, it will be general search behavior (e.g., approach to the food port) (Timberlake 1994). Similar sequences happen in sexual conditioning (Domjan 1994).

Defense is also organized in distinct phases along a predatory imminence scale that is anchored by safety at one end and predator attack at the other (Fanselow and Lester 1988; Fanselow 1994). Again different behaviors are appropriate at different points on the continuum. The rat forages at night to reduce the possibility of encountering a predator, but if a predator is encountered it freezes to reduce the likelihood of attack. However, if the predator makes contact the rat stops freezing and makes vigorous attempts at escape. We have suggested that in fear conditioning the CR is one step lower than the response to the US (Fanselow 1989). A shock, which models painful contact with the predator, produces as a UR a vigorous activity burst. However, a CS paired with shock produces freezing. When a rat lives in a context and receives very infrequent presentations of shock, rather than freeze it adjusts its meal patterns. This approach is called a functional behavior systems approach and it offers considerable power in explaining the CR–UR relationship (Timberlake and Fanselow 1994).

## A MODERN DEFINITION OF PAVLOVIAN CONDITIONING

The common definition of Pavlovian conditioning, that via repeated pairings of a neutral stimulus with a stimulus that elicits a reflex the neutral stimulus acquires the ability to elicit that the reflex, is neither accurate nor reflective of the richness of Pavlovian conditioning. Rather, Pavlovian conditioning is the way we learn about dependent relationships between stimuli. As Bolles and Fanselow (1980) stated, "the heart of Pavlovian conditioning . . . is the change in meaning of the CS; a once neutral cue becomes significant for the animal because it serves as a signal for the US." CRs are not limited to replicas of a reflex elicited by the US, but are functional sets of behaviors that facilitate adaptive responding in the face of that US. Our modern definition of Pavlovian conditioning is "the process whereby experience with a conditional re-

lationship between stimuli bestows these stimuli with the ability to promote adaptive behavior patterns that did not occur before the experience." A CR is any response that can be directly attributed to that conditional relationship.

In the above definition, Pavlovian conditioning is considered a process and not a mechanism. Conditioning is embedded in the neural systems that evolved for very different functions (e.g., defense, reproduction, feeding). There is no Pavlovian learning system per se; rather, because of the adaptive value of anticipating events, Pavlovian conditioning appears to have evolved independently within each of these systems. At a process level, each type of Pavlovian conditioning has general similarities. For example, CSs condition better when they precede the US. The best predictors gain associative strength at the cost of other potential predictors. Process models such as Rescorla–Wagner and Pearce–Hall apply very generally. But there are specific differences. Eyeblink, fear, and taste conditioning have their own timeframes and tolerate different delays between CS and US. Some types of learning, such as taste aversion and fear conditioning, are exceedingly rapidly learned, perhaps because a one-time mistake in these domains has dire evolutionary consequences. However, Pavlovian conditioning of specific motor behaviors are often slow, perhaps because to be effective they must be highly refined and well timed. This becomes clear when one focuses on the brain circuits that support conditioning. There is little overlap in the circuitry of functionally distinct classes of conditioning. Thus, at a mechanistic level each type of conditioning needs to be considered on its own. In the remainder of this review, we provide a bit more detail about two such examples: fear conditioning and appetitive conditioning.

## IN-DEPTH LOOK AT TWO FORMS OF CONDITIONING

### Fear Conditioning

The laboratory study of fear conditioning began with Watson and Rayner's famous "Little Albert" demonstration in which a young child

Cite this article as *Cold Spring Harb Perspect Biol* doi: 10.1101/cshperspect.a021717

learned to fear an initially attractive white rat that was paired with a disturbing loud noise (Watson and Rayner 1920). During much of the last century, fear conditioning was used as a way to examine learned motivation in rats (e.g., Miller 1948). Much of this work used fear CSs to motivate avoidance responses. However, the avoidance literature led to few advances in our understanding of fear per se. Furthermore, the rules of reinforcement seemed to depend more on the particular avoidance response investigated than any general reinforcement process (Bolles et al. 1966). Indeed, rats often seemed incapable of learning avoidance responses, even when those responses occurred contiguous with the putative reinforcement (Bolles 1969). All of this changed when Bolles (1970) argued that what happens during fear learning is an activation of a defensive behavior system that functioned to limit behaviors to those that evolved to protect animals from danger, particularly predation. His species-specific defense reaction theory focused research on fear as an investigation of defensive behavior and that refocusing spurred on the detailed understanding of fear we have today.

The use of fear conditioning as a tool for understanding Pavlovian conditioning dramatically increased when Annau and Kamin (1961) developed a convenient metric for fear learning. They used Estes and Skinner's (1941) conditioned suppression task but simply suggested that one could quantify suppression as a ratio between CS and pre-CS responding. Conditioned suppression in rats, using the Annau–Kamin suppression ratio, along with eyeblink conditioning in rabbits (Gormezano and More 1964; Thompson 1988), dominated Pavlovian conditioning research for the next 20 years. Following experiments conducted in Bolles' laboratory (Fanselow and Bolles 1979; Bouton and Bolles 1980; Sigmundi et al. 1980), freezing gradually replaced suppression as the dominant way of assessing fear learning (Anagnostaras et al. 2010).

Why do rodents freeze in fear conditioning experiments? This follows directly from the recognition of fear as the activation of the functional behavior system serving defense. Fear

conditioning activates one particular phase of defense, the postencounter phase when a predator has been detected, but is not on the verge of contact (Fanselow and Lester 1988). Freezing is effective at this point for two reasons: (1) stationary prey are more difficult to detect than moving prey, and (2) for many predators, the releasing stimulus for attack is movement. Several other things go on while the rodent freezes, heart rate changes, blood pressure increases, and breathing becomes shallow and rapid. Pain sensitivity is also decreased. If freezing fails to avoid attack the rat will burst into a protean frenzy akin to panic. Through freezing, the rat readies itself for such an activity burst and this preparation can be measured as a potentiated startle response to a loud noise. All of these responses can and have been used to measure conditional fear. Such measures have identified a critical descending circuit for fear learning.

### The Descending Fear Circuit

The BLA complex is the hub of the fear circuit (Fanselow and LeDoux 1999). This frontotemporal cortical region consists of the lateral, basolateral, and basomedial nuclei and receives input from the thalamus and from other cortical regions (Swanson and Petrovich 1998). Importantly, both CS and US information converges on single neurons within the region (Romanski et al. 1993) promoting $N$-methyl-D-aspartate receptor (NMDA)-dependent synaptic plasticity (Miserendino et al. 1990; Fanselow and Kim 1994; Maren and Fanselow 1995). The resulting long-term potentiation (LTP) is what gives the CS the ability to activate defensive behaviors. Indeed, BLA plasticity is sufficient to produce fear learning. Using a mouse with a global knockout of the Creb (adenosine $3',5'$-monophosphate response element-binding protein), which is needed for LTP, Han and colleagues (2007) found they could rescue fear conditioning by replacing Creb in the BLA. This should not be taken to mean that plasticity in other regions is not normally involved in fear conditioning; blocking RNA synthesis, also needed for LTP in regions upstream of the BLA also attenuates fear conditioning (Parsons et al. 2006; Helmstetter et al. 2008). At least some of this upstream plasticity depends on ascending BLA input (Maren et al. 2001; Talk et al. 2004).

Three glutamatergic outputs from the BLA drive fear responding. One consists of projections to the nearby striatal-like central nucleus (CeN) and another is to clusters of GABAergic cells lying in the capsule between the BLA and CeN (Paré et al. 2004). These intercalated cells (ITC) in turn project to the CeN. In addition, the BLA projects to several of the bed nuclei of the stria terminalis (BNST). Output from the BNST drives sustained fear responses, while the CeN drives more short-lived responses. So, for example, fear responses to long CSs and contextual cues depend more on the BNST, and those to discrete, brief CSs depend more on the CeN (Davis et al. 1997; Waddell et al. 2006).

The medial portion of the CeN sends GABAergic projections to the regions responsible for individual fear behaviors, such as the periaqueductal gray (vPAG) (freezing, analgesia, vocalization), hypothalamus (hormonal responses and hypertension), dorsal motor nucleus of the vagus (heart rate), and nucleus reticularis pontis caudalis (potentiated startle) (Davis 1989). GABAergic circuitry within the CeN regulates this output (Haubensak et al. 2010). For example, the lateral nucleus does not project to the medial region of CeN that contains projections neurons. Rather, it activates cells in the lateral portion of CeN. This region consists of opposing fear on and off cells that contact the medial CeN's projection neurons.

The medial CeN generates two important CRs, freezing and analgesia, by projecting to the ventral portion of the vPAG (Fanselow 1991). Within the vPAG the freezing and analgesic CRs have different neurochemical coding as the analgesia, but not freezing, is blocked by opioid antagonists (Helmstetter and Fanselow 1987). The vPAG in turn projects to the rostral ventral medulla and then to motor neurons in the ventral horn of the spinal cord to produce freezing and to the dorsal horn of the spinal cord to produce analgesia by inhibiting ascending pain information (Morgan et al. 2008).

## Negative Feedback and Error Correction

Given that the analgesia-producing descending opioid circuitry is engaged as a fear CR, as the CR builds, the reinforcing ability of a painful US will be reduced. Thus, endogenous opioids provide critical negative feedback regulation of fear learning that is responsible for Rescorla–Wagner type effects. The earliest evidence for this comes from the finding that the opioid antagonist naloxone prevents a phenomenon called preference for signaled shock (Fanselow 1979). If given a choice between two environments that deliver identical shock except that in one the shock is preceded by a CS and the other in which it is unsignaled, rats chose the signaled shock environment (Lockard 1963). This occurs because the tone overshadows context conditioning in the signaled case, but in the unsignaled situation all the associative strength goes to the context (Fanselow 1980b). Thus, the rat is choosing to go to the least frightening context. Naloxone, by antagonizing the negative feedback mechanism, prevents this from happening. Naloxone also prevents conventional overshadowing of a light CS by a tone CS (Zelikowsky and Fanselow 2010). Kamin blocking is also prevented by systemic (Fanselow and Bolles 1979) and intra-PAG administration of opioid antagonists (Cole and McNally 2007).

The negative feedback regulation of fear is perhaps most clearly illustrated by the effects of opioid antagonists on simple learning curves. The Rescorla–Wagner model predicts that the asymptote of learning should depend on US intensity (Rescorla and Wagner 1972) and the empirical evidence for this is unequivocal (Annau and Kamin 1961; Young and Fanselow 1992). Functionally, this makes sense; a mild threat should not produce overwhelming fear, but a significant threat should. When opioid antagonists are given during learning asymptotic levels become high and undifferentiated by shock intensity (Fanselow 1981; Young and Fanselow 1992). Indeed, a shock intensity that is normally barely able to condition detectable levels of fear will, under the antagonist, result in the same asymptote as a maximally effective US.

Cite this article as *Cold Spring Harb Perspect Biol* doi: 10.1101/cshperspect.a021717

Opioid antagonists, to some extent, also prevent changes in associative strength that are caused by negative $\Delta V$ values such as in extinction (McNally et al. 2004a) and overexpectation (McNally et al. 2004b). Thus, in fear conditioning, an opioid negative feedback circuit accounts for the same wealth of phenomena as the Rescorla–Wagner model. In strong support of this notion, there are neurons in both the vPAG and BLA that respond as if they encode the critical error signal (i.e., $\Delta V$) (Johansen et al. 2010).

### Mechanistic Models

Knowledge of this circuit should allow us to advance from process models, such as Rescorla–Wagner and Pearce–Hall, to mechanistic models that incorporate what we know about the circuitry and synaptic plasticity. Toward this goal, Krasne and colleagues (2011) have proposed a computational model that uses LTP rules at afferents onto BLA excitatory and inhibitory neurons from the prefrontal cortex, thalamus, and hippocampus that drive both freezing and a negative regulation of fear learning at the PAG. The model predicts the phenomena anticipated by process models such as blocking and extinction. However, the model also predicts some of the phenomena that have been difficult to accommodate in process models, notably renewal of extinguished fear. The model suggests that this is accomplished by potentiation of inhibitory neurons some of which encode conjunctions of context, CS, and extinction. Such conjunctive information is communicated to the BLA as connections between the BLA, prefrontal cortex (extinction processing), and hippocampus (context processing) are necessary for renewal (Orsini et al. 2011). The model requires that extinction promotes LTP of these neurons, and extinction is, indeed, blocked by intra-amygdala application of NMDA-antagonists (Falls et al. 1992). Finally, the BLA contains neurons whose increased activity during extinction coincides with the reduction of activity in neurons that were potentiated by fear acquisition (Herry et al. 2008). Indeed, extinction in a novel context recruits a unique population of context-dependent neurons in the BLA (Orsini et al. 2013).

### Appetitive Conditioning

Appetitive Pavlovian conditioning controls a large majority of our reward-related behavior and is most often studied the laboratory with a discrete tone or light CS and food US in hungry rodents. The CS can then come to elicit a variety of CRs. These include specific consummatory reactions (mouth movements related to consummation of the specific reward) (Grill and Norgren 1978), conditioned approach to the food source (i.e., goal approach or "goal tracking"), and, if the CS is visual and localizable (Cleland and Davey 1983), conditioned approach to the stimulus itself (i.e., "sign tracking" or autoshaping) (Brown and Jenkins 1968; Boakes 1977). All of these CRs are sensitive to posttraining revaluation of the US, suggesting that, at least to some extent (see Holland 2008), they are guided by a cognitive image of the specific predicted reward (Holland and Rescorla 1975; Cleland and Davey 1982; Berridge 1991). Food-predictive CSs can also serve to reinforce instrumental behavior (i.e., conditioned reinforcement), invigorate a nonselective range of instrumental actions (i.e., general PIT) and induce conditioned locomotor activation (Estes 1948; Rescorla and Solomon 1967; Lovibond 1983; Dickinson and Dawson 1987). All of these conditioned behaviors are insensitive to US devaluation (Rescorla 1994; Holland 2004; Parkinson et al. 2005), suggesting that they are not guided by retrieval of an identity-specific mental representation of the reward. Rather these Pavlovian "incentive motivational" effects of the US result because the CS acquires general motivational value (Konorski 1967; Dickinson and Dawson 1987) via a connection with the more general (e.g., caloric or fluidic properties). These are not mutually exclusive; the same food-predictive CS can both induce goal approach CRs and invigorate instrumental activity (PIT). Moreover, although goal- and CS-approach conditioning responses are sensitive to US devaluation, these effects are smaller than the complete rejection of the devalued food that results from such procedures. This suggests that Pavlovian conditioning can engender multiple associative processes and may, therefore, engage a variety of levels of neural processing. Below, we consider the neural mechanisms responsible for the development of appetitive Pavlovian CRs and evaluate how these differ depending on the form of the response (and presumably also the form of learning).

## Dopamine and Reward-Predication Error

The activity of dopamine neurons in the ventral tegmental area (VTA) and substantia nigra (SNc) displays many properties of a Rescorla–Wagner/temporal difference reward-prediction error signal. Unexpected reward delivery results in a phasic increase in dopamine cell activity, but when the reward is expected based on the presence of a CS, it no longer elicits such activity; rather the unexpected presentation of the CS induces the phasic response (Schultz 2002). If a conditioned reward is unexpectedly delivered, it elicits a phasic increase in dopamine cell activity (Hollerman and Schultz 1998). Both omission of an expected reward or presentation of an aversive event will induce a phasic pause in tonic dopamine cell activity (Tobler et al. 2003; Ungless et al. 2004). These seminal findings have been replicated in dopaminergic cells identified by optogenetic "phototagging" (Cohen et al. 2012). Moreover, prediction error-like phasic dopamine release has been detected in striatal terminal regions (Day et al. 2007; Roitman et al. 2008; Brown et al. 2011; Wassum et al. 2012; Hart et al. 2014).

Dopamine has also been causally linked to reward-prediction error. A pretrained CS will not only block acquisition of the CR to a novel stimulus, but will also prevent the burst firing in dopamine neurons that would typically occur during learning (Waelti et al. 2001). Such blocking will be prevented if a reward-prediction error-like dopamine signal is artificially induced at the time of reward delivery with optogenetic activation of VTA dopamine cells (Steinberg et al. 2013). These and related studies demonstrate that phasic dopamine release can convey a short-latency, phasic reward signal indicating the difference between actual and predicted rewards important for driving learning.

That phasic mesolimbic dopamine release acts as a reward-prediction error signal selectively when the CS becomes a "motivational magnet" to elicit sign tracking (Berridge and Robinson 2003) suggests that phasic dopamine, at least in the nucleus accumbens (NAc), may have a role beyond a passive learning process. In support of this, phasic mesolimbic dopamine release is both necessary for and tracks the ability of a food-paired stimulus to PIT (Wassum et al. 2011a, 2013). Moreover, enhancing the activity of dopamine release specifically in the NAc will enhance both PIT (Wyvell and Berridge 2000) and sign tracking (Peciña and Berridge 2013). Phasic mesolimbic dopamine is, therefore, also involved in the incentive motivational impact of rewards and related stimuli that contribute to the online performance of CRs (for further review, see Berridge 2007).

The above evidence suggests that phasic mesolimbic dopamine release may serve a dual role in Pavlovian conditioned responding, involved in both reward prediction error-mediated acquisition of motivational value to reward-predictive stimuli and the online contribution of this Pavlovian incentive motivation to conditioned responding and reward seeking. This latter response-invigorating function is consistent with the physiological effects of phasic dopamine on striatal projection neurons (SPNs). These GABAergic SPNs can be divided into two projection systems: a direct path to the basal ganglia output nuclei and an indirect path to these output nuclei. Direct pathway activation triggers behavior by disinhibiting motor control areas (Deniau and Chevalier 1985; Freeze et al. 2013; Goldberg et al. 2013). These direct pathway SPNs selectively express the D1, but not D2, dopamine receptors (Surmeier et al. 2007) that have a high affinity for phasic (rather than tonic) dopamine release (Corbit and Janak 2010). Because of coupling to $G_{\alpha s/olf}$ G-proteins that stimulate cyclic AMP (cAMP) and the activity of protein kinase A (PKA), D1 receptor activation leads to increased SPN excitability (for review, see Gerfen and Surmeier 2011).

Therefore, high-concentration dopamine surges activate D1Rs in the direct path, increasing the excitability of these SPNs and, thereby, transiently disinhibiting motor output nuclei.

### Circuitry for General Motivational Value

The mesolimbic and nigrastriatal dopamine pathways are only part of the appetitive Pavlovian conditioning circuit. Indeed, dopamine release can be modulated at both the cell body and terminal fields by varied inputs. The CeN is a large component of this circuit. This region projects to both the VTA and SNc (Gonzales and Chesselet 1990; Fudge and Haber 2000), which innervate the NAc and dorsal striatum, respectively, and these projections can have an indirect excitatory effect on a subpopulation of dopamine neurons through inhibition of local GABAergic interneurons (Rouillard and Freeman 1995; Chuhma et al. 2011). The CeN is required to learn from (negative) reward-prediction error (Holland and Gallagher 1993; Holland et al. 2001; Haney et al. 2010) and CeN-SNc-projecting neurons will become activated after learning by reward-prediction error, suggesting that communication between these structures relates to reward-prediction error-mediated learning (Lee et al. 2010). An intact CeN is also required for the acquisition of conditioned-orienting (Gallagher et al. 1990) and sign-tracking responding (Parkinson et al. 2000), but is not required for more specific consummatory CRs (Chang et al. 2012). Similarly, the CeN is required for general, but not outcome-specific PIT (Hall et al. 2001; Holland and Gallagher 2003; Corbit and Balleine 2005). These data suggest that the CeN is required for the acquisition of general motivational value to a CS, but not for acquisition or use of a detailed S–O association. In further support of this hypothesis, rats with CeN lesions are both able to acquire Pavlovian conditioned food port (i.e., goal) approach responding and to flexibly adjust such responding to the current value of the US (Hatfield et al. 1996).

In addition to the CeN, the lateral habenula (LHb) may also play a role in appetitive Pavlovian conditioning by influencing phasic striatal dopamine signaling. The LHb projects to both the VTA and SNc (Herkenham and Nauta 1979), and stimulation of this structure inhibits the activity of dopamine neurons in these regions (Christoph et al. 1986; Ji and Shepard 2007; Matsumoto and Hikosaka 2007). Interestingly, LHb neurons show a firing pattern opposite to a reward-prediction error signal (Matsumoto and Hikosaka 2007), suggesting a potential inhibitory influence over dopamine-mediated reward-prediction error signals. Of course, the circuitry surrounding this influence is more complex than this (for review, see Hikosaka 2010).

### Circuitry of Outcome Representations

Pretraining lesions of the orbitofrontal cortex (OFC), BLA, NAc shell and core, and mediodorsal thalamus (MD) do not prevent the acquisition of Pavlovian goal approach responding, but do render this behavior insensitive to US devaluation. Because sensitivity to US devaluation requires a rather detailed reward representation, these regions comprise a circuitry important for the encoding of details in the acquired S–O association and/or the use of this information to guide conditioned responding.

Evidence suggests that the OFC is uniquely important for the acquisition and integration of information about the specific identifying features of US for both Pavlovian learning and performance. Pre- and posttraining OFC lesions result in Pavlovian-conditioned responding that is insensitive to devaluation (Pickens et al. 2003; Ostlund and Balleine 2007). The OFC is not required for choosing between valued and devalued rewards when they are present, supporting a primary role in the use of a specific reward representation. Moreover, OFC lesions abolish outcome-specific PIT (Ostlund and Balleine 2007), such that reward-predicative stimuli are unable to retrieve a detailed cognitive representation of the US and are, therefore, unable to bias action selection. Similar findings with US devaluation procedures have been found in humans (Gottfried et al. 2003) and in nonhuman primates (Murray et al. 2007; for further review, see McDannald et al. 2014a).

OFC neurons can fire in response to appetitive CSs and in anticipation of the predicted reward (Schoenbaum et al. 2003) and this encoding depends on BLA input. The converse is also true; associative encoding in BLA neurons relies on OFC function (Saddoris et al. 2005). The OFC and BLA therefore function in a circuit vital for encoding and retrieving reward-specific representations in Pavlovian S–O associations. Indeed, as with the OFC, pretraining BLA lesions result in CRs that are insensitive to US devaluation (Hatfield et al. 1996). The BLA is also vital for specific PIT (Corbit and Balleine 2005). Recent evidence suggests that rapid glutamate signaling within the BLA mediates this effect and encodes specific reward representations (Wassum 2014). The BLA's role in representing outcome-specific information is limited to motivationally significant events. An intact BLA is not required to represent the outcome-specific aspects of neutral events (Dwyer and Killcross 2006). Correlates of "valueless" reward representations have been identified in the OFC (McDannald et al 2014b), suggesting that the BLA may incorporate value in to outcome-specific representations sent from the OFC. Indeed, the BLA is required for attaching motivational significance to rewards themselves (Parkes and Balleine 2013), an effect that relies on μ-opioid receptor activation (Wassum et al. 2009, 2011b).

The BLA and CeN make distinct functional contributions to Pavlovian conditioning. The BLA and CeN are arranged partly in series and much work from aversive Pavlovian conditioning proposes their serial function (Fendt and Fanselow 1999; LeDoux 2000). However, the CeN and BLA each possess independent input and output that allow them to also act in parallel (Balleine and Killcross 2006).

The BLA receives excitatory projections from the MD (van Vulpen and Verwer 1989), which itself is required for Pavlovian conditioning responses to be modified on the basis of posttraining changes in the US value (Mitchell et al. 2007; Izquierdo and Murray 2010). Although, pretraining MD lesions disrupt the sensitivity of Pavlovian CRs to US devaluation, posttraining lesions do not. The MD is, how-ever, recruited posttraining when previous associations need to be suppressed to allow new associations to be formed (Pickens 2008). Posttraining MD lesions will also disrupt outcome-specific PIT (Ostlund and Balleine 2007).

Last, the taste-processing cortices are a prime candidate for US representations, at least for food USs. Indeed, food-paired CSs can activate the gustatory region of the insular cortex (GC) in both rodents (Dardou et al 2006, 2007) and humans (Veldhuizen et al. 2007; Small et al. 2008). Importantly, these CSs activate the very same neuronal ensembles in the GC that were activated by the food US itself, providing a neural substrate of CS retrieval of a reward representation (Saddoris et al. 2009).

It is, perhaps, not surprising that all of the above structures either directly or indirectly make connections with the striatum. Both the NAc core and shell are required for Pavlovian CRs to be sensitive to outcome devaluation (Singh et al. 2010). Interestingly, as in the amygdala, an NAc dissociation exists in the encoded Pavlovian associations driving PIT. The NAc core is required for the general, but not specific component of PIT, whereas the opposite is true of the NAc shell. A similar dissociation exists in the dorsolateral (DLS) and dorsomedial striatum (DMS), respectively (Corbit and Janak 2007). Although the contribution of the DLS may be more vital for the instrumental component of PIT, evidence does suggest that the posterior DMS is critical for the formation of S–O (Corbit and Janak 2010). If, as mentioned, these striatal structures influence motor output during Pavlovian conditioning, these regions are likely key integration sites where both specific (OFC, BLA, MD, GC) and general (CeN and maybe LHb) information is integrated to influence Pavlovian conditioned responding. How this is achieved is still a matter of intense interest. Of course, the circuitry described here is not complete. The ventral pallidum (Smith et al. 2009), anterior cingulate cortex (Cardinal et al. 2003), and hippocampus (Ito et al. 2005; Gilboa et al. 2014), to name a few, have all been implicated in Pavlovian conditioning and more work is needed to fully delineate the entire circuit.

Cite this article as *Cold Spring Harb Perspect Biol* doi: 10.1101/cshperspect.a021717

## CONCLUSIONS

Pavlovian conditioning is the process underlying how the brain represents relationships between environmental stimuli. The circuits that serve conditioning are information-processing circuits that correct their errors in prediction and allocate associative strength to the best predictors of significant events. Formation of those associations depends on conditional relationships between the relevant stimuli and is not a result of simple pairing. Through those associations, CSs can exert behavioral change either directly by promoting specific behaviors (S–R associations), or indirectly, by activating representations of the events they predict (S–O associations). In the latter, CSs can activate a general representation of the hedonic/motivational aspects of a US and by so doing promote or inhibit alternate classes of behavior. But they can also activate a representation of the specific sensory aspects of the US, through which they promote behaviors directed at obtaining, or avoiding, specific outcomes. Thus, Pavlovian stimuli are crucial to many, if not most, of the behaviors in which vertebrate animals partake.

Pavlovian conditioning is a functional process. By function we mean that CS-elicited behaviors directly impact biological fitness. By process we mean that a single circuit or mechanism does not mediate conditioning. Rather, the mechanisms of conditioning are imbedded in circuits that serve specific adaptive functions such as feeding, reproduction, and defense. However, likely through convergent evolution, there is considerable generality in the rules that govern conditioning. Thus, there is considerable power and generality in models of this process such as those of Rescorla and Wagner (1972) and Pearce and Hall (1980). To the extent that we know the specifics of the circuit, there appears to be a lot of generality of how the circuits are wired, although the specifics vary. One commonality seems to be a negative feedback loop by which the CS activates a circuit that dampens the reinforcing US input. This feedback regulates the amount of learning, serving an adaptive function of keeping the level of conditioning within a maximally functional range. Perhaps Rescorla (1988) said it best when he stated, "Pavlovian conditioning, it is not what you think."

## REFERENCES

Anagnostaras SG, Wood SC, Shuman T, Cai DJ, Leduc AD, Zurn KR, Zurn JB, Sage JR, Herrera GM. 2010. Automated assessment of Pavlovian conditioned freezing and shock reactivity in mice using the video freeze system. *Front Behav Neurosci* **4:** 158.

Annau Z, Kamin LJ. 1961. The conditioned emotional response as a function of intensity of the US. *J Comp Physiol Psychol* **54:** 428–432.

Balleine B. 1994. Asymmetrical interactions between thirst and hunger in Pavlovian-instrumental transfer. *Q J Exp Psychol B* **47:** 211–231.

Balleine BW, Killcross S. 2006. Parallel incentive processing: An integrated view of amygdala function. *Trends Neurosci* **29:** 272–279.

Berridge KC. 1991. Modulation of taste affect by hunger, caloric satiety, and sensory-specific satiety in the rat. *Appetite* **16:** 103–120.

Berridge KC. 2007. The debate over dopamine's role in reward: The case for incentive salience. *Psychopharmacology* **191:** 391–431.

Berridge KC, Robinson TE. 2003. Parsing reward. *Trends Neurosci* **26:** 507–513.

Betts S, Brandon S, Wagner A. 1996. Dissociation of the blocking of conditioned eyeblink and conditioned fear following a shift in US locus. *Anim Learn Behav* **24:** 459–470.

Boakes RA. 1977. Performance on learning to associate a stimulus with positive reinforcement, in operant-Pavlovian interactions. In *Operant-Pavlovian interactions* (ed. Davis H, Hurwitz HMB), pp. 67–97. Erlbaum, Hillsdale, NJ.

Bolles R. 1972. Reinforcement, expectancy, and learning. *Psychol Rev* **79:** 394–409.

Bolles RC, Fanselow MS. 1980. A perceptual-defensive-recuperative model of fear and pain. *Behav Brain Sci* **3:** 291–301.

Bolles RC, Zeigler HP. 1967. *Theory of motivation.* Harper & Row, New York.

Bolles RC, Stokes LW, Younger MS. 1966. Does CS termination reinforce avoidance behavior? *J Comp Physiol Psychol* **62:** 201–207.

Bouton ME. 1993. Context, time, and memory retrieval in the interference paradigms of Pavlovian learning. *Psychol Bull* **114:** 80–99.

Bouton ME, Bolles RC. 1980. Conditioned fear assessed by freezing and by the suppression of three different baselines. *Anim Learn Behav* **8:** 429–434.

Bray S, Rangel A, Shimojo S, Balleine B, O'Doherty JP. 2008. The neural mechanisms underlying the influence of Pavlovian cues on human decision making. *J Neurosci* **28:** 5861–5866.

Brogden WJ. 1939. Sensory pre-conditioning. *J Exp Psychol* **25:** 323–332.

Brown PL, Jenkins HM. 1968. Auto-shaping of the pigeons key-peck. *J Exp Anal Behav* **11:** 1–8.

Brown HD, McCutcheon JE, Cone JJ, Ragozzino ME, Roitman MF. 2011. Primary food reward and reward-predictive stimuli evoke different patterns of phasic dopamine signaling throughout the striatum. *Eur J Neurosci* **34:** 1997–2006.

Bush RR, Mosteller F. 1955. *Stochastic models for learning.* Wiley, Oxford.

Canli T, Donegan NH. 1995. Conditioned diminution of the unconditioned response in rabbit eyeblink conditioning: Identifying neural substrates in the cerebellum and brainstem. *Behav Neurosci* **109:** 874–892.

Cardinal RN, Parkinson JA, Marbini HD, Toner AJ, Bussey TJ, Robbins TW, Everitt BJ. 2003. Role of the anterior cingulate cortex in the control over behavior by Pavlovian conditioned stimuli in rats. *Behav Neurosci* **117:** 566–587.

Chang SE, Wheeler DS, Holland PC. 2012. Effects of lesions of the amygdala central nucleus on autoshaped lever pressing. *Brain Res* **1450:** 49–56.

Christoph GR, Leonzio RJ, Wilcox KS. 1986. Stimulation of the lateral habenula inhibits dopamine-containing neurons in the substantia nigra and ventral tegmental area of the rat. *J Neurosci* **6:** 613–619.

Chuhma N, Tanaka KF, Hen R, Rayport S. 2011. Functional connectome of the striatal medium spiny neuron. *J Neurosci* **31:** 1183–1192.

Cleland GG, Davey GC. 1982. The effects of satiation and reinforcer develuation on signal-centered behavior in the rat. *Learn Motiv* **13:** 343–360.

Cleland GG, Davey GC. 1983. Autoshaping in the rat: The effects of localizable visual and auditory signals for food. *J Exp Anal Behav* **40:** 47–56.

Cohen JY, Haesler S, Vong L, Lowell BB, Uchida N. 2012. Neuron-type-specific signals for reward and punishment in the ventral tegmental area. *Nature* **482:** 85–88.

Cole S, McNally GP. 2007. Opioid receptors mediate direct predictive fear learning: Evidence from one-trial blocking. *Learn Mem* **14:** 229–235.

Collier G, Hirsch E, Hamlin PH. 1972. The ecological determinants of reinforcement in the rat. *Physiol Behav* **9:** 705–716.

Colwill RM, Motzkin DK. 1994. Encoding of the unconditioned stimulus in Pavlovian conditioning. *Anim Learn Behav* **22:** 384–394.

Corbit LH, Balleine BW. 2005. Double dissociation of basolateral and central amygdala lesions on the general and outcome-specific forms of Pavlovian-instrumental transfer. *J Neurosci* **25:** 962–970.

Corbit LH, Janak PH. 2007. Inactivation of the lateral but not medial dorsal striatum eliminates the excitatory impact of Pavlovian stimuli on instrumental responding. *J Neurosci* **27:** 13977–13981.

Corbit LH, Janak PH. 2010. Posterior dorsomedial striatum is critical for both selective instrumental and Pavlovian reward learning. *Eur J Neurosci* **31:** 1312–1321.

Corbit LH, Janak PH, Balleine BW. 2007. General and outcome-specific forms of Pavlovian-instrumental transfer: The effect of shifts in motivational state and inactivation of the ventral tegmental area. *Eur J Neurosci* **26:** 3141–3149.

Dardou D, Datiche F, Cattarelli M. 2006. Fos and Egr1 expression in the rat brain in response to olfactory cue after taste-potentiated odor aversion retrieval. *Learn Mem* **13:** 150–160.

Dardou D, Datiche F, Cattarelli M. 2007. Does taste or odor activate the same brain networks after retrieval of taste potentiated odor aversion? *Neurobiol Learn Mem* **88:** 186–197.

Davis M. 1989. Neural systems involved in fear-potentiated startle. *Ann NY Acad Sci* **563:** 165–183.

Davis M, Walker DL, Lee Y. 1997. Amygdala and bed nucleus of the stria terminalis: Differential roles in fear and anxiety measured with the acoustic startle reflex. *Philos Trans R Soc Lond B Biol Sci* **352:** 1675–1687.

Day JJ, Roitman MF, Wightman RM, Carelli RM. 2007. Associative learning mediates dynamic shifts in dopamine signaling in the nucleus accumbens. *Nat Neurosci* **10:** 1020–1028.

Dayan P, Berridge KC. 2014. Model-based and model-free Pavlovian reward learning: Revaluation, revision, and revelation. *Cogn Affect Behav Neurosci* **14:** 473–492.

Delamater AR. 2012. On the nature of CS and US representations in Pavlovian learning. *Learn Behav* **40:** 1–23.

Delamater AR, Oakeshott S. 2007. Learning about multiple attributes of reward in Pavlovian conditioning. *Ann NY Acad Sci* **1104:** 1–20.

Deniau JM, Chevalier G. 1985. Disinhibition as a basic process in the expression of striatal functions. II: The striato-nigral influence on thalamocortical cells of the ventromedial thalamic nucleus. *Brain Res* **334:** 227–233.

Dickinson A, Balleine BW. 2002. The role of learning in the operation of motivational systems. In *Learning, motivation and emotion, volume 3 of Steven's handbook of experimental psychology* (ed. Gallistel CR), pp. 497–533. Wiley, New York.

Dickinson A, Dawson GR. 1987. Pavlovian processes in the motivational control of instrumental performance. *Q J Exp Psychol (Hove)* **39:** 201–213.

Domjan M. 1994. Formulation of a behavior system for sexual conditioning. *Psychon Bull Rev* **1:** 421–428.

Domjan M, O'Vary D, Greene P. 1988. Conditioning of appetitive and consummatory sexual behavior in male Japanese quail. *J Exp Anal Behav* **50:** 505–519.

Dwyer DM, Killcross S. 2006. Lesions of the basolateral amygdala disrupt conditioning based on the retrieved representations of motivationally significant events. *J Neurosci* **26:** 8305–8309.

Eikelboom R, Stewart J. 1982. Conditioning of drug-induced physiological responses. *Psychol Rev* **89:** 507–528.

Estes W. 1948. Discriminative conditioning. II: Effects of a Pavlovian conditioned stimulus upon a subsequently established operant response. *J Exp Psychol* **38:** 173–177.

Estes WK, Skinner BF. 1941. Some quantitative properties of anxiety. *J Exp Psychol* **29:** 392–400.

Falls WA, Miserendino MJ, Davis M. 1992. Extinction of fear-potentiated startle: Blockade by infusion of an NMDA antagonist into the amygdala. *J Neurosci* **12:** 854–863.

Fanselow MS. 1979. Naloxone attenuates rats' preference for signaled shock. *Physiol Psychol* **7**: 70–74.

Fanselow MS. 1980a. Conditional and unconditional components of post-shock freezing. *Pavlov J Biol Sci* **15**: 177–182.

Fanselow MS. 1980b. Signaled shock-free periods and preference for signaled shock. *J Exp Psychol Anim Behavior Processes* **6**: 65–80.

Fanselow MS. 1981. Naloxone and Pavlovian fear conditioning. *Learn Motiv* **12**: 398–419.

Fanselow MS. 1982. The postshock activity burst. *Anim Learn Behav* **10**: 448–454.

Fanselow MS. 1984. Opiate modulation of the active and inactive components of the postshock reaction: Parallels between naloxone pretreatment and shock intensity. *Behav Neurosci* **98**: 269–277.

Fanselow MS. 1986a. Conditioned fear-induced opiate analgesia: A competing motivational state theory of stress analgesia. *Ann NY Acad Sci* **467**: 40–54.

Fanselow MS. 1986b. Associative vs topographical accounts of the immediate shock-freezing deficit in rats: Implications for the response selection rules governing species-specific defensive reactions. *Learn Motiv* **17**: 16–39.

Fanselow MS. 1989. The adaptive function of conditioned defensive behavior: An ecological approach to Pavlovian stimulus substitution theory. In *Ethoexperimental approaches to the study of behavior* (ed. Blanchard RJ, et al.), pp. 151–166. Kluwer, Boston.

Fanselow MS. 1991. The midbrain periaqueductal gray as a coordinator of action in response to fear and anxiety. In *The midbrain periaqueductal gray matter: Functional, anatomical and immunohistochemical organization* (ed. Depaulis A, Bandler R), pp. 151–173. Plenum, New York.

Fanselow MS. 1994. Neural organization of the defensive behavior system responsible for fear. *Psychon Bull Rev* **1**: 429–438.

Fanselow MS. 2010. From contextual fear to a dynamic view of memory systems. *Trends Cogn Sci* **14**: 7–15.

Fanselow MS, Baackes MP. 1982. Conditioned fear-induced opiate analgesia on the formalin test: Evidence for two aversive motivational systems. *Learn Motiv* **13**: 200–221.

Fanselow MS, Bolles RC. 1979. Triggering of the endorphin analgesic reaction by a cue previously associated with shock: Reversal by naloxone. *Bull Psychon Soc* **14**: 88–90.

Fanselow MS, Kim JJ. 1994. Acquisition of contextual Pavlovian fear conditioning is blocked by application of an NMDA receptor antagonist D,L-2-amino-5-phosphonovaleric acid to the basolateral amygdala. *Behav Neurosci* **108**: 210–212.

Fanselow MS, LeDoux JE. 1999. Why we think plasticity underlying Pavlovian fear conditioning occurs in the basolateral amygdala. *Neuron* **23**: 229–232.

Fanselow MS, Lester LS. 1988. A functional behavioristic approach to aversively motivated behavior: Predatory imminence as a determinant of the topography of defensive behavior. In *Evolution and learning* (ed. Bolles RC, Beecher MD), pp. 185–211. Erlbaum, Hillsdale, NJ.

Fanselow MS, Landeira-Fernandez J, DeCola JP, Kim JJ. 1994. The immediate-shock deficit and postshock analgesia: Implications for the relationship between the analgesic CR and UR. *Anim Learn Behav* **22**: 72–76.

Fendt M, Fanselow MS. 1999. The neuroanatomical and neurochemical basis of conditioned fear. *Neurosci Biobehav Rev* **23**: 743–760.

Flaherty CF, Becker HC. 1984. Influence of conditioned stimulus context on hyperglycemic conditioned responses. *Physiol Behav* **33**: 587–593.

Freeze BS, Kravitz AV, Hammack N, Berke JD, Kreitzer AC. 2013. Control of basal ganglia output by direct and indirect pathway projection neurons. *J Neurosci* **33**: 18531–18539.

Fudge JL, Haber SN. 2000. The central nucleus of the amygdala projection to dopamine subpopulations in primates. *Neuroscience* **97**: 479–494.

Gallagher M, Graham PW, Holland PC. 1990. The amygdala central nucleus and appetitive Pavlovian conditioning: Lesions impair one class of conditioned behavior. *J Neurosci* **10**: 1906–1911.

Gerfen CR, Surmeier DJ. 2011. Modulation of striatal projection systems by dopamine. *Annu Rev Neurosci* **34**: 441–466.

Gilboa A, Sekeres M, Moscovitch M, Winocur G. 2014. Higher-order conditioning is impaired by hippocampal lesions. *Curr Biol* **24**: 2202–2207.

Goldberg JH, Farries MA, Fee MS. 2013. Basal ganglia output to the thalamus: Still a paradox. *Trends Neurosci* **36**: 695–705.

Gonzales C, Chesselet MF. 1990. Amygdalonigral pathway: An anterograde study in the rat with *Phaseolus vulgaris* leucoagglutinin (PHA-L). *J Comp Neurol* **297**: 182–200.

Gormezano I, More JW. 1964. Yoked comparisons of contingent and noncontingent US presentations in human eyelid conditioning. *Psychon Sci* **1**: 231–232.

Gottfried JA, O'Doherty J, Dolan RJ. 2003. Encoding predictive reward value in human amygdala and orbitofrontal cortex. *Science* **301**: 1104–1107.

Graham JM, Desjardins C. 1980. Classical conditioning: Induction of luteinizing hormone and testosterone secretion in anticipation of sexual activity. *Science* **210**: 1039–1041.

Grill HJ, Norgren R. 1978. The taste reactivity test. I: Mimetic responses to gustatory stimuli in neurologically normal rats. *Brain Res* **143**: 263–279.

Hall J, Parkinson JA, Connor TM, Dickinson A, Everitt BJ. 2001. Involvement of the central nucleus of the amygdala and nucleus accumbens core in mediating Pavlovian influences on instrumental behaviour. *Eur J Neurosci* **13**: 1984–1992.

Han JH, Kushner SA, Yiu AP, Cole CJ, Matynia A, Brown RA, Neve RL, Guzowski JF, Silva AJ, Josselyn SA. 2007. Neuronal competition and selection during memory formation. *Science* **316**: 457–460.

Haney RZ, Calu DJ, Takahashi YK, Hughes BW, Schoenbaum G. 2010. Inactivation of the central but not the basolateral nucleus of the amygdala disrupts learning in response to overexpectation of reward. *J Neurosci* **30**: 2911–2917.

Hart AS, Rutledge RB, Glimcher PW, Phillips PE. 2014. Phasic dopamine release in the rat nucleus accumbens symmetrically encodes a reward prediction error term. *J Neurosci* **34**: 698–704.

Hatfield T, Han JS, Conley M, Gallagher M, Holland P. 1996. Neurotoxic lesions of basolateral, but not central, amygdala interfere with Pavlovian second-order conditioning and reinforcer devaluation effects. *J Neurosci* **16:** 5256–5265.

Haubensak W, Kunwar PS, Cai H, Ciocchi S, Wall NR, Ponnusamy R, Biag J, Dong HW, Deisseroth K, Callaway EM, et al. 2010. Genetic dissection of an amygdala microcircuit that gates conditioned fear. *Nature* **468:** 270–276.

Helmstetter FJ, Fanselow MS. 1987. Effects of naltrexone on learning and performance of conditional fear-induced freezing and opioid analgesia. *Physiol Behav* **39:** 501–505.

Helmstetter FJ, Parsons RG, Gafford GM. 2008. Macromolecular synthesis, distributed synaptic plasticity, and fear conditioning. *Neurobiol Learn Mem* **89:** 324–337.

Herkenham M, Nauta WJ. 1979. Efferent connections of the habenular nuclei in the rat. *J Comp Neurol* **187:** 19–47.

Herry C, Ciocchi S, Senn V, Demmou L, Müller C, Lüthi A. 2008. Switching on and off fear by distinct neuronal circuits. *Nature* **454:** 600–606.

Hikosaka O. 2010. The habenula: From stress evasion to value-based decision-making. *Nat Rev Neurosci* **11:** 503–513.

Holland PC. 1977. Conditioned stimulus as a determinant of the form of the Pavlovian conditioned response. *J Exp Psychol Anim Behav Process* **3:** 77–104.

Holland PC. 1979a. Differential effects of omission contingencies on various components of Pavlovian appetitive responding in rats. *J Exp Psychol* **17:** 1680–1694.

Holland PC. 1979b. The effects of qualitative and quantitative variation in the US on individual components of Pavlovian appetitive conditioned behavior in rats. *Anim Learn Behav* **7:** 424–432.

Holland PC. 1981. The effects of satiation after first- and second-order appetitive conditioning in rats. *Pavlov J Biol Sci* **16:** 18–24.

Holland PC. 2004. Relations between Pavlovian-instrumental transfer and reinforcer devaluation. *J Exp Psychol Anim Behav Process* **30:** 104–117.

Holland PC. 2008. Cognitive versus stimulus-response theories of learning. *Learn Behav* **36:** 227–241.

Holland PC, Gallagher M. 1993. Effects of amygdala central nucleus lesions on blocking and unblocking. *Behav Neurosci* **107:** 235–245.

Holland PC, Gallagher M. 2003. Double dissociation of the effects of lesions of basolateral and central amygdala on conditioned stimulus-potentiated feeding and Pavlovian-instrumental transfer. *Eur J Neurosci* **17:** 1680–1694.

Holland PC, Rescorla RA. 1975. The effect of two ways of devaluing the unconditioned stimulus after first- and second-order appetitive conditioning. *J Exp Psychol Anim Behav Process* **1:** 355–363.

Holland PC, Straub JJ. 1979. Differential effects of two ways of devaluing the unconditioned stimulus after Pavlovian appetitive conditioning. *J Exp Psychol Anim Behav Process* **5:** 65–78.

Holland PC, Chik Y, Zhang Q. 2001. Inhibitory learning tests of conditioned stimulus associability in rats with lesions of the amygdala central nucleus. *Behav Neurosci* **115:** 1154–1158.

Hollerman JR, Schultz W. 1998. Dopamine neurons report an error in the temporal prediction of reward during learning. *Nat Neurosci* **1:** 304–309.

Hollis KL, Cadieux EL, Colbert MM. 1989. The biological function of Pavlovian conditioning: A mechanism for mating success in the blue gourami (*Trichogaster trichopterus*). *J Comp Psychol* **103:** 115–121.

Hull CL. 1942. Conditioning: Outline of a systematic theory of learning. In *The forty-first yearbook of the National Society for the Study of Education. Part II: The psychology of learning* (ed. Henry NB), pp. 61–95. University of Chicago Press, Chicago.

Hull C. 1943. *Principles of behavior.* Appleton, New York.

Ito R, Everitt BJ, Robbins TW. 2005. The hippocampus and appetitive Pavlovian conditioning: Effects of excitotoxic hippocampal lesions on conditioned locomotor activity and autoshaping. *Hippocampus* **15:** 713–721

Izquierdo A, Murray EA. 2010. Functional interaction of medial mediodorsal thalamic nucleus but not nucleus accumbens with amygdala and orbital prefrontal cortex is essential for adaptive response selection after reinforcer devaluation. *J Neurosci* **30:** 661–669.

Ji H, Shepard PD. 2007. Lateral habenula stimulation inhibits rat midbrain dopamine neurons through a GABA_A receptor-mediated mechanism. *J Neurosci* **27:** 6923–6930.

Johansen JP, Tarpley JW, LeDoux JE, Blair HT. 2010. Neural substrates for expectation-modulated fear learning in the amygdala and periaqueductal gray. *Nat Neurosci* **13:** 979–986.

Kamin LJ. 1969. Predictability, surprise, attention, and conditioning. In *Punishment and aversive behavior* (ed. Campbell BA, Church RM), pp. 279–296. Appleton-Century-Crofts, New York.

Kamin LJ, Gaioni SJ. 1974. Compound conditioned emotional response conditioning with differentially salient elements in rats. *J Comp Physiol Psychol* **87:** 591–597.

Kiernan MJ, Westbrook RF, Cranney J. 1995. Immediate shock, passive avoidance, and potentiated startle: Implications for the unconditioned response to shock. *Anim Learn Behav* **23:** 22–30.

Kim JJ, Krupa DJ, Thompson RF. 1998. Inhibitory cerebello-olivary projections and blocking effect in classical conditioning. *Science* **279:** 570–573.

Koffka K. 1935. *The principles of Gestalt psychology.* Harcourt Brace, New York.

Kohler W. 1940. *Dynamics in psychology.* Liveright, New York.

Konorski J. 1948. *Conditioned reflexes and neuron organization.* Cambridge University Press, New York.

Konorski J. 1967. *Integrative activity of the brain: An interdisciplinary approach.* University of Chicago Press, Chicago.

Konorski J, Miller S. 1937. On two types of conditioned reflex. *J Gen Psychol* **16:** 264–272.

Krasne FB, Fanselow MS, Zelikowsky M. 2011. Design of a neurally plausible model of fear learning. *Front Behav Neurosci* **5:** 41.

Kremer EF. 1978. The Rescorla–Wagner model: Losses in associative strength in compound conditioned stimuli. *J Exp Psychol Anim Behav Process* **4:** 22–36.

Kruse JM, Overmier JB, Konz WA, Rokke E. 1983. Pavlovian conditioned stimulus effects upon instrumental choice behavior are reinforcer specific. *Learn Motiv* **14:** 165–181.

LeDoux JE. 2000. The amygdala and emotion: A view through fear. In *The Amygdala: A functional analysis* (ed. Aggleton JP), pp. 289–310. Oxford University Press, Oxford.

Lee HJ, Gallagher M, Holland PC. 2010. The central amygdala projection to the substantia nigra reflects prediction error information in appetitive conditioning. *Learn Mem* **17:** 531–538.

Lewin K. 1936. *Principles of topological psychology.* McGraw-Hill, New York.

Lockard JS. 1963. Choice of a warning signal or no warning signal in an unavoidable shock situation. *J Comp Physiol Psychol* **56:** 65–80.

Lovibond PF. 1983. Facilitation of instrumental behavior by a Pavlovian appetitive conditioned stimulus. *J Exp Psychol Anim Behav Process* **9:** 225–247.

Lubow RE, Moore AU. 1959. Latent inhibition: The effect of nonreinforced pre-exposure to the conditional stimulus. *J Comp Physiol Psychol* **52:** 415–419.

Mackintosh NJ. 1975. A theory of attention: Variations in the associability of stimuli with reinforcement. *Psychol Rev* **82**.

Mackintosh NJ. 1976. Overshadowing and stimulus intensity. *Anim Learn Behav* **4:** 186–192.

Malvaez M, Greenfield VY, Wang AS, Yorita AM, Feng L, Linker KE, Monbouquette HG, Wassum KM. 2015. Basolateral amygdala rapid glutamate release encodes an outcome-specific representation vital for reward-predictive cues to selectively invigorate reward-seeking actions. *Sci Rep* **5:** 12511.

Maren S, Fanselow MS. 1995. Synaptic plasticity in the basolateral amygdala induced by hippocampal formation stimulation in vivo. *J Neurosci* **15:** 7548–7564.

Maren S, Yap SA, Goosens KA. 2001. The amygdala is essential for the development of neuronal plasticity in the medial geniculate nucleus during auditory fear conditioning in rats. *J Neurosci* **21:** RC135.

Matsumoto M, Hikosaka O. 2007. Lateral habenula as a source of negative reward signals in dopamine neurons. *Nature* **447:** 1111–1115.

Matthews RN, Domjan M, Ramsey M, Crews D. 2007. Learning effects on sperm competition and reproductive fitness. *Psychol Sci* **18:** 758–762.

McDannald MA, Jones JL, Takahashi YK, Schoenbaum G. 2014a. Learning theory: A driving force in understanding orbitofrontal function. *Neurobiol Learn Mem* **108:** 22–27.

McDannald MA, Esber GR, Wegener MA, Wied HM, Liu TL, Stalnaker TA, Jones JL, Trageser J, Schoenbaum G. 2014b. Orbitofrontal neurons acquire responses to "valueless" Pavlovian cues during unblocking. *eLife* **3:** e02653.

McNally GP, Pigg M, Weidemann G. 2004a. Opioid receptors in the midbrain periaqueductal gray regulate extinction of Pavlovian fear conditioning. *J Neurosci* **24:** 6912–6919.

McNally GP, Pigg M, Weidemann G. 2004b. Blocking, unblocking, and overexpectation of fear: A role for opioid receptors in the regulation of Pavlovian association formation. *Behav Neurosci* **118:** 111–120.

Miller NE. 1948. Studies of fear as an acquirable drive fear as motivation and fear-reduction as reinforcement in the learning of new responses. *J Exp Psychol* **38:** 89–101.

Miserendino MJ, Sananes CB, Melia KR, Davis M. 1990. Blocking of acquisition but not expression of conditioned fear-potentiated startle by NMDA antagonists in the amygdala. *Nature* **345:** 716–718.

Mitchell AS, Browning PG, Baxter MG. 2007. Neurotoxic lesions of the medial mediodorsal nucleus of the thalamus disrupt reinforcer devaluation effects in rhesus monkeys. *J Neurosci* **27:** 11289–11295.

Morgan MM, Whitier KL, Hegarty DM, Aicher SA. 2008. Periaqueductal gray neurons project to spinally projecting GABAergic neurons in the rostral ventromedial medulla. *Pain* **140:** 376–386.

Murray EA, O'Doherty JP, Schoenbaum G. 2007. What we know and do not know about the functions of the orbitofrontal cortex after 20 years of cross-species studies. *J Neurosci* **27:** 8166–8169.

Orsini CA, Kim JH, Knapska E, Maren S. 2011. Hippocampal and prefrontal projections to the basal amygdala mediate contextual regulation of fear after extinction. *J Neurosci* **31:** 17269–17277.

Orsini CA, Yan C, Maren S. 2013. Ensemble coding of context-dependent fear memory in the amygdala. *Front Behav Neurosci* **7:** 199.

Ostlund SB, Balleine BW. 2007. Orbitofrontal cortex mediates outcome encoding in Pavlovian but not instrumental conditioning. *J Neurosci* **27:** 4819–4825.

Ostlund SB, Balleine BW. 2008. Differential involvement of the basolateral amygdala and mediodorsal thalamus in instrumental action selection. *J Neurosci* **28:** 4398–4405.

Paré D, Quirk GJ, Ledoux JE. 2004. New vistas on amygdala networks in conditioned fear. *J Neurophysiol* **92:** 1–9.

Parkes SL, Balleine BW. 2013. Incentive memory: Evidence the basolateral amygdala encodes and the insular cortex retrieves outcome values to guide choice between goal-directed actions. *J Neurosci* **33:** 8753–8763.

Parkinson JA, Robbins TW, Everitt BJ. 2000. Dissociable roles of the central and basolateral amygdala in appetitive emotional learning. *Eur J Neurosci* **12:** 405–413.

Parkinson JA, Roberts AC, Everitt BJ, Di Ciano P. 2005. Acquisition of instrumental conditioned reinforcement is resistant to the devaluation of the unconditioned stimulus. *Q J Exp Psychol B* **58:** 19–30.

Parsons RG, Riedner BA, Gafford GM, Helmstetter FJ. 2006. The formation of auditory fear memory requires the synthesis of protein and mRNA in the auditory thalamus. *Neuroscience* **141:** 1163–1170.

Pavlov IP. 1927. *Conditioned Reflexes, an investigation of the psychological activity of the cerebral cortex.* Oxford University Press, New York.

Pavlov I. 1932. The reply of a physiologist to psychologists. *Psychol Rev* **39:** 91–127.

Pearce JM, Hall G. 1980. A model for Pavlovian learning: Variations in the effectiveness of conditioned but not of unconditioned stimuli. *Psychol Rev* **87**.

Peciña S, Berridge KC. 2013. Dopamine or opioid stimulation of nucleus accumbens similarly amplify cue-triggered "wanting" for reward: Entire core and medial shell mapped as substrates for PIT enhancement. *Eur J Neurosci* 37: 1529–1540.

Pickens CL. 2008. A limited role for mediodorsal thalamus in devaluation tasks. *Behav Neurosci* 122: 659–676.

Pickens CL, Holland PC. 2004. Conditioning and cognition. *Neurosci Biobehav Rev* 28: 651–661.

Pickens CL, Saddoris MP, Setlow B, Gallagher M, Holland PC, Schoenbaum G. 2003. Different roles for orbitofrontal cortex and basolateral amygdala in a reinforcer devaluation task. *J Neurosci* 23: 11078–11084.

Rescorla RA. 1968. Probability of shock in the presence and absence of CS in fear conditioning. *J Comp Physiol Psychol* 66: 1–5.

Rescorla RA. 1971. Summation and retardation tests of latent inhibition. *J Comp Physiol Psychol* 75: 77–81.

Rescorla R. 1973a. *Mechanisms of formation and inhibition of conditioned reflex*. Academy of Sciences of the U.S.S.R., Moscow.

Rescorla RA. 1973b. Effect of US habituation following conditioning. *J Comp Physiol Psychol* 82: 137–143.

Rescorla RA. 1974. Effect of inflation of the unconditioned stimulus value following conditioning. *J Comp Physiol Psychol* 86: 101–106.

Rescorla RA. 1980. Simultaneous and successive associations in sensory preconditioning. *J Exp Psychol Anim Behav Process* 6: 207–216.

Rescorla RA. 1988. Pavlovian conditioning. It's not what you think it is. *Am Psychol* 43: 151.

Rescorla RA. 1994. Transfer of instrumental control mediated by a devalued outcome. *Anim Learn Behav* 22: 27–33.

Rescorla RA, Solomon RL. 1967. Two-process learning theory: Relationships between Pavlovian conditioning and instrumental learning. *Psychol Rev* 74: 151–182.

Rescorla RA, Wagner AR. 1972. A theory of Pavlovian conditioning: Variations in the effectiveness of reinforcement and non-reinforcement. In *Classical conditioning II* (ed. Black AH, Prokasy WF), pp. 64–99. Appleton-Century Crofts, New York.

Rizley RC, Rescorla RA. 1972. Associations in second-order conditioning and sensory preconditioning. *J Comp Physiol Psychol* 81: 1–11.

Robinson TE, Berridge KC. 1993. The neural basis of drug craving: An incentive-sensitization theory of addiction. *Brain Res Brain Res Rev* 18: 247–291.

Robinson MJ, Berridge KC. 2013. Instant transformation of learned repulsion into motivational "wanting." *Curr Biol* 23: 282–289.

Roitman MF, Wheeler RA, Wightman RM, Carelli RM. 2008. Real-time chemical responses in the nucleus accumbens differentiate rewarding and aversive stimuli. *Nat Neurosci* 11: 1376–1377.

Romanski LM, Clugnet MC, Bordi F, LeDoux JE. 1993. Somatosensory and auditory convergence in the lateral nucleus of the amygdala. *Behav Neurosci* 107: 444–450.

Rouillard C, Freeman AS. 1995. Effects of electrical stimulation of the central nucleus of the amygdala on the in

vivo electrophysiological activity of rat nigral dopaminergic neurons. *Synapse* 21: 348–356.

Rozeboom WW. 1958. What is learned? An empirical enigma. *Psychol Rev* 65: 22–33.

Saddoris MP, Gallagher M, Schoenbaum G. 2005. Rapid associative encoding in basolateral amygdala depends on connections with orbitofrontal cortex. *Neuron* 46: 321–331.

Saddoris MP, Holland PC, Gallagher M. 2009. Associatively learned representations of taste outcomes activate taste-encoding neural ensembles in gustatory cortex. *J Neurosci* 29: 15386–15396.

Schoenbaum G, Setlow B, Saddoris MP, Gallagher M. 2003. Encoding predicted outcome and acquired value in orbitofrontal cortex during cue sampling depends upon input from basolateral amygdala. *Neuron* 39: 855–867.

Schultz W. 2002. Getting formal with dopamine and reward. *Neuron* 36: 241–263.

Siegel S. 1975a. Evidence from rats that morphine tolerance is a learned response. *J Comp Physiol Psychol* 89: 498–506.

Siegel S. 1975b. Conditioning insulin effects. *J Comp Physiol Psychol* 89: 89–99.

Siegel S. 1991. Tolerance: Role of conditioning processes. *NIDA Res Monogr* 1991: 213–229.

Sigmundi RA, Bouton ME, Bolles RC. 1980. Conditioned freezing in the rat as a function of shock intensity and CS modality. *Bull Psychon Soc* 15: 254–256.

Singh T, McDannald MA, Haney RZ, Cerri DH, Schoenbaum G. 2010. Nucleus accumbens core and shell are necessary for reinforcer devaluation effects on Pavlovian conditioned responding. *Front Integr Neurosci* 4: 7126.

Skinner BF. 1938. *The behavior of organisms: An experimental analysis*. Appleton-Century, Oxford.

Skinner BF. 1950. Are theories of learning necessary? *Psychol Rev* 57: 193–216.

Small DM, Veldhuizen MG, Felsted J, Mak YE, McGlone F. 2008. Separable substrates for anticipatory and consummatory food chemosensation. *Neuron* 57: 786–797.

Smith KS, Tindell AJ, Aldridge JW, Berridge KC. 2009. Ventral pallidum roles in reward and motivation. *Behav Brain Res* 196: 155–167.

Spence KW. 1956. *Behavior theory and conditioning*. Yale University Press, New Haven.

Steinberg EE, Keiflin R, Boivin JR, Witten IB, Deisseroth K, Janak PH. 2013. A causal link between prediction errors, dopamine neurons and learning. *Nat Neurosci* 16: 966–973.

Surmeier DJ, Ding J, Day M, Wang Z, Shen W. 2007. D1 and D2 dopamine-receptor modulation of striatal glutamatergic signaling in striatal medium spiny neurons. *Trends Neurosci* 30: 228–235.

Sutherland NS, Mackintosh NJ. 1971. *Mechanisms of animal discrimination learning*. Academic, London.

Sutton RS. 1988. Learning to predict by methods of temporal differences. *Machine Learning* 3: 9–44.

Sutton R, Barto A. 1990. Time-derivative models of Pavlovian reinforcement. In *Learning and computational neuroscience: Foundations of adaptive networks* (ed. Gabriel M, Moore J), pp. 497–537. MIT Press, Boston.

Swanson LW, Petrovich GD. 1998. What is the amygdala? *Trends Neurosci* **21**: 323–331.

Talk A, Kashef A, Gabriel M. 2004. Effects of conditioning during amygdalar inactivation on training-induced neuronal plasticity in the medial geniculate nucleus and cingulate cortex in rabbits (*Oryctolagus cuniculus*). *Behav Neurosci* **118**: 944–955.

Thompson RF. 1988. The neural basis of basic associative learning of discrete behavioral responses. *Trends Neurosci* **11**: 152–155.

Thorndike EL. 1898. Animal intelligence: An experimental study of the associative processes in animals. *Psychol Rev* **2**(Suppl): i–109.

Timberlake W. 1994. Behavior systems, associationism, and Pavlovian conditioning. *Psychon Bull Rev* **1**: 405–420.

Timberlake W, Fanselow MS. 1994. Symposium on behavior systems: Learning, neurophysiology, and development. *Psychon Bull Rev* **1**: 403–404.

Tobler PN, Dickinson A, Schultz W. 2003. Coding of predicted reward omission by dopamine neurons in a conditioned inhibition paradigm. *J Neurosci* **23**: 10402–10410.

Tolman EC. 1949. There is more than one kind of learning. *Psychol Rev* **56**: 144–155.

Tolman EC. 1951. *Purposive behavior in animals and men*. University of California Press, Berkeley, CA.

Ungless MA, Magill PJ, Bolam JP. 2004. Uniform inhibition of dopamine neurons in the ventral tegmental area by aversive stimuli. *Science* **303**: 2040–2042.

Vandercar DH, Schneiderman N. 1967. Interstimulus interval functions in different response systems during classical conditioning. *Psychon Sci* **9**: 9–10.

van Vulpen EH, Verwer RW. 1989. Organization of projections from the mediodorsal nucleus of the thalamus to the basolateral complex of the amygdala in the rat. *Brain Res* **500**: 389–394.

Veldhuizen MG, Bender G, Constable RT, Small DM. 2007. Trying to detect taste in a tasteless solution: Modulation of early gustatory cortex by attention to taste. *Chem Senses* **32**: 569–581.

Waddell J, Morris RW, Bouton ME. 2006. Effects of bed nucleus of the stria terminalis lesions on conditioned anxiety: Aversive conditioning with long-duration conditional stimuli and reinstatement of extinguished fear. *Behav Neurosci* **120**: 324–336.

Waelti P, Dickinson A, Schultz W. 2001. Dopamine responses comply with basic assumptions of formal learning theory. *Nature* **412**: 43–48.

Wagner AR, Brandon SE. 1989. Evolution of a structured connectionist model of Pavlovian conditioning (AESOP). In *Contemporary learning theories: Pavlovian conditioning and the status of traditional learning theory* (ed. Klein SB, Mowrer RR), pp. 149–189. Erlbaum, Hillsdale, NJ.

Wagner AR, Brandon SE. 2001. A componential theory of Pavlovian conditioning. In *Handbook of contemporary learning theories* (ed. Mower RR, Klein SB), pp. 23–64. Erlbaum, Mahwah, NJ.

Wagner AR, Logan FA, Haberlandt K, Price T. 1968. Stimulus selection in animal discrimination learning. *J Exp Psychol* **76**: 171–180.

Wassum KM, Cely IC, Balleine BW, Maidment NT. 2009. Distinct opioid circuits determine the palatability and the desirability of rewarding events. *Proc Natl Acad Sci* **106**: 12512–12517.

Wassum KM, Ostlund SB, Balleine BW, Maidment NT. 2011a. Differential dependence of Pavlovian incentive motivation and instrumental incentive learning processes on dopamine signaling. *Learn Mem* **18**: 475–483.

Wassum KM, Cely IC, Balleine BW, Maiment NT. 2011b. μ Opioid receptor activation in the basolateral amygdala mediates the learning of increases but not decreases in the incentive value of a food reward. *J Neurosci* **31**: 1583–1599.

Wassum KM, Ostlund SB, Maidment NT. 2012. Phasic mesolimbic dopamine signaling precedes and predicts performance of a self-initiated action sequence task. *Biol Psychiatry* **71**: 846–854.

Wassum KM, Ostlund SB, Loewinger GC, Maidment NT. 2013. Phasic mesolimbic dopamine release tracks reward seeking during expression of Pavlovian-to-instrumental transfer. *Biol Psychiatry* **73**: 747–755.

Watson JB. 1916. The place of the conditioned-reflex in psychology. *Psychol Rev* **23**: 89.

Watson JB, Rayner R. 1920. Conditioned emotional reactions. *Am Psychol* **15**: 313–317.

Williams BA. 1994. Conditioned reinforcement: Experimental and theoretical issues. *Behav Anal* **17**: 261–285.

Woods SC, Makous W, Hutton RA. 1969. Temporal parameters of conditioned hypoglycemia. *J Comp Physiol Psychol* **69**: 301–307.

Wyvell CL, Berridge KC. 2000. Intra-accumbens amphetamine increases the conditioned incentive salience of sucrose reward: Enhancement of reward "wanting" without enhanced "liking" or response reinforcement. *J Neurosci* **20**: 8122–8130.

Young SL, Fanselow MS. 1992. Associative regulation of Pavlovian fear conditioning: Unconditional stimulus intensity, incentive shifts, and latent inhibition. *J Exp Psychol Anim Behav Process* **18**: 400–413.

Zamble E, Hadad GM, Mitchell JB, Cutmore TR. 1985. Pavlovian conditioning of sexual arousal: First- and second-order effects. *J Exp Psychol Anim Behav Process* **11**: 598–610.

Zelikowsky M, Fanselow MS. 2010. Opioid regulation of Pavlovian overshadowing in fear conditioning. *Behav Neurosci* **124**: 510–519.

Zener K, McCurdy HG. 1939. Analysis of motivational factors in conditioned behavior. I: The differential effect of changes in hunger upon conditioned, unconditioned, and spontaneous salivary secretion. *J Psychol* **8**: 321–350.

Zentall TR, Hogan DE. 1975. Key pecking in pigeons produced by pairing keylight with inaccessible grain. *J Exp Anal Behav* **23**: 199–206.

# Molecular Genetic Strategies in the Study of Corticohippocampal Circuits

Christopher C. Angelakos and Ted Abel

Department of Biology, University of Pennsylvania, Philadelphia, Pennsylvania 19104-6018

*Correspondence:* abele@sas.upenn.edu

The first reproductively viable genetically modified mice were created in 1982 by Richard Palmiter and Ralph Brinster (Palmiter RD, Brinster RL, Hammer RE, Trumbauer ME, Rosenfeld MG, Birnberg NC, Evans RM. 1982. Dramatic growth of mice that develop from eggs microinjected with metallothionein-growth hormone fusion genes. *Nature* 300: 611–615). In the subsequent 30 plus years, numerous groundbreaking technical advancements in genetic manipulation have paved the way for improved spatially and temporally targeted research. Molecular genetic studies have been especially useful for probing the molecules and circuits underlying how organisms learn and remember—one of the most interesting and intensively investigated questions in neuroscience research. Here, we discuss selected genetic tools, focusing on corticohippocampal circuits and their implications for understanding learning and memory.

Genetic approaches can be subdivided into two main categories—forward genetic and reverse genetic dissection. Forward genetic analysis involves observing a phenotype, and then identifying the responsible gene(s). This approach has been especially successful in *Drosophila* research, leading to the discovery of genes important for memory encoding, such as *dunce* and *rutabaga* (Dudai et al. 1976; Livingstone et al. 1984). However, forward genetic dissection has been difficult to implement in rodent research owing to the complexity of rodent behaviors (Takahashi et al. 2008).

In contrast, reverse genetic analysis—manipulating gene(s) and observing phenotype—has been immensely successful for studying learning and memory in mice, the closest phylogenetic relative to humans on which pow-

erful genetic manipulation can be conferred. A series of studies by Martin Evans, Mario Capecchi, and Oliver Smithies in the 1980s led to the development of the first "knockout" mouse in 1989 (Evans and Kaufman 1981; Bradley et al. 1984; Smithies et al. 1985; Doetschman et al. 1987; Thomas and Capecchi 1987; Mansour et al. 1988). Knockout mice have a gene deleted by injecting a nonfunctional homolog of the gene into mouse embryonic stem cells. The inactivated gene then replaces the endogenous gene through homologous recombination. These stem cells are then implanted into the embryo of another mouse strain, and the resulting offspring, lacking the targeted gene, are selected and bred. For example, knocking out α-calcium-calmodulin-dependent protein kinase II (α-CaMKII), a protein preferentially expressed in excitatory forebrain neurons, results in deficits in spatial memory and long-term potentiation (LTP), the presumptive cellular correlate of learning (Silva et al. 1992a,b). Knockout technologies have also been used to show the importance of posttranslational DNA modifications for learning and memory. For example, knocking out *Gadd45b*, a gene involved in activity-induced DNA methylation, results in abnormal dendritic development of newborn neurons in the adult brain, deficient long-term contextual fear memory, and the reduction of genes necessary for adult neurogenesis (Ma et al. 2009; Leach et al. 2012). Moreover, knockout of the cyclic adenosine monophosphate (cAMP) response element-binding (CREB) protein (CBP), which can act as a histone acetyltransferase, impairs long-term fear and object-recognition memory and well as LTP (Alarcón et al. 2004). Conversely, knocking out histone deacetylase 2 (HDAC2), an enzyme that removes posttranslational amino-terminal acetyl residues on histones, increases fear memory and LTP (Guan et al. 2009).

In contrast to knockout mice, transgenic (random integration) and knockin (targeted insertion) mice have a gene added into their genome rather than deleted. Knockin technology may be used to render an endogenous gene inactive by introducing a nonfunctional variant of the gene, such as one containing a point mutation in the functional domain. In contrast to knockouts, the knockin approach spares other domains of the protein, such as regulatory regions, and may assuage some of the difficulties in interpreting complex phenotypes resulting from knocking out a gene of interest. For example, knockin techniques revealed that the functional domain of the transcriptional coactivator CBP is necessary for long-term memory storage (Wood et al. 2005, 2006).

Although knockout and knockin techniques are reproducible and may serve as a good model of human genetic disorders, they suffer from many disadvantages. Namely, the earliest versions of these methodologies had no spatial or temporal sensitivity. A gene may have differential functions between development and adulthood, and deleting a gene may result in nonphysiological compensatory mechanisms during development. The NR1 subunit of the *N*-methyl-D-aspartate (NMDA) receptor is important for spatial memory (Tsien et al. 1996b). However, these findings would not have been possible without spatially restricted genetic deletion, as NR1 knockout animals die neonatally (Forrest et al. 1994). Additionally, mice having a dominant-negative form of CREB show dwarfism and hypoplasia in the anterior pituitary (Struthers et al. 1991), whereas targeted deletion of the α and δ isoforms of CREB show deficits in long-term, but not short-term memory (Bourtchuladze et al. 1994). The differences in these phenotypes can be explained by strong up-regulation and compensation by β CREB, which is normally found in very low levels and is likely to have a relatively minor role in wild-type (WT) animals (Blendy et al. 1996; Gass et al. 1998).

These findings clearly show the limitations of conventional knockout, knockin, and transgenic genetic manipulation. To circumvent these issues, more spatially and temporally restricted genetic technologies had to be developed. Here, we will discuss some of the major recent technical advancements in the field of molecular genetics, including the discovery of spatially restricted promoters, as well as the advent of viral, pharmacological, and optical manipulations. In the context of these approaches, we will highlight evidence for the importance of corticohippocampal circuitry for various forms of learning and memory.

## SPATIALLY RESTRICTED PROMOTERS

The targeting of proteins expressed selectively in certain cell types or brain regions has allowed for more spatially restricted genetic manipulation. By driving expression of a transgene using one (or more) of these promoters, one can probe the functional significance of specific brain regions or even cell types without harsh, nonselective lesions or pharmacological inactivation. Numerous brain region–specific promoters exist, including a minimal Msx1 promoter for craniofacial tissues (Takahashi et al. 1997), Drd1a for striatonigral neurons and Drd2 for striato-

Cite this article as *Cold Spring Harb Perspect Biol* doi: 10.1101/cshperspect.a021725

pallidal neurons (Gerfen et al. 1990), and Etv1 for layer 5 of the neocortex, the main output layer of the cortex (Yoneshima et al. 2006). Examples of cell-type-specific promoters include the neuron-specific endolase (NSE) (Forss-Petter et al. 1990), the astrocyte-specific glial fibrillary acidic protein (GFAP) (Brenner et al. 1994), and the Purkinje cell and retinal bipolar neuron-specific L7/Pcp2 (Oberdick et al. 1990).

One especially useful brain region–specific promoter for the field of learning and memory is the α-CaMKII promoter, which is preferentially expressed in adult forebrain neurons, including throughout the hippocampus (Mayford et al. 1996a). α-CaMKII protein is not only spatially restricted to brain regions known to be important for learning and memory, but it is also temporally restricted in that its expression levels are 20- to 60-fold higher at 3–4 wk postnatal than during development (Sugiura and Yamauchi 1992). This temporal window allows researchers to probe learning and memory, specifically during time points when synaptic connections underlying higher order cognitive functions are strengthening, while also avoiding many of the aforementioned developmental issues of genetic manipulation. Under control of the α-CaMKII promoter, researchers have manipulated and investigated molecules important for learning and memory, such as the regulatory subunit of protein kinase A (PKA) (Abel et al. 1997), brain-derived neurotrophic factor (BDNF) (Cunha et al. 2009), and cyclin-dependent kinase 5 (Cdk5) (Su et al. 2013). Additionally, researchers have used the α-CaMKII promoter in concert with other genetic tools, such as Cre recombinase (Tsien et al. 1996a), the tetracycline transactivator system (tTA) (Mayford et al. 1996b), and pharmacogenetic manipulation (Cao et al. 2008), to probe memory function. These techniques will be discussed in detail below.

## DOMINANT-NEGATIVE INHIBITION

First suggested as a research tool by Hershkowitz in 1987, dominant-negative mutations involve using a mutated, nonfunctional variant of a protein of interest to probe the protein's func-tion (Hershkowitz 1987). Because the mutation is "dominant," this manipulation overrides the endogenous version of the protein. Examples of dominant-negative mutation strategies include mutating the catalytic subunit of an enzyme while leaving the localization sequence intact, or mutating a protein that must combine into a multimer to function. Dominant-negative inhibition is especially useful for situations in which different molecules can activate the same signaling pathway, or when different isoforms of a protein exist. Where deleting a gene using conventional knockout strategies might allow for compensatory mechanisms, dominant-negative inhibition may still render the protein or pathway inactive.

In 1997, Abel and colleagues in the Kandel laboratory used dominant-negative techniques to generate a catalytically inactive form of the regulatory subunit of PKA, known as R(AB) (Abel et al. 1997). PKA is cAMP dependent, but has two cAMP-binding sites. Previous studies knocking out each of the individual cAMP subunits (RIβ and Cβ$_1$) failed to alter the activity of PKA, likely because of compensation by the other cAMP subunit (Brandon et al. 1995; Qi et al. 1996). However, the dominant-negative mutation was able to render both subunits inactive simultaneously. R(AB) mice showed deficient long-term contextual fear memory, decreased late phase LTP (L-LTP), and impaired spatial memory on a Morris water maze probe trial task. However, short-term contextual fear memory and early phase LTP (E-LTP) were normal in these animals (Abel et al. 1997). Collectively, these experiments revealed the necessity of PKA for consolidation of short-term memory into protein synthesis–dependent long-term memory.

More recently, dominant-negative transgenic mice driven by the α-CaMKII promoter have been used to show the importance of proteins involved in transcription for the formation of long-term memories. The transcriptional coactivators CBP and p300, for example, interact with histone acetyltransferases (HATs) and numerous transcription factors known to be important for long-term memory formation (Vo and Goodman 2001). Mice expressing dominant-negative CBP in $\alpha$-CaMKII$^+$ cells show impairments in a long-term visual-paired comparison task, the Morris water maze, and late-phase LTP (Korzus et al. 2004; Wood et al. 2005). Meanwhile, mice expressing dominant-negative p300 show deficits in long-term contextual fear and object-recognition memory (Oliveira et al. 2007). Additionally, mice expressing a dominant-negative $Ca^{2+}$/calmodulin selectively in excitatory forebrain nuclei have decreased gene expression and deficient long-term contextual fear memory, novel object recognition, and Morris water maze performance (Limbäck-Stokin et al. 2004).

## Cre/loxP SYSTEM

The Cre/loxP system is used, naturally, by the P1 bacteriophage to circularize DNA and aid in the replication process. In 1988, this system was engineered into mammalian cells by Sauer and Henderson (1988), allowing for cell-type or site-specific deletion of a single mammalian gene. LoxP sites can be inserted into embryonic stem cells via homologous recombination such that two loxP sites "flank" one or more exons encoding a targeted gene. Crossing a flanked mouse line with a mouse line expressing Cre recombinase under control by a cell-type or site-specific promoter deletes the targeted gene wherever the promoter being used to drive Cre recombinase is expressed. Because of disparate sites of Cre transgene integration into the genome, different Cre mouse lines created under control of the same promoter will have slightly different Cre expression patterns. This characteristic of Cre recombination has allowed the Tonegawa laboratory to create mouse lines with specific deletions in the CA1, CA3, or dentate gyrus (DG) subregions of the hippocampus proper (Fig. 1). For example, driving Cre recombinase expression with the $\alpha$-CaMKII promoter produced several founder lines of transgenic mice with varied patterns of Cre expression. Surprisingly, one founder line showed expression of Cre recombinase seemingly selective for area CA1 of the hippocampus, allowing for hippocampus subregion–specific genetic manipulation for the first time (Tsien et al. 1996a).

Because NMDA receptors (NMDARs) were known to be crucial for LTP in the hippocampus (Bliss and Collingridge 1993), the Tonegawa laboratory decided to use this mouse line to selectively delete the NMDA receptor 1 gene, which encodes the NR1 subunit of NMDARs, from CA1 pyramidal cells. These mice did not show LTP in CA1 and showed impaired spatial memory, whereas nonspatial forms of memory remained intact (Fig. 1C–E) (Tsien et al. 1996b). CA1 pyramidal cells, called "place cells," are known to preferentially fire when an animal is in a certain location in space (O'Keefe and Dostrovsky 1971; see Moser et al. 2015), leading to the theory that these cells encode spatial information by making up an internalized geographical representation of an animal's environment. Using the same CA1 NMDAR1 deleted mouse line, Wilson's laboratory showed that the spatial specificity of the firing patterns of individual CA1 place cells was significantly decreased, further corroborating the importance of region CA1 for representing spatial memory (McHugh et al. 1996). However, more recent studies in a mouse line using a genetic silencer to conditionally delete NMDARs in adult CA1 neurons reveal normal acquisition and storage of spatial memory (Bannerman et al. 2012). This and other evidence suggests that the spatial memory deficits seen using the purported CA1-specific Cre founder line may in part be caused by spread of Cre expression (and, thus, NMDAR deletion) to overlying cortical neurons (see Hoeffer et al. 2008; Wiltgen et al. 2010).

Using the same principles of differential Cre expression among founder lines, the Tonegawa laboratory then drove Cre recombinase with the kainite KA1 promoter, which is preferentially expressed in hippocampus regions CA3 and DG (Wisden and Seeburg 1993). One founder line possessed CA3-specific Cre recombination. Deleting the NR1 subunit of the NMDAR in CA3 produced mice that could acquire and retrieve spatial memories as well as WT animals, but suffered from significant deficits in retrieving these memories if presented with only a subset of the original visual cues (Fig. 1G,H). Moreover, these mice had normal CA1 place fields under full-cue task conditions, but de-

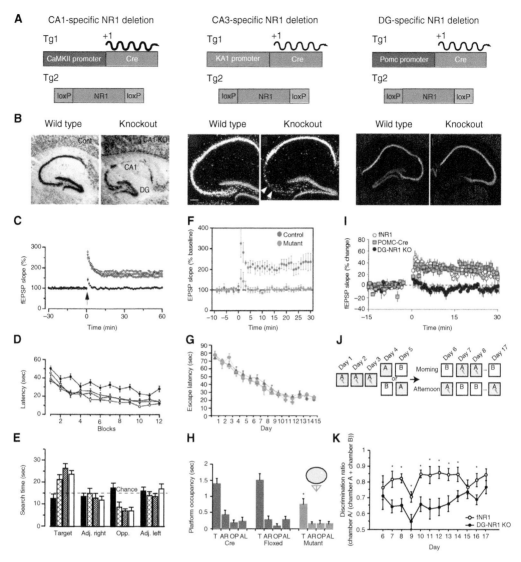

**Figure 1.** The deletion of *N*-methyl-D-aspartate (NMDA) receptor NR1 subunit in specific subregions of the hippocampus. (*A*) Loss of NMDA receptor function in specific hippocampal areas can be achieved by using specific promoters to drive Cre recombinase. (*B*) In situ hybridization of *NMDAR1* mRNA from wild types (WTs) (*left* panels) and subregion-specific NR1 knockout (KO) mice (*right* panels). (*C*) CA1-restricted deletion of the NR1 subunit impairs long-term potentiation (LTP) in the Schaffer collateral pathway (open symbols, WTs; closed circles, mutant). (*D*) CA1-specific knockout mice are impaired during training in the spatial version of the Morris water maze (open symbols, WTs; closed circles, mutant), and (*E*) during the probe test (black bars indicate mutants). (*F*) CA3-specific deletion of the NR1 subunit blocks LTP at the recurrent commissural/associational synapses. (*G*) CA3-specific knockout does not affect training in the spatial version of the Morris water maze, (*H*) but results in a reduced preference for the target quadrant during a probe trial with partial cues, indicating an impairment in pattern completion. (*I*) Perforant path LTP is lost as a consequence of ablation of the NR1 subunit in the dentate gyrus. (*J*) A design to test context discrimination in WT and dentate gyrus-specific NMDA receptor knockout mice. For the first 3 days of conditioning, mice visited only chamber A and each day received a single foot shock. Freezing was measured once in chamber A and once in chamber B over the subsequent 2 days. During days 6–17, mice visited each chamber daily (receiving a shock in one of the two), and freezing was assessed during the first 3 min in each chamber. (*K*) Dentate gyrus (DG)-restricted ablation of NMDA receptors results in impaired pattern separation indicated by a reduced discrimination ratio between a shocked context (context A) and a nonshocked context (context B). adj., Adjacent; opp., opposite. (From Havekes and Abel 2009; reprinted, with permission, from Elsevier © 2009; original sources for data are Tsien et al. 1996b; Nakazawa et al. 2002; McHugh et al. 2007.)

creased place cell specificity under partial-cue conditions (Nakazawa et al. 2002). Thus, in agreement with earlier behavioral work (Moser et al. 2015), region CA3 of the hippocampus seems to play an important role in pattern completion—that is, the ability to retrieve a complete memory from an incomplete set of cues. Subsequent experiments showed that these mice also have impaired one-trial memory on a delayed matching-to-place (DMP) version of the water maze (Steele and Morris 1999), in which the location of the escape platform is altered each day (Nakazawa et al. 2003).

Analogous methods were used to obtain DG-specific Cre recombination. The proopiomelanocortin (PomC) promoter is expressed in the dentate gyrus granule cell layer, arcuate nucleus of the hypothalamus, and the nucleus of the solitary tract (Balthasar et al. 2004). McHugh and colleagues (McHugh et al. 2007) used the PomC promoter in conjunction with the Cre/loxP system to generate a transgenic line with DG-specific NMDAR receptor NR1 deletion. These mice performed normally on a contextual fear-conditioning task, but were impaired at differentiating between two similar contexts (Fig. 1J,K). These findings, along with earlier behavioral work, indicate that the DG is important for pattern separation—the ability to discriminate between similar memories in time and space.

Apart from these examples, Cre/loxP deletion in adult forebrain excitatory neurons using the α-CaMKII promoter has revealed the importance of numerous molecules for learning and memory. For example, BDNF and the neural cell–adhesion molecule (NCAM) have both been shown to be important for LTP in region CA1 of the hippocampus (Xu et al. 2000; Bukalo et al. 2004). Using this methodology, the Abel laboratory deleted the transcriptional coactivator p300 from the adult forebrain, establishing its importance for long-term novel object recognition and long-term contextual fear memory (Oliveira et al. 2011).

The Cre/loxP system has allowed for precise cell-type or brain region–specific genetic manipulation. Despite this advantage, Cre recombination does not allow for temporal ma-

nipulation apart from using promoters that are expressed selectively at different times of development. Our memories have both spatial and temporal components. Without being able to genetically assess the temporal dynamics of memory, we cannot efficiently proceed to address some of the fundamental questions of the molecules and circuitry underlying learning and memory. Thus, development of a technology allowing for inducible genetic manipulation was needed to probe these questions. The tTA allows researchers to do just that.

## TETRACYCLINE TRANSACTIVATOR SYSTEM

The Tn10-specified tetracycline-resistance operon of Escherichia coli was genetically modified in 1992 by Gossen and Bujard (1992) to temporally control gene expression in the HeLa cell line. In E. coli, the tetracycline repressor protein (tetR) binds to its operator, which suppresses the expression of resistance genes. In the presence of the antibiotic tetracycline, tetR dissociates from its operator, allowing for the expression of resistance proteins and antibiotic resistance. Gossen and Bujard fused the tetR protein with the activation domain of the herpes simplex virus, creating a version of tetR called tTA that was functional in eukaryotic cells. When tTA binds to the tetracycline operator (tetO), the transcription of a transgene of interest takes place. In the presence of tetracycline or doxycycline (Dox), which can be administered through a mouse's food or water, the tetracycline system is suppressed, allowing for temporal control of transgene expression.

In 1996, Mark Mayford in the Kandel laboratory used the tTA system to reversibly express a $Ca^{2+}$-independent, constitutively active CaMKII in a forebrain-specific manner (Mayford et al. 1996b). This was achieved by crossing mice expressing tTA under control of the α-CaMKII promoter with mice expressing a dominant active α-CaMKII (CaMKII-Asp$^{286}$) driven by the tetO promoter (Fig. 2). The bitransgenic offspring showed deficits in spatial memory, and LTP in the θ frequency range (10 Hz) was eliminated in CA1 of the hippocampus. Interesting-

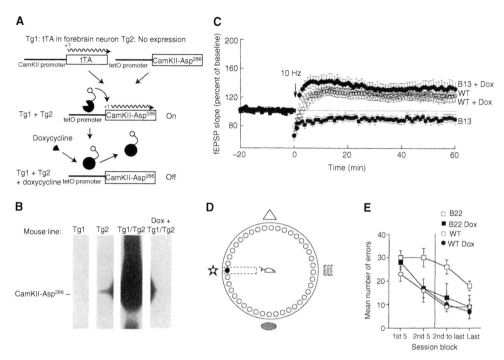

**Figure 2.** Regulation of the α-calcium-calmodulin-dependent protein kinase II (CaMKII) transgene with the tetracycline transactivator (tTA) system reversibly alters long-term potentiation (LTP) and spatial memory. (*A*) Strategy used to obtain forebrain-specific doxycycline (Dox)-regulated transgene expression. Two independent lines of transgenic mice are obtained, and the two transgenes are introduced into a single mouse through mating. (*B*) Quantification by RT-PCR Southern blot of CaMKII-Asp$^{286}$ expression from the tetO promoter (Tg1, mouse carrying only the CaMKII promoter-tTA transgene; Tg2, mouse carrying only the tetO-CaMKII-Asp$^{286}$ transgene; Tg1/Tg2, double transgenic mouse carrying both the CaMKII promoter-tTA transgene and tetO-CaMKII-Asp$^{286}$ transgenes; Tg1/Tg2 + Dox, double transgenic mouse treated with doxycycline [2 mg/ml] plus 5% sucrose in the drinking water for 4 wk). (*C*) Double transgenic mice (B13) fail to potentiate following stimulation at 10 Hz for 1.5 min. Doxycycline treatment reversed the defect in B13 mice. (*D*) The Barnes circular maze. In this spatial memory task, mice are required to use distal cues to find the location of the escape hole. (*E*) Mean number of errors across session blocks composed of five sessions. Double transgenic mice (B22) show impaired performance on the Barnes maze. This impairment is reversed after doxycycline treatment. WT, Wild type. (From Mayford et al. 1996b; reprinted, with permission, from The American Association for the Advancement of Science © 1996.)

ly, when the transgenic CaMKII-Asp$^{286}$ was suppressed by administering Dox through the drinking water, the memory and LTP impairments were restored to normal (Fig. 2C–E) (Mayford et al. 1996b). This seminal study showed unequivocally that these memory deficits were exclusively caused by adult neuronal activity, separate of any developmental effects.

One advantage of the tTA system over previously available genetic strategies is its ability to dissociate different stages of memory temporally for the first time. For example, Isiegas and colleagues (Isiegas et al. 2006) used the tetracy-cline system to temporally control dominant-negative R(AB) expression in adult forebrain neurons. By keeping mice on Dox during development and contextual fear conditioning, Isiegas et al. were then able to elucidate the role of PKA selectively in fear extinction by removing Dox before initiating a fear extinction paradigm. They found that inhibition of PKA facilitates fear extinction, suggesting that PKA may help protect against memory erasure.

The temporal-resolution advantages of the tetracycline system are best revealed when combined with other spatially specific genetic manipulations, such as the aforementioned hippocampus subregion–specific knockouts. By combining the tTA system with the predominantly CA1 Cre/loxP deletion of the NR1 subunit of the NMDA receptor, Shimizu et al. (Shimizu et al. 2000) assessed the role of the NMDAR on long-term memory consolidation. Here, a Dox-on system was used such that administration of Dox knocked out NMDARs in CA1, whereas mice off Dox expressed normal NMDAR function. Mice off Dox learned a 7-day Morris water maze hidden-platform task as well as control animals. However, when Dox was administered after training (during consolidation) to knock out CA1-NMDARs, the mutant mice showed memory deficits on a retrieval probe trial, as assessed by increased latency to locate the hidden platform. Analogous results were found when removing NMDARs in CA1 during consolidation of long-term contextual fear memory. These results showed that NMDARs in CA1 are important for the consolidation of short-term hippocampus-dependent memories into long-term memories.

Recently, the Tonegawa laboratory used the tTA system in combination with the Cre/loxP system to temporally restrict synaptic transmission selectively in region CA3 of the hippocampus. The tetanus toxin (TeTX) light chain specifically cleaves synaptobrevin 2, which is part of the vesicle-associated membrane protein (VAMP) family necessary for vesicle release and synaptic transmission. By driving TeTX specifically to CA3 of the hippocampus using the kainate KA1 promoter, the Tonegawa laboratory was able to block synaptic transmission from CA3 to CA1. Controlling this manipulation with the tTA system, they found that CA3-TeTX mice had memory impairments in one-trial novel contextual fear conditioning and deficient spatial tuning of CA1 place cells. Incremental spatial memory acquisition and recall in the Morris water maze was normal. Additionally, these mice had memory impairments in a contextual fear paradigm testing pattern-completion-based recall in which mice are first pre-exposed to the context without footshock before the training day. If administered Dox to repress the TeTX transgene, the CA3-TeTX mice

performed similarly to controls, demonstrating in a reversible manner the importance for CA3 → CA1 synapses for one-trial memory, spatial representation, and pattern completion (Nakashiba et al. 2008). Subsequent analysis of this mouse line revealed the importance of CA3 output for consolidation of long-term contextual fear memories. Using Dox regulation of transgene expression to block CA3 synaptic transmission during the consolidation phase of memory, contextual fear memory was impaired 6 wk later. However, when CA3 output was normal during consolidation but blocked during retrieval, the mice showed no memory deficits, suggesting that CA3 → CA1 synaptic activity is important for the consolidation, but not for the retrieval of long-term contextual memories (Nakashiba et al. 2009).

Next, the Tonegawa laboratory used the same principles to generate DG-TeTX mice under control of the tTA system. The DG of the hippocampus is one of two areas of the brain (along with the subventricular zone) known to generate new neurons into adulthood, a process known as adult neurogenesis (Altman and Das 1965; Eriksson et al. 1998). Using the tTA system, the Tonegawa laboratory was able to block DG → CA3 synaptic transmission selectively in developmentally derived DG neurons, while sparing adult-born DG neurons. This manipulation produced normal pattern separation as probed by contextual fear discrimination between disparate contexts, but reduced pattern completion on the preexposure-dependent contextual fear-conditioning paradigm. When young, adult-born DG cells were specifically ablated using X-ray radiation, the mice had deficits in pattern separation of dissimilar contexts. Thus, adult-born DG neurons appear to be necessary for pattern completion, whereas older, developmentally derived DG neurons appear necessary for normal pattern separation (Nakashiba et al. 2012).

Moving out of the hippocampus, the oxidation resistance 1 (oxr1) promoter was recently used to generate a mouse line expressing Cre recombination selectively in superficial layer III of the medial entorhinal cortex (MEC). The MEC has two disparate projections into

 Cite this article as *Cold Spring Harb Perspect Biol* doi: 10.1101/cshperspect.a021725

the hippocampus. One, known as the trisysnaptic pathway, originates from layer II of the MEC and projects to the DG and then ultimately to CA3 and CA1. The other, the monosynaptic pathway, projects from layer III of the MEC directly to CA1. Using the tTA system, a pOxr1-Cre line, and TeTX driven by the α-CaMKII promoter, the Tonegawa laboratory was able to selectively and reversibly block synaptic transmission of the monosynaptic pathway. Doing so produced spatial working memory deficits in the DMP Morris water maze task as well as a delayed nonmatching-to-place T maze task. The mutants were also impaired in the acquisition, but not the retrieval phase of cued trace fear conditioning in which a tone was paired temporally with a footshock. Together, these studies suggest that the monosynaptic pathway from MEC layer III → CA1 is important for temporal associations and spatial working memory (Suh et al. 2011).

The tTA system has also been used to assess the effects of glia cells on learning and memory. In the 1990s, it was discovered that activation of astrocytic glutamate receptors (GluRs) increases $Ca^{2+}$ levels in astrocytes, producing astrocytic transmitter release, called "gliotransmission" (Cornell-Bell et al. 1990; Araque et al. 1998). In 2005, Phil Haydon's laboratory generated a mouse line expressing dominant-negative N-ethylmaleimide-sensitive factor attachment protein receptor (dn-SNARE) selectively in astrocytes under temporal control of the tTA system, allowing for attenuation of gliotransmission via disruption of the vesicular release machinery. Using this model, it was discovered that gliotransmission alters both basal neuronal synaptic transmission and late-phase LTP (Pascual et al. 2005). Later studies with dn-SNARE mice showed that these mice are protected against the intense sleep drive and hippocampus-dependent object-recognition memory impairments normally observed following sleep deprivation (Halassa et al. 2009; Florian et al. 2011).

## PHARMACOGENETICS

Although the tTA system allows for inducible and reversible genetic manipulation, the time course of transgene expression/suppression after Dox administration/removal is slow, taking 3–4 wk (Mayford et al. 1996b). One method developed to improve on this issue was made by engineering an inducible form of Cre recombinase and fusing it to a mutated human estrogen receptor (ER) (Metzger et al. 1995). The ER is naturally localized to the cytoplasm by chaperone proteins. Injecting tamoxifen disrupts the interaction between the fusion protein (CreERT2) and chaperone proteins and causes the ER to relocate to the nucleus, initiating Cre recombination of the floxed gene.

This technique was used by Imayoshi et al. (2008) to study neurogenesis in the DG. In one mouse line, CreERT2 was driven by the Nestin promoter, which is expressed in immature progenitor cells. In a second mouse line, a fragment of the diphtheria toxin was knocked into a noncoding region of the mature neuron-specific enolase 2. In this second mouse line, a STOP codon was flanked by lox P sites. Thus, crossing the two lines (coupled with administration of tamoxifin) produced Cre recombination specifically in newly differentiated mature neurons expressing enolase 2. This manipulation allowed for expression of Diphtheria toxin, specifically killing adult-born neurons. Eliminating neurogenesis in the DG resulted in spatial memory deficits on the Morris water maze as well as deficits in hippocampus-dependent contextual fear conditioning.

A complimentary study, using the CreERT2 system to enhance neurogenesis, was performed in the laboratory of René Hen. Here, a mouse line expressing CreERT2 under control of the Nestin promoter was crossed with a mouse line containing a floxed *Bax* gene in neural stem cells (Sahay et al. 2011). BAX is proapoptotic (Oltval et al. 1993), and tamoxifin-dependent Cre deletion of BAX results in decreased cell death of neural stem cells. This manipulation leads to a threefold increase of adult-born neurons in the dentate gyrus 6 wk after tamoxifin injection. Behaviorally, these mice displayed normal object recognition, spatial memory, and contextual fear conditioning. Interestingly, these mice had enhanced discrimination of two similar contexts during contextual fear conditioning, lending further credence to the importance of the DG for pattern separation (Sahay et al. 2011).

Although the CreERT2 system allows for more precise temporal genetic control, tamoxifin is an endogenous antagonist of the ER. Thus, tamoxifin administration affects both the engineered CreERT2 receptor and endogenous ERs, confounding phenotypic interpretations. Moreover, larger quantities of tamoxifin and/or CreERT2 expression have been shown to be toxic (Higashi et al. 2009). Recent advancements have used destabilized forms of Cre (Sando et al. 2013) or synthetic G protein–coupled receptors (GPCRs) that are activated exclusively by synthetic ligands. The first attempt at developing a designer receptor exclusively activated by designer drugs (DREADD), was performed by Strader and colleagues in 1991 by developing a mutated form of the β2-adrenergic receptor through site-directed mutagenesis that was not responsive to its endogenous ligand (Strader et al. 1991). However, this receptor had low affinity for its synthetic ligand, making it impractical for in vivo usage.

An improvement on this methodology was developed and used by the Abel laboratory to incorporate the *Aplysia* $G_s$-coupled octopamine receptor in mouse excitatory forebrain neurons under control of the tTA system (Fig. 3A) (Isiegas et al. 2008). The octopamine receptor is activated by octopamine, which is found predominately in invertebrates but only in trace levels in mammals (Berry et al. 2004). Injection of octopamine produces rapid increases in cAMP. Using this manipulation, the Abel laboratory found that activating cAMP levels with octopamine increases hippocampus LTP and enhances the consolidation of contextual fear memory (Fig. 3B–D) (Isiegas et al. 2008). In a recent follow-up study, targeted delivery and activation of the octopamine receptor exclusively in hippocampus excitatory neurons during sleep deprivation prevented the object location memory impairments typically observed following sleep loss (Havekes et al. 2014). Together, these studies show the importance of cAMP during both wake and sleep for the consolidation of hippocampus-dependent memories.

Recent technical advancements have resulted in the creation of a wealth of new DREADDs capable for use in vivo. For example, a synthetic $G_q$-coupled human $M_3$ muscarinic DREADD ($hM_3D_q$) was developed to be selectively activated by the pharmacologically inert drug clozapine-N-oxide (CNO) (Alexander et al. 2009). Mayford and colleagues used a promoter for the transcription factor c-fos to drive expression of this receptor, along with the tTA transgene, to activated neurons. Activated neurons were labeled with the $hM_3D_q$ receptor during exploration of a novel context (context A). Mice were then injected with CNO and fear conditioned in a second context (context B), activating the same population of neurons that were activated in context A. This manipulation resulted in activation of competing neural ensembles, resulting in a hybrid memory between context A and context B (Garner et al. 2012).

## VIRAL/OPTOGENETIC APPROACHES

Viruses operate by infecting host cells and incorporating their genetic material into the host's genome. By modifying viruses to remove the harmful components that hijack the host's cellular machinery and instead inserting promoters and transgenes of interest, researchers have been able to develop viral vectors that allow for spatially precise genetic control. A viral construct typically includes a cell-type-specific promoter, a transgene of interest, and a marker, such as a fluorescent protein. Injecting the viral vector into the brain allows for cell-type and brain region–specific delivery of the viral genetic material, termed transduction.

The use of viral vectors has proved fruitful for investigating the roles of corticohippocampal circuitry in learning and memory. In one study, the transcription factor myocyte enhancer factor 2 (MEF2), which negatively regulates the formation of dendritic spines (Flavell et al. 2006), was found to be important for the consolidation of contextual fear memory in the anterior cingulate cortex (ACC). MEF2-dependent transcription was increased in layer 2/3 of the ACC at various time points after contextual fear conditioning using a replication-defective

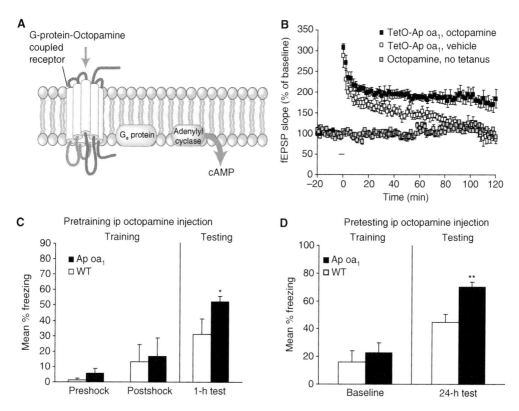

**Figure 3.** Activation of the octopamine receptor in mouse forebrain neurons enhances long-term potentiation (LTP) and memory. (*A*) *Aplysia* octopamine receptor (Ap oa$_1$) is a G-protein-coupled receptor whose activation by octopamine stimulates adenylyl cyclase activity, which in turn synthesizes cyclic adenosine monophosphate (cAMP). (*B*) Hippocampal one-train LTP (100 Hz for 1 sec) is potentiated and lengthened in Ap oa$_1$ mice after octopamine application. Arrow indicates when the potentiation stimulus was delivered. (*C*) Training and 24-h long-term memory test during contextual fear conditioning in Ap oa$_1$ transgenic mice and wild-type (WT) littermates injected intraperitoneally with octopamine 30 min before training. No differences in freezing behavior are observed between Ap oa$_1$ transgenic mice and WT littermates before and after the shock during training. In contrast, Ap oa$_1$ transgenic mice show a significant increase in freezing behavior when reexposed to the fear-conditioned context 24 h after training. (*D*) Training and 24-h long-term memory testing during fear conditioning in Ap oa$_1$ transgenic mice and WT littermates receiving an intraperitoneal injection of octopamine 30 min before the retrieval session. Ap oa$_1$ transgenic mice and WT mice show similar levels of freezing during the training session (baseline). However, Ap oa$_1$ transgenic mice show significantly increased freezing behavior compared with WT mice during the contextual fear memory test 24 h after training. (Image reprinted from Isiegas et al. 2008 under the U.S. Fair Use Guidelines available from the Copyright Office at the Library of Congress.)

herpes simplex virus. Mice were injected 1 d or 42 d after contextual fear conditioning and memory was tested 1 wk later. Mice with enhanced MEF2-dependent transcription 1 d after training had decreased fear memory as well as decreased spine density in ACC, whereas mice injected 42 d after training did not (Vetere et al. 2011). These results suggest that MEF2 acts

in a time-dependent manner to negatively regulate memory consolidation and spine formation in layer 2/3 of the ACC. They also show the involvement of the neocortex, in addition to the hippocampus, for the consolidation of contextual fear memories.

A recent study by Lovett-Barron et al. (2014) combined pharmacogenetics, the Cre-LoxP system, and adeno-associated viruses (AAV) to inactivate somatostatin-expressing (SOM$^+$) interneurons in CA1 and assess their role in two forms of contextual fear conditioning. Neurons expressing the Cl$^-$ channel PSAM$^{L141F}$-GlyR are inactivated for 15–20 min on administration of the synthetic ligand PSAM$^{89}$. PSAM$^{L141F}$-GlyR was packaged in an AAV and injected bilaterally into CA1 of Som-Cre mice. Systemic administration of PSAM$^{89}$ during conditioning impaired memory in both head-fixed water-licking suppression contextual fear conditioning (hf-CFC) and freely moving contextual fear conditioning, suggesting that inhibitory circuits comprised of SOM$^+$ interneurons in CA1 are necessary for multiple forms of contextual fear conditioning. Inactivating parvalbumin-expressing interneurons in CA1, however, did not impair contextual fear memory. Using a separate AAV to express the Ca$^{2+}$ indicator GCaMP and two-photon imaging, the researchers also showed that SOM$^+$ interneurons in CA1 are preferentially activated during hf-CFC.

As shown in previous studies, the spatial precision of viral techniques is most useful when used in combination with other genetic methodologies. For example, Warburton et al. used an adenovirus to inject a dominant-negative form of CREB (dn-CREB) into the perirhinal cortex. dn-CREB in the perirhinal cortex-impaired long-term object-recognition memory and perirhinal LTP (Warburton et al. 2005). However, perhaps the most fascinating combinatorial approach involves using the spatial precision of viral vectors with the recently developed, temporally precise technique known as optogenetics. Previously mentioned techniques allow for precise genetic manipulation on the course of minutes to hours. However, neurons fire on the order of milliseconds. The development and optimization of optogenetics over the last decade has allowed for millisecond control of neural activity in a brain region or cell-type-specific manner. Optogenetics involves the use of light-activated ion channels known as opsins, found naturally in certain bacteria or algae, and engineering them into mammalian cells to confer control of neuronal excitability on optical pre-

sentation. Channelrhodopsin-2 (ChR2) is one such light-gated proton channel found naturally in the green alga *Chlamydomonas reinhardtii*. In 2003, Georg Nagel and Ernst Bamberg expressed and characterized ChR2 in *Xenopus* oocytes and mammalian cell cultures, producing reliable depolarizing currents on activation with 450-nm blue light (Nagel et al. 2003). In 2005, Ed Boyden and Karl Deisseroth collaborated with Nagel and Bamberg and bioengineered ChR2 into cultured mammalian neurons for the first time. They found that ChR2 can produce single neuronal spikes or long trains of depolarization depending on the interval of light stimulation (Boyden et al. 2005).

In recent years, the technique of optogenetics has been applied to the field of learning and memory to study circuitry and populations of cell types with millisecond precision. Pharmacologic and lesion studies indicate that the hippocampus is crucial for recent memories, but, over time, these memories become hippocampus independent, relying instead on the neocortex (i.e., Kim and Fanslow 1992; Maren et al. 1997). To test these ideas with more precise temporal control, Deisseroth's laboratory engineered an enhanced halorhodopsin chloride channel from the archaea *Natronomonas pharaoni* (eNpHR3.1), which acts as an optogenetic inhibitor on stimulation with 680 nm red light. CaMKIIα:eNpHR3.1 was incorporated into a lentiviral vector and injected into hippocampus CA1 (Fig. 4A,B). Mice were then trained on contextual fear conditioning. Surprisingly, optic inhibition of CA1 during the 5-min test session impaired retrieval of the remote memory 28 d after training (Fig. 4C). However, if the inhibition interval was prolonged to 30 min, remote memory was undisturbed (Fig. 4D) (Goshen et al. 2011). These findings suggest that the time course of hippocampus inhibition is important—with longer periods of inhibition, as with lesion studies, there may be sufficient time for cortical compensatory circuits to take over. Along these lines, optogenetic inhibition delivered into the anterior cingulate cortex (ACC) impaired remote, but not recent contextual fear memories (Fig. 4E,F) (Goshen et al. 2011).

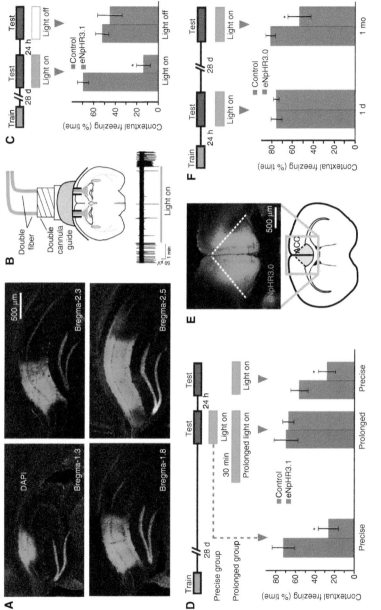

**Figure 4.** Optogenetic inactivation of CA1 impairs recent (24 h) long-term memories, whereas optogenetic inactivation of anterior cingulate cortex (ACC) impairs remote (28 d) memory. (*A*) Double lentiviral injection resulted in eNpHR3.1 expression in CA1 only. (*B*) Bilateral in vivo light administration to CA1 (*top*). Illumination of CA1 neurons in eNpHR3.1-expressing mice resulted in a reversible reduction in spiking frequency without affecting average spike amplitude. A representative optrode recording trace is shown (*bottom*). (*C*) CA1 optogenetic inhibition prevented remote memory. This disruption was reversible, as when the same mice were reintroduced to the conditioning context with no illumination, they showed intact fear responses. Optogenetic inactivation of CA1 had no impact on cued fear memory, which is amygdala dependent (not shown). (*D*) CA1 optogenetic inhibition prevented remote fear memory only when light was administered precisely during testing (precise group, *left*), but not when the light was on continuously for 30 min before (as well as during) the test (prolonged group, *middle*). When the prolonged group mice were retested the next day with precise light, their recall was disrupted (prolonged group, *middle*). (*E*) eNpHR3.0 expression in the ACC. (*F*) Precise light administration resulted in inhibition of remote but not recent memory recall. (From Goshen et al. 2011; reprinted, with permission, from Elsevier © 2011.)

Optogenetic techniques have also been used recently to study the roles of previously difficult-to-study regions and cell types of corticohippocampal circuitry. The Tonegawa laboratory used optogenetic stimulation of DG mossy fibers in concert with electrode recording from hippocampus region CA2, an area between CA3 and CA1 comprised of large pyramidal cells, to discover a previously unknown circuit connection from DG → CA2. These techniques in combination with CA2-specific markers and tracing studies revealed a trisynaptic circuit from DG → CA2 → CA1, the function of which remains to be explored (Kohara et al. 2014). In a separate experiment, the Tonegawa laboratory identified a subset of excitatory cells in entorhinal cortex layer II, which they termed "island cells." Island cells project to CA1 and inhibit the excitatory projections from medial EC layer III onto CA1 via feedforward inhibition. Island cells are positive for the Wfs1 promoter and using a Wfs1-Cre line, channelrhodopsin-2 was virally targeted to EC layer II island cells. These island cells were then optically stimulated during trace fear conditioning (TFC), a hippocampus-dependent task in which a delay (trace) is incorporated between the presentation of a tone and a foot shock. Stimulating island cells during TFC resulted in memory impairments both during training and testing. Similar results were observed by optically inhibiting EC layer III inputs to CA1 using the pOxr1-Cre line described earlier. These freezing deficits were not present if the trace interval was removed, suggesting that projections from EC layer II and EC layer III to CA1 are both important for temporal association memory (Kitamura et al. 2014). Optogenetics has also recently been used to label active neurons during contextual fear conditioning by driving expression of ChR2 to DG neurons expressing the c-fos promoter. Mice were placed in a novel context A and active neurons were labeled with ChR2. This population of neurons was then optically activated during fear conditioning in context B. When mice were placed back into context A, they showed enhance freezing even though they were never shocked in context A, in effect creating a false memory representation (Ramirez et al. 2013).

## CONCLUDING REMARKS

The use of molecular-genetic techniques to study the corticohippocampal circuitry underlying learning and memory has exploded in the three decades since Palmiter and colleagues developed the first knockout mouse strain (Palmiter et al. 1982). Knockin and dominant-negative approaches have revealed the functionality of disparate subdomains of proteins. The Cre LoxP system, along with the discovery of spatially restricted promoters, have allowed for brain region and cell-type-specific genetic manipulation. The tetracycline system, pharmacogenetics, viral vectors, and optogenetics have made inducible and reversible genetic studies possible (Table 1). When used in combination, these techniques have permitted great spatial and temporal control, leading to some of the fascinating findings reviewed herein.

Still, much improvement in genetic technology is needed. Despite the multitude of genetic tools at researchers' disposal, better spatial and temporal control is needed. One way to improve specificity is through the discovery of new spatially and temporally restricted promoters. For example, no known promoters exist that specifically target the perirhinal cortex, which projects to entorhinal cortex, receives projections from CA1, and plays an important role in memory consolidation (Apergis-Schoute et al. 2007; Furtak et al. 2007). Additionally, promoter expression may change over the course of development, or with neural activity. For instance, the transcription factor *zif268* is important for the induction and maintenance of LTP, and is induced in CA1 by contextual fear conditioning and spatial memory (i.e., Cole et al. 1989; Guzowski et al. 2001; Hall et al. 2001; Jones et al. 2001). However, it remains difficult to modulate *zif268* with the temporal precision necessary to parse apart its effects on the different stages of memory.

One novel and exciting gene editing technology involves using clustered regularly interspaced short palindromic repeats (CRISPRs) found in bacteria in association with CRISPR associated (cas) genes, such as *cas9*. The CRISPR/cas9 system delivers cas9 proteins,

**Table 1.** Summary table of genetic techniques outlined in this review and their properties

| Genetic technique | Regulator | Inducible? | Promoter-specific expression? | Method of genome insertion |
|---|---|---|---|---|
| *Vector delivery techniques* | | | | |
| Conventional knockout | None | No | No | Targeted mutation through homologous recombination |
| Knockin | None | In combination with an inducible technique | Yes | Targeted mutation through homologous recombination |
| Transgenesis | None | In combination with an inducible technique | Yes | Random insertion into genome |
| Viral vector | None | In combination with an inducible technique | Yes | Targeted, random, or episomal |
| *Genetic manipulation techniques* | | | | |
| Dominant negative | None | In combination with an inducible technique | Yes | Depends on the vector delivery technique chosen |
| Cre/loxP System | Promoter | Can be (i.e., CreERT2, DD-Cre) | Yes | Depends on the vector delivery technique chosen |
| Tetracycline system | Doxycycline | Yes | Yes | Depends on the vector delivery technique chosen |
| DREADDs | Synthetic ligand | Yes | Yes | Depends on the vector delivery technique chosen |
| Optogenetics | Light | Yes | Yes | Depends on the vector delivery technique chosen |
| CRISPR/CAS9 | Cas9 endonuclease + guide RNAs | Yes | Yes | Depends on the vector delivery technique chosen |

along with a guide RNA, to any predetermined location in the genome. This system cuts DNA wherever the researcher desires, allowing for precise addition, subtraction, or modification of genes (Pennisi 2013). This exciting technology is still in its infancy, but should substantially aid in the speed and accuracy of producing future innovative genetically modified mouse lines.

Another point that must be considered in rodent genetic research is the impact of genetic background and environment on memory studies. Studies using common mouse strains, such C57BL/6J, CBA/J, DBA/2J, 129/SvJ, and BALB/cByJ, show differences in both LTP and performance on various spatial and nonspatial hippocampus-dependent memory tasks (Wehner and Silva 1996; Nguyen et al. 2000). The environment also plays a large role on the interpretation of results. Crabbe and colleagues tested eight commonly used mouse strains on several different behaviors in three different lab-

oratories. Despite keeping all other variables as consistent as possible, including the age of the mice and time of day of testing, the researchers found marked differences in behavioral performance between laboratories (Crabbe et al. 1999). Additionally, a recent study indicated that the sex of the experimenter may even differentially alter stress levels in rodents (Sorge et al. 2014). Studies such as these reveal the careful contemplation that must be taken before designing experiments involving mouse behavior.

Finally, an important step in the interpretation of genetic research is the ability to manipulate complete behavioral circuits as opposed to individual genes. Only then can the impact of corticohippocampal circuitry on complex learning and memory behaviors be more comprehensively addressed. The human brain is amazingly complex, consisting of billons of neurons with countless more synaptic connections. Probing one gene, one cell type, or one brain region only scratches the surface of the mechanisms underlying behavior. Although the task to study complete behavioral circuits is an arduous one, strategically combining many of the techniques discussed in this review may be the best current approach to begin to tackle this question. However, new genetic techniques are also needed. Exciting new circuit-based approaches are constantly being developed and applied to the field of learning and memory (see Brown et al. 2015; Jercog et al. 2015; Mayford and Reijmers 2015). Only 25 years have passed since the development of the first knockout mouse in 1989. Surely the next 25 years will bring about a vastly improved array of new genetic research tools—perhaps capable of addressing in more complete detail the corticohippocampal circuitry underlying learning and memory.

## MOUSE DATABASES

### Mouse Lines Available

Mutant Mouse Regional Resource Centers (MMRRC): www.mmrrc.org
Jackson Laboratory: www.jaxmice.jax.org
International Mouse Strain Resource (IMSR): www.findmice.org

### Mouse Genome Browsers

UCSC Genome Browser: www.genome.ucsc.edu
Jackson Laboratory: www.informatics.jax.org

## ACKNOWLEDGMENTS

We thank Sarah Ferri, Shane Poplawski, and Rolf Hansen III for comments on a previous version of this manuscript.

## REFERENCES

*Reference is also in this collection.

Abel T, Nguyen PV, Barad M, Deuel TA, Kandel ER, Bourtchouladze R. 1997. Genetic demonstration of a role for PKA in the late phase of LTP and in hippocampus-based long-term memory. Cell 88: 615–626.

Alarcón JM, Malleret G, Touzani K, Vronskaya S, Ishii S, Kandel ER, Barco A. 2004. Chromatin acetylation, memory, and LTP are impaired in CBP$^{+/-}$ mice: A model for the cognitive deficit in Rubinstein–Taybi syndrome and its amelioration. Neuron 42: 947–959.

Alexander GM, Rogan SC, Abbas AI, Armbruster BN, Pei Y, Allen JA, Nonneman RJ, Hatmann J, Moy SS, Nicolelis MA, et al. 2009. Remote control of neuronal activity in transgenic mice expressing evolved G protein-coupled receptors. Neuron 63: 27–39.

Altman J, Das GD. 1965. Autoradiographic and histological evidence of postnatal hippocampal neurogenesis in rats. J Comp Neurol 124: 319–335.

Apergis-Schoute J, Pinto A, Paré D. 2007. Muscarinic control of long-range GABAergic inhibition within the rhinal cortices. J Neurosci 27: 4061–4071.

Araque A, Parpura V, Sanzgiri RP, Haydon PG. 1998. Glutamate-dependent astrocyte modulation of synaptic transmission between cultured hippocampal neurons. Eur J Neurosci 10: 2129–2142.

Balthasar N, Coppari R, McMinn J, Liu SM, Lee CE, Tang V, Kenny CD, McGovern RA, Chua SC, Elmquist JK, et al. 2004. Leptin receptor signaling in POMC neurons is required for normal body weight homeostasis. Neuron 42: 983–991.

Bannerman DM, Bus T, Taylor A, Sanderson DJ, Schwarz I, Jensen V, Hvalby O, Rawlins JN, Seeburg PH, Sprengel R. 2012. Dissecting spatial knowledge from spatial choice by hippocampal NMDA receptor deletion. Nat Neurosci 15: 1153–1159.

Berry MD. 2004. Mammalian central nervous system trace amines. Pharmacologic amphetamines, physiologic neuromodulators. J Neurochem 90: 257–271.

Blendy JA, Kaestner KH, Schmid W, Gass P, Schutz G. 1996. Targeting of the CREB gene leads to up-regulation of a novel CREB mRNA isoform. EMBO J 15: 1098–1106.

Bliss TV, Collingridge GL. 1993. A synaptic model of memory: Long-term potentiation in the hippocampus. Nature 361: 31–39.

Bourtchuladze R, Frenguelli B, Blendy J, Cioffi D, Schutz G, Silva AJ. 1994. Deficient long-term memory in mice with a targeted mutation of the cAMP-responsive element-binding protein. Cell 79: 59–68.

Boyden ES, Zhang F, Bamberg E, Nagel G, Deisseroth K. 2005. Millisecond-timescale, genetically targeted optical control of neural activity. Nat Neurosci 8: 1263–1268.

Bradley A, Evans M, Kaufman MH, Robertson E. 1984. Formation of germ-line chimaeras from embryo-derived teratocarcinoma cell lines. Nature 309: 255–256.

Brandon EP, Zhuo M, Huang Y-Y, Ming Q, Gerhold KA, Burton KA, Kandel ER, McKnight GS, Idzerda RL. 1995. Hippocampal long-term depression and depotentiation are defective in mice carrying a targeted disruption of the gene encoding the RIβ subunit of cAMP-dependent protein kinase. Proc Natl Acad Sci 92: 8852–8855.

Brenner M, Kisseberth WC, Su Y, Besnard F, Messing A. 1994. GFAP promoter directs astrocyte-specific expression in transgenic mice. J Neurosci 14: 1030–1037.

* Brown TI, Staresina BP, Wagner AD. 2015. Noninvasive functional and anatomical imaging of the human medial temporal lobe. Cold Spring Harb Perspect Biol 7: a021840.

Bukalo O, Fentrop N, Lee AY, Salmen B, Law JW, Wotjak CT, Schweizer M, Dityatev A, Schachner M. 2004. Condition-

Cite this article as Cold Spring Harb Perspect Biol doi: 10.1101/cshperspect.a021725

al ablation of the neural cell adhesion molecule reduces precision of spatial learning, long-term potentiation, and depression in the CA1 subfield of mouse hippocampus. *J Neurosci* **24:** 1565–1577.

Cao X, Wang H, Mei B, An S, Yin L, Wang LP, Tsien JZ. 2008. Inducible and selective erasure of memories in the mouse brain via chemical-genetic manipulation. *Neuron* **60:** 353–366.

Cole AJ, Saffen DW, Baraban JM, Worley PF. 1989. Rapid increase of an immediate early gene messenger RNA in hippocampal neurons by synaptic NMDA receptor activation. *Nature* **340:** 474–476.

Cornell-Bell AH, Finkbeiner SM, Cooper MS, Smith SJ. 1990. Glutamate induces calcium waves in cultured astrocytes: Long-range glial signaling. *Science* **247:** 470–473.

Crabbe JC, Wahlsten D, Dudek BC. 1999. Genetics of mouse behavior: Interactions with laboratory environment. *Science* **284:** 1670–1672.

Cunha C, Angelucci A, D'Antoni A, Dobrossy MD, Dunnett SB, Berardi N, Brambilla R. 2009. Brain-derived neurotrophic factor (BDNF) overexpression in the forebrain results in learning and memory impairments. *Neurobiol Dis* **33:** 358–368.

Doetschman T, Gregg RG, Maeda N, Hooper ML, Melton DW, Thompson S, Smithies O. 1987. Targeted correction of a mutant HPRT gene in mouse embryonic stem cells. *Nature* **330:** 576–578.

Dudai Y, Jan Y-N, Byers D, Quinn W, Benzer S. 1976. *dunce*, a mutant of *Drosophila melanogaster* deficient in learning. *Proc Natl Acad Sci* **73:** 1684–88.

Eriksson PS, Perfilieva E, Björk-Eriksson T, Alborn AM, Nordborg C, Peterson DA, Gage FH. 1998. Neurogenesis in the adult human hippocampus. *Nat Med* **4:** 1313–1317.

Evans MJ, Kaufman MH. 1981. Establishment in culture of pluripotential cells from mouse embryos. *Nature* **292:** 154–156.

Flavell SW, Cowan CW, Kim TK, Greer PL, Lin Y, Paradis S, Griffith EC, Hu LS, Chen C, Greenberg ME. 2006. Activity-dependent regulation of MEF2 transcription factors suppresses excitatory synapse number. *Science* **311:** 1008–1012.

Florian CD, Vecsey CG, Halassa MM, Haydon PG, Abel T. 2011. Astrocyte-derived adenosine and A1 receptor activity contribute to sleep loss-induced deficits in hippocampal synaptic plasticity and memory in mice. *J Neurosci* **31:** 6956–6962.

Forrest D, Yuzaki M, Soares HD, Ng L, Luk DC, Sheng M, Stewart CL, Morgan JI, Connor JA, Curran T. 1994. Targeted disruption of NMDA receptor 1 gene abolishes NMDA response and results in neonatal death. *Neuron* **13:** 325–338.

Forss-Petter S, Danielson PE, Catsicas S, Battenberg E, Price J, Nerenberg M, Sutcliffe JG. 1990. Transgenic mice expressing β-galactosidase in mature neurons under neuron-specific enolase promoter control. *Neuron* **5:** 187–197.

Furtak SC, Wei SM, Agster KL, Burwell RD. 2007. Functional neuroanatomy of the parahippocampal region in the rat: The perirhinal and postrhinal cortices. *Hippocampus* **17:** 709–722.

Garner AR, Rowland DC, Hwang SY, Baumgaertel K, Roth BL, Kentros C, Mayford M. 2012. Generation of a synthetic memory trace. *Science* **335:** 1513–1516.

Gass P, Wolfer DP, Balschun D, Rudolph D, Frey U, Lipp HP, Schutz G. 1998. Deficits in memory tasks of mice with CREB mutations depend on gene dosage. *Learn Mem* **5:** 274–288.

Gerfen CR, Engber TM, Mahan LC, Susel Z, Chase TN, Monsma FJ, Sibley DR. 1990. D1 and D2 dopamine receptor regulated gene expression of striatonigral and striatopallidal neurons. *Science* **250:** 1429–1432.

Goshen I, Brodsky M, Prakash R, Wallace J, Gradinaru V, Ramakrishnan C, Deisseroth K. 2011. Dynamics of retrieval strategies for remote memories. *Cell* **147:** 678–689.

Gossen M, Bujard H. 1992. Tight control of gene expression in mammalian cells by tetracycline-responsive promoters. *Proc Natl Acad Sci* **89:** 5547–5551.

Guan JS, Haggarty SJ, Giacometti E, Dannenberg JH, Joseph N, Gao J, Nieland T, Zhou Y, Wang X, Mazitschek R, et al. 2009. HDAC2 negatively regulates memory formation and synaptic plasticity. *Nature* **459:** 55–60.

Guzowski JF, Setlow B, Wagner EK, McGaugh JL. 2001. Experience-dependent gene expression in the rat hippocampus after spatial learning: A comparison of the immediate-early genes *Arc*, *c-fos*, and *zif268*. *J Neurosci* **21:** 5089–5098.

Halassa MM, Florian C, Fellin T, Munoz JR, Lee SY, Abel T, Haydon PG, Frank MG. 2009. Astrocytic modulation of sleep homeostasis and cognitive consequences of sleep loss. *Neuron* **61:** 213–219.

Hall J, Thomas KL, Everitt BJ. 2001. Cellular imaging of *zif268* expression in the hippocampus and amygdala during contextual and cued fear memory retrieval: Selective activation of hippocampal CA1 neurons during the recall of contextual memories. *J Neurosci* **21:** 2186–2193.

Havekes R, Abel T. 2009. Genetic dissection of neural circuits and behavior in *Mus musculus*. *Adv Genet* **65:** 1–38.

Havekes R, Bruinenberg VM, Tudor JC, Ferri SL, Baumann A, Meerlo P, Abel T. 2014. Transiently increasing cAMP levels selectively in hippocampal excitatory neurons during sleep deprivation prevents memory deficits caused by sleep loss. *J Neurosci* **34:** 15715–15721.

Herskowitz I. 1987. Functional inactivation of genes by dominant negative mutations. *Nature* **329:** 219–222.

Higashi AY, Ikawa T, Muramatsu M, Economides AN, Niwa A, Okuda T, Murphy AJ, Rojas J, Heike T, Nakahata T, et al. 2009. Direct hematological toxicity and illegitimate chromosomal recombination caused by the systemic activation of CreER$^{T2}$. *J Immunol* **182:** 5633–5640.

Hoeffer CA, Tang W, Wong H, Santillan A, Patterson RJ, Martinez LA, Tejada-Simon MV, Paylor R, Hamilton SL, Klann E. 2008. Removal of FKBP12 enhances mTOR-Raptor interactions, LTP, memory, and perseverative/repetitive behavior. *Neuron* **60:** 832–845.

Imayoshi I, Sakamoto M, Ohtsuka T, Takao K, Miyakawa T, Yamaguchi M, Mori K, Ikeda T, Itohara S, Kageyama R. 2008. Roles of continuous neurogenesis in the structural and functional integrity of the adult forebrain. *Nat Neurosci* **11:** 1153–1161.

Isiegas C, Park A, Kandel ER, Abel T, Lattal KM. 2006. Transgenic inhibition of neuronal protein kinase A activity facilitates fear extinction. *J Neurosci* **26:** 12700–12707.

Isiegas C, McDonough C, Huang T, Havekes R, Fabian S, Wu LJ, Xu H, Zhao MG, Kim JI, Lee YS, et al. 2008. A novel conditional genetic system reveals that increasing neuronal cAMP enhances memory and retrieval. *J Neurosci* **28:** 6220–6230.

* Jercog P, Rogerson T, Schnitzer MJ. 2015. Large-scale fluorescence calcium imaging methods for studies of long-term memory in behaving mammals. *Cold Spring Harb Perspect Biol* doi: 10.1101/cshperspect.a021824.

Jones MW, Errington ML, French PJ, Fine A, Bliss TVP, Garel S, Charnay P, Bozon B, Laroche S, Davis S. 2001. A requirement for the immediate early gene *Zif268* in the expression of late LTP and long-term memories. *Nat Neurosci* **4:** 289–296.

Kim JJ, Fanselow MS. 1992. Modality-specific retrograde amnesia of fear. *Science* **256:** 675–677.

Kitamura T, Pignatelli M, Suh J, Kohara K, Yoshiki A, Abe K, Tonegawa S. 2014. Island cells control temporal association memory. *Science* **343:** 896–901.

Kohara K, Pignatelli M, Rivest AJ, Jung HY, Kitamura T, Suh J, Frank D, Kajikawa K, Mise N, Obata Y, et al. 2014. Cell type-specific genetic and optogenetic tools reveal hippocampal CA2 circuits. *Nat Neurosci* **17:** 269–279.

Korzus E, Rosenfeld MG, Mayford M. 2004. CBP histone acetyltransferase activity is a critical component of memory consolidation. *Neuron* **42:** 961–972.

Leach PT, Poplawski SG, Kenney JW, Hoffman B, Liebermann DA, Abel T, Gould TJ. 2012. *Gadd45b* knockout mice exhibit selective deficits in hippocampus-dependent long-term memory. *Learn Mem* **19:** 319–324.

Limbäck-Stokin K, Korzus E, Nagaoka-Yasuda R, Mayford M. 2004. Nuclear calcium/calmodulin regulates memory consolidation. *J Neurosci* **24:** 10858–10867.

Livingstone MS, Sziber PP, Quinn WG. 1984. Loss of calcium/calmodulin responsiveness in adenylate cyclase of *rutabaga*, a *Drosophila* learning mutant. *Cell* **137:** 205–215.

Lovett-Barron M, Kaifosh P, Kheirbek MA, Danielson N, Zaremba JD, Reardon TR, Turi GF, Zemelman Hen R, Losonczy A. 2014. Dendritic inhibition in the hippocampus supports fear learning. *Science* **343:** 857–863.

Ma D, Jang M-H, Guo JU, Kitabatake Y, Chang M-L, Pow-Anpongkul N, Flavell RA, Lu B, Ming G-L, Song H. 2009. Neuronal activity–induced Gadd45b promotes epigenetic DNA demethylation and adult neurogenesis. *Science* **323:** 1074–1077.

Mansour SL, Thomas KR, Capecchi MR. 1988. Disruption of the proto-oncogene int-2 in mouse embryo-derived stem cells: A general strategy for targeting mutations to non-selectable genes. *Nature* **336:** 348–352.

Maren S, Aharonov G, Fanselow MS. 1997. Neurotoxic lesions of the dorsal hippocampus and Pavlovian fear conditioning in rats. *Behav Brain Res* **88:** 261–274.

* Mayford M, Reijmers L. 2015. Exploring memory representations with activity-based genetics. *Cold Spring Harb Perspect Biol* doi: 10.1101/cshperspect.a021832.

Mayford M, Baranes D, Podsypanina K, Kandel ER. 1996a. The 3′-untranslated region of CaMKIIα is a *cis*-acting signal for the localization and translation of mRNA in dendrites. *Proc Natl Acad Sci* **93:** 13250–13255.

Mayford M, Bach ME, Huang YY, Wang L, Hawkins RD, Kandel ER. 1996b. Control of memory formation through regulated expression of a CaMKII transgene. *Science* **274:** 1678–1683.

McHugh TJ, Blum KI, Tsien JZ, Tonegawa S, Wilson MA. 1996. Impaired hippocampal representation of space in CA1-specific NMDAR1 knockout mice. *Cell* **87:** 1339–1349.

McHugh TJ, Jones MW, Quinn JJ, Balthasar N, Coppari R, Elmquist JK, Lowell BB, Fanselow MS, Wilson MA, Tonegawa S. 2007. Dentate gyrus NMDA receptors mediate rapid pattern separation in the hippocampal network. *Science* **317:** 94–99.

Metzger D, Clifford J, Chiba H, Chambon P. 1995. Conditional site-specific recombination in mammalian cells using a ligand-dependent chimeric Cre recombinase. *Proc Natl Acad Sci* **92:** 6991–6995.

* Moser M-B, Rowland DC, Moser EI. 2015. Place cells, grid cells, and memory. *Cold Spring Harb Perspect Biol* **7:** a021808.

Nagel G, Szellas T, Huhn W, Kateriya S, Adeishvili N, Berthold P, Ollig D, Hegemann P, Bamberg E. 2003. Channelrhodopsin-2, a directly light-gated cation-selective membrane channel. *Proc Natl Acad Sci* **100:** 13940–13945.

Nakashiba T, Young JZ, McHugh TJ, Buhl DL, Tonegawa S. 2008. Transgenic inhibition of synaptic transmission reveals role of CA3 output in hippocampal learning. *Science* **319:** 1260–1264.

Nakashiba T, Buhl DL, McHugh TJ, Tonegawa S. 2009. Hippocampal CA3 output is crucial for ripple-associated reactivation and consolidation of memory. *Neuron* **62:** 781–787.

Nakashiba T, Cushman JD, Pelkey KA, Renaudineau S, Buhl DL, McHugh TJ, Rodriguez Barrera V, Chittajallu R, Iwamoto KS, McBain CJ, et al. 2012. Young dentate granule cells mediate pattern separation, whereas old granule cells facilitate pattern completion. *Cell* **149:** 188–201.

Nakazawa K, Quirk MC, Chitwood RA, Watanabe M, Yeckel MF, Sun LD, Kato A, Carr CA, Johnston D, Wilson MA, et al. 2002. Requirement for hippocampal CA3 NMDA receptors in associative memory recall. *Science* **297:** 211–218.

Nakazawa K, Sun LD, Quirk MC, Rondi-Reig L, Wilson MA, Tonegawa S. 2003. Hippocampal CA3 NMDA receptors are crucial for memory acquisition of one-time experience. *Neuron* **38:** 305–315.

Nguyen PV, Abel T, Kandel ER, Bourtchouladze R. 2000. Strain-dependent differences in LTP and hippocampus-dependent memory in inbred mice. *Learn Mem* **7:** 170–179.

Oberdick J, Smeyne RJ, Mann JR, Zackson S, Morgan JI. 1990. A promoter that drives transgene expression in cerebellar Purkinje and retinal bipolar neurons. *Science* **248:** 223–226.

O'Keefe J, Dostrovsky J. 1971. The hippocampus as a spatial map. Preliminary evidence from unit activity in the freely-moving rat. *Brain Res* **34:** 171–175.

Cite this article as *Cold Spring Harb Perspect Biol* doi: 10.1101/cshperspect.a021725

Oliveira AM, Wood MA, McDonough CB, Abel T. 2007. Transgenic mice expressing an inhibitory truncated form of p300 exhibit long-term memory deficits. *Learn Mem* **14:** 564–572.

Oliveira AM, Estévez MA, Hawk JD, Grimes S, Brindle PK, Abel T. 2011. Subregion-specific p300 conditional knock-out mice exhibit long-term memory impairments. *Learn Mem* **18:** 161–169.

Oltval ZN, Milliman CL, Korsmeyer SJ. 1993. Bcl-2 heterodimerizes in vivo with a conserved homolog, Bax, that accelerates programed cell death. *Cell* **74:** 609–619.

Palmiter RD, Brinster RL, Hammer RE, Trumbauer ME, Rosenfeld MG, Birnberg NC, Evans RM. 1982. Dramatic growth of mice that develop from eggs microinjected with metallothionein-growth hormone fusion genes. *Nature* **300:** 611–615.

Pascual O, Casper KB, Kubera C, Zhang J, Revilla-Sanchez R, Sul JY, Takano H, Moss SJ, McCarthy K, Haydon PG. 2005. Astrocytic purinergic signaling coordinates synaptic networks. *Science* **310:** 113–116.

Pennisi E. 2013. The CRISPR craze. *Science* **341:** 833–836.

Qi M, Zhuo M, Skålheggm BS, Brandon EP, Kandel ER, McKnight GS, Idzerda RL. 1996. Impaired hippocampal plasticity in mice lacking the C$\beta_1$ catalytic subunit of cAMP-dependent protein kinase. *Proc Natl Acad Sci* **93:** 1571–1576.

Ramirez S, Liu X, Lin PA, Suh J, Pignatelli M, Redondo RL, Ryan TJ, Tonegawa S. 2013. Creating a false memory in the hippocampus. *Science* **341:** 387–391.

Sahay A, Scobie KN, Hill AS, O'Carroll CM, Kheirbek MA, Burghardt NS, Fenton AA, Dranovsky A, Hen R. 2011. Increasing adult hippocampal neurogenesis is sufficient to improve pattern separation. *Nature* **472:** 466–470.

Sando R III, Baumgaertel K, Pieraut S, Torabi-Rander N, Wandless TJ, Mayford M, Maximov A. 2013. Inducible control of gene expression with destabilized Cre. *Nat Methods* **10:** 1085–1088.

Sauer B, Henderson N. 1988. Site-specific DNA recombination in mammalian cells by the Cre recombinase of bacteriophage P1. *Proc Natl Acad* **85:** 5166–5170.

Shimizu E, Tang YP, Rampon C, Tsien JZ. 2000. NMDA receptor-dependent synaptic reinforcement as a crucial process for memory consolidation. *Science* **290:** 1170–1174.

Silva AJ, Stevens CF, Tonegawa S, Wang Y. 1992a. Deficient hippocampal long-term potentiation in α-calcium-calmodulin kinase II mutant mice. *Science* **257:** 201–206.

Silva AJ, Paylor R, Wehner JM, Tonegawa S. 1992b. Impaired spatial learning in α-calcium-calmodulin kinase II mutant mice. *Science* **257:** 206–211.

Smithies O, Gregg RG, Boggs SS, Koralewski MA, Kucherlapati RS. 1985. Insertion of DNA sequences into the human chromosomal β-globin locus by homologous recombination. *Nature* **317:** 230–234.

Sorge RE, Martin LJ, Isbester KA, Sotocinal SG, Rosen S, Tuttle AH, Wieskopf JS, Acland EL, Dokova A, Kadoura B, et al. 2014. Olfactory exposure to males, including men, causes stress and related analgesia in rodents. *Nat Methods* **11:** 629–632.

Steele RJ, Morris RGM. 1999. Delay-dependent impairment of a matching-to-place task with chronic and intrahippocampal infusion of the NMDA-antagonist D-AP5. *Hippocampus* **9:** 118–136.

Strader CD, Gaffney T, Sugg EE, Candelore MR, Keys R, Patchett AA, Dixon RA. 1991. Allele-specific activation of genetically engineered receptors. *J Biol Chem* **266:** 5–8.

Struthers RS, Vale WW, Arias C, Sawchenko PE, Montminy MR. 1991. Somatotroph hypoplasia and dwarfism in transgenic mice expressing a non-phosphorylatable CREB mutant. *Nature* **350:** 622–624.

Su SC, Rudenko A, Cho S, Tsai LH. 2013. Forebrain-specific deletion of Cdk5 in pyramidal neurons results in mania-like behavior and cognitive impairment. *Neurobiol Learn Mem* **105:** 54–62.

Sugiura H, Yamauchi T. 1992. Developmental changes in the levels of $Ca^{2+}$/calmodulin-dependent protein kinase II α and β proteins in soluble and particulate fractions of the rat brain. *Brain Res* **593:** 97–104.

Suh J, Rivest AJ, Nakashiba T, Tominaga T, Tonegawa S. 2011. Entorhinal cortex layer III input to the hippocampus is crucial for temporal association memory. *Science* **334:** 1415–1420.

Takahashi T, Guron C, Shetty S, Matsui H, Raghow R. 1997. A minimal murine *Msx-1* gene promoter. *J Biol Chem* **272:** 22667–22678.

Takahashi JS, Shimomura K, Kumar V. 2008. Searching for genes underlying behavior: Lessons from circadian rhythms. *Science* **322:** 909–912.

Thomas KR, Capecchi MR. 1987. Site-directed mutagenesis by gene targeting in mouse embryo-derived stem cells. *Cell* **51:** 503–512.

Tsien JZ, Chen DF, Gerber D, Tom C, Mercer EH, Anderson DJ, Mayford M, Kandel ER, Tonegawa S. 1996a. Subregion- and cell-type-restricted gene knockout in mouse brain. *Cell* **87:** 1317–1326.

Tsien JZ, Huerta PT, Tonegawa S. 1996b. The essential role of hippocampal CA1 NMDA receptor-dependent synaptic plasticity in spatial memory. *Cell* **87:** 1327–1338.

Vetere G, Restivo L, Cole CJ, Ross PJ, Ammassari-Teule M, Josselyn SA, Frankland PW. 2011. Spine growth in the anterior cingulate cortex is necessary for the consolidation of contextual fear memory. *Proc Natl Acad Sci* **108:** 8456–8460.

Vo N, Goodman RH. 2001. CREB-binding protein and p300 in transcriptional regulation. *J Biol Chem* **276:** 13505–13508.

Warburton EC, Glover CP, Massey PV, Wan H, Johnson B, Bienemann A, Deuschle U, Kew JN, Aggleton JP, Bashir ZI, et al. 2005. cAMP responsive element-binding protein phosphorylation is necessary for perirhinal long-term potentiation and recognition memory. *J Neurosci* **25:** 6296–6303.

Wehner JM, Silva A. 1996. Importance of strain differences in evaluations of learning and memory processes in null mutants. *Ment Retard Dev Disabil Res Rev* **2:** 243–248.

Wiltgen BJ, Royle GA, Gray EE, Abdipranoto A, Thangthaeng N, Jacobs N, Saab F, Tonegawa S, Heinemann SF, O'Dell TJ, et al. 2010. A role for calcium-permeable

AMPA receptors in synaptic plasticity and learning. *PLoS ONE* **5:** e12818.

Wisden W, Seeburg PH. 1993. A complex mosaic of high-affinity kainate receptors in rat brain. *J Neurosci* **13:** 3582–3598.

Wood MA, Kaplan MP, Park A, Blanchard EJ, Oliveira AM, Lombardi TL, Abel T. 2005. Transgenic mice expressing a truncated form of CREB-binding protein (CBP) exhibit deficits in hippocampal synaptic plasticity and memory storage. *Learn Mem* **12:** 111–119.

Wood MA, Attner MA, Oliveira AM, Brindle PK, Abel T. 2006. A transcription factor-binding domain of the coactivator CBP is essential for long-term memory and the expression of specific target genes. *Learn Mem* **13:** 609–617.

Xu B, Gottschalk W, Chow A, Wilson RI, Schnell E, Zang K, Wang D, Nicoll A, Lu B, Reichardt LF. 2000. The role of brain-derived neurotrophic factor receptors in the mature hippocampus: Modulation of long-term potentiation through a presynaptic mechanism involving TrkB. *J Neurosci* **20:** 6888–6897.

Yoneshima H, Yamasaki S, Voelker CC, Molnar Z, Christophe E, Audinat E, Takemoto M, Nishiwaki M, Tsuji S, Fujita I, et al. 2006. Er81 is expressed in a subpopulation of layer 5 neurons in rodent and primate neocortices. *Neuroscience* **137:** 401–412.

Cite this article as *Cold Spring Harb Perspect Biol* doi: 10.1101/cshperspect.a021725

# The Corticohippocampal Circuit, Synaptic Plasticity, and Memory

Jayeeta Basu[1] and Steven A. Siegelbaum[2,3,4]

[1]Department of Neuroscience and Physiology, NYU Neuroscience Institute, New York University School of Medicine, New York, New York 10016

[2]Kavli Institute for Brain Science, Columbia University, New York, New York 10032

[3]Department of Neuroscience, Columbia University, New York, New York 10032

[4]Department of Pharmacology, Columbia University, New York, New York 10032

*Correspondence:* sas8@columbia.edu

Synaptic plasticity serves as a cellular substrate for information storage in the central nervous system. The entorhinal cortex (EC) and hippocampus are interconnected brain areas supporting basic cognitive functions important for the formation and retrieval of declarative memories. Here, we discuss how information flow in the EC–hippocampal loop is organized through circuit design. We highlight recently identified corticohippocampal and intrahippocampal connections and how these long-range and local microcircuits contribute to learning. This review also describes various forms of activity-dependent mechanisms that change the strength of corticohippocampal synaptic transmission. A key point to emerge from these studies is that patterned activity and interaction of coincident inputs gives rise to associational plasticity and long-term regulation of information flow. Finally, we offer insights about how learning-related synaptic plasticity within the corticohippocampal circuit during sensory experiences may enable adaptive behaviors for encoding spatial, episodic, social, and contextual memories.

Ever since the description by Scoville and Milner (1957) of the profound anterograde amnesia in patient H.M. following bilateral temporal lobe resection, the hippocampus and surrounding temporal lobe structures have been extensively studied for their role in memory storage (Squire and Wixted 2011). Fifteen years after this initial finding, our understanding of the neurophysiological bases of hippocampal function were greatly enhanced by two breakthroughs: Bliss and Lomo's (1973) finding of activity-dependent long-term potentiation (LTP) of synaptic transmission in the hippo-

campus, and the discovery of hippocampal place cells, neurons that encode the spatial position of an animal, by O'Keefe and Dostrovsky (1971). These discoveries stimulated a number of subsequent advances in our understanding of various forms of long-term synaptic plasticity at different stages of information processing in the corticohippocampal circuit, and how such plasticity contributes to memory storage and spatial representation.

Changes in synaptic efficacy ultimately act by altering the flow of information through neural circuits. Thus, a deeper understanding of how synaptic plasticity may subserve the encoding of memory requires a detailed knowledge of the paths of information flow through the corticohippocampal circuit, and how neural activity alters such information processing. Because there have been a number of recent excellent reviews on the importance of various forms of hippocampal synaptic plasticity in learning and memory (Morris et al. 2013; Bannerman et al. 2014), we largely focus on recent studies elucidating new features of the corticohippocampal circuit and new forms of plasticity that are tuned to the dynamics of this circuit.

## INFORMATION FLOW THROUGH THE CORTICOHIPPOCAMPAL CIRCUIT

The hippocampus is important for both spatial and nonspatial forms of declarative or explicit memory (Squire et al. 2004), our knowledge of people, places, things, and events. In addition to encoding spatial information (Burgess and O'Keefe 1996), hippocampal neurons may also encode time during episodic events (Pastalkova et al. 2008; Kraus et al. 2013; Macdonald et al. 2013). How does the hippocampus encode and store these diverse memories? Although a definitive answer is lacking, our knowledge of the corticohippocampal circuit has greatly expanded in recent years, giving us a new appreciation for the multiple pathways by which information is processed in the hippocampus, an important step to achieving a circuit-level understanding of how this brain region stores memories.

Neurons wire up during development to form circuits that provide the architectural framework for information flow in the brain, enabling one brain area to influence another. Learning requires the association of information from different coactive brain areas during sensory experiences or the reprocessing of internal representations. Such associations result in plastic changes in synaptic and cell-wide functions that enable the formation of preferentially connected cell assemblies. At the level of a single neuron, this could occur through integration and association of inputs that coincide in time or space to influence the neuron's output.

As initially defined by Lorente de Nó (1934), the hippocampal region is composed of several subregions, including dentate gyrus (DG), and the CA3, CA2, and CA1 regions of the hippocampus proper. In the rat brain, there are estimated to be ∼1,000,000 DG granule neurons, 300,000 CA3 pyramidal neurons, ∼30,000 CA2 pyramidal neurons, and 300,000 CA1 pyramidal neurons (Amaral and Witter 1989). In addition to these excitatory, principal neurons, there are many classes of inhibitory neurons in the hippocampus, although the total number of these inhibitory interneurons is only about 10% to 20% that of the principal cells. In CA1 alone, more than 20 types of GABAergic interneurons, have been classified according to their morphology, location, molecular and electrophysiological properties, and synaptic targets (Klausberger and Somogyi 2008). A key goal in hippocampal research is to gain an understanding as to how these different subregions process their inputs during learning to generate an output contributing to distinct aspects of memory encoding.

### The Corticohippocampal Circuit I: Classical Pathways

A typical CA1 pyramidal neuron in the rat receives a total of ∼30,000 glutamatergic synaptic inputs distributed throughout its dendritic tree. In addition, it receives ∼1700 GABAergic inputs (Megias et al. 2001). The major source of glutamatergic input to hippocampus comes from the entorhinal cortex (EC), a polymodal sensory association area that conveys both nonspatial sensory information (from the lateral entorhinal cortex [LEC]) and spatial information (from the medial entorhinal cortex [MEC]). This sensory information is then processed within the hippocampus by several parallel circuits, ultimately leaving the hippocampus through CA1, the major output pathway (van Strien et al. 2009).

One striking feature of certain neurons in layers II and III of EC (EC LII and LIII), termed grid cells, is that they show spatially tuned firing patterns consisting of a hexagonally spaced array of grid-like firing fields that cover the extent of a two-dimensional environment (Fyhn et al. 2004; Hafting et al. 2005; Buzsaki and Moser 2013). Convergent input from a number of EC grid cells with slightly shifted firing fields has been proposed to produce an interference-like pattern that gives rise to the single well-defined place field of a hippocampal place cell (O'Keefe and Burgess 2005; Solstad et al. 2006; Bush et al. 2014).

Information from the superficial layers of EC reaches CA1 through both direct and indirect pathways (Fig. 1) (van Strien et al. 2009). The most well-characterized route is the indirect stream of information flow through the trisynaptic path. In this glutamatergic circuit, EC LII stellate cells send excitatory projections through the perforant path (PP) to granule cells of the DG, whose mossy fiber projections excite CA3 pyramidal neurons, which in turn excite CA1 pyramidal cells through the Schaffer collat-

eral (SC) pathway (EC LII → DG → CA3 → CA1). In addition to the trisynaptic path, CA1 pyramidal neurons also receive a direct glutamatergic projection from EC LIII pyramidal neurons through the temporoammonic or PP (PP, EC LIII → CA1).

These direct and indirect corticohippocampal inputs target distinct regions of the CA1 pyramidal neuron dendritic tree, with the SC inputs of the indirect pathway forming synapses on more "proximal" regions of CA1 pyramidal neuron apical dendrites in a layer of CA1 known as stratum radiatum (SR). In contrast, the direct PP inputs from EC form synapses on the "distal" regions of CA1 pyramidal neuron apical dendrites in a layer known as stratum lacunosum molecular (SLM). As a result of their distinct dendritic locations, these two inputs provide different levels of excitatory drive, due in part to differential attenuation by dendritic

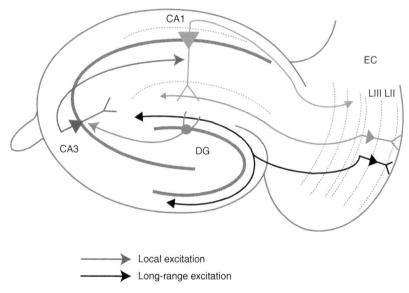

Figure 1. The classical corticohippocampal glutamatergic circuits. The classical corticohippocampal circuit comprises of glutamatergic input from the superficial entorhinal cortex (EC) layers (LII and LIII) to CA1 pyramidal neurons via the trisynaptic and monosynaptic paths; hippocampal back projections to the deep layers of EC complete the loop. Sensory signals drive the perforant path (PP, purple) inputs from EC LIII pyramidal neurons to distal CA1 pyramidal neuron dendrites (light blue). Activated EC LII pyramidal neurons send inputs to dentate gyrus (DG, black), which sends mossy fiber axons (dark green) to CA3 and then CA3 feeds onto CA1 neurons through Schaffer collateral (SC, dark red) excitatory inputs. A major output of the hippocampus arises from CA1 pyramidal neurons, which project to lateral ventricles (LVs) of EC. There is a 15-20 ms timing delay for transmission of information from EC LII to CA1 through the trisynaptic path compared with that from EC LIII to CA1 via the monosynaptic path.

cable properties, with the PP synapses of the direct pathway providing weak excitation at the soma compared with the strong excitation provided by the SC inputs of the indirect pathway. These two classes of glutamatergic inputs also activate a number of different classes of GABAergic interneurons that inhibit CA1 output in a feedforward manner (Fig. 2).

## The Corticohippocampal Circuit
## II: Neoclassical Pathways

More recent results show that, in addition to their classical inputs from CA3 and EC, CA1 neurons also receive strong excitatory input from the CA2 region of the hippocampus (CA2 → CA1) (Chevaleyre and Siegelbaum 2010), a relatively small area nestled between CA3 and CA1 (Fig. 3). Although first identified by Lorente de Nó in 1934, the CA2 region has remained relatively unexplored, due in part to its small size and transitional location between

CA3 and CA1. However, recent studies using genetic approaches clearly show the unique molecular identity of CA2 pyramidal neurons (Lein et al. 2005; Hitti and Siegelbaum 2014; Kohara et al. 2014) and suggest the importance of the CA2 region in certain hippocampal functions (Piskorowski and Chevaleyre 2012).

Similar to CA1 neurons, CA2 pyramidal neurons receive both direct input from LII EC neurons (EC LII → CA2) onto their distal dendrites (Cui et al. 2013; Hitti and Siegelbaum 2014) and indirect input from the CA3 SC pathway (EC LII → DG → CA3 → CA2) onto their proximal dendrites (Chevaleyre and Siegelbaum 2010). In addition, CA2 also receives weaker mossy fiber excitatory input directly from DG granule cells (EC LII → DG → CA2) (Kohara et al. 2014).

In striking contrast to CA1, CA2 pyramidal neurons are excited much more strongly by their direct EC input compared with their SC input (Chevaleyre and Siegelbaum 2010). Fur-

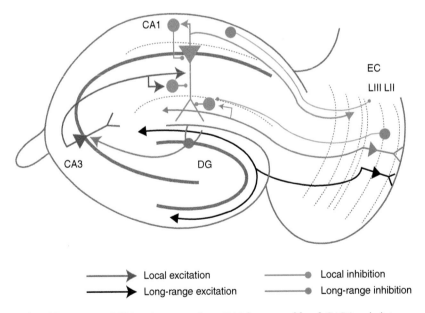

Figure 2. Local and long-range GABAergic connections. CA1 has several local GABAergic interneurons (red). These target the CA1 pyramidal neuron soma, axon, and dendrite to modulate pyramidal neuron activity in a domain-specific manner. Schaffer collateral (SC)- and entorhinal cortex (EC)-associated inhibitory microcircuits provide feedforward inhibition, whereas feedback inhibition is recruited recurrently when the CA1 pyramidal neuron fires an action potential. Long-range inhibitory projections (green) from the EC provide direct inhibition preferentially to local interneurons (INs) in CA1. Long-range projections from GABAergic neurons in stratum oriens (SO) of hippocampus to layer II/III (LII/LIII) of the EC have also been described.

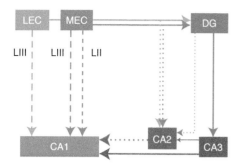

**Figure 3.** The updated corticohippocampal circuit. This circuit diagram integrates the recently discovered glutamatergic inputs from the entorhinal cortex (EC) to the hippocampus as well as connections within the hippocampus. In addition to the classical trisynaptic (EC layer II [LII] → dentate gyrus [DG] → CA3 → CA1, solid line) and monosynaptic (EC layer III (LIII) → CA1, large-dashed lines) pathways of information flow, CA1 also receives monosynaptic projections from LII of the medial entorhinal cortex (MEC) (small dashed line) and CA2 receives direct inputs from LII of both MEC and lateral entorhinal cortex (LEC) (dotted lines). Within the hippocampus, CA2 sends prominent inputs to CA1, targeting dendritic domains (stratum oriens [SO]/stratum radiatum [SR]) (red) that overlap with the CA3 → CA1 inputs. CA2 also receives weak inputs from DG and CA3. The thickness of the arrowed lines emphasizes the strength of the input connection.

thermore, a CA2 pyramidal neuron forms a strong multiquantal synaptic connection with a CA1 pyramidal neuron, eliciting an excitatory postsynaptic potential (EPSP) that is significantly larger than the uniquantal response at the synaptic connection between a CA3 and CA1 pyramidal neuron (Chevaleyre and Siegelbaum 2010). As a result, CA2 mediates a powerful disynaptic circuit directly linking cortical input to hippocampal output (EC → CA2 → CA1), which operates in parallel to the trisynaptic path. Although CA2, in principle, can participate in a quadrisynaptic path linking EC to CA1 (EC → DG → CA3 → CA2 → CA1), strong feedforward inhibition from CA3 to CA2 normally limits efficient information flow through this pathway.

More recent evidence suggests the existence of an additional direct pathway from EC to CA1 (Fig. 3) involving projections from LII pyramidal neurons of MEC (EC LII → CA1) (Kitamura et al. 2014). These inputs also target the CA1 pyramidal neuron distal dendrites but are weaker compared with the EC LIII inputs. In addition, the EC LII projections provide strong excitatory drive to local CA1 interneurons, thus recruiting pronounced feedforward inhibition onto CA1 pyramidal neurons. In addition to the LII and LIII excitatory inputs to CA1, MEC also sends direct GABAergic inputs to all fields of the hippocampus (Fig. 2) (Melzer et al. 2012). These inhibitory projections likely originate in LII/III of the EC and show a bias toward innervating hippocampal GABAergic interneurons near the SR/SLM border in CA1.

## Potential Function of Parallel Corticohippocampal Streams

What is the purpose of the parallel streams of corticohippocampal information flow? Do the various hippocampal subfields (DG, CA1, CA2, and CA3) act as semi-independent parallel information-processing units? Does the information received by these neighboring subfields differ in its cellular origin or in its quality? Or do the hippocampal subfields act primarily as serial processing stations that are transforming cortical input in discrete and successive steps?

The fact that CA1 integrates a direct representation of spatial and nonspatial sensory information (originating in the MEC and LEC) with information that has been processed through hippocampal disynaptic and trisynaptic circuits has led to the idea that CA1 performs a comparison between the immediate spatio-sensory context and stored mnemonic information. However, the nature of the comparison is not known.

One suggestion is that CA1 acts as a novelty detector, comparing stored memory-related information in DG and CA3 with ongoing direct sensory representations from the EC (Lisman and Grace 2005; Duncan et al. 2012). The strong feedforward inhibition triggered by the direct EC inputs may serve as an inhibitory gate on information flow arriving from the SC inputs

J. Basu and S.A. Siegelbaum

by means of the action of dopaminergic novelty signals (Lisman and Grace 2005). Another suggestion is that the direct inputs may provide instructive signals for assessing the salience of information flow through the trisynaptic path using a timing-dependent plasticity rule, which is tuned to the delay-line architecture of the hippocampal circuit (Dudman et al. 2007). Because of the two extra synapses and conduction and integration delays, information flowing through the trisynaptic path arrives at CA1 neurons some 15 to 20 msec after the arrival of information through the direct EC inputs (Yeckel and Berger 1990). Such a circuit design allows for integration and comparison of temporally coordinated inputs that may contribute to the encoding of sequential episodic events (Dudman et al. 2007; Ahmed and Mehta 2009; Mizuseki et al. 2009; Basu et al. 2013).

## The Corticohippocampal Circuit III: Outputs of the Hippocampus

CA1 pyramidal neurons provide the major output from the hippocampus, sending projections to a number of brain regions, including the neighboring subiculum, perirhinal cortex, prefrontal cortex, and amygdala (van Groen and Wyss 1990). A small fraction of pyramidal neurons from dorsomedial CA1 also project to the restrosplenial cortex (Wyss and Van Groen 1992). One particularly strong output goes back to the EC, in which CA1 axons excite layer V pyramidal cells, which in turn send excitatory feedback input to the EC LII/III, thereby completing an EC → hippocampus → EC loop (Figs. 1 and 3) (Naber et al. 2001). In addition to excitatory pyramidal neuronal outputs, certain GABAergic neurons in specific layers of CA1 project directly to retrohippocampal and cortical areas. These include long-range GABAergic projections from somatostatin (SOM) and mGluR1a expressing inhibitory neurons in stratum oriens (SO) to the subiculum and medial septum (MS) (Jinno et al. 2007; Fuentealba et al. 2008). GABAergic cells located in the border of SR and SLM (often expressing muscarinic AChRs or mGluR1as) project to retrosplenial cortex and indusium gresium

(Jinno et al. 2007). There are also SOM-expressing GABAergic neurons in SO of CA1 and hilus of DG that send direct projections to the superficial layers of MEC (Fig. 2) and striatum (Melzer et al. 2012). The long-range inhibitory projections are often highly myelinated (Jinno et al. 2007) and predominantly target local GABAergic interneurons at the projection sites (Melzer et al. 2012). These properties have led to the suggestion that long-range inhibitory projections may be important for coordinating the timing between the hippocampus and its cortical targets (Buzsaki and Chrobak 1995) and could serve a disinhibitory role (Caputi et al. 2013).

At present, there are conflicting data as to whether CA2 neurons project outside of the hippocampus. One study using a cell-type-specific rabies virus retrograde tracing strategy reported that CA2 pyramidal neurons send projections to EC LII neurons (Rowland et al. 2013), the source of the direct cortical input to CA2 (Hitti and Siegelbaum 2014; Kohara et al. 2014). CA2 has also been reported to project to the supramammillary nucleus (Cui et al. 2013), a hypothalamic region long known to provide strong input to CA2 (Haglund et al. 1984; Vertes 1992; Magloczky et al. 1994; Ochiishi et al. 1999; Kiss et al. 2000).

## The Corticohippocampal Circuit IV: Heterogeneity within the CA1 Pyramidal Neuron Population

Although many studies treat CA1 pyramidal neurons as a uniform population, there is increasing evidence for heterogeneity along each of the three spatial axes of the hippocampus: the septotemporal (or dorsoventral) longitudinal axis, the proximodistal transverse axis (CA2 to subiculum), and the deep superficial radial axis in the stratum pyramidale (SP) cell body layer (with deep referring to pyramidal neurons closer to SO and superficial referring to pyramidal neurons closer to SR).

Along the transverse axis, proximal CA1 neurons (closer to the CA2 border) receive direct input primarily from the MEC, whereas distal CA1 neurons (closer to the border with

subiculum).

136  Cite this article as *Cold Spring Harb Perspect Biol* doi: 10.1101/cshperspect.a021733

subiculum) receive direct input primarily from the LEC (Fig. 3) (Ishizuka et al. 1990; Witter and Amaral 1991). This topographical arrangement is reversed in subiculum. Pyramidal neurons in CA1 and subiculum also show a strong proximodistal gradient from regular firing in CA1 to prominent burst firing in subiculum (and CA3) (Jarsky et al. 2008; Kim and Spruston 2012).

Marked differences in pyramidal neuron-firing properties along the transverse axis have also been recorded in vivo. Thus, distal CA1 pyramidal neurons, which receive largely non-spatial input from the LEC, show dispersed firing during spatial navigation and tend to have multiple place fields (Henriksen et al. 2010) and display increased 20–40 Hz coupling with the LEC during spatial associational learning behavior (Igarashi et al. 2014). As expected, the firing of proximal CA1 pyramidal neurons are more strongly coupled to MEC neuron firing at theta frequencies and show greater spatial modulation and more compact place fields compared with their distal counterparts (Henriksen et al. 2010).

Since the initial studies of Lorente de Nó (1934), it has been suggested that there may be two separate sublayers of CA1 pyramidal neurons along the radial axis: a relatively tight superficial layer (closer to SR) and a broader more dispersed deep layer (closer to SO) (Slomianka et al. 2011). Molecular evidence supporting this view comes from the finding that the superficial neurons, but not deep pyramidal neurons, are enriched in calbindin and zinc (Baimbridge and Miller 1982; Baimbridge et al. 1982; Dong et al. 2008; Slomianka et al. 2011). During development, expression of Sox5 specifies pyramidal neurons in the deep layers, whereas the coexpression of Satb2 and Zbtb20 may determine the fate of superficial neurons (Nielsen et al. 2010; Xie et al. 2010). Morphologically, the deep neurons have fewer oblique dendrites in SR and more extensive branching in SO than do superficial neurons (Bannister and Larkman 1995b). The two sublayers also have distinct input–output connectivity, with deep neurons receiving preferential input from CA2 pyramidal neurons (Kohara et al. 2014) and local inhibitory parvalbumin-positive basket cells (Lee

et al. 2014), the latter distinction first reported by Lorente de Nó (1934).

Importantly, the two layers show different functional properties in vitro and during behavior. The presence in superficial neurons of calbindin, which buffers calcium and interferes with N-methyl-D-aspartate receptor (NMDAR)-dependent LTP, may lead to plasticity differences in the two populations (Arai et al. 1994; Bannister and Larkman 1995a,b). The superficial cells show a larger influence of the hyperpolarization-activated cation current, Ih, which contributes to the resting integrative properties of the neurons (Jarsky et al. 2008). In vivo, the deep neurons show a greater tendency for spatial tuning during navigation as well as stronger modulation during slow wave sleep (Mizuseki et al. 2011).

Finally, there are functional and structural differences along the longitudinal axis of the hippocampus, including differences in protein expression along the dorsoventral axis that actually define three distinct regions: dorsal, intermediate, and ventral hippocampus (Dong et al. 2010). In addition, ventral hippocampal CA1 pyramidal neurons are more excitable than dorsal neurons, perhaps resulting from a stronger expression of the HCN1 cation channel in the ventral hippocampus (Dougherty et al. 2013).

Dorsal hippocampal place cells show more precise place fields compared with ventral hippocampal place cells, whose place fields are more diffuse (Jung et al. 1994), although this may not degrade the ability of the ventral hippocampus to encode spatial position because of the higher ventral cell-firing rates (Keinath et al. 2014). This difference in place-field size corresponds to a dorsoventral gradient of increasing grid field size and spacing in MEC LII grid cells (Brun et al. 2008), which show a topographic dorsoventral projection to the hippocampus (van Strien et al. 2009). Dorsal and ventral hippocampi also differ in their outputs. Ventral hippocampus projects more strongly to prefrontal cortex and amygdala compared with dorsal hippocampus (Ishikawa and Nakamura 2006). Moreover, these differences reflect the distinct behavioral roles of these two regions, with dorsal hippocampus more important for spatial and contextual memory and ventral hippocampus more important for emotional- and anxiety-related behaviors (Fanselow and Dong 2010).

## ROLE OF DISTINCT REGIONS OF THE CORTICOHIPPOCAMPAL CIRCUIT IN LEARNING AND MEMORY

Studies using both conventional lesions and sophisticated genetic approaches have identified specific roles of distinct corticohippocampal circuit elements in learning and memory (Fig. 4), and have begun to identify the role of specific regions in disease. CA1, which provides the major output of the hippocampus, is essential for most, if not all, forms of hippocampal-dependent memory as CA1 lesions in both humans and other mammalian species, including rodents, leads to severe memory impairment (Zola-Morgan et al. 1986; Squire 2004). At present, the role of the hippocampus and CA1 in memory recall is somewhat unclear, as certain patients with hippocampal damage (including H.M.), as well as lesioned animals, are able to recall remote memories formed before hippocampal damage (Squire 2004). However, other patients with hippocampal CA1 lesions have both severe retrograde and anterograde loss of episodic memory (Bartsch et al. 2011).

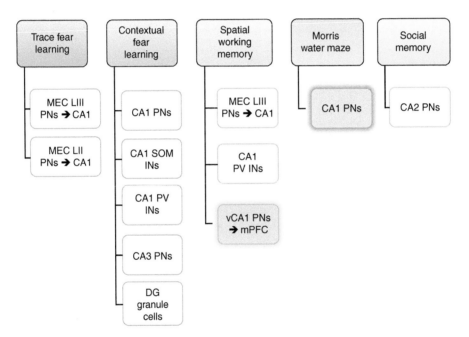

Figure 4. Cellular and circuit correlates of behavioral learning. Genetic and classical lesions have elucidated how the different pathways and cell populations comprising the corticohippocampal circuit support various forms of associational learning and declarative memory functions. All subfields, except the grayed boxes, show results from behavioral experiments involving genetically targeted cell type or input specific functional manipulations. Gray boxes indicate physical or chemical lesion-based findings. Both the medial entorhinal cortex (MEC) layer II (LII) (Kitamura et al. 2014) and layer III (LIII) (Suh et al. 2011) glutamatergic inputs to CA1 are used for trace fear conditioning (TFC). CA1 pyramidal neurons (PNs) (Goshen et al. 2011), CA3 pyramidal neurons (Nakashiba et al. 2008), dentate gyrus (DG), granule cells (Nakashiba et al. 2012; Kheirbek et al. 2013), parvalbumin (PV) interneurons (INs) (Donato et al. 2013), and somatostatin (SOM) interneurons (Lovett-Barron et al. 2014) have all been found to be involved in contextual fear-learning behavior. Spatial working memory requires activity in the MEC LIII pyramidal neurons projections to CA1 (Suh et al. 2011; Yamamoto et al. 2014), CA1 PV interneurons (Murray et al. 2011), and the ventral hippocampal projections to the prefrontal cortex (Wang and Cai 2006).

Moreover, a number of studies have shown the activation of the hippocampus, including CA1, during memory recall (Rugg and Vilberg 2013). Furthermore, optogenetic experiments have shown that temporally precise and transient inactivation of dorsal CA1 pyramidal neurons in rodents markedly impairs recall (Goshen et al. 2011). This suggests that long-term inactivation of the hippocampus through genetic or physical lesions may result in a variable degree of compensatory changes in other brain regions that enable recall in some individuals but not in others.

One surprising finding is that lesions of dorsal CA1 lead to a disruption of grid cell spatial tuning patterns within the EC, which provides the major input to the hippocampus (Bonnevie et al. 2013). In addition, hippocampal place-cell activity can precede the appearance of well-defined EC grid-cell firing during early postnatal development (Langston et al. 2010; Wills et al. 2010). These results suggest that feedback from the hippocampus may be necessary for optimal grid-field formation. As "border cells" that fire when an animal reaches the edges of an environment develop in the EC before grid cells, it has been proposed that these border cells may provide the input that gives rise to hippocampal place cells (Bjerknes et al. 2014). Furthermore, removal of CA1 inputs to the EC converts grid cells into head direction cells (Bonnevie et al. 2013).

In contrast to the profound changes produced by lesions in CA1, lesions of DG appear to cause more subtle changes in memory performance. Although such lesions do not disrupt a basic form of contextual fear conditioning, they do result in an impaired ability to distinguish between closely related environments, a process termed pattern separation (Leutgeb et al. 2007; Sahay et al. 2011; Nakashiba et al. 2012), the ability to distinguish between closely related environments. Interestingly, DG is a prominent site of adult neurogenesis (Drew et al. 2013) and the newborn neurons appear preferentially involved in pattern separation (Nakashiba et al. 2012). DG is also affected preferentially during age-related memory loss (Small et al. 2011; Pavlopoulos et al. 2013),

whereas Alzheimer's disease initially targets the EC (Braak and Braak 1985). Interestingly, a recent study (Kheirbek et al. 2013) using optogenetic activation and silencing of granule cells shows that dorsal DG is important for encoding but not retrieval of contextual fear memories. Genetic knockout studies suggest that CA3 is important for pattern completion (Nakazawa et al. 2002), the ability to recall a memory from partial cues, and one-trial forms of contextual learning (Fig. 4), in which a strong aversive stimulus results in rapid memory formation without the usual need for repeated trials and spatial reference memory.

The primary role of the trisynaptic path in memory formation and spatial encoding was called into question by physical (Brun et al. 2002) and genetic (Nakashiba et al. 2008) lesion studies, which showed that removal of CA3 input to CA1 had surprisingly little effect on spatial reference memory performance in the Morris water maze (MWM), or on the rate of CA1 neuron place-cell firing in vivo. To explain these findings, it was postulated that the direct EC LIII (and possibly EC LII) inputs to CA1 might be sufficient to drive normal CA1 firing rates to support hippocampus-dependent memory.

In support of the importance of the direct EC LIII inputs in spatial memory, chemical and electrolytic lesions of the MEC inputs to CA1 were found to lead to a degradation of place-field precision (Brun et al. 2008) and deficits in consolidation of spatial long-term memory (Remondes and Schuman 2004). However, highly selective genetic lesions or optogenetic silencing of glutamatergic MEC LIII inputs to CA1 impaired trace fear conditioning (TFC), a form of temporal association memory, and spatial working memory, but did not perturb spatial reference memory, contextual fear conditioning, or place-cell firing (Fig. 4) (Suh et al. 2011; Kitamura et al. 2014; Yamamoto et al. 2014). The recently identified EC LII pyramidal neuron projections to CA1 are also involved in trace fear learning but have an opposite role to EC LIII inputs, acting to inhibit fear memory by recruiting a population of dendritic targeting CA1 inhibitory neurons (Kitamura et al. 2014).

The suggestion that the direct EC inputs to CA1 can compensate for loss of CA3 inputs is difficult to reconcile with the finding that the direct EC inputs to CA1 provide only a weak excitatory drive and so cannot elicit CA1 output, except when stimulated in high-frequency bursts (Jarsky et al. 2005). One alternative explanation is that the residual function of CA1 pyramidal neurons and memory task performance following CA3 lesions is maintained by activity through the disynaptic pathway mediated by CA2. However, when output from CA2 pyramidal neurons was silenced using a Cre-dependent viral vector to express tetanus toxin in the Amigo2-Cre mouse line, there was little change in spatial reference memory (MWM), object or odor recognition (including social odors), or context memory (contextual fear conditioning). Surprisingly, inactivation of CA2 did cause a profound deficit in social memory, the ability of an animal to recognize and remember individual members of its species (Fig. 4) (Hitti and Siegelbaum 2014). The role of the hippocampus in social memory in humans is well illustrated by the case of H.M., who could no longer form memories of new individuals following temporal lobe resection (Corkin 2002).

The finding that spatial and contextual fear conditioning are intact following lesioning of all three major classes of inputs to CA1 is somewhat surprising, and could indicate that one or two input pathways may be able to maintain memory storage capacity in the absence of the third pathway. Perhaps this explains why there are the parallel processing routes for information flow from the EC to hippocampus that are sufficient but not necessary to sustain the critical functions of the hippocampus, namely, spatial and contextual encoding. Another possibility is that long-term lesions and cell-type-specific genetic ablations may lead to compensatory mechanisms that result in strengthening or redistribution of functional gain of otherwise weaker pathways, thus enabling them to sustain the function of the lesioned circuit. Studies using temporally precise inactivation of pairs of hippocampal regions and inputs may help determine whether different regions do indeed provide such compensation.

## EXTRAHIPPOCAMPAL AND NEUROMODULATORY TUNING OF CORTICOHIPPOCAMPAL CIRCUITS DURING LEARNING AND MEMORY

There are several neuromodulatory inputs to the various subfields of the hippocampus that strongly influence synaptic transmission and activity in the corticohippocampal circuit and contribute significantly to learning behaviors. These projections often target specific subpopulations of GABAergic interneurons in addition to their glutamatergic counterparts. Furthermore, there are substantial differences in the innervation patterns and behaviorally triggered activity of such inputs.

Aminergic modulatory inputs from midbrain and brain stem can markedly alter the function of various circuit elements and modulate memory formation. The importance of dopamine (DA) in regulating short-term synaptic transmission, long-term plasticity, and learning and memory has received a great deal of attention (Jay 2003; Lisman and Grace 2005). Studies using bath application of dopaminergic agonists generally report a strong suppression of the direct EC input to CA1, with little effect on the SC input (Lisman and Grace 2005; Ito and Schuman 2007).

Other studies show that DA facilitates the induction of a late phase of LTP at the SC → CA1 pyramidal neuron synapses by activation of D1/D5 receptors and an increase in cAMP levels both in vitro (Huang and Kandel 1995) and in vivo (Lemon and Manahan-Vaughan 2006). Moreover, DA has been found to be important for both hippocampal memory formation (Wilkerson and Levin 1999; Rossato et al. 2009; Bethus et al. 2010) and the stability of place-field representations (Kentros et al. 1998). Such results have led to the suggestion that DA release in response to novelty acts as a reward signal to enhance memory storage by increasing the relative influence of the hippocampal SC inputs to CA1 versus the EC inputs (Lisman and Grace 2005).

Cholinergic inputs also play a key role in regulating hippocampal function and memory. A recent study (Lovett-Barron et al. 2014) found that cholinergic input from the MS activates a class of SOM-positive dendritic targeting interneurons during the unconditioned aversive stimulus in a contextual fear-conditioning task. Interfering with this circuit-based mechanism by silencing dendrite-targeting inhibition in CA1 during presentation of the unconditioned stimulus impaired fear learning (Fig. 4). The increased inhibition at CA1 pyramidal neuron distal dendrites acts to suppress any coincident excitatory inputs conveyed by the EC direct pathway. The investigators postulate that such a mechanism may be useful for preventing cortical sensory information representing the aversive stimuli from becoming confounded with the representation of the context, enabling recall of the context alone to trigger a fear response.

Another extrahippocampal modulatory input to CA1 comes from the thalamic nucleus reuniens (NuRe), which receives input from the prefrontal cortex. Anatomical and electrophysiological studies show that the NuRe sends excitatory inputs to the SLM region of CA1, where it innervates both local interneurons and the distal dendrites of pyramidal neurons (Wouterlood et al. 1990; Hirayasu and Wada 1992; Dolleman-Van der Weel et al. 1997, 2009). Xu and Südhof (2013) recently found that this pathway is important for specificity of contextual representations by preventing the generalization of fear memory, based on experiments in which the NuRe projections were selectively silenced.

The hippocampal CA2 region also receives neuromodulatory inputs from various midbrain nuclei in addition to its EC LII and intrahippocampal inputs. These include the reciprocal projections between CA2 and the supramammillary nucleus (SUM), MS, diagonal band of Broca (DBB) as well as afferent vasopressinergic projections from paraventricular nuclei of the hypothalamus to CA2 (Cui et al. 2013; Hitti and Siegelbaum 2014). In fact, the strong input from the hypothalamic nuclei as well as a high level of expression of the arginine vasopressin receptor AVPR1b likely contributes to the ability of the CA2 region to participate in a specific circuit for the storage and recall of social memory (Young et al. 2006; Hitti and Siegelbaum 2014). A global knockout of

AVPR1b (Wersinger et al. 2002) shows decreased social memory, the ability to remember interactions with conspecifics, and decreased temporal memory for event order (DeVito et al. 2009). Whereas AVPR1b is widely expressed (albeit at lower levels) throughout the brain, including in hypothalamus, the deficit in socially motivated aggressive behavior in the knockout mouse was partially rescued by selective expression of AVPR1b in dorsal CA2 using spatially, but not genetically, targeted viral injections (Pagani et al. 2014). It will be of interest to determine whether AVPR1b expression in CA2 is also required for formation of social memory (Hitti and Siegelbaum 2014).

## LONG-TERM SYNAPTIC PLASTICITY IN THE CORTICOHIPPOCAMPAL CIRCUIT IN LEARNING AND MEMORY

One of the striking features of synapses in the central nervous system, especially those in the corticohippocampal circuit, is the extent to which their strength can be regulated for prolonged periods of time by different patterns of synaptic activity, a process termed activity-dependent synaptic plasticity. In some instances, the plasticity is homosynaptic in that activity within a given synaptic pathway leads to altered strength of synaptic communication at the same synapses that were activated (Fig. 5A,B). Other examples of synaptic plasticity are heterosynaptic in that activity in one synaptic pathway influences the function of another pathway (Fig. 5C). Although there is a great deal of correlative evidence linking forms of plasticity to hippocampal-dependent memory formation, the precise role of plasticity mechanisms in learning and memory formation remains controversial (see reviews Morris 2013; Bannerman et al. 2014). This complexity likely reflects the fact that there are multiple forms of plasticity that differ in the pattern of activity required for their induction, the molecular mechanisms for both their induction and expression, and the duration of the plastic changes (Fig. 5).

### Homosynaptic Plasticity

The importance of homosynaptic forms of activity-dependent plasticity was first postulated

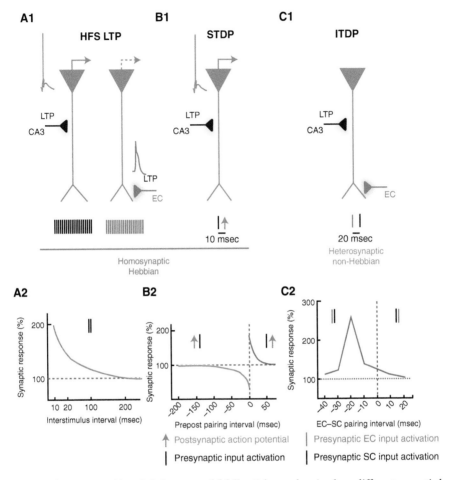

**Figure 5.** Synaptic learning motifs and their temporal fidelity. Scheme showing how different synaptic learning rules emerge in the CA1 microcircuit from temporally patterned activity of synaptic inputs (CA3 to CA1 Schaffer collateral [SC] inputs in black; entorhinal cortex [EC] to CA1 inputs in purple) and the postsynaptic CA1 pyramidal neuron (blue). (*A1*) High-frequency stimulation long-term potentiation (HFS LTP) is a homosynaptic form of synaptic learning in which strong tetanic stimulation (200 pulses at 20–100 Hz) of the CA3 or EC inputs can strengthen the synaptic output of that specific active input pathway. HFS LTP is Hebbian in that its induction requires Schaffer collateral (SC) input stimulation to evoke somatic spikes, or perforant path (PP) input stimulation to trigger dendritic spikes in the postsynaptic CA1 pyramidal neurons. (*A2*) Plot depicting the frequency dependence of SC HFS LTP. Postsynaptic excitatory response recorded in CA1 pyramidal neurons plotted as a function of the interstimulus interval for tetanic stimulation of the CA3–CA1 SC inputs. Preinduction baseline synaptic response is 100%. (Adapted, with values, from Thomas et al. 1996; Aihara et al. 1997; Zakharenko et al. 2003; Alarcon et al. 2004.) (*B1*) Spike-timing-dependent plasticity (STDP) is induced in CA1 pyramidal neurons by temporally precise pairing of synaptic inputs from CA3 with a postsynaptic spike triggered by injecting a brief current step in the soma. The pairing is typically repeated 50–100 times at 10 Hz. (*B2*) LTP is induced when the presynaptic input precedes the postsynaptic spike by 5–20 msec, whereas long-term depression (LTD) prevails when the pairing sequence is reversed (postsynaptic spike before the presynaptic input). (−) Timing intervals indicate pre- before postsynaptic pairing. (Adapted from data in Bi and Poo 1998; Debanne et al. 1998; Nishiyama et al. 2000.) (*C1*) Input-timing-dependent plasticity (ITDP) is induced when EC and SC inputs are stimulated 20 msec apart (EC before SC) at subthreshold strengths (hence, non-Hebbian) for 90 sec at a 1 Hz frequency. ITDP is expressed in the CA1 pyramidal neuron as a long-term potentiation (LTP) of the SC- mediated postsynaptic depolarization without a change in the PP-evoked response (hence, heterosynaptic). (*C2*) Induction of ITDP is finely tuned to the 20 msec pairing interval, even a 10 msec deviation from this preferred timing interval is ineffective. (−) Timing intervals indicate EC before SC input pairing. (Adapted from data in Basu et al. 2013.)

by Donald Hebb (1949) on theoretical grounds as a mechanism for forming neural assemblies. According to Hebb:

> When an axon of cell A is near enough to excite a cell B and repeatedly or persistently takes part in firing it, some growth process or metabolic change takes place in one or both cells such that A's efficiency, as one of the cells firing B, is increased.

Hebb's idea was that this type of synaptic learning rule would provide for the wiring of neuronal assemblies with common response properties. For example, it could explain how a linear array of neighboring retinal ganglion cells and their lateral geniculate targets, which both have circular receptive fields, are able to connect to a common cortical neuron in primary visual cortex to generate typical response selectivity to oriented bars of light. However, Hebbian plasticity is also ideally suited for the formation of neural ensembles that encode a given memory.

### Hebbian LTP

LTP of synaptic transmission represents the classic example of a Hebbian synaptic learning rule. Bliss and Lomo induced LTP by a brief, strong tetanic stimulation of the PP inputs to DG in anesthetized rabbits, which produced a long-lasting enhancement in the strength of excitatory synaptic transmission from PP to DG that lasted for hours to days (Bliss and Lomo 1973). Following its initial discovery, LTP was subsequently found to be inducible at nearly all stages of hippocampal synaptic transmission.

At most synapses, LTP follows Hebb's synaptic learning rules in that it requires strong synaptic activity that is able to drive spike firing in the postsynaptic cells (Bliss and Collingridge 1993; Bliss et al. 2014). The one exception is mossy fiber LTP from DG granules cells to CA3 pyramidal neurons, which is found in most studies to require only presynaptic activity with no requirement for action potential firing in the postsynaptic neuron (Nicoll and Malenka 1995). Tetanic stimulation of the SCs from CA3 pyramidal neurons also induces Hebbian LTP at the synapses these axons make onto other

CA3 pyramidal neurons (recurrent collaterals) as well as at their synapses with CA1 pyramidal neurons (Fig. 5A). In contrast, tetanic high-frequency stimulation (HFS) does not normally induce LTP in the classical Hebbian sense at the SC synapses onto CA2 pyramidal neurons (Zhao et al. 2007) because of their strong expression of the G-protein regulatory protein RGS14 (Lee et al. 2010) and enhanced $Ca^{2+}$ pump activity (Simons et al. 2009). However, a form of LTP was recently described at SC → CA2 synapses that relies on suppression of feedforward inhibition (see below) (Piskorowski and Chevaleyre 2013). Finally, LTP can also be induced by tetanic stimulation of the direct EC inputs to CA1 and CA2, although the extent of LTP onto CA1 is typically quite small. Thus, HFS LTP is a widespread form of homosynaptic activity-dependent plasticity found at all excitatory synapses throughout the hippocampal circuit.

Our understanding of the synaptic and molecular mechanisms underlying LTP has progressed greatly since its initial discovery. However, a number of fundamental questions concerning the basic properties of LTP, as well as the precise role that different forms of LTP play in learning and memory, remain unanswered.

The key characteristics of activity-dependent Hebbian LTP are established by the properties of the NMDARs, whose activation is critical for the induction of LTP at many synapses, including the SC → CA1 synapses (Collingridge et al. 1983). Fast excitatory synapses rely on two classes of ionotropic glutamate receptors, the NMDARs and α-amino-3-hydroxy-5-methyl-4-isoxazolepropionic acid receptors (AMPARs). NMDARs differ from most ionotropic ligand-gated channels (including the AMPARs), in that they require membrane depolarization in addition to glutamate neurotransmitter to function. At typical negative resting potentials, the pore of the NMDAR is blocked by an $Mg^{2+}$ ion. As a result, basal synaptic transmission normally relies on activation of the AMPARs. However, when presynaptic activity is coupled with strong postsynaptic depolarization, such as during tetanic stimulation, the NMDARs are relieved from their Mg$^{2+}$ blockade through electrostatic repulsion. NMDARs are also distinguished from most AMPARs by their high permeability to Ca$^{2+}$, which acts as a second messenger inside cells to activate a number of downstream signaling cascades. As a result, NMDAR activation during strong synaptic stimulation causes a rise in intracellular Ca$^{2+}$ levels, which leads to activation of the Ca$^{2+}$-calmodulin-dependent protein kinase II (CaMKII), a step critical for the induction of LTP (Malenka et al. 1989; Malinow et al. 1989).

A major question that has dominated the field is whether LTP results from a presynaptic change, involving an increase in glutamate release, or a postsynaptic change, resulting from an increase in the postsynaptic response to glutamate, or a coordinated change in presynaptic and postsynaptic properties. This has led to a lively controversy in the field that remains to date. A number of investigators state the case that LTP is largely postsynaptic, resulting from the insertion of AMPARs into the postsynaptic membrane (Huganir and Nicoll 2013; Nicoll and Roche 2013). Others argue that LTP is largely presynaptic (Enoki et al. 2009) or is a mixture of distinct pre and postsynaptic processes (Bliss et al. 2014).

Much of the controversy likely results from the fact that LTP is not a unitary phenomenon (Mayford et al. 2012). Rather a neuron can express multiple forms of long-lasting synaptic plasticity that differ in the pattern of activity needed for their induction, underlying molecular mechanism, time scale of onset and duration, and role in learning and memory (Fig. 5). The behaviorally linked activity state and the coupled molecular tuning of upstream inputs and downstream targets will also determine the ability of a particular circuit to serve as a substrate for plasticity, a process termed metaplasticity (Abraham and Tate 1997), and thereby allow the modulation of information flow. One clear indication of this diversity is provided by the observation that certain forms of LTP that appear to have a presynaptic locus of expression do not require activation of NMDARs but are primarily mediated by Ca$^{2+}$ influx through voltage-gated calcium channels (Grover and

Teyler 1990). Another important finding is that deletion of the GluR1 AMPA receptor subunit blocks LTP induced by tetanic stimulation but only partially inhibits LTP induced by theta burst stimulation (Hoffman et al. 2002) and has little effect on synaptic potentiation during spike-timing-dependent plasticity (STDP) (Frey et al. 2009), a form of plasticity induced by the pairing of an EPSP with a postsynaptic action potential (Markram et al. 1997).

Another important distinction among different forms of synaptic plasticity lies in the time course and duration of the change in synaptic efficacy. LTP in the hippocampus has both early and late phases. For example, stimulation of SC inputs using one train of tetanic stimulation (at 100 Hz for 1 sec) produces an early phase of potentiation (E-LTP) that lasts for 1 to 3 h and does not require protein synthesis. Stimulation with four or more identical trains spaced several minutes apart recruits a late phase of potentiation (L-LTP), which can last for 24 h or more (Huang and Kandel 1994). Unlike E-LTP, L-LTP requires new protein synthesis and also depends on the activation of protein kinase A (PKA) (Frey et al. 1993; Abel et al. 1997) through the action of modulatory transmitters, such as DA (Huang and Kandel 1995). The role of L-LTP has been investigated genetically using mice that express a mutant gene that blocks the catalytic subunit of PKA, or carry an inhibitory mutation in the CREB-1 gene. Both lines of mice have a serious defect in long-term spatial memory and similar defects in LTP. The early phase is normal but the late phase is blocked, providing evidence linking the phases of LTP to the phases of memory storage (Silva et al. 1992; Abel et al. 1997; Bourtchouladze et al. 1998). Theoretical studies suggest that the expression of different temporal phases of LTP enables the stability of long-term memory traces as new memories are encoded through ongoing plastic changes in synaptic function (Fusi et al. 2005).

## Long-Term Depression

The presence of LTP raises the prospect that synapses may become saturated during the lifetime of an animal, and thus no longer able to encode new memories. This potential problem is averted by the antagonistic processes of depotentiation and long-term depression (LTD) (Bear and Abraham 1996), which can be induced by prolonged periods (5–15 min) of low frequency stimulation (1–5 Hz). Depotentiation refers to the reversal of LTP seen when low frequency stimulation is delivered shortly after induction of LTP. LTD is induced by low frequency stimulation in the absence of prior induction of LTP. Similar to LTP, there are also different forms of LTD. One prominent form requires activation of NMDARs and an influx of postsynaptic $Ca^{2+}$, similar to NMDAR-dependent LTP (Bear and Abraham 1996). Another form of LTD requires activation of metabotropic glutamate receptors (Luscher and Huber 2010).

How can $Ca^{2+}$ influx through NMDARs differentially trigger LTP versus LTD? The answer seems to depend on the magnitude and duration of the postsynaptic $Ca^{2+}$ signal, with a large elevation in $Ca^{2+}$ triggering LTP and a more long-lasting but low $Ca^{2+}$ signal resulting in LTD (Neveu and Zucker 1996). These differences may reflect the distinct molecular mechanisms of LTP versus LTD, with LTP requiring activation of CaMKII, whereas LTD requires activation of the calcium-calmodulin-dependent phosphatase calcineurin (Mulkey et al. 1994), which is activated by lower levels of $Ca^{2+}$.

## Spike-Timing-Dependent Plasticity

The relevance of LTP for learning and memory has been questioned because of the nonphysiological, prolonged high-frequency tetanic stimulation required for its induction. However, a form of long-lasting synaptic potentiation can be induced by more physiologically relevant patterns of stimulation through STDP, involving the pairing of a relatively weak presynaptic stimulus with the firing of a postsynaptic action potential. Moreover the potentiation during STDP shows a strict dependence on the timing of the EPSP and spike (Fig. 5B) (Magee and Johnston 1997; Markram et al. 1997). If the postsynaptic spike follows the presynaptic ac-

tion potential by 10 msec, a synapse is maximally potentiated, conforming to a Hebbian learning rule. In contrast, if the spike precedes the presynaptic stimulus, the synapse becomes depressed. If the presynaptic and postsynaptic cells fire action potentials separated by 40 msec or more, there is no change in the synaptic strength. This timing dependence is consistent with the properties of the NMDARs, where the firing of a postsynaptic spike following, but not preceding, glutamate release will be able to relieve $Mg^{2+}$ block, thereby enabling $Ca^{2+}$ influx and the induction of plasticity. Thus, STDP provides a mechanism for strengthening or weakening synapses depending on the correlation or anticorrelation between presynaptic and postsynaptic firing. Similar to LTP and LTD, STDP also requires activation of NMDARs. However, as mentioned above, STDP is likely to recruit distinct downstream mechanisms from tetanus-induced LTP.

## Role of LTP and LTD in Learning and Memory

The properties of NMDA-dependent synaptic plasticity provide an attractive cellular mechanism for forming learned associations. However, it has been significantly more difficult to prove direct causative links between LTP and behavior. The first correlation between LTP and spatial memory was provided by the demonstration that pharmacological blockade of NMDARs prevents spatial reference memory formation, assessed by the MWM (Morris et al. 1986). Importantly, blockade of the NMDARs did not impair the ability of an animal to learn to swim to the platform when it is visible, a task that does not require the hippocampus. Pharmacological blockade of NMDARs also decreases the stability of place fields across two recording sessions separated by 24 h (Kentros et al. 1998).

Subsequent studies showed that NMDAR blockade does not prevent memory formation in the water maze if rats are pretrained on the task in a different spatial setting (Bannerman et al. 1995; Saucier and Cain 1995). This suggests that NMDAR-independent mechanisms can suffice for learning spatial associations under some conditions. Whether the spatial learning occurs through non-NMDAR-dependent forms of LTP or through another mechanism is not known.

Genetic evidence linking LTP in the CA1 region of the hippocampus with learning was again provided by experiments that rely on manipulation of NMDAR function. Mice lacking an essential NMDAR subunit (NR1) in the CA1 region (Tsien et al. 1996) show a loss of tetanus-induced LTP at the SC → CA1 pathway and have a profound deficit in the MWM. However, subsequent NR1 knockout-based studies have raised questions about these initial conclusions (Bannerman et al. 2014). First, the initial NR1 deletion mouse line was found to have a loss of NR1 outside the hippocampus. Furthermore, a second transgenic mouse line in which deletion of NR1 was indeed restricted to DG and CA1 showed no deficit in spatial reference memory, whereas spatial working memory performance was impaired (Bannerman et al. 2012). Interestingly, these mice did show difficulties in relearning a new location of the MWM platform after initial training. This result has led to the suggestion that hippocampal NMDAR-dependent LTP is necessary for resolving conflicts between stored information and the current sensory context rather than for encoding paired associative memories (Bannerman et al. 2014).

It is also important to realize that the establishment of a link between NMDARs and certain hippocampal-dependent memory tasks does not necessarily imply that the memory formation results from NMDAR-dependent LTP. This is because NMDARs subserve a number of functions apart from the induction of LTP, including generation of a late phase of the glutamatergic EPSP and regenerative dendritic voltage signals termed NMDAR spikes (Golding et al. 1999, 2002; Schiller et al. 2000). Conversely, the finding that pharmacological blockade or genetic deletion of the NMDAR does not alter learning and memory in certain tasks suggests a potential role of non-NMDAR-dependent forms of LTP.

An important line of evidence linking LTP to hippocampal learning and memory comes from in vivo extracellular recording experiments during learning behaviors. For example, one study (Whitlock et al. 2006) found that one-trial inhibitory avoidance learning produces an enhancement in synaptic responses evoked by electrical stimulation of SC → CA1 pyramidal neuron inputs. Moreover, learning behaviors brought about the same changes in AMPAR phosphorylation and membrane trafficking as seen during induction of LTP by tetanic stimulation.

Although less thoroughly examined than LTP, several lines of evidence suggest the importance of LTD in both learning and its extinction. Pharmacological blockade of NMDAR-dependent hippocampal LTD was found to impair consolidation of long-term spatial memory in the MWM (Ge et al. 2010). Furthermore, Kemp and Manahan-Vaughan (2004) showed an enhanced induction of CA1 LTD by low-frequency stimulation when rats explored novel objects in a new environment. Links between mGluR-dependent hippocampal LTD and object-place learning was provided by a recent study (Di Prisco et al. 2014) in mice carrying a phosphorylation-deficient mutation of the translational factor EF2$\alpha$.

Some of the best evidence linking LTP and LTD to learning and memory comes from studies in amygdala, in which the relatively simple circuitry and its role in well-established behavioral paradigms provide an important experimental advantage (McNally et al. 2011). Early studies showed that cued fear conditioning, in which animals learn to associate a tone with a shock, led to an LTP-like enhancement in the synaptic response both to the auditory conditioned stimulus recorded from the amygdala in vivo (Rogan et al. 1997) and to electrical stimulation of thalamic input to the amygdala in an in vitro slice preparation (McKernan and Shinnick-Gallagher 1997). Cued fear conditioning also occluded subsequent induction of LTP at corticoamygdala synapses in acute amygdala slices (Tsvetkov et al. 2002). Interestingly, a learned safety response in which an auditory stimulus signals the absence of aversive stimuli led to an LTD-like decrease in the synaptic response of lateral amygdala to the auditory cue (Rogan et al. 2005). Recent in vivo experiments using optogenetics to directly stimulate the corticothalamic auditory inputs to amygdala show that induction of LTD can suppress fear conditioning memory and that the subsequent induction of LTP can restore the fearful memory (Nabavi et al. 2014), providing some of the strongest evidence linking memory to plastic changes that enhance or diminish synaptic responses.

## Inhibitory Circuits in Plasticity, Learning, and Memory

In addition to plastic changes at the excitatory synapses onto CA1 pyramidal neurons, a number of studies have now described activity-dependent plastic changes in the inhibitory synapses these neurons receive. Several recent studies have begun to address the role of inhibition and inhibitory plasticity in learning and memory, as discussed next (see also the recent reviews by Kullmann et al. 2012; Wester and McBain 2014).

One prominent form of activity-dependent plasticity of inhibitory synaptic transmission is mediated by the endocannabinoid-signaling pathway (Castillo et al. 2012; Younts and Castillo 2014). Although endocannabinoids have been reported to contribute to LTD and LTP at excitatory SC synapses in some experimental paradigms (Ohno-Shosaku et al. 2002; Peterfi et al. 2012), the most prominent action of these signaling molecules involves the suppression of inhibition. This results from the presynaptic inhibition of GABA release caused by the binding of endocannabinoids to G-protein-coupled CB1 receptors present on presynaptic terminals of cholecystokinin-positive (CCK$^+$) inhibitory neurons. Endocannabinoids were first found to mediate a short-lasting suppression of inhibition observed following strong postsynaptic depolarization of the CA1 pyramidal neuron (Wilson and Nicoll 2001). Later studies found that tetanic or theta burst stimulation of the SC pathway could induce an endocannabinoid-dependent LTD of GABA release (Katona et al. 1999; Chevaleyre and Castillo 2003; Freund and Katona 2007). Interestingly, in addition to these stronger stimulation paradigms (Chevaleyre and Castillo 2003, 2004), endocannabinoid CB1R-mediated synaptic modulation can

be induced by coincident recruitment of presynaptic inputs with subthreshold depolarizations and activation of mGluRs (Hashimoto-dani et al. 2007; Basu et al. 2013). Such endocannabinoid-mediated plasticity of inhibition can have a long-lasting effect to enhance the output of the excitatory CA1 circuit (see below) (Chevaleyre and Castillo 2004; Zhu and Lovinger 2007; Basu et al. 2013; Younts et al. 2013).

Several studies have used pharmacological and genetic approaches to address the role of CB1Rs in hippocampal-dependent memory behaviors. Unconditional deletion of the CB1R in mice results in heightened freezing responses to hippocampal-dependent CFC and an increased overgeneralization of fear memories to a strong unconditioned aversive stimulus. This behavioral effect was accompanied by an increase of LTP at the PP inputs to DG (Jacob et al. 2012). Evidence suggesting the importance of hippocampal CB1Rs comes from the finding that specific knockdown of these receptors in pyramidal cells and interneurons of mouse dorsal hippocampus both impairs associative learning during hippocampal-dependent trace eyeblink conditioning and reduces HFS-induced SC LTP (Madronal et al. 2012).

One surprising finding comes from a study showing that selective deletion of the CB1R from astrocytes impairs both in vivo LTD induced by the cannabinoid agonist THC as well as performance in a spatial working memory version of a MWM task (Han et al. 2012). This form of LTD also requires activation of NR2B subunit containing NMDARs and down-regulation of surface AMPARs. In contrast, selective deletion of the CB1R from cortical and hippocampal glutamatergic or GABAergic neurons produced little change in working memory or in vivo endocannabinoid-dependent LTD.

As described above, tetanic stimulation normally fails to elicit classical homosynaptic LTP at glutamatergic SC–CA2 pyramidal neuron synapses (Caruana et al. 2012). However, when inhibition is intact, high-frequency 10 Hz or theta-burst stimulation of CA3 inputs to CA2 can induce potentiation of information flow in this pathway. Such activity leads to a δ-opioid-dependent LTD of GABAergic feedforward inhibition mediated by parvalbumin (PV) interneurons. The decreased inhibition results in a long-term enhancement of the ability of SC stimulation to excite the CA2 pyramidal neurons (Piskorowski and Chevaleyre 2013).

Neuromodulatory tuning of PV interneurons in the hippocampal CA1 region also occurs through the actions of oxytocin (Owen et al. 2013). This hormone enhances the spontaneous firing of bistratified and basket PV interneurons, leading to the suppression of spontaneous spiking activity in CA1 pyramidal neurons. At the same time, oxytocin decreases feedforward inhibition onto CA1 pyramidal neurons in response to activation of the SC inputs, likely a result of short-term synaptic depression of GABA release caused by the elevated spontaneous firing rate. The reduction in feedforward inhibition enhances the firing of synaptically evoked action potentials in CA1 pyramidal neurons. These dual actions of oxytocin to inhibit spontaneous pyramidal neuron firing, while enhancing net evoked synaptic excitation, greatly increase the signal to noise ratio in information transfer through the trisynaptic circuit.

Genetic silencing of PV interneurons in dorsal hippocampus results in a deficit in spatial working memory but does not impair long-term spatial memory (Fig. 4) (Murray et al. 2011). Although PV interneurons were silenced by targeted injections into the CA1 region of a viral vector expressing tetanus toxin, some of the behavioral changes may have resulted from expression of the toxin in the neighboring CA2 region, which has an unusually dense population of PV interneurons. Silencing PV interneurons also increases the firing rates of place cells within the place field and shifted spike timing in relation to spatially modulated theta phase; however, it does not affect their place-field size (Royer et al. 2012).

The oriens-lacunosum moleculare (OLM) subpopulation of SOM-expressing interneurons has also been shown to powerfully regulate memory encoding. These neurons are located in SO where they are recruited in a recurrent fashion by CA1 pyramidal neuron output and by subcortical cholinergic inputs. The OLM neu-

rons send their axons to SLM where they powerfully inhibit the distal dendrites of CA1 pyramidal neurons, thereby suppressing the excitatory effects of the direct EC inputs to CA1. Activation of the OLM neurons by cholinergic inputs in response to a fearful stimulus (foot shock) has been found to be important for encoding of contextual fear conditioning (Lovett-Barron et al. 2014).

## Corticohippocampal Heterosynaptic Plasticity

LTP and STDP are nonsupervised, homosynaptic Hebbian learning rules in which activity in a single class of excitatory synaptic inputs alters the efficacy of those same inputs. In contrast, cerebellar LTD (Ito 2001) represents a heterosynaptic supervised learning rule, in which activity in one set of excitatory synapses (climbing fibers), is thought to provide an error signal that induces plasticity (LTD) at another set of excitatory synapses (parallel fibers), which plays an important role in certain forms of motor learning. Might the complex and convergent set of cortical and hippocampal inputs to a CA1 pyramidal neuron also support activity-dependent heterosynaptic learning rules?

The fact that CA1 pyramidal neurons receive both weak direct sensory input from the EC and strong processed or mnemonic input from CA3 has led a number of groups to investigate the possible function of this dual input. Strong paired activation of the EC and SC inputs antagonizes the induction of HFS LTP at the SC inputs because the EC inputs produce a strong inhibitory response in the CA1 pyramidal neurons (Levy et al. 1998; Remondes and Schuman 2004). Other groups have found that stimulation of the SC pathway before PP activation can potentiate the propagation of the EC EPSP to the soma (Jarsky et al. 2005; Ang et al. 2006).

In contrast to the suppressive effect of EC input stimulation on SC LTP noted above, the pairing of a brief 100 Hz burst of four EC stimuli with a single SC stimulus at theta frequency (5 Hz) causes a small long-lasting potentiation in the SC pathway (25% increase) when the SC stimulus occurs within 70 msec of the end of the burst (Judge and Hasselmo 2004). Theta burst stimulation of the EC pathway alone can produce a slow, small (20%–40%) potentiation of the SC pathway, with no change in the EC synaptic response (Han and Heinemann 2013). Simultaneous theta burst stimulation of the entorhinal cortex and SC inputs results in dendritic plateau potentials and a 30% to 60% LTP of the PP EPSP (Takahashi and Magee 2009).

One particularly interesting feature of the direct and trisynaptic corticohippocampal inputs to CA1 pyramidal neurons is that they are organized in a delay line architecture, in which information carried by the direct entorhinal cortex inputs arrive at CA1 pyramidal neurons some 15–20 msec before the arrival of information propagated through the trisynaptic path (Yeckel and Berger 1990). Our laboratory (Dudman et al. 2007) examined whether such a delay-line architecture might be used to implement a timing-dependent synaptic learning rule. The paired activation of the entorhinal cortex and SC inputs at a 20 msec delay interval (PP before SC) in hippocampal slices induces a surprisingly large (100%–300%) potentiation of the SC synaptic response, without altering the entorhinal cortex-evoked response. In contrast, pairings at slightly different intervals (10 or 30 msec) or reversed timing (SC before entorhinal cortex) produced little or no long-lasting potentiation (Fig. 5C) (Dudman et al. 2007; Basu et al. 2013). This phenomenon was termed input timing-dependent plasticity (ITDP), (Dudman et al. 2007), by analogy to STDP.

ITDP shares key features with SC HFS LTP in that it requires $Ca^{2+}$ influx through NMDARs and some postsynaptic depolarization (Dudman et al. 2007; Basu et al. 2013). It differs from LTP in that it does not require a large postsynaptic depolarization or CA1 action potential output. ITDP also requires activation of the mGluR1a metabotropic glutamate receptor and $IP_3$ receptor-dependent $Ca^{2+}$ release from internal stores (Dudman et al. 2007; Basu et al. 2013).

What is responsible for the large enhancement in synaptic transmission observed during ITDP? One of the unusual properties of ITDP is

that it can be robustly induced when inhibition is intact. Indeed blockade of GABA receptors greatly reduces the magnitude of ITDP, indicating the importance of inhibitory synaptic transmission (Xu et al. 2012; Basu et al. 2013). Because the PSP generated in a CA1 pyramidal neuron in response to SC stimulation is determined by the overlapping SC EPSP and the feedforward IPSP, the enhanced depolarizing response during ITDP could result, in principle, from either an increase in the EPSP or a decrease in the feedforward IPSP. Our laboratory (Basu et al. 2013) approached this question by examining the EPSP and IPSP separately and found that ITDP results from both a long-lasting potentiation of the EPSP (E-LTP) and a long-lasting depression of the IPSP (I-LTD). Moreover, I-LTD was found to result from a selective decrease in perisomatic inhibition from CCK-expressing interneurons mediated by the release of endocannabinoids and the activation of CB1 G-protein-coupled receptors.

The fact that ITDP is highly tuned to the timing delay for propagation of information through the trisynaptic versus direct paths to CA1 pyramidal neurons suggests that it might be useful for assessing the salience of mnemonic information processed through DG and CA3 to the immediate sensory context encoded by EC inputs. A role for ITDP in learning and memory is consistent with the deficit in learned temporal associations seen on inactivation of LIII neurons in the MEC (Suh et al. 2011; Kitamura et al. 2014) and with the learning defects seen on deletion of the CB1 receptors described above. A possible role for ITDP in spatial coding is also suggested by its dependence on CCK inhibitory basket cells, which fire synchronously during spatially tuned theta activity at a phase that just precedes place-cell firing as the animal enters the corresponding place field (Klausberger et al. 2005). Endocannabinoid-dependent modulation of feedforward inhibition mediated by CCK interneurons may be a plausible way for the CA1 microcircuit to generate higher contrast for functionally linked pyramidal neuron ensembles in a use-dependent manner, such as during contextual or spatial coding. During sensory experience-driven associative learning,

ITDP could be useful for the assignment of weights to previously stored hippocampal representations based on the online cortical sensory information stream.

## SUMMARY

One of the most difficult problems in linking synaptic plasticity mechanisms to declarative memory is the sparse and distributed nature of the functionally linked circuits. How the various neural representations of the environment are modified with learning, the location of the critical sites of plasticity, and how these modified circuits are recruited to alter motor output during a behavioral memory task are still largely unclear. This makes interpreting the effects of a single type of pharmacological manipulation quite difficult. However, recent genetic-based approaches allowing the marking of neural assemblies activated by learning paradigms and reactivated during memory recall (Garner et al. 2012; Liu et al. 2012; Ramirez et al. 2013; Denny et al. 2014) offer an exciting approach for the identification and spatiotemporal dissection of the specific circuit elements that are involved in formation of specific memories.

Given that different forms of synaptic plasticity are tightly tuned to the temporal patterns of activation, defined circuit elements that participate in the induction or expression of specific forms of synaptic plasticity could be optogenetically manipulated during behavior to simulate how such temporal codes contribute to learning (Nabavi et al. 2014). Cell-type as well as compartment-specific in vivo functional imaging during learning behaviors will also provide glimpses of cellular dynamics, synapse strengthening or weakening as memories are formed in the live animal (Holtmaat et al. 2006; Lai et al. 2012; Donato et al. 2013; Ziv et al. 2013; Grienberger et al. 2014; Lovett-Barron et al. 2014). We expect that such advances will greatly enhance our understanding of how changes in information flow through the corticohippocampal circuit through both homosynaptic and heterosynaptic plasticity mechanisms

contribute to hippocampal-dependent learning and memory.

## REFERENCES

Abel T, Nguyen PV, Barad M, Deuel TA, Kandel ER, Bourtchouladze R. 1997. Genetic demonstration of a role for PKA in the late phase of LTP and in hippocampus-based long-term memory. *Cell* **88**: 615–626.

Abraham WC, Tate WP. 1997. Metaplasticity: A new vista across the field of synaptic plasticity. *Prog Neurobiol* **52**: 303–323.

Ahmed OJ, Mehta MR. 2009. The hippocampal rate code: Anatomy, physiology and theory. *Trends Neurosci* **32**: 329–338.

Aihara T, Tsukada M, Crair MC, Shinomoto S. 1997. Stimulus-dependent induction of long-term potentiation in CA1 area of the hippocampus: Experiment and model. *Hippocampus* **7**: 416–426.

Alarcon JM, Hodgman R, Theis M, Huang YS, Kandel ER, Richter JD. 2004. Selective modulation of some forms of Schaffer collateral-CA1 synaptic plasticity in mice with a disruption of the *CPEB-1* gene. *Learn Mem* **11**: 318–327.

Amaral DG, Witter MP. 1989. The three-dimensional organization of the hippocampal formation: A review of anatomical data. *Neuroscience* **31**: 571–591.

Ang CW, Carlson GC, Coulter DA. 2006. Massive and specific dysregulation of direct cortical input to the hippocampus in temporal lobe epilepsy. *J Neurosci* **26**: 11850–11856.

Arai A, Black J, Lynch G. 1994. Origins of the variations in long-term potentiation between synapses in the basal versus apical dendrites of hippocampal neurons. *Hippocampus* **4**: 1–9.

Baimbridge KG, Miller JJ. 1982. Immunohistochemical localization of calcium-binding protein in the cerebellum, hippocampal formation and olfactory bulb of the rat. *Brain Res* **245**: 223–229.

Baimbridge KG, Miller JJ, Parkes CO. 1982. Calcium-binding protein distribution in the rat brain. *Brain Res* **239**: 519–525.

Bannerman DM, Good MA, Butcher SP, Ramsay M, Morris RG. 1995. Distinct components of spatial learning revealed by prior training and NMDA receptor blockade. *Nature* **378**: 182–186.

Bannerman DM, Bus T, Taylor A, Sanderson DJ, Schwarz I, Jensen V, Hvalby Ø, Rawlins JN, Seeburg PH, Sprengel R. 2012. Dissecting spatial knowledge from spatial choice by hippocampal NMDA receptor deletion. *Nat Neurosci* **15**: 1153–1159.

Bannerman DM, Sprengel R, Sanderson DJ, McHugh SB, Rawlins JN, Monyer H, Seeburg PH. 2014. Hippocampal synaptic plasticity, spatial memory and anxiety. *Nat Rev Neurosci* **15**: 181–192.

Bannister NJ, Larkman AU. 1995a. Dendritic morphology of CA1 pyramidal neurones from the rat hippocampus. I: Branching patterns. *J Comp Neurol* **360**: 150–160.

Bannister NJ, Larkman AU. 1995b. Dendritic morphology of CA1 pyramidal neurones from the rat hippocam-

pus. II: Spine distributions. *J Comp Neurol* **360:** 161–171.

Bartsch T, Döhring J, Rohr A, Jansen O, Deuschl G. 2011. CA1 neurons in the human hippocampus are critical for autobiographical memory, mental time travel, and autonoetic consciousness. *Proc Natl Acad Sci* **108:** 17562–17567.

Basu J, Srinivas KV, Cheung SK, Taniguchi H, Huang ZJ, Siegelbaum SA. 2013. A cortico-hippocampal learning rule shapes inhibitory microcircuit activity to enhance hippocampal information flow. *Neuron* **79:** 1208–1221.

Bear MF, Abraham WC. 1996. Long-term depression in hippocampus. *Annu Rev Neurosci* **19:** 437–462.

Bethus I, Tse D, Morris RG. 2010. Dopamine and memory: Modulation of the persistence of memory for novel hippocampal NMDA receptor-dependent paired associates. *J Neurosci* **30:** 1610–1618.

Bi GQ, Poo MM. 1998. Synaptic modifications in cultured hippocampal neurons: Dependence on spike timing, synaptic strength, and postsynaptic cell type. *J Neurosci* **18:** 10464–10472.

Bjerknes TL, Moser EI, Moser MB. 2014. Representation of geometric borders in the developing rat. *Neuron* **82:** 71–78.

Bliss TV, Collingridge GL. 1993. A synaptic model of memory: Long-term potentiation in the hippocampus. *Nature* **361:** 31–39.

Bliss TV, Lomo T. 1973. Long-lasting potentiation of synaptic transmission in the dentate area of the anaesthetized rabbit following stimulation of the perforant path. *J Physiol* **232:** 331–356.

Bliss TV, Collingridge GL, Morris RG. 2014. Synaptic plasticity in health and disease: Introduction and overview. *Philos Trans R Soc Lond B Biol Sci* **369:** 20130129.

Bonnevie T, Dunn B, Fyhn M, Hafting T, Derdikman D, Kubie JL, Roudi Y, Moser EI, Moser MB. 2013. Grid cells require excitatory drive from the hippocampus. *Nat Neurosci* **16:** 309–317.

Bourtchouladze R, Abel T, Berman N, Gordon R, Lapidus K, Kandel ER. 1998. Different training procedures recruit either one or two critical periods for contextual memory consolidation, each of which requires protein synthesis and PKA. *Learn Mem* **5:** 365–374.

Braak H, Braak E. 1985. On areas of transition between entorhinal allocortex and temporal isocortex in the human brain. Normal morphology and lamina-specific pathology in Alzheimer's disease. *Acta Neuropathol* **68:** 325–332.

Brun VH, Otnass MK, Molden S, Steffenach HA, Witter MP, Moser MB, Moser EI. 2002. Place cells and place recognition maintained by direct entorhinal-hippocampal circuitry. *Science* **296:** 2243–2246.

Brun VH, Leutgeb S, Wu HQ, Schwarcz R, Witter MP, Moser EI, Moser MB. 2008. Impaired spatial representation in CA1 after lesion of direct input from entorhinal cortex. *Neuron* **57:** 290–302.

Burgess N, O'Keefe J. 1996. Neuronal computations underlying the firing of place cells and their role in navigation. *Hippocampus* **6:** 749–762.

Bush D, Barry C, Burgess N. 2014. What do grid cells contribute to place cell firing? *Trends Neurosci* **37:** 136–145.

Buzsaki G, Chrobak JJ. 1995. Temporal structure in spatially organized neuronal ensembles: A role for interneuronal networks. *Curr Opin Neurobiol* **5:** 504–510.

Buzsaki G, Moser EI. 2013. Memory, navigation and theta rhythm in the hippocampal-entorhinal system. *Nat Neurosci* **16:** 130–138.

Caputi A, Melzer S, Michael M, Monyer H. 2013. The long and short of GABAergic neurons. *Curr Opin Neurobiol* **23:** 179–186.

Caruana DA, Alexander GM, Dudek SM. 2012. New insights into the regulation of synaptic plasticity from an unexpected place: Hippocampal area CA2. *Learn Mem* **19:** 391–400.

Castillo PE, Younts TJ, Chavez AE, Hashimotodani Y. 2012. Endocannabinoid signaling and synaptic function. *Neuron* **76:** 70–81.

Chevaleyre V, Castillo PE. 2003. Heterosynaptic LTD of hippocampal GABAergic synapses: A novel role of endocannabinoids in regulating excitability. *Neuron* **38:** 461–472.

Chevaleyre V, Castillo PE. 2004. Endocannabinoid-mediated metaplasticity in the hippocampus. *Neuron* **43:** 871–881.

Chevaleyre V, Siegelbaum SA. 2010. Strong CA2 pyramidal neuron synapses define a powerful disynaptic cortico-hippocampal loop. *Neuron* **66:** 560–572.

Collingridge GL, Kehl SJ, McLennan H. 1983. Excitatory amino acids in synaptic transmission in the Schaffer collateral-commissural pathway of the rat hippocampus. *J Physiol* **334:** 33–46.

Corkin S. 2002. What's new with the amnesic patient H.M.? *Nat Rev Neurosci* **3:** 153–160.

Cui Z, Gerfen CR, Young WS III. 2013. Hypothalamic and other connections with dorsal CA2 area of the mouse hippocampus. *J Comp Neurol* **521:** 1844–1866.

Debanne D, Gahwiler BH, Thompson SM. 1998. Long-term synaptic plasticity between pairs of individual CA3 pyramidal cells in rat hippocampal slice cultures. *J Physiol* **507:** 237–247.

Denny CA, Kheirbek MA, Alba EL, Tanaka KF, Brachman RA, Laughman KB, Tomm NK, Turi GF, Losonczy A, Hen R. 2014. Hippocampal memory traces are differentially modulated by experience, time, and adult neurogenesis. *Neuron* **83:** 189–201.

DeVito LM, Konigsberg R, Lykken C, Sauvage M, Young WS III, Eichenbaum H. 2009. Vasopressin 1b receptor knockout impairs memory for temporal order. *J Neurosci* **29:** 2676–2683.

Di Prisco GV, Huang W, Buffington SA, Hsu CC, Bonnen PE, Placzek AN, Sidrauski C, Krnjevic K, Kaufman RJ, Walter P, et al. 2014. Translational control of mGluR-dependent long-term depression and object-place learning by eIF2α. *Nat Neurosci* **17:** 1073–1082.

Dolleman-Van der Weel MJ, Lopes da Silva FH, Witter MP. 1997. Nucleus reuniens thalami modulates activity in hippocampal field CA1 through excitatory and inhibitory mechanisms. *J Neurosci* **17:** 5640–5650.

Dolleman-van der Weel MJ, Morris RG, Witter MP. 2009. Neurotoxic lesions of the thalamic reuniens or mediodorsal nucleus in rats affect non-mnemonic aspects of watermaze learning. *Brain Struct Funct* **213:** 329–342.

Donato F, Rompani SB, Caroni P. 2013. Parvalbumin-expressing basket-cell network plasticity induced by experience regulates adult learning. *Nature* **504:** 272–276.

Dong Z, Han H, Cao J, Zhang X, Xu L. 2008. Coincident activity of converging pathways enables simultaneous long-term potentiation and long-term depression in hippocampal CA1 network in vivo. *PLoS ONE* **3:** e2848.

Dong S, Rogan SC, Roth BL. 2010. Directed molecular evolution of DREADDs: A generic approach to creating next-generation RASSLs. *Nat Protoc* **5:** 561–573.

Dougherty KA, Nicholson DA, Diaz L, Buss EW, Neuman KM, Chetkovich DM, Johnston D. 2013. Differential expression of HCN subunits alters voltage-dependent gating of h-channels in CA1 pyramidal neurons from dorsal and ventral hippocampus. *J Neurophysiol* **109:** 1940–1953.

Drew LJ, Fusi S, Hen R. 2013. Adult neurogenesis in the mammalian hippocampus: Why the dentate gyrus? *Learn Mem* **20:** 710–729.

Dudman JT, Tsay D, Siegelbaum SA. 2007. A role for synaptic inputs at distal dendrites: Instructive signals for hippocampal long-term plasticity. *Neuron* **56:** 866–879.

Duncan K, Ketz N, Inati SJ, Davachi L. 2012. Evidence for area CA1 as a match/mismatch detector: A high-resolution fMRI study of the human hippocampus. *Hippocampus* **22:** 389–398.

Enoki R, Hu YL, Hamilton D, Fine A. 2009. Expression of long-term plasticity at individual synapses in hippocampus is graded, bidirectional, and mainly presynaptic: Optical quantal analysis. *Neuron* **62:** 242–253.

Fanselow MS, Dong HW. 2010. Are the dorsal and ventral hippocampus functionally distinct structures? *Neuron* **65:** 7–19.

Freund TF, Katona I. 2007. Perisomatic inhibition. *Neuron* **56:** 33–42.

Frey U, Huang YY, Kandel ER. 1993. Effects of cAMP simulate a late stage of LTP in hippocampal CA1 neurons. *Science* **260:** 1661–1664.

Frey MC, Sprengel R, Nevian T. 2009. Activity pattern-dependent long-term potentiation in neocortex and hippocampus of GluA1 (GluR-A) subunit-deficient mice. *J Neurosci* **29:** 5587–5596.

Fuentealba P, Begum R, Capogna M, Jinno S, Marton LF, Csicsvari J, Thomson A, Somogyi P, Klausberger T. 2008. Ivy cells: A population of nitric-oxide-producing, slow-spiking GABAergic neurons and their involvement in hippocampal network activity. *Neuron* **57:** 917–929.

Fusi S, Drew PJ, Abbott LF. 2005. Cascade models of synaptically stored memories. *Neuron* **45:** 599–611.

Fyhn M, Molden S, Witter MP, Moser EI, Moser MB. 2004. Spatial representation in the entorhinal cortex. *Science* **305:** 1258–1264.

Garner AR, Rowland DC, Hwang SY, Baumgaertel K, Roth BL, Kentros C, Mayford M. 2012. Generation of a synthetic memory trace. *Science* **335:** 1513–1516.

Ge Y, Dong Z, Bagot RC, Howland JG, Phillips AG, Wong TP, Wang YT. 2010. Hippocampal long-term depression is required for the consolidation of spatial memory. *Proc Natl Acad Sci* **107:** 16697–16702.

Golding NL, Jung HY, Mickus T, Spruston N. 1999. Dendritic calcium spike initiation and repolarization are controlled by distinct potassium channel subtypes in CA1 pyramidal neurons. *J Neurosci* **19:** 8789–8798.

Golding NL, Staff NP, Spruston N. 2002. Dendritic spikes as a mechanism for cooperative long-term potentiation. *Nature* **418:** 326–331.

Goshen I, Brodsky M, Prakash R, Wallace J, Gradinaru V, Ramakrishnan C, Deisseroth K. 2011. Dynamics of retrieval strategies for remote memories. *Cell* **147:** 678–689.

Grienberger C, Chen X, Konnerth A. 2014. NMDA receptor-dependent multidendrite Ca$^{2+}$ spikes required for hippocampal burst firing in vivo. *Neuron* **81:** 1274–1281.

Grover LM, Teyler TJ. 1990. Two components of long-term potentiation induced by different patterns of afferent activation. *Nature* **347:** 477–479.

Hafting T, Fyhn M, Molden S, Moser MB, Moser EI. 2005. Microstructure of a spatial map in the entorhinal cortex. *Nature* **436:** 801–806.

Haglund L, Swanson LW, Kohler C. 1984. The projection of the supramammillary nucleus to the hippocampal formation: An immunohistochemical and anterograde transport study with the lectin PHA-L in the rat. *J Comp Neurol* **229:** 171–185.

Han EB, Heinemann SF. 2013. Distal dendritic inputs control neuronal activity by heterosynaptic potentiation of proximal inputs. *J Neurosci* **33:** 1314–1325.

Han J, Kesner P, Metna-Laurent M, Duan T, Xu L, Georges F, Koehl M, Abrous DN, Mendizabal-Zubiaga J, Grandes P, et al. 2012. Acute cannabinoids impair working memory through astroglial CB1 receptor modulation of hippocampal LTD. *Cell* **148:** 1039–1050.

Hashimotodani Y, Ohno-Shosaku T, Kano M. 2007. Ca$^{2+}$-assisted receptor-driven endocannabinoid release: Mechanisms that associate presynaptic and postsynaptic activities. *Curr Opin Neurobiol* **17:** 360–365.

Hebb DO. 1949. *The organization of behavior*. Wiley, New York.

Henriksen EJ, Colgin LL, Barnes CA, Witter MP, Moser MB, Moser EI. 2010. Spatial representation along the proximodistal axis of CA1. *Neuron* **68:** 127–137.

Hirayasu Y, Wada JA. 1992. N-methyl-D-aspartate injection into the massa intermedia facilitates development of limbic kindling in rats. *Epilepsia* **33:** 965–970.

Hitti FL, Siegelbaum SA. 2014. The hippocampal CA2 region is essential for social memory. *Nature* **508:** 88–92.

Hoffman DA, Sprengel R, Sakmann B. 2002. Molecular dissection of hippocampal theta-burst pairing potentiation. *Proc Natl Acad Sci* **99:** 7740–7745.

Holtmaat A, Wilbrecht L, Knott GW, Welker E, Svoboda K. 2006. Experience-dependent and cell-type-specific spine growth in the neocortex. *Nature* **441:** 979–983.

Huang YY, Kandel ER. 1994. Recruitment of long-lasting and protein kinase A-dependent long-term potentiation in the CA1 region of hippocampus requires repeated tetanization. *Learn Mem* **1:** 74–82.

Huang YY, Kandel ER. 1995. D1/D5 receptor agonists induce a protein synthesis-dependent late potentiation in the CA1 region of the hippocampus. *Proc Natl Acad Sci* **92:** 2446–2450.

Huganir RL, Nicoll RA. 2013. AMPARs and synaptic plasticity: The last 25 years. *Neuron* **80:** 704–717.

Igarashi KM, Lu L, Colgin LL, Moser MB, Moser EI. 2014. Coordination of entorhinal-hippocampal ensemble activity during associative learning. *Nature* **510**: 143–147.

Ishikawa A, Nakamura S. 2006. Ventral hippocampal neurons project axons simultaneously to the medial prefrontal cortex and amygdala in the rat. *J Neurophysiol* **96**: 2134–2138.

Ishizuka N, Weber J, Amaral DG. 1990. Organization of intrahippocampal projections originating from CA3 pyramidal cells in the rat. *J Comp Neurol* **295**: 580–623.

Ito M. 2001. Cerebellar long-term depression: Characterization, signal transduction, and functional roles. *Physiol Rev* **81**: 1143–1195.

Ito HT, Schuman EM. 2007. Frequency-dependent gating of synaptic transmission and plasticity by dopamine. *Front Neural Circuits* **1**: 1.

Jacob W, Marsch R, Marsicano G, Lutz B, Wotjak CT. 2012. Cannabinoid CB1 receptor deficiency increases contextual fear memory under highly aversive conditions and long-term potentiation in vivo. *Neurobiol Learn Mem* **98**: 47–55.

Jarsky T, Roxin A, Kath WL, Spruston N. 2005. Conditional dendritic spike propagation following distal synaptic activation of hippocampal CA1 pyramidal neurons. *Nat Neurosci* **8**: 1667–1676.

Jarsky T, Mady R, Kennedy B, Spruston N. 2008. Distribution of bursting neurons in the CA1 region and the subiculum of the rat hippocampus. *J Comp Neurol* **506**: 535–547.

Jay TM. 2003. Dopamine: A potential substrate for synaptic plasticity and memory mechanisms. *Prog Neurobiol* **69**: 375–390.

Jinno S, Klausberger T, Marton LF, Dalezios Y, Roberts JD, Fuentealba P, Bushong EA, Henze D, Buzsaki G, Somogyi P. 2007. Neuronal diversity in GABAergic long-range projections from the hippocampus. *J Neurosci* **27**: 8790–8804.

Judge SJ, Hasselmo ME. 2004. Theta rhythmic stimulation of stratum lacunosum-moleculare in rat hippocampus contributes to associative LTP at a phase offset in stratum radiatum. *J Neurophysiol* **92**: 1615–1624.

Jung MW, Wiener SI, McNaughton BL. 1994. Comparison of spatial firing characteristics of units in dorsal and ventral hippocampus of the rat. *J Neurosci* **14**: 7347–7356.

Katona I, Sperlagh B, Sik A, Kafalvi A, Vizi ES, Mackie K, Freund TF. 1999. Presynaptically located CB1 cannabinoid receptors regulate GABA release from axon terminals of specific hippocampal interneurons. *J Neurosci* **19**: 4544–4558.

Keinath AT, Wang ME, Wann EG, Yuan RK, Dudman JT, Muzzio IA. 2014. Precise spatial coding is preserved along the longitudinal hippocampal axis. *Hippocampus* **24**: 1533–1543.

Kemp A, Manahan-Vaughan D. 2004. Hippocampal long-term depression and long-term potentiation encode different aspects of novelty acquisition. *Proc Natl Acad Sci* **101**: 8192–8197.

Kentros C, Hargreaves E, Hawkins RD, Kandel ER, Shapiro M, Muller RV. 1998. Abolition of long-term stability of new hippocampal place cell maps by NMDA receptor blockade. *Science* **280**: 2121–2126.

Kheirbek MA, Drew LJ, Burghardt NS, Costantini DO, Tannenholz L, Ahmari SE, Zeng H, Fenton AA, Hen R. 2013. Differential control of learning and anxiety along the dorsoventral axis of the dentate gyrus. *Neuron* **77**: 955–968.

Kim Y, Spruston N. 2012. Target-specific output patterns are predicted by the distribution of regular-spiking and bursting pyramidal neurons in the subiculum. *Hippocampus* **22**: 693–706.

Kiss J, Csaki A, Bokor H, Shanabrough M, Leranth C. 2000. The supramammillo-hippocampal and supramammillo-septal glutamatergic/aspartatergic projections in the rat: A combined [³H] D-aspartate autoradiographic and immunohistochemical study. *Neuroscience* **97**: 657–669.

Kitamura T, Pignatelli M, Suh J, Kohara K, Yoshiki A, Abe K, Tonegawa S. 2014. Island cells control temporal association memory. *Science* **343**: 896–901.

Klausberger T, Somogyi P. 2008. Neuronal diversity and temporal dynamics: The unity of hippocampal circuit operations. *Science* **321**: 53–57.

Klausberger T, Marton LF, O'Neill J, Huck JH, Dalezios Y, Fuentealba P, Suen WY, Papp E, Kaneko T, Watanabe M, et al. 2005. Complementary roles of cholecystokinin- and parvalbumin-expressing GABAergic neurons in hippocampal network oscillations. *J Neurosci* **25**: 9782–9793.

Kohara K, Pignatelli M, Rivest AJ, Jung HY, Kitamura T, Suh J, Frank D, Kajikawa K, Mise N, Obata Y, et al. 2014. Cell type-specific genetic and optogenetic tools reveal hippocampal CA2 circuits. *Nat Neurosci* **17**: 269–279.

Kraus BJ, Robinson RJ II, White JA, Eichenbaum H, Hasselmo ME. 2013. Hippocampal "time cells": Time versus path integration. *Neuron* **78**: 1090–1101.

Kullmann DM, Moreau AW, Bakiri Y, Nicholson E. 2012. Plasticity of inhibition. *Neuron* **75**: 951–962.

Lai CS, Franke TF, Gan WB. 2012. Opposite effects of fear conditioning and extinction on dendritic spine remodelling. *Nature* **483**: 87–91.

Langston RF, Ainge JA, Couey JJ, Canto CB, Bjerknes TL, Witter MP, Moser EI, Moser MB. 2010. Development of the spatial representation system in the rat. *Science* **328**: 1576–1580.

Lee SE, Simons SB, Heldt SA, Zhao M, Schroeder JP, Vellano CP, Cowan DP, Ramineni S, Yates CK, Feng Y, et al. 2010. RGS14 is a natural suppressor of both synaptic plasticity in CA2 neurons and hippocampal-based learning and memory. *Proc Natl Acad Sci* **107**: 16994–16998.

Lee SH, Marchionni I, Bezaire M, Varga C, Danielson N, Lovett-Barron M, Losonczy A, Soltesz I. 2014. Parvalbumin-positive basket cells differentiate among hippocampal pyramidal cells. *Neuron* **82**: 1129–1144.

Lein ES, Callaway EM, Albright TD, Gage FH. 2005. Redefining the boundaries of the hippocampal CA2 subfield in the mouse using gene expression and 3-dimensional reconstruction. *J Comp Neurol* **485**: 1–10.

Lemon N, Manahan-Vaughan D. 2006. Dopamine $D_1/D_5$ receptors gate the acquisition of novel information through hippocampal long-term potentiation and long-term depression. *J Neurosci* **26**: 7723–7729.

Leutgeb JK, Leutgeb S, Moser MB, Moser EI. 2007. Pattern separation in the dentate gyrus and CA3 of the hippocampus. *Science* **315**: 961–966.

Levy WB, Desmond NL, Zhang DX. 1998. Perforant path activation modulates the induction of long-term potentiation of the Schaffer collateral–hippocampal CA1 response: Theoretical and experimental analyses. *Learn Mem* **4:** 510–518.

Lisman JE, Grace AA. 2005. The hippocampal-VTA loop: Controlling the entry of information into long-term memory. *Neuron* **46:** 703–713.

Liu X, Ramirez S, Pang PT, Puryear CB, Govindarajan A, Deisseroth K, Tonegawa S. 2012. Optogenetic stimulation of a hippocampal engram activates fear memory recall. *Nature* **484:** 381–385.

Lorente de Nó R. 1934. Studies on the structure of the cerebral cortex. II: Continuation of the study of the ammonic system. *J Psychol Neurol* **46:** 113–177.

Lovett-Barron M, Kaifosh P, Kheirbek MA, Danielson N, Zaremba JD, Reardon TR, Turi GF, Hen R, Zemelman BV, Losonczy A. 2014. Dendritic inhibition in the hippocampus supports fear learning. *Science* **343:** 857–863.

Luscher C, Huber KM. 2010. Group 1 mGluR-dependent synaptic long-term depression: Mechanisms and implications for circuitry and disease. *Neuron* **65:** 445–459.

Macdonald CJ, Carrow S, Place R, Eichenbaum H. 2013. Distinct hippocampal time cell sequences represent odor memories in immobilized rats. *J Neurosci* **33:** 14607–14616.

Madronal N, Gruart A, Valverde O, Espadas I, Moratalla R, Delgado-Garcia JM. 2012. Involvement of cannabinoid CB1 receptor in associative learning and in hippocampal CA3-CA1 synaptic plasticity. *Cereb Cortex* **22:** 550–566.

Magee JC, Johnston D. 1997. A synaptically controlled, associative signal for Hebbian plasticity in hippocampal neurons. *Science* **275:** 209–213.

Magloczky Z, Acsady L, Freund TF. 1994. Principal cells are the postsynaptic targets of supramammillary afferents in the hippocampus of the rat. *Hippocampus* **4:** 322–334.

Malenka RC, Kauer JA, Perkel DJ, Mauk MD, Kelly PT, Nicoll RA, Waxham MN. 1989. An essential role for postsynaptic calmodulin and protein kinase activity in long-term potentiation. *Nature* **340:** 554–557.

Malinow R, Schulman H, Tsien RW. 1989. Inhibition of postsynaptic PKC or CaMKII blocks induction but not expression of LTP. *Science* **245:** 862–866.

Markram H, Lubke J, Frotscher M, Sakmann B. 1997. Regulation of synaptic efficacy by coincidence of postsynaptic APs and EPSPs. *Science* **275:** 213–215.

Mayford M, Siegelbaum SA, Kandel ER. 2012. Synapses and memory storage. *Cold Spring Harb Perspect Biol* **4:** a005751.

McKernan MG, Shinnick-Gallagher P. 1997. Fear conditioning induces a lasting potentiation of synaptic currents in vitro. *Nature* **390:** 607–611.

McNally GP, Johansen JP, Blair HT. 2011. Placing prediction into the fear circuit. *Trends Neurosci* **34:** 283–292.

Megias A, Martinez-Senac MM, Delgado J, Saborido A. 2001. Regulation of transverse tubule ecto-ATPase activity in chicken skeletal muscle. *Biochem J* **353:** 521–529.

Melzer S, Michael M, Caputi A, Eliava M, Fuchs EC, Whittington MA, Monyer H. 2012. Long-range-projecting GABAergic neurons modulate inhibition in hippocampus and entorhinal cortex. *Science* **335:** 1506–1510.

Mizuseki K, Sirota A, Pastalkova E, Buzsaki G. 2009. Theta oscillations provide temporal windows for local circuit computation in the entorhinal-hippocampal loop. *Neuron* **64:** 267–280.

Mizuseki K, Diba K, Pastalkova E, Buzsaki G. 2011. Hippocampal CA1 pyramidal cells form functionally distinct sublayers. *Nat Neurosci* **14:** 1174–1181.

Morris RG. 2013. NMDA receptors and memory encoding. *Neuropharmacology* **74:** 32–40.

Morris RG, Hagan JJ, Rawlins JN. 1986. Allocentric spatial learning by hippocampectomised rats: A further test of the "spatial mapping" and "working memory" theories of hippocampal function. *Q J Exp Psychol B* **38:** 365–395.

Morris RG, Steele RJ, Bell JE, Martin SJ. 2013. N-methyl-D-aspartate receptors, learning and memory: Chronic intraventricular infusion of the NMDA receptor antagonist D-AP5 interacts directly with the neural mechanisms of spatial learning. *Eur J Neurosci* **37:** 700–717.

Mulkey RM, Endo S, Shenolikar S, Malenka RC. 1994. Involvement of a calcineurin/inhibitor-1 phosphatase cascade in hippocampal long-term depression. *Nature* **369:** 486–488.

Murray AJ, Sauer JF, Riedel G, McClure C, Ansel L, Cheyne L, Bartos M, Wisden W, Wulff P. 2011. Parvalbumin-positive CA1 interneurons are required for spatial working but not for reference memory. *Nat Neurosci* **14:** 297–299.

Nabavi S, Fox R, Proulx CD, Lin JY, Tsien RY, Malinow R. 2014. Engineering a memory with LTD and LTP. *Nature* **511:** 348–352.

Naber PA, Lopes da Silva FH, Witter MP. 2001. Reciprocal connections between the entorhinal cortex and hippocampal fields CA1 and the subiculum are in register with the projections from CA1 to the subiculum. *Hippocampus* **11:** 99–104.

Nakashiba T, Young JZ, McHugh TJ, Buhl DL, Tonegawa S. 2008. Transgenic inhibition of synaptic transmission reveals role of CA3 output in hippocampal learning. *Science* **319:** 1260–1264.

Nakashiba T, Cushman JD, Pelkey KA, Renaudineau S, Buhl DL, McHugh TJ, Rodriguez Barrera V, Chittajallu R, Iwamoto KS, McBain CJ, et al. 2012. Young dentate granule cells mediate pattern separation, whereas old granule cells facilitate pattern completion. *Cell* **149:** 188–201.

Nakazawa K, Quirk MC, Chitwood RA, Watanabe M, Yeckel MF, Sun LD, Kato A, Carr CA, Johnston D, Wilson MA, et al. 2002. Requirement for hippocampal CA3 NMDA receptors in associative memory recall. *Science* **297:** 211–218.

Neveu D, Zucker RS. 1996. Postsynaptic levels of $[Ca^{2+}]_i$ needed to trigger LTD and LTP. *Neuron* **16:** 619–629.

Nicoll RA, Malenka RC. 1995. Contrasting properties of two forms of long-term potentiation in the hippocampus. *Nature* **377:** 115–118.

Nicoll RA, Roche KW. 2013. Long-term potentiation: Peeling the onion. *Neuropharmacology* **74:** 18–22.

Nielsen JV, Blom JB, Noraberg J, Jensen NA. 2010. Zbtb20-induced CA1 pyramidal neuron development and area enlargement in the cerebral midline cortex of mice. *Cereb Cortex* **20:** 1904–1914.

Nishiyama M, Hong K, Mikoshiba K, Poo MM, Kato K. 2000. Calcium stores regulate the polarity and input specificity of synaptic modification. *Nature* **408:** 584–588.

Ochiishi T, Saitoh Y, Yukawa A, Saji M, Ren Y, Shirao T, Miyamoto H, Nakata H, Sekino Y. 1999. High level of adenosine A1 receptor-like immunoreactivity in the CA2/CA3a region of the adult rat hippocampus. *Neuroscience* **93:** 955–967.

Ohno-Shosaku T, Tsubokawa H, Mizushima I, Yoneda N, Zimmer A, Kano M. 2002. Presynaptic cannabinoid sensitivity is a major determinant of depolarization-induced retrograde suppression at hippocampal synapses. *J Neurosci* **22:** 3864–3872.

O'Keefe J, Burgess N. 2005. Dual phase and rate coding in hippocampal place cells: Theoretical significance and relationship to entorhinal grid cells. *Hippocampus* **15:** 853–866.

O'Keefe J, Dostrovsky J. 1971. The hippocampus as a spatial map. Preliminary evidence from unit activity in the freely moving rat. *Brain Res* **34:** 171–175.

Owen SF, Tuncdemir SN, Bader PL, Tirko NN, Fishell G, Tsien RW. 2013. Oxytocin enhances hippocampal spike transmission by modulating fast-spiking interneurons. *Nature* **500:** 458–462.

Pagani JH, Zhao M, Cui Z, Williams Avram SK, Caruana DA, Dudek SM, Young WS. 2014. Role of the vasopressin 1b receptor in rodent aggressive behavior and synaptic plasticity in hippocampal area CA2. *Mol Psychiatry* **20:** 490–499.

Pastalkova E, Itskov V, Amarasingham A, Buzsaki G. 2008. Internally generated cell assembly sequences in the rat hippocampus. *Science* **321:** 1322–1327.

Pavlopoulos E, Jones S, Kosmidis S, Close M, Kim C, Kovalerchik O, Small SA, Kandel ER. 2013. Molecular mechanism for age-related memory loss: The histone-binding protein RbAp48. *Sci Transl Med* **5:** 200ra115.

Peterfi Z, Urban GM, Papp OI, Nemeth B, Monyer H, Szabo G, Erdelyi F, Mackie K, Freund TF, Hajos N, et al. 2012. Endocannabinoid-mediated long-term depression of afferent excitatory synapses in hippocampal pyramidal cells and GABAergic interneurons. *J Neurosci* **32:** 14448–14463.

Piskorowski RA, Chevaleyre V. 2012. Synaptic integration by different dendritic compartments of hippocampal CA1 and CA2 pyramidal neurons. *Cell Mol Life Sci* **69:** 75–88.

Piskorowski RA, Chevaleyre V. 2013. δ-Opioid receptors mediate unique plasticity onto parvalbumin-expressing interneurons in area CA2 of the hippocampus. *J Neurosci* **33:** 14567–14578.

Ramirez S, Liu X, Lin PA, Suh J, Pignatelli M, Redondo RL, Ryan TJ, Tonegawa S. 2013. Creating a false memory in the hippocampus. *Science* **341:** 387–391.

Remondes M, Schuman EM. 2004. Role for a cortical input to hippocampal area CA1 in the consolidation of a long-term memory. *Nature* **431:** 699–703.

Rogan MT, Staubli UV, LeDoux JE. 1997. Fear conditioning induces associative long-term potentiation in the amygdala. *Nature* **390:** 604–607.

Rogan MT, Leon KS, Perez DL, Kandel ER. 2005. Distinct neural signatures for safety and danger in the amygdala and striatum of the mouse. *Neuron* **46:** 309–320.

Rossato JI, Bevilaqua LR, Izquierdo I, Medina JH, Cammarota M. 2009. Dopamine controls persistence of long-term memory storage. *Science* **325:** 1017–1020.

Rowland DC, Weible AP, Wickersham IR, Wu H, Mayford M, Witter MP, Kentros CG. 2013. Transgenically targeted rabies virus demonstrates a major monosynaptic projection from hippocampal area CA2 to medial entorhinal layer II neurons. *J Neurosci* **33:** 14889–14898.

Royer S, Zemelman BV, Losonczy A, Kim J, Chance F, Magee JC, Buzsaki G. 2012. Control of timing, rate and bursts of hippocampal place cells by dendritic and somatic inhibition. *Nat Neurosci* **15:** 769–775.

Rugg MD, Vilberg KL. 2013. Brain networks underlying episodic memory retrieval. *Curr Opin Neurobiol* **23:** 255–260.

Sahay A, Scobie KN, Hill AS, O'Carroll CM, Kheirbek MA, Burghardt NS, Fenton AA, Dranovsky A, Hen R. 2011. Increasing adult hippocampal neurogenesis is sufficient to improve pattern separation. *Nature* **472:** 466–470.

Saucier D, Cain DP. 1995. Spatial learning without NMDA receptor-dependent long-term potentiation. *Nature* **378:** 186–189.

Schiller J, Major G, Koester HJ, Schiller Y. 2000. NMDA spikes in basal dendrites of cortical pyramidal neurons. *Nature* **404:** 285–289.

Scoville WB, Milner B. 1957. Loss of recent memory after bilateral hippocampal lesions. *J Neurol Neurosurg Psychiatry* **20:** 11–21.

Silva AJ, Paylor R, Wehner JM, Tonegawa S. 1992. Impaired spatial learning in α-calcium-calmodulin kinase II mutant mice. *Science* **257:** 206–211.

Simons SB, Escobedo Y, Yasuda R, Dudek SM. 2009. Regional differences in hippocampal calcium handling provide a cellular mechanism for limiting plasticity. *Proc Natl Acad Sci* **106:** 14080–14084.

Slomianka L, Amrein I, Knuesel I, Sorensen JC, Wolfer DP. 2011. Hippocampal pyramidal cells: The reemergence of cortical lamination. *Brain Struct Funct* **216:** 301–317.

Small SA, Schobel SA, Buxton RB, Witter MP, Barnes CA. 2011. A pathophysiological framework of hippocampal dysfunction in ageing and disease. *Nat Rev Neurosci* **12:** 585–601.

Solstad T, Moser EI, Einevoll GT. 2006. From grid cells to place cells: A mathematical model. *Hippocampus* **16:** 1026–1031.

Squire LR. 2004. Memory systems of the brain: A brief history and current perspective. *Neurobiol Learn Mem* **82:** 171–177.

Squire LR, Wixted JT. 2011. The cognitive neuroscience of human memory since H.M. *Annu Rev Neurosci* **34:** 259–288.

Squire LR, Stark CE, Clark RE. 2004. The medial temporal lobe. *Annu Rev Neurosci* **27:** 279–306.

Suh J, Rivest AJ, Nakashiba T, Tominaga T, Tonegawa S. 2011. Entorhinal cortex layer III input to the hippocampus is crucial for temporal association memory. *Science* **334:** 1415–1420.

Takahashi H, Magee JC. 2009. Pathway interactions and synaptic plasticity in the dendritic tuft regions of CA1 pyramidal neurons. *Neuron* **62:** 102–111.

Thomas MJ, Moody TD, Makhinson M, O'Dell TJ. 1996. Activity-dependent β-adrenergic modulation of low frequency stimulation induced LTP in the hippocampal CA1 region. *Neuron* **17:** 475–482.

Tsien JZ, Huerta PT, Tonegawa S. 1996. The essential role of hippocampal CA1 NMDA receptor-dependent synaptic plasticity in spatial memory. *Cell* **87:** 1327–1338.

Tsvetkov E, Carlezon WA, Benes FM, Kandel ER, Bolshakov VY. 2002. Fear conditioning occludes LTP-induced presynaptic enhancement of synaptic transmission in the cortical pathway to the lateral amygdala. *Neuron* **34:** 289–300.

Van Groen T, Wyss JM. 1990. Extrinsic projections from area CA1 of the rat hippocampus: Olfactory, cortical, subcortical, and bilateral hippocampal formation projections. *J Comp Neurol* **302:** 515–528.

Van Strien NM, Cappaert NL, Witter MP. 2009. The anatomy of memory: An interactive overview of the parahippocampal-hippocampal network. *Nat Rev Neurosci* **10:** 272–282.

Vertes RP. 1992. PHA-L analysis of projections from the supramammillary nucleus in the rat. *J Comp Neurol* **326:** 595–622.

Wang GW, Cai JX. 2006. Disconnection of the hippocampal-prefrontal cortical circuits impairs spatial working memory performance in rats. *Behav Brain Res* **175:** 329–336.

Wersinger SR, Ginns EI, O'Carroll AM, Lolait SJ, Young WS III. 2002. Vasopressin V1b receptor knockout reduces aggressive behavior in male mice. *Mol Psychiatry* **7:** 975–984.

Wester JC, McBain CJ. 2014. Behavioral state-dependent modulation of distinct interneuron subtypes and consequences for circuit function. *Curr Opin Neurobiol* **29C:** 118–125.

Whitlock JR, Heynen AJ, Shuler MG, Bear MF. 2006. Learning induces long-term potentiation in the hippocampus. *Science* **313:** 1093–1097.

Wilkerson A, Levin ED. 1999. Ventral hippocampal dopamine D1 and D2 systems and spatial working memory in rats. *Neuroscience* **89:** 743–749.

Wills TJ, Cacucci F, Burgess N, O'Keefe J. 2010. Development of the hippocampal cognitive map in preweanling rats. *Science* **328:** 1573–1576.

Wilson RI, Nicoll RA. 2001. Endogenous cannabinoids mediate retrograde signalling at hippocampal synapses. *Nature* **410:** 588–592.

Witter MP, Amaral DG. 1991. Entorhinal cortex of the monkey: V. Projections to the dentate gyrus, hippocampus, and subicular complex. *J Comp Neurol* **307:** 437–459.

Wouterlood FG, Saldana E, Witter MP. 1990. Projection from the nucleus reuniens thalami to the hippocampal region: Light and electron microscopic tracing study in the rat with the anterograde tracer *Phaseolus vulgaris*-leucoagglutinin. *J Comp Neurol* **296:** 179–203.

Wyss JM, Van Groen T. 1992. Connections between the retrosplenial cortex and the hippocampal formation in the rat: A review. *Hippocampus* **2:** 1–11.

Xie Z, Ma X, Ji W, Zhou G, Lu Y, Xiang Z, Wang YX, Zhang L, Hu Y, Ding YQ, et al. 2010. Zbtb20 is essential for the specification of CA1 field identity in the developing hippocampus. *Proc Natl Acad Sci* **107:** 6510–6515.

Xu W, Sudhof TC. 2013. A neural circuit for memory specificity and generalization. *Science* **339:** 1290–1295.

Xu JY, Zhang J, Chen C. 2012. Long-lasting potentiation of hippocampal synaptic transmission by direct cortical input is mediated via endocannabinoids. *J Physiol* **590:** 2305–2315.

Yamamoto J, Suh J, Takeuchi D, Tonegawa S. 2014. Successful execution of working memory linked to synchronized high-frequency γ oscillations. *Cell* **157:** 845–857.

Yeckel MF, Berger TW. 1990. Feedforward excitation of the hippocampus by afferents from the entorhinal cortex: Redefinition of the role of the trisynaptic pathway. *Proc Natl Acad Sci* **87:** 5832–5836.

Young WS, Li J, Wersinger SR, Palkovits M. 2006. The vasopressin 1b receptor is prominent in the hippocampal area CA2 where it is unaffected by restraint stress or adrenalectomy. *Neuroscience* **143:** 1031–1039.

Younts TJ, Castillo PE. 2014. Endogenous cannabinoid signaling at inhibitory interneurons. *Curr Opin Neurobiol* **26:** 42–50.

Younts TJ, Chevaleyre V, Castillo PE. 2013. CA1 pyramidal cell theta-burst firing triggers endocannabinoid-mediated long-term depression at both somatic and dendritic inhibitory synapses. *J Neurosci* **33:** 13743–13757.

Zakharenko SS, Patterson SL, Dragatsis I, Zeitlin SO, Siegelbaum SA, Kandel ER, Morozov A. 2003. Presynaptic BDNF required for a presynaptic but not postsynaptic component of LTP at hippocampal CA1-CA3 synapses. *Neuron* **39:** 975–990.

Zhao M, Choi YS, Obrietan K, Dudek SM. 2007. Synaptic plasticity (and the lack thereof) in hippocampal CA2 neurons. *J Neurosci* **27:** 12025–12032.

Zhu PJ, Lovinger DM. 2007. Persistent synaptic activity produces long-lasting enhancement of endocannabinoid modulation and alters long-term synaptic plasticity. *J Neurophysiol* **97:** 4386–4389.

Ziv Y, Burns LD, Cocker ED, Hamel EO, Ghosh KK, Kitch LJ, El Gamal A, Schnitzer MJ. 2013. Long-term dynamics of CA1 hippocampal place codes. *Nat Neurosci* **16:** 264–266.

Zola-Morgan S, Squire LR, Amaral DG. 1986. Human amnesia and the medial temporal region: Enduring memory impairment following a bilateral lesion limited to field CA1 of the hippocampus. *J Neurosci* **6:** 2950–2967.

Cite this article as *Cold Spring Harb Perspect Biol* doi: 10.1101/cshperspect.a021733

# The Regulation of Transcription in Memory Consolidation

## Cristina M. Alberini[1] and Eric R. Kandel[2,3,4,5,6]

[1]Center for Neural Science, New York University, New York, New York 10003

[2]Zuckerman Mind Brain Behavior Institute, New York State Psychiatric Institute, New York, New York 10032

[3]Department of Neuroscience, New York State Psychiatric Institute, New York, New York 10032

[4]Kavli Institute for Brain Science, New York State Psychiatric Institute, New York, New York 10032

[5]Howard Hughes Medical Institute, New York State Psychiatric Institute, New York, New York 10032

[6]College of Physicians and Surgeons of Columbia University, New York State Psychiatric Institute, New York, New York 10032

*Correspondence:* ca60@nyu.edu; erk5@columbia.edu

De novo transcription of DNA is a fundamental requirement for the formation of long-term memory. It is required during both consolidation and reconsolidation, the posttraining and postreactivation phases that change the state of the memory from a fragile into a stable and long-lasting form. Transcription generates both mRNAs that are translated into proteins, which are necessary for the growth of new synaptic connections, as well as noncoding RNA transcripts that have regulatory or effector roles in gene expression. The result is a cascade of events that ultimately leads to structural changes in the neurons that mediate long-term memory storage. The de novo transcription, critical for synaptic plasticity and memory formation, is orchestrated by chromatin and epigenetic modifications. The complexity of transcription regulation, its temporal progression, and the effectors produced all contribute to the flexibility and persistence of long-term memory formation. In this article, we provide an overview of the mechanisms contributing to this transcriptional regulation underlying long-term memory formation.

The ability to form long-term memories and to store them for periods ranging from days to weeks to a whole lifetime is one of the brain functions most critical for adaptation and survival. Without the ability to store information about our experiences for the long term, our lives would be a series of disconnected fragments. Memories shape our character and, thus, contribute to every aspect of our individuality. The process of long-term memory formation is complex and is accompanied by long-lasting structural modification in the brain.

Long-term memories do not form immediately after learning but develop with time. They are initially fragile, but through a process of stabilization, known as memory consolidation, they become resistant to disruption (Bailey et al. 1996; McGaugh 2000; Dudai 2012). The biological mechanisms underlying consolidation start with a rapid phase of de novo gene expression, known as cellular or molecular consolidation. This begins at the onset of training and is a fundamental signature of long-term memory formation found in numerous species and in both explicit and implicit types of memories. Consolidation is manifest not only in behavior, but also in the cellular and molecular mechanisms contributing to long-term synaptic plasticity (Kandel 2001, 2014; Alberini 2009). Although it was long believed that molecular consolidation is completed rapidly, within a few hours, it recently emerged that in vivo it continues for at least 24 h, a temporal window in which circadian rhythms and sleep may make an important contribution (Eckel-Mahan and Storm 2009; Wang et al. 2011; Tononi and Cirelli 2014).

For example, in the hippocampus, a brain region critical for the formation of long-term contextual, spatial, and episodic memories, the gene expression–dependent phase necessary for the consolidation of inhibitory avoidance memory in rats lasts for more than 24 h, and seems to be completed by 48 h after training (Taubenfeld et al. 2001b; Alberini 2009; Bekinschtein et al. 2014). In addition, through a subsequent process called system consolidation, the initial critical role of the hippocampus continues for up to weeks in mice and even up to years in humans, although, over time, it can become dispensable (Squire et al. 2004; Wiltgen and Tanaka 2013). As a result, memory loss can still occur weeks after training in animals, and even years in humans, when the hippocampus is either inactivated or ablated. At the end of this phase of system consolidation, memories become insensitive to disruption by either pharmacological or molecular manipulations or hippocampal disruption/inactivation, and are therefore considered consolidated at the system level. System consolidation is primarily found in hippocampus-based explicit memories. Implicit memories, such as emotional Pavlovian associations that require the amygdala (e.g., cued fear conditioning) undergo molecular consolidation, but they are not known to undergo system consolidation like the hippocampal-dependent ones. However, it is important to keep in mind that, although the distinction between implicit and

explicit memory can be seen and studied in laboratory settings in which conditions can be controlled, different implicit and explicit memory processes generally interact to form long-term memories in real life (Phelps 2004).

It was long believed that the molecular/cellular phase of memory consolidation occurs only once, following training. Recently, however, we have learned that when memories that have become resistant to molecular interferences, hence, consolidated with respect to the molecular consolidation process are reactivated by a retrieval event, they can, as a result, become temporarily sensitive to disruption. During the first few hours after reactivation, de novo transcription and translation are required as is the case during consolidation. Only over time do the memories regain their stability and resilience. Because of the similarities with consolidation, this postreactivation process of stabilization is known as memory reconsolidation (Sara 2000; Alberini 2011; Nader and Einarsson 2010). The reconsolidation of different types of memories shows different temporal boundaries. Although hippocampal-based memories undergo a temporal gradient of stabilization for postretrieval interference and seem to become at one point resilient (Milekic and Alberini 2002; Suzuki et al. 2004; Frankland et al. 2006; Graff et al. 2014), amygdala-based memories can reconsolidate for a long time after training (Debiec et al. 2002, 2006). The reasons for this difference may very well reside in the different processing and mechanisms of the distinct memory systems. It is thought that hippocampal–cortical system consolidation may explain the temporal window during which these memories can undergo reconsolidation (Alberini 2011).

A fundamental biological mechanism for both consolidation and reconsolidation is de novo gene expression, and both processes recruit several overlapping mechanisms, including transcription factors and regulators. Thus, important questions to be understood are: How can DNA transcription promote memory stabilization? Are these changes in gene expression transient or do they last for weeks, months, or even years paralleling memory storage? This central issue has attracted the attention of many neuroscientists in the last two decades and has been investigated in in vitro models and a variety of behavioral paradigms in invertebrates as well as in mammals. These studies have asked: What regulatory mechanisms of transcription are involved in long-term memory formation? Can these changes in gene expression by themselves explain the long persistence of memories? What are the genes transcribed in response to the experience and what are their functions? Are these molecular mechanisms important therapeutic targets for treating memory disorders?

In this review, we summarize some of the current answers to these questions. Given the large number of studies, we will only be able to describe examples of (1) the major classes of transcription factors that play a critical role in both posttraining and postretrieval transcriptional regulation; (2) the target sequences regulated, including effector genes; and (3) the noncoding RNAs that have recently been found to regulate transcription and that can lead to chromatin, DNA, and RNA modifications that act in concert with transcription factors to regulate transcription important for memory consolidation.

## MEMORY CONSOLIDATION AND RECONSOLIDATION REQUIRE TRANSCRIPTION

Transcription, the first step of gene expression, is the mechanism that copies a sequence of DNA into RNAs. It is a complex process that requires the concerted action of protein and RNA complexes that together dictate the expression of target genes. It is estimated that 5%–10% of the expressed sequences in the human genome encode for transcription regulators, which indicates the importance and complexity of regulating transcription. Transcription regulators include DNA-binding proteins that dictate the rate of gene transcription and are commonly known as transcription factors, cofactors that interact with transcription factors, chromatin regulators, the general transcription machinery, and their regulators. The complexity of transcription regulation implies the versatility, selec-

tivity, and flexibility of the process, which, in fact, governs all cellular functions.

Studies using inhibitors of mRNA transcription in a variety of species ranging from invertebrates to mammals have shown that memory consolidation requires the synthesis of mRNAs and their translation into proteins, and that these transcriptional events are a fundamental and evolutionarily conserved mechanism of long-term memory formation (Brink et al. 1966; Agranoff 1967; Squire and Barondes 1970; Thut and Lindell 1974; Wetzel et al. 1976; Nestler 1993; Pedreira et al. 1996). As described earlier, memory consolidation recruits transcription and translation at multiple phases during an initial and limited temporal window. For example, in the rat hippocampus, a key region for explicit memory formation, at least two periods of transcription are needed to establish a long-term inhibitory avoidance memory. The first period of transcription occurs at about the time of training and the second occurs around 3–6 h later (Quevedo et al. 1999; Igaz et al. 2002). As mentioned earlier, the requirement for this initial transcription in the hippocampus continues for more than 24 h but ends by 48 h after training (Taubenfeld et al. 2001b; Garcia-Osta et al. 2006; Bekinschtein et al. 2007; Chen et al. 2011).

Another phase of transcription is required during reconsolidation. Although the underlying mechanisms are much less understood, it has been found that memory reactivation reinitiates a phase of gene expression, as revealed by the amnesia caused when either mRNA synthesis or the function of transcription factors is inhibited after retrieval (Sangha et al. 2003; Suzuki et al. 2004; Da Silva et al. 2008; Maddox et al. 2010; Cheval et al. 2012; Arguello et al. 2013). As in consolidation, the necessity of mRNA transcription in reconsolidation has been observed in many species ranging from invertebrates like *Lymnaea stagnalis* to mammals (Sangha et al. 2003; Suzuki et al. 2004; Merlo et al. 2005; Arguello et al. 2013; Veyrac et al. 2014), indicating its general and evolutionarily conserved role in the fragile phases of memory.

The important role of transcription has also been confirmed in cellular mechanisms contributing to long-term memory formation. These cellular mechanisms include long-lasting changes of the strength of synaptic connections in long-term facilitation (LTF) in the invertebrate *Aplysia californica*, and long-term potentiation (LTP) and long-term depression (LTD) in mammalian brain cells (Lynch 2004), thus strengthening the conclusion that transcription and gene expression are essential and general mechanisms necessary for stabilizing functions supported by long-term plasticity.

## CLASSES OF TRANSCRIPTION FACTORS INVOLVED IN LONG-TERM MEMORY CONSOLIDATION

Since the initial studies of the 1990s on the identification of transcription factors required for long-term plasticity and memory, it has emerged that one of the gene expression pathways required across species, types of memories, and memory systems for long-term plasticity and memory consolidation is that activated by cAMP-dependent mechanisms and mediated by members of the family known as cAMP-response element-binding proteins (CREB) (Dash et al. 1990; Bourtchuladze et al. 1994; Yin et al. 1994; Bartsch et al. 1995; Silva et al. 1998; Scott et al. 2002; Yin and Tully 2006; Alberini 2009; Kandel 2012). In the invertebrates *A. californica* and *Drosophila melanogaster*, the activation of the cascade cAMP-protein kinase A (PKA)-CREB is critical for plasticity and memory formation (Yin et al. 1994; Kandel 2012; but see Perazzona 2004). Specifically, cAMP-PKA activation initiates short-term synaptic changes that subsequently link via nuclear translocation of PKA, ERK, and perhaps other kinases to the activation and recruitment of CREB proteins and gene transcription (Bacskai et al. 1993; Martin et al. 1997; Ch'ng et al. 2012). Most of these mechanisms are conserved in the mammalian brain where the CREB-dependent pathway has also been shown to be necessary for long-term memory formation and long-term synaptic plasticity (Benito and Barco 2010; Barco and Marie 2011). Moreover, the overexpression of CREB promotes long-term memory storage from protocols that otherwise only induce

short-term memory indicating its proactive role (Josselyn et al. 2001; Barco et al. 2002; Josselyn and Nguyen 2005; Viosca et al. 2009; Gruart et al. 2012).

Although CREB represents one of the earliest identified transcription factors required for long-term memory formation, transcription regulation is a complex mechanism that involves the interactions of several transcription factors that can activate or inhibit transcription, cofactors, and general transcription proteins as well as chromatin-modifying proteins. In fact, CREB, like many other transcription factors, is expressed in many cell types throughout the organism, is regulated by several intracellular pathways, and is involved in several processes through different protein/chromatin complexes. CREB refers to the activator isoform, but the CREB family of transcription factors includes several members (such as CREB-2 and activating transcription factor [ATF-4]) that can act as inhibitors of transcription. Thus, its specific contribution to long-term memory formation is defined by the orchestrated regulation of the context in which CREB functions.

Hence, CREB, although essential, is one of several transcriptional events required for memory consolidation and reconsolidation. We next turn to examples of other transcription factors belonging to different families found to critically mediate memory consolidation and reconsolidation.

One gene controlled by and downstream from CREB activation in the context of learning or long-term plasticity is the CCAAT enhancer–binding protein (C/EBP), an immediate early gene (IEG) whose disruption or overexpression, like that of CREB, blocks or promotes long-term synaptic plasticity and long-term memory consolidation, respectively (Alberini et al. 1994; Lee at al. 2001; Taubenfeld et al. 2001a,b; Arguello et al. 2013). This indicates an intimate functional link between the two families of transcription regulators. One important aspect of this functional link is that via C/EBP, CREB controls a transcriptional cascade (Goelet et al. 1986; Alberini et al. 1994).

The biological implication of the contribution of a cascade of gene expression is that it governs a complex cellular function through a controlled and specific amplification of the initial signal. The result is a stable and long-lasting functional change that, at the same time, maintains flexibility and dynamism. Although the transcription factors of the cascade confer specificity through the ensemble of regulated target genes, chromatin and DNA modifications maintain the changes, as we will explain below (Guan et al. 2002; Levenson and Sweatt 2005). Additional signaling regulation can add, eliminate, or change the controllers of the cascades, thus turning on and/or off all of their downstream events, hence reversing or modifying the functional state of the cell. In fact, through specificity and cooperativity, gene-expression cascades lead to precise concerted actions.

Other transcription factors regulated as IEGs include the c-Fos and the zinc-finger protein Zif268 (also known as early growth response protein [EGR]-1). The transcription of these transcription factors is induced by activity following learning and play a necessary role in long-term memory formation (Guzowski 2002). Zif268 in particular is required in a variety of brain regions for consolidation and reconsolidation of different forms of explicit memories (Veyrac et al. 2014).

Understanding the transcriptional events underlying long-term plasticity and memory formation also provides important tools for asking further molecular questions. For example, in addition to experiments of knockout/knockdown or functional blockade of IEGs, which leads to the identification of their role in plasticity and memory, the detection of activity-induced expression of c-Fos and Zif268 can be used as a survey of activity patterns elicited by learning, retrieval, or any behavioral response of interest. Furthermore, the IEG regulatory elements (e.g., promoter regions) can be used to build readouts of activity-dependent responses. For instance, constructs can be engineered using a c-Fos promoter placed in front of the tetracycline transactivator (tTA or TET-off), which is known to drive the induction of tTA during high-level neural activity. tTA is a transcription factor that can be blocked by the antibiotic doxycycline (Dox), but, when ex-

pressed, it drives the transcription of genes controlled by a tetO promoter. The presence of a second transgene carrying a tetO-linked reporter, like the somato-axonal marker tau-lacZ together with elements of stabilization, has been used to reveal c-fos promoter–driven active cellular networks. This approach (developed by Mayford and colleagues) has been used to reveal the map of neural networks activated by and responding to experience and processing of representations (Reijmers et al. 2007). The induction of c-fos during learning can also be utilized to selectively express a receptor that can regulate activity when desired. With this method, artificial memory traces can be created. For example, by artificially activating an ensemble of cells in a given context (e.g., context B), which earlier had been activated by the exposure to a different context (e.g., context A), one can create a hybrid, artificial memory representation (Garner et al. 2012; Ramirez et al. 2013). These studies show the importance of understanding and using the transcriptional mechanisms underlying memory formation and storage to develop novel strategies that can be useful in research as well as translational applications.

In addition to these classical IEGs, other classes of transcription factors play critical roles in long-term memory. These include the nuclear factor-κ light-chain enhancer of activated B cells (NF-κB), members of the families nuclear receptor 4a (NR4a), serum response factor (SRF), and neuronal Per-Arnt-Sim (PAS) homology factor 4 (NPAS4), just to mention a few. NF-κB, expressed in both neurons and glia, is induced by LTP and by learning tasks like water maze, novel object recognition, and contextual fear conditioning. Its knockout results in memory impairment indicating that it plays a critical role in memory-related synaptic plasticity (Kaltschmidt et al. 2006; Romano et al. 2006; Ahn et al. 2008; Crampton and O'Keeffe 2013; Snow et al. 2013). An interesting feature of NF-κB is its synaptic localization, which implies that it plays a dual role in long-term memory, first as a signaling molecule at the synapse and second as a transcriptional regulator on translocation into the nucleus (Romano et al. 2006). In line with the functional requirement in memory consolidation and reconsolidation, primary functions targeted by family members of the NF-κB are the growth and morphological changes of axonal and dendritic arbors in several regions of the developing and mature central nervous system (CNS) (de la Fuente et al. 2011; Gutierrez and Davies 2011).

The nuclear receptor (NR) superfamily of transcription factors includes ligand-activated transcription factors implicated in cell differentiation, development, proliferation, and metabolism. They contain a zinc-finger DNA-binding domain and a carboxy-terminal ligand-binding domain. The expression of some members of this superfamily, including the NR4a family of orphan receptors, increases in the hippocampus immediately after learning, and their function is necessary for hippocampus-dependent contextual fear and object recognition memory as well as the transcriptional-dependent LTP (Bridi and Abel 2013). Notably, the level of NR4a increases following treatments that inhibit histone deacetylase (HDAC) (Hawk et al. 2012), a modification of histones that favors gene expression and promotes memory enhancement, as discussed below. Blocking NR4a signaling blocks the HDAC inhibitor-mediated memory enhancement suggesting that the Nr4a gene family significantly contributes to memory consolidation and enhancement (Pena de Ortiz et al. 2000; Hawk and Abel 2011; Hawk et al. 2012; Bridi and Abel 2013).

SRF, like CREB, is a major controller of IEG expression associated with actin-mediated contractile and motile cell functions (Knöll and Nordheim 2009). In adult brain, SRF is required for the acquisition of novel contextual information and for hippocampal-dependent spatial memory, as well as for LTP and LTD (Ramanan et al. 2005; Etkin et al. 2006). SRF, expressed mostly in neurons but not in glia, regulates activity-dependent gene expression, neuronal precursor cell migration, and morphological differentiation in both the developing and adult neurons. Its downstream signal activation recruits MAPKs, CaM kinases, and Rho/actin signaling cascades. SRF is required for long-lasting cellular changes because it modulates actin microfilament dynamics and associated mor-

phological functions. Thus, SRF seems to be positioned to couple the initial neural activation with the structural cell and synaptic modifications required for long-term maintenance of synaptic connections. SRF target genes include c-fos, Egr1, Egr2, and SRF itself as well as actin cytoskeletal genes (e.g., Acta1, Actb, Actg2) (Knöll and Nordheim 2009). Like CREB, SRF can recruit different cofactors and binding proteins, which can lead to either activation or repression of target genes; but, unlike CREB, it seems not to be involved in cell survival or apoptosis.

Although many transcription factors, such as CREB, c-Fos, Zif268, and NF-κB, are expressed throughout the brain in a variety of different cell populations, other transcription factors involved in memory formation appear to have expression that is restricted to a subpopulation of cells. One example is NPAS4, a b-helix–loop–helix–PAS transcription factor induced by neuronal activity, which on heterodimerization with ARNT2 regulates genes involved in inhibitory synapse formation. Both the expression and activity of NPAS4 are tightly coupled to neuronal activity; neuronal depolarization, ischemia, seizure, and learning, all rapidly and transiently induce the expression of NPAS4, which, in turn, regulates the expression of genes involved in increasing the number of inhibitory synapses, thus maintaining homeostasis of neuronal activity (Lin et al. 2008; Kim et al. 2010). Because of its role in supporting inhibitory synapse formation, deletion of NPAS4, not surprisingly, results in glutamate neurotoxicity and neurodegeneration, hyperactivity, seizures, anxiety, and cognitive impairments. Some of these phenotypes are reminiscent of those found in autism and schizophrenia (Lin et al. 2008; Coutellier et al. 2012). Interestingly, conditional deletion of NPAS4 selectively in the CA3 region of the hippocampus in adult mice impairs contextual memory formation (Ramamoorthi et al. 2011). Furthermore, environmental experience leads to expression of NPAS4 in hippocampal pyramidal neurons, which promotes an increase of inhibitory synapses on the cell soma but a decrease in the number of inhibitory synapses on the apical dendrites. This differential regulation of somatic and apical dendritic inhibition may allow compartmental integration or plasticity (Bloodgood et al. 2013).

These are only a few examples of transcription factors belonging to different families critically implicated in memory consolidation and reconsolidation. They provide a flavor of how complex the transcription regulation responding to experience and mediating memory consolidation and reconsolidation is. Each of them in different combinatorial complexes regulates distinct sets of target genes that also remain differentially expressed over time. Because of their role in regulating activity-induced gene expression, the transcription factors involved in various brain functions, including learning and memory, are also implicated in a number of neuropsychiatric and neurodegenerative diseases.

## CHROMATIN STRUCTURAL ALTERATIONS ASSOCIATED WITH LONG-TERM MEMORY

Transcription factors are key regulators of transcription, but they can only function if the appropriate DNA regulatory sequences are accessible to them. The gatekeepers of these regulatory sequences are the histone proteins around which the DNA is wrapped and that mediate compaction or relaxation of DNA sequences. In the nucleus of eukaryotic cells, the DNA complexes with histone proteins (also known as chromatin) form compact structures called nucleosomes, which are similar to beads on a necklace. Histone proteins have small tails that, by protruding from the nucleosome, offer themselves for the addition of specific marks on individual amino acids. The chromatin changes, which include histone posttranslational modifications, chromatin remodeling, and histone variant exchange, produce a unique combination or code that controls the way the DNA is packaged, hence available for transcription. Tight packaging inhibits DNA accessibility to the transcription machinery, whereas loose packaging allows DNA sequences to be accessible for transcription. In addition to chromatin changes, there are also chemical modifications of the

DNA itself, which together are referred to as "epigenetic changes" that regulate the availability and temporal duration of gene expression.

The discovery that the requirement for de novo gene expression is necessary for memory consolidation and reconsolidation first suggested that epigenetic changes must play essential roles. As morphological changes in synaptic structures have been found to correlate with the persistence of memory retention (*Aplysia*, mammals, songbirds), it is thought that the persistence of patterns of gene expression, hence the organization of chromatin and DNA modifications, orchestrate and control information retention by regulating transcription that translates into synaptic structural changes.

Until around the year 2000, chromatin modification had been studied primarily in the context of development and differentiation. Despite extensive studies of transcription in the brain, little was known about whether external events that affect transcription modulate chromatin structure in neurons.

Swank and Sweatt (2001) were the first to suggest that histone modifications may play a role in regulating gene expression associated with long-term memory formation.

The demonstration that histone acetylation was indeed a critical step for the de novo gene expression required for long-term plasticity mechanisms came in 2002 from Guan et al. who explored chromatin structure and protein–DNA interaction in *Aplysia* neuronal cultures in the context of learning-related synaptic plasticity using chromatin immunoprecipitation. The investigators focused on the chromatin around the promoter of C/EBP, which, as we have seen, is an early response gene downstream from CREB-1 with several CRE elements in the promoter region. C/EBP is rapidly induced during the formation of long-term memory and its induction is critical for long-term synaptic plasticity and memory (Alberini et al. 1994). Preventing the induction of C/EBP blocks LTF, whereas overexpression of C/EBP enhances LTF.

Guan et al. (2002) found that when *Aplysia* was exposed to repeated pulses of serotonin (5-HT), a protocol that induces long-term memory as a result of CREB and C/EBP expression, the

CREB-binding protein (CBP), which is capable of binding to CREB1, was recruited to the promoter of C/EBP to form a CREB1-CBP complex. In addition to CREB1, a small amount of CREB2—an inhibitor of CREB1—is bound to the promoter in the untreated state, and this small amount of CREB-2 decreases further after 5-HT exposure. This decrease in CREB2 most likely represents a displacement of CREB2 from the promoter after serotonin induction. With the induction of the C/EBP, the TATA-box-binding protein is also recruited to the promoter.

Guan et al. (2002) next asked whether the induction of C/EBP involves regulation of histone acetylation. They found that, indeed, exposure to repeated pulses of serotonin increased the acetylation of both histone H3 and H4 at the C/EBP promoter. Unlike H3, there was a strong basal acetylation of H4 in untreated animals. Both histone H3 and H4 have several lysine residues that can be acetylated. Guan et al. (2002) found that the acetylation and deacetylation of histones at the C/EBP promoter correlated with the induction and the termination of C/EBP expression.

Guan et al. (2002) next went on to ask: How are the excitatory and inhibitory inputs on a simple neuron integrated into a coherent output? Although the question of synaptic integration has been much studied, little was known about how neurons sum up opposing signals for long-term synaptic plasticity and memory. To address this question, they studied the same *Aplysia* sensory neurons that undergo LTF in response to serotonin. These neurons also undergo long-term synaptic depression in response to the peptide transmitter FMRFamide. Each of these transmitters produces synapse-specific actions when applied to one set of terminals and not the other. But when experimenters applied to the sensory neurons simultaneous pulses of the facilitating transmitter serotonin at one of the set of terminals while applying inhibitory transmitter FMRFamide at the other set of terminals, long-term synaptic depression dominated and shut off LTF, centrally preventing it from being expressed.

They next used chromatin immunoprecipitation assays and found that, although seroto-

nin induces the transcription of C/EBP through CREB1 activation and CBP recruitment with the consequent increase of histone acetylation, FMRF leads to CREB1 displacement by CREB2 and the recruitment of histone deacetylase 5 (HDAC5). When the two transmitters are applied together, facilitation is blocked and CREB2 and HDAC5 displace CREB1-CBP thereby deacetylating histones (Fig. 1). These studies show that long-term integration of spatially separate inhibitory and excitatory synaptic inputs occurs in the nucleus and is achieved by regulating chromatin structure and gene induction bidirectionally. Inhibitory inputs dominate in long-term integration by overriding the effects of facilitatory inputs on histone acetylation and gene induction. Thus, a neuron integrates opposite inputs at several levels: on the cell membrane to determine short-term response and in the nucleus to determine long-term response.

In subsequent years, many studies have investigated the mechanisms of chromatin and DNA modifications that, together with the recruitment of specific transcription complexes, regulate gene expression in the brain, and particularly in plasticity, learning and memory, reward, and cognitive processes in general (for reviews, see Maze et al. 2013; Peixoto and Abel 2013). Below, we limit our description to a few examples of chromatin regulators that have been found to critically control gene expression in memory formation through the action of some of the transcription factors that we have described above as essential in memory consolidation.

Because the consolidation and reconsolidation of different types of memories use different memory systems and are mediated by different brain regions, it is likely that different patterns of chromatin and DNA modifications occur in a cell-specific manner and differentially evolve over time. It follows that, to understand memory consolidation at the level of gene expression, it is necessary not only to identify the learning-induced pattern of expressed genes but also the learning-induced combinatorial interaction and assembly of transcription factors, cofactors, chromatin modifications, and remodeling as well as DNA modifications that control gene

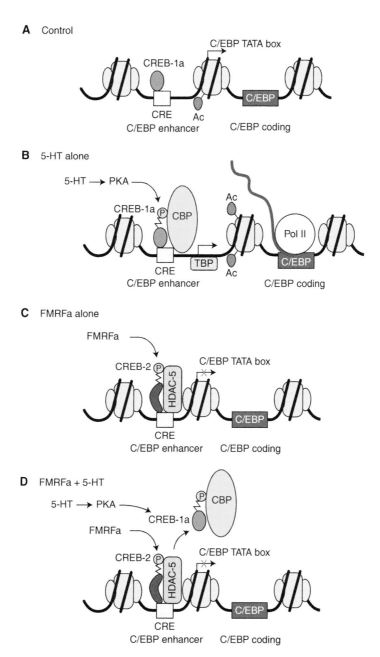

**Figure 1.** How 5-HT and FMRFa bidirectionally regulate chromatin changes leading to c/ebp transcription. (*A*) At the basal level, CREB1a resides on the C/EBP promoter; some lysine residues of histones are acetylated. (*B*) 5-HT, through PKA, phosphorylates CREB1 that binds to the C/EBP promoter. Phosphorylated CREB1 then forms a complex with CBP at the promoter. CBP then acetylates lysine residues of the histones (for example, K8 of H4). Acetylation modulates chromatin structure, enabling the transcription machinery to bind and regulate gene expression. (*C*) FMRFa activates CREB2, which displaces CREB1 from the C/EBP promoter. HDAC5 is then recruited to deacetylate histones. As a result, the gene is repressed. (*D*) If the neuron is exposed to both FMRFa and 5-HT, CREB1a is replaced by CREB2 at the promoter even though it might still be phosphorylated through the 5-HT-PKA pathway, and HDAC5 is then recruited to deacetylate histones, blocking gene induction (from Guan et al. 2002).

expression in a comprehensive way rather than in isolation. Moreover, as memory consolidation and reconsolidation evolve and change with time, allowing memory storage to be dynamic, the temporal evolution of the underlying molecular changes needs to be understood to comprehend how memory consolidation, persistence, and storage occur.

Although this complexity has thus far prevented the elucidation of a comprehensive picture, specific types of chromatin changes that in the brain play a critical role in memory formation and storage have been identified. Removing or adding acetyl groups from histones, through the action of HDACs or histone acetyltransferases (HATs), respectively, modifies the histones wrapping around the DNA, thus influencing gene expression. In rodent brain, changes in histones acetylation occur in response to protocols that induce late LTP (L-LTP), a form of long-term plasticity that, like memory consolidation, requires transcription (Weaver et al. 2004; Levenson and Sweatt 2006); and HDAC inhibitors promote memory formation and enhancement (Rudenko and Tsai 2014). Although the identification of the specific mechanisms that mediate the memory-enhancing effect of HDAC inhibitors is still in progress, HDAC2 seems to have an important role (Guan et al. 2009; Morris et al. 2013). Furthermore, and once again evolutionarily conserved, the recruitment of the HAT CBP controls the gene expression critical for plasticity and memory consolidation and enhancement (Alarcon et al. 2004; Korzus et al. 2004; Wood et al. 2006). As promoting CREB activation in CBP mutant mice rescues their memory impairment phenotypes, it follows that, as discussed above, CREB-dependent gene expression indeed plays a primary and proactive role in long-term memory formation (Bourtchouladze et al. 2003). Clinical data also underscore this conclusion, as mutations of CBP, or the related protein p300, are associated with devastating cognitive impairments, like that of the Rubinstein–Taybi syndrome. In animal models of Rubinstein–Taybi and in cell lines derived from Rubinstein–Taybi syndrome patients, cognitive impairments and histone acetylation defects can be ameliorated by HDAC inhibitors (Alarcon et al. 2004; Lopez-Atalaya et al. 2012; Park et al. 2014).

Genetic mutations of mice have shown that most classes of HDAC are involved in memory formation and its accompanying regulation of the IEG. For example, the class I HDAC1 and 3 bidirectionally modulate memory retention and regulate the expression of IEGs like c-Fos, NR4a, and Zif268. On the other hand, another class I HDAC, the HDAC2, acts as a negative regulator of memory formation by binding to the promoter regions of numerous regulatory genes, including Zif268, CREB, CBP, and effector genes like Homer1, Arc, GLUA1/2, NR2A/2B, Nrx1/3, Shank3, and PSD95 (Guan et al. 2009).

In addition to chromatin changes, long-term memory formation also depends on DNA methyltransferases (Miller and Sweatt 2007), indicating that chemical DNA modifications are also fundamental for the process (see Jarome and Lubin 2013, Zovkic et al. 2013, and Rudenko and Tsai 2014 for more information).

## NONCODING RNAs IN THE REGULATION OF TRANSCRIPTION

mRNAs are not the only target of transcription and chromatin regulation in memory consolidation and reconsolidation. Noncoding RNAs, like micro-RNAs (miRNAs), PIWI-interacting RNAs (piRNAs), and long noncoding RNAs (ncRNAs), are also targeted, and their expression in turn regulates transcriptional and posttranscriptional mechanisms. To obtain an understanding of how chromatin structure might be regulated, Rajasethupathy et al. (2012) performed a systematic screen of small RNAs in *Aplysia*. They found that there were not one but two classes of small RNAs regulated by neural activity: miRNAs and piRNAs. MiRNAs are a class of conserved 20–23-nucleotide noncoding RNAs that, in turn, critically contribute to transcriptional and posttranscriptional regulation of gene expression and depend on the RNAi machinery for maturation and function. Rajasethupathy et al. (2012) identified 170 distinct miRNAs: nine of these were enriched in the brain and several were down-regulated by sero-

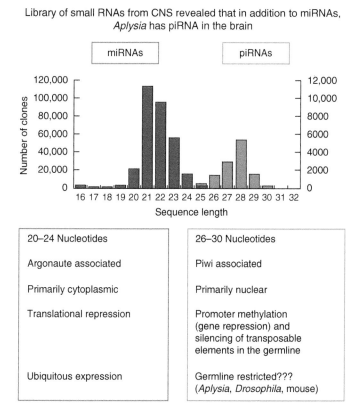

Figure 2. Two classes of small RNAs in *Aplysia* central nervous system (CNS). A size histogram of the cloned small RNAs revealed two populations, and further characterization confirmed the new class of sequences (blue) to be piRNAs.

tonin, the modulatory transmitter released during learning (Fig. 2).

Characterization of these neural-enriched miRNAs revealed that miRNA 124, the most abundant and conserved brain-specific miRNA, was exclusively present presynaptically in the sensory-to-motor synapse, where it constrained serotonin-induced synaptic facilitation through the regulation of the mRNA of the transcription factor CREB1. The activation of serotonin inhibits miRNA 124, thereby disinhibiting the translation of CREB1 making possible long-term transcription.

But in addition to miRNAs, Rajasethupathy made the surprising discovery of a second class of small noncoding RNA molecules, piRNAs, 28 to 32 nucleotides in length previously thought to be restricted to germ cells (Fig. 2). One of these piRNAs—piRNA-F—increased in re-

sponse to serotonin and led to the methylation of the promoter of CREB2 and to its silencing. Thus, serotonin regulates both piRNAs and miRNAs in a coordinated fashion, illustrating the integrative interactive action of small RNAs on the transcriptional level. Serotonin inhibits miRNA 124 rapidly and facilitates the activation of CREB1, which begins the process of memory consolidation, whereas piRNA-F, also activated by serotonin but with a delay, leads to methylation and thus repression of the promoter CREB2, allowing CREB1 to be active for a longer period of time, thereby establishing a stable long-term change in the sensory neuron that consolidates memory and puts it into the long-term phase (Fig. 3).

Examples of miRNA regulation have also been found in mammalian memory consolidation and plasticity. For example, the class III

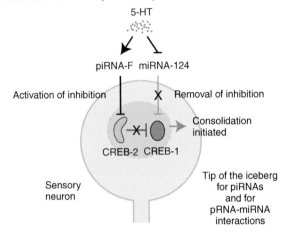

A New view of the consolidation switch

Coordinated action on transcription alters protein level of CREB-1 and CREB-2

**Figure 3.** Epigenetic regulation of the transcriptional switch. 5-HT inhibits miRNA-124 and thus facilitates the activation of CREB-1, which begins the process of memory consolidation, whereas piRNA, also activated by 5-HT but with a delay, methylates and thus represses the promoter of CREB-2, allowing CREB-1 to be active for a longer period of time.

NAD-dependent deacetylase SIRT1 limits the transcription of the miRNA 134, which represses the expression of critical plasticity/memory-related genes, such as *creb* and *bdnf*, hence promoting and regulating memory retention (Gao et al. 2010). Overexpression or inhibition of miRNA-132 in forebrain neurons, respectively, increases or reduces spine density (Hansen et al. 2010; Magill et al. 2010) and miRNA-132 accumulates in response to activity (Nudelman et al. 2010). Furthermore, miRNA-132 also inhibits methyl-CpG-binding protein 2 (MeCP2), a regulator of transcription that binds to methylated DNA (Klein et al. 2007). Long ncRNAs can target different aspects of transcription regulation by modulating transcriptional activators or repressors, different components of the transcription reaction, such as RNA polymerase (RNAP) II, and even the DNA structure (Goodrich and Kugel 2006). NcRNAs modulate the function of transcription factors by several mechanisms, including by functioning themselves as coregulators, modifying transcription factor activity, or regulating the association and activity of coregulators. Hence, it is speculated that ncRNAs, which together with transcription factors, chromatin, and DNA modifica-

tions, finely control gene expression in complex eukaryotes and may have a critical role in regulating gene expression in memory consolidation and reconsolidation (Mercer et al. 2008).

## AN OVERALL VIEW: THE FUNCTIONAL OUTCOME OF TRANSCRIPTION IN MEMORY STABILIZATION AND STORAGE

As mentioned earlier, understanding transcription underlying complex brain functions, such as memory consolidation and reconsolidation, requires the characterization of the complexity of the underlying mechanisms, including modification of DNA and chromatin, activation or inhibition of the expression of transcription factors, formation of diverse DNA-binding complexes that regulate gene expression, transcription regulation of target genes, which include those translated into proteins and regulatory sequences, such as miRNAs and ncRNAs, and further downstream changes, including transport, translation, and activation of the effector genes.

What is the role of this complexity in the stabilization processes of consolidation and reconsolidation? Why is transcription regulation required for memory stabilization? Given the complexity of memory storage, a clear answer is not yet available. Most information we have relates to single pathways, cell types, or restricted temporal windows. However, we have attempted to provide some answers given the knowledge that is available.

First, de novo gene expression is required for memory stabilization because it provides a controlled mechanism for inducing cellular changes that can persist for a long time, in fact, establishing a pattern of gene expression governed by the epigenome determines cellular changes that last as long as the epigenomic and transcription factor activation are maintained. Second, it offers specificity, as specific changes can occur through the selection of programs of gene expression. For example, in line with the accepted view of long-term synaptic plasticity, if the number and/or type of synaptic contacts change in response to learning, and this synaptic network constitutes the site where information is stored, then gene expression would be able to provide all the proteins, and only those, necessary to build and sustain those long-lasting synaptic changes. Hence, de novo gene expression is indeed an ideal underlying mechanism, because it offers regulated changes and stability and at the same time a vast repertoire of combinatorial possibilities for rapid responses to the changing environment. In other words, transcription regulation can provide a mechanism and explanation for the dynamic nature of memory storage. The complexity of the gene expression repertoire also allows for integration of information. Third, gene expression is different in different cell types (e.g., inhibitory versus excitatory neurons/synapses or versus astrocyte, microglia, oligodendocytes), and together the cell-specific patterns of gene expression can produce distinct functional integration patterns. These combinatorial patterns of gene expression translate into cellular functions, which then result in the cross talk among different brain areas that underlie memory consolidation, reconsolidation, and storage. Hence, transcription and gene-expression mechanisms can also explain how information can be stored in a system-wide dynamic network. It has, in fact, become increasingly

clear that memory storage is not a fixed function, but rather is a very dynamic process, whereby consolidation and reconsolidation together with possibly other retrieval-dependent processes and multiple trace consolidation/reconsolidation play a role in maintaining and changing information over time (Alberini 2011; Inda et al. 2011; Alberini et al. 2012).

System-wide molecular changes clearly must involve numerous forms of modification, cooperation, and regulations. Thus, one can ask: What types of cellular functions reflect the outcome of the transcription required for memory stabilization? A reasonable answer is that, given that memory consolidation and reconsolidation are the result of a network involving several brain regions, it is likely that the underlying transcriptional mechanism regulates both general as well as brain region- or cell-specific events. The field of transcription regulation in long-term memory is relatively new and, as a first step, in the last 20–25 yr, has mainly focused on identifying general, common mechanisms of cellular and synaptic plasticity. Here we will mention two examples of these general mechanisms that are mediated by CREB: the regulation of excitability and activation of growth pathways and related structural changes.

Overexpression of CREB in neurons promotes their preferential recruitment into a fear memory trace (Han et al. 2007). Specifically, if neurons in the amygdala of mice overexpress CREB, they are more likely to be activated, as revealed by IEG expression, following fear conditioning testing. This suggested that the CREB-overexpressing neurons are predisposed to participate in the memory trace and possibly that the system is not hardwired at the level of individual neurons but that certain molecular states of neurons, such as those promoted by CREB overexpression, favor a neuron's recruitment into the memory trace. Furthermore, selective ablation of the CREB-overexpressing neurons disrupts the fear memory, whereas ablation of a similar number of random neurons in the amygdala has no effect, demonstrating their functional role in memory formation and storage (Han et al. 2009). One of the general functions promoted by CREB expression seems to be neuronal excitability (Lopez de Armentia et al. 2007; Viosca et al. 2009; Zhou et al. 2009; Tong et al. 2010; Gruart et al. 2012) and, like the overexpression of CREB, increased excitability promotes cellular recruitment (allocation) in long-term memory (Rogerson et al. 2014) leading to the conclusion that CREB-dependent gene expression governs the circuitry recruited in memory formation and storage by increasing neuronal excitability.

Another general function regulated by CREB-dependent pathways that may contribute to cellular changes required for memory stabilization is growth (Persengiev and Green 2003). Growth pathways are activated in response to experience and play a key role in neurons during long-term memory formation. CREB is required to regulate the expression of growth factors, such as the brain-derived neurotrophic factor (BDNF), which plays a central role in the induction and maintenance of long-term plasticity and long-term memory and is a critical mediator of synaptic structural changes (Lu et al. 2008). Hence, a plausible hypothesis is that transcription regulated by CREBs controls growth as a fundamental outcome of long-term plasticity and memory (Alberini 2009; Kida 2012; Finsterwald and Alberini 2013; Nestler 2013).

These are only two examples thus far characterized, but given the critical role of de novo transcription in long-term memory, it is important that it be fully understood. Despite the great progress made in the last 25 yr, the understanding of transcription regulation in long-term memory is clearly in its infancy. A number of questions remain to be addressed including: What is the contribution of transcription regulation in long-term memory from each cell type and brain region? How does transcription alter each cell involved? Which are the end products? Are they general or selective in different cells? What is general or selective in different types of memories? How does previous experience change the learning-dependent transcriptional regulation? The answers to these questions will come from future studies and this knowledge will also be critical for a better understanding of memory disorders.

## REFERENCES

Abel T, Havekes R, Saletin JM, Walker MP. 2013. Sleep, plasticity and memory from molecules to whole-brain networks. *Curr Biol* 23: R774–R788.

Agranoff BW. 1967. Agents that block memory. In *The neurosciences: A study program*. Rockefeller University Press, New York.

Ahn HJ, Hernandez CM, Levenson JM, Lubin FD, Liou HC, Sweatt. 2008. c-Rel, an NF-κB family transcription factor, is required for hippocampal long-term synaptic plasticity and memory formation. *Learn Mem* 15: 539–549.

Alarcón JM, Malleret G, Touzani K, Vronskaya S, Ishii S, Kandel ER, Barco A. 2004. Chromatin acetylation, memory, and LTP are impaired in CBP$^{+/-}$ mice: A model for the cognitive deficit in Rubinstein–Taybi syndrome and its amelioration. *Neuron* 42: 947–959.

Alberini CM. 2009. Transcription factors in long-term memory and synaptic plasticity. *Physiol Rev* 89: 121–145.

Alberini CM. 2011. The role of reconsolidation and the dynamic process of long-term memory formation and storage. *Front Behav Neurosci* 5: 12.

Alberini CM, Ghirardi M, Metz R, Kandel ER. 1994. C/EBP is an immediate-early gene required for the consolidation of long-term facilitation in *Aplysia*. *Cell* 76: 1099–1114.

Alberini CM, Johnson SA, Ye X. 2012. Mechanisms and functions of memory reconsolidation: Lingering consolidation and the dynamic memory trace. In *Memory reconsolidation* (ed. Alberini CM). Academic, New York.

Arguello AA, Ye X, Bozdagi O, Pollonini G, Tronel S, Bambah-Mukku D, Huntley GW, Platano D, Alberini CM. 2013. CCAAT enhancer binding protein δ plays an essential role in memory consolidation and reconsolidation. *J Neurosci* 33: 3646–3658.

Bacskai BJ, Hochner B, Mahaut-Smith M, Adams SR, Kaang B-K, Kandel ER, Tsien RY. 1993. Spatially resolved dynamics of cAMP and protein kinase A subunits in *Aplysia* sensory neurons. *Science* 260: 222–226.

Bailey CH, Bartsch D, Kandel ER. 1996. Toward a molecular definition of long-term memory storage. *Proc Natl Acad Sci* 93: 13445–13452.

Barco A, Marie H. 2011. Genetic approaches to investigate the role of CREB in neuronal plasticity and memory. *Mol Neurobiol* 44: 330–349.

Barco A, Alarcon JM, Kandel ER. 2002. Expression of constitutively active CREB protein facilitates the late phase of long-term potentiation by enhancing synaptic capture. *Cell* 108: 689–703

Bartsch D, Ghirardi M, Skehel PA, Karl KA, Herder SP, Chen M, Bailey CH, Kandel ER. 1995. Aplysia CREB2 represses long-term facilitation: Relief of repression converts transient facilitation into long-term functional and structural change. *Cell* 83: 979–992.

Bekinschtein P, Cammarota M, Igaz LM, Bevilaqua LR, Izquierdo I, Medina JH. 2007. Persistence of long-term memory storage requires a late protein synthesis- and BDNF- dependent phase in the hippocampus. *Neuron* 53: 261–277.

Bekinschtein P, Cammarota M, Medina JH. 2014. BDNF and memory processing. *Neuropharmacology* 76: 677–683.

Benito E, Barco A. 2010. CREB's control of intrinsic and synaptic plasticity: Implications for CREB-dependent memory models. *Trends Neurosci* **33**: 230–240.

Bloodgood BL, Sharma N, Browne HA, Trepman AZ, Greenberg ME. 2013. The activity-dependent transcription factor NPAS4 regulates domain-specific inhibition. *Nature* **503**: 121–125.

Bourtchuladze R, Frenguelli B, Blendy J, Cioffi D, Schutz G, Silva AJ. 1994. Deficient long-term memory in mice with a targeted mutation of the cAMP-responsive element-binding protein. *Cell*. **79**: 59–68.

Bourtchouladze R, Lidge R, Catapano R, Stanley J, Gossweiler S, Romashko D, Scott R, Tully T. 2003. A mouse model of Rubinstein–Taybi syndrome: Defective long-term memory is ameliorated by inhibitors of phosphodiesterase 4. *Proc Natl Acad Sci* **100**: 10518–10522.

Bridi MS, Abel T. 2013. The NR4A orphan nuclear receptors mediate transcription-dependent hippocampal synaptic plasticity. *Neurobiol Learn Mem* **105**: 151–158.

Brink JJ, Davis RE, Agranoff BW. 1966. Effects of puromycin, acetoxycycloheximide and actinomycin D on protein synthesis in goldfish brain. *J Neurochem* **13**: 889–896.

Chen DY, Stern SA, Garcia-Osta A, Saunier-Rebori B, Pollonini G, Bambah-Mukku D, Blitzer RD, Alberini CM. 2011. A critical role for IGF-II in memory consolidation and enhancement. *Nature* **469**: 491–497.

Cheval H, Chagneau C, Levasseur G, Veyrac A, Faucon-Biguet N, Laroche S, Davis S. 2012. Distinctive features of Egr transcription factor regulation and DNA binding activity in CA1 of the hippocampus in synaptic plasticity and consolidation and reconsolidation of fear memory. *Hippocampus* **22**: 631–642.

Ch'ng TH, Uzgil B, Lin P, Avliyakulov NK, O'Dell TJ, Martin KC. 2012. Activity-dependent transport of the transcriptional coactivator CRTC1 from synapse to nucleus. *Cell* **150**: 207–221.

Coutellier L, Beraki S, Ardestani PM, Saw NL, Shamloo M. 2012. Npas4: A neuronal transcription factor with a key role in social and cognitive functions relevant to developmental disorders. *PLoS ONE* **7**: e46604.

Crampton SJ, O'Keeffe GW. 2013. NF-κB: Emerging roles in hippocampal development and function. *Int J Biochem Cell Biol* **45**: 1821–1824.

Dash PK, Hochner B, Kandel ER. 1990. Injection of the cAMP-responsive element into the nucleus of *Aplysia* sensory neurons blocks long-term facilitation. *Nature* **345**: 718–721.

Da Silva WC, Bonini JS, Bevilaqua LR, Medina JH, Izquierdo I, Cammarota M. 2008. Inhibition of mRNA synthesis in the hippocampus impairs consolidation and reconsolidation of spatial memory. *Hippocampus* **18**: 29–39.

Debiec J, LeDoux JE. 2006. Noradrenergic signaling in the amygdala contributes to the reconsolidation of fear memory: Treatment implications for PTSD. *Ann NY Acad Sci* **1071**: 521–524.

Debiec J, LeDoux JE, Nader K. 2002. Cellular and systems reconsolidation in the hippocampus. *Neuron* **36**: 527–538.

de la Fuente V, Freudenthal R, Romano A. 2011. Reconsolidation or extinction: Transcription factor switch in the determination of memory course after retrieval. *J Neurosci* **31**: 5562–5573.

Dudai Y. 2012. The restless engram: Consolidations never end. *Annu Rev Neurosci* **35**: 227–247.

Eckel-Mahan KL, Storm DR. 2009. Circadian rhythms and memory: Not so simple as cogs and gears. *EMBO Rep* **10**: 584–591.

Etkin A, Alarcón JM, Weisberg SP, Touzani K, Huang YY, Nordheim A, Kandel ER. 2006. A role in learning for SRF: Deletion in the adult forebrain disrupts LTD and the formation of an immediate memory of a novel context. *Neuron* **50**: 127–143.

Finsterwald C, Alberini CM. 2013. Stress and glucocorticoid receptor-dependent mechanisms in long-term memory: From adaptive responses to psychopathologies. *Neurobiol Learn Mem* **112**: 17–29.

Frankland PW, Ding HK, Takahashi E, Suzuki A, Kida S, Silva AJ. 2006. Stability of recent and remote contextual fear memory. *Learn Mem* **13**: 451–457.

Gao J, Wang WY, Mao YW, Gräff J, Guan JS, Pan L, Mak G, Kim D, Su SC, Tsai LH. 2010. A novel pathway regulates memory and plasticity via SIRT1 and miR-134. *Nature* **466**: 1105–1109.

Garcia-Osta A, Tsokas P, Pollonini G, Landau EM, Blitzer R, Alberini CM. 2006. MuSK expressed in the brain mediates cholinergic responses, synaptic plasticity, and memory formation. *J Neurosci* **26**: 7919–7932.

Garner AR, Rowland DC, Hwang SY, Baumgaertel K, Roth BL, Kentros C, Mayford M. 2012. Generation of a synthetic memory trace. *Science* **335**: 1513–1516.

Goelet P, Castellucci VF, Schacher S, Kandel ER. 1986. The long and the short of long-term memory: A molecular framework. *Nature* **322**: 419–422.

Goodrich JA, Kugel JF. 2006. Non-coding-RNA regulators of RNA polymerase II transcription. *Nat Rev Mol Cell Biol* **7**: 612–616.

Gräff J, Joseph NF, Horn ME, Samiei A, Meng J, Seo J, Rei D, Bero AW, Phan TX, Wagner F, et al. 2014. Epigenetic priming of memory updating during reconsolidation to attenuate remote fear memories. *Cell* **156**: 261–276.

Gruart A, Benito E, Delgado-García JM, Barco A. 2012. Enhanced cAMP response element-binding protein activity increases neuronal excitability, hippocampal long-term potentiation, and classical eyeblink conditioning in alert behaving mice. *J Neurosci* **32**: 17431–17441.

Guan Z, Giustetto M, Lomvardas S, Kim JH, Miniaci MC, Schwartz JH, Thanos D, Kandel ER. 2002. Integration of long-term-memory-related synaptic plasticity involves bidirectional regulation of gene expression and chromatin structure. *Cell* **111**: 483–493.

Guan JS, Haggarty SJ, Giacometti E, Dannenberg JH, Joseph N, Gao J, Nieland TJ, Zhou Y, Wang X, Mazitschek R, et al. 2009. HDAC2 negatively regulates memory formation and synaptic plasticity. *Nature* **459**: 55–60.

Gutierrez H, Davies AM. 2011. Regulation of neural process growth, elaboration and structural plasticity by NF-κB. *Trends Neurosci* **34**: 316–325.

Guzowski JF. 2002. Insights into immediate-early gene function in hippocampal memory consolidation using antisense oligonucleotide and fluorescent imaging approaches. *Hippocampus* **12**: 86–104.

Han JH, Kushner SA, Yiu AP, Cole CJ, Matynia A, Brown RA, Neve RL, Guzowski JF, Silva AJ, Josselyn SA. 2007. Neuronal competition and selection during memory formation. *Science* **316:** 457–460.

Han JH, Kushner SA, Yiu AP, Hsiang HL, Buch T, Waisman A, Bontempi B, Neve RL, Frankland PW, Josselyn SA. 2009. Selective erasure of a fear memory. *Science* **323:** 1492–1496.

Hansen KF, Sakamoto K, Wayman GA, Impey S, Obrietan K. 2010. Transgenic miR132 alters neuronal spine density and impairs novel object recognition memory. *PLoS ONE* **5:** e15497.

Hawk JD, Abel T. 2011. The role of NR4A transcription factors in memory formation. *Brain Res Bull* **85:** 21–29.

Hawk JD, Bookout AL, Poplawski SG, Bridi M, Rao AJ, Sulewski ME, Kroener BT, Mangelsdorf DJ, Abel T. 2012. NR4A nuclear receptors support memory enhancement by histone deacetylase inhibitors. *J Clin Invest* **122:** 3593–3602.

Igaz LM, Vianna MR, Medina JH, Izquierdo I. 2002. Two time periods of hippocampal mRNA synthesis are required for memory consolidation of fear-motivated learning. *J Neurosci* **22:** 6781–6789.

Inda MC, Muravieva EV, Alberini CM. 2011. Memory retrieval and the passage of time: From reconsolidation and strengthening to extinction. *J Neurosci* **31:** 1635–1643.

Jarome TJ, Lubin FD. 2013. Histone lysine methylation: Critical regulator of memory and behavior. *Rev Neurosci* **24:** 375–387.

Josselyn SA, Nguyen PV. 2005. CREB, synapses and memory disorders: Past progress and future challenges. *Curr Drug Targets CNS Neurol Disord* **4:** 481–497.

Josselyn SA, Shi C, Carlezon WA Jr, Neve RL, Nestler EJ, Davis M. 2001. Long-term memory is facilitated by cAMP response element-binding protein overexpression in the amygdala. *J Neurosci* **21:** 2404–2412.

Kaltschmidt B, Ndiaye D, Korte M, Pothion S, Arbibe L, Prüllage M, Pfeiffer J, Lindecke A, Staiger V, Israël A, et al. 2006. NF-κB regulates spatial memory formation and synaptic plasticity through protein kinase A/CREB signaling. *Mol Cell Biol* **26:** 2936–2946.

Kandel ER. 2001. The molecular biology of memory storage: A dialog between genes and synapses. *Biosci Rep 2001* **21:** 565–611.

Kandel ER. 2012. The molecular biology of memory: cAMP, PKA, CRE, CREB-1, CREB-2, and CPEB. *Mol Brain* **5:** 14.

Kandel ER, Dudai Y, Mayford MR. 2014. The molecular and systems biology of memory. *Cell* **157:** 163–186.

Kida S. 2012. A functional role for CREB as a positive regulator of memory formation and LTP. *Exp Neurobiol* **21:** 136–140.

Kim TK, Hemberg M, Gray JM, Costa AM, Bear DM, Wu J, Harmin DA, Laptewicz M, Barbara-Haley K, Kuersten S, et al. 2010. Widespread transcription at neuronal activity-regulated enhancers. *Nature* **465:** 182–187.

Klein ME, Lioy DT, Ma L, Impey S, Mandel G, Goodman RH. 2007. Homeostatic regulation of MeCP2 expression by a CREB-induced microRNA. *Nat Neurosci* **10:** 1513–1514.

Knöll B, Nordheim A. 2009. Functional versatility of transcription factors in the nervous system: The SRF paradigm. *Trends Neurosci* **32:** 432–442.

Korzus E, Rosenfeld MG, Mayford M. 2004. CBP histone acetyltransferase activity is a critical component of memory consolidation. *Neuron* **42:** 961–972.

Lee JA, Kim HK, Kim KH, Han JH, Lee YS, Lim CS, Chang DJ, Kubo T, Kaang BK. 2001. Overexpression of and RNA interference with the CCAAT enhancer-binding protein on long-term facilitation of *Aplysia* sensory to motor synapses. *Learn Mem* **8:** 220–226.

Levenson JM, Sweatt JD. 2005. Epigenetic mechanisms in memory formation. *Nat Rev Neurosci* **6:** 108–118.

Levenson JM, Sweatt JD. 2006. Epigenetic mechanisms: A common theme in vertebrate and invertebrate memory formation. *Cell Mol Life Sci* **63:** 1009–1016.

Lin Y, Bloodgood BL, Hauser JL, Lapan AD, Koon AC, Kim TK, Hu LS, Malik AN, Greenberg ME. 2008. Activity-dependent regulation of inhibitory synapse development by Npas4. *Nature* **455:** 1198–1204.

Lopez de Armentia M, Jancic D, Olivares R, Alarcon JM, Kandel ER, Barco A. 2007. cAMP response element-binding protein-mediated gene expression increases the intrinsic excitability of CA1 pyramidal neurons. *J Neurosci* **27:** 13909–13918.

Lopez-Atalaya JP, Gervasini C, Mottadelli F, Spena S, Piccione M, Scarano G, Selicorni A, Barco A, Larizza L. 2012. Histone acetylation deficits in lymphoblastoid cell lines from patients with Rubinstein–Taybi syndrome. *J Med Genet* **49:** 66–74.

Lu Y, Christian K, Lu B. 2008. BDNF: A key regulator for protein synthesis-dependent LTP and long-term memory? *Neurobiol Learn Mem* **89:** 312–323.

Lynch MA. 2004. Long-term potentiation and memory. *Physiol Rev* **84:** 87–136.

Maddox SA, Monsey MS, Schafe GE. 2010. Early growth response gene 1 (Egr-1) is required for new and reactivated fear memories in the lateral amygdala. *Learn Mem* **18:** 24–38.

Magill ST, Cambronne XA, Luikart BW, Lioy DT, Leighton BH, Westbrook GL, Mandel G, Goodman RH. 2010. microRNA-132 regulates dendritic growth and arborization of newborn neurons in the adult hippocampus. *Proc Natl Acad Sci* **107:** 20382–20387.

Martin KC, Michael D, Rose JC, Barad M, Casadio A, Zhu H, Kandel ER. 1997. MAP kinase translocates into the nucleus of the presynaptic cell and is required for long-term facilitation in *Aplysia*. *Neuron* **18:** 899–912.

Maze I, Noh KM, Allis CD. 2013. Histone regulation in the CNS: Basic principles of epigenetic plasticity. *Neuropsychopharmacology* **38:** 3–22.

McGaugh JL. 2000. Memory—A century of consolidation. *Science* **287:** 248–251.

Mercer TR, Dinger ME, Mariani J, Kosik KS, Mehler MF, Mattick JS. 2008. Noncoding RNAs in long-term memory formation. *Neuroscientist* **14:** 434–445.

Merlo E, Freudenthal R, Maldonado H, Romano A. 2005. Activation of the transcription factor NF-κB by retrieval is required for long-term memory reconsolidation. *Learn Mem* **12:** 23–29.

Cite this article as *Cold Spring Harb Perspect Biol* doi: 10.1101/cshperspect.a021741

Milekic MH, Alberini CM. 2002. Temporally graded requirement for protein synthesis following memory reactivation. *Neuron* **36**: 521–525.

Miller CA, Sweatt JD. 2007. Covalent modification of DNA regulates memory formation. *Neuron* **53**: 857–869.

Morris MJ, Mahgoub M, Na ES, Pranav H, Monteggia LM. 2013. Loss of histone deacetylase 2 improves working memory and accelerates extinction learning. *J Neurosci* **33**: 6401–6411.

Nader K, Einarsson EO. 2010. Memory reconsolidation: An update. *Ann NY Acad Sci* **1191**: 27–41.

Nestler EJ. 1993. Cellular responses to chronic treatment with drugs of abuse. *Crit Rev Neurobiol* **7**: 23–39.

Nestler EJ. 2013. Cellular basis of memory for addiction. *Dialogues Clin Neurosci* **15**: 431–443.

Nudelman AS, DiRocco DP, Lambert TJ, Garelick MG, Le J, Nathanson NM, Storm DR. 2010. Neuronal activity rapidly induces transcription of the CREB-regulated microRNA-132, in vivo. *Hippocampus* **20**: 492–498.

Park E, Kim Y, Ryu H, Kowall NW, Lee J, Ryu H. 2014. Epigenetic mechanisms of Rubinstein–Taybi syndrome. *Neuromolecular Med* **16**: 16–24.

Pedreira ME, Dimant B, Maldonado H. 1996. Inhibitors of protein and RNA synthesis block context memory and long-term habituation in the crab *Chasmagnathus*. *Pharmacol Biochem Behav* **54**: 611–617.

Peixoto L, Abel T. 2013. The role of histone acetylation in memory formation and cognitive impairments. *Neuropsychopharmacology* **38**: 62–76.

Peña de Ortiz S, Maldonado-Vlaar CS, Carrasquillo Y. 2000. Hippocampal expression of the orphan nuclear receptor gene hzf-3/nurr1 during spatial discrimination learning. *Neurobiol Learn Mem* **74**: 161–178.

Perazzona B, Isabel G, Preat T, Davis RL. 2004. The role of cAMP response element-binding protein in *Drosophila* long-term memory. *J Neurosci* **24**: 8823–8828.

Persengiev SP, Green MR. 2003. The role of ATF/CREB family members in cell growth, survival and apoptosis. *Apoptosis* **8**: 225–228.

Phelps EA. 2004. Human emotion and memory: Interactions of the amygdala and hippocampal complex. *Curr Opin Neurobiol* **14**: 198–202.

Quevedo J, Vianna MR, Roesler R, de-Paris F, Izquierdo I, Rose SP. 1999. Two time windows of anisomycin-induced amnesia for inhibitory avoidance training in rats: Protection from amnesia by pretraining but not pre-exposure to the task apparatus. *Learn Mem* **6**: 600–607.

Rajasethupathy P, Antonov I, Sheridan R, Frey S, Sander C, Tuschl T, Kandel ER. 2012. A role for neuronal piRNAs in the epigenetic control of memory-related synaptic plasticity. *Cell* **149**: 693–707.

Ramamoorthi K, Fropf R, Belfort GM, Fitzmaurice HL, McKinney RM, Neve RL, Otto T, Lin Y. 2011. Npas4 regulates a transcriptional program in CA3 required for contextual memory formation. *Science* **334**: 1669–1675.

Ramanan N, Shen Y, Sarsfield S, Lemberger T, Schütz G, Linden DJ, Ginty DD. 2005. SRF mediates activity-induced gene expression and synaptic plasticity but not neuronal viability. *Nat Neurosci* **8**: 759–767.

Ramirez S, Liu X, Lin PA, Suh J, Pignatelli M, Redondo RL, Ryan TJ, Tonegawa S. 2013. Creating a false memory in the hippocampus. *Science* **341**: 387–391.

Reijmers LG, Perkins BL, Matsuo N, Mayford M. 2007. Localization of a stable neural correlate of associative memory. *Science* **317**: 1230–1233.

Rogerson T, Cai DJ, Frank A, Sano Y, Shobe J, Lopez-Aranda MF, Silva AJ. 2014. Synaptic tagging during memory allocation. *Nat Rev Neurosci* **15**: 157–169.

Romano A, Freudenthal R, Merlo E, Routtenberg A. 2006. Evolutionarily conserved role of the NF-κB transcription factor in neural plasticity and memory. *Eur J Neurosci* **24**: 1507–1516.

Rudenko A, Tsai LH. 2014. Epigenetic modifications in the nervous system and their impact upon cognitive impairments. *Neuropharmacology* **80**: 70–82.

Sangha S, Scheibenstock A, Lukowiak K. 2003. Reconsolidation of a long-term memory in *Lymnaea* requires new protein and RNA synthesis and the soma of right pedal dorsal 1. *J Neurosci* **23**: 8034–8040.

Sara SJ. 2000. Retrieval and reconsolidation: Toward a neurobiology of remembering. *Learn Mem* **7**: 73–84.

Scott R, Bourtchuladze R, Gossweiler S, Dubnau J, Tully T. 2002. CREB and the discovery of cognitive enhancers. *J Mol Neurosci* **19**: 171–177.

Silva AJ, Kogan JH, Frankland PW, Kida S. 1998. CREB and memory. *Annu Rev Neurosci* **21**: 127–148.

Snow WM, Stoesz BM, Kelly DM, Albensi BC. 2013. Roles for NF-κB and gene targets of NF-κB in synaptic plasticity, memory, and navigation. *Mol Neurobiol* **49**: 757–770.

Squire LR, Barondes S. 1970. Actinomycin-D: Effects on memory at different times after training. *Nature* **225**: 649–650.

Squire LR, Stark CE, Clark RE. 2004. The medial temporal lobe. *Annu Rev Neurosci* **27**: 279–306.

Suzuki A, Josselyn SA, Frankland PW, Masushige S, Silva AJ, Kida S. 2004. Memory reconsolidation and extinction have distinct temporal and biochemical signatures. *J Neurosci* **24**: 4787–4795.

Swank MW, Sweatt JD. 2001. Increased histone acetyltransferase and lysine acetyltransferase activity and biphasic activation of the ERK/RSK cascade in insular cortex during novel taste learning. *J Neurosci* **21**: 3383–3391.

Taubenfeld SM, Wiig KA, Monti B, Dolan B, Pollonini G, Alberini CM. 2001a. Fornix-dependent induction of hippocampal CCAAT enhancer-binding protein β and δ co-localizes with phosphorylated cAMP response element-binding protein and accompanies long-term memory consolidation. *J Neurosci* **21**: 84–91.

Taubenfeld SM, Milekic MH, Monti B, Alberini CM. 2001b. The consolidation of new but not reactivated memory requires hippocampal C/EBPβ. *Nat Neurosci* **4**: 813–818.

Thut PD, Lindell TJ. 1974. α-Amanitin inhibition of mouse brain form II ribonucleic acid polymerase and passive avoidance retention. *Mol Pharmacol* **10**: 146–154.

Tong H, Steinert JR, Robinson SW, Chernova T, Read DJ, Oliver DL, Forsythe ID. 2010. Regulation of Kv channel expression and neuronal excitability in rat medial nucleus of the trapezoid body maintained in organotypic culture. *J Physiol* **588:** 1451–1468.

Tononi G, Cirelli C. 2014. Sleep and the price of plasticity: From synaptic and cellular homeostasis to memory consolidation and integration. *Neuron* **81:** 12–34.

Veyrac A, Besnard A, Caboche J, Davis S, Laroche S. 2014. The transcription factor zif268/egr1, brain plasticity, and memory. *Prog Mol Biol Transl Sci* **122:** 89–129.

Viosca J, Lopez de Armentia M, Jancic D, Barco A. 2009. Enhanced CREB-dependent gene expression increases the excitability of neurons in the basal amygdala and primes the consolidation of contextual and cued fear memory. *Learn Mem* **16:** 193–197

Wang G, Grone B, Colas D, Appelbaum L, Mourrain P. 2011. Synaptic plasticity in sleep: Learning, homeostasis and disease. *Trends Neurosci* **34:** 452–463.

Weaver IC, Cervoni N, Champagne FA, D'Alessio AC, Sharma S, Seckl JR, Dymov S, Szyf M, Meaney MJ. 2004. Epigenetic programming by maternal behavior. *Nat Neurosci* **7:** 847–854.

Wetzel W, Ott T, Matthies H. 1976. Is actinomycin D suitable for the investigation of memory processes? *Pharmacol Biochem Behav* **4:** 515–519.

Wiltgen BJ, Tanaka KZ. 2013. Systems consolidation and the content of memory. *Neurobiol Learn Mem* **106:** 365–371.

Wood MA, Hawk JD, Abel T. 2006. Combinatorial chromatin modifications and memory storage: A code for memory? *Learn Mem* **13:** 241–244.

Yin JC, Tully T. 1996. CREB and the formation of long-term memory. *Curr Opin Neurobiol* **6:** 264–268.

Yin JCP, Wallach JS, Del Vecchio M, Wilder EL, Zhou H, Quinn WG, Tully T. 1994. Induction of a dominant-negative CREB transgene specifically blocks long-term memory in *Drosophila melanogaster. Cell* **79:** 49–58.

Zhou Y1, Won J, Karlsson MG, Zhou M, Rogerson T, Balaji J, Neve R, Poirazi P, Silva AJ. 2009. CREB regulates excitability and the allocation of memory to subsets of neurons in the amygdala. *Nat Neurosci* **12:** 1438–1443.

Zovkic IB, Guzman-Karlsson MC, Sweatt JD. 2013. Epigenetic regulation of memory formation and maintenance. *Learn Mem* **20:** 61–74.

# Structural Components of Synaptic Plasticity and Memory Consolidation

Craig H. Bailey[1,2,3], Eric R. Kandel[1,2,3,4], and Kristen M. Harris[5]

[1]Department of Neuroscience, College of Physicians and Surgeons of Columbia University, New York, New York 10027

[2]New York State Psychiatric Institute, New York, New York 10032

[3]Kavli Institute for Brain Sciences, New York, New York 10032

[4]Howard Hughes Medical Institute, Chevy Chase, Maryland 20815-6789

[5]Department of Neuroscience, Center for Learning and Memory, Institute for Neuroscience, The University of Texas at Austin, Austin, Texas 78712-0805

*Correspondence:* chb1@columbia.edu

Consolidation of implicit memory in the invertebrate *Aplysia* and explicit memory in the mammalian hippocampus are associated with remodeling and growth of preexisting synapses and the formation of new synapses. Here, we compare and contrast structural components of the synaptic plasticity that underlies these two distinct forms of memory. In both cases, the structural changes involve time-dependent processes. Thus, some modifications are transient and may contribute to early formative stages of long-term memory, whereas others are more stable, longer lasting, and likely to confer persistence to memory storage. In addition, we explore the possibility that trans-synaptic signaling mechanisms governing de novo synapse formation during development can be reused in the adult for the purposes of structural synaptic plasticity and memory storage. Finally, we discuss how these mechanisms set in motion structural rearrangements that prepare a synapse to strengthen the same memory and, perhaps, to allow it to take part in other memories as a basis for understanding how their anatomical representation results in the enhanced expression and storage of memories in the brain.

Santiago Ramón y Cajal (1894) used the insights provided by his remarkable light microscopic observations of neurons selectively stained with the Golgi method to propose the first cellular theory of memory storage as an anatomical change in the functional connections between nerve cells, later called synapses (Sherrington 1897). For most of the last century, chemical synapses were thought to convey information in only one direction—from the presynaptic to the postsynaptic neuron. It now is clear that synaptic transmission is a bidirectional and self-modifiable form of cell–cell communication (Peters et al. 1976; Jessell and Kandel 1993). This appreciation of reciprocal signaling between pre- and postsynaptic elements is consistent with other forms of intercellular communication and provides a conceptual framework for understanding memory-induced changes in the structure of the synapse. Indeed, an increasing body of evidence suggests that trans-synaptic signaling and coordinated recruitment of pre- and postsynaptic mechanisms underlie consolidation of both implicit and explicit forms of memory storage (Marrone 2005; Hawkins et al. 2006; Bailey et al. 2008).

Studies in a variety of systems have found that molecular mechanisms of consolidation and long-term storage of memory begin at the level of the synapse. Existing proteins are modified, signals are sent back to the nucleus so that specific genes are expressed, and gene products are transported back to the synapse where the local synthesis of new protein is triggered to allow for the remodeling, addition, and elimination of synapses (Bailey and Kandel 1985; Bailey et al. 1996; Kandel 2001; Bourne and Harris 2008, 2012). These structural components of synaptic plasticity are thought to represent a cellular change that contributes to both implicit and explicit memory consolidation (Greenough and Bailey 1988; Bailey and Kandel 1993; Bailey et al. 2005; Bourne and Harris 2008, 2012). The association between alterations in the structure and/or number of synapses and memory storage has led to numerous studies regarding the signaling pathways that might couple molecular changes to structural changes. In addition, parallel homeostatic mechanisms have been identified that can trigger synaptic scaling, which serves to stabilize the strengthened synapses while weakening or eliminating other synapses, thus providing specificity during memory consolidation (Bourne and Harris 2011; Schacher and Hu 2014).

In this review, we compare and contrast structural changes at the synapse during both implicit and explicit memory consolidation, as well as the molecular signaling pathways that initiate the learning-induced structural changes versus those that serve to maintain these changes over time. Toward that end, we will focus on two experimental model systems and several prototypic forms of synaptic plasticity that we have worked on and that have been extensively studied as representative examples of memory storage: long-term habituation and sensitization of

the gill-withdrawal reflex in *Aplysia*. These are examples of implicit memory consolidation and hippocampal-based long-term potentiation (LTP) and long-term depression (LTD), as candidate mechanisms for the synaptic plasticity underlying explicit memory storage in mammals. These will serve as useful points of comparison to consider similarities, differences, and still-existing limitations in our understanding of the functional significance of the structural synaptic plasticity recruited during the consolidation of both implicit and explicit forms of memory.

## STRUCTURAL CHANGES AND CONSOLIDATION OF IMPLICIT BEHAVIORAL MEMORY

Structural mechanisms contributing to implicit memory storage have been most extensively studied for sensitization of the gill-withdrawal reflex in *Aplysia* (Kandel 2009). Sensitization is an elementary form of nonassociative learning, a form of learned fear, by which an animal learns about the properties of a single noxious stimulus. When a light touch is applied to the siphon of an *Aplysia*, it responds by withdrawing its gill and siphon. This response is enhanced when the animal is given a noxious, sensitizing stimulus, such as a mild shock to its tail. The memory for sensitization of the withdrawal reflex is graded: a single tail shock produces short-term sensitization that lasts for minutes, whereas five repeated tail shocks given at spaced intervals produce long-term sensitization that lasts for up to several weeks (Frost et al. 1985). Both short- and long-term sensitization lead to enhanced transmission at a critical synaptic locus: the monosynaptic connection between identified mechanoreceptor sensory neurons and their follower cells.

In the early 1980s, studies in *Aplysia* began to explore the structural changes that underlie memory consolidation: the transition from short- to long-term sensitization. By combining selective intracellular-labeling techniques with the analysis of serial thin sections and transmission electron microscopy (TEM), complete 3D reconstructions of unequivocally identified sensory neuron synapses were quantitatively analyzed from both control and behaviorally modified animals (Fig. 1).

The storage of long-term memory for sensitization (lasting several weeks) was accompanied by two classes of structural changes at identified synapses between the sensory neurons and their target neurons: (1) a remodeling of the preexisting presynaptic compartment leading to an increase in the number, size, and vesicle complement of the active zones (regions modified for transmitter release) of sensory neurons from sensitized animals compared

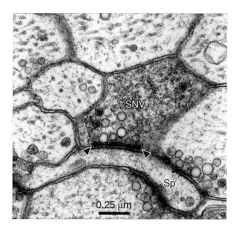

Figure 1. Fine structure of an identified sensory neuron presynaptic varicosity. A thin section containing a sensory neuron varicosity (SNV) labeled with horseradish peroxidase (HRP) (Bailey et al. 1979) is shown. The density of the HRP reaction product allows one to clearly distinguish the labeled sensory neuron profile from unlabeled profiles while still being able to visualize the intracellular contents of the identified varicosity. This portion of the identified sensory neuron presynaptic compartment contains three dense core vesicles and a population of electron-lucent vesicles, some of which cluster at the electron-dense specializations that define the active zone (between arrow heads). In this thin section, the labeled sensory neuron presynaptic varicosity forms a synaptic contact with an unlabeled postsynaptic dendritic spine (Sp) of a follower neuron. By combining this selective intracellular-labeling technique with the analysis of serial thin sections and transmission EM, complete 3D reconstructions of active zone morphology (number, size, and vesicle complement) in unequivocally identified sensory neuron synapses were quantitatively analyzed from both control and behaviorally modified animals. (Unpublished electron micrograph courtesy of Mary Chen and Craig Bailey.)

with untrained controls (Bailey and Chen 1983, 1988b), and (2) a more expansive growth process that led to a twofold increase in the number of synaptic varicosities (boutons), as well as an enlargement of each neuron's synaptic arbor when compared with sensory neurons from untrained animals (Fig. 2) (Bailey and Chen 1988a). Moreover, in control animals, ~60% of the fully reconstructed varicosities lacked a structurally detectable active zone (Bailey and Chen 1983). The extent to which learning can convert these nascent and presynaptically silent synapses into mature and functionally competent synaptic connections is discussed below.

By comparing the time course for each class of morphological change with the behavioral duration of the memory, Bailey and Chen (1989) found that only the increases in the number of varicosities and active zones, which persisted unchanged for at least 1 week and were partially reversed at the end of 3 weeks, paralleled the time course of behavioral memory storage and, thus, could contribute to the retention of long-term sensitization. These results directly correlate a change in the structure of an identified synapse to a long-lasting behavioral memory and suggest that the morphological alterations could represent an anatomical substrate for memory consolidation. The learning-induced growth of new sensory neuron synapses in the abdominal ganglion that accompanies long-term sensitization of the gill-withdrawal reflex also was found to occur in subsequent behavioral studies of sensitization in the pleural ganglion mediating the tail-siphon withdrawal reflex in *Aplysia* (Wainwright et al. 2002).

In addition to long-term sensitization, Bailey and Chen (1983, 1988a) also examined, in the same studies, the structural correlates of long-term habituation. Unlike what they observed following long-term sensitization, this behavioral form of persistent synaptic depression was associated with decreases in the number, size, and vesicle complement of sensory neuron active zones, as well as a 35% reduction in the total number of synapses that the sensory neurons make on their follower cells when compared with sensory neurons from untrained animals. Thus, long-term behavioral modifica-

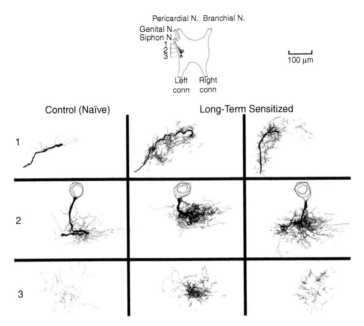

**Figure 2.** Learning-related growth of the sensory neuron synaptic arbor induced by long-term sensitization in *Aplysia*. Serial 3D reconstructions of identified sensory neurons labeled with horseradish peroxidase (HRP) from long-term sensitized and control animals are shown. Total extent of the synaptic neuropil arbors of sensory neurons from one control (untrained) and two long-term sensitized animals are shown. In each case, the rostral (row 3) to caudal (row 1) extent of the arbor is divided roughly into thirds. Each panel was produced by the superimposition of camera lucida tracings of all HRP-labeled processes present in 17 consecutive slab-thick Epon sections and represents a linear segment through the ganglion of roughly 340 μm. For each composite, ventral is *up*, dorsal is *down*, lateral is to the *left*, and medial is to the *right*. By examining images across each row (rows 1, 2, and 3), the viewer is comparing similar regions of each sensory neuron. In all cases, the synaptic arbor of long-term sensitized cells is markedly expanded compared with cells from control (untrained) animals, and parallels the concomitant twofold increase in the total number of sensory neuron presynaptic varicosities. (From Bailey and Chen 1988a; modified, with permission.)

tions in *Aplysia* not only can induce the growth of new synaptic connections, but also the pruning of preexisting connections.

This bidirectional structural remodeling of the same synapse following opposing forms of learning, in turn, provided some insights into how the anatomical representations of enduring memories might be accomplished at the more complex systems level. In the mammalian brain, each memory is likely to be distributed and embedded in many synaptic connections. Clearly, the brain cannot accommodate the storage of such a large number of memories by constant growth of new synaptic connections alone. The studies on long-term habituation in *Aplysia* provide an experimental foundation for an alternative hypothesis, that is, although initial long-

term storage may be dependent on a growth process, the brain appears to have the ability to reorganize and refine this representation in an experience-dependent fashion by pruning old or inappropriate synapses, thus reducing the total number of synapses required to carry each memory over time. A corollary of this would be the prediction that as a memory is strengthened over time, which is thought to occur with retrieval and recall, no new synapses would form, but rather there is an increase of signal-to-noise as the appropriate synapses are enlarged and strengthened, whereas the inappropriate synapses are eliminated (see, for example, Xu et al. 2009; Yang et al. 2009; Bourne and Harris 2011).

These initial studies in *Aplysia* showed that learning-induced structural changes occur at the level of specific identified synapses known to be critically involved in the behavioral modification providing direct evidence supporting Ramón y Cajal's prescient suggestions that synaptic connections between neurons are not immutable, but are modified by learning and may serve as key components of memory expression and storage. Moreover, the growth of new synapses may represent a stable component required for the consolidation of memory storage and raises the possibility that the persistence of the long-term process might be achieved, at least in part, because of the relative stability of these changes in synaptic structure (Bailey and Chen 1990; Bailey 1991; Bailey and Kandel 2008a).

## IMPLICIT MEMORY MECHANISMS CAN BE RECONSTITUTED IN CULTURED *APLYSIA* NEURONS

The simplicity of the neuronal circuit underlying sensitization, including direct monosynaptic connections between identified mechanoreceptor sensory neurons and their follower cells (Castellucci et al. 1970), has allowed reduction of the analysis of the short- and long-term memory for sensitization to the cell and molecular level. This monosynaptic sensory to motor neuron connection, which is glutamatergic, can be reconstituted in dissociated cell culture and reproduces what is observed during behavioral training by replacing tail shocks with brief applications of serotonin (5-HT), a modulatory transmitter normally released by sensitizing stimuli in the intact animal (Montarolo et al. 1986; Marinesco and Carew 2002). A single, brief application of 5-HT produces a short-term change in synaptic effectiveness (short-term facilitation [STF]), whereas repeated and spaced applications produce changes in synaptic strength that can last for more than a week (long-term facilitation [LTF]).

The molecular changes associated with STF and LTF differ fundamentally in at least two ways. First, the long-term but not the short-term changes require the activation of transcription and new protein synthesis (Schwartz et al. 1971; Montarolo et al. 1986; Castellucci et al. 1989). Second, as we have just seen at the

behavioral level, the long-term but not the short-term processes involve the growth of new sensory-to-motor-neuron synapses, which, when reconstituted in dissociated cell culture, are induced by five repeated applications of 5-HT and depend on transcription and translation (Bailey et al. 1992b) as well as the presence of an appropriate target cell similar to the synapse formation that occurs during development (Glanzman et al. 1990).

## REMODELING AND ACTIVATION OF PREEXISTING SILENT SYNAPSES DURING LTF

Kim et al. (2003) followed remodeling and growth at the same specific synaptic varicosities continuously over time and examined the functional contribution of these presynaptic structural changes to different time-dependent phases of facilitation. Live time-lapse confocal imaging was performed on sensory neurons containing the whole cell marker *Alexa-594*, and the presynaptic marker proteins *synaptophysin-eGFP* and *synapto-PHluorin* (*synPH*), which monitor changes in synaptic vesicle distributions and active transmitter-release sites, respectively. The results showed that initially, when a sensory neuron was cocultured with its postsynaptic motor neuron L7, ~12% of the presynaptic varicosities that were labeled with *Alexa-594* lacked *synaptophysin-eGFP* and *synPH* labeling and, thus, were not competent to release transmitter. Repeated pulses of 5-HT induced a rapid activation of these silent presynaptic terminals through the filling of preexisting empty (nascent) varicosities with synaptic vesicles and active zone material. This filling and unsilencing of preexisting sensory neuron varicosities began at 0.5 h after exposure to the five pulses of 5-HT, was completed within 3–6 h, and accounted for ~32% of the newly activated synapses present at 24 h. Thus, the rapid activation of silent presynaptic varicosities suggests that, in addition to its role in LTF, this remodeling of preexisting nascent synapses may also contribute to the intermediate phases of synaptic plasticity and implicit memory storage (Fig. 3) (Ghirardi et al. 1995; Mauelshagen et al. 1996; Sutton et al. 2001).

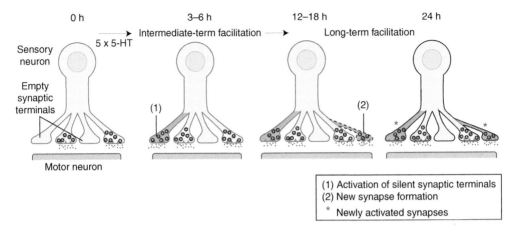

**Figure 3.** Time course and functional contribution of two distinct presynaptic structural changes associated with intermediate-term facilitation and long-term facilitation (LTF) in *Aplysia*. Repeated pulses of 5-HT in sensory to motor neuron cocultures trigger two distinct classes of learning-related presynaptic structural changes: (1) the rapid filling of synaptic vesicles and active zone material to preexisting silent sensory neuron varicosities (3–6 h), and (2) the slower generation of new sensory neuron synaptic varicosities (12–18 h). The resultant newly filled and newly formed varicosities are functionally competent (capable of evoked transmitter release) and contribute to the synaptic enhancement that underlies LTF measured at 24 h. The rapid filling and activation of silent presynaptic terminals at 3 h suggests that, in addition to its role in LTF, this remodeling of preexisting varicosities may also contribute to the intermediate phase of synaptic plasticity. Red triangles represent transmitter-release sites (active zones). (From Kim et al. 2003; modified, with permission.)

## LEARNING-RELATED ADDITION OF NEW FUNCTIONAL SYNAPSES DURING LTF

LTF also is accompanied by a second class of learning-related presynaptic structural change: a slower generation of new and functionally effective sensory neuron varicosities. Time-lapse imaging revealed that new sensory neuron varicosities began to form 12–18 h after exposure to five pulses of 5-HT and accounted for 68% of newly activated synapses at 24 h (Kim et al. 2003).

How are these new varicosities formed? The 5-HT-induced recruitment of synaptic vesicles and active zone material to a preexisting varicosity leads directly to both an enrichment of these presynaptic constituents, as well as to an overall increase in the size of the varicosity. The presynaptic remodeling and growth is followed by the apparent division or splitting of a subset of these preexisting varicosities (Hatada et al. 2000; Kim et al. 2003; Udo et al. 2005). This dynamic process may lead to the budding off of components of the active zone and cognate synaptic vesicle cluster from each preexisting

presynaptic compartment, similar to the creation of "orphan-release sites" in mammalian cultures (Ziv and Garner 2004), which could then serve as nucleation loci to seed the subsequent differentiation and establishment of new presynaptic varicosities (Bailey and Kandel 2008b).

These findings, the first to be made on individually identified presynaptic varicosities, suggest that the duration of changes in synaptic effectiveness that accompany different phases of memory storage may be reflected by the differential regulation of two fundamentally disparate forms of presynaptic compartment: (1) nascent, silent varicosities that can be rapidly and reversibly remodeled into active transmitter-release sites, and (2) mature, more stable, and functionally competent varicosities that, following long-term training, may undergo a process of fission to form new stable synaptic contacts.

These morphological findings, in turn, raised the question: What are the cellular and molecular mechanisms responsible for in these two distinct classes of learning-related presynaptic structural change?

Cite this article as *Cold Spring Harb Perspect Biol* doi: 10.1101/cshperspect.a021758

## INITIAL STEPS OF LEARNING-RELATED SYNAPTIC GROWTH IN *APLYSIA*

### Spontaneous Transmitter Release and Trans-Synaptic Recruitment of Pre- and Postsynaptic Mechanisms

Similar to synaptogenesis during development (McAllister 2007), the growth of new synaptic connections induced by learning in the adult requires the participation of both pre- and postsynaptic components of the synapse. In *Aplysia*, a newly discovered intermediate phase of memory initiates structural remodeling in preexisting synapses, which, in turn, serves as an early step contributing to the synaptic growth during the long-term phase and, therefore, requires participation of both pre- and postsynaptic components of the synapse, although not transcription (Ghirardi et al. 1995; Kim et al. 2003). Jin et al. (2012a,b) found that application of protein kinase A (PKA), which initiates the intermediate phase, leads to an increase in spontaneous transmitter release from the presynaptic sensory neuron and provides the critical trans-synaptic signal for recruitment of the molecular machinery of the postsynaptic motor neuron and subsequent remodeling of preexisting synapses, which represent the initial steps of synaptic growth. The spontaneous release is regulated by an *Aplysia* neurotrophin (ApNT) ligand (Kassabov et al. 2013) released by the presynaptic neuron that contributes, in a PKA-dependent manner, to intermediate-term facilitation by enhancing spontaneous transmitter release (Hawkins et al. 2012) and inducing growth. ApNT does so, in part, by contributing an autocrine signal to the presynaptic sensory neurons via its cognate Trk autoreceptors. Spontaneous release activates postsynaptic metabotropic glutamate receptors (mGluR5), which increase IP3 production, causing release of calcium from intracellular stores, which leads to the insertion of new α-amino-3-hydroxy-5-methyl-4-isoxazolepropionic acid (AMPA) receptors (Jin et al. 2012a,b) and the first phase of remodeling in the postsynaptic neuron. Blocking the postsynaptic $Ca^{2+}$ signal blocks postsynaptic participation and growth.

### Remodeling of the Presynaptic Actin Network

The 5-HT-induced enrichment of synaptic vesicle proteins and recruitment of active zone components in both preexisting and newly formed sensory neuron synapses during LTF in cultured *Aplysia* neurons involve an activity-dependent rearrangement of the presynaptic actin cytoskeleton (Udo et al. 2005; see also Hatada et al. 2000). Application of toxin B, a general inhibitor of the Rho family of proteins, blocks 5-HT-induced LTF, as well as growth of new synapses in sensorimotor neuron coculture. Moreover, repeated pulses of 5-HT selectively induce the spatial and temporal regulation of the activity of one of the small Rho families of GTPases, Cdc42, at a subset of sensory neuron presynaptic varicosities. The activation of ApCdc42 induced by 5-HT is dependent on both the phosphoinositide-3-kinase (PI3K) and phospholipase C (PLC) pathways and, in turn, recruits the downstream effectors p21-activated kinase (PAK) and neuronal Wiskott–Aldrich syndrome protein (N-WASP) to regulate and remodel the presynaptic actin network.

### Three Types of Cell-Adhesion Molecule-Mediated Trans-Synaptic Interactions

De novo synapse formation during development requires specific trans-synaptic protein interactions. This is also true for learning-induced synaptic growth in *Aplysia*. These trans-synaptic interactions, which reflect a second, later stage in synaptic growth—the generation of new functionally competent varicosities (Kim et al. 2003)—involve at least three types of cell-adhesion interactions. The selective 5-HT-induced, clathrin-mediated internalization of the transmembrane isoform of an immunoglobulin-related cell-adhesion molecule in *Aplysia* (apCAM) in the presynaptic sensory neuron is thought to be a preliminary and permissive step for the expression of LTF and synaptic growth (Fig. 4) (Bailey et al. 1992a, 1997; Mayford et al. 1992; Han et al. 2004). Down syndrome cell-adhesion molecule (Dscam) is required both pre- and postsynaptically for clustering of AMPA receptors and the emergence of new synaptic connections (Li et al. 2009). In addition,

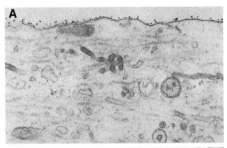

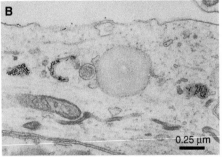

0.25 µm

**Figure 4.** Differential down-regulation of the glycosyl-phosphoinositol (GPI)-linked versus transmembrane isoforms of cell-adhesion molecule in *Aplysia* (apCAM): the role in learning-related synaptic growth. Some of the first evidence for a role of cell-adhesion molecules (CAMs) during learning and memory came from studies of an immunoglobulin-related CAM in *Aplysia*, designated apCAM, which is homologous to neural cell–adhesion molecules (NCAMs) in vertebrates and Fasciclin II in *Drosophila*. To determine the fate of the two isoforms of apCAM in learning-related synaptic growth, gold-conjugated epitope-tagged constructs of either the transmembrane (TM) or GPI-linked isoforms were overexpressed in *Aplysia* sensory neurons. (*A*) Neurite of a sensory neuron expressing the GPI-linked isoform of apCAM following a 1-h exposure to 5-HT. Note, virtually all of the gold complexes (black dots) remain on the surface membrane with none inside despite a robust 5-HT-induced activation of the endosomal pathway leading to significant accumulations of internal membranous profiles. (*B*) Neurite of a sensory neuron expressing the transmembrane isoform of apCAM following a 1-h exposure to 5-HT. In contrast to the lack of down-regulation of the GPI-linked isoform, 5-HT has a dramatic effect on the transmembrane isoform of apCAM, removing most of it from the surface membrane, resulting in heavy accumulations of gold complexes within presumptive endocytic compartments. This 5-HT-induced, clathrin-mediated selective internalization of the transmembrane isoform of apCAM in the presynaptic sensory neuron leads to: (1) defasciculation, a process that

neurexin (presynaptic) and neuroligin (postsynaptic) are required for both LTF and the associated synaptic growth induced by serotonin. Interestingly, introduction into the motor neuron of the R451C mutation of neuroligin-3, which is linked to autism, interrupts transsynaptic signaling and blocks both intermediate-term facilitation and LTF (Choi et al. 2011).

## Signaling from the Synapse to the Nucleus

Studies in *Aplysia*-cultured neurons also have explored how signals from the synapse are sent to the nucleus and how activity at the synapse informs the nucleus to alter transcription. Earlier work had shown that repeated pulses of 5-HT activate PKA, which recruits mitogen-associated protein kinase (MAPK), and both translocate to the nucleus where they phosphorylate transcription factors and activate gene expression required for the induction of long-term memory (Bacskai et al. 1993; Martin et al. 1997b). In more recent studies, Lee et al. (2007, 2012) found that the repeated pulses of serotonin required to induce LTF and activate PKA, in turn, phosphorylate CAM-associated protein (CAMAP), a transcriptional regulator that is tethered to the synapse via the cell-adhesion molecule (CAM), apCAM. Phosphorylation of CAMAP dissociates it from apCAM, leading to the internalization of apCAM described above and also the translocation of CAMAP from the synapse to the nucleus of sensory neurons, where it contributes to activating CREB1 and

destabilizes adhesive contacts normally inhibiting synaptic growth, (2) endocytic activation that results in a redistribution of membrane components to sites in which new synapses form, and, finally, (3) the normal expression of long-term facilitation (LTF) and synaptic growth. These findings also suggest that previously established connections might remain intact following exposure to 5-HT because they would be held in place by the adhesive, homophilic interactions of the GPI-linked isoforms, and the process of outgrowth from sensory neuron axons would be initiated by down-regulation of the transmembrane form at extrasynaptic sites of membrane apposition. (From Bailey et al. 1997; modified, with permission.)

ApC/EBP-mediated transcription (Alberini et al. 1994) required for the initiation of synaptic growth and LTF. This retrograde signaling also removes the inhibition of microRNA 124, thereby enhancing the activation of CREB-1 and leading to activation of piRNA-F, which methylates and shuts off the promoter of CREB-2, the repressor gene, for >24 h, allowing the action of CREB to be prolonged (Rajasethupathy et al. 2009, 2012).

## Coordinated Transport from the Cell Body to the Synapse

Puthanveettil et al. (2008) considered how anterograde signaling and the gene products, required for the initiation of synaptic growth, move from the cell body of the sensory neuron to its presynaptic terminals, and from the cell body of the motor neuron to its postsynaptic dendritic spines. The induction of LTF and synaptic growth requires up-regulation of the molecular motor kinesin heavy chain (KHC), which mediates fast axonal transport of organelles, messenger RNAs (mRNAs), and proteins in a microtubule- and ATP-dependent manner. Kinesins are rapidly up-regulated in both pre- and postsynaptic neurons by five pulses of 5-HT. Moreover, inhibition of ApKHC1 in either the pre- or postsynaptic neuron blocks induction of LTF, whereas up-regulation of KHC in the presynaptic neurons alone is sufficient for the induction of LTF. The mRNA and protein cargo associated with ApKHC includes neurexin and neuroligin involved in de novo synapse formation and piccolo and bassoon proteins required for formation and stabilization of the presynaptic active zone. These data support the idea that the building blocks important for the final stages of new synapse formation induced by learning need to be transported in a coordinated fashion from the cell body to the synapses.

## STABILIZATION OF NEW SYNAPSES DURING LTF IN *APLYSIA*

Studies of synapse-specific long-term plasticity in *Aplysia* first suggested the molecular mecha-

nisms underlying the initiation of LTF and synaptic growth are likely to differ from those required for their long-term maintenance (Martin et al. 1997a; Casadio et al. 1999; Si et al. 2003). Induction of changes in synaptic function and structure, measured 24 h after 5-HT treatment, requires only nuclear transcription and somatic translation, whereas persistence of these synaptic modifications, measured at 72 h, requires, in addition, local protein synthesis at the synapse (Casadio et al. 1999).

To determine the role of local protein synthesis and its time window in stabilization of learning-related synaptic growth and persistence of LTF, Miniaci et al. (2008) used the modified *Aplysia* culture system, consisting of a single bifurcated sensory neuron contacting two spatially separated motor neurons (Martin et al. 1997a). Local application of emetine, an inhibitor of protein synthesis, to one set of sensorimotor neuron synapses following five pulses of 5-HT blocked LTF when given at either 24 h or 48 h, but had no effect when applied at 72 h after 5-HT. The inhibition of local protein synthesis at 24 h led to a selective retraction of newly formed varicosities induced by 5-HT when compared with preexisting varicosities (Fig. 5). This late phase of local protein synthesis is importantly regulated by the *Aplysia* homolog of cytoplasmic polyadenylation element-binding protein (ApCPEB), which promotes translational activation (Si et al. 2003). Local application of a specific TAT-antisense (TAT-AS) oligonucleotide to ApCPEB 24 h after repeated pulses of 5-HT blocked the stable maintenance of both LTF and synaptic growth (Fig. 6).

Combined, these results defined a temporally distinct and local phase of stabilization, indicating that the consolidation process for learning-related synaptic growth extends to ~72 h. During this time, 5-HT-induced newly formed varicosities are labile and require sustained CPEB-dependent local protein synthesis to acquire the more stable properties of mature varicosities (for additional self-sustaining, molecular modifications that may lead to the long-term maintenance of structural changes and memory storage, see also Bailey et al. 2004 and Si and Kandel 2015).

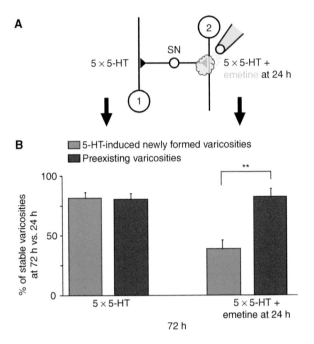

**Figure 5.** Local perfusion of emetine at 24 h leads to a selective retraction of 5-HT-induced newly formed sensory neuron varicosities. (*A*) Diagram of a single bifurcated sensory neuron (SN) in contact with two spatially separated L7 gill-motor neurons (1 and 2) and experimental protocol. (*B*) To assess the dynamic properties of the 5-HT-induced newly formed varicosities, their stability was compared under two different experimental conditions: 5-HT (*left*) and 5-HT + emetine (*right*). Culture dishes containing the bifurcated sensory neuron–motor neuron preparation were treated with five pulses of 5-HT at time 0 and 24 h later, one of the two branches was perfused locally with emetine. Each individual fluorescently labeled 5-HT-induced newly formed and preexisting varicosity was imaged at 24 h and then the exact target field was reimaged to determine the presence or absence of the same individual varicosities at 72 h. The number of 5-HT-induced newly formed and preexisting varicosities that were present at 72 h were compared with the number of varicosities in the same respective class observed at 24 h. At the branch that only received 5-HT, 81.3% of the 5-HT-induced newly formed varicosities (red, *left*) and 80.3% of the preexisting varicosities (blue, *left*) were maintained at 72 h when compared with 24 h. In contrast, at the branch that received emetine 24 h after 5-HT treatment, only 38.1% of the 5-HT-induced newly formed varicosities (red, *right*) were maintained at 72 h versus 81.63% of the preexisting varicosities (blue, *right*). In both cases, the 5-HT-induced new varicosities represent varicosities that formed between 0 and 24 h and remained stable at 72 h. Each histogram illustrates the mean percentage ( ± SEM) of identified varicosities maintained at 72 h compared with 24 h. The selective retraction of 5-HT-induced newly formed varicosities induced by local application of emetine shows that during the stabilization phase this population of learning-related varicosities is significantly more labile and sensitive to disruption than the population of preexisting sensory neuron varicosities. (From Miniaci et al. 2008; modified, with permission.)

## LTP AS A MODEL SYNAPTIC MECHANISM CONTRIBUTING TO CONSOLIDATION OF EXPLICIT MEMORY STORAGE IN THE MAMMALIAN BRAIN

As is the case with implicit memory storage in invertebrates, explicit, hippocampal-based memory is also stored by means of structural changes at the synapse. Much of the work on synaptic plasticity as a cellular mechanism of hippocampal learning and memory has been performed using the model systems of LTP and LTD (Bliss et al. 2013). LTP is a persistent increase in synaptic strength induced by brief high-frequency stimulation, whereas LTD is a persistent decrease in synaptic strength induced

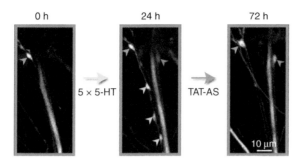

**Figure 6.** A late phase of sustained cytoplasmic polyadenylation element-binding (CPEB) protein-dependent local protein synthesis is required to stabilize learning-related synaptic growth. A specific *Aplysia* CPEB (ApCPEB) antisense oligonucleotide covalently coupled to an 11-amino-acid peptide derived from the HIV-TAT protein (TAT-antisense [TAT-AS]) was locally perfused to one branch of the bifurcated sensorimotor neuron culture preparation for 30 min at 24 h after 5-HT treatment. This antisense oligo has previously been shown to lead to the depletion of ApCPEB messenger RNA (mRNA) and to a selective decrease in the level of CPEB protein (Si et al. 2003). Local perfusion of the TAT-AS selectively reduced the number of 5-HT-induced newly formed varicosities maintained at 72 h compared with preexisting varicosities, similar to what was observed with the local perfusion of emetine (see Fig. 5). This figure contains confocal images of green fluorescent protein (GFP)-labeled sensory neuron presynaptic varicosities in contact with the postsynaptic motor neuron L7 (not labeled), and illustrates the results of three imaging sessions of a representative example of the entire sensory neuron–motor neuron synaptic field. Before the application of 5-HT, a single preexisting sensory neuron varicosity is present (green arrowhead) in this field of view. After repeated applications of 5-HT for 24 h, four newly formed sensory neuron varicosities (one red and three yellow arrowheads) are present along with the single preexisting varicosity seen at time 0. The local perfusion of TAT-AS at 24 h to this synaptic area induces the selective pruning of three newly formed varicosities (yellow arrowheads) without affecting the preexisting varicosity (green arrowhead). The red arrowhead represents the only 5-HT-induced newly formed varicosity in this field that is maintained at 72 h. (From Miniaci et al. 2008; modified, with permission.)

by longer episodes of low-frequency stimulation. Two paradigms, tetanic stimulation and θ-burst stimulation (TBS), have been commonly used to investigate how long LTP can last. Tetanic stimulation, involving three or more episodes of 100 pulses (100 Hz) delivered at 10-min intervals, saturates LTP and last for many hours in mature hippocampal slices in vitro (Huang and Kandel 1994; Frey et al. 1995). TBS provides a more natural paradigm, resembling firing patterns of hippocampal pyramidal cells in vivo (Buzsaki et al. 1987; Staubli and Lynch 1987; Abraham and Huggett 1997; Nguyen and Kandel 1997; Morgan and Teyler 2001; Buzsaki 2002; Leinekugel et al. 2002; Hyman et al. 2003; Raymond and Redman 2006; Mohns and Blumberg 2008). TBS, producing maximal LTP, consists of eight trains delivered at 30-s intervals with each train being 10 bursts at 5 Hz of four pulses at 100 Hz (Abraham and Huggett 1997). Tetanic stimulation induces LTP primarily through the activation of $N$-methyl-$\mathrm{D}$-aspartate receptors (NMDARs), whereas TBS engages multiple induction mechanisms, including activation of NMDARs and voltage-gated calcium channels, back-propagating action potentials, release of calcium from intracellular stores, as well as release of brain-derived neurotrophic factor (BDNF) (Buzsaki et al. 1987; Staubli and Lynch 1987; Abraham and Huggett 1997; Nguyen and Kandel 1997; Morgan and Teyler 2001; Buzsaki 2002; Hyman et al. 2003; Raymond and Redman 2006). Like tetanic stimulation, TBS also produces LTP that lasts for >3 h and has a late phase that is protein synthesis dependent (Nguyen and Kandel 1997; Kelleher et al. 2004; Martin 2004; Yang et al. 2008).

Evidence that LTP and learning share mechanisms comes from work showing that they occlude one another, namely, that a strong learning experience before testing for LTP results in less LTP and, conversely, inducing LTP in vivo can occlude subsequent learning (Barnes et al. 1994; Moser et al. 1998; Habib et al. 2013; Takeuchi et al. 2014). LTP and LTD interact along a sliding scale in which the more saturated with potentiation that a set of synapses becomes, the more resistant they are to additional potentiation (Abraham and Bear 1996; Abraham et al. 2001). On the contrary, the more depressed a population of synapses becomes, the more likely that subsequent stimulation will reverse the depression. The properties of LTP have been shown to depend not only on the induction protocols, but also on the time of day, and the age, strain, and species of the animal (Harris and Teyler 1983; Diana et al. 1994; Manahan-Vaughan and Schwegler 2011; Bowden et al. 2012; Cao and Harris 2012).

Hence, differences in structural outcomes might arise when induction is by glutamate uncaging at individual spines versus chemical, tetanic, or TBS of multiple synapses. Similarly, results may differ between young (prepubescent or cultured) versus more mature hippocampal neurons. Because more is known about the structural correlates of LTP, we focus here on the structural components of synaptic plasticity associated with LTP and recognize that although some commonalities are beginning to emerge, additional research will be needed to determine whether a uniform theory of structural plasticity underlying LTP and explicit memory can be applied across paradigms.

## DENDRITIC SPINES IN THE MAMMALIAN BRAIN

The major focus of the structural plasticity studies in the hippocampus has been the dendritic spines: the postsynaptic receptive surface area of the synapse. Dendritic spines are protrusions with diverse lengths and shapes that stud the surface of many neurons throughout the brain and are the major sites of excitatory synapses. This diversity allows spines to increase the total postsynaptic surface area and, thus, more synaptic connections can form in a compact volume of neuropil than if the same synapses had to line up along a more uniform dendritic shaft (Harris and Kater 1994). Hippocampal den-

dritic spines can vary up to 100-fold in their dimensions and most of their volume is concentrated in a bulbous head, which is connected to the dendritic shaft through a constricted neck of low volume (Fig. 7) (Harris et al. 1992). A thickened postsynaptic density (PSD), characteristic of excitatory synapses, occupies the head of a dendritic spine. Isolated postsynaptic densities have been found to contain numerous proteins, including receptors, ion channels, scaffolding proteins, enzymatic signaling molecules, cytoskeletal elements and motor proteins, exocytic and endocytic trafficking proteins, and CAMs (Kennedy 2000; Sheng and Hoogenraad 2007; Harris and Weinberg 2012). Larger spines tend to have larger, more irregularly shaped synapses with a higher density of glutamate receptors (Matsuzaki et al. 2001; Nicholson et al. 2006).

Larger spines are also more likely to contain smooth endoplasmic reticulum, which regulates calcium and integral membrane protein trafficking (Spacek and Harris 1997; Cui-Wang et al. 2012). In especially large spines, the smooth endoplasmic reticulum forms a spine apparatus, which has Golgi-like functions for posttranslational modification of proteins (Fig. 7) (Spacek and Harris 1997; Pierce et al. 2000; Horton et al. 2005). Larger spines are also more likely to contain polyribosomes, which mediate local protein synthesis (Steward and Schuman 2001; Ostroff et al. 2002; Bourne et al. 2007), and endosomal compartments, which serve local recycling of receptors and membrane management during developmental spine outgrowth and learning-related synaptic plasticity (Cooney et al. 2002; Park et al. 2006). Larger dendritic spines and PSDs are associated with presynaptic axonal boutons, which contain more synaptic vesicles (Harris and Stevens 1989; Lisman and Harris 1993; Harris and Sultan 1995; Shepherd and Harris 1998; Sorra et al. 2006; Bourne et al. 2013). Larger dendritic spines and synapses are also more likely to be associated with perisynaptic astroglial processes (Ventura and Harris 1999; Witcher et al. 2007), which support synapse formation and stabilization, as well as synapse elimination (Clarke and Barres 2013).

These features suggest that larger spines might produce a larger response to glutamate,

Cite this article as *Cold Spring Harb Perspect Biol* doi: 10.1101/cshperspect.a021758

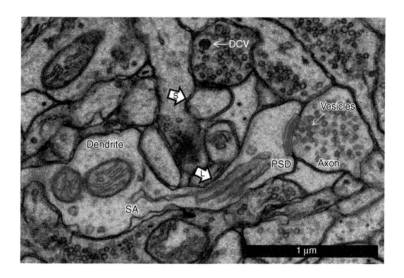

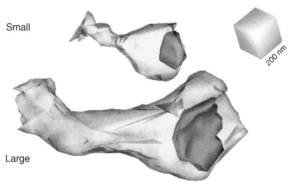

Figure 7. Small and large dendritic spines and associated structures in the mature rat hippocampus. These spines are from the middle of stratum radiatum of area CA1 of a perfusion-fixed preparation. (*Top*) Electron micrograph (EM) illustrating a small (S) and large (L) dendritic spine, the postsynaptic density (PSD, red) of the large spine, presynaptic axon (green) and vesicles it contains, as well as the perisynaptic astroglial processes (light blue). The presynaptic axon of the small spine also contains a small dense-core vesicle, which is usually associated with transport packets involved in delivering presynaptic active zone proteins to growing synapses. (*Bottom*) These two spines (yellow) are illustrated in 3D reconstructions with their associated PSDs at the same scale as in the top EMs. DCV, dense-core vesicle; SA, sample area.

released from the presynaptic terminal acting on it, and give rise to local modulation of intracellular calcium, receptor trafficking and recycling, protein translation and degradation, or interaction with perisynaptic astroglia. However, it is rare that any one spine contains all of these features (Cooney et al. 2002). Interestingly, even in the mature hippocampus, >75% of all spines are small dendritic spines with head diameters of <0.6 μm. These small spines are more prone to rapid formation and elimination

depending on age and the conditions of activation (Bourne and Harris 2007, 2011; Macdougall and Fine 2014).

## ANALYSIS OF STRUCTURAL PLASTICITY ON DENDRITIC SPINES

Many studies show that dendritic spine structure is dynamic both under normative conditions in vivo and in response to conditions of synaptic plasticity, which could contribute to learning and memory (reviewed in Yuste and Bonhoeffer 2001; Alvarez and Sabatini 2007; Bourne and Harris, 2007, 2008; Rogerson et al. 2014). For example, spatial training (Moser et al. 1997) and exposure to enriched environments (Kozorovitskiy et al. 2005) alters spine number in the hippocampus, and hippocampal-dependent associative learning has been associated with an increase in large dendritic spines sharing the same presynaptic axonal boutons (Geinisman et al. 2001). Hippocampal dendritic spines are also sensitive to estrogens. As a result, overectomized and estrogen-deprived female rats or postmenopausal primates show both cognitive decline and loss of dendritic spines in key cortical areas and/or the hippocampus, both of which are reversed with estrogen-replacement therapy (Foy et al. 2010; Bailey et al. 2011). There are many examples suggesting that different dendritic spines are responsive during different stages and forms of learning and memory. For example, hippocampal dendritic spines seem to be more sensitive during early stages of learning, increasing in number shortly after fear conditioning, whereas cortical neurons appear to acquire more spines later (Restivo et al. 2009). With fear conditioning, spines in the prefrontal association cortex are eliminated, whereas extinction of fear conditioning results in spine formation on the same pyramidal cell dendritic branches (Lai et al. 2012). Hippocampal dendritic spines respond similarly, with neurons active during fear conditioning having fewer dendritic spines (Sanders et al. 2012), and AMPA receptors are preferentially recruited to large hippocampal dendritic spines during fear conditioning (Matsuo et al. 2008). Importantly, these studies provide evidence that the spine remodeling is specific to the synaptic circuits that were active during learning, although they do not rule out involvement of other circuits that were not imaged. Spine shape and number are not necessarily dependable predictors of synapse size, location, or composition (Fiala et al. 1998; Toni et al. 2007; Bock et al. 2011; Shu et al. 2011). A more reliable assessment requires nanoscale 3D reconstruction from serial section EM, which allows one to understand how changes

in structure affect synaptic connectivity and function (Harlow et al. 2001; Denk and Horstmann 2004; Coggan et al. 2005; Toni et al. 2007; Lichtman and Sanes 2008; Meinertzhagen et al. 2009; Cardona et al. 2010; Mishchenko et al. 2010; Ostroff et al. 2010; Bock et al. 2011; Helmstaedter et al. 2011; Bourne and Harris 2012; Cardona 2013; Lu et al. 2013; Wilke et al. 2013). Live imaging with two-photon microscopy also has revealed rapid, activity-dependent turnover of spines, which is common in the neocortex (and, presumably, the hippocampus) during development, but as an animal matures, more of the spines begin to stabilize (Alvarez and Sabatini 2007; Holtmaat and Svoboda 2009). This form of imaging has also revealed dynamic changes in the shapes of individual dendritic spines during the uncaging of glutamate at single spines (Matsuzaki et al. 2001; Kasai et al. 2010, 2004).

Estimates of dendritic spine size and dynamics from live imaging can provide a reasonable first approximation of synapse size in the mature hippocampus because serial section EM reconstruction reveals that spine volume correlates with synaptic area (Harris and Stevens 1987). Thus, spine dynamics readily distinguish stable from unstable spines; however, interpretation of the effect on synaptic connectivity is complicated because of the fact that, during development, excitatory synapses often occur directly on the dendritic shafts of immature but rarely on the shafts of mature spiny hippocampal dendrites (Fiala et al. 1998). In addition, many hippocampal CA1 spines, with apparently mature shapes, can form multiple synapses with different presynaptic axons during development, but multisynaptic CA1 spines are extremely rare in the normal mature hippocampus (Harris et al. 1992; Fiala et al. 1998; Sorra and Harris 2000). In other brain regions, such as the neocortex and thalamus, excitatory, inhibitory, and neuromodulatory synapses can all occur on the same dendritic spine (Spacek and Lieberman 1974; Van Horn et al. 2000).

Finally, crucial subcellular components (such as polyribosomes, smooth endoplasmic reticulum, mitochondria, microtubules, perisynaptic astroglial processes, and presynaptic dense-core vesicles) occur at only a small fraction of dendritic spines. Retrospective EM combines light (two-photon) and EM and promises new understanding, although refinement is needed because the reaction products currently used to track the dendrites can obscure synapses and subcellular organelles (Zito et al. 1999; Knott et al. 2006; Nagerl et al. 2007). Despite these caveats, a review of the literature is beginning to reveal a variety of structural components that underlie the initial (5–30 min), intermediate (~1–2 h), and enduring phases of LTP (reported to last up to a year in vivo [Abraham et al. 2002]), with interesting parallels to learning in the hippocampus (Frey and Morris 1997; Reymann and Frey 2007) and *Aplysia*, as discussed above.

## STRUCTURAL SYNAPTIC PLASTICITY OCCURRING DURING LTP IN THE IMMATURE AND MATURE BRAIN

Based on molecular, neurophysiological, and structural analyses, the properties of LTP lasting >3 h are substantially different from those mediating the first hour of potentiation. Within minutes following the induction of LTP, silent synapses, which are commonly found in the developing nervous system, undergo activation by the insertion or functional modification of AMPA receptors (AMPARs) (Edwards 1991; Isaac et al. 1995; Liao et al. 1995; Durand et al. 1996; Petralia et al. 1999; Malinow et al. 2000; Malinow and Malenka 2002; Groc et al. 2006; Hanse et al. 2009; Macdougall and Fine 2014). Initially, potentiation (or depression) can be sustained by these changes in glutamate receptor properties and composition, but longer-lasting potentiation (or depression) involves structural alterations in spines and synapses in both the "immature" (Engert and Bonhoeffer 1999; Maletic-Savatic et al. 1999; Ostroff et al. 2002; Lang et al. 2004; Matsuzaki et al. 2004; Nagerl et al. 2004, 2007; Kopec et al. 2006) and "mature" hippocampus (Van Harreveld and Fifkova 1975; Trommald et al. 1996; Chen et al. 2004; Nagerl et al. 2004; Popov et al. 2004; Zhou et al. 2004; Stewart et al. 2005; Bourne et al. 2007).

Comparison of results from producing LTP in acute slices from P15 and the adult Long–Evans rat hippocampus revealed interesting differences, even in such basic findings as spine number and synapse size (Fig. 8). Representative 3D reconstructions illustrate that P15 dendrites are much less spiny than adult (P55–71) dendrites (Fig. 8A). During LTP, P15 dendrites add spines and synapses, whereas adult dendrites have fewer spines and synapses (Fig. 8B). In contrast, the small added synapses decreased average synapse size at P15, whereas those in the adult were, on average, larger than during control stimulation (Fig. 8C). Despite these dramatic changes in spine numbers and synapse sizes during LTP, the summed area of synaptic input along the length of dendrites was not altered by LTP at either age. This finding suggests synaptic resources were redistributed to support more spines at P15 and larger synapses in the adults. Interestingly, at P15, total synaptic input has only reached about one third the adult value, which might explain why spine formation predominates during LTP at young ages, whereas spine growth and stabilization predominates in adults. Thus, in the adult hippocampus, control stimulation produces more, smaller, and, presumably, less-effective synapses, whereas LTP results in fewer, larger, and, presumably, more effective synapses (Fig. 8E). These observations are consistent with the hypothesis that synaptic scaling and heterosynaptic competition regulate total synaptic input on a neuron such that limited resources are redistributed to strengthened synapses (Turrigiano and Nelson 2004; Turrigiano 2007; Bourne and Harris 2008; Nelson and Turrigiano 2008; Fiete et al. 2010).

## MOLECULAR MECHANISMS UNDERLYING THE STRUCTURAL CHANGES THAT ACCOMPANY SYNAPTIC PLASTICITY PRODUCED BY LTP

The last two decades have seen a large number of studies using labeled molecules to track their effects with light microscopy on the structural integrity of spines and synapses, largely through the modulation of actin filaments, scaffolding

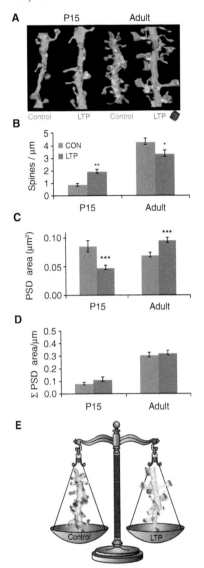

**Figure 8.** Age differences in the structural correlates of long-term potentiation (LTP) in acute rat hippocampal slices. (*A*) 3D electron microscopy (EM) of representative dendrites that received control (CON) stimulation versus induction of LTP by θ-burst stimulation (TBS). (*B*) Opposite effects of TBS on spine density and (*C*) synaptic surface area (postsynaptic density [PSD]) at P15 versus adult (P60–70) dendrites. (*D*) Yet, the summed surface area of the synapses per micron length of dendrite was unchanged by LTP at either age. These graphs also illustrate that neither spine density nor summed synapse area has reached adult levels by P15. (*E*) Thus, as illustrated for adult dendrites, either a dendritic segment supports more, smaller, and presumably less-effective synapses or more, larger, and presumably more-effective synapses.

proteins, receptors, and other growth-promoting or -reducing factors at the synapse (Bonhoeffer and Yuste 2002; Ouyang et al. 2005; Alvarez et al. 2007; Sfakianos et al. 2007; Bourne and Harris 2008; Steiner et al. 2008; Loebrich and Nedivi 2009; Budnik and Salinas 2011). Despite dramatic structural plasticity, some synapses show remarkable tenacity (Minerbi et al. 2009), lasting as long as some memories (Xu et al. 2009; Yang et al. 2009). The profound changes in dendritic and synaptic structure and function are also associated with changes in ion channel and receptor density, which are developmentally regulated and are dependent on dendrite caliber and distance from the soma (Maletic-Savatic et al. 1995; Kang et al. 1996; Miyashita and Kubo 1997; Hsia et al. 1998; Magee et al. 1998; Petralia et al. 1999; Rongo and Kaplan 1999; Sans et al. 2000; Molnar et al. 2002; Frick et al. 2003; Bender et al. 2007; Gasparini et al. 2007; Stuart et al. 2008). Recent experiments and computational models suggest that dendritic segments, rather than individual dendritic spines, might be the "minimal units" of synaptic plasticity (Poirazi et al. 2003; Govindarajan et al. 2006, 2011; Losonczy and Magee 2006; Harvey et al. 2008).

In the developing hippocampus, nascent synapses and surface specializations have distinct PSDs, but no presynaptic vesicles (Vaughn 1989; Fiala et al. 1998; Ahmari and Smith 2002). Live imaging and retrospective EM from hippocampal cultures has revealed that small dense-core vesicles (DCVs), which carry active zone proteins, are transported to and inserted at nascent synapses, which soon thereafter become functional (Buchanan et al. 1989; Ahmari et al. 2000; Zhai et al. 2001; Shapira and others 2003; Sabo et al. 2006; Tao-Cheng 2007; Zampighi et al. 2008; for review, see Ziv and Garner 2004). This is similar to what is found in *Aplysia* sensory to motor neuron cocultures, in which time-lapse imaging suggests that rapid activation also turns nascent or silent presynaptic varicosities into active transmitter-releasing sites (Kim et al. 2003). Recent work in mature rat hippocampal slices suggests that the recruitment of presynaptic vesicles to nascent zones of preexisting synapses facilitates a rapid activation of silent synaptic regions during LTP (Bell et al. 2014).

Both nascent and active zones of mature hippocampal synapses have a distinct PSD, but unlike the active zone, the presynaptic side of a nascent zone lacks synaptic vesicles (Fig. 9A–D) (Spacek and Harris 1998; Bell et al. 2014). Immunogold labeling has revealed glutamate receptors at the edges of cultured hippocampal synapses (Nair et al. 2013) and in nascent zones of mature hippocampal synapses

(Bell et al. 2014). However, stochastic modeling suggests that falloff in glutamate concentration in the synaptic cleft reduces the probability of glutamate receptor activation from 0.4 at the center of a release site to 0.1 just 200 nm away (Franks et al. 2002, 2003). The average distance from vesicles docked at active zones to adjacent nascent zones was $\sim$200 nm; hence, the conversion of nascent zones to functional active zones via recruitment of presynaptic vesicles may constitute the initial phase of LTP (Bell et al. 2014).

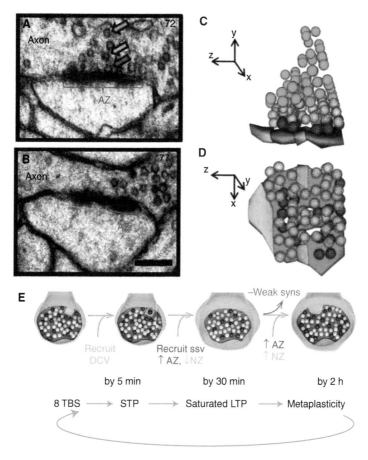

**Figure 9.** Plasticity of synaptic nascent zones at the edges of synapses from the mature rat hippocampus. (*A–D*) Electron micrographs (EMs) and 3DEM through representative sections of a synapse to distinguish active zones (AZ, red) from nascent zones (NZ, aqua). Synaptic vesicles are colorized to distinguish docked vesicles (dark blue) from vesicles in a pool within 94 nm of the presynaptic membrane (light purple) from the reserve pool (green). NZs had no presynaptic vesicles located within 94 nm perpendicular to them. (*E*) Model of the sequence of morphological changes associated with different times following the induction of long-term potentiation (LTP) by theta-burst stimulation (TBS), which could participate in the preparation of synapses for subsequent augmentation of LTP. DCV, dense-core vesicle; syns, synapses; ssv, small synaptic vesicle; STP, short-term potentiation.

This conversion could be facilitated by the insertion of DCVs at existing nascent zones, as DCVs moved into more presynaptic boutons by 5 min following the induction of LTP, and by 30 min, DCV frequency had returned to control levels, as additional presynaptic vesicles were recruited to nascent zones (Bell et al. 2014). By 2 h, there were fewer small dendritic spines relative to control stimulation in the same slices (Bourne and Harris 2011), and both nascent and active zones were enlarged, potentially, in preparation for synapses to undergo further plasticity (Fig. 9E) (Cao and Harris 2012; Bell et al. 2014).

Support for the hypothesis that DCVs are involved in the initial stages of structural synaptic plasticity comes from analysis of their composition and movements. In addition to active zone proteins, DCVs also transport CAMs (Zhai et al. 2001). CAMs provide bidirectional signaling and coordinated recruitment of pre- and postsynaptic proteins and receptors (Benson et al. 2000; Sytnyk et al. 2002; Li and Sheng 2003; Scheiffele 2003; Ziv and Garner 2004; Waites et al. 2005; Akins and Biederer 2006; Benson and Huntley 2010). DCVs contain cadherins (Zhai et al. 2001), which cluster at the edges of synapses (Fannon and Colman 1996; Uchida et al. 1996; Elste and Benson 2006), regulate AMPAR trafficking (Zhai et al. 2001; Nuriya and Huganir 2006; Saglietti et al. 2007), and contribute to the stabilization of enhanced synaptic efficacy during LTP (Bozdagi et al. 2000, 2010; Tanaka et al. 2000; Mendez et al. 2010). DCVs could transport other presynaptic CAMs that might play a role in nascent zone conversion. For example, presynaptic neurexin-1β (Nrx-1β) has two postsynaptic partners, neuroligin-1 (NLG-1) and postsynaptic leucine-rich repeat transmembrane protein 2 (LRRTM2). This extracellular binding modulates presynaptic vesicle release and promotes synapse initiation and stabilization together with N-cadherin (Ichtchenko et al. 1995; Song et al. 1999; Scheiffele et al. 2000; Dean et al. 2003; Graf et al. 2004; Futai et al. 2007; Heine et al. 2008; Sudhof 2008; deWit et al. 2009; Linhoff et al. 2009; Wittenmayer et al. 2009; Stan et al. 2010; Soler-Llavina et al. 2013). Furthermore, the Nrx-1β/NLG-1

complex binds with PSD-95, Stargazin, and other proteins that reduce AMPAR diffusion (Irie et al. 1997; Barrow et al. 2009; Mondin et al. 2011; Giannone et al. 2013). Presynaptic ephrin-B might also participate in nascent zone conversion, as its extracellular binding to postsynaptic EphB receptors has been implicated in the recruitment of presynaptic vesicles, NMDARs, and AMPARs to synapses during maturation and plasticity (Henkemeyer et al. 2003; Kayser et al. 2006; Lim et al. 2008; Klein 2009; Lai and Ip 2009; Nolt et al. 2011; Murata and Constantine-Paton 2013). Whether DCV-transported proteins are engaged in nascent zone conversion and growth at mature hippocampal synapses remains to be determined. However, the aforementioned results from the mature hippocampus provide further links between the early phase of LTP and the remodeling of synapses via regulation of apCAMs during LTF in *Aplysia* and neural cell–adhesion molecules (NCAMs) during hippocampal learning (Senkov et al. 2006).

Protein synthesis is elevated during periods of synaptogenesis (Phillips et al. 1990; Sebeo et al. 2009), and spines with polyribosomes have enlarged synapses by 2 h after the induction of LTP following tetanic stimulation in the developing (Ostroff et al. 2002) and mature hippocampus (Bourne et al. 2007). Endosomes and smooth endoplasmic reticulum also play key roles in LTP; however, <20% of all dendritic spines contain polyribosomes, endosomes, or smooth endoplasmic reticulum (Spacek and Harris 1997; Cooney et al. 2002; Park et al. 2006, 2008). Even within the dendritic shaft, a single polyribosome or sorting endosome appears to serve 10-20 different dendritic spines (Cooney et al. 2002). This sparse distribution of core structures could be critical in determining where structural plasticity can occur along dendrites.

## SYNAPSE GROWTH, METAPLASTICITY, AND THE ADVANTAGE OF SPACED LEARNING

Some patterns of stimulation have no direct effect on synaptic strength, but instead modulate the subsequent expression of plasticity.

This phenomenon is known as metaplasticity (Huang et al. 1992; Abraham and Tate 1997; Young and Nguyen 2005). Recently, there has been a surge of interest in testing the effects of spacing episodes of LTP induction as a model for understanding mechanisms of spaced as opposed to distributed learning (Lynch and Gall 2013; Lynch et al. 2013; Wang et al. 2014). Training that is spaced over time produces stronger and longer memories than massed learning, and the efficacy of memory is dependent on the interval between repetitions (Ebbinghaus 1885; Fields 2005). Similarly, if episodes of TBS that initially saturate LTP are spaced by 1 h, more LTP can be induced (Kramar et al. 2012). Interestingly, the number of TBS episodes required to saturate initial LTP, as well as the delay needed between episodes to allow enhanced LTP, is age, strain, and species specific (Cao and Harris 2012). As the prior discussion illustrates, soon after induction, both pre- and postsynaptic processes are recruited to support synapse growth during the later phase of LTP in the mature hippocampus. However, the magnitude of LTP from the first saturating induction was stable. This observation suggests that the growth and formation of nascent zones is a form of metaplasticity because they form without influencing existing synaptic function, but instead they provide a substrate for subsequent LTP (Fig. 9E).

## AN OVERALL VIEW AND FUTURE DIRECTIONS

Perhaps the most striking finding in the cell biology of memory is that the consolidation and long-term storage of memory involves transcription in the nucleus and structural changes at the synapse (Bailey and Kandel 2009). These structural components of learning-related synaptic plasticity can be grouped into two general categories: (1) remodeling and enlargement of preexisting synapses, and (2) alterations in the number of synapses, including both the addition and elimination of synaptic connections (Bailey and Kandel 1993, 2004; Bourne and Harris 2007, 2008).

Studies in *Aplysia* and the hippocampus have provided evidence that activity-dependent remodeling of preexisting synapses and changes in the number of synapses might play an important role in the expression and storage of information at both the level of individual synaptic connections, as well as in more complex neuronal networks by modulating the activity of the neural network in which this structural plasticity occurs. In both cases, some structural modifications are transient and may contribute to early formative stages of long-term memory, whereas others are more stable, longer lasting, and may confer persistence to the expression of memory storage.

The role of structural synaptic plasticity in memory consolidation raises several questions central to an understanding of how memories are stored in the brain. First, there is the issue of causality versus correlation. Are the structural changes at synapses a consequence of learning, or are they a correlate of learning, or perhaps a purely homeostatic response, or a cellular preparation of new computational space? Second, are memories stored over time in the same synapses? Or are they distributed such that, over time, they can be stored in different synapses so that the system can be efficiently degraded without affecting performance? For the consolidation and persistence of long-term memory, the evidence is quite clear. The same synapses that grow out seem to carry the memory storage. For reconsolidation, the evidence is less clear. There is now evidence that the memory becomes distributed with time, and that the memory can be stored in different synapses of the same neuron so the memory at the systems level can be efficiently degraded without affecting performance. However, reconsolidation can only be activated for a short period of time, usually a few days to a few weeks; thus, the ability to render the memory labile has a limited time window (see Alberini and Kandel 2015). Finally, recent studies suggest the possibility that the long-term memory may not be stored in the synapse, but rather in nuclear programs within the soma. According to this hypothesis, the synaptic changes (both functional and structural) would represent how the storage of each memory is expressed (Chen et al. 2014). Answers to these questions are still being examined in a variety of memory systems and will provide a more refined understanding of the family of mechanisms that contribute to memory consolidation.

For example, we know that consolidation of explicit memory in mammals at the systems level involves redistribution of the information over new circuits, particularly in the neocortex (Dudai 2012). How is the structural plasticity at the level of individual synapses modified and, perhaps, reorganized to reconfigure the redistributed activity in more expansive neuronal networks following this transfer to the systems level in the cortex?

In vivo imaging reveals subsets of dendritic spines and presynaptic axonal boutons remain highly dynamic in the adult neocortex (Grutzendler et al. 2002; Holtmaat et al. 2005; De Paola et al. 2006; Majewska et al. 2006; Stettler et al. 2006; Lee et al. 2008; for review, see Holtmaat and Svoboda 2009; Hübener and Bonhoeffer 2010). Moreover, recent results show that dramatic spine remodeling, including the formation and stabilization of new spines, can be correlated with the degree of behavioral training and can occur in relevant cortical areas (Xu et al. 2009; Yang et al. 2009; Moczulska et al. 2013).

Although a number of technical hurdles remain, the continuing improvements in optical and molecular approaches raise hope that the ability to visualize, in real time, the synaptic changes that mediate the flow and storage of information in specific neural circuits will come to fruition in the not-too-distant future (Hübener and Bonhoeffer 2010; Mayford et al. 2012). When combined with retrospective 3D reconstruction from serial section EM of identified synapses, it also should be possible to reveal the fundamental underlying structural and molecular mechanisms of long-term memory expression and storage in complex circuits in different regions of the brain.

## REFERENCES

*Reference is also in this collection.

Abraham WC, Bear MF. 1996. Metaplasticity: The plasticity of synaptic plasticity. Trends Neurosci 19: 126–130.

Abraham WC, Huggett A. 1997. Induction and reversal of long-term potentiation by repeated high-frequency stimulation in rat hippocampal slices. Hippocampus 7: 137–145.

Abraham WC, Tate WP. 1997. Metaplasticity: A new vista across the field of synaptic plasticity. Prog Neurobiol 52: 303–323.

Abraham WC, Mason-Parker SE, Bear MF, Webb S, Tate WP. 2001. Heterosynaptic metaplasticity in the hippocampus in vivo: A BCM-like modifiable threshold for LTP. Proc Natl Acad Sci 98: 10924–10929.

Abraham WC, Logan B, Greenwood JM, Dragunow M. 2002. Induction and experience-dependent consolidation of stable long-term potentiation lasting months in the hippocampus. J Neurosci 22: 9626–9634.

Ahmari SE, Smith SJ. 2002. Knowing a nascent synapse when you see it. Neuron 34: 333–336.

Ahmari SE, Buchanan J, Smith SJ. 2000. Assembly of presynaptic active zones from cytoplasmic transport packets. Nat Neurosci 3: 445–451.

Akins MR, Biederer T. 2006. Cell-cell interactions in synaptogenesis. Curr Opin Neurobiol 16: 83–89.

* Alberini CM, Kandel ER. 2015. The regulation of transcription in memory consolidation. Cold Spring Harb Perspect Biol 7: a021741.

Alberini CM, Ghirardi M, Metz R, Kandel ER. 1994. C/EBP is an immediate-early gene required for the consolidation of long-term facilitation in Aplysia. Cell 76: 1099–1114.

Alvarez VA, Sabatini BL. 2007. Anatomical and physiological plasticity of dendritic spines. Annu Rev Neurosci 30: 79–97.

Alvarez VA, Ridenour DA, Sabatini BL. 2007. Distinct structural and ionotropic roles of NMDA receptors in controlling spine and synapse stability. J Neurosci 27: 7365–7376.

Bacskai BJ, Hochner B, Mahaut-Smith M, Adams SR, Kaang BK, Kandel ER, Tsien RY. 1993. Spatially resolved dynamics of cAMP and protein kinase A subunits in Aplysia sensory neurons. Science 260: 222–226.

Bailey CH. 1991. Morphological basis of short- and long-term memory in Aplysia. In Perspectives on cognitive neuroscience (ed. Weingartner H, Lister R), pp. 76–92. Oxford University Press, New York.

Bailey CH, Chen M. 1983. Morphological basis of long-term habituation and sensitization in Aplysia. Science 220: 91–93.

Bailey CH, Chen M. 1988a. Long-term memory in Aplysia modulates the total number of varicosities of single identified sensory neurons. Proc Natl Acad Sci 85: 2373–2377.

Bailey CH, Chen M. 1988b. Long-term sensitization in Aplysia increases the number of presynaptic contacts onto the identified gill motor neuron L7. Proc Natl Acad Sci 85: 9356–9359.

Bailey CH, Chen M. 1989. Time course of structural changes at identified sensory neuron synapses during long-term sensitization in Aplysia. J Neurosci 9: 1774–1780.

Bailey CH, Chen M. 1990. Morphological alterations at identified sensory neuron synapses during long-term sensitization in Aplysia. In The biology of memory, Vol. 23. Symposium Medicum Hoechst (ed. Squire L, Lindenlaub E), pp. 135–153. Schattauer, Stuttgart, Germany.

Bailey CH, Kandel ER. 1985. Molecular approaches to the study of short-term and long-term memory. In Functions *of the brain* (ed. Coen CW), pp. 98–129. Clarendon, Oxford.

Bailey CH, Kandel ER. 1993. Structural changes accompanying memory storage. *Annu Rev Physiol* **55:** 397–426.

Bailey CH, Kandel ER. 2004. Synaptic growth and the persistence of long-term memory: A molecular perspective. In *The new cognitive neurosciences* (ed. Gazzaniga MS), Vol. III, pp. 647–664. MIT Press, Cambridge, MA.

Bailey CH, Kandel ER. 2008a. Synaptic remodeling, synaptic growth and the storage of long-term memory in *Aplysia*. In *The essence of memory, progress in brain research* (ed. Sossin W, et al.), pp. 179–198. Elsevier, Amsterdam.

Bailey CH, Kandel ER. 2008b. Activity-dependent remodeling of presynaptic boutons. In *New encyclopedia of neuroscience* (ed. Squire L), pp. 67–74. Elsevier, Amsterdam.

Bailey CH, Kandel ER. 2009. Synaptic and cellular basis of learning. In *Handbook of neuroscience for behavioral sciences* (ed. Cacioppo JT, Berntson GG), Vol. 1, Chap. 27, pp. 528–551. Wiley, New York.

Bailey CH, Thompson EB, Castellucci VF, Kandel ER. 1979. Ultrastructure of the synapses of sensory neurons that mediate the gill-withdrawal reflex in *Aplysia*. *J Neurocytol* **8:** 415–444.

Bailey CH, Chen M, Keller F, Kandel ER. 1992a. Serotonin-mediated endocytosis of apCAM: An early step of learning-related synaptic growth in *Aplysia*. *Science* **25:** 645–649.

Bailey CH, Montarolo PG, Chen M, Kandel ER, Schacher S. 1992b. Inhibitors of protein and RNA synthesis block the structural changes that accompany long-term heterosynaptic plasticity in the sensory neurons of *Aplysia*. *Neuron* **9:** 749–758.

Bailey CH, Bartsch D, Kandel ER. 1996. Toward a molecular definition of long-term memory storage. *Proc Natl Acad Sci* **93:** 13445–13452.

Bailey CH, Kaang BK, Chen M, Marin C, Lim A, Kandel ER. 1997. Mutation in the phosphorylation sites of MAP kinase blocks learning-related internalization of apCAM in *Aplysia* sensory neurons. *Neuron* **18:** 913–924.

Bailey CH, Kandel ER, Si K. 2004. The persistence of long-term memory: A molecular approach to self-sustaining changes in learning-induced synaptic growth. *Neuron* **44:** 49–57.

Bailey CH, Kandel ER, Si K, Choi YB. 2005. Toward a molecular biology of learning-related synaptic growth in *Aplysia*. *Cell Sci Rev* **2:** 27–57.

Bailey CH, Barco A, Hawkins RD, Kandel ER. 2008. Molecular studies of learning and memory in *Aplysia* and the hippocampus: A comparative analysis of implicit and explicit memory storage. In *Learning and memory: Comprehensive reference, Vol 4. Molecular mechanisms of memory* (ed. Byrne JH), pp. 11–29. Elsevier, Amsterdam.

Bailey ME, Wang AC, Hao J, Janssen WG, Hara Y, Dumitriu D, Hof PR, Morrison JH. 2011. Interactive effects of age and estrogen on cortical neurons: Implications for cognitive aging. *Neuroscience* **191:** 148–158.

Barnes CA, Jung MW, McNaughton BL, Korol DL, Andreasson K, Worley PF. 1994. LTP saturation and spatial learning disruption: Effects of task variables and saturation levels. *J Neurosci* **14:** 5793–5806.

Barrow SL, Constable JR, Clark E, El-Sabeawy F, McAllister AK, Washbourne P. 2009. Neuroligin1: A cell adhesion molecule that recruits PSD-95 and NMDA receptors by distinct mechanisms during synaptogenesis. *Neural Dev* **4:** 17.

Bell ME, Bourne JN, Chirillo MA, Harris KM. 2014. Conversion of nascent to active zones as synapses enlarge during long-term potentiation in mature hippocampus. *J Comp Neurol* **522:** 3861–3884.

Bender RA, Kirschstein T, Kretz O, Brewster AL, Richichi C, Ruschenschmidt C, Shigemoto R, Beck H, Frotscher M, Baram TZ. 2007. Localization of HCN1 channels to presynaptic compartments: Novel plasticity that may contribute to hippocampal maturation. *J Neurosci* **27:** 4697–4706.

Benson DL, Huntley GW. 2010. Building and remodeling synapses. *Hippocampus* **22:** 954–968.

Benson DL, Schnapp LM, Shapiro L, Huntley GW. 2000. Making memories stick: Cell-adhesion molecules in synaptic plasticity. *Trends Cell Biol* **10:** 473–482.

Bliss TV, Collingridge GL, Morris RG. 2013. Synaptic plasticity in health and disease: Introduction and overview. *Philos Trans R Soc Lond B Biol Sci* **369:** 20130129.

Bock DD, Lee WC, Kerlin AM, Andermann ML, Hood G, Wetzel AW, Yurgenson S, Soucy ER, Kim HS, Reid RC. 2011. Network anatomy and in vivo physiology of visual cortical neurons. *Nature* **471:** 177–182.

Bonhoeffer T, Yuste R. 2002. Spine motility. Phenomenology, mechanisms, and function. *Neuron* **35:** 1019–1027.

Bourne J, Harris KM. 2007. Do thin spines learn to be mushroom spines that remember? *Curr Opin Neurobiol* **17:** 381–386.

Bourne JN, Harris KM. 2008. Balancing structure and function at hippocampal dendritic spines. *Annu Rev Neurosci* **31:** 47–67.

Bourne JN, Harris KM. 2011. Coordination of size and number of excitatory and inhibitory synapses results in a balanced structural plasticity along mature hippocampal CA1 dendrites during LTP. *Hippocampus* **21:** 354–373.

Bourne JN, Harris KM. 2012. Nanoscale analysis of structural synaptic plasticity. *Curr Opin Neurobiol* **22:** 372–382.

Bourne JN, Sorra KE, Hurlburt J, Harris KM. 2007. Polyribosomes are increased in spines of CA1 dendrites 2 h after the induction of LTP in mature rat hippocampal slices. *Hippocampus* **17:** 1–4.

Bourne JN, Chirillo MA, Harris KM. 2013. Presynaptic ultrastructural plasticity along CA3 → CA1 axons during LTP in mature hippocampus. *J Comp Neurol* **521:** 3898–3912.

Bowden JB, Abraham WC, Harris KM. 2012. Differential effects of strain, circadian cycle, and stimulation pattern on LTP and concurrent LTD in the dentate gyrus of freely moving rats. *Hippocampus* **22:** 1363–1370.

Bozdagi O, Shan W, Tanaka H, Benson DL, Huntley GW. 2000. Increasing numbers of synaptic puncta during late-phase LTP: N-cadherin is synthesized, recruited to synaptic sites, and required for potentiation. *Neuron* **28:** 245–259.

Bozdagi O, Wang XB, Nikitczuk JS, Anderson TR, Bloss EB, Radice GL, Zhou Q, Benson DL, Huntley GW. 2010. Persistence of coordinated long-term potentiation and dendritic spine enlargement at mature hippocampal CA1 synapses requires N-cadherin. *J Neurosci* **30:** 9984–9989.

Buchanan J, Sun YA, Poo MM. 1989. Studies of nerve-muscle interactions in *Xenopus* cell culture: Fine structure of early functional contacts. *J Neurosci* **9:** 1540–1554.

Budnik V, Salinas PC. 2011. Wnt signaling during synaptic development and plasticity. *Curr Opin Neurobiol* **21:** 151–159.

Buzsaki G. 2002. θ Oscillations in the hippocampus. *Neuron* **33:** 325–340.

Buzsaki G, Haas HL, Anderson EG. 1987. Long-term potentiation induced by physiologically relevant stimulus patterns. *Brain Res* **435:** 331–333.

Cao G, Harris KM. 2012. Developmental regulation of the late phase of long-term potentiation (L-LTP) and metaplasticity in hippocampal area CA1 of the rat. *J Neurophysiol* **107:** 902–912.

Cardona A. 2013. Towards semi-automatic reconstruction of neural circuits. *Neuroinformatics* **11:** 31–33.

Cardona A, Saalfeld S, Preibisch S, Schmid B, Cheng A, Pulokas J, Tomancak P, Hartenstein V. 2010. An integrated micro- and macroarchitectural analysis of the *Drosophila* brain by computer-assisted serial section electron microscopy. *PLoS Biol* **8:** e1000502.

Casadio A, Martin KC, Giustetto M, Zhu H, Chen M, Bartsch D, Bailey CH, Kandel ER. 1999. A transient, neuron-wide form of CREB-mediated long-term facilitation can be stabilized at specific synapses by local protein synthesis. *Cell* **99:** 221–237.

Castellucci V, Pinsker H, Kupfermann I, Kandel ER. 1970. Neuronal mechanisms of habituation and dishabituation of the gill-withdrawal reflex in *Aplysia. Science* **167:** 1745–1748.

Castellucci VF, Blumenfeld H, Goelet P, Kandel ER. 1989. Inhibitor of protein synthesis blocks long-term behavioral sensitization in the isolated gill-withdrawal reflex of *Aplysia. Science* **220:** 91–93.

Chen YC, Bourne J, Pieribone VA, Fitzsimonds RM. 2004. The role of actin in the regulation of dendritic spine morphology and bidirectional synaptic plasticity. *NeuroReport* **15:** 829–832.

Chen S, Cai D, Pearce K, Sun PYW, Roberts AC, Glanzman DL. 2014. Reinstatement of long-term memory following erasure of its behavioral and synaptic expression in *Aplysia. eLife* **3:** e03896.

Choi Y-B, Li H-L, Kassabov SR, Jin I, Puthanveettil SV, Karl KA, Lu Y, Kim J-H, Bailey CH, Kandel ER. 2011. Neurexin-Neuroligin transsynaptic interaction mediates learning-related synaptic remodeling and long-term facilitation in *Aplysia. Neuron* **70:** 1–14.

Clarke LE, Barres BA. 2013. Emerging roles of astrocytes in neural circuit development. *Nat Rev Neurosci* **14:** 311–321.

Coggan JS, Bartol TM, Esquenazi E, Stiles JR, Lamont S, Martone ME, Berg DK, Ellisman MH, Sejnowski TJ. 2005. Evidence for ectopic neurotransmission at a neuronal synapse. *Science* **309:** 446–451.

Cooney JR, Hurlburt JL, Selig DK, Harris KM, Fiala JC. 2002. Endosomal compartments serve multiple hippocampal dendritic spines from a widespread rather than a local store of recycling membrane. *J Neurosci* **22:** 2215–2224.

Cui-Wang T, Hanus C, Cui T, Helton T, Bourne J, Watson D, Harris KM, Ehlers MD. 2012. Local zones of endoplasmic reticulum complexity confine cargo in neuronal dendrites. *Cell* **148:** 309–321.

Dean C, Scholl FG, Choih J, DeMaria S, Berger J, Isacoff E, Scheiffele P. 2003. Neurexin mediates the assembly of presynaptic terminals. *Nat Neurosci* **6:** 708–716.

Denk W, Horstmann H. 2004. Serial block-face scanning electron microscopy to reconstruct three-dimensional tissue nanostructure. *PLoS Biol* **2:** e329.

De Paola V, Holtmaat A, Knott G, Song S, Wilbrecht L, Caroni P, Svoboda K. 2006. Cell type-specific structural plasticity of axonal branches and boutons in the adult neocortex. *Neuron* **49:** 861–875.

deWit J, Sylwestrak E, O'Sullivan ML, Otto S, Tiglio K, Savas JN, Yates JR III, Comoletti D, Taylor P, Ghosh A. 2009. LRRTM2 interacts with Neurexin1 and regulates excitatory synapse formation. *Neuron* **64:** 799–806.

Diana G, Domenici MR, Loizzo A, Scotti de Carolis A, Sagratella S. 1994. Age and strain differences in rat place learning and hippocampal dentate gyrus frequency-potentiation. *Neurosci Lett* **171:** 113–116.

Dudai Y. 2012. The restless engram: Consolidations never end. *Annu Rev Neurosci* **35:** 227–247.

Durand GM, Kovalchuk Y, Konnerth A. 1996. Long-term potentiation and functional synapse induction in developing hippocampus. *Nature* **381:** 71–75.

Ebbinghaus H. 1885. *Über das gedächnis: Untersuchungen zur experimentellen psychologie* [*Memory: A contribution to experimental psychology*]. Veit, Leipzig, Germany.

Edwards F. 1991. Neurobiology. LTP is a long-term problem. *Nature* **350:** 271–272.

Elste AM, Benson DL. 2006. Structural basis for developmentally regulated changes in cadherin function at synapses. *J Comp Neurol* **495:** 324–335.

Engert F, Bonhoeffer T. 1999. Dendritic spine changes associated with hippocampal long-term synaptic plasticity. *Nature* **399:** 66–70.

Fannon AM, Colman DR. 1996. A model for central synaptic junctional complex formation based on the differential adhesive specificities of the cadherins. *Neuron* **17:** 423–434.

Fiala JC, Feinberg M, Popov V, Harris KM. 1998. Synaptogenesis via dendritic filopodia in developing hippocampal area CA1. *J Neurosci* **18:** 8900–8911.

Fields RD. 2005. Making memories stick. *Sci Am* **292:** 75–81.

Fiete IR, Senn W, Wang CZ, Hahnloser RH. 2010. Spike-time-dependent plasticity and heterosynaptic competition organize networks to produce long scale-free sequences of neural activity. *Neuron* **65:** 563–576.

Foy MR, Baudry M, Akopian GK, Thompson RF. 2010. Regulation of hippocampal synaptic plasticity by estrogen and progesterone. *Vitam Horm* **82:** 219–239.

Cite this article as *Cold Spring Harb Perspect Biol* doi: 10.1101/cshperspect.a021758

Franks KM, Bartol TM Jr, Sejnowski TJ. 2002. A Monte Carlo model reveals independent signaling at central glutamatergic synapses. *Biophys J* **83:** 2333–2348.

Franks KM, Stevens CF, Sejnowski TJ. 2003. Independent sources of quantal variability at single glutamatergic synapses. *J Neurosci* **23:** 3186–3195.

Frey U, Morris RG. 1997. Synaptic tagging and long-term potentiation. *Nature* **385:** 533–536.

Frey U, Schollmeier K, Reymann KG, Seidenbecher T. 1995. Asymptotic hippocampal long-term potentiation in rats does not preclude additional potentiation at later phases. *Neuroscience* **67:** 799–807.

Frick A, Magee J, Koester HJ, Migliore M, Johnston D. 2003. Normalization of $Ca^{2+}$ signals by small oblique dendrites of CA1 pyramidal neurons. *J Neurosci* **23:** 3243–3250.

Frost WN, Castellucci VF, Hawkins RD, Kandel ER. 1985. Monosynaptic connections made by the sensory neurons of the gill- and siphon-withdrawal reflex in *Aplysia* participates in the storage of long-term memory for sensitization. *Proc Natl Acad Sci* **82:** 8266–8269.

Futai K, Kim MJ, Hashikawa T, Scheiffele P, Sheng M, Hayashi Y. 2007. Retrograde modulation of presynaptic release probability through signaling mediated by PSD-95-neuroligin. *Nat Neurosci* **10:** 186–195.

Gasparini S, Losonczy A, Chen X, Johnston D, Magee JC. 2007. Associative pairing enhances action potential backpropagation in radial oblique branches of CA1 pyramidal neurons. *J Physiol* **580:** 787–800.

Geinisman Y, Berry RW, Disterhoft JF, Power JM, van der Zee EA. 2001. Associative learning elicits the formation of multiple-synapse boutons. *J Neurosci* **21:** 5568–5573.

Ghirardi M, Montarolo PG, Kandel ER. 1995. A novel intermediate stage in the transition between short- and long-term facilitation in the sensory to motor neuron synapse of *Aplysia*. *Neuron* **14:** 413–420.

Giannone G, Mondin M, Grillo-Bosch D, Tessier B, Saint-Michel E, Czondor K, Sainlos M, Choquet D, Thoumine O. 2013. Neurexin-1β binding to neuroligin-1 triggers the preferential recruitment of PSD-95 versus gephyrin through tyrosine phosphorylation of neuroligin-1. *Cell Rep* **3:** 1996–2007.

Glanzman DL, Kandel ER, Schacher S. 1990. Target-dependent structural changes accompanying long-term synaptic facilitation in *Aplysia* neurons. *Science* **249:** 779–802.

Govindarajan A, Kelleher RJ, Tonegawa S. 2006. A clustered plasticity model of long-term memory engrams. *Nat Rev Neurosci* **7:** 575–583.

Govindarajan A, Israely I, Huang SY, Tonegawa S. 2011. The dendritic branch is the preferred integrative unit for protein synthesis-dependent LTP. *Neuron* **69:** 132–146.

Graf ER, Zhang X, Jin SX, Linhoff MW, Craig AM. 2004. Neurexins induce differentiation of GABA and glutamate postsynaptic specializations via neuroligins. *Cell* **119:** 1013–1026.

Greenough WT, Bailey CH. 1988. The anatomy of a memory: Convergence of results across a diversity of tests. *Trends Neurosci* **11:** 142–147.

Groc L, Gustafsson B, Hanse E. 2006. AMPA signalling in nascent glutamatergic synapses: There and not there! *Trends Neurosci* **29:** 132–139.

Grutzendler J, Kasthuri N, Gan WB. 2002. Long-term dendritic spine stability in the adult cortex. *Nature* **420:** 812–816.

Habib D, Tsui CK, Rosen LG, Dringenberg HC. 2013. Occlusion of low-frequency-induced, heterosynaptic long-term potentiation in the rat hippocampus in vivo following spatial training. *Cereb Cortex* **24:** 3090–3096.

Han JH, Lim YS, Kandel ER, Kaang BK. 2004. Role of *Aplysia* cell adhesion molecules during 5-HT-induced long-term functional and structural changes. *Learn Mem* **11:** 421–435.

Hanse E, Taira T, Lauri S, Groc L. 2009. Glutamate synapse in developing brain: An integrative perspective beyond the silent state. *Trends Neurosci* **32:** 532–537.

Harlow ML, Ress D, Stoschek A, Marshall RM, McMahan UJ. 2001. The architecture of active zone material at the frog's neuromuscular junction. *Nature* **409:** 479–484.

Harris KM, Kater SB. 1994. Dendritic spines: Cellular specializations imparting both stability and flexibility to synaptic function. *Annu Rev Neurosci* **17:** 341–371.

Harris KM, Stevens JK. 1987. Study of dendritic spines by serial electron microscopy and three-dimensional reconstructions. *Neurol Neurobiol* **37:** 179–199.

Harris KM, Stevens JK. 1989. Dendritic spines of CA1 pyramidal cells in the rat hippocampus: Serial electron microscopy with reference to their biophysical characteristics. *J Neurosci* **9:** 2982–2997.

Harris KM, Sultan P. 1995. Variation in the number, location and size of synaptic vesicles provides an anatomical basis for the nonuniform probability of release at hippocampal CA1 synapses. *Neuropharmacology* **34:** 1387–1395.

Harris KM, Teyler TJ. 1983. Age differences in a circadian influence on hippocampal LTP. *Brain Res* **261:** 69–73.

Harris KM, Weinberg RJ. 2012. Ultrastructure of synapses in the mammalian brain. *Cold Spring Harb Perspect Biol* **4:** a005587.

Harris KM, Jensen FE, Tsao B. 1992. Three-dimensional structure of dendritic spines and synapses in rat hippocampus (CA1) at postnatal day 15 and adult ages: Implications for the maturation of synaptic physiology and long-term potentiation. *J Neurosci* **12:** 2685–2705.

Harvey CD, Yasuda R, Zhong H, Svoboda K. 2008. The spread of Ras activity triggered by activation of a single dendritic spine. *Science* **321:** 136–140.

Hatada Y, Wu F, Sun ZY, Schacher S, Goldberg DJ. 2000. Presynaptic morphological changes associated with long-term synaptic facilitation are triggered by actin polymerization at preexisting varicosities. *J Neurosci* **20:** RC82.

Hawkins RD, Kandel ER, Bailey CH. 2006. Molecular mechanisms of memory storage in *Aplysia*. *Biol Bull* **210:** 174–191.

Hawkins RD, Kandel ER, Jin I. 2012. Possible roles of spontaneous transmitter release in homeostasis and growth related plasticity at *Aplysia* sensory-motor neuron synapses. In *Molluscan neuroscience in the genomic era: From gastropods to cephalopods*, Abstract 294.06. Scripps Research Institute, Jupiter, FL.

Heine M, Thoumine O, Mondin M, Tessier B, Giannone G, Choquet D. 2008. Activity-independent and subunit-specific recruitment of functional AMPA receptors at neurexin/neuroligin contacts. *Proc Natl Acad Sci* **105**: 20947–20952.

Helmstaedter M, Briggman KL, Denk W. 2011. High-accuracy neurite reconstruction for high-throughput neuroanatomy. *Nat Neurosci* **14**: 1081–1088.

Henkemeyer M, Itkis OS, Ngo M, Hickmott PW, Ethell IM. 2003. Multiple EphB receptor tyrosine kinases shape dendritic spines in the hippocampus. *J Cell Biol* **163**: 1313–1326.

Holtmaat A, Svoboda K. 2009. Experience-dependent structural synaptic plasticity in the mammalian brain. *Nat Rev Neurosci* **10**: 647–658.

Holtmaat AJ, Trachtenberg JT, Wilbrecht L, Shepherd GM, Zhang X, Knott GW, Svoboda K. 2005. Transient and persistent dendritic spines in the neocortex in vivo. *Neuron* **45**: 279–291.

Horton AC, Racz B, Monson EE, Lin AL, Weinberg RJ, Ehlers MD. 2005. Polarized secretory trafficking directs cargo for asymmetric dendrite growth and morphogenesis. *Neuron* **48**: 757–771.

Hsia AY, Malenka RC, Nicoll RA. 1998. Development of excitatory circuitry in the hippocampus. *J Neurophysiol* **79**: 2013–2024.

Huang YY, Kandel ER. 1994. Recruitment of long-lasting and protein kinase A-dependent long-term potentiation in the CA1 region of hippampus requires repeated tetanization. *Learn Mem* **1**: 74–82.

Huang YY, Colino A, Selig DK, Malenka RC. 1992. The influence of prior synaptic activity on the induction of long-term potentiation. *Science* **255**: 730–733.

Hübener M, Bonhoeffer T. 2010. Searching for engrams. *Neuron* **67**: 363–371.

Hyman JM, Wyble BP, Goyal V, Rossi CA, Hasselmo ME. 2003. Stimulation in hippocampal region CA1 in behaving rats yields long-term potentiation when delivered to the peak of θ and long-term depression when delivered to the trough. *J Neurosci* **23**: 11725–11731.

Ichtchenko K, Hata Y, Nguyen T, Ullrich B, Missler M, Moomaw C, Sudhof TC. 1995. Neuroligin 1: A splice site-specific ligand for β-neurexins. *Cell* **81**: 435–443.

Irie M, Hata Y, Takeuchi M, Ichtchenko K, Toyoda A, Hirao K, Takai Y, Rosahl TW, Sudhof TC. 1997. Binding of neuroligins to PSD-95. *Science* **277**: 1511–1515.

Isaac JT, Nicoll RA, Malenka RC. 1995. Evidence for silent synapses: Implications for the expression of LTP. *Neuron* **15**: 427–434.

Jessell TM, Kandel ER. 1993. Synaptic transmission: A bidirectional and self-modifiable form of cell–cell communication. *Cell* **72** (Suppl): 1–30.

Jin I, Puthanveettil S, Udo H, Karl K, Kandel ER, Hawkins RD. 2012a. Spontaneous transmitter release is critical for the induction of long-term and intermediate-term facilitation in *Aplysia*. *Proc Natl Acad Sci* **109**: 9131–9136.

Jin I, Udo H, Rayman JB, Puthanveettil S, Kandel ER, Hawkins RD. 2012b. Spontaneous transmitter release recruits postsynaptic mechanisms of long-term and intermediate-term facilitation in *Aplysia*. *Proc Natl Acad Sci* **109**: 9137–9142.

Kandel ER. 2001. The molecular biology of memory storage: A dialogue between genes and synapses. *Science* **294**: 1030–1038.

Kandel ER. 2009. The biology of memory: A forty-year perspective. *J Neurosci* **29**: 12748–12756.

Kang J, Huguenard JR, Prince DA. 1996. Development of BK channels in neocortical pyramidal neurons. *J Neurophysiol* **76**: 188–198.

Kasai H, Fukuda M, Watanabe S, Hayashi-Takagi A, Noguchi J. 2010. Structural dynamics of dendritic spines in memory and cognition. *Trends Neurosci* **33**: 121–129.

Kassabov SR, Choi YB, Karl KA, Vishwasrao HD, Bailey CH, Kandel ER. 2013. A single *Aplysia* neurotrophin mediates synaptic facilitation via differentially processed isoforms secreted as mature or precursor forms. *Cell Rep* **3**: 1–15.

Kayser MS, McClelland AC, Hughes EG, Dalva MB. 2006. Intracellular and transsynaptic regulation of glutamatergic synaptogenesis by EphB receptors. *J Neurosci* **26**: 12152–12164.

Kelleher RJ, Govindarajan A, Tonegawa S. 2004. Translational regulatory mechanisms in persistent forms of synaptic plasticity. *Neuron* **44**: 59–73.

Kennedy MB. 2000. Signal-processing machines at the postsynaptic density. *Science* **290**: 750–754.

Kim JH, Udo H, Li HL, Youn TY, Chen M, Kandel ER, Bailey CH. 2003. Presynaptic activation of silent synapses and growth of new synapses contribute to intermediate and long-term facilitation in *Aplysia*. *Neuron* **40**: 151–165.

Klein R. 2009. Bidirectional modulation of synaptic functions by Eph/ephrin signaling. *Nat Neurosci* **12**: 15–20.

Knott GW, Holtmaat A, Wilbrecht L, Welker E, Svoboda K. 2006. Spine growth precedes synapse formation in the adult neocortex in vivo. *Nat Neurosci* **9**: 1117–1124.

Kopec CD, Li B, Wei W, Boehm J, Malinow R. 2006. Glutamate receptor exocytosis and spine enlargement during chemically induced long-term potentiation. *J Neurosci* **26**: 2000–2009.

Kozorovitskiy Y, Gross CG, Kopil C, Battaglia L, McBreen M, Stranahan AM, Gould E. 2005. Experience induces structural and biochemical changes in the adult primate brain. *Proc Natl Acad Sci* **102**: 17478–17482.

Kramar EA, Babayan AH, Gavin CF, Cox CD, Jafari M, Gall CM, Rumbaugh G, Lynch G. 2012. Synaptic evidence for the efficacy of spaced learning. *Proc Natl Acad Sci* **109**: 5121–5126.

Lai KO, Ip NY. 2009. Synapse development and plasticity: Roles of ephrin/Eph receptor signaling. *Curr Opin Neurobiol* **19**: 275–283.

Lai CS, Franke TF, Gan WB. 2012. Opposite effects of fear conditioning and extinction on dendritic spine remodelling. *Nature* **483**: 87–91.

Lang C, Barco A, Zablow L, Kandel ER, Siegelbaum SA, Zakharenko SS. 2004. Transient expansion of synaptically connected dendritic spines upon induction of hippocampal long-term potentiation. *Proc Natl Acad Sci* **101**: 16665–16670.

Lee SH, Lim CS, Park H, Lee JA, Han JH, Kim H, Cheang YH, Lee SH, Lee YS, Ko HG, et al. 2007. Nuclear translocation of CAM-associated protein activates transcription for long-term facilitation in *Aplysia*. *Cell* **129**: 801–812.

Lee WC, Chen JL, Huang H, Leslie JH, Amitai Y, So PT, Nedivi E. 2008. A dynamic zone defines interneuron remodeling in the adult neocortex. *Proc Natl Acad Sci* **105:** 19968–19973.

Lee SH, Shim J, Choi SL, Lee N, Lee CH, Bailey CH, Kandel ER, Jang DJ, Kaang BK. 2012. Learning-related synaptic growth mediated by internalization of *Aplysia* cell adhesion molecule is controlled by membrane phosphatidylinositol 4,5-bisphosphate synthetic pathway. *J Neuroscience* **32:** 16296–16305.

Leinekugel X, Khazipov R, Cannon R, Hirase H, Ben-Ari Y, Buzsaki G. 2002. Correlated bursts of activity in the neonatal hippocampus in vivo. *Science* **296:** 2049–2052.

Li Z, Sheng M. 2003. Some assembly required: The development of neuronal synapses. *Nat Rev Mol Cell Biol* **4:** 833–841.

Li H-L, Huang BS, Vishwasrao H, Sutedja N, Chen W, Jin I, Hawkins RD, Bailey CH, Kandel ER. 2009. Dscam mediates remodeling of glutamate receptors in *Aplysia* during de novo and learning-related synapse formation. *Neuron* **61:** 527–540.

Liao D, Hessler NA, Malinow R. 1995. Activation of postsynaptically silent synapses during pairing-induced LTP in CA1 region of hippocampal slice. *Nature* **375:** 400–404.

Lichtman JW, Sanes JR. 2008. Ome sweet ome: What can the genome tell us about the connectome? *Curr Opin Neurobiol* **18:** 346–353.

Lim BK, Matsuda N, Poo MM. 2008. Ephrin-B reverse signaling promotes structural and functional synaptic maturation in vivo. *Nat Neurosci* **11:** 160–169.

Linhoff MW, Lauren J, Cassidy RM, Dobie FA, Takahashi H, Nygaard HB, Airaksinen MS, Strittmatter SM, Craig AM. 2009. An unbiased expression screen for synaptogenic proteins identifies the LRRTM protein family as synaptic organizers. *Neuron* **61:** 734–749.

Lisman J, Harris KM. 1993. Quantal analysis and synaptic anatomy—Integrating two views of hippocampal plasticity. *Trends Neurosci* **16:** 141–147.

Loebrich S, Nedivi E. 2009. The function of activity-regulated genes in the nervous system. *Physiol Rev* **89:** 1079–1103.

Losonczy A, Magee JC. 2006. Integrative properties of radial oblique dendrites in hippocampal CA1 pyramidal neurons. *Neuron* **50:** 291–307.

Lu W, Bushong EA, Shih TP, Ellisman MH, Nicoll RA. 2013. The cell-autonomous role of excitatory synaptic transmission in the regulation of neuronal structure and function. *Neuron* **78:** 433–439.

Lynch G, Gall CM. 2013. Mechanism based approaches for rescuing and enhancing cognition. *Front Neurosci* **7:** 143.

Lynch G, Kramar EA, Babayan AH, Rumbaugh G, Gall CM. 2013. Differences between synaptic plasticity thresholds result in new timing rules for maximizing long-term potentiation. *Neuropharmacology* **64:** 27–36.

Macdougall MJ, Fine A. 2014. The expression of long-term potentiation: Reconciling the priests and the positivists. *Philos Trans R Soc Lond B Biol Sci* **369:** 20130135.

Magee J, Hoffman D, Colbert C, Johnston D. 1998. Electrical and calcium signaling in dendrites of hippocampal pyramidal neurons. *Annu Rev Physiol* **60:** 327–346.

Majewska AK, Newton JR, Sur M. 2006. Remodeling of synaptic structure in sensory cortical areas in vivo. *J Neurosci* **26:** 3021–3029.

Maletic-Savatic M, Lenn NJ, Trimmer JS. 1995. Differential spatiotemporal expression of $K^+$ channel polypeptides in rat hippocampal neurons developing in situ and in vitro. *J Neurosci* **15:** 3840–3851.

Maletic-Savatic M, Malinow R, Svoboda K. 1999. Rapid dendritic morphogenesis in CA1 hippocampal dendrites induced by synaptic activity. *Science* **283:** 1923–1927.

Malinow R, Malenka RC. 2002. AMPA receptor trafficking and synaptic plasticity. *Annu Rev Neurosci* **25:** 103–126.

Malinow R, Mainen ZF, Hayashi Y. 2000. LTP mechanisms: From silence to four-lane traffic. *Curr Opin Neurobiol* **10:** 352–357.

Manahan-Vaughan D, Schwegler H. 2011. Strain-dependent variations in spatial learning and in hippocampal synaptic plasticity in the dentate gyrus of freely behaving rats. *Front Behav Neurosci* **5:** 7.

Marinesco S, Carew TJ. 2002. Serotonin release evoked by tail nerve stimulation in the CNS of *Aplysia*: Characterization and relationship to heterosynaptic plasticity. *J Neurosci* **22:** 2299–2312.

Marrone DF. 2005. The morphology of bi-directional experience-dependent cortical plasticity: A meta-analysis. *Brain Res Brain Res Rev* **50:** 100–113.

Martin KC. 2004. Local protein synthesis during axon guidance and synaptic plasticity. *Curr Opin Neurobiol* **14:** 305–310.

Martin EC, Casadio A, Zhu H, Yaping E, Rose J, Chen M, Bailey CH, Kandel ER. 1997a. Synapse-specific long-term facilitation of *Aplysia* sensory somatic synapses: A function for local protein synthesis memory storage. *Cell* **91:** 927–938.

Martin KC, Michael D, Rose JC, Barad M, Casadio A, Zhu H, Kandel ER. 1997b. MAP kinase translocates into the nucleus of the presynaptic cell and is required for long-term facilitation in *Aplysia*. *Neuron* **18:** 899–912.

Matsuo N, Reijmers L, Mayford M. 2008. Spine-type-specific recruitment of newly synthesized AMPA receptors with learning. *Science* **319:** 1104–1107.

Matsuzaki M, Ellis-Davies GC, Nemoto T, Miyashita Y, Iino M, Kasai H. 2001. Dendritic spine geometry is critical for AMPA receptor expression in hippocampal CA1 pyramidal neurons. *Nat Neurosci* **4:** 1086–1092.

Matsuzaki M, Honkura N, Ellis-Davies GC, Kasai H. 2004. Structural basis of long-term potentiation in single dendritic spines. *Nature* **429:** 761–766.

Mauelshagen J, Parker GR, Carew TJ. 1996. Dynamics of induction and expression of long-term synaptic facilitation in *Aplysia*. *J Neurosci* **16:** 7099–7108.

Mayford M, Barzilai A, Keller F, Schacher S, Kandel ER. 1992. Modulation of an NCAM-related adhesion molecule with long-term synaptic plasticity in *Aplysia*. *Science* **256:** 638–644.

Mayford M, Siegelbaum SA, Kandel ER. 2012. Synapses and memory storage. *Cold Spring Harb Perspect Biol* **4:** a005751.

McAllister AM. 2007. Dynamic aspects of CNS synapse formation. *Ann Rev Neurosci* **30:** 425–450.

Meinertzhagen IA, Takemura SY, Lu Z, Huang S, Gao S, Ting CY, Lee CH. 2009. From form to function: The ways to know a neuron. *J Neurogenet* **23**: 68–77.

Mendez P, De Roo M, Poglia L, Klauser P, Muller D. 2010. N-cadherin mediates plasticity-induced long-term spine stabilization. *J Cell Biol* **189**: 589–600.

Minerbi A, Kahana R, Goldfeld L, Kaufman M, Marom S, Ziv NE. 2009. Long-term relationships between synaptic tenacity, synaptic remodeling, and network activity. *PLoS Biol* **7**: e1000136.

Miniaci MC, Kim J-H, Puthenveettil S, Si K, Zhu H, Kandel ER, Bailey CH. 2008. Sustained CPEB-dependent local protein synthesis is required to stabilize synaptic growth for persistence of long-term facilitation in *Aplysia*. *Neuron* **59**: 1024–1036.

Mishchenko Y, Hu T, Spacek J, Mendenhall J, Harris KM, Chklovskii DB. 2010. Ultrastructural analysis of hippocampal neuropil from the connectomics perspective. *Neuron* **67**: 1009–1020.

Miyashita T, Kubo Y. 1997. Localization and developmental changes of the expression of two inward rectifying $K^+$-channel proteins in the rat brain. *Brain Res* **750**: 251–263.

Moczulska KE, Tinter-Thiede J, Peter M, Ushakova L, Wernle T, Bathellier B, Rumpel S. 2013. Dynamics of dendritic spines in the mouse auditory cortex during memory formation and memory recall. *Proc Natl Acad Sci* **110**: 18315–18320.

Mohns EJ, Blumberg MS. 2008. Synchronous bursts of neuronal activity in the developing hippocampus: Modulation by active sleep and association with emerging gamma and θ rhythms. *J Neurosci* **28**: 10134–10144.

Molnar E, Pickard L, Duckworth JK. 2002. Developmental changes in ionotropic glutamate receptors: Lessons from hippocampal synapses. *Neuroscientist* **8**: 143–153.

Mondin M, Labrousse V, Hosy E, Heine M, Tessier B, Levet F, Poujol C, Blanchet C, Choquet D, Thoumine O. 2011. Neurexin-neuroligin adhesions capture surface-diffusing AMPA receptors through PSD-95 scaffolds. *J Neurosci* **31**: 13500–13515.

Montarolo PG, Goelet P, Castellucci VF, Morgan J, Kandel ER, Schacher S. 1986. A critical period for macromolecular synthesis in long-term heterosynaptic facilitation in *Aplysia*. *Science* **234**: 1249–1254.

Morgan SL, Teyler TJ. 2001. Electrical stimuli patterned after the θ-rhythm induce multiple forms of LTP. *J Neurophysiol* **86**: 1289–1296.

Moser MB, Trommald M, Egeland T, Andersen P. 1997. Spatial training in a complex environment and isolation alter the spine distribution differently in rat CA1 pyramidal cells. *J Comp Neurol* **380**: 373–381.

Moser EI, Krobert KA, Moser MB, Morris RG. 1998. Impaired spatial learning after saturation of long-term potentiation. *Science* **281**: 2038–2042.

Murata Y, Constantine-Paton M. 2013. Postsynaptic density scaffold SAP102 regulates cortical synapse development through EphB and PAK signaling pathway. *J Neurosci* **33**: 5040–5052.

Nagerl UV, Eberhorn N, Cambridge SB, Bonhoeffer T. 2004. Bidirectional activity-dependent morphological plasticity in hippocampal neurons. *Neuron* **44**: 759–767.

Nagerl UV, Kostinger G, Anderson JC, Martin KA, Bonhoeffer T. 2007. Protracted synaptogenesis after activity-dependent spinogenesis in hippocampal neurons. *J Neurosci* **27**: 8149–8156.

Nair D, Hosy E, Petersen JD, Constals A, Giannone G, Choquet D, Sibarita JB. 2013. Super-resolution imaging reveals that AMPA receptors inside synapses are dynamically organized in nanodomains regulated by PSD95. *J Neurosci* **33**: 13204–13224.

Nelson SB, Turrigiano GG. 2008. Strength through diversity. *Neuron* **60**: 477–482.

Nguyen PV, Kandel ER. 1997. Brief θ-burst stimulation induces a transcription-dependent late phase of LTP requiring cAMP in area CA1 of the mouse hippocampus. *Learn Mem* **4**: 230–243.

Nicholson DA, Trana R, Katz Y, Kath WL, Spruston N, Geinisman Y. 2006. Distance-dependent differences in synapse number and AMPA receptor expression in hippocampal CA1 pyramidal neurons. *Neuron* **50**: 431–442.

Nolt MJ, Lin Y, Hruska M, Murphy J, Sheffler-Colins SI, Kayser MS, Passer J, Bennett MV, Zukin RS, Dalva MB. 2011. EphB controls NMDA receptor function and synaptic targeting in a subunit-specific manner. *J Neurosci* **31**: 5353–5364.

Nuriya M, Huganir RL. 2006. Regulation of AMPA receptor trafficking by N-cadherin. *J Neurochem* **97**: 652–661.

Ostroff LE, Fiala JC, Allwardt B, Harris KM. 2002. Polyribosomes redistribute from dendritic shafts into spines with enlarged synapses during LTP in developing rat hippocampal slices. *Neuron* **35**: 535–545.

Ostroff LE, Cain CK, Bedont J, Monfils MH, Ledoux JE. 2010. Fear and safety learning differentially affect synapse size and dendritic translation in the lateral amygdala. *Proc Natl Acad Sci* **107**: 9418–9423.

Ouyang Y, Wong M, Capani F, Rensing N, Lee CS, Liu Q, Neusch C, Martone ME, Wu JY, Yamada K, et al. 2005. Transient decrease in F-actin may be necessary for translocation of proteins into dendritic spines. *Eur J Neurosci* **22**: 2995–3005.

Park M, Salgado JM, Ostroff L, Helton TD, Robinson CG, Harris KM, Ehlers MD. 2006. Plasticity-induced growth of dendritic spines by exocytic trafficking from recycling endosomes. *Neuron* **52**: 817–830.

Park MK, Choi YM, Kang YK, Petersen OH. 2008. The endoplasmic reticulum as an integrator of multiple dendritic events. *Neuroscientist* **14**: 68–77.

Peters A, Palay SL, Webster D. 1976. *The fine structure of the nervous system: The neurons and supporting cells.* W.B. Saunders, Philadelphia.

Petralia RS, Esteban JA, Wang YX, Partridge JG, Zhao HM, Wenthold RJ, Malinow R. 1999. Selective acquisition of AMPA receptors over postnatal development suggests a molecular basis for silent synapses. *Nat Neurosci* **2**: 31–36.

Phillips LL, Pollack AE, Steward O. 1990. Protein synthesis in the neuropil of the rat dentate gyrus during synapse development. *J Neurosci* **26**: 474–482.

Pierce JP, van Leyen K, McCarthy JB. 2000. Translocation machinery for synthesis of integral membrane and secretory proteins in dendritic spines. *Nat Neurosci* **3**: 311–313.

Cite this article as *Cold Spring Harb Perspect Biol* doi: 10.1101/cshperspect.a021758

Poirazi P, Brannon T, Mel BW. 2003. Arithmetic of subthreshold synaptic summation in a model CA1 pyramidal cell. *Neuron* **37:** 977–987.

Popov VI, Davies HA, Rogachevsky VV, Patrushev IV, Errington ML, Gabbott PL, Bliss TV, Stewart MG. 2004. Remodelling of synaptic morphology but unchanged synaptic density during late phase long-term potentiation (LTP): A serial section electron micrograph study in the dentate gyrus in the anaesthetised rat. *Neurosci* **128:** 251–262.

Puthanveettil SV, Monje FJ, Miniaci MC, Choi YB, Karl KA, Khandros E, Gawinowicz MA, Sheetz MP, Kandel ER. 2008. A new component in synaptic plasticity: Upregulation of kinesin in the neurons of the gill-withdrawal reflex. *Cell* **135:** 960–973.

Rajasethupathy P, Fiumara F, Sheridan R, Betel D, Puthanveettil SV, James J, Russo JJ, Sander C, Tuschl T, Kandel ER. 2009. Characterization of small RNAs in *Aplysia* reveals a role for MIR-124 in constraining long-term synaptic plasticity through CREB. *Neuron* **66:** 803–817.

Rajasethupathy P, Antonov I, Sheridan R, Frey S, Sander C, Tuschl T, Kandel ER. 2012. A role for neuronal piRNAs in the epigenetic control of memory-related synaptic plasticity. *Cell* **149:** 693–707.

Ramón y Cajal S. 1894. Croonian lecture. La fine structure des centres nerveux. *Proc R Soc London* **55:** 444–468; translated in De Felipe J, Jones EG. 1988. *Cajal on the cerebral cortex. An annotated translation of the complete writings*, pp. 83–88. Oxford University Press, New York.

Raymond CR, Redman SJ. 2006. Spatial segregation of neuronal calcium signals encodes different forms of LTP in rat hippocampus. *J Physiol* **570:** 97–111.

Restivo L, Vetere G, Bontempi B, Ammassari-Teule M. 2009. The formation of recent and remote memory is associated with time-dependent formation of dendritic spines in the hippocampus and anterior cingulate cortex. *J Neurosci* **29:** 8206–8214.

Reymann KG, Frey JU. 2007. The late maintenance of hippocampal LTP: Requirements, phases, "synaptic tagging," "late-associativity" and implications. *Neuropharmacology* **52:** 24–40.

Rogerson T, Cai DJ, Frank A, Sano Y, Shobe J, Lopez-Aranda MF, Silva AJ. 2014. Synaptic tagging during memory allocation. *Nat Rev Neurosci* **15:** 157–169.

Rongo C, Kaplan JM. 1999. CaMKII regulates the density of central glutamatergic synapses in vivo. *Nature* **402:** 195–199.

Sabo SL, Gomes RA, McAllister AK. 2006. Formation of presynaptic terminals at predefined sites along axons. *J Neurosci* **26:** 10813–10825.

Saglietti L, Dequidt C, Kamieniarz K, Rousset MC, Valnegri P, Thoumine O, Beretta F, Fagni L, Choquet D, Sala C, et al. 2007. Extracellular interactions between GluR2 and N-cadherin in spine regulation. *Neuron* **54:** 461–477.

Sanders J, Cowansage K, Baumgärtel K, Mayford M. 2012. Elimination of dendritic spines with long-term memory is specific to active circuits. *J Neurosci* **32:** 12570–12578.

Sans N, Petralia RS, Wang YX, Blahos J, Hell JW, Wenthold RJ. 2000. A developmental change in NMDA receptor-associated proteins at hippocampal synapses. *J Neurosci* **20:** 1260–1271.

Schacher S, Hu JY. 2014. The less things change, the more they are different: Contributions of long-term synaptic plasticity and homeostasis to memory. *Learn Mem* **21:** 128–134.

Scheiffele P. 2003. Cell-cell signaling during synapse formation in the CNS. *Annu Rev Neurosci* **26:** 485–508.

Scheiffele P, Fan J, Choih J, Fetter R, Serafini T. 2000. Neuroligin expressed in nonneuronal cells triggers presynaptic development in contacting axons. *Cell* **101:** 657–669.

Schwartz H, Castellucci VF, Kandel ER. 1971. Functions of identified neurons and synapses in abdominal ganglion of *Aplysia* in absence of protein synthesis. *J Neurophysiol* **34:** 9639–9653.

Sebeo J, Hsiao K, Bozdagi O, Dumitriu D, Ge Y, Zhou Q, Benson DL. 2009. Requirement for protein synthesis at developing synapses. *J Neurosci* **29:** 9778–9793.

Senkov O, Sun M, Weinhold B, Gerardy-Schahn R, Schachner M, Dityatev A. 2006. Polysialylated neural cell adhesion molecule is involved in induction of long-term potentiation and memory acquisition and consolidation in a fear-conditioning paradigm. *J Neurosci* **26:** 10888–109898.

Sfakianos MK, Eisman A, Gourley SL, Bradley WD, Scheetz AJ, Settleman J, Taylor JR, Greer CA, Williamson A, Koleske AJ. 2007. Inhibition of Rho via Arg and p190RhoGAP in the postnatal mouse hippocampus regulates dendritic spine maturation, synapse and dendrite stability, and behavior. *J Neurosci* **27:** 10982–10992.

Shapira M, Zhai RG, Dresbach T, Bresler T, Torres VI, Gundelfinger ED, Ziv NE, Garner CC. 2003. Unitary assembly of presynaptic active zones from Piccolo-Bassoon transport vesicles. *Neuron* **38:** 237–252.

Sheng M, Hoogenraad CC. 2007. The postsynaptic architecture of excitatory synapses: A more quantitative view. *Annu Rev Biochem* **76:** 823–847.

Shepherd GM, Harris KM. 1998. Three-dimensional structure and composition of CA3 → CA1 axons in rat hippocampal slices: Implications for presynaptic connectivity and compartmentalization. *J Neurosci* **18:** 8300–8310.

Sherrington CS. 1897. The central nervous system. In *A textbook of physiology*, 7th ed. (ed. Foster M), Vol. 3. MacMillan, London.

Shu X, Lev-Ram V, Deerinck TJ, Qi Y, Ramko EB, Davidson MW, Jin Y, Ellisman MH, Tsien RY. 2011. A genetically encoded tag for correlated light and electron microscopy of intact cells, tissues, and organisms. *PLoS Biol* **9:** e1001041.

* Si K, Kandel ER. 2015. Long-term maintenance. *Cold Spring Harb Perspect Biol* doi: 10.1101/cshperspect.a021774.

Si K, Giustetto M, Etkin A, Hsu R, Janisiewicz AM, Miniaci MC, Kim JH, Zhu H, Kandel ER. 2003. A neuronal isoform of CPEB regulates local protein synthesis and stabilizes synapse-specific long-term facilitation in *Aplysia*. *Cell* **115:** 893–904.

Soler-Llavina GJ, Arstikaitis P, Morishita W, Ahmad M, Sudhof TC, Malenka RC. 2013. Leucine-rich repeat transmembrane proteins are essential for maintenance of long-term potentiation. *Neuron* **79:** 439–446.

Song JY, Ichtchenko K, Sudhof TC, Brose N. 1999. Neuroligin 1 is a postsynaptic cell-adhesion molecule of excitatory synapses. *Proc Natl Acad Sci* **96:** 1100–1105.

Sorra KE, Harris KM. 2000. Overview on the structure, composition, function, development, and plasticity of hippocampal dendritic spines. *Hippocampus* **10**: 501–511.

Sorra KE, Mishra A, Kirov SA, Harris KM. 2006. Dense core vesicles resemble active-zone transport vesicles and are diminished following synaptogenesis in mature hippocampal slices. *Neurosci* **141**: 2097–2106.

Spacek J, Harris KM. 1997. Three-dimensional organization of smooth endoplasmic reticulum in hippocampal CA1 dendrites and dendritic spines of the immature and mature rat. *J Neurosci* **17**: 190–203.

Spacek J, Harris KM. 1998. Three-dimensional organization of cell adhesion junctions at synapses and dendritic spines in area CA1 of the rat hippocampus. *J Comp Neurol* **393**: 58–68.

Spacek J, Lieberman AR. 1974. Ultrastructure and three-dimensional organization of synaptic glomeruli in rat somatosensory thalamus. *J Anat* **117**: 487–516.

Stan A, Pielarski KN, Brigadski T, Wittenmayer N, Fedorchenko O, Gohla A, Lessmann V, Dresbach T, Gottmann K. 2010. Essential cooperation of N-cadherin and neuroligin-1 in the transsynaptic control of vesicle accumulation. *Proc Natl Acad Sci* **107**: 11116–11121.

Staubli U, Lynch G. 1987. Stable hippocampal long-term potentiation elicited by "θ" pattern stimulation. *Brain Res* **435**: 227–234.

Steiner P, Higley MJ, Xu W, Czervionke BL, Malenka RC, Sabatini BL. 2008. Destabilization of the postsynaptic density by PSD-95 serine 73 phosphorylation inhibits spine growth and synaptic plasticity. *Neuron* **60**: 788–802.

Stettler DD, Yamahachi H, Li W, Denk W, Gilbert CD. 2006. Axons and synaptic boutons are highly dynamic in adult visual cortex. *Neuron* **49**: 877–887.

Steward O, Schuman EM. 2001. Protein synthesis at synaptic sites on dendrites. *Annu Rev Neurosci* **24**: 299–325.

Stewart MG, Medvedev NI, Popov VI, Schoepfer R, Davies HA, Murphy K, Dallerac GM, Kraev IV, Rodriguez JJ. 2005. Chemically induced long-term potentiation increases the number of perforated and complex postsynaptic densities but does not alter dendritic spine volume in CA1 of adult mouse hippocampal slices. *Eur J Neurosci* **21**: 3368–3378.

Stuart G, Spruston N, Hausser M. 2008. *Dendrites*, 2nd ed. Oxford University Press, New York.

Sudhof TC. 2008. Neuroligins and neurexins link synaptic function to cognitive disease. *Nature* **455**: 903–911.

Sutton MA, Masters SE, Bagnall MW, Carew TJ. 2001. Molecular mechanisms underlying a unique intermediate phase of memory in *Aplysia*. *Neuron* **31**: 143–154.

Sytnyk V, Leshchyns'ka I, Delling M, Dityateva G, Dityatev A, Schachner M. 2002. Neural cell adhesion molecule promotes accumulation of TGN organelles at sites of neuron-to-neuron contacts. *J Cell Biol* **159**: 649–661.

Takeuchi T, Duszkiewicz AJ, Morris RG. 2014. The synaptic plasticity and memory hypothesis: Encoding, storage and persistence. *Philos Trans R Soc Lond B Biol Sci* **369**: 20130288.

Tanaka H, Shan W, Phillips GR, Arndt K, Bozdagi O, Shapiro L, Huntley GW, Benson DL, Colman DR. 2000. Molecular modification of N-cadherin in response to synaptic activity. *Neuron* **25**: 93–107.

Toni N, Teng EM, Bushong EA, Aimone JB, Zhao C, Consiglio A, van Praag H, Martone ME, Ellisman MH, Gage FH. 2007. Synapse formation on neurons born in the adult hippocampus. *Nat Neurosci* **10**: 727–734.

Trommald M, Hulleberg G, Andersen P. 1996. Long-term potentiation is associated with new excitatory spine synapses on rat dentate granule cells. *Learn Mem* **3**: 218–228.

Turrigiano G. 2007. Homeostatic signaling: The positive side of negative feedback. *Curr Opin Neurobiol* **17**: 318–324.

Turrigiano GG, Nelson SB. 2004. Homeostatic plasticity in the developing nervous system. *Nat Rev Neurosci* **5**: 97–107.

Uchida N, Honjo Y, Johnson KR, Wheelock MJ, Takeichi M. 1996. The catenin/cadherin adhesion system is localized in synaptic junctions bordering transmitter release zones. *J Cell Biol* **135**: 767–779.

Udo H, Jin I, Kim J-H, Li H-L, Youn T, Hawkins RD, Kandel ER, Bailey CH. 2005. Serotonin-induced regulation of the actin network for learning-related synaptic growth requires CdC42, N-WASP and PAK in *Aplysia* sensory neurons. *Neuron* **45**: 887–901.

Van Harreveld A, Fifkova E. 1975. Swelling of dendritic spines in the fascia dentata after stimulation of the perforant fibers as a mechanism of post-tetanic potentiation. *Exp Neurol* **49**: 736–749.

Van Horn SC, Erisir A, Sherman SM. 2000. Relative distribution of synapses in the A-laminae of the lateral geniculate nucleus of the cat. *J Comp Neurol* **416**: 509–520.

Vaughn JE. 1989. Fine structure of synaptogenesis in the vertebrate central nervous system. *Synapse* **3**: 255–285.

Ventura R, Harris KM. 1999. Three-dimensional relationships between hippocampal synapses and astrocytes. *J Neurosci* **19**: 6897–6906.

Wainwright ML, Zhang H, Byrne JH, Cleary LJ. 2002. Localized neuronal outgrowth induced by long-term sensitization training in *Aplysia*. *J Neurosci* **22**: 4132–4141.

Waites CL, Craig AM, Garner CC. 2005. Mechanisms of vertebrate synaptogenesis. *Annu Rev Neurosci* **28**: 251–274.

Wang Y, Zhu G, Briz V, Hsu YT, Bi X, Baudry M. 2014. A molecular brake controls the magnitude of long-term potentiation. *Nat Commun* **5**: 3051.

Wilke SA, Antonios JK, Bushong EA, Badkoobehi A, Malek E, Hwang M, Terada M, Ellisman MH, Ghosh A. 2013. Deconstructing complexity: Serial block-face electron microscopic analysis of the hippocampal mossy fiber synapse. *J Neurosci* **33**: 507–522.

Witcher MR, Kirov SA, Harris KM. 2007. Plasticity of perisynaptic astroglia during synaptogenesis in the mature rat hippocampus. *Glia* **55**: 13–23.

Wittenmayer N, Korber C, Liu H, Kremer T, Varoqueaux F, Chapman ER, Brose N, Kuner T, Dresbach T. 2009. Postsynaptic Neuroligin1 regulates presynaptic maturation. *Proc Natl Acad Sci* **106**: 13564–13569.

Xu T, Yu X, Perlik AJ, Tobin WF, Zweig JA, Tennant K, Jones T, Zuo Y. 2009. Rapid formation and selective stabilization of synapses for enduring motor memories. *Nature* **462**: 915–919.

Yang Y, Wang XB, Frerking M, Zhou Q. 2008. Spine expansion and stabilization associated with long-term potentiation. *J Neurosci* **28**: 5740–5751.

Yang G, Pan F, Gan WB. 2009. Stably maintained dendritic spines are associated with lifelong memories. *Nature* **462**: 920–924.

Young JZ, Nguyen PV. 2005. Homosynaptic and heterosynaptic inhibition of synaptic tagging and capture of long-term potentiation by previous synaptic activity. *J Neurosci* **25**: 7221–7231.

Yuste R, Bonhoeffer T. 2001. Morphological changes in dendritic spines associated with long-term synaptic plasticity. *Annu Rev Neurosci* **24**: 1071–1089.

Zhai RG, Vardinon-Friedman H, Cases-Langhoff C, Becker B, Gundelfinger ED, Ziv NE, Garner CC. 2001. Assembling the presynaptic active zone: A characterization of an active one precursor vesicle. *Neuron* **29**: 131–143.

Zhou Q, Homma KJ, Poo MM. 2004. Shrinkage of dendritic spines associated with long-term depression of hippocampal synapses. *Neuron* **44**: 749–757.

Zito K, Parnas D, Fetter RD, Isacoff EY, Goodman CS. 1999. Watching a synapse grow: Noninvasive confocal imaging of synaptic growth in *Drosophila. Neuron* **22**: 719–729.

Ziv NE, Garner CC. 2004. Cellular mechanisms of presynaptic assembly. *Nat Rev Neurosci* **5**: 385–399.

# Memory Consolidation

Larry R. Squire[1,2,3], Lisa Genzel[4], John T. Wixted[3], and Richard G. Morris[4]

[1]VA San Diego Healthcare System, San Diego, California 92161

[2]Departments of Psychiatry and Neurosciences, University of California, San Diego, La Jolla, California 92093

[3]Department of Psychology, University of California, San Diego, La Jolla, California 92093

[4]Centre for Cognitive and Neural Systems, The University of Edinburgh, Edinburgh EH8 9JZ, United Kingdom

*Correspondence:* lsquire@ucsd.edu

Conscious memory for a new experience is initially dependent on information stored in both the hippocampus and neocortex. Systems consolidation is the process by which the hippocampus guides the reorganization of the information stored in the neocortex such that it eventually becomes independent of the hippocampus. Early evidence for systems consolidation was provided by studies of retrograde amnesia, which found that damage to the hippocampus-impaired memories formed in the recent past, but typically spared memories formed in the more remote past. Systems consolidation has been found to occur for both episodic and semantic memories and for both spatial and nonspatial memories, although empirical inconsistencies and theoretical disagreements remain about these issues. Recent work has begun to characterize the neural mechanisms that underlie the dialogue between the hippocampus and neocortex (e.g., "neural replay," which occurs during sharp wave ripple activity). New work has also identified variables, such as the amount of preexisting knowledge, that affect the rate of consolidation. The increasing use of molecular genetic tools (e.g., optogenetics) can be expected to further improve understanding of the neural mechanisms underlying consolidation.

Memory consolidation refers to the process by which a temporary, labile memory is transformed into a more stable, long-lasting form. Memory consolidation was first proposed in 1900 (Müller and Pilzecker 1900; Lechner et al. 1999) to account for the phenomenon of retroactive interference in humans, that is, the finding that learned material remains vulnerable to interference for a period of time after learning. Support for consolidation was already available in the facts of retrograde amnesia, especially as outlined in the earlier writings of Ribot (1881). The key observation was that recent memories are more vulnerable to injury or disease than remote memories, and the significance of this finding for consolidation was immediately appreciated.

> In normal memory a process of organization is continually going on—a physical process of organization and a psychological process of repetition and association. In order that ideas may become a part of permanent memory, time must elapse for these processes of organization to be completed. (Burnham 1903, p. 132)

It is useful to note that the term consolidation has different contemporary usages that derive from the same historical sources. For example, the term is commonly used to describe events at the synaptic/cellular level (e.g., protein synthesis), which stabilize synaptic plasticity within hours after learning. In contrast, systems consolidation, which is the primary focus of this review, refers to gradual reorganization of the brain systems that support memory, a process that occurs within long-term memory itself (Squire and Alvarez 1995; Dudai and Morris 2000; Dudai 2012).

Systems consolidation is typically, and accurately, described as the process by which memories, initially dependent on the hippocampus, are reorganized as time passes. By this process, the hippocampus gradually becomes less important for storage and retrieval, and a more permanent memory develops in distributed regions of the neocortex. The idea is not that memory is literally transferred from the hippocampus to the neocortex, for information is encoded in the neocortex as well as in hippocampus at the time of learning. The idea is that gradual changes in the neocortex, beginning at the time of learning, establish stable long-term memory by increasing the complexity, distribution, and connectivity among multiple cortical regions. Recent findings have enriched this perspective by emphasizing the dynamic nature of long-term memory (Dudai and Morris 2013). Memory is reconstructive and vulnerable to error, as in false remembering (Schacter and Dodson 2001). Also, under some conditions, long-term memory can transiently return to a labile state (and then gradually stabilize), a phenomenon termed reconsolidation (Nader et al. 2000; Sara 2000; Alberini 2005). In addition, the rate of consolidation can be influenced by the amount of prior knowledge that is available about the material to be learned (Tse et al. 2007; van Kesteren et al. 2012).

Neurocomputational models of consolidation (McClelland et al. 1995; McClelland 2013) describe how the acquisition of new knowledge might proceed and suggest a purpose for consolidation. As originally described, elements of information are first stored in a fast-learning hippocampal system. This information directs the training of a "slow learning" neocortex, whereby the hippocampus gradually guides the development of connections between the multiple cortical regions that are active at the time of learning and that represent the memory. Training of the neocortex by the hippocampus (termed "interleaved" training) allows new information to be assimilated into neocortical networks with a minimum of interference. In simulations (McClelland et al. 1995), rapid learning of new information, which was inconsistent with prior knowledge, was shown to cause interference and disrupt previously established representations ("catastrophic interference"). The gradual incorporation of information into the neocortex during consolidation avoids this problem. In a recent revision of this framework (McClelland 2013), neocortical learning is characterized, not so much as fast or slow, but as dependent on prior knowledge. If the information to be learned is consistent with prior knowledge, neocortical learning can be more rapid.

This review considers several types of evidence that illuminate the nature of the consolidation process: studies of retrograde amnesia in memory-impaired patients, studies of healthy volunteers with neuroimaging, studies of sleep and memory, studies of experimental animals, both with lesions or other interventions, and studies that track neural activity as time passes after learning.

## STUDIES OF RETROGRADE AMNESIA IN MEMORY-IMPAIRED PATIENTS

One useful source of information about memory consolidation comes from studies in memory-impaired patients of memory for past news events or other public facts. In an early study of memory for television programs that had broadcast for only a single season, patients receiving electroconvulsive therapy for depressive illness showed temporally limited retrograde amnesia extending back 1–3 years (Squire et al. 1975). This phenomenon was slow to be related to neuroanatomy because of the need to test patients with well-characterized lesions who developed a memory impairment at a known time. In one of the first studies to meet these requirements, six memory-impaired patients with bilateral damage limited to the hippocampus were given a test of 250 news events covering 50 years (Manns et al. 2003). The patients showed a similar graded memory loss extending just a few years into the premorbid period (for two additional patients, see Kapur and Brooks 1999). Similar results have been obtained in neuroimaging studies of hippocampal activity, for example, when volunteers recalled past news events that had occurred 1–30 years ago (Smith and Squire 2009). These results suggest a consolidation process whereby the human hippocampus can be needed to support memory for factual information (semantic memory) for as long as a few years after learning, but is not needed after that time.

Patients with damage that not only includes the hippocampus but larger regions of the medial temporal (and sometimes lateral temporal) lobes as well can have severe retrograde memory loss covering decades (Bayley et al. 2006; Bright et al. 2006). Nevertheless, patients with medial temporal lobe lesions can have considerable sparing of premorbid memory (e.g., recognition of famous faces: patient H.M. in Marslen-Wilson and Teuber 1975; spatial knowledge of the childhood environment: patient E.P. in Teng and Squire 1999). These findings show that the brain regions damaged in such patients, for example, the entorhinal, perirhinal, and parahippocampal cortices, although important for new learning, are not similarly important for recollecting the past. The implication is that information initially requires the integrity of medial temporal lobe structures, but is reorganized as time passes so as to depend much less (or not at all) on these same structures.

Another useful source of evidence about memory consolidation comes from studies of autobiographical memory for past events (episodic memory). The study of autobiographical memory presents unique challenges because it depends on the analysis of spoken narratives that are often difficult to corroborate and on elaborate scoring methods that can be difficult to duplicate across laboratories. The findings

from studies of autobiographical memory are mixed. In several studies, patients with restricted hippocampal lesions, or larger medial temporal lobe lesions, successfully recalled detailed memories from early life (Eslinger 1998; Bayley et al. 2003; Buchanan et al. 2005; Kirwan et al. 2008; Kopelman and Bright 2012). Squire and Bayley (2007) offer additional discussion of single-case reports. Whether assessed by counting the details in tape-recorded narratives, rating the quality of narratives, or scoring responses to a few standardized questions (on a 0–9 scale), patients performed like controls (Fig. 1). Recall of episodes from the recent past was impaired (see also Thaiss and Petrides 2008). Yet, other patients produced fewer details about past events, even events from early life (Rosenbaum et al. 2005; Steinvorth et al. 2005; Race et al. 2011).

Why are the findings mixed? A lingering and challenging issue concerns the locus and extent of brain damage and the possibility that damage outside the medial temporal lobe contributes to impaired performance. Lesions that extend lateral to the medial temporal lobe, for example,

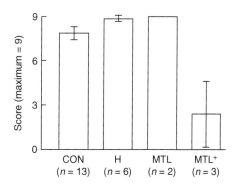

**Figure 1.** Performance on the remote memory (childhood) portion of the autobiographical memory interview (Kopelman et al. 1989; three questions; maximum score = 9) by controls (CON), patients with circumscribed hippocampal (H) lesions, patients with large medial temporal lobe (MTL) lesions, and patients with medial temporal lobe lesions plus additional damage in the neocortex (MTL⁺). Brackets show standard error of the mean. Patients with damage limited to the MTL have the capacity to recall remote events, but this capacity is diminished when the damage extends into the neocortex.

could impair remote memory by damaging regions thought to be repositories of long-term memory. In one of the studies reporting an extensive deficit in autobiographical memory (Rosenbaum et al. 2005), all four patients had significant damage to the posterior temporal cortex. Although the deficit was attributed to hippocampal damage, the data show that the best predictor of retrograde memory performance was the volume of the posterior temporal cortex and the cingulate cortex. The findings seem as consistent with the idea that an extended network of regions supports remote autobiographical recollection (Svoboda et al. 2006) as with the idea that the medial temporal lobe itself is especially critical. In another study of eight patients (Race et al. 2011), quantitative magnetic resonance imaging (MRI) data were available for only four and, of these, two had damage to lateral temporal cortex.

The fate of past autobiographical memories has special significance for ideas about consolidation. Multiple-trace theory (Moscovitch et al. 2006), later elaborated as the transformation hypothesis (Winocur and Moscovitch 2011), holds that memories retaining contextual details (such as episodic memories) remain dependent on the hippocampus so long as they persist. The consolidation of memory into the neocortex is thought to involve a loss of time and place, contextual information, and a transition to more gist-based, fact-like semantic memory. By this account, detailed context information about past autobiographical events does not consolidate into the neocortex. Note that, if the hippocampus supports the capacity for episodic recollection during the full lifetime of a memory, then hippocampal damage should not only reduce the number of details that can be produced (as has sometimes been reported), but should also result in fact-based narratives that reflect semantic knowledge and lack episodic (time and place) information.

Tulving (1985) introduced a method that can be used to test this idea. He argued that it is possible to recover information about past events from either episodic or semantic memory. In the former case, the phenomenal experience is remembering (becoming aware of one's own past). In the latter case, the experience is knowing, as when one is simply convinced that something is true without appreciating when or where the knowledge was acquired.

The method then involves asking people when they recall a past event whether they "remember" the occurrence of the event, or whether they simply "know" in some other way that it occurred. The so-called "remember–know" procedure has been used extensively in studies of learning and memory and has also been used to assess the quality of retrieved memories from the past. This procedure is well suited to address the question of interest. When memory-impaired patients with restricted medial temporal lobe lesions recollect an event from the remote past, do they report that they "remember" or that they simply "know" about the event? It took a long time to test this question experimentally. In the first formal test, three patients with hippocampal lesions and two with large medial temporal lobe lesions recalled events from their early life and, then, were asked to decide whether they "remembered" the incident or simply "knew" that it had happened (Bayley et al. 2005). The patients produced as many details as controls and also rated most of their memories as "remember" (87.1% vs. 80.3% for controls). Thus, these patients reported that they were "remembering" and retrieving from episodic memory. The results suggest that episodic retrieval from the remote past is possible after medial temporal lobe damage and that such retrieval is independent of the medial temporal lobe.

There are patients who cannot "remember" the past at all, but these are not patients with damage limited to the hippocampus. These are patients with significant damage outside the medial temporal lobe whose narratives are not just short on detail, but also lack any episodic content. Their knowledge of the past seems impersonal like their other knowledge of the world (patient S.S. in Cermak and O'Connor 1983; patient K.C. in Rosenbaum et al. 2005; patient G.T. in Bayley et al. 2005). They know a few facts about their own past, but cannot recall a single incident or event. For example, patient K.C. is described as follows:

The most striking fact about K.C.'s amnesia is that he cannot recollect a single personal happening or event from his life. He does not remember any incidents from all the years preceding his accident, nor can he remember any of the normally highly memorable things that have happened to him. (Tulving et al. 1988, p. 9)

If episodic memory depends on the hippocampus for as long as memory persists, then patients with restricted hippocampal lesions should be as empty of a personal past as patient K.C. Yet, a condition so severe has not, to our knowledge, been described for patients with restricted hippocampal lesions. The available evidence suggests that autobiographical memories consolidate and ultimately depend on a distributed network of cortical regions.

## NEUROIMAGING STUDIES OF NORMAL MEMORY

Consolidation has also been explored in healthy volunteers using neuroimaging methods like positron emission tomography (PET) or functional magnetic resonance imaging (fMRI). Neuroimaging studies can establish whether or not a particular structure (e.g., the hippocampus, medial prefrontal cortex, or a network of structures) is active when recent and remote memories are retrieved, but this method does not indicate whether a structure is necessary for retrieval. Thus, a temporal gradient of hippocampal activation (e.g., greater activation for recent than for remote memories) might reflect a decreasing dependence on the hippocampus as memories age, but such a result might also reflect differences in the extent to which memories of different ages are relearned and reencoded as they are recollected. Considerations such as these point to the complexities of interpreting neuroimaging data and may help to explain why the relevant neuroimaging literature is decidedly mixed.

In neuroimaging studies, participants are often tested for their semantic knowledge of public events that occurred at various times in the recent and remote past. With this approach, some studies have found greater activity in the medial temporal lobe during the recollection of recent memories compared with remote memories (Haist et al. 2001; Douville et al. 2005; Smith and Squire 2009), but other studies have found no difference (e.g., Maguire et al. 2001; Maguire and Frith 2003; Bernard et al. 2004). The latter results (like all null results) are difficult to interpret and may reflect a failure to detect a true difference. If so, the available results may not be as contradictory as they appear. More work is needed to settle this issue.

With respect to episodic (autobiographical) memories, several approaches have been used, and the literature is similarly mixed. In one typical design, a prescan interview is conducted to elicit autobiographical memories from the recent and remote past, and these memories are then queried during scanning (e.g., Niki and Luo 2002; Piefke et al. 2003; Bonnici et al. 2012, 2013; Söderlund et al. 2012). Studies using this design sometimes found greater medial temporal lobe activity in association with recent recollections compared with remote recollections (i.e., evidence of a temporal gradient), but sometimes there was no evidence for a temporal gradient. However, a recognized limitation of the prescan interview is, that later in the scanner, participants may retrieve what they had just recollected about their recent and remote memories. If so, all the memories are effectively recent, a circumstance that would work against finding a temporal gradient (Cabeza and St Jacques 2007).

An alternative design involves cueing subjects in the scanner for recent and remote autobiographical memories without a prescan interview (e.g., Gilboa et al. 2004; Rekkas and Constable 2005; Viard et al. 2010). In one study (Rekkas and Constable 2005), volunteers took a prescan tour of a campus (to create recent memories) and then, in the scanner, recalled either recent events from the campus tour or remote events that had occurred during elementary school years. A temporal gradient of hippocampal activity was observed, but the level of activity was "higher" for remote memories than for recent memories (a result not anticipated by any account of memory consolidation). In fMRI studies, it is often difficult to know whether measured activity reflects reencoding or retrieval (Stark and Okado 2003). One possibility is that interesting and personally significant memories from the remote past might have stimulated greater reencoding activity in the hippocampus than less interesting, and less significant, memories from the recent past.

A different approach is to use a prospective design that affords experimental control over the memories from different time periods. In this design, participants learn similar materials at multiple different time points before scanning. A number of recent neuroimaging studies have used this approach (Takashima et al. 2006, 2009; Yamashita et al. 2009; Furman et al. 2012; Harand et al. 2012). For example, in one study (Takashima et al. 2009), participants memorized two sets of face-location associations; one was studied 24 h before testing (remote memories) and the other studied 15 min before testing (recent memories). Activity in the hippocampus decreased (and activity in the neocortex increased) as a function of time after learning. Concomitantly, functional connectivity between the hippocampus and cortical areas decreased over time, whereas connectivity within the cortical areas increased. This temporal gradient is shorter than what is typically observed in lesion studies, but the findings are nevertheless in agreement with the idea that the hippocampus becomes less important for memory as time passes. We shall consider the speed of consolidation in more detail in a later section on animal studies.

Furman et al. (2012), also using a prospective design, tested memory of documentary film clips. Hippocampal activity declined as time passed over a period of months when memory was tested by recognition (an indication of memory consolidation), but it remained stable across time when memory was tested by recall. A complementary increase in activity in cortical areas as a function of time was not observed; instead, cortical activity decreased as well.

A concern that could be raised about prospective studies is that, by the time memory is tested in the scanner, many older memories will have been forgotten. As a result, the surviving remote memories are relatively durable and are being compared with a mixture of durable and less durable recent memories. One elegant prospective memory study addressed this issue (Yamashita et al. 2009). Activity in the hippocampus and temporal neocortex was monitored as participants recalled two sets of paired-associate figures (fractal images), which they had memorized at two different times. One set had been studied 8 weeks before testing (remote memories), and the other had been studied just before testing (recent memories). Overall memory accuracy at the time of the test was equated for these two conditions by providing more study time for the items studied 8 weeks earlier. The results were that a region in the right hippocampus was more active during retrieval of new memories than old memories, whereas in the left temporal neocortex, the opposite pattern was observed. These results are consistent with a decreasing role of the hippocampus and an increasing role of the neocortex as memories age across a period of 50 days.

Imaging studies have also been used to explore how and under what conditions consolidation occurs—a process that presumably involves some relatively long-lasting communication between the hippocampus and the neocortex. One proposal for how this could be accomplished is through the phenomenon of "neural replay," which refers to the spontaneous recurrence of hippocampal activity that occurred originally during learning. Neural replay has most often been observed in rats during "non"–rapid eye movement [NREM]) slow-wave sleep (Wilson and McNaughton 1994), but something akin to neural replay seems to occur in humans as well. For example, a study of regional cerebral blood flow found that hippocampal areas that were active during route learning in a virtual environment (a hippocampus-dependent, spatial learning task) were active again during subsequent slow-wave sleep (Peigneux et al. 2004). Moreover, the degree of activation during slow-wave sleep correlated with performance on the task the next day. In a conceptually related study (Rasch et al. 2007), cueing recently formed odor-card associations by odor reexposure during slow-wave sleep, but not during rapid eye movement (REM) sleep, increased hippocampal activity (as measured by fMRI), and resulted in less forgetting after sleep. In both studies, hippocampal reactivation, which presumably initiated hippocampal–neocortical dialogue, occurred within hours of the learning episode.

Other studies have found that the replay-like activity observed during NREM sleep can also occur during quiet wakefulness, suggesting that consolidation-related processes may proceed whenever the hippocampus is not otherwise engaged in encoding activity (Karlsson and Frank 2009; Mednick et al. 2011). For example, following paired-associates learning (objects or scenes paired with faces), functional connectivity between the hippocampus and a portion of the lateral occipital complex increased during a postlearning rest period (Tambini et al. 2010). Moreover, the strength of this effect predicted subsequent memory performance. The specific patterns of activity associated with encoding experiences can also recur during subsequent offline rest periods. In one study (Tambini and Davachi 2013), hippocampal activity patterns persisted into postencoding rest periods, and the persistence of some of these patterns correlated with later memory performance for the material presented during the study. In related work, an object-scene paired-associates learning task was followed by a 2-min delay, followed by a cued-recall test (Staresina et al. 2013). Successful recall of individual study events was predicted by the degree to which those events elicited similar patterns of activation in the entorhinal cortex during the encoding and delay periods. Inasmuch as these studies have involved activity occurring rather soon after learning, it will be useful to evaluate the possible role of rehearsal, either intentional or spontaneous.

If neural replay is related to memory consolidation, it should be possible to find evidence of replay at longer time intervals after learning (i.e., across the portion of long-term memory during which consolidation is thought to occur). In a study of trace eyeblink conditioning in rats, activity in the medial prefrontal cortex, selective for the acquired association, developed over a period of several weeks and in the absence of continued training (Takehara-Nishiuchi and McNaughton 2008). In this case, training initi-

ated gradual processes that developed within long-term memory. These findings are consistent with the idea that the encoding of a memory can be followed by replay-like activity during subsequent offline periods. The extended hippocampal–cortical communication resulting from replay is thought to lead gradually to a memory that is represented in the neocortex and independent of the hippocampus.

## ANIMAL STUDIES OF MEMORY CONSOLIDATION

Animal research on memory consolidation has the same starting point as human work: establishing whether and how memory stabilizes with the passage of time. It differs in allowing for invasive experiments using interventions, such as lesions, physiological monitoring and recording, and molecular techniques. Key empirical issues include (1) Are temporal gradients of retrograde amnesia reliably observed after comparable experimental interventions? (2) Is there evidence of time-dependent changes in physiological function that relate to or mediate aspects of memory consolidation? (3) Can contemporary molecular-genetic techniques be used to shed new light on the systems issues concerning hippocampal–neocortical dialogue?

### Are Temporal Gradients of Retrograde Amnesia Reliably Observed in Animals?

Beginning more than 20 years ago, studies in nonhuman primates, rodents, and other species confirmed the existence of a temporal gradient of amnesia in animals. Thus, monkeys learned multiple object discrimination tasks at five different intervals before surgery (Fig. 2A). After lesions of the hippocampal formation, they remembered better the problems learned 12 weeks before the lesion than problems learned just before the lesion (Zola-Morgan and Squire 1990). A study of the social transmission of food preference task in rats yielded a similar but shorter temporal gradient (Winocur 1990). In context fear conditioning, rats given hippocampal lesions 28 days after conditioning still displayed fear of the chamber in which condi-

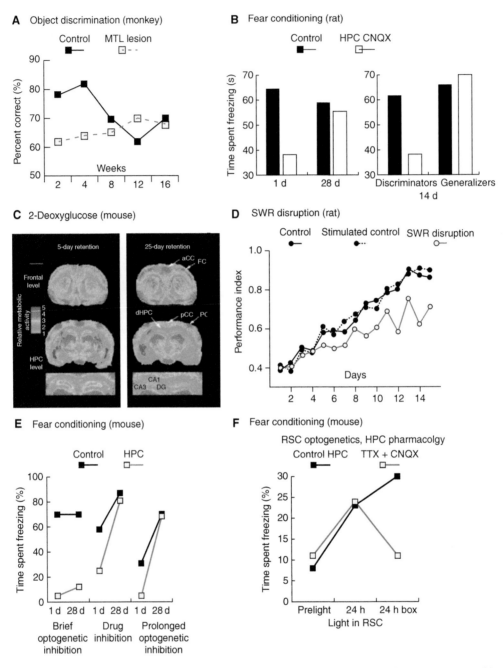

Figure 2. Animal studies revealing temporal gradients and other characteristics of cortical memory measured in a variety of different tasks. (*A*) Object discrimination learning in monkeys shows a temporal gradient over a period of 12 weeks (based on Zola-Morgan and Squire 1990). (*B*) Context fear conditioning sometimes shows a temporal gradient, but does not always do so. In a study in which a temporal gradient was observed in animals tested during pharmacological inhibition of the hippocampus (*left* panel), animals that successfully discriminated two testing contexts show a loss of memory with hippocampal inhibition, whereas animals that generalized do not (*right* panel based on Figs. 2 and 5 of Wiltgen et al. 2010). (*C*) Glucose uptake measured using radiolabeled 2-deoxyglucose shows a time-dependent increase in the cortex 25 days after radial-maze learning in mice (based on Bontempi et al. 1999). (*Legend continues on following page.*)

tioning had occurred (indexed as "freezing"), whereas those trained only 1 day before the lesion did not (Kim and Fanselow 1992). Similarly, in a within-subjects procedure, rats froze in the context in which fear conditioning had occurred 50 days before a hippocampal lesion, but not in a different context in which fear conditioning had occurred just before the lesion (Anagnostaras et al. 1999).

These studies made two important points. First, their prospective designs enabled behavioral training to be appropriately timed in relation to lesions that could be both complete and limited to a structure of interest. Second, the findings indicated that it was not simply that a lesion after training caused a deficit in memory. The key finding was that a group with a long time interval between training and surgery performed paradoxically better than a group with a shorter training–surgery interval. Discussions of retrograde amnesia have emphasized the importance of this feature of the data (Squire 1992). Indeed, the finding that remote memory can be better than recent memory after a lesion is usually considered to be the gold standard for showing memory consolidation.

However, not all studies of retrograde amnesia in animals have yielded a gradient, for example, in context fear conditioning. Thus, one group has repeatedly failed to find a temporal gradient (Lehmann et al. 2007; Sutherland et al. 2008; Sparks et al. 2011). Another group found no temporal gradient despite systematic manipulation of potentially relevant parameters: lesion method, lesion size, single trial training, and massed and spaced multitrial training (Broadbent and Clark 2013).

Early studies in the water maze also did not find a temporal gradient (Bolhuis 1994; Sutherland et al. 2001). Our two groups investigated this in detail. Varying the task to a dry-land version, called the "oasis" maze, revealed no sparing of remote memory after radio frequency lesions of the hippocampus (Clark et al, 2005a). Extended training of the water maze task before the lesion, in an effort to minimize floor effects, also did not reveal a gradient (Clark et al. 2005b). One possibility was that this impairment in remote memory was a retrieval deficit rather than a true loss of consolidated memory. To address this, a reminder procedure was used in two water maze studies involving neurotoxic lesions (de Hoz et al. 2004; Martin et al. 2005). The Atlantis platform procedure provided the reminder. The platform is near the bottom of the water maze during a probe trial, but becomes available for escape after 60 sec. In this way, a probe test can be performed and the animal is "reminded" of the correct location. Performance in a second probe test 1 h later was significantly better, indicating that reminding can work. However, this effect occurred only in partially lesioned rats that had been trained shortly before the lesion. Furthermore, successful reminding of memory was not observed in remote memory groups trained long before the lesion. Thus, a temporal gradient can appear in the water maze once access to memory is improved but in the direction opposite to that predicted by standard consolidation theory.

However, as navigation through space involves the hippocampal formation in rodents (O'Keefe and Nadel 1978; Moser et al. 2008), the failure to see intact memory could be be-

**Figure 2.** (*Continued*) (*D*) Detection and shutdown of hippocampal sharp-wave ripples (SWR), a candidate mediator of consolidation, slows learning of a spatial radial-arm maze task (based on Girardeau et al. 2009). (*E*) Comparison of short and prolonged optogenetic (halorhodopsin)-induced inhibition of the hippocampus. There is an unexpected effect of brief hippocampal inhibition after 28 days in a training paradigm that shows a temporal gradient with more prolonged inhibition (based on Goshen et al. 2011). (*F*) Optogenetic activation of neurons in the retrosplenial cortex (RSC) labeled with channelrhodopsin via c-fos activation during context fear conditioning is sufficient to elicit a freezing response, bypassing the need for hippocampal binding during the early stage of systems consolidation. Even when the hippocampus (HPC) was inactivated by tetrodotoxin (TTX) and 6-cyano-7-nitroquinoxaline-2,3-dione (CNQX), direct optogenetic activation of RSC elicited greater freezing 24 h after conditioning (*middle*) than during preconditioning (*left*). In the absence of optogenetic activation (*right*), hippocampal activity was essential to reactivate memory so soon after conditioning (based on Fig. 4E of Cowensage et al. 2014). MTL, medial temporal lobe.

cause of a secondary disruption of the expression of memory, even for a memory that had been consolidated in the cortex. It may be, for this reason, that the reminding effect was observed in animals with partial neurotoxic lesions with intact fibers of passage. If so, reducing the navigational demands by arranging for swimming and escape within a circular corridor or "annulus" might reveal intact memory in animals with larger lesions. However, such a finding was also not observed (Clark et al. 2007). Thus, something about allocentric spatial memory tasks, at least in the water maze, appears to be different from the tasks discussed earlier, possibly including some long-term storage of spatial information in the hippocampus. This difference has recently been driven home by a study using a within-subjects design. In the same experimental subjects, a temporal gradient was found for context fear conditioning but not for spatial memory (Winocur et al. 2013). Why the gradient is reliably seen in some studies of context fear conditioning but not in others, nor in the water maze, is an issue of current interest (Winocur et al. 2013).

It has been shown that the hippocampus has to be active at the time of retrieval for the expression of both recent and remote memory in rodents (Liang et al. 1994; Broadbent et al. 2006), and this fact points to the need for a different approach. One would like to inactivate the hippocampus reversibly *during* the putative memory consolidation process, but allow it to work normally *during* learning and, later, at recall. One relevant study used a chronic reversible blockade of GluR1-5 receptors in the dorsal hippocampus (for 7 days, beginning 1 or 5 days after learning), with blockade confirmed metabolically using 2-deoxyglucose (2-DG). Memory retention was tested 16 days after training with the hippocampus, once again, working normally (Riedel et al. 1999). A deficit in the consolidation of memory was observed whether the inhibition was begun 1 day after training or after a delay of 5 days. These findings suggest that "turning off" the hippocampus for 7 days *after* learning, despite normal function during encoding and retrieval, does cause retrograde amnesia. Although this study lacked a long-

delay condition before the onset of hippocampal inhibition (which might then show spatial memory to be intact), it confirms the idea that water maze learning may involve posttraining consolidation just as the standard model predicts. However, it is unclear whether the pharmacological intervention affects only the dialogue between the hippocampus and neocortex for stabilization of cortical traces (Alvarez and Squire 1994), or might also affect memory traces within the hippocampus itself (Nadel and Moscovitch 1997). A further difficulty is that a study using continuous posttraining infusions of either lidocaine or CNQX for 7 days did not replicate the Riedel et al. (1999) findings (Broadbent et al. 2010). However, a later study using postlearning transection of the temporoammonic path from the entorhinal cortex to area CA1 of the hippocampus did reveal poor memory tested 28 days later (Remondes and Schuman 2004), which was interpreted as interrupting hippocampal–neocortical dialogue. That is, after learning, ongoing cortical input conveyed by the temporoammonic path is required to consolidate long-term spatial memory. Follow-up studies need to be conducted that look not only at the impact of sustained and reversible inactivation of the hippocampus, but also at relevant structures of the neocortex (retrosplenial, anterior cingulate, medial prefrontal cortex) that might be engaged in systems consolidation.

For the present, it is clear that after appropriately timed lesions, a temporal gradient favoring better remote memory is seen too often to discount, including in a spatial paired-associate task to be discussed below (Tse et al. 2007), but it is unclear why it is not always observed with either permanent lesions or temporary inactivation. The next section identifies some relevant factors.

## Interpreting Behavioral Measures and Cognitive Factors

Human studies are characterized by distinct ways of measuring memory—from simple measures of percent correct to more subjective measures, such as the remember–know procedure.

 Cite this article as *Cold Spring Harb Perspect Biol* doi: 10.1101/cshperspect.a021766

It is unclear how to apply such subtle qualitative distinctions to animals who cannot verbally report a specific event. In a typical task, it is simply the apparent strength of a memory that is being tested. Recent animal research has endeavored to determine whether different temporal gradients of retrograde amnesia are associated with different training protocols and/or whether representations in memory change qualitatively with the passage of time. What is emerging is evidence that features of a test protocol can matter and that the temporal gradient of retrograde amnesia is not a fixed entity.

One important variable is whether the information being learned is completely novel or can be related to previously acquired knowledge. How easily new information can be assimilated into a neocortical knowledge structure, such as a "schema," may depend on the extent to which subjects have an available framework of prior knowledge relevant to the new information being learned (Bransford 1979; McClelland 2013; Ghosh and Gilboa 2014). Rapid assimilation of information into a schema should speed up the time course of systems consolidation. This idea was investigated in rats by first training them to learn multiple paired associates. The associations involved the flavor/odor of a food reward buried in a sand well and the spatial location in an "event arena" where the food could be found (Tse et al. 2007). The use of spatial location as one member of a paired associate enabled the animals to build a map or "schema" indicating what food was where. Learning was slow, ∼6 weeks, but the animals could eventually use the taste or smell of the food to direct their digging at the location where the corresponding food could be found. The initial learning of paired associates was impaired by neurotoxic hippocampal lesions given before training, indicating that this learning is "hippocampus dependent." Critically, when normal animals were trained and acquired a schema, and lesions were made at different times after the introduction of new paired associates, a temporal gradient of retrograde amnesia was observed. Specifically, when lesions were made only 3 h after training, later memory was poor. However, when lesions were made 2 days

after training, the animals showed good memory. Thus, a rapid temporal gradient was observed (unlike in the water maze), reflecting the apparent assimilation of new information into the previously trained neocortical schema.

These results raise the question: What is different about context fear conditioning, the water maze, and the event arena? Temporal gradients are seen in the first and third of these, but not usually in the second. Spatial navigation is important in the water maze, but not in context fear conditioning, and not so important in the event arena because the animal can decide whether or not to dig for food at each sand well after it gets to it. Another difference between these three tasks, to which Tse et al. (2007) drew particular attention, is that in studies to date, animals entering context fear conditioning and water maze experiments are "experimentally naïve." That is, they have not been trained on other tasks, nor do they have a history of prior learning like adult humans. The implicit supposition has been that this exceptional degree of control is a desirable feature of animal protocols, but this may be incorrect. The results from Tse et al. (2007) raise the possibility that the temporal parameters of consolidation are not a biological given (such as the interaction of fast [hippocampal] and slow [cortical] learning systems discussed by McClelland et al. 1995), but can be influenced by cognitive factors, that is, the time course of consolidation is not fixed, and that this is the case is an important topic of current study (Tse et al. 2007). The paired-associate protocol may produce a relatively short temporal gradient of retrograde amnesia because the animals do not remember the individual events associated with a paired-associate learning trial, but can nevertheless modify a preexisting semantic memory in the cortex (Fig. 3B). An additional justification for this second point is that new computational modeling indicates that past experience may make the extraction of statistical regularities easier and faster (McClelland 2013).

It is possible that this same idea may illuminate context fear conditioning studies in some circumstances. For example, following up earlier work (Riccio et al. 1984), new studies of fear

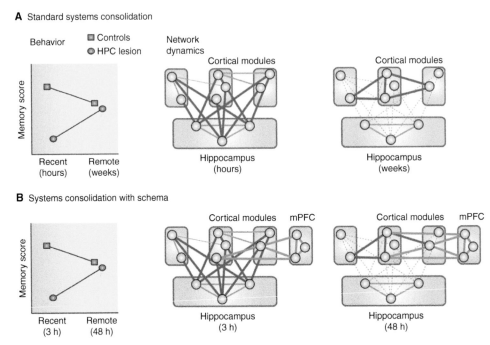

**Figure 3.** Hypothetical models of hippocampal–neocortical interactions during memory consolidation. (*A*) The standard model supposes that information is stored simultaneously in the hippocampus and in multiple cortical modules during learning and that, after learning, the hippocampal formation guides a process by which cortical modules are gradually bound together over time. This process is considered to be slow, occurring across weeks, months, or even longer (based on Frankland and Bontempi 2005). (*B*) In situations in which prior knowledge is available and, thus, cortical modules are already connected at the start of learning, a similar hippocampal–neocortical-binding process takes place. However, this process may involve the assimilation of new information into an existing "schema" rather than the slower process of creating intercortical connectivity (based on van Kesteren et al. 2012). HPC, hippocampus; mPFC, medial prefrontal cortex.

conditioning have shown that the ability to discriminate between the training context (context A) and a novel context (context B) diminishes over the course of a month (Wiltgen and Silva 2007; Wiltgen et al. 2010). Importantly, these studies revealed a temporal gradient of retrograde amnesia when mice were tested in the training context (context A) 1 or 28 days after training (hippocampal inactivation impaired performance after 1 day, but not after 28 days). What is the relationship between the loss of discriminability between contexts A and B across 28 days and the evidence for consolidation over the same time period? One possibility is that consolidation occurs against a background of normal forgetting. Alternatively, the decline in discriminability might mean that consolidation is a time-dependent process that

gradually extracts statistical regularities while permitting details to be lost.

A study relevant to these issues assessed fear conditioning 14 days after training. A significant negative correlation was found between an individual animal's discrimination ability (trained vs. novel context) and the extent to which the animal showed freezing in the "novel" context (Wiltgen et al. 2010). There was no correlation between discrimination ability and the extent of freezing in the "trained" context. In addition, pharmacological inactivation of the hippocampus after 14 days impaired memory for the training context in animals that discriminated between contexts, but not in animals that failed to discriminate (Fig. 2B). One interpretation of this result is that, if a detailed contextual ("episodic-like") memory is available, whether immediately or long after training, the hippocampus is engaged and required. Moreover, if only a "semantic" or "gist" memory is available, which does not discriminate between contexts, memory can be supported by the cortex alone.

An alternative way to understand these results is that they may reflect group differences in the rate of consolidation. Mice were separated into two groups based on their ability to discriminate context A from context B. Accordingly, the two groups could have differed from each other in many other ways as well (notwithstanding the finding that performance in context A was similar for the two groups). Indeed, the amount of gist or detailed memory that was available may not be relevant. The high-discrimination group might take longer to consolidate than the low-discrimination group and, even after that point is reached, the animals in the high-discrimination group might still be able to discriminate between contexts. This result would imply that even memory for detail can consolidate. This analysis predicts that if a much longer period of posttraining consolidation were allowed, enabling the high-discriminating group to fully consolidate, hippocampal inactivation would have no effect despite successful context discrimination.

Another study of how memory changes over time involved use of the water maze (Richards et al. 2014). Mice were trained on a delayed matching-to-place protocol in which the escape platform moved location each day. On each day, there was a rapid decline in escape latency as each new location was learned. In one version of this task, the platform was located in various places in the north quadrant of the pool twice as often as in various places in the east quadrant. When tested 1 day after completing 8 days of training, mice searched primarily in the most recently trained location. However, when tested after 30 days, mice tended to search twice as often in the north quadrant as in the east quadrant. That is, after 30 days, search was driven less by any single day's training, or even by the last day's training, but rather by the cumulative statistical distribution of training experience across days. When mice had to learn a new escape location

either 1 day or 30 days after training, they could remember this new location better if it conflicted with the distribution of locations learned 30 days earlier than if it conflicted with the distribution of locations learned only 1 day earlier.

The nature of this stabilization of pattern memory or gist memory over time remains unclear. The investigators propose that the finding is unlikely to be caused by the degradation of memories over time or to memory strength being simply a function of relative recency. Instead, like Wiltgen et al. (2010), they favor the view that consolidation entails an active process of extracting the gist or pattern from what was learned, although forgetting individual instances (including the most recent instance). Whereas this perspective does provide an accurate description of the data, it is unclear whether a qualitative process/mechanism of gist extraction is, in practice, actually required. Successive training days to platforms in distinct locations will create multiple rapidly declining memory traces whose residual strength over time may not decline to zero. Instead, these traces could gradually summate to result in particular areas of the water maze having different associative strengths. This summation process could occur during posttraining consolidation and result in the time-dependent appearance of an apparent gist memory even in the absence of a specific pattern extraction process.

Taken together, the studies summarized in this section indicate that temporal gradients of retrograde amnesia can be observed, but there is presently uncertainty about both the time course of the gradients and the factors that mediate them. The possibility that there are qualitative changes in the character of memory during consolidation is currently an active area of research.

## MONITORING METABOLIC, IMMEDIATE EARLY GENE, AND NEUROPHYSIOLOGICAL SIGNATURES OF CONSOLIDATION AT VARIOUS TIMES AFTER TRAINING

Monitoring physiological changes over time offers a different window on consolidation processes. One strategy has been to look at region-specific alterations of 2-DG uptake or immediate early gene (IEG) activation in the hippocampus and neocortex at various times after training. One study trained mice in an eight-arm radial-maze task (Olton et al. 1979) in which some arms were rewarded and others were never rewarded (Bontempi et al. 1999). Increased 2-DG uptake in the hippocampus was found when testing occurred shortly after training. Importantly, there was a time-dependent decrease in hippocampal uptake when testing was delayed for several weeks, and there was a corresponding increase in 2-DG uptake in the neocortex (Fig. 2C). Follow-up work found similar temporal patterns in the expression levels of IEGs, such as c-fos (Frankland et al. 2004; Maviel et al. 2004). Direct transfer of information from the hippocampus to the cortex is not thought to occur in consolidation. However, the process of cortical stabilization may require guidance, whether the stabilization involves synaptogenesis or persistent changes in synaptic strength in the cortex. One suggestion for this process is the idea of a "tagging" process at neocortical synapses (Lesburgueres et al. 2011). Tagging refers to a local, posttranslational change in cortical neurons that occurs in parallel with memory encoding.

Using a social transmission of food-preference paradigm in which a demonstrator animal "teaches" an observer animal which foods are safe to eat, these investigators explored the idea that neurons might undergo a "tagging process" in the orbitofrontal cortex at the time of memory encoding. This process would help to ensure that the gradual, hippocampus-driven rewiring of cortical networks, which ultimately support long-term memory, is directed appropriately. Using a combination of lesions and pharmacological interventions, these studies established that tagging was amino-3-(5-methyl-3-oxo-1,2-oxazol-4-yl) propanoic acid (AMPA)- and N-methyl-D-aspartate receptor–dependent, information-specific, and capable of modulating remote memory persistence by affecting the temporal dynamics of hippocampal–cortical interactions. The concept emerging from these studies is in line with standard consolidation theory. Specifically, the concept is

that neocortical tags are set early at the time of memory encoding and time-dependent changes develop across time as dynamic interactions occur between the hippocampus and neocortex.

When animals were tested for their memory of context fear conditioning at 1 and 30 days after training, correlations in the expression of the IEG c-fos were observed across brain areas (Wheeler et al. 2013). Specifically, the absolute levels of IEG expression in up to 80 brain areas revealed a striking change in the correlation matrix between brain areas 1 and 30 days after training. These results indicate time-dependent changes in cortical networks over this time period. This "functional connectome" is strongly suggestive of a time-dependent reorganization and stabilization of neocortical networks. However, given the uncertainty about which factors influence temporal gradients of behavioral memory (Broadbent and Clark 2013), caution is also appropriate in the case of IEG measures. That time-dependent changes are observed in IEG expression is clear. Less clear is what these changes correspond to in terms of underlying memory processes. For example, c-fos activation likely reflects increased neural activity at the time of retrieval, but this activation does not necessarily measure retrieval itself. Although there was no opportunity for "new" fear learning during the retrieval tests (because reexposure to the training context was not accompanied by additional presentations of the unconditional shock stimulus), it is nevertheless possible that retrieval could be associated with new encoding (of the retrieval test itself, for example). That is, what is procedurally a retrieval paradigm (Wheeler et al. 2013) might reflect gene expression associated with retrieval-related encoding.

IEG expression in the hippocampus and cortex has also been measured as animals express a learned schema in the event arena or are required to assimilate new information (Tse et al. 2011). In that study, two IEGs, zif-268 and Arc, were measured and found to be upregulated throughout consolidation-relevant midline regions of the neocortex at the time of memory encoding, broadly similar to the areas identified in earlier work at long time periods

Cite this article as Cold Spring Harb Perspect Biol doi: 10.1101/cshperspect.a021766 after learning (Frankland and Bontempi 2005). The magnitude of the increase for these IEGs was related to how readily the new information would later be expressed, that is, how easily it would be assimilated into the existing and activated knowledge structure if the full period of consolidation had been allowed to take place. These findings support the possibility, raised earlier (Lesburgueres et al. 2011), that cortical tagging helps promote the neocortical consolidation process. They are also consistent with the idea, derived from studies of amygdala-dependent fear conditioning, that neurons are recruited into a memory trace as a function of neuronal excitability (Yiu et al. 2014).

Another and distinct window onto putative consolidation mechanisms comes from neurophysiological recordings from relevant neuronal ensembles of the hippocampus and neocortex during sleep. It has been proposed that consolidation occurs specifically during NREM sleep (Jenkins and Dallenbach 1924; Marr 1971; Smith 1996). Two mechanisms have since been proposed. One is "replay," the reactivation of patterns of network activity that had occurred during previous experience and that are thought to lead to potentiation of relevant synaptic connections in the cortex. Replay starts in the hippocampus and propagates to the cortex, with reprocessing there to extract statistical overlap from different encoding episodes (Wilson and McNaughton 1994; Diekelmann and Born 2010; Battaglia et al. 2012; Genzel et al. 2014). The other suggested mechanism for consolidation during sleep is "downscaling"— "with sleep homeostatically but nonspecifically regulating synaptic weights to improve the signal-to-noise ratio of memory traces" (Tononi and Cirelli 2006, 2014). The combined "push–pull" action of replay on the one hand ("push" equals potentiating "important" traces) and downscaling on the other ("pull" equals weakening irrelevant traces) may together aid the construction and updating of memory networks in the cortex (Diekelmann and Born 2010; Lewis and Durrant 2011; Genzel et al. 2014). Sleep replay seems to be a widespread phenomenon requiring participation and cooperation among many different brain areas, whereas downscaling is thought to be a more local process, which is locally initiated and regulated.

What mechanisms are responsible for memory traces becoming consolidated during sleep? To initiate replay in the hippocampus, a slow oscillation starts in the prefrontal cortex, consisting of an alternation between states of generalized cortical excitation and depolarized membrane potentials (UP states) and generalized states of relative neuronal silence (DOWN states). This oscillation, as seen in intracranial recordings of patients being evaluated for epileptic surgery, travels across the brain to the medial temporal lobe (Nir et al. 2011), where it is followed by sharp wave ripples in the hippocampus. During sharp wave ripples, 30% of hippocampal neurons increase their firing rate, and the replay of sequential firing of neurons that encode, for example, previous spatial experiences occurs in a temporally compressed form (Wilson and McNaughton 1994). Hippocampal replay was first observed in the rat when place cells were seen to fire during sleep in the same sequence as they fired on a linear track when the rat was actually running (Wilson and McNaughton 1994). Since then, many other interesting attributes of sharp wave ripples and replay have been reported. Replay occurs 7–10 times faster than the original experience (Euston et al. 2007) and seems to be homeostatically regulated. The increased appearance of sharp wave ripples after their disruption is seen only after a significant learning period (Girardeau et al. 2014). Further, the type of neural network being activated during replay, that is, the specific memory, can be recognized by the typical morphology of the ripple (Reichinnek et al. 2010), and longer memories seem to be replayed across multiple sharp wave ripples (Davidson et al. 2009). By recording neuronal firing patterns in the prefrontal cortex at choice points in a maze and during sleep, it has been shown that hippocampal replay during sleep (but not during the awake state) is directly communicated to the prefrontal cortex. As a result, there is simultaneous expression in the hippocampus and prefrontal cortex of learning-associated neural firing patterns (Peyrache et al. 2009). Evidence for replay has also been found in other brain areas, including the striatum, motor cortex, and visual cortex (Euston et al. 2007; Ji and Wilson 2007; Aton et al. 2009; Lansink et al. 2009; Yang et al. 2014). Additionally, if sharp-wave ripples are followed by neocortical spindles, secondary larger waves of increased firing rates can be observed in the cortex (Wierzynski et al. 2009). At the same time, by comparing unit firing in the cortex and hippocampus during spindles, it was observed in one study that the cortex became functionally deafferented from hippocampal inputs during sleep spindles (Peyrache et al. 2011). Spindles may also play an additional but separate role in local cortical processing during sleep, which is independent of the hippocampal replay mechanism (Andrillon et al. 2011).

As discussed earlier, recent studies in humans have tried to increase the probability of replay during sleep by cueing recently learned memories via smell or sound associations (Rasch et al. 2007; Rudoy et al. 2009). In a related study in rats, two running tracks were associated with distinct sounds. In this case, cueing replay during sleep did not increase the absolute amount of replay. Instead, cueing biased replay toward the cued content at the cost of the noncued content (Bendor and Wilson 2012).

Interestingly, replay is observed not only during sleep, but can occur during waking as well. Intervention studies have addressed whether there are different functions of replay in sleep versus awake states by interrupting sharp wave ripples with electrical stimulation whenever they occur (Fig. 2D). This work has hinted at the possibility that replay during sleep is important for consolidation processes, whereas awake replay is more associated with spatial working memory and navigational planning (Girardeau et al. 2009; Ego-Stengel and Wilson 2010; Jadhav et al. 2012; Pfeiffer and Foster 2013). However, both awake and sleep replay do seem to contribute to later memory performance (Dupret et al. 2010), with awake replay, perhaps, serving to initially stabilize memory but with sleep replay operating in association with systems consolidation.

To summarize, there are time-dependent changes in physiological function that relate to or mediate aspects of systems memory consolidation, as now shown with 2-DG, immediate early gene mapping, and electrophysiological recordings. New technologies, notably optical recording, offer the opportunity to study these changes dynamically in large numbers of neurons. This approach might require a head-fixed, virtual-reality paradigm that could be used over the time periods after learning when consolidation occurs.

## THE USE OF MOLECULAR–GENETIC TECHNIQUES TO ILLUMINATE MECHANISMS OF HIPPOCAMPAL–NEOCORTICAL DIALOGUE

New approaches are starting to use elegant inducible and reversible genetic interventions to examine systems consolidation. The focus here is not on the specific molecular mechanisms of consolidation (discussed elsewhere in this collection), but on the use of molecular engineering approaches to illuminate the dynamics of systems consolidation. Two studies illustrate the approach.

In one study, optogenetic inhibition of the hippocampus was deployed to examine the interplay of the hippocampus and neocortex, emphasizing the opportunity that this technique offers for precise temporal intervention (Goshen et al. 2011). Halorhodopsin-induced inhibition of area CA1 of the hippocampus reduced the frequency of cell firing and blocked the acquisition and retrieval of contextual fear conditioning. Brief optogenetic inhibition (for 5 min) of hippocampus consistently interfered with retrieval (after 28 days, 9 weeks, and 12 weeks). In contrast, more extended pharmacological and optogenetic inhibition spared remote memory, but impaired recent memory, revealing a temporal gradient (Fig. 2E). Brief inhibition of CA1 resulted in a decreased c-fos expression at recall in both CA1 and the anterior cingulate cortex (ACC), whereas extended inhibition decreased expression in CA1, but increased expression in ACC. These results could mean that prolonged inhibition allows for compensatory activity to develop and that this compensatory activity supports remote memory performance. An alternative is that prolonged inhibition allows nonspecific effects of the inhibition to wear off and remote memory to be successfully expressed.

The concept of rapid cortical tagging (Lesburgueres et al. 2011; Tse et al. 2011) suggests that, even if the hippocampus is normally engaged for a short period for "binding" disparate cortical networks during memory encoding and the early stages of consolidation, a memory trace of some kind is rapidly formed in the cortex. The cortical trace in this case corresponds to the episodic memory-like trace that was formed during the initial experience and not to a gist-memory trace. This trace may ordinarily require co-occurrent activity in the hippocampus for behavioral expression during the consolidation period, but this involvement of the hippocampus might be mimicked optogenetically. That is, if the role of the hippocampus during the consolidation period is to engage cortical neurons selectively during the stabilization process, this function of the hippocampus might be achieved optogenetically by selectively activating those cortical neurons that are involved in the memory.

A relevant step forward has used context fear conditioning and tet-TAG mice in which c-fos is used to drive the insertion of a channelrhodopsin-like protein into learning-activated cells (Cowensage et al. 2014). A subset of neurons in the retrosplenial cortex (RSC) was tagged in this way during context fear conditioning. Subsequently, it was possible to activate just this subset of RSC cells optogenetically (despite diffuse light activation of all cells) because only those neurons were activated that had been firing at the time of cortical memory encoding and had, therefore, expressed c-fos. The mice showed elevated freezing in response to the light, an observation made even more secure through the use of a context discrimination procedure (reminiscent of Wiltgen's work cited earlier). Increased freezing was also observed in response to light activation of the RSC network when the hippocampus was simultaneously inactivated with lidocaine (Fig. 2F). Thus, although the hippocampus may ordinarily serve to guide the process of neocortical stabilization, its role can be bypassed if another (exogenous) method is used to activate the appropriate subset of neocortical neurons. Whereas the method of bypassing the hippocampus is artificial, this is an important proof-of-principle study establishing that memory traces in the cortex are sufficient for memory expression.

## CONCLUSIONS

Evidence for memory consolidation has accumulated in the laboratory and the clinic for more than 100 years. Yet, quantitative studies of retrograde amnesia began only in the 1970s, and the idea that consolidation involves a dialogue between the hippocampus and neocortex is even more recent. Information is stored initially in both the hippocampus and neocortex, and the hippocampus then guides a gradual process of reorganization and stabilization whereby information in the neocortex eventually becomes independent of the hippocampus. This is the so-called standard model of memory consolidation depicted in Figure 3A. It is now known that the rate of this process depends on the extent to which new information can be related to preexisting knowledge, such as networks of connected neurons called "schemas" (Fig. 3B). Consolidation occurs for both facts and events and for both spatial and nonspatial information, although it appears to be masked in tasks that depend prominently on spatial navigation (e.g., the water maze in rodents), as the integrity of the hippocampal formation is required for memory expression. Molecular genetic tools, including optogenetics and metabolic markers, are now being used to explore further the mechanisms by which consolidation occurs.

## ACKNOWLEDGMENTS

This work is supported by the Medical Research Service of the Department of Veterans Affairs (L.R.S.), The National Institute of Mental Health (NIMH) Grant 24600 (L.R.S.), an Advanced Investigator Grant (Project 268800-NEUROSCHEMA) from the European Research Council (R.G.M.), and a postdoctoral fellowship from Society-in-Science (L.G.).

## REFERENCES

Alberini CM. 2005. Mechanisms of memory stabilization: Are consolidation and reconsolidation similar or distinct processes? *Trends Neurosci* **28:** 51–56.

Alvarez P, Squire LR. 1994. Memory consolidation and the medial temporal lobe: A simple network model. *Proc Natl Acad Sci* **91:** 7041–7045.

Anagnostaras SG, Maren S, Fanselow MS. 1999. Temporally graded retrograde amnesia of contextual fear after hippocampal damage in rats: Within-subjects examination. *J Neurosci* **19:** 1106–1114.

Andrillon T, Nir Y, Staba RJ, Ferrarelli F, Cirelli C, Tononi G, Fried I. 2011. Sleep spindles in humans: Insights from intracranial EEG and unit recordings. *J Neurosci* **31:** 17821–17834.

Aton SJ, Seibt J, Dumoulin M, Jha SK, Steinmetz N, Coleman T, Naidoo N, Frank MG. 2009. Mechanisms of sleep-dependent consolidation of cortical plasticity. *Neuron* **61:** 454–466.

Battaglia FP, Borensztajn, Bod GR. 2012. Structured cognition and neural systems: From rats to language. *Neurosci Biobehav Rev* **36:** 1626–1639.

Bayley PJ, Hopkins RO, Squire LR. 2003. Successful recollection of remote autobiographical memories by amnesic patients with medial temporal lobe lesions. *Neuron* **37:** 135–144.

Bayley PJ, Gold JJ, Hopkins RO, Squire LR. 2005. The neuroanatomy of remote memory. *Neuron* **46:** 799–810.

Bayley PJ, Hopkins RO, Squire LR. 2006. The fate of old memories after medial temporal lobe damage. *J Neurosci* **26:** 13311–13317.

Bendor D, Wilson MA. 2012. Biasing the content of hippocampal replay during sleep. *Nat Neurosci* **15:** 1439–1444.

Bernard FA, Bullmore ET, Graham KS, Thompson SA, Hodges JR, Fletcher PC. 2004. The hippocampal region is involved in successful recognition of both remote and recent famous faces. *Neuroimage* **22:** 1704–1714.

Bolhuis J, Stewart CA, Forrest EM. 1994. Retrograde amnesia and memory reactivation in rats with ibotenate lesions to the hippocampus or subiculum. *Q J Exp Psychol* **47:** 129–150.

Bonnici HM, Chadwick MJ, Lutti A, Hassabis D, Weiskopf N, Maguire EA. 2012. Detecting representations of recent and remote autobiographical memories in vmPFC and hippocampus. *J Neurosci* **32:** 16982–16991.

Bonnici HM, Chadwick MJ, Maguire EA. 2013. Representations of recent and remote autobiographical memories in hippocampal subfields. *Hippocampus* **23:** 849–854.

Bontempi B, Laurent-Demir C, Destrade C, Jaffard R. 1999. Time-dependent reorganization of brain circuitry underlying long-term memory storage. *Nature* **400:** 671–675.

Bransford JD. 1979. *Human cognition: Learning, understanding and remembering.* Wadsworth, Belmont, CA.

Bright P, Buckman JR, Fradera A, Yoshimasu H, Colchester ACF, Kopelman MD. 2006. Retrograde amnesia in patients with hippocampal, medial temporal, temporal lobe, or frontal pathology. *Learn Mem* **13:** 545–557.

Broadbent NJ, Clark RE. 2013. Remote context fear conditioning remains hippocampus-dependent irrespective of training protocol, training-surgery interval, lesion size, and lesion method. *Neurobiol Learn Mem* **106:** 300–308.

Broadbent NJ, Squire LR, Clark RE. 2006. Reversible hippocampal lesions disrupt water maze performance during both recent and remote memory tests. *Learn Mem* **13:** 187–191.

Broadbent NJ, Squire LR, Clark RE. 2010. Sustained dorsal hippocampal activity is not obligatory for either the maintenance or retrieval of long-term spatial memory. *Hippocampus* **20:** 1366–1375.

Buchanan TW, Tranel D, Adolphs R. 2005. Emotional autobiographical memories in amnesic patients with medial temporal lobe damage. *J Neurosci* **25:** 3151–3160.

Burnham WH. 1903. Retroactive amnesia: Illustrative cases and a tentative explanation. *Am J Psychol* **14:** 382–396.

Cabeza R, St Jacques P. 2007. Functional neuroimaging of autobiographical memory. *Trends Cogn Sci* **11:** 219–227.

Cermak LS, O'Connor M. 1983. The anterograde and retrograde amnesia in a patient with herpes encephalitis. *Neuropsychologia* **39:** 213–224.

Clark RE, Broadbent NJ, Squire LR. 2005a. Hippocampus and remote spatial memory in rats. *Hippocampus* **15:** 260–272.

Clark RE, Broadbent NJ, Squire LR. 2005b. Impaired remote spatial memory after hippocampal lesions despite extensive training beginning early in life. *Hippocampus* **15:** 340–346.

Clark RE, Broadbent, Squire LR. 2007. The hippocampus and spatial memory: Findings with a novel modification of the water maze. *J Neurosci* **27:** 6647–6654.

Cowensage KK, Shuman T, Dillingham BC, Chang A, Golshani P, Mayford M. 2014. Direct reactivation of a coherent neocortical memory of context. *Neuron* **84:** 432–441.

Davidson TJ, Kloosterman F, Wilson MA. 2009. Hippocampal replay of extended experience. *Neuron* **63:** 497–507.

de Hoz L, Martin S, Morris RG. 2004. Forgetting, reminding, and remembering: The retrieval of lost spatial memory. *PLoS Biol* **2:** E225.

Diekelmann S, Born J. 2010. The memory function of sleep. *Nat Rev Neurosci* **11:** 114–126.

Douville K, Woodard JL, Seidenberg M, Miller SK, Leveroni CL, Nielson KA, Franczak M, Antuono P, Rao SM. 2005. Medial temporal lobe activity for recognition of recent and remote famous names: An event related fMRI study. *Neuropsychologia* **43:** 693–703.

Dudai Y. 2012. The restless engram: Consolidations never end. *Ann Rev Neurosci* **35:** 227–247.

Dudai Y, Morris RGM. 2000. To consolidate or not to consolidate: What are the questions? In *Brain, perception, memory advances in cognitive sciences* (ed. Bulhuis JJ), pp. 149–162. Oxford University Press, Oxford.

Dudai Y, Morris RGM. 2013. Memorable trends. *Neuron* **80:** 742–750.

Dupret D, O'Neill J, Pleydell-Bouverie B, Csicsvari J. 2010. The reorganization and reactivation of hippocampal maps predict spatial memory performance. *Nat Neurosci* **13:** 995–1002.

Ego-Stengel V, Wilson MA. 2010. Disruption of ripple-associated hippocampal activity during rest impairs spatial learning in the rat. *Hippocampus* **20:** 1–10.

Eslinger PJ. 1998. Autobiographical memory after temporal lobe lesions. *Neurocase* **4**: 481–495.

Euston DR, Tatsuno M, McNaughton BL. 2007. Fast-forward playback of recent memory sequences in prefrontal cortex during sleep. *Science* **318**: 1147–1150.

Frankland PW, Bontempi B. 2005. The organization of recent and remote memories. *Nat Rev Neurosci* **6**: 119–130.

Frankland PW, Bontempi B, Talton LE, Kaczmarek L, Silva AJ. 2004. The involvement of the anterior cingulate cortex in remote contextual fear memory. *Science* **304**: 881–883.

Furman O, Mendelsohn A, Dudai Y. 2012. The episodic engram transformed: Time reduces retrieval-related brain activity but correlates it with memory accuracy. *Learn Mem* **19**: 575–587.

Genzel L, Kroes MC, Dresler M, Battaglia FP. 2014. Light sleep versus slow wave sleep in memory consolidation: A question of global versus local processes? *Trends Neurosci* **37**: 10–19.

Ghosh VE, Gilboa A. 2014. What is a memory schema? A historical perspective on current neuroscience literature. *Neuropsychologia* **53**: 104–114.

Gilboa A, Winocur G, Grady CL, Hevenor SJ, Moscovitch M. 2004. Remembering our past: Functional neuroanatomy of recollection of recent and very remote personal events. *Cereb Cortex* **14**: 1214–1225.

Girardeau GA, Benchenane A, Wiener SI, Buzsaki G, Zugaro MB. 2009. Selective suppression of hippocampal ripples impairs spatial memory. *Nat Neurosci* **12**: 1222–1223.

Girardeau G, Cei A, Zugaro M. 2014. Learning-induced plasticity regulates hippocampal sharp wave-ripple drive. *J Neurosci* **34**: 5176–5183.

Goshen I, Brodsky M, Prakash R, Wallace J, Gradinaru V, Ramakrishnan C, Deisseroth K. 2011. Dynamics of retrieval strategies for remote memories. *Cell* **147**: 678–689.

Haist F, Bowden Gore J, Mao H. 2001. Consolidation of human memory over decades revealed by functional magnetic resonance imaging. *Nat Neurosci* **4**: 1139–1145.

Harand C, Bertran F, La Joie R, Landeau B, Mézenge F, Desgranges B, Peigneux P. 2012. The hippocampus remains activated over the long term for the retrieval of truly episodic memories. *PLoS ONE* **7**: e43495.

Jadhav SP, Kemere C, German PW, Frank LM. 2012. Awake hippocampal sharp-wave ripples support spatial memory. *Science* **336**: 1454–1458.

Jenkins JG, Dallenbach KM. 1924. Oblivescence during sleep and waking. *Am J Psychol* **35**: 605–612.

Ji D, Wilson MA. 2007. Coordinated memory replay in the visual cortex and hippocampus during sleep. *Nat Neurosci* **10**: 100–107.

Kapur N, Brooks DJ. 1999. Temporally specific retrograde amnesia in two cases of discrete bilateral hippocampal pathology. *Hippocampus* **9**: 247–254.

Karlsson MP, Frank LM. 2009. Awake replay of remote experiences in the hippocampus. *Nat Neurosci* **12**: 913–918.

Kim JJ, Fanselow MS. 1992. Modality-specific retrograde amnesia of fear. *Science* **256**: 675–677.

Kirwan CB, Bayley PJ, Galvan VV, Squire LR. 2008. Detailed recollection of remote autobiographical memory after damage to the medial temporal lobe. *Proc Natl Acad Sci* **105**: 2676–2680.

Kopelman MD, Bright P. 2012. On remembering and forgetting our autobiographical pasts: Retrograde amnesia and Andrew Mayes's contribution to neuropsychological method. *Neuropsychologia* **50**: 2961–2972.

Kopelman MD, Wilson BA, Baddeley AD. 1989. The autobiographical memory interview: A new assessment of autobiographical and personal semantic memory in amnesic patients. *J Clin Exp Neuropsy* **5**: 724–744.

Lansink CS, Goltstein PM, Lankelma JV, McNaughton BL, Pennartz CM. 2009. Hippocampus leads ventral striatum in replay of place-reward information. *PLoS Biol* **7**: e1000173.

Lechner HA, Squire LR, Byrne JH. 1999. 100 years of consolidation—Remembering Müller and Pilzecker. *Learn Mem* **2**: 77–87.

Lehmann H, Lacanilao S, Sutherland RJ. 2007. Complete or partial hippocampal damage produces equivalent retrograde amnesia for remote contextual fear memories. *Eur J Neurosci* **25**: 1278–1286.

Lesburgueres E, Gobbo O, Alaux-Cantin S, Hambucken A, Trifilieff P, Bontempi B. 2011. Early tagging of cortical networks is required for the formation of enduring associative memory. *Science* **331**: 924–928.

Lewis PA, Durrant SJ. 2011. Overlapping memory replay during sleep builds cognitive schemata. *Trends Cogn Sci* **15**: 343–351.

Liang KC, Hon W, Tyan YM, Liao WL. 1994. Involvement of hippocampal NMDA and AMPA receptors in acquisition, formation and retrieval of spatial memory in the Morris water maze. *Chin J Physiol* **37**: 201–212.

Maguire EA, Frith CD. 2003. Lateral asymmetry in the hippocampal response to the remoteness of autobiographical memories. *J Neurosci* **23**: 5302–5307.

Maguire EA, Henson RNA, Mummery CJ, Frith CD. 2001. Activity in prefrontal cortex, not hippocampus, varies parametrically with the increasing remoteness of memories. *Neuroreport* **12**: 441–444.

Manns JR, Hopkins RO, Squire LR. 2003. Semantic memory and the human hippocampus. *Neuron* **37**: 127–133.

Marr D. 1971. Simple memory: A theory for archicortex. *Phil Trans R Soc Lond B Biol Sci* **262**: 23–81.

Marslen-Wilson WD, Teuber HL. 1975. Memory for remote events in anterograde amnesia: Recognition of public figures from news photographs. *Neuropsychologia* **13**: 353–364.

Martin SJ, de Hoz L, Morris RG. 2005. Retrograde amnesia: Neither partial nor complete hippocampal lesions in rats result in preferential sparing of remote spatial memory, even after reminding. *Neuropsychologia* **43**: 609–624.

Maviel T, Durkin TP, Menzaghi F, Bontempi B. 2004. Sites of neocortical reorganization critical for remote spatial memory. *Science* **305**: 96–99.

McClelland JL. 2013. Incorporating rapid neocortical learning of new schema-consistent information into complementary learning systems theory. *J Exp Psychol Gen* **142**: 1190–1210.

McClelland JL, McNaughton BL, O'Reilly RC. 1995. Why there are complementary learning systems in the hippocampus and neocortex: Insights from the successes and failures of connectionist models of learning and memory. *Psychol Rev* **102:** 419–457.

Mednick SC, Cai DJ, Shuman T, Anagnostaras S, Wixted JT. 2011. An opportunistic theory of cellular and systems consolidation. *Trends in Neurosci* **34:** 504–514.

Moscovitch M, Nadel L, Winocur G, Gilboa A, Rosenbaum RS. 2006. The cognitive neuroscience of remote episodic, semantic and spatial memory. *Curr Opin Neurobiol* **16:** 179–190.

Moser EI, Kropff E, Moser MB. 2008. Place cells, grid cells, and the brain's spatial representation system. *Annu Rev Neurosci* **31:** 69–89.

Müller GE, Pilzecker A. 1900. Experimentelle Beiträge zur Lehre vom Gedächtnis. [Experimental contributions to the science of memory]. *Z Psychol Ergänzungsband* **1:** 1–300.

Nadel L, Moscovitch M. 1997. Memory consolidation, retrograde amnesia and the hippocampal complex. *Curr Opin Neurobiol* **7:** 217–227.

Nader K, Schafe GE, Le Doux JE. 2000. Fear memories require protein synthesis in the amygdala for reconsolidation after retrieval. *Nature* **406:** 722–726.

Niki K, Luo J. 2002. An fMRI study on the time-limited role of the medial temporal lobe and long-term topographical autobiographic memory. *J Cog Neurosci* **14:** 500–507.

Nir Y, Staba RJ, Andrillon T, Vyazovskiy VV, Cirelli C, Fried I, Tononi G. 2011. Regional slow waves and spindles in human sleep. *Neuron* **70:** 153–169.

O'Keefe J, Nadel L. 1978. *The hippocampus as a cognitive map.* Clarendon, Oxford.

Olton DS, Becker JT, Handelmann GE. 1979. Hippocampus, space, and memory. *Brain Behav Sci* **2:** 313–365.

Peigneux P, Laureys S, Fuchs S, Collette F, Perrin F, Reggers J, Phillips C, Degueldre C, Del Fiore G, Aerts J, et al. 2004. Are spatial memories strengthened in the human hippocampus during slow wave sleep? *Neuron* **44:** 535–545.

Peyrache A, Khamassi M, Benchenane K, Wiener SI, Battaglia FP. 2009. Replay of rule-learning related neural patterns in the prefrontal cortex during sleep. *Nat Neurosci* **12:** 919–926.

Peyrache A, Battaglia FP, Destexhe A. 2011. Inhibition recruitment in prefrontal cortex during sleep spindles and gating of hippocampal inputs. *Proc Natl Acad Sci* **108:** 17207–17212.

Pfeiffer BE, Foster DJ. 2013. Hippocampal place-cell sequences depict future paths to remembered goals. *Nature* **497:** 74–79.

Piefke M, Weiss P, Ailes K, Markowitsch H, Fink G. 2003. Differential remoteness and emotional tone modulate the neural correlates of autobiographical memory. *Brain* **126:** 650–668.

Race E, Keane MM, Verfaellie M. 2011. Medial temporal lobe damage causes deficits in episodic memory and episodic future thinking not attributable to deficits in narrative construction. *J Neurosci* **31:** 10262–10269.

Rasch B, Buchel C, Gais S, Born J. 2007. Odor cues during slow-wave sleep prompt declarative memory consolidation. *Science* **315:** 1426–1429.

Reichinnek S, Kunsting T, Draguhn A, Both M. 2010. Field potential signature of distinct multicellular activity patterns in the mouse hippocampus. *J Neurosci* **30:** 15441–15449.

Riedel G, Micheau J, Lam AM, Roloff EL, Martin SJ, Bridge H, de Hoz L, Poeschel B, McCulloch J, Morris RG. 1999. Reversible neural inactivation reveals hippocampal participation in several memory processes. *Nat Neurosci* **2:** 898–905.

Rekkas PV, Constable RT. 2005. Evidence that autobiographic memory retrieval does not become independent of the hippocampus: An fMRI study contrasting very recent with remote events. *J Cogn Neurosci* **17:** 1950–1961.

Remondes M, Schuman EM. 2004. Role for a cortical input to hippocampal area CA1 in the consolidation of a long-term memory. *Nature* **431:** 699–703.

Ribot T. 1881. *Les maladies de la memoire* [*Diseases of memory*]. Appleton-Century-Crofts, New York.

Riccio DC, Richardson RD, Ebner L. 1984. Memory retrieval deficits based upon altered contextual cues: A paradox. *Psychol Bull* **96:** 152–165.

Richards BA, Xia F, Santoro A, Husse J, Woodin MA, Josselyn SA, Frankland PW. 2014. Patterns across multiple memories are identified over time. *Nat Neurosci* **17:** 981–986.

Rosenbaum RS, Kohler S, Schacter DL, Moscovitch M, Westmacott R, Black SE, Gao F, Tulving E. 2005. The case of K.C.: Contributions of a memory-impaired person to memory theory. *Neuropsychologia* **43:** 989–1021.

Rudoy JD, Voss JL, Westerberg CE, Paller KA. 2009. Strengthening individual memories by reactivating them during sleep. *Science* **326:** 1079.

Sara SJ. 2000. Retrieval and reconsolidation: Toward a neurobiology of remembering. *Learn Mem* **7:** 73–84.

Schacter DL, Dodson CS. 2001. Misattribution, false recognition and the sins of memory. *Philos Trans R Soc London B Biol Sci* **356:** 1385–1393.

Smith C. 1996. Sleep states, memory processes and synaptic plasticity. *Behav Brain Res* **78:** 49–56.

Smith CN, Squire LR. 2009. Medial temporal lobe activity during retrieval of semantic memory is related to the age of the memory. *J Neurosci* **29:** 930–938.

Söderlund H, Moscovitch M, Kumar N, Mandic M, Levine B. 2012. As time goes by: Hippocampal connectivity changes with remoteness of autobiographical memory retrieval. *Hippocampus* **22:** 670–679.

Sparks FT, Lehmann H, Hernandez K, Sutherland RJ. 2011. Suppression of neurotoxic lesion-induced seizure activity: Evidence for a permanent role for the hippocampus in contextual memory. *PLoS ONE* **6:** e27426.

Spooner R, Thomson A, Hall J, Morris RG, Salter SH. 1994. The Atlantis platform: A new design and further developments of Buresova's on-demand platform for the water maze. *Learn Mem* **1:** 203–211.

Squire LR. 1992. Memory and the hippocampus: A synthesis from findings with rats, monkeys, and humans. *Psychol Rev* **99:** 195–231.

Squire LR, Alvarez P. 1995. Retrograde amnesia and memory consolidation: A neurobiological perspective. *Curr Opin Neurobiol* **5:** 169–177.

Squire LR, Bayley PJ. 2007. The neuroscience of remote memory. *Curr Opin Neurobiol* **17:** 185–196.

Squire LR, Slater PC, Chace PM. 1975. Retrograde amnesia: Temporal gradient in very long-term memory following electroconvulsive therapy. *Science* **187**: 77–79.

Squire LR, Clark RE, Knowlton BJ. 2001. Retrograde amnesia. *Hippocampus* **11**: 50–55.

Staresina BP, Alin A, Kriegeskorte N, Henson RN. 2013. Awake reactivation predicts memory in humans. *Proc Natl Acad Sci* **110**: 21159–21164.

Stark CEL, Okado Y. 2003. Making memories without trying: Medial temporal lobe activity associated with incidental memory formation during recognition. *J Neurosci* **23**: 6748–6753.

Steinvorth S, Levine B, Corkin S. 2005. Medial temporal lobe structures are needed to re-experience remote autobiographical memories: Evidence from H.M. and W.R. *Neuropsychologia* **43**: 479–496.

Sutherland RJ, Weisend MP, Mumby D, Astur RS, Hanlon FM, Koerner A, Thomas MJ, Wu Y, Moses SN, Cole C, Hamilton DA, Hoesing JM. 2001. Retrograde amnesia after hippocampal damage: Recent vs. remote memories in two tasks. *Hippocampus* **11**: 27–42.

Sutherland RJ, O'Brien J, Lehmann H. 2008. Absence of systems consolidation of fear memories after dorsal, ventral, or complete hippocampal damage. *Hippocampus* **18**: 710–718.

Svoboda E, McKinnon MC, Levine B. 2006. The functional neuroanatomy of autobiographical memory: A meta-analysis. *Neuropsychologia* **44**: 2189–2208.

Takashima A, Petersson KM, Rutters F, Tendolkar I, Jensen O, Zwarts MJ, McNaughton BL, Fernández G. 2006. Declarative memory consolidation in humans: A prospective functional magnetic resonance imaging study. *Proc Natl Acad Sci* **103**: 756–761.

Takashima A, Nieuwenhuis ILC, Jensen O, Talamini LM, Rijpkema M, Fernández G. 2009. Shift from hippocampal to neocortical centered retrieval network with consolidation. *J Neurosci* **29**: 10087–10093.

Takehara-Nishiuchi K, McNaughton BL. 2008. Spontaneous changes of neocortical code for associative memory during consolidation. *Science* **322**: 960–963.

Tambini A, Davachi L. 2013. Persistence of hippocampal multivoxel patterns into postencoding rest is related to memory. *Proc Natl Acad Sci* **110**: 19591–19596.

Tambini A, Ketz N, Davachi L. 2010. Enhanced brain correlations during rest are related to memory for recent experiences. *Neuron* **65**: 280–290.

Teng E, Squire LR. 1999. Memory for places learned long ago is intact after hippocampal damage. *Nature* **400**: 675–677.

Thaiss L, Petrides M. 2008. Autobiographical memory of the recent past following frontal cortex or temporal lobe excisions. *Eur J Neurosci* **28**: 829–840.

Tononi G, Cirelli C. 2006. Sleep function and synaptic homeostasis. *Sleep Med Rev* **10**: 49–62.

Tononi G, Cirelli C. 2014. Sleep and the price of plasticity: From synaptic and cellular homeostasis to memory consolidation and integration. *Neuron* **81**: 12–34.

Tse D, Langston RF, Kakeyama M, Bethus I, Spooner PA, Wood ER, Witter MP, Morris RG. 2007. Schemas and memory consolidation. *Science* **316**: 76–82.

Tse D, Takeuchi T, Kakeyama T, Kajii Y, Okuno H, Tohyama C, Bito H, Morris RG. 2011. Schema-dependent gene activation and memory encoding in neocortex. *Science* **333**: 891–895.

Tulving E. 1985. Memory and consciousness. *Can Psychol* **26**: 1–12.

Tulving E, Schacter DL, McLachlan DR, Moscovitch M. 1988. Priming of semantic autobiographical knowledge: A case study of retrograde amnesia. *Brain Cogn* **8**: 3–20.

van Kesteren MT, Ruiter DJ, Fernandez G, Henson RN. 2012. How schema and novelty augment memory formation. *Trends Neurosci* **35**: 211–219.

Viard A, Lebreton K, Chételat G, Desgranges B, Landeau B, Young A, De La Saette V, Eustache F, Piolino P. 2010. Patterns of hippocampal–neocortical interactions in the retrieval of episodic autobiographical memories across the entire life-span of aged adults. *Hippocampus* **20**: 153–165.

Wheeler AL, Teixeira CM, Wang AH, Xiong X, Kovacevic N, Lerch JP, McIntosh AR, Parkinson J, Frankland PW. 2013. Identification of a functional connectome for long-term fear memory in mice. *PLoS Comput Biol* **9**: e1002853.

Wierzynski CM, Lubenov EV, Gu M, Siapas AG. 2009. State-dependent spike-timing relationships between hippocampal and prefrontal circuits during sleep. *Neuron* **61**: 587–596.

Wilson MA, McNaughton BL. 1994. Reactivation of hippocampal ensemble memories during sleep. *Science* **265**: 676–679.

Wiltgen BJ, Silva AJ. 2007. Memory for context becomes less specific with time. *Learn Mem* **14**: 313–317.

Wiltgen BJ, Zhou M, Cai Y, Balaji J, Karlsson MG, Parivash SN, Li W, Silva AJ. 2010. The hippocampus plays a selective role in the retrieval of detailed contextual memories. *Curr Biol* **20**: 1336–1344.

Winocur G. 1990. Anterograde and retrograde amnesia in rats with dorsal hippocampal or dorsomedial thalamic lesions. *Behav Brain Res* **38**: 145–154.

Winocur G, Moscovitch M. 2011. Memory transformation and systems consolidation. *J Int Neuropsychol Soc* **17**: 766–780.

Winocur G, Sekeres MJ, Binns MA, Moscovitch M. 2013. Hippocampal lesions produce both nongraded and temporally graded retrograde amnesia in the same rat. *Hippocampus* **23**: 330–341.

Yamashita K, Hirose S, Kunimatsu A, Aoki S, Chikazoe J, Jimura K, Masutani Y, Abe O, Ohtomo K, Miyashita Y, et al. 2009. Formation of long-term memory representation in human temporal cortex related to pictorial paired associates. *J Neurosci* **29**: 10335–10340.

Yang G, Lai C, Cichon J, Ma L, Li W, Gan W. 2014. Sleep promotes branch-specific formation of dendritic spines after learning. *Science* **344**: 1173–1178.

Yiu AP, Mercaldo V, Yan C, Richards B, Rashid AJ, Hsiang HL, Pressey J, Mahadevan V, Tran MM, Kushner SA, et al. 2014. Neurons are recruited to a memory trace based on relative neuronal excitability immediately before training. *Neuron* **83**: 722–735.

Zola-Morgan S, Squire LR. 1990. The primate hippocampal formation: Evidence for a time-limited role in memory storage. *Science* **250**: 288–290.

# The Role of Functional Prion-Like Proteins in the Persistence of Memory

Kausik Si[1,2] and Eric R. Kandel[3,4,5,6]

[1]Stowers Institute for Medical Research, Kansas City, Missouri 64113

[2]Department of Physiology, School of Medicine, University of Kansas Medical Center, Kansas City, Kansas 66160

[3]Howard Hughes Medical Institute, Chevy Chase, Maryland 20815-6789

[4]Departments of Neuroscience and Psychiatry, College of Physicians and Surgeons of Columbia University, New York, New York 10027

[5]Zuckerman Mind Brain Behavior Institute, New York State Psychiatric Institute, New York, New York 10032

[6]Kavli Institute for Brain Sciences, New York, New York 10032

*Correspondence:* ksi@stowers.org

Prions are a self-templating amyloidogenic state of normal cellular proteins, such as prion protein (PrP). They have been identified as the pathogenic agents, contributing to a number of diseases of the nervous system. However, the discovery that the neuronal RNA-binding protein, cytoplasmic polyadenylation element-binding protein (CPEB), has a prion-like state that is involved in the stabilization of memory raised the possibility that prion-like proteins can serve normal physiological functions in the nervous system. Here, we review recent experimental evidence of prion-like properties of neuronal CPEB in various organisms and propose a model of how the prion-like state may stabilize memory.

Prions are proteinaceous infectious agents that were discovered in the 1980s by Stanley Prusiner while studying Creutzfeldt–Jakob disease (Prusiner 1982). Prusiner and colleagues showed them to be an amyloidogenic, self-perpetuating, forms of a normal cellular protein, termed prion protein or PrP. Prp in its self-perpetuating state kills cells. Prusiner and colleagues found that PrPs exist in at least two conformations: monomeric and aggregated (Fig. 1). The transition among these forms occurs spontaneously and only the aggregated conformation is pathogenic. Soon, PrPs were found to contribute to other neurodegenerative disorders in people, including kuru, transmissible spongiform encephalopathies, as well as bovine spongiform encephalopathy in cows (Prusiner 1994; Aguzzi and Weissmann 1998).

There is now a growing consensus that similar prion-like, self-templating mechanisms underlie a variety of neurodegenerative disorders, including amyotrophic lateral sclerosis, Alzheimer's disease, Parkinson's disease, and Huntington's disease (Polymenidou and Cleveland 2012).

Not all prions, however, appear to be disease causing. Fungal prions, for instance, are nontoxic, and some may even be beneficial to the

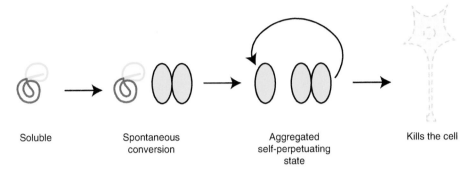

**Figure 1.** Pathogenic prions exist in two states (soluble and aggregated and self-perpetuating). The conversion from the soluble to the aggregated form is spontaneous and the aggregated, self-perpetuating form is often toxic and kills the cell.

cells that harbor them (Wickner 1994; Shorter and Lindquist 2005; Crow and Li 2011). In 2003, Si and Kandel serendipitously discovered a prion-like protein in multicellular eukaryotes—the nervous system of the marine snail *Aplysia*—whose aggregated and self-perpetuating form contributes to the maintenance of long-term changes in synaptic efficacy. This functional prion-like protein differs from pathogenic prions in two important ways: (1) The conversion to the prion-like state is regulated by a physiological signal, and (2) the aggregated form has an identified physiological function (Fig. 2). Recent identification of new functional prion-like proteins in various organisms, including human, supports the idea that non-

pathogenic prions may perform a wide range of biologically meaningful roles (Coustou et al. 1997; Eaglestone et al. 1999; True and Lindquist 2000; Ishimaru et al. 2003; True et al. 2004; Hou et al. 2011; Jarosz et al. 2014).

In this review, we focus on functional prion-like proteins in the brain and specifically on the prion-like properties of the cytoplasmic polyadenylation element-binding protein (CPEB), and examine how the prion-like state can control protein synthesis at the synapse and, thereby, synaptic plasticity and long-lasting memory. We anticipate the studies of CPEB would also provide some generalizable concepts as to how prion-based protein switches in multicellular eukaryotes may work.

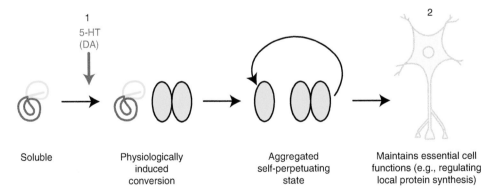

**Figure 2.** "Functional" prion: memory. "Functional" prions differ from conventional prions in two ways. First, the conversion is triggered by a physiological signal, and second, the aggregated, self-perpetrating forms have a physiological function. 5-HT, Serotonin; DA, dopamine.

 Cite this article as *Cold Spring Harb Perspect Biol* doi: 10.1101/cshperspect.a021774

## A METASTABLE PROTEIN CONFORMATION SWITCH FOR SYNAPSE-SPECIFIC LOCAL PROTEIN SYNTHESIS

### Synapses Are Independent Units of Change

A single neuron can have up to $10^3$ synapses. These are the units of information storage for short-term memory. Because long-term memory storage requires gene expression, which takes place in the nucleus, one might expect long-term memory to involve cell-wide changes. To explore whether individual synapses can also serve as units for long-term memory lasting days and weeks, Martin and colleagues (Martin et al. 1997a; Casadio et al. 1999) performed experiments in which serotonin (5-HT), the modulatory transmitter released during training for learned fear in *Aplysia*, was applied locally to one of the two branches of the bifurcating sensory neurons that innervate two separate motor neurons. These experiments, as well as parallel experiments by Frey and Morris in the rodent hippocampus (Frey and Morris 1997), show that individual synapses can be independently modified to give rise to long-term synaptic plasticity, which persists for $>24$ h (Fig. 3). Thus, long-term facilitation (LTF) and its associated synaptic changes are synapse-specific. Moreover, this synapse specificity requires the action of the transcription factor cyclic adenosine-3-monophosphate (cAMP)-response element-binding protein 1 (CREB-1). Blocking CREB-1 expression blocks the synapse-specific facilitation. These findings imply that signals are sent not

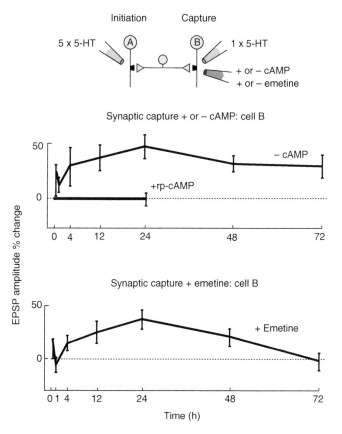

Figure 3. There are two molecular components to the "synaptic" mark (Martin et al. 1997a,b). (*Top*) Activation of the cyclic adenosine-3-monophosphate (cAMP)-dependent protein kinase is required to have any capture at all. (*Middle, bottom*) Local protein synthesis is necessary for the maintenance of long-term facilitation (LTF) beyond 24 h. (From Casadio et al. 1999; with permission, from Elsevier © 1999.) 5-HT, Serotonin; EPSP, excitatory postsynaptic potential.

only from the synapse back to the nucleus (Martin et al. 1997a; Lee et al. 2007), but also from the nucleus to specific synapses.

## The "Synaptic Capture" Hypothesis

How does CREB-dependent transcription in the cell body, which is accessible to all synapses, confer synapse-specific LTF? To explain how this specificity can be achieved in a biologically economical way, despite the massive number of synapses made by a single neuron, Martin et al. (1997a) and Frey and Morris (1997) proposed the "synaptic capture" hypothesis. This hypothesis, also referred as "synaptic tagging,"

proposed two possibilities that are not mutually exclusive. First, the products of gene expression are delivered to all of the synapses throughout the cell, but are functionally incorporated in only those synapses that have been molecularly altered or "tagged" by previous synaptic activity. Second, the products of activated transcription and translation are selectively delivered to the synapse that has been "tagged" by synaptic activity. The model of a "synaptic tag," whereby gene products are delivered to all synapses but are only used at synapses that have been tagged, has been supported by a number of studies both in *Aplysia* (Figs. 3 and 4) (Martin et al. 1997b; Casadio et al. 1999) and in the rodent hippo-

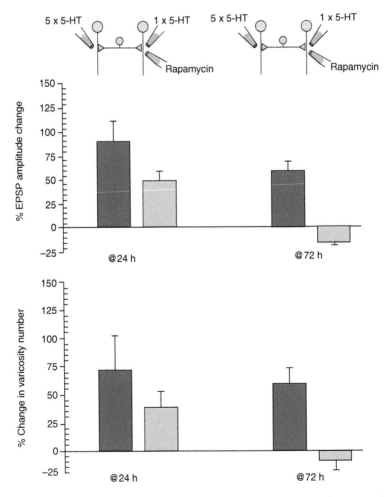

**Figure 4.** Rapamycin blocks the maintenance of growth at 72 h. (From Casadio et al. 1999; with permission, from Elsevier © 1999.) 5-HT, Serotonin; EPSP, excitatory postsynaptic potential.

campus (Frey and Morris 1997, 1998; Barco et al. 2002; Dudek and Fields 2002).

## MOLECULAR MECHANISMS OF SYNAPTIC CAPTURE

This raised the question: How is a synapse tagged? The clues as to how synapses are "tagged" came from the observations that the synapse-specific LTF requires more than the activation of CREB-driven gene transcription in the nucleus (Casadio et al. 1999). Injection of phosphorylated CREB-1 into the cell body of the sensory neuron gives rise to LTF at all the synapses. This facilitation, however, is not maintained beyond 24–48 h, nor is it accompanied by synaptic growth. Maintenance and growth is achieved only if the synapse is also simultaneously activated by a single pulse of 5-HT.

By applying various chemical inhibitors to the synapse, Martin et al. (1997a) found two distinct components of tagging in *Aplysia*. The first component initiates long-term synaptic plasticity and growth, and requires the cAMP-dependent protein kinase A (PKA). The second component stabilizes and maintains long-term functional and structural changes at the synapse and requires, in addition to protein synthesis

in the cell body, local protein synthesis at the synapse (Figs. 4 and 5) (Martin et al. 1997a,b).

## The Control of Local Protein Synthesis at the Marked Synapse

That local protein synthesis is required specifically for the persistent phase of functional and structural change of the synapse, suggested that some critical proteins are either limiting or absent in an unstimulated synapse and are made locally at the tagged synapse. Because messenger RNAs (mRNAs) are made in the cell body, and then shipped to all synapses, it suggested to Si et al. (2003a) that these mRNAs are likely to be dormant before they reach the tagged synapse. If that were true, the tag could be a regulator of translation that is capable of activating dormant mRNAs.

A search for regulators of protein synthesis in neurons that can confer both spatial and temporal restrictions led to the identification of CPEB protein (Si et al. 2003a). CPEB is a family of RNA-binding protein first discovered by Hake and Richter in *Xenopus* oocytes (Hake and Richter 1994). Richter and colleagues have found that oocyte maternal RNA is silent until translationally activated by CPEB (Fig. 5) and

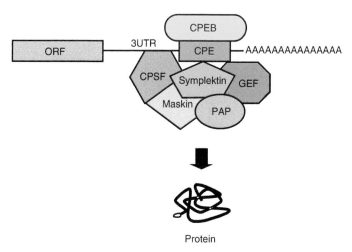

**Figure 5.** The cytoplasmic polyadenylation element binding protein (CPEB) binds to the cytoplasmic polyadenylation element (CPE) and activates dormant messenger RNA (mRNA). ORF, Open reading frame; UTR, untranslated region; CPSF, cleavage/polyadenylation specificity factor; GEF, guanine nucleotide exchange factor; PAP, poly(A) polymerase.

CPEB activity is modulated by external signals, such as hormone progesterone.

In a search for CPEB in *Aplysia*, Si et al. (2003a) found that *Aplysia* neurons had, in addition to the developmental isoform described by Richter, a new isoform of CPEB (ApCPEB) with distinct properties. ApCPEB had three features that made it an attractive candidate for a synapse-specific tag for stabilization: (1) in response to 5-HT, the stimulus that produces the tag, the amount of ApCPEB increases only in the stimulated synapse (Fig. 6); (2) ApCPEB regulates translation of the target mRNA and some of these mRNA targets are involved in cellular growth; and, finally, (3) blocking of ApCPEB at a tagged (active) synapse prevented the maintenance without affecting the initiation of LTF (Si et al. 2003a,b). Intriguingly, synapses in which LTF has already been induced, blocking ApCPEB blocks the persistence of that facilitated state (Miniaci et al. 2008). Thus, the maintenance of LTF over a period of days is a dynamic process requiring continuous presence of CPEB and protein synthesis (Si et al. 2003a).

## A Prion-Like Mechanism of CPEB Regulates Local Protein Synthesis

Because proteins have a relatively short half-life compared with the duration of memory, structural changes at the synaptic level have been postulated to confer stability to the memory, and it was implicitly assumed that the requirement for activity-dependent molecular changes was transient. However, studies by Martin et al. and others have suggested that the maintenance of learning-related structural alterations requires ongoing macromolecular synthesis (Martin et al. 1997a; Kandel 2001). This posed a problem because there is constant turnover of proteins at the synapse. A potential solution to this problem of molecular turnover is either to have proteins with an unusually long half-life or proteins that participate in a self-sustaining biochemical reaction. In 1984, Crick first addressed the possibility of a self-sustained molecular alteration as the basis of long-term memory storage using protein phosphorylation as a candidate mechanism (Crick 1984).

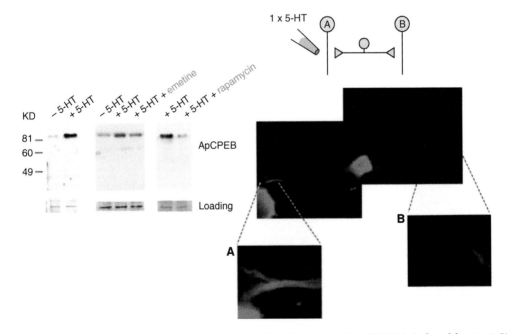

**Figure 6.** *Aplysia* cytoplasmic polyadenylation element-binding protein (ApCPEB) is induced by 1× 5-HT (serotonin) in the process of the sensory neuron. (From Si et al. 2003; reprinted, with permission, from the author.) KD, Knockdown.

One plausible candidate for a self-sustaining biochemical reaction came from analysis of the amino acid sequence of ApCPEBs. In examining the sequence, Si et al. made the surprising observation that the neuronal isoform of ApCPEB has a glutamine and asparagine (Q- and N)-rich amino-terminal domain reminiscent of well-characterized prions in yeast (Fig. 7). Almost 50% of the first 150 amino acids were glutamine or asparagine. A search of the protein sequence database revealed putative homologs of the *Aplysia* neuronal CPEB in *Drosophila*, mouse, and human, with amino-terminal extensions of similar character.

An array of assays developed for studying prions in yeast revealed that ApCPEB could exist in yeast in two distinct conformational states, very much like other yeast PrPs. One of these states is monomeric and incapable of self-perpetuation. The other state is aggregated and stably inherited across generations in yeast, although occasional switches to the monomeric state occur (Si et al. 2003b; Heinrich and Lindquist 2011).

Si et al. (2010) went on to explore the conformational state of ApCPEB in *Aplysia* neu-

rons. They found that when ApCPEB was overexpressed in *Aplysia* sensory neurons, it formed punctate structures that were amyloid-like in nature, a common characteristic of all known prions and prion-like proteins. These punctate structures were caused by self-assembly of the ApCPEB protein and once formed they could recruit newly synthesized protein, a feature necessary for self-sustenance. Intriguingly, application of five pulses of 5-HT, which produces LTF, increased the number of puncta suggesting that the aggregation of ApCPEB could be regulated by modulators of synaptic activity. Importantly, injection of an antibody that selectively binds the aggregated form of ApCPEB did not prevent the initiation of LTF but selectively blocked its maintenance beyond 24 h (Fig. 7).

## microRNA-22 Gates Long-Term Heterosynaptic Plasticity in *Aplysia* through Presynaptic Regulation of CPEB and Downstream Targets

It soon became apparent that the aggregation of ApCPEB is critical for maintenance and can be regulated by 5-HT. This raised the ques-

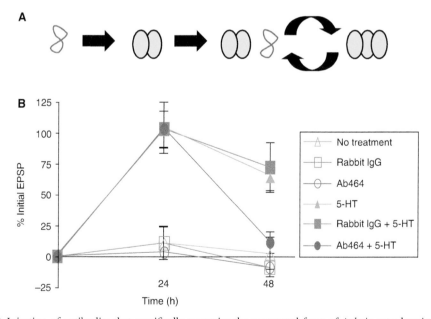

**Figure 7.** Injection of antibodies that specifically recognize the aggregated form of *Aplysia* cytoplasmic polyadenylation element-binding protein (ApCPEB) (Ab464) blocks persistence of long-term facilitation (LTF). 5-HT, Serotonin; EPSP, excitatory postsynaptic potential. (From Si et al. 2010; reprinted, with permission, from the author.)

. Si and E.R. Kandel

tion: How does the heterosynaptic modulator 5-HT—which induces learning-related long-term plasticity of sensorimotor neuron synapses—determine the initial up-regulation, and the consequent conversion of CPEB into its aggregated state in the presynaptic sensory neurons? In search of a mechanism whereby 5-HT up-regulates the levels of ApCPEB in sensory neurons, Fiumara et al. (2015) identified an *Aplysia* microRNA 22 (miR-22) that is expressed in sensory neurons. MiR-22 is down-regulated by 5-HT and inhibition of miR-22 increases CPEB levels and enhances LTF, mimicking the effect of 5-HT-dependent up-regulation of CPEB. Conversely, the presynaptic overexpression of miR-22 impairs LTF, mimicking the physiological effect of CPEB down-regulation.

Fiumara et al. next asked, how does miR-22 effect ApCPEB aggregation? They found, interestingly, that miR-22 overexpression also indirectly down-regulates presynaptic expression of *Aplysia* protein kinase C (aPKC), most likely via its effect on CPEB, because aPKC 3′UTR (untranslated region) contains several cytoplasmic polyadenylation elements (CPEs). Earlier studies in *Drosophila* also identified the *Drosophila* aPKC mRNA as a target for Orb2, the ortholog of *Aplysia* CPEB (Mastushita-Sakai et al. 2010). The aPKC/M has an established function in synaptic plasticity in the postsynaptic compartment, both in vertebrates and in *Aplysia* (Sacktor 2011; Bougie et al. 2012). Using genetic approaches based on the selective presynaptic overexpression of wild-type and dominant negative forms of the kinase, Fiumara et al. found that aPKC also acts presynaptically in the maintenance of LTF. Taken together, these results support a model in which the presynaptic 5-HT-dependent down-regulation of miR-22 promotes the up-regulation and activation of CPEB and downstream targets, including an atypical PKC. As discussed below in the context of *Drosophila* Orb2, phosphorylation may modulate aggregation of CPEB. Therefore, the activation of aPKC may act synergistically with CPEB or directly affect CPEB to maintain learning-related long-term synaptic facilitation.

An activity-dependent prion-like switch controlled by 5-HT acting through miR-22 could, therefore, serve as a mechanism to maintain a self-sustained activated molecular state. According to this model (Fig. 11), ApCPEB in the sensory neuron has at least two conformational states: (1) a recessive monomeric state, in which ApCPEB is inactive or acts as a repressor of translation; and (2) a dominant, self-sustaining, active multimeric state. In a naïve synapse, the basal level of ApCPEB is low and the protein is in the monomeric state. An increase in the amount of ApCPEB induced by inhibition of miR-22 by 5-HT, results in the conversion of ApCPEB from the monomeric to an aggregated prion-like state, which might be more active or be devoid of the inhibitory function of the basal state. Once the prion state is established at an activated synapse, dormant mRNAs, made in the cell body and distributed globally to all synapses, can be activated only locally, through the local activation of ApCPEB. Because the activated ApCPEB can be self-perpetuating, it can contribute to a self-sustaining synapse-specific long-term molecular change and provide a mechanism for the stabilization of learning-related synaptic modification and growth, and contribute to the persistence of memory storage (Fig. 8).

In addition to regulating the level of expression, are there other means for regulating the autocatalytic, self-perpetuating chemical reaction resulting in synaptic enhancement? One possibility is that molecular chaperones might regulate the conformational states of ApCPEB. Indeed, Si et al. (2003b) found that the heat shock protein HSP104 is capable of reversing the active form of ApCPEB in yeast. A second mechanism of regulation is through the restriction of the action of ApCPEB to specific synapses. If ApCPEB in the active state could escape from a newly potentiated synapse, the entire cellular ApCPEB would then be converted to an active state, thereby leading to erroneous potentiation of all synapses. Thus, there must be a mechanism that effectively restricts the activated ApCPEB at the potentiated synapse. One possibility is that the polymerization of ApCPEB creates a diffusion barrier restricting

ite this article as *Cold Spring Harb Perspect Biol* doi: 10.1101/cshperspect.a021774

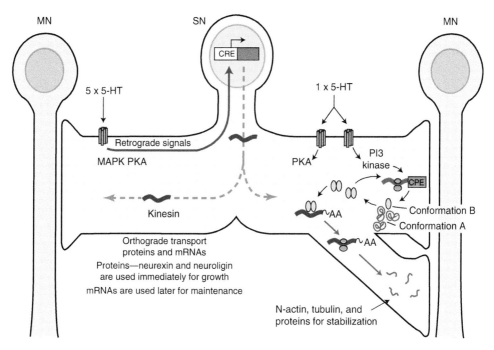

**Figure 8.** A model for the initiation and persistence of long-term memory storage. A single pulse of serotonin (5-HT) to sensory neuron (SN) and motor neuron (MN) synapses recruits the short-term process, and this requires protein kinase A (PKA) activity. PI3 kinase also becomes activated and stimulates cytoplasmic poly-adenylation element binding protein (CPEB)-dependent translation. Five pulses of 5-HT activate cyclic aden-osine-3-monophosphate (cAMP)-response element-binding protein 1 (CREB-1) in the nucleus through the PKA–mitogen-activated protein kinase (PKA–MAPK) pathway and facilitate enhanced, kinesin-mediated fast axonal transport of proteins, mRNAs, and organelles to synapses. These activation steps are critical for the initiation of long-term facilitation (LTF). RNAs transported by kinesin may be used for persistence. During the persistence phase, CPEB at the stimulated synapse activates polyadenylation of cytoplasmic polyadenylation element (CPE)-containing RNAs through a prion-like mechanism for the self-perpetuation of synapse-specific memory storage. (From Puthanveettil and Kandel 2011; reprinted, with permission, from the authors.)

the protein at the activated synapse. Compartmentalization of activated ApCPEB could also be facilitated through interactions of its polymers with other cellular components, including those that are translated on ApCPEB activation. For example, the actin cytoskeleton is involved in the formation, propagation, and intracellular localization of the yeast prion [$PSI^+$] (Ganusova et al. 2006), and oligomeric ApCPEB has been found to activate translation of actin mRNAs (Si et al. 2003a).

## ApCPEB Homolog Also Exists in *Drosophila*

The proposed critical role and mode of action of CPEB in the maintenance of long-term plasticity and hence memory is not restricted to *Aply-*

*sia*. The *Drosophila* CPEB family is comprised of two genes: Orb1 and Orb2. Orb1 is an ortholog of mammalian CPEB1 and Orb2 belongs to the mammalian CPEB2–3 family (Keleman et al. 2007). All family members share common RNA binding motifs in the carboxy-terminal end but differ significantly in the amino-terminal end, suggesting functional differences among isoforms. Similar to the ApCPEB, *Drosophila* CPEB, Orb2, carries a prion-like domain at the amino-terminal end and the Orb2 prion domain can functionally substitute the canonical prion domains of yeast prion Sup35 (R Hervás, L Li, and M Carrión-Vázquez, unpubl.).

Similar to ApCPEB, Orb2 has two distinct physical states in the adult fly brain: a monomeric state and a stable sodium dodecyl sulfate (SDS)-resistant amyloid-like oligomeric state. The oligomers are formed at physiological concentrations of Orb2 protein and stimulation of behaviorally relevant neurons, such as the Kenyon cells of the mushroom body or octopamine or dopamine-responsive neurons, increase the level of the oligomeric Orb2 (Krüttner et al. 2012; Majumdar et al. 2012). Considering the evolutionary distance between the *Aplysia* CPEB and *Drosophila* Orb2, these observations suggested that the amyloidogenic oligomers of Orb2/CPEB may act to stabilize activity-dependent changes in synaptic efficacy across species.

## Role of Prion-Like State of *Drosophila* Orb2 in the Persistence of Memory

Is the prion-like conversion of *Drosophila* Orb2 important for long-lasting memory? Keleman et al. (2007) found that when the Q-rich 80 amino acids from the prion-like domain of *Drosophila* Orb2 were removed, the flies had no problem in forming short-term male courtship-suppression memory; however, the flies failed to form any long-term memory. These observations suggested that the prion-like domain is important for long-term memory. To test whether the prion-like conversion is necessary for long-term memory, Majumdar et al. (2012) created a specific mutant variant of Orb2. The *Drosophila* Orb2 has two protein isoforms: Orb2A and Orb2B. Both isoforms contain an RNA-binding domain, a zinc finger domain, and an amino-terminal prion-like domain, but amino acid sequences upstream of Q-rich sequence are different. The short Orb2A isoform has eight and the longer isoform Orb2B has 162 amino acids upstream of the prion-like domain. These two isoforms also differ in their biophysical properties. Orb2A forms amyloids more efficiently than Orb2B both in vitro and in vivo. Although present at a very low level, deletion of Orb2A prevented Orb2B oligomerization and in the adult brain Orb2A and Orb2B forms hetero-oligomers. These observations suggested that the rare Orb2A protein either acts directly as a seed to induce activity-dependent oligomerization of the abundant

Orb2B protein or Orb2A oligomerization indirectly affects oligomerization of Orb2B. The seeded oligomerization of Orb2 was further supported by Krüttner et al. (2012), who observed that the prion-like domain of Orb2A is sufficient for long-term memory as long as there is full-length Orb2B protein. Majumdar et al. (2012) performed a random mutagenesis screen and identified a number of mutations in the eight amino acids that are unique to the Orb2A isoform that prevented Orb2A oligomerization. One of these point mutations in Orb2A, Orb2AF5>Y5, reduced activity-dependent amyloid-like oligomerization of Orb2 when introduced into the fly. Notably, flies carrying this mutation showed a very specific memory deficit (Fig. 9). In two different behavioral paradigms, male courtship-suppression memory and appetitive associative memory, the memory score of mutant flies were similar to the memory score of wild type a day after training. However, unlike wild type flies, the mutants memory begins to decay at 2 d and by 3 d there was no measureable memory. The loss of memory in the Orb2 mutant that cannot form Orb2 amyloid is reminiscent of the loss of synaptic facilitation in sensorimotor neuron synapse on inhibition of the ApCPEB oligomers via antibody.

The seeded prion-like conversion also suggested that the amount and localization of Orb2A protein is likely to be a key determinant of when and where amyloid-like conversion would occur. Therefore, a key question in the regulation of Orb2 prion-like conversion becomes: how is the expression of the of Orb2A protein regulated? In an effort to understand the regulation of Orb2 prion-like conversion, White-Grindley et al. (2014) noticed that Orb2A protein has a very short half-life (~1 h) consistent with its low abundance. Transducer of Erb2 or Tob, a previously known regulator of SMAD (small body size and mothers against decapentaplegic)-dependent transcription (Yoshida et al. 2000, 2003) and CPEB-mediated translation (Hosoda et al. 2011) associates with both Orb2A and Orb2B, but increases the half-life of only Orb2A. Both Orb2 and Tob are phosphoproteins and PP2A controls the phosphorylation status of Orb2A and Orb2B. Stimulation

Cite this article as *Cold Spring Harb Perspect Biol* doi: 10.1101/cshperspect.a021774

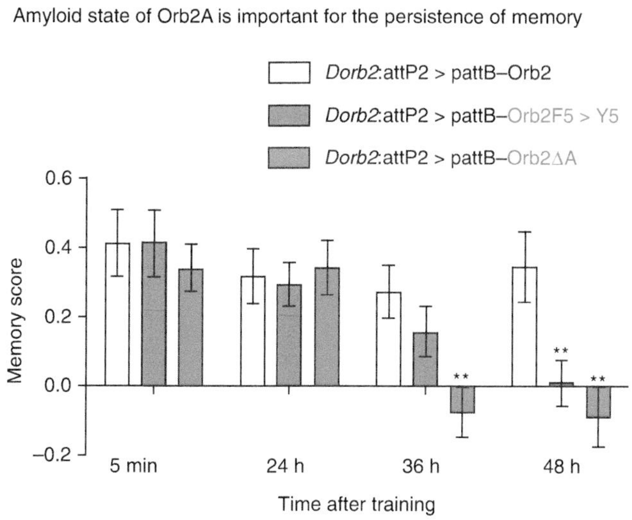

Amyloid state of Orb2A is important for the persistence of memory

Figure 9. Orb2A is required for stable memory formation. The Orb2A mutations were introduced into a genomic fragment, and the mutant flies were generated by introducing the genomic fragments into an orb2 null background. Deletion of Orb2A or a mutation that prevents Orb2A oligomerization (F5 > Y5) selectively interferes with persistence of male courtship-suppression memory beyond a day.

with tyramine or activation of mushroom body neurons enhances the association of Tob with Orb2A and Tob promotes Orb2 phosphorylation by recruiting Lim kinase (LimK). Phosphorylation destabilizes Orb2 associated Tob, whereas it increases the monomeric Orb2A protein level.

Protein phosphatase 2A (PP2A), an autocatalytic phosphatase, is known to act as a bidirectional switch in activity-dependent changes in synaptic activity (Mulkey et al. 1993; Kikuchi et al. 2003; Belmeguenai and Hansel 2005; Pi and Lisman 2008). PP2A activity is down-regulated on induction of long-term potentiation (LTP) of hippocampal CA1 synapses and up-regulated during long-term depression (LTD) (Pi and Lisman 2008). Similarly, LimK, which is synthesized locally at the synapse (Schratt et al. 2006) in response to synaptic activation, is also critical for long-term changes in synaptic activity and synaptic growth (Meng et al. 2002).

These observations suggested a model, in which in the basal state synaptic PP2A keeps the available Orb2A in a un- or hypophosphorylated and thereby unstable state. On stimulation the Tob protein that is constitutive-

ly present at the synapse binds to and stabilizes the un- or hypophosphorylated Orb2A and recruits the activated LimK to the Tob–Orb2 complex, allowing Orb2 phosphorylation. Concomitant decreases in PP2A activity and phosphorylation by other kinases increases Orb2A half-life. The increase in Orb2A level as well as change in phosphorylation may induce conformational changes in Orb2A, which allows Orb2A to act as a seed. Alternatively, accumulation and oligomerization of Orb2A creates an environment that is conducive to overall Orb2 oligomerization. Phosphorylation of Orb2 may allow integration of various signaling pathways allowing Orb2 prion-like conversion to act as an activity-dependent molecular "switch."

## EXPLICIT MEMORY STORAGE IN MAMMALS AND THE CYTOPLASMIC POLYADENYLATION ELEMENT BINDING PROTEIN 3

### CPEB3 Is a Prion-Like Protein

Is a prion-like state of CPEB, which is important for LTF in *Aplysia* and long-term memory in *Drosophila*, also important for synaptic plasticity and various forms of memory in mammals?

In mouse and in human, there are four distinct CPEB genes: CPEB1 to CPEB4 (Theis et al. 2003). Although all CPEB isoforms in mammals share a similar RNA-binding domain, they differ significantly in the amino-terminal domain. Among these isoforms CPEB1, is the most well characterized and is involved in some form of LTP (Alarcon et al. 2004). Among the other CPEB genes, CPEB2 and CPEB3 have Q/N-rich regions reminiscent of the ApCPEB and Orb2 amino-terminal domain (Fig. 10) (Theis et al. 2003). More recent studies (Chen and Huang 2012) have revealed that two small aggregation-prone regions are also present in CPEB2 (Fig. 10), but the protein has not yet been studied from a functional point of view. Only the contribution of the amino-terminal domain of CPEB3 has so far been functionally interrogated at the molecular and behavioral level.

To explore the potential prion-like property of CPEB3, Stephan et al. (2015) expressed CPEB3 in yeast and found that it displayed the two essential features of a prion-like protein: (1) it forms amyloidogenic aggregate, and (2) aggregates are heritable across cell division. Interestingly, consistent with these findings, Krüttner et al. (2012) recently found that the CPEB3 Q-rich amino terminus can functionally substi-

tute the Orb2A prion-like domain in *Drosophila*. Like ApCPEB, Orb2A has been found to behave like a prion in yeast (Si et al. 2003a) and to form amyloid-like aggregates that mediate the persistence of memory (Krüttner et al. 2012; Majumdar et al. 2012). Moreover, deletion analysis of the prion domain of CPEB3 revealed a novel, tripartite organization consisting of two Q/N-rich aggregation-promoting domains surrounding a regulatory module that mediates an interaction between CPEB3 and the actin cytoskeleton. Although the functional significance of the interaction between CPEB3 and actin is ongoing, nonetheless these data provide evidence that CPEB3 has the capacity to undergo a prion-like conversion.

## THE ROLE FOR CPEB3 AGGREGATES IN LOCAL PROTEIN SYNTHESIS

Does a prion-like conversion of CPEB3 evident in yeast also occur in the brain and, if so, what function does it serve? Fioriti et al. (2015) found that in the basal state CPEB3 binds to and represses the translation of its target mRNAs in the brain, such as the α-amino-3-hydroxy-5-methyl-4-isoxazolepropionic acid (AMPA) receptor subunits GluA1 and GluA2 (Huang

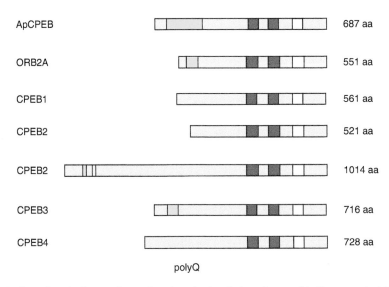

Figure 10. Mice have four isoforms of cytoplasmic polyadenylation element-binding protein (CPEB): CPEB1 to CPEB4. The amino-terminal domain of CPEB3 most resembles the prion-like domain of *Aplysia* CPEB (ApCPEB). aa, Amino acid.

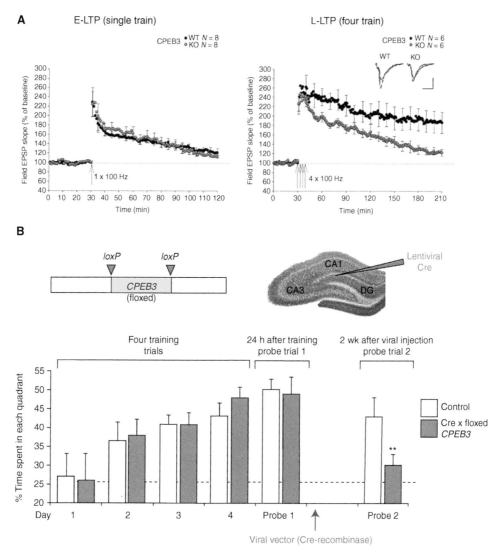

**Figure 11.** Cytoplasmic polyadenylation element-binding protein 3 (CPEB3) conditional knockout mice show impaired (*A*) maintenance of dopamine-dependent synaptic plasticity, and (*B*) long-term memory. LTP, long-term potentiation; WT, wild type; KO, knockout; DG, dentate gyrus.

et al. 2006; Pavlopoulos et al. 2011). In turn, CPEB3 promotes the translation of the AMPA receptor (AMPAR) following monoubiquitination by the ubiquitin ligase neuralized (Pavlopoulos et al. 2011). Together, these data suggest that CPEB3 can act as a repressor in the basal state and can be converted to an activator in aggregation by posttranslational modification.

This raised the question: How does CPEB3 switch from one state to the other from being a repressor to being an activator of translation?

Similarly, to what has been observed in *Aplysia* and *Drosophila*, Fioriti et al. found that mouse CPEB3 form aggregates on synaptic activation in culture as well as performing a behavioral task in vivo. Moreover the dual role in translation, the switch from repression to activation, is correlated with change of CPEB3 from a soluble to an aggregated form. The propensity of CPEB3 to form aggregates derive from its amino terminus domain, which comprises, as we have seen, two regions rich in glutamine and a low complexity sequence, which is predicted to be poorly structured and to form aggregates (Fiumara et al. 2010).

## The Persistence of Synaptic Plasticity and Memory Storage Requires CPEB3-Mediated Protein Synthesis in the Hippocampus

To determine the role of CPEB3 in the persistence of synaptic plasticity and memory, Fioriti et al. (2015) generated a conditional knockout strain of CPEB3 and surveyed the contribution of CPEB3 to the maintenance of memory. They found that CPEB3-mediated protein synthesis is required for maintenance, but not for memory acquisition (Fig. 11). The memory deficit is observed in two different behavioral paradigms, spatial object recognition and the Morris water maze task, suggesting that CPEB3-mediated processes are required for different types of hippocampal-based spatial learning tasks. Fioriti et al. (2015) also found that CPEB3 loses its ability to maintain long-term synaptic plasticity and long-term memory if its prion-like amino terminus domain is deleted. Fioriti et al., therefore, proposes that, like the *Aplysia* CPEB and the *Drosophila* Orb2A, CPEB3 can sustain the persistence of memory through a stimulus-induced conformation change, which causes protein aggregation and a change in function that allows enhanced translation of CPEB3 target mRNAs, such as the AMPAR subunits GluA1 and GluA2.

These results provide the first evidence for a prion-like mechanism to sustain memory in the mouse brain during consolidation and maintenance.

## SUMOylation Inactivates, whereas Ubiquitination Activates CPEB3

The dominant nature of the CPEB3 aggregates characteristic of other CPEB-related functional prions led Drisaldi and colleagues (2015) to search for inhibitory constraints that might be important in regulating aggregate formation. They found that small-ubiquitin-like modifier or SUMOylation of CPEB3 acts on an inhibitory constraint. In its basal state, CPEB3 is SUMOylated in hippocampal neurons and in its SU-

MOylated form CPEB3 is monomeric and acts as a repressor of translation. Following neuronal stimulation, CPEB3 is converted into an active form, which is associated with a decrease in SUMOylation and an increase of aggregation. A chimeric CPEB3 protein fused to SUMO prevents the protein from aggregating and from activating the translation of target mRNAs. These findings suggest a model whereby SUMO regulates translation of mRNAs and synaptic plasticity by modulating the aggregation of CPEB3

Because SUMOylation keeps it in an inactive state, what activates CPEB3? Pavlopoulos et al. found that CPEB3 is activated by Neuralized1, an E3 ubiquitin ligase (Pavlopoulos et al. 2011). CPEB3 interacts with Neuralized1 in dendrites of adult hippocampal neurons. In mice overexpressing Neuralized1, specifically in the forebrain, the levels of monomeric CPEB3 are increased in the hippocampus, whereas CPEB1 and CPEB4 are unaffected. Pavlopoulos et al. found that CPEB3 interacts with Neuralized1 via its amino-terminal, prion-like domain, and that this interaction leads to the monoubiquitination and consequent activation of CPEB3. Strikingly, overexpression of Neuralized1 activates CPEB3 in cultured hippocampal neurons (Fig. 12).

These results suggest a model whereby Neuralized1-mediated ubiquination facilitates hippocampal plasticity and hippocampal-dependent memory storage by modulating the activity of CPEB3 and CPEB3-dependent protein synthesis. In response to synaptic activity, the protein levels of Neuralized1 are increased, leading to the ubiquination and activation of CPEB3, and consequent production of synaptic components critical for the formation of new functional synaptic connections. Because CPEB3 can be SUMOylated as well as ubiquitinated the relationship between these two posttranslational modification is of interest.

Finally, although the evidence for a plausible role of prion-like conversion of neuronal CPEB in long-lasting memory is growing, still several important questions remain unanswered. Does persistence of memory require continued presence of the prion-like state? Does decay of memory coincide with the dis-

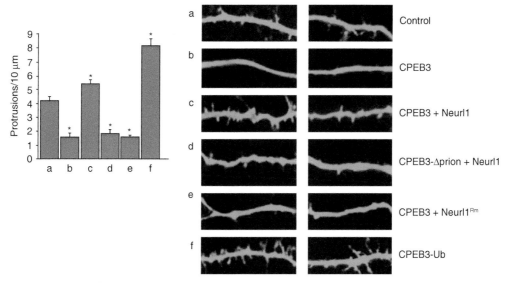

Figure 12. The modulation of cytoplasmic polyadenylation element-binding protein 3 (CPEB3) by Neuralized1 (Neurl1) and ubiquitin alters the number of spines in cultured hippocampal neurons. Modulation is blocked by removal of the prion-like or ubiquitin ligase domains. Dendrites of cultured hippocampal neurons expressing enhanced green fluorescent protein (EGFP) alone (control) or EGFP and the indicated proteins are shown. The averaged density of spines is also shown. (From Pavlopoulos et al. 2011; reprinted, with permission, from Elsevier © 2011.)

appearance of the prion-like state, and can a transient memory be stabilized by artificial recruitment of the prion-like state? Does the prion-like state maintain altered state of protein synthesis in subset of synapses for the entire duration of a memory? What are the biochemical consequences of prion-like conversion at the translational level?

## Functional Prions Are Likely to Have a Distinctive Structure

A particular fascinating observation to emerge from these studies is that various species CPEB in *Aplysia*, *Drosophila*, and mice can form functional prionogenic aggregates in the mature adult neurons that support memory, whereas other prions and amyloids formed by a number of other proteins in the nervous system causes cognitive deficit and neurodegeneration. How can these two diametrically opposite outcomes be reconciled? The simplest and most likely answer is some inherent structural differences between functional and toxic amyloids and prions. However, the specific structural and molecular features, if any, that distinguish functional

prions from pathological prions are currently not known.

Conventional prions have been found to have a β-sheet-rich structure, and the structural transition from soluble to aggregated forms occurs through uncontrolled structural transitions and consequent misfolding and aggregation. With functional prions, such as *Aplysia* CPEB, *Drosophila* Orb2, and CPEB3, the conversion from one state to another is regulated by physiological signals. To search for a difference between the two classes of prions—pathogenic and functional—Fiumara et al. (2010) searched for other types of structures, and found that in addition to β-sheet there are ApCPEB coiled-coil α helices that can mediate prion-like oligomerization. However, unlike β-sheets, coiled-coils are responsive to environmental signal and, therefore, regulatable (see Fiumara et al. 2010).

Raveendra et al. (2013) next performed more in depth structural studies on ApCPEB using functional nuclear magnetic resonance (NMR). Consistent with studies of Fiumara et al., these functional NMR studies revealed that the prion-like domain is not solely composed of β-sheet but also has a novel mixed structure containing helical and random coil stretches. This mixed structure view is consistent with bioinformatics and mutagenesis studies (Fiumara et al. 2010) that predicted that the glutamine rich prion domain of ApCPEB has a propensity to form α-helical coiled-coil structures. This "mixed structure model" (Fig. 13) has a plausible advantage in that it might allow the β-sheet to form a fiber axis scaffold. In turn, this would allow the carboxy-terminal domains to be stacked together yet be exposed and free to bind mRNA on the surface of the β-sheet fiber axis scaffold. This mixed structure model would allow the coordinated translation of the population of interrelated mRNAs required for the stabilization of synaptic growth.

## An Overall View: Functional Prions in Perspective

The realization that protein conformational switches could provide a means for inheritance of phenotypes dates back >20 years (Wickner 1994). Although prions were initially discovered as infectious proteinaceous agents that are associated with a class of fatal degenerative diseases of the mammalian brain, the discovery of fungal

prions—which are not associated with disease—first suggested that the effects of prion mechanisms on cellular physiology could be viewed in a different light. Fungal prions as epigenetic determinants alter a range of cellular processes, including metabolism and gene expression (Tompa and Friedrich 1998; Eaglestone et al. 1999; True and Lindquist 2000; True et al. 2004; Halfmann et al. 2012). These changes lead to a variety of prion-associated phenotypes. The data we review here provides one of the early examples of the existence of a prion-like protein conformational switch in the brain that may instead of causing loss of memory allow stabilization of memory. The mechanistic similarities between prion-like propagation of CPEBs in snails, flies, and mammals suggest that prions are not a biological anomaly, but, instead, could perhaps embody a ubiquitous regulatory mechanism. Indeed, functional prion-like protein or other self-assembling proteins have now been found in other species including human (Hou et al. 2011). Therefore, it is tempting to speculate that in the nervous system there might be other proteins in addition to CPEB that serve normal physiological functions at the prion state and perhaps the preponderance of amyloid-based disease in the nervous

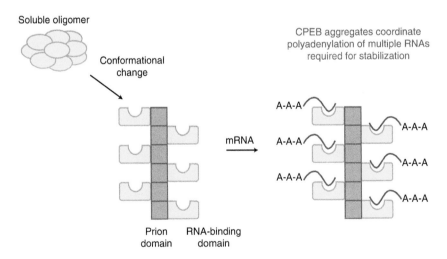

**Figure 13.** A mixed structure model for prion-like aggregation of *Aplysia* cytoplasmic polyadenylation element-binding protein (ApCPEB) and protein synthesis. The prion domains are stacked together, leaving the RNA-binding domain exposed, and free to bind messenger RNAs (mRNAs) on the surface of the β-sheet fiber-axis scaffold. This can allow the coordinated translation of the population of interrelated mRNAs required for stabilization of synaptic growth (based on Raveendra et al. 2013).

system is linked to the presence of functional prions in the nervous system.

## REFERENCES

Aguzzi A, Weissmann C. 1998. Prion diseases. *Haemophilia* **4:** 619–627.

Alarcon JM, Hodgman R, Theis M, Huang YS, Kandel ER, Richter JD. 2004. Selective modulation of some forms of schaffer collateral-CA1 synaptic plasticity in mice with a disruption of the *CPEB-1* gene. *Learn Mem* **11:** 318–327.

Barco A, Alarcon JM, Kandel ER. 2002. Expression of constitutively active CREB protein facilitates the late phase of long-term potentiation by enhancing synaptic capture. *Cell* **108:** 689–703.

Belmeguenai A, Hansel C. 2005. A role for protein phosphatases 1, 2A, and 2B in cerebellar long-term potentiation. *J Neurosci* **25:** 10768–10772.

Bougie JK, Cai D, Hastings M, Farah CA, Chen S, Fan X, McCamphill PK, Glanzman DL, Sossin WS. 2012. Serotonin-induced cleavage of the atypical protein kinase C Apl III in *Aplysia*. *J Neurosci* **32:** 14630–14640.

Casadio A, Martin KC, Giustetto M, Zhu H, Chen M, Bartsch D, Bailey CH, Kandel ER. 1999. A transient, neuron-wide form of CREB-mediated long-term facilitation can be stabilized at specific synapses by local protein synthesis. *Cell* **99:** 221–237.

Chen PJ, Huang YS. 2012. CPEB2–eEF2 interaction impedes HIF-1α RNA translation. *EMBO J* **31:** 959–971.

Coustou V, Deleu C, Saupe S, Beguent J. 1997. The protein product of the *het-s* heterokaryon incompatibility gene of the fungus *Podospora anserina* behaves as a prion analog. *Proc Natl Acad Sci* **94:** 9773–9778.

Crick F. 1984. Memory and molecular turnover. *Nature* **312:** 101.

Crow ET, Li L. 2011. Newly identified prions in budding yeast, and their possible functions. *Semin Cell Dev Biol* **22:** 452–459.

Drisaldi B, Colnaghi L, Fioriti L, Rao N, Myers C, Snyder AM, Metzger DJ, Tarasoff J, Konstantinov E, Fraser P, et al. 2015. SUMOylation is an inhibitory constraint that regulates the prion-like aggregation and activity of CPEB3. *Cell Rep* **11:** 1694–1702.

Dudek SM, Fields RD. 2002. Somatic action potentials are sufficient for late-phase LTP-related cell signaling. *Proc Natl Acad Sci* **99:** 3962–3967.

Eaglestone SS, Cox BS, Tuite MF. 1999. Translation termination efficiency can be regulated in *Saccharomyces cerevisiae* by environmental stress through a prion-mediated mechanism. *EMBO J* **18:** 1974–1981.

Fioriti L, Myers C, Huang YY, Li X, Stephan J, Trifilieff P, Kosmidis S, Drisaldi B, Pavlopoulos E, Kandel ER. 2015. The persistence of hippocampal-based memory requires protein synthesis mediated by the prion-like protein CPEB3. *Neuron* **86:** 1433–1448.

Fiumara F, Fioriti L, Kandel ER, Hendrickson WA. 2010. Essential role of coiled coils for aggregation and activity of Q/N-rich prions and PolyQ proteins. *Cell* **143:** 1121–1135.

Fiumara F, Rajasethupathy P, Antonov I, Kosmidis S, Sossin W, Kandel E. 2015. MicroRNA-22 gates long-term heterosynaptic plasticity in *Aplysia* through presynaptic regulation of CPEB and downstream targets. *Cell Rep* **11:** 1866–1875.

Frey U, Morris RG. 1997. Synaptic tagging and long-term potentiation. *Nature* **385:** 533–536.

Frey U, Morris RG. 1998. Weak before strong: Dissociating synaptic tagging and plasticity-factor accounts of late-LTP. *Neuropharmacology* **37:** 545–552.

Ganusova EE, Ozolins LN, Bhagat S, Newnam GP, Wegrzyn RD, Sherman MY, Chernoff YO. 2006 Modulation of prion formation, aggregation, and toxicity by the actin cytoskeleton in yeast. *Mol Cell Biol* **26:** 617–629.

Hake LE, Richter JD. 1994. CPEB is a specificity factor that mediates cytoplasmic polyadenylation during *Xenopus oocyte* maturation. *Cell* **79:** 617–627.

Halfmann R, Jarosz DF, Jones SK, Chang A, Lancaster AK, Lindquist S. 2012. Prions are a common mechanism for phenotypic inheritance in wild yeasts. *Nature* **482:** 363–368.

Heinrich SU, Lindquist S. 2011. Protein-only mechanism induces self-perpetuating changes in the activity of neuronal *Aplysia* cytoplasmic polyadenylation element-binding protein (CPEB). *Proc Natl Acad Sci* **108:** 2999–3004.

Hosoda N, Funakoshi Y, Hirasawa M, Yamagishi R, Asano Y, Miyagawa R, Ogami K, Tsujimoto M, Hoshino S. 2011. Anti-proliferative protein Tob negatively regulates CPEB3 target by recruiting Caf1 deadenylase. *EMBO J* **30:** 1311–1323.

Hou F, Sun L, Zheng H, Skaug B, Jiang QX, Chen ZJ. 2011. MAVS forms functional prion-like aggregates to activate and propagate antiviral innate immune response. *Cell* **146:** 448–461.

Huang YS, Kan MC, Lin CL, Richter JD. 2006. CPEB3 and CPEB4 in neurons: Analysis of RNA-binding specificity and translational control of AMPA receptor GluR2 mRNA. *EMBO J* **25:** 4865–4876.

Ishimaru D, Andrade LR, Teixeira LS, Quesado PA, Maiolino LM, Lopez PM, Cordeiro Y, Costa LT, Heckl WM, Weissmüller G, et al. 2003. Fibrillar aggregates of the tumor suppressor p53 core domain. *Biochemistry* **42:** 9022–9027.

Jarosz DF, Lancaster AK, Brown JC, Lindquist S. 2014. An evolutionarily conserved prion-like element converts wild fungi from metabolic specialists to generalists. *Cell* **58:** 1072–1082.

Kandel ER. 2001. The molecular biology of memory storage: A dialogue between genes and synapses. *Science* **294:** 1113–1120.

Keleman K, Krüttner S, Alenius M, Dickson BJ. 2007. Function of the *Drosophila* CPEB protein Orb2 in long-term courtship memory. *Nat Neurosci* **10:** 1587–1593.

Kikuchi S, Fujimoto K, Kitagawa N, Fuchikawa T, Abe M, Oka K, Takei K, Tomita M. 2003. Kinetic simulation of signal transduction system in hippocampal long-term potentiation with dynamic modeling of protein phosphatase 2A. *Neural Netw* **16:** 1389–1398.

Krüttner S, Stepien B, Noordermeer JN, Mommaas MA, Mechtler K, Dickson BJ, Keleman K. 2012. *Drosophila*

CPEB Orb2A mediates memory independent of its RNA-binding domain. *Neuron* **76**: 383–395.

Lee SH, Lim CS, Park H, Lee JA, Han JH, Kim H, Cheang YH, Lee SH, Lee YS, Ko HG, et al. 2007. Nuclear translocation of CAM-associated protein activates transcription for long-term facilitation in *Aplysia*. *Cell* **129**: 801–812.

Majumdar A, Cesario WC, White-Grindley E, Jiang H, Ren F, Khan MR, Li L, Choi EM, Kannan K, Guo F, et al. 2012. Critical role of amyloid-like oligomers of *Drosophila* Orb2 in the persistence of memory. *Cell* **148**: 515–529.

Martin KC, Casadio A, Zhu H, Yaping E, Rose JC, Chen M, Bailey CH, Kandel ER. 1997a. Synapse-specific, long-term facilitation of *Aplysia* sensory to motor synapses: A function for local protein synthesis in memory storage. *Cell* **91**: 927–938.

Martin KC, Michael D, Rose JC, Barad M, Casadio A, Zhu H, Kandel ER. 1997b. MAP kinase translocates into the nucleus of the presynaptic cell and is required for long-term facilitation in *Aplysia*. *Neuron* **18**: 899–912.

Mastushita-Sakai T, White-Grindley E, Samuelson J, Seidel C, Si K. 2010. *Drosophila* Orb2 targets genes involved in neuronal growth, synapse formation, and protein turnover. *Proc Natl Acad Sci* **107**: 11987–11992.

Meng Y, Zhang Y, Tregoubov V, Janus C, Cruz L, Jackson M, Lu WY, MacDonald JF, Wang JY, Falls DL, et al. 2002. Abnormal spine morphology and enhanced LTP in LIMK-1 knockout mice. *Neuron* **35**: 121–133.

Miniaci MC, Kim JH, Puthanveettil SV, Si K, Zhu H, Kandel ER, Bailey CH. 2008. Sustained CPEB-dependent local protein synthesis is required to stabilize synaptic growth for persistence of long-term facilitation in *Aplysia*. *Neuron* **59**: 1024–1036.

Mulkey RM, Herron CE, Malenka RC. 1993. An essential role for protein phosphatases in hippocampal long-term depression. *Science* **261**: 1051–1055.

Pavlopoulos E, Trifilieff P, Chevaleyre V, Fioriti L, Zairis S, Pagano A, Malleret G, Kandel ER. 2011. Neuralized1 activates CPEB3: A function of non-proteolytic ubiquitin in synaptic plasticity and memory storage. *Cell* **147**: 1369–1383.

Pi HJ, Lisman JE. 2008. Coupled phosphatase and kinase switches produce the tristability required for long-term potentiation and long-term depression. *J Neurosci* **28**: 13132–13138.

Polymenidou M, Cleveland DW. 2012. Prion-like spread of protein aggregates in neurodegeneration. *J Exp Med* **209**: 889–893.

Prusiner SB. 1982. Novel proteinaceous infectious particles cause scrapie. *Science* **216**: 136–144.

Prusiner SB. 1994. Biology and genetics of prion diseases. *Annu Rev Microbiol* **48**: 655–686.

Puthanveettil S, Kandel ER. 2011. Molecular mechanisms for the initiation and maintenance of long-term memory storage. In *Two faces of evil: Cancer and neurodegeneration* (ed. Curran T, Christen Y), pp. 143–160. Springer, New York.

Raveendra BL, Siemer AB, Puthanveettil SV, Hendrickson WA, Kandel ER, McDermott AE. 2013. Characterization of prion-like conformational changes of the neuronal isoform of *Aplysia* CPEB. *Nat Struct Mol Biol* **20**: 495–501.

Sacktor TC. 2011. How does PKMz maintain long-term memory? *Nat Rev Neurosci* **12**: 9–15.

Schratt GM, Tuebing F, Nigh EA, Kane CG, Sabatini ME, Kiebler M, Greenberg ME. 2006. A brain-specific microRNA regulates dendritic spine development. *Nature* **439**: 283–289.

Shorter J, Lindquist S. 2005. Prions as adaptive conduits of memory and inheritance. *Nat Rev Genet* **6**: 435–450.

Si K, Giustetto M, Etkin A, Hsu R, Janisiewicz AM, Miniaci MC, Kim JH, Zhu H, Kandel ER. 2003a. A neuronal isoform of CPEB regulates local protein synthesis and stabilizes synapse specific long-term facilitation in *Aplysia*. *Cell* **115**: 893–904.

Si K, Lindquist S, Kandel ER. 2003b. A neuronal isoform of the *Aplysia* CPEB has prion-like properties. *Cell* **115**: 879–891.

Si K, Choi YB, White-Grindley E, Majumdar A, Kandel ER. 2010. *Aplysia* CPEB can form prion-like multimers in sensory neurons that contribute to long-term facilitation. *Cell* **140**: 421–435.

Stephan JS, Fioriti L, Lamba N, Derkatch IL, Kandel ER. 2015. The CPEB3 protein important in memory persistence is a functional prion that interacts with the actin cytoskeleton. *Cell Rep* **11**: 1772–1785.

Theis M, Si K, Kandel ER. 2003. Two previously undescribed members of the mouse CPEB family of genes and their inducible expression in the principal cell layers of the hippocampus. *Proc Natl Acad Sci* **100**: 9602–9607.

Tompa P, Friedrich P. 1998. Prion proteins as memory molecules: A hypothesis. *Neuroscience* **86**: 1037–1043.

True HL, Lindquist SL. 2000. A yeast prion provides a mechanism for genetic variation and phenotypic diversity. *Nature* **407**: 477–483.

True HL, Berlin I, Lindquist SL. 2004 Epigenetic regulation of translation reveals hidden genetic variation to produce complex traits. *Nature* **431**: 184–187.

White-Grindley E, Li L, Mohammad Khan R, Ren F, Saraf A, Florens L, Si K. 2014. Contribution of Orb2A stability in regulated amyloid-like oligomerization of *Drosophila* Orb2. *PLoS Biol* **12**: e1001786.

Wickner RB. 1994. [URE3] as an altered *URE2* protein: Evidence for a prion analog in *Saccharomyces cerevisiae*. *Science* **264**: 566–569.

Yoshida Y, Tanaka S, Umemori H, Minowa O, Usui M, Ikematsu N, Hosoda E, Imamura T, Kuno J, Yamashita T, et al. 2000. Negative regulation of BMP/Smad signaling by Tob in osteoblasts. *Cell* **103**: 1085–1097.

Yoshida Y, von Bubnoff A, Ikematsu N, Blitz IL, Tsuzuku JK, Yoshida EH, Umemori H, Miyazono K, Yamamoto T, Cho KW. 2003. Tob proteins enhance inhibitory Smad-receptor interactions to repress BMP signaling. *Mech Dev* **120**: 629–637.

# Reconsolidation and the Dynamic Nature of Memory

Karim Nader

Psychology Department, McGill University, Montréal, Quebec H3A 1B1, Canada

*Correspondence:* karim.nader@mcgill.ca

Memory reconsolidation is the process in which reactivated long-term memory (LTM) becomes transiently sensitive to amnesic agents that are effective at consolidation. The phenomenon was first described more than 50 years ago but did not fit the dominant paradigm that posited that consolidation takes place only once per LTM item. Research on reconsolidation was revitalized only more than a decade ago with the demonstration of reconsolidation in a well-defined behavioral protocol (auditory fear conditioning in the rat) subserved by an identified brain circuit (basolateral amygdala). Since then, reconsolidation has been shown in many studies over a range of species, tasks, and amnesic agents, and cellular and molecular correlates of reconsolidation have also been identified. In this review, I will first define the evidence on which reconsolidation is based, and proceed to discuss some of the conceptual issues facing the field in determining when reconsolidation does and does not occur. Last, I will refer to the potential clinical implications of reconsolidation.

Learning and memory are commonly depicted as going through a set of phases. There is the learning or encoding phase, in which information is acquired, by stabilization phase, in which specific mechanisms are engaged to stabilize initially unstable new information (referred to as synaptic consolidation) (Glickman 1961; McGaugh 1966), the "storage" or maintenance phase, during which other mechanisms are involved to maintain the memory, and the retrieval phase, in which specific mechanisms permit a memory to be retrieved (Miller and Springer 1973; Spear 1973). For a long time, from a neurobiological perspective, only acquisition and memory stabilization (Martin et al. 2000; Kandel 2001; Dudai 2004) were considered to be active phases, in the sense that neurons had to perform certain computations or synthesize new RNA and proteins for these

phases of memory processing to be performed successfully. After acquisition and stabilization, all other phases were implicitly thought by many to be passive readout of changes in the circuits mediating the long-term memory (LTM). However, the picture has now changed and the maintenance of memory is portrayed as an active process. One of the reasons for this change is the demonstration that a consolidated LTM can become susceptible to disruption and restoration, a process termed "reconsolidation" (Spear 1973; Nader et al. 2000; Sara 2000). There are now detailed molecular and cellular models of this time-dependent active memory phase.

This review will first describe the logic of the findings that brought the existence of the consolidation process to light. I will then describe how we concluded that a consolidated memory undergoes reconsolidation in a well-defined behavioral protocol (auditory fear conditioning in the rat). I will then refer to the range of species, tasks, and treatments in which reconsolidation have been reported. One aspect of reconsolidation that has attracted experimental attention involves the finding that there seem to be conditions that facilitate, inhibit, or even prevent reconsolidation from occurring. I present an approach that could help to identify such conditions. Last, I will discuss potential clinical implications of reconsolidation.

## CONSOLIDATION: THE DOMINANT MODEL OF MEMORY STORAGE

Consolidation is the time-dependent stabilization of newly acquired memory (Fig. 1Ai) (Ebbinghaus 1885; Müller and Pilzecker 1900; Glickman 1961; McGaugh 1966, Dudai 2004). At the synaptic level of analysis, this process, referred to as synaptic consolidation, is thought to be a universal property of neurons that subserve memory formation.

The evidence for the consolidation process has been derived from various lines of evidence demonstrating the presence of a postacquisition time interval during which new memories are labile and sensitive to challenges (Fig. 1Ai). First, performance can be impaired by amnesic treatments, such as electroconvulsive shock (Duncan 1949), protein synthesis inhibitors (PSIs) (Flexner et al. 1965), or by new learning (Gordon and Spear 1973). Second, retention can be enhanced by administration of certain compounds, such as strychnine (McGaugh and Krivanek 1970). Crucially, these manipulations are only effective when administered shortly after new learning, but not when given after a few hours. These types of results led to the conclusion that memory exists in two states. When susceptible to enhancement or impairment, memory resides in a labile state, but if it is insensitive to these treatments, memory is stable and, by definition, consolidated (McGaugh 1966; Dudai 2004).

This same logic was used by Schafe and colleagues to test for the existence of a consolidation process in the lateral and basal amygdala (LBA) for auditory fear memory. When an amnesic agent, the protein-synthesis inhibitor anisomycin, is infused into the LBA, shortly after training, short-term memory (STM) is intact but LTM is impaired (Fig. 1Aii) (Schafe and LeDoux 2000); however, LTM remains intact when the infusion is delayed for 6 h. This pattern of results conforms to the operational definition of consolidation in the sense that the aspect of fear-conditioning memory that requires protein synthesis within the LBA is consolidated within at most 6 h after learning. In addition, we assume that the experimental manipulation-induced amnesia for those computations that the LBA supposedly mediates, that is, the association between the conditioned stimulus (the tone; CS) and the unconditioned stimulus (the foot shock; US) (Rodrigues et al. 2009).

One of the tenets of the cellular consolidation model is that learning induces changes in synaptic efficacy, suggesting that the physiological "unit" of cellular consolidation is the synapse. Two main candidate cellular mechanisms that were postulated to implement these changes are long-term potentiation (LTP) and long-term depression (LTD) (Malenka and Nicoll 1999; Martin et al. 2000). In parallel to the distinction of STM and LTM, with the latter being consolidated by a protein-synthesis-dependent process, LTP is also divided into an early transient phase (E-LTP) and a stabilized, RNA- and protein-synthesis-dependent late phase (L-LTP) (Goelet et al. 1986).

## BEHAVIORAL EVIDENCE FOR A RECONSOLIDATION PROCESS

The existence of a reconsolidation process in the LBA for consolidated, that is, long-term auditory fear memory was identified in a study (Nader et al. 2000) that in logic and design followed those for consolidation as described in Schafe and LeDoux (2000). One day after conditioning, at a time when, according to the results from the consolidation study, memory should be stabilized and immune to the amnesic agent, we reminded animals of the conditioning session by exposing them again to the CS, that is, the tone (Nader et al. 2000). Anisomycin, at the same dose, concentration, and rate as in the

Cite this article as *Cold Spring Harb Perspect Biol* doi: 10.1101/cshperspect.a021782

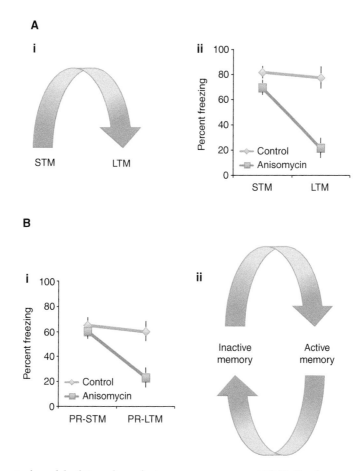

**Figure 1.** Conceptual model of time-dependent memory processes. (*Ai*) Textbook account of consolidation demonstrating that memories consolidate over time into long-term memory (LTM). The critical point is to show that once a memory is in LTM, it is thought to remain fixed or permanent (Glickman 1961; McGaugh 1966). (*Aii*) A typical demonstration of a consolidation blockade (Schafe and LeDoux 2000)—intact short-term memory (STM) and impaired LTM, a pattern of impairment that defines a consolidation impairment (Dudai 2004; McGaugh 2004). (*Bi*) A typical demonstration of a reconsolidation blockade. Intact postreactivation STM (PR-STM) and impaired LTM (PR-LTM), meeting the definitions for a consolidation blockade (Dudai 2004; McGaugh 2004). (*Bii*) An alternate model of memory that incorporates the findings of consolidation and reconsolidation datasets proposed by Lewis (1979). Consolidation theory cannot explain the reconsolidation dataset. New and reactivated memories are in an "active state" and then over time they stabilize and exist in an "inactive memory state." When a memory in an inactive memory state is remembered it returns to an active memory state.

consolidation study (Schafe and LeDoux 2000) was then infused into the LBA either immediately or 6 h after the memory retrieval. When anisomycin was administered immediately, anisomycin-treated animals show intact postreactivation STM (PR-STM) but impaired PR-LTM (Fig. 1Bi), a pattern of results that is identical to what is found when blocking consolidation (Fig. 1Aii) (Schafe and LeDoux 2000). However,

if the postreactivation infusion was delayed by 6 h, anisomycin had no effect, demonstrating that the reactivation-induced lability was transient. Importantly, animals that were not reminded before anisomycin infusions had intact memory.

Staying strictly within the commonly accepted consolidation framework, and applying only the rationale on which this framework is based, the following four conclusions can be drawn from the results of these experiments. First, the observation that the memory was insensitive to anisomycin when it was not reactivated shows that it was "consolidated" 24 h after training—at least with regard to the specific amnesic treatment applied. Second, that only the reactivated memory was sensitive to disruption shows that memory was in a labile state after reactivation. Third, the observation that the anisomycin-treated animals showed intact STM and impaired LTM after reactivation implies that a consolidation-like process is triggered by reactivation. And finally, given the amnesic treatment was ineffective 6 h after reactivation, this postreactivation restabilization process is, like consolidation, a time-dependent process. Taken together, these four conclusions yield the interpretation that reactivation of a consolidated memory returns it again to a labile state from which the memory has to undergo stabilization (i.e., reconsolidate) over time (Nader et al. 2000).

Consolidation and reconsolidation are, thus, both deduced from the evidence of a transient period of instability. In the case of consolidation, this window is initiated after acquisition of new information. In the case of reconsolidation, it is initiated after reactivation of an existing, consolidated memory representation. As is the case for consolidation, only during the reconsolidation phase can memory be enhanced by "memory enhancers" (Gordon 1977b; Rodriguez et al. 1993, 1999; Horne et al. 1997), or impaired by either amnesic treatments (Misanin et al. 1968) or interference produced by new learning (Gordon 1977a). These treatments are ineffective when reconsolidation is complete, which is also the case for consolidation.

The term "reconsolidation" was originally introduced in the context of a discussion on memory retrieval. Spear (1973) asked "... how will the dynamic aspects of memory be handled, that is, with successive learning trials or related successive experiences does the entire memory reconsolidate anew or merely the new information?" As a consequence of the perceived inability of the consolidation hypothesis to account for reconsolidation, new memory models were

developed that treated new and reactivated consolidated memories in similar ways (Fig. 1Bii) (Spear 1973; Lewis 1979).

In recent years, reconsolidation has been shown across a variety of species, tasks, and amnesic treatments (Table 1). In light of this evidence, it was therefore postulated that reconsolidation represents a fundamental memory process (Nader and Hardt 2009). One of the most striking findings in this literature is a study by Lee (2008), who devised specific tools to block consolidation or reconsolidation mechanisms. Most students of memory would assume that presenting additional learning trials to a consolidated memory would engage consolidation mechanisms, which will make the memory stronger. However, the evidence from the aforementioned study suggests that a memory has to undergo reconsolidation to be strengthened. Moreover, memory strengthening by new learning was mediated by reconsolidation and not consolidation mechanisms. This evidence suggests that a recently acquired memory will be mediated by consolidation mechanisms within a time window of $\sim 5$ h. However, for the rest of the memory's lifetime, the memory will engage reconsolidation mechanisms (Lee 2009).

## ALTERNATIVE INTERPRETATIONS

Reconsolidation, as we discussed above, has been defined by applying the very standards that define consolidation. Therefore, certain nonspecific interpretations of the reconsolidation hypothesis pose the same challenges to the consolidation hypothesis, a consequence that is rarely acknowledged. The complexity of the data poses a problem for alternative interpretations, which should not merely provide new explanations for the reconsolidation dataset, but need to allow for predictions that are different from those offered by the reconsolidation model. For this reason, we will not address all the previous alternative interpretations here. A detailed discussion of these alternative interpretations including facilitation of extinction, transient retrieval impairment, nonspecific effects, state-dependent learning, and new learning is presented in Nader and Hardt (2009).

Cite this article as *Cold Spring Harb Perspect Biol* doi: 10.1101/cshperspect.a021782

**Table 1.** Some of the paradigms in which reconsolidation has been reported

| | |
|---|---|
| Experimental paradigm | Habituation (Rose and Rankin 2006) |
| | Auditory fear conditioning (Nader et al. 2000) |
| | Contextual fear conditioning (Debiec et al. 2002) |
| | Instrumental learning (Sangha et al. 2003; but see Hernandez and Kelley 2004) |
| | Inhibitory avoidance (Anokhin et al. 2002; Milekic and Alberini 2002) |
| | Conditioned aversion learning (Eisenberg et al. 2003) |
| | Motor sequence learning (Walker et al. 2003) |
| | Incentive learning (Wang et al. 2005) |
| | Object recognition (Kelly et al. 2003) |
| | Spatial memory (Suzuki et al. 2004; Morris et al. 2006) |
| | Memory for drug reward (Lee et al. 2005; Miller and Marshall 2005; Valjent et al. 2006) |
| | Episodic memory (Hupbach et al. 2007) |
| Treatment | Protein-synthesis inhibition (Nader et al. 2000) |
| | RNA synthesis inhibition (Sangha et al. 2003) |
| | Inhibition of kinase activity (Kelly et al. 2003; Duvarci et al. 2005) |
| | Anesthesia (Eisenberg et al. 2003) |
| | Protein-knockout mice (Bozon et al. 2003) |
| | Antisense (Taubenfeld et al. 2001; Lee et al. 2004) |
| | Inducible knockout mice (Kida et al. 2002) |
| | Receptor antagonists (Przybyslawski et al. 1999; Debiec and Ledoux 2004; Suzuki et al. 2004) |
| | Interference by new learning (Walker et al. 2003; Hupbach et al. 2007) |
| | Potentiated reconsolidation by increase in kinase activity (Tronson et al. 2006) |
| Species | *Aplysia* (Cai et al. 2012; Lee et al. 2012) |
| | Nematodes (Rose and Rankin 2006) |
| | Honeybees (Stollhoff et al. 2005) |
| | Snails (Sangha et al. 2003) |
| | Sea slugs (Child et al. 2003) |
| | Fish (Eisenberg et al. 2003) |
| | Crabs (Pedreira et al. 2002) |
| | Chicks (Anokhin et al. 2002) |
| | Mice (Kida et al. 2002) |
| | Rats (Nader et al. 2000), rat pups (Gruest et al. 2004) |
| | Humans (Walker et al. 2003; Hupbach et al. 2007; Kindt et al. 2009; Schiller et al. 2010) |

Examples from various experimental paradigms, treatments, and species for studies reporting evidence for a reconsolidation process since the year 2000.

## EVIDENCE FOR RECONSOLIDATION ACROSS LEVELS OF ANALYSIS

Evidence for reconsolidation does not come solely from the behavioral level of analysis. A cellular phenomenon akin to reconsolidation was shown for L-LTP (Fonseca et al. 2006). In this study, the investigators report that when anisomycin is added 2 h after the induction of L-LTP, it has no effect on L-LTP maintenance. If, however, the potentiated pathway is reactivated by administering test pulses that inhibit protein synthesis, the potentiation is intact shortly after reactivation but becomes impaired over time.

This suggests that reactivation of stabilized L-LTP returns its substrate to a labile state, in which it can be disrupted by inhibiting protein synthesis. Other evidence includes reports that reconsolidation blockade reverses increases in field potentials induced by fear conditioning in the lateral amygdala (LA) in intact animals (Doyere et al. 2007). In summary, this evidence suggests the presence of a cellular correlate of the behaviorally shown reconsolidation impairment.

More recently, two papers using the classic paradigm of *Aplysia* to study sensitization and long-term facilitation (LTF) reported that reconsolidation affects these kinds of processes.

Indeed, when reconsolidation was blocked, the sensory-motor synaptic enhancement typically observed after LTF was reversed (Cai et al. 2012; Lee et al. 2012).

At the molecular level, interfering with reconsolidation can, in a time-dependent manner, remove molecular correlates of memory induced by learning and subsequent consolidation. Miller and Marshall (2005) showed that place-preference learning activates the extracellular signal-regulated kinase (ERK) in the nucleus accumbens. Blocking the activated ERK in the nucleus accumbens after reactivation results in intact PR-STM but impaired PR-LTM. In these amnesic animals, this also leads to the absence of ERK and its downstream transcription factors in the nucleus accumbens (see also Valjent et al. 2006, who show a reduction in ERK and GluA1 phosphorylation using a similar procedure). Studying mechanisms of long-term habituation in *Caenorhabditis elegans*, Rose and Rankin (2006) showed that administering heat-shock or the non-*N*-methyl-D-aspartate (NMDA) glutamatergic antagonist, DMQX, after reactivation of a consolidated memory dramatically returns expression of α-amino-3-hydroxy-5-methyl-4-isoxazolepropionic (AMPA) receptors in the mechanosensory neuron to a level typical for naïve animals (Rose and Rankin 2006). Importantly, the reconsolidation effects in all of these studies were contingent on memory reactivation—in the absence of a reminder the amnesic treatments were ineffective.

Another study described the biochemical process that destabilizes a consolidated memory and the subsequent reconsolidation process at the level of postsynaptic AMPA receptors. Learning is thought to lead to AMPA receptor trafficking: calcium-permeable AMPA receptors are inserted into the postsynaptic density (PSD), then over time replaced by calcium-impermeable receptors (Rumpel et al. 2005). Hong et al. (2013) asked what the AMPA receptor dynamics would be when a memory is destabilized and then reconsolidated. They found that memory destabilization is associated with calcium-permeable AMPA receptors. Indeed, blocking the introduction of calcium-perme-

able AMPA receptors into the PSD prevented the memory from being apparently unstored. Thus, they found that the replacement of calcium-permeable AMPA receptors by calcium-impermeable AMPA receptors mediated the process of reconsolidation.

These studies are only examples of the dataset that provides strong evidence for the existence of a transient postreactivation period of memory plasticity, that is, memory reconsolidation, on the behavioral, physiological, and molecular levels of analysis.

## CAN MECHANISMS MEDIATING PRESYNAPTIC PLASTICITY UNDERGO RECONSOLIDATION?

Both post- and presynaptic mechanisms are posited to contribute to synaptic plasticity and memory (Finnie and Nader 2012; Kandel et al. 2014). One theory on the locus of memory posits that presynaptic changes are critical for LTM and L-LTP (Bliss and Collingridge 1993). These presynaptic changes are thought to increase the probability of vesicle release.

In studies that examined cellular or molecular correlates of consolidation or reconsolidation, blocking the respective memory processes were reported to reverse the learning-induced molecular/cellular correlates. For example, Bailey and Kandel (1993) reported that the blockade of consolidation in *Aplysia* with a protein-synthesis inhibitor prevented the increase in the number of synapses following sensitization to the point where this number of synapses was comparable to the level of synapses in naïve animals. The same pattern of results has been shown in reconsolidation studies, as can be seen in the previous section.

Tsvetkov et al. (2002) have shown that auditory fear conditioning induces predominantly presynaptic enhancements in both inputs to the LA thought to mediate fear learning. Recently, this group assessed what would happen to these learning-induced presynaptic enhancements after blocking reconsolidation with rapamycin, a protein-synthesis inhibitor. They reported that these presynaptic enhancements were not reduced, but that a reduction in postsynaptic AMPA receptors correlated with the behavioral impairments (Li et al. 2013). This finding suggests that the postsynaptic mechanisms detect how much potential exists on the presynaptic terminals and adjusts the postsynaptic AMPA receptors accordingly.

There are two theoretical implications of these findings for reconsolidation. First, perhaps, presynaptic mechanisms of long-term plasticity are independent of reconsolidation. This would entail that only the postsynaptic mechanisms of LTM could be susceptible to reconsolidation blockade. The second possibility is that presynaptic mechanisms are affected by reconsolidation, but the amnesic treatment used, a PSI, was not appropriate to target the presynaptic mechanisms mediating reconsolidation. We know that presynaptic enhancements are not affected by PSIs. Therefore, a tool transiently challenging the mechanisms mediating long-term presynaptic efficacy would be needed to test this hypothesis.

## RECONSOLIDATION IS NOT UNIVERSAL

The fact that memory reconsolidation has been found across levels of analysis does not imply that reconsolidation is universal, that is, observed under any circumstance. Another variation of the theme that reconsolidation is not a universal property of memory is the concept of constraints on this phenomenon, or "boundary conditions." These are conditions of a physiological, environmental, or psychological nature, in which memory that normally would reconsolidate no longer does. Several boundary conditions have been proposed, such as the dominance of the association over behavior (Eisenberg et al. 2003), competition with extinction (Eisenberg et al. 2003; Pedreira and Maldonado 2003; Suzuki et al. 2004), memory age (Milekic and Alberini 2002; Eisenberg and Dudai 2004; Suzuki et al. 2004), predictability of the reactivation stimulus (Pedreira et al. 2004; Morris et al. 2006), and training intensity (Suzuki et al. 2004). Others, however, have not identified similar boundary conditions in other protocols for extinction (Stollhoff et al. 2005; Duvarci et al. 2006), old memories (Debiec et al. 2002; Lee et al. 2005), predictability of the reactivation

stimulus (Pedreira et al. 2002; Bozon et al. 2003; Sangha et al. 2003; Valjent et al. 2006), or strength of training (Debiec et al. 2002; Lee et al. 2005). Whether additional parameters moderate boundary conditions remains to be seen.

The observed inconsistencies in the identification of the boundary conditions might be caused by the absence of agreed-on, standard experimental parameters required to test the presence of such boundary conditions. For example, if memory disruption is not observed within a set of experimental parameters, then it is concluded that the memory does not undergo reconsolidation under those conditions. A number of reports, however, have shown that a memory may undergo reconsolidation only under specific reactivation conditions (De Vietti and Holiday 1972; Bozon et al. 2003; Suzuki et al. 2004). The implication of these findings is that it is difficult to conclude, based on behavioral studies, that a memory never undergoes reconsolidation. Therefore, the question remains whether the negative effects on which the boundary conditions are based imply that a given memory never undergoes reconsolidation under those inferred conditions, or that same memory is still capable of undergoing reconsolidation with a different reactivation protocol that still includes the same inferred boundary conditions (Figs. 2 and 3). Given that the parameter space of possible reactivation procedures is essentially infinite, a generalized boundary condition may be difficult to prove at the behavioral level. This may explain part of the inconsistency in the field of boundary conditions of reconsolidation (Dreyfuss et al. 2009).

An understanding of how boundary conditions are mediated across levels of analysis is critical because targeting reconsolidation of traumatic memories has been proposed to be a potential treatment for posttraumatic stress disorder (PTSD) (Przybyslawski and Sara 1997; Debiec et al. 2002; Schiller et al. 2010). Specifically, blocking the reconsolidation of traumatic memories might weaken the long-term maintenance of these traumatic memories, in turn, reducing PTSD pathology. However, if strong aversive experiences act as boundary conditions on reconsolidation (Suzuki et al. 2004), then

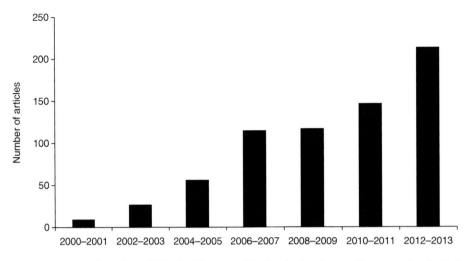

Figure 2. The number of articles published with reconsolidation in the abstract (based on data in Nader et al. 2000).

this would suggest that the traumatic memories in PTSD patients may be resistant to undergoing reconsolidation thereby negating reconsolidation as a potential therapeutic target. Therefore, understanding boundary conditions, such as strength of training, is critical to ensure we know whether it is possible to target reconsolidation of very strong fear memories and, if so, what the optimal conditions are to allow an extremely strong fear memory to undergo reconsolidation.

To this end, Wang et al. (2009) found that strong auditory training produced memories that initially did not undergo reconsolidation but they did so over time on the order of 1 mo. This suggests that the boundary condition induced by strong training is transient (Fig. 4A). This in itself is striking, as the implicit assumption is that once a memory stops undergoing reconsolidation it will never begin again. This was the first demonstration that a putative boundary condition could be transient (Wang et al. 2009).

Wang et al. (2009) hypothesized that one principle that could mediate boundary conditions is to down-regulate the mechanisms that allow memories to undergo reconsolidation (Fig. 5). What could the molecular mechanism be that inhibit reconsolidation of strong memories for up to 30 d after training in the LBA?

Ben Mamou et al. (2006) showed the NMDA receptor antagonists for the NR2B subunits are necessary in reactivation-induced destabilization, but that this destabilization does not get expressed at the behavioral level. Specifically, prereactivation administration of ifenprodil (an NR2B antagonist) prevented the memory from being impaired by postreactivation anisomycin; however, the ifenprodil itself had no effect on the expression of freezing (Fig. 6). New strong memories show similar properties: normal expression of freezing during reactivation but insensitivity to postreactivation anisomycin. The investigators reasoned that strong training may down-regulate NR2B expression in the LBA, thereby making the memory insensitive to postreactivation anisomycin infusions but capable of being expressed normally. The investigators suggest that NR2B expression in the LBA should be reduced under conditions when memories did not undergo reconsolidation but should remain normal when memories underwent reconsolidation (Fig. 7). That was indeed observed. NR2B levels were normal when the memory underwent reconsolidation, but drastically reduced under the conditions in which the memory did not undergo reconsolidation (Fig. 4B). The reduction was subunit-selective, with NR1 subunits constant at all time points.

Cite this article as *Cold Spring Harb Perspect Biol* doi: 10.1101/cshperspect.a021782

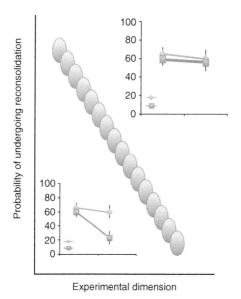

Figure 3. Possible functions describing the constraints on reconsolidation. It is still an open question whether the functions are linear or exponent. Different experimental conditions may produce different functions. The experimental space to the *left* of the curve is determined by examples in which the memory undergoes reconsolidation as shown in the schematic behavioral impairment. The evidence for constraints on reconsolidation is derived from negative findings as shown in the schematic on the *right*. That is a logical limitation of the behavioral approach to this issue. Therefore, we suggested that a complementary approach to help resolve this issue would be to identify a molecular correlate for the absence of reconsolidation. This would act as positive evidence for the existence of the constraint.

The suggested role of the NR2B subunits in regulating when fear memory in the LBA will undergo reconsolidation may not generalize to all memory systems or types of memory. Currently, there are several studies that have examined the mechanisms involved in transforming a consolidated memory to a labile state. Although the NR2B subunit is critical for memories to return to a labile state within the LBA for fear conditioning (Ben Mamou et al. 2006), NMDA receptors in the hippocampus and within the amygdala for appetitive memories are thought to play a direct role in the restabilization process following the reactivation of the memory (Milton et al. 2008; Suzuki et al. 2008).

In the hippocampus, voltage-gated calcium channels (VGCCs) (Suzuki et al. 2008) and protein degradation (Lee et al. 2008) are critical for a memory to return to a labile state.

## DOES RECONSOLIDATION IMPLY RECAPITULATION OF CONSOLIDATION

An important but somewhat neglected aspect of this debate is that the protocols used to study reconsolidation are different from those used to study consolidation, which renders comparison of results problematic. For example, in auditory fear conditioning, both CS and US are presented, leading to activation of afferents that relay auditory and pain information to the amygdala. Neurons that are thought to be the site of plasticity in the LBA are proposed to receive concurrent activation by these afferents (Blair et al. 2001). As a consequence, a series of second messenger systems are activated that are thought to lead to transcription and translation of proteins required for consolidation (Maren 2001; Schafe et al. 2001). In reconsolidation studies, however, typically only the CS is presented to reactive and induce plasticity in consolidated memory. Thus, consolidation studies examine the neurobiological changes after a CS and US are presented together, although reconsolidation studies examine neurobiological changes that happen after presentation of a CS alone. For this reason, at the brain systems/circuits and molecular level, consolidation and reconsolidation must be different, as only the former directly activates the pathways that relay US information to the amygdala, which are not directly activated in reconsolidation studies. Therefore, the demonstration of differences in brain regions or circuits mediating consolidation and reconsolidation may be rather trivial (Nader et al. 2005). It remains unclear which of the reported differences between consolidation and reconsolidation reflect genuine differences between the two processes as opposed to differences in the protocols used to induce them. A study in which differences between reconsolidation and consolidation were not attributable to differences in the protocols is the first to shed some light on this issue (Lee et al. 2004). These

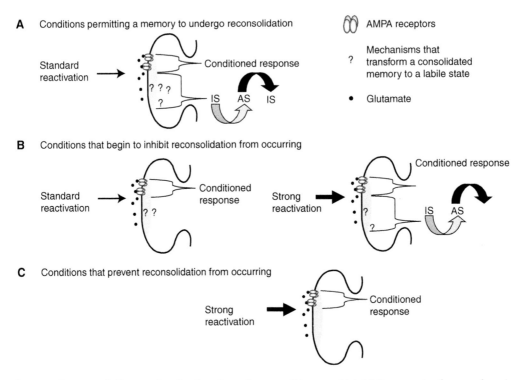

**Figure 4.** Conceptual diagram showing how boundary conditions could inhibit memories from undergoing reconsolidation across memory types and memory systems. (*A*) Under experimental conditions when a memory undergoes reconsolidation, the mechanisms allowing a memory to be transformed from a consolidated to a labile active state (AS), must be present and functional at the synapse ("?" in the figure). These mechanisms, of course, will involve more than surface receptors and will likely include a number of molecular processes that have yet to be identified. (*B*) Experimental conditions that begin to inhibit memories from undergoing reconsolidation may lead to a partial reduction in a mechanism that is critical for the induction of reconsolidation. The partial reduction might be sufficient to prevent the induction of reconsolidation when a standard protocol is used. However, there may still be sufficient amounts of this mechanism to permit the memory to undergo reconsolidation when a stronger reactivation is used. (*C*) Under conditions in which the memory does not undergo reconsolidation, a boundary condition, a necessary mechanism for the induction of reconsolidation, is reduced to the point that alternative reactivation protocols cannot induce the memory to undergo reconsolidation. AMPA, α-amino-3-hydroxy-5-methyl-4-isoxazolepropionic; IS, inactive state.

investigators reported a double dissociation, separating the mechanisms mediating consolidation from those that mediate reconsolidation (Lee et al. 2004; see also von Hertzen and Giese 2005).

## POTENTIAL CLINICAL IMPLICATIONS

The finding that consolidated memories return to a labile state and have to be restored has significant potential implications for a number of clinical conditions, such as PTSD, addiction,

obsessive-compulsive disorder (OCD), or delusions/hallucinations. An understanding of the mechanisms mediating reconsolidation could provide the basis for developing new or refining old therapeutic tools to successfully manage some of these conditions. As an example of how this could be applied, imagine a patient with PTSD whose symptoms were resistant to therapy. A new way of treating this condition could be to reactivate the patient's traumatic memory and block its reconsolidation. Theoretically, this may lead to a rapid "cure."

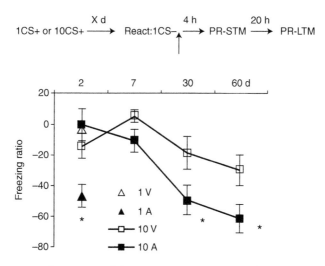

**Figure 5.** Strong memories undergo reconsolidation at 30 and 60, but not 7 d after training. The *top* panel of each subfigure represents the behavior protocol. Separate groups of animals were lateral and basal amygdala (LBA)-cannulated and trained with 10-tone foot-shock pairings. The memory was reactivated at 7, 30, or 60 d after training. The freezing ration was computed as postreactivation long-term memory (PR-LTM)—PR short-term memory (STM)/(PR-STM)x*100%. Intra-LBA anisomycin infusion impaired the PR-LTM only when the strong memory was reactivated at 30- and 60-d after training. The asterisks (*) indicate significant group differences. (From Wang et al. 2009; reproduced, with permission, from the authors.)

Early evidence suggests that this may work. Rubin et al. (1969) and Rubin (1976) treated patients suffering from either hallucinations, delusions, major depression, or OCD with electroconvulsive therapy (ECT). In contrast to other studies that administered ECT when the subjects were anaesthetized, Rubin and colleagues kept the patients awake and directed them to focus on the objects of their compulsions or hallucinations. This experimental procedure was thought to reactivate the neural mechanisms mediating those memories when the ECT was delivered. All of the subjects were reportedly "cured" of their condition, even though some had had up to 30 previous ECT treatments while under anesthesia. The majority remained symptom-free for the 2-yr period between the treatment and the publication of the manuscript. The fact that ECT was effective only when the memories were presumed to be reactivated, but not when the memory reactivation was omitted (i.e., when the patient was anesthetized), suggests, in principle, that reconsolidation occurs in humans. Furthermore, this study provides evidence that the possibility of

curing someone by removing a memory in a single session may not be so remote.

More recent candidate treatments tend to be less intrusive than ECT (Kroes et al. 2014). For example, β-adrenergic antagonists, such as propranolol, have few side effects and were reported to block reconsolidation of aversive and appetitive memories preferentially stored in the amygdala. However, behaviorally updating or extinguishing the reactivated memory during the time window of reconsolidation in the absence of drug treatment, or the use of sensory distractors, may also be used to block reconsolidation in potential treatment protocols in humans (Schiller et al. 2010; Schiller and Phelps 2011; Gray and Liotta 2012).

The first attempt to target reconsolidation with a β-blocker in patients with enduring PTSD symptoms reported a reduction in the strength of traumatic memories after a 15-min intervention (Brunet et al. 2008). It is important to note that some of these patients had been suffering from these PTSD symptoms for close to 30 yr. Furthermore, it is remarkable that a single reactivation seems to cause an old and

K. Nader

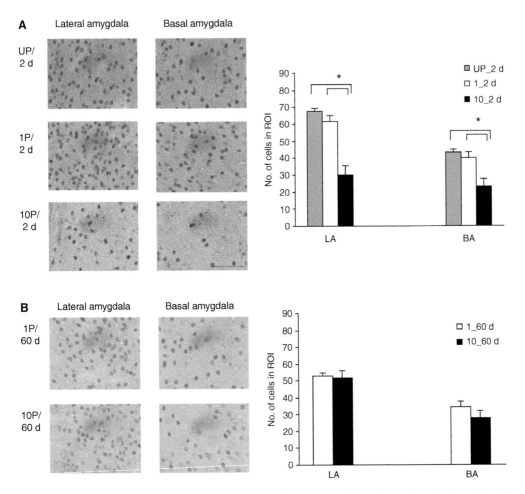

Figure 6. NR2B-subunit levels, assessed by immunohistochemistry (IHC), are inversely related to the ability of the strong memories to undergo reconsolidation over time. (*A*) Animals received 10-tone foot-shock pairings (10P), one pairing (1P), or one foot shock followed by an unpaired tone (UP). They were killed 2 d after training, a time when the memory does not undergo reconsolidation, and their brains were later processed for IHC. The *left* panel represents the actual staining in regions of interest (ROI) in lateral and basal amygdala (LBA) in individual groups (*n* = 4/group). The graph shows the quantification of NR2B-positive cell numbers in each ROI. Although 1P and UP animals showed similar level of NR2B-immuno-stained cells, 10P animals showed significantly less stained cells in either lateral amygdala (LA) or basal amygdala (BA). The asterisks (*) indicate significant group differences. (*B*) Animals received either 10P or 1P. They were killed 60 d after training, a time when the memory does undergo reconsolidation, and their brains were later processed for IHC. Both groups show similar level of NR2B-positive cells in LA and BA. All pictures in the *left* panel are in the same scale. Each data point is represented as mean ± S.E.M. Scale bar, 80 μm.

consolidated memory to become labile again. Drug craving (Xue et al. 2012; Saladin et al. 2013) and PTSD (Brunet et al. 2008; Menzies 2012) are two clinical conditions, in which it has been reported that targeting their underlying maintaining mechanisms through reconsolida-tion can lead to some clinical improvement (Nader et al. 2013). However, the results of other studies indicate that the practical usefulness of blocking reconsolidation in the treatment of trauma requires further exploration (Spring et al. 2015; Wood et al. 2015).

Cite this article as *Cold Spring Harb Perspect Biol* doi: 10.1101/cshperspect.a021782

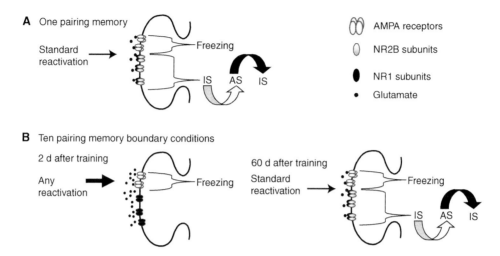

**Figure 7.** A diagram hypothesizing how strength of auditory fear memory can influence a mechanism that allows reconsolidation to occur at the lateral and basal amygdala (LBA) synapses. (*A*) After one pairing, NR2B containing *N*-methyl-D-aspartate (NMDA) receptors are activated for the memory to go from a consolidated inactive state (IS) to a labile active state (AS) (gray curved arrow). The labile memory then undergoes reconsolidation and return to IS (black curved arrow). Independently of this mechanism, α-amino-3-hydroxy-5-methyl-4-isoxazolepropionic (AMPA) receptors are critical for the expression of freezing behavior. After one pairing, there are sufficient NR2B subunits at the synapse for the memory to undergo reconsolidation when standard reactivation procedures are used. (*B*) After 10 pairings, via hippocampus-mediated mechanisms (not shown), levels of the NR2B, but not NR1, are reduced 2 d after training. NR2B reduction eliminates a necessary mechanism for the memory to undergo reconsolidation. Hence, reactivation by standard or alternative protocols should be unable to induce reconsolidation. Meanwhile, the existing AMPA receptors will mediate normal expression of freezing. NR2B level returns to normal 60 d after training. Thus, standard reactivation procedures are now sufficient to induce reconsolidation.

## CONCLUSION

Reconsolidation indicates that the maintenance of memory over time is active rather than passive. The evidence for reconsolidation comes from a spectrum of species, amnesic agents, and reinforcers, spanning all levels of analysis from the molecular and physiological to the behavioral levels, thereby suggesting that that it is a fundamental property of memory.

Reconsolidation remains a topic of intensive research. One area of investigation that is being studied involves the identification of boundary conditions in reconsolidation. I have suggested that identifying a molecular or cellular indicator when memories undergo reconsolidation is a complementary approach that can alleviate some of the problems inherent in attempts to identify boundary conditions on reconsolidation.

There is a growing interest in utilizing reconsolidation blockage as a potential therapeutic tool in several clinical conditions, most importantly PTSD. The use of reconsolidation protocols for clinical purposes must, however, await further exploration.

## ACKNOWLEDGMENTS

K.N. is supported by the Canadian Institutes of Health Research, Natural Sciences and Engineering Research Council of Canada, Canadian Foundation for Innovation, The Volkswagen, Alfred P. Sloan, and EJLB (Echo Foundation). K.N. is the James McGill Chair.

## REFERENCES

Anokhin KV, Tiunova AA, Rose SP. 2002. Reminder effects—Reconsolidation or retrieval deficit? Pharmacological dissection with protein synthesis inhibitors following reminder for a passive-avoidance task in young chicks. *Eur J Neurosci* **15**: 1759–1765.

Bailey CH, Kandel ER. 1993. Structural changes accompanying memory storage. *Annu Rev Physiol* **55**: 397–426.

Ben Mamou C, Gamache K, Nader K. 2006. NMDA receptors are critical for unleashing consolidated auditory fear memories. *Nat Neurosci* **9**: 1237–1239.

Blair HT, Schafe GE, Bauer EP, Rodrigues SM, LeDoux JE. 2001. Synaptic plasticity in the lateral amygdala: A cellular hypothesis of fear conditioning. *Learn Mem* **8**: 229–242.

Bliss TV, Collingridge GL. 1993. A synaptic model of memory: Long-term potentiation in the hippocampus. *Nature* **361**: 31–39.

Bozon B, Davis S, Laroche S. 2003. A requirement for the immediate early gene *zif268* in reconsolidation of recognition memory after retrieval. *Neuron* **40**: 695–701.

Brunet A, Orr SP, Tremblay J, Robertson K, Nader K, Pitman RK. 2008. Effect of post-retrieval propranolol on psychophysiologic responding during subsequent script-driven traumatic imagery in post-traumatic stress disorder. *J Psychiatry Res* **42**: 503–506.

Cai D, Pearce K, Chen S, Glanzman DL. 2012. Reconsolidation of long-term memory in *Aplysia*. *Curr Biol* **22**: 1783–1788.

Child FM, Epstein HT, Kuzirian AM, Alkon DL. 2003. Memory reconsolidation in hermissenda. *Biol Bull* **205**: 218–219.

Debiec J, Ledoux JE. 2004. Disruption of reconsolidation but not consolidation of auditory fear conditioning by noradrenergic blockade in the amygdala. *Neuroscience* **129**: 267–272.

Debiec J, LeDoux JE, Nader K. 2002. Cellular and systems reconsolidation in the hippocampus. *Neuron* **36**: 527–538.

De Vietti T, Holiday JH. 1972. Retrograde amnesia produced by electroconvulsive shock after reactivation of a consolidated memory trace: A replication. *Psychon Sci* **29**: 137–138.

Doyere V, Debiec J, Monfils MH, Schafe GE, LeDoux JE. 2007. Synapse-specific reconsolidation of distinct fear memories in the lateral amygdala. *Nat Neurosci* **10**: 414–416.

Dreyfuss JL, Regatieri CV, Jarrouge TR, Cavalheiro RP, Sampaio LO, Nader HB. 2009. Heparan sulfate proteoglycans: Structure, protein interactions and cell signaling. *An Acad Bras Cienc* **81**: 409–429.

Dudai Y. 2004. The neurobiology of consolidations, or, how stable is the engram? *Annu Rev Psychol* **55**: 51–86.

Duncan CP. 1949. The retroactive effect of electroconvulsive shock. *J Comp Physiol Psychol* **42**: 32–44.

Duvarci S, Nader K, Ledoux JE. 2005. Activation of extracellular signal-regulated kinase–mitogen-activated protein kinase cascade in the amygdala is required for memory reconsolidation of auditory fear conditioning. *Eur J Neurosci* **21**: 283–289.

Duvarci S, Mamou CB, Nader K. 2006. Extinction is not a sufficient condition to prevent fear memories from undergoing reconsolidation in the basolateral amygdala. *Eur J Neurosci* **24**: 249–260.

Ebbinghaus M. 1885. *Über das Gedächtnis* (ed. K. Buehler). Leipzig, Germany.

Eisenberg M, Dudai Y. 2004. Reconsolidation of fresh, remote, and extinguished fear memory in medaka: Old fears don't die. *Eur J Neurosci* **20**: 3397–3403

Eisenberg M, Kobilo T, Berman DE, Dudai Y. 2003. Stability of retrieved memory: Inverse correlation with trace dominance. *Science* **301**: 1102–1104.

Finnie PS, Nader K. 2012. The role of metaplasticity mechanisms in regulating memory destabilization and reconsolidation. *Neurosci Biobehav Rev* **36**: 1667–1707.

Flexner LB, Flexner JB, Stellar E. 1965. Memory and cerebral protein synthesis in mice as affected by graded amounts of puromycin. *Exp Neurol* **13**: 264–272.

Fonseca R, Nagerl UV, Bonhoeffer T. 2006. Neuronal activity determines the protein synthesis dependence of long-term potentiation. *Nat Neurosci* **9**: 478–480.

Glickman S. 1961. Perseverative neural processes and consolidation of the memory trace. *Psychol Bull* **58**: 218–233.

Goelet P, Castellucci VF, Schacher S, Kandel ER. 1986. The long and short of long-term memory—A molecular framework. *Nature* **322**: 419–422.

Gordon WC. 1977a. Similarities of recently acquired and reactivated memories in interference. *Am J Psychol* **90**: 231–242.

Gordon WC. 1977b. Susceptibility of a reactivated memory to the effects of strychnine: A time-dependent phenomenon. *Physiol Behav* **18**: 95–99.

Gordon WC, Spear NE. 1973. Effect of reactivation of a previously acquired memory on the interaction between memories in the rat. *J Exp Psychol* **99**: 349–355.

Gray RM, Liotta RF. 2012. PTSD: Extinction, reconsolidation, and the visual-kinesthetic dissociation protocol. *Traumatology* **18**: 3–16.

Gruest N, Richer P, Hars B. 2004. Memory consolidation and reconsolidation in the rat pup require protein synthesis. *J Neurosci* **24**: 10488–10492.

Hernandez PJ, Kelley AE. 2004. Long-term memory for instrumental responses does not undergo protein synthesis-dependent reconsolidation upon retrieval. *Learn Mem* **11**: 748–754.

Hong I, Kim J, Lee S, Ko HG, Nader K, Kaang BK, Tsien RW, Choi S. 2013. AMPA receptor exchange underlies transient memory destabilization on retrieval. *Proc Natl Acad Sci* **110**: 8218–8223.

Horne CA, Rodriguez WA, Wright TP, Padilla JL. 1997. Time-dependent effects of fructose on the modulation of a reactivated memory. *Prog Neuropsychopharmacol Biol Psychiatry* **21**: 649–658.

Hupbach A, Gomez R, Hardt O, Nadel L. 2007. Reconsolidation of episodic memories: A subtle reminder triggers integration of new information. *Learn Mem* **14**: 47–53.

Kandel ER. 2001. The molecular biology of memory storage: A dialogue between genes and synapses. *Science* **294**: 1030–1038.

Kandel ER, Dudai Y, Mayford MR. 2014. The molecular and systems biology of memory. *Cell* **157**: 163–186.

Kelly A, Laroche S, Davis S. 2003. Activation of mitogen-activated protein kinase/extracellular signal-regulated kinase in hippocampal circuitry is required for consolidation and reconsolidation of recognition memory. *J Neurosci* **23:** 5354–5360.

Kida S, Josselyn SA, de Ortiz SP, Kogan JH, Chevere I, Masushige S, Silva AJ. 2002. CREB required for the stability of new and reactivated fear memories. *Nat Neurosci* **5:** 348–355.

Kindt M, Soeter M, Vervliet B. 2009. Beyond extinction: Erasing human fear responses and preventing the return of fear. *Nat Neurosci* **12:** 256–258.

Kroes MC, Tendolkar I, van Wingen GA, van Waarde JA, Strange BA, Fernandez G. 2014. An electroconvulsive therapy procedure impairs reconsolidation of episodic memories in humans. *Nat Neurosci* **17:** 204–206.

Lee JL. 2008. Memory reconsolidation mediates the strengthening of memories by additional learning. *Nat Neurosci* **11:** 1264–1266.

Lee JL. 2009. Reconsolidation: Maintaining memory relevance. *Trends Neurosci* **32:** 413–420.

Lee JL, Everitt BJ, Thomas KL. 2004. Independent cellular processes for hippocampal memory consolidation and reconsolidation. *Science* **304:** 839–843.

Lee JL, Di Ciano P, Thomas KL, Everitt BJ. 2005. Disrupting reconsolidation of drug memories reduces cocaine-seeking behavior. *Neuron* **47:** 795–801.

Lee SH, Choi JH, Lee N, Lee HR, Kim JI, Yu NK, Choi SL, Lee SH, Kim H, Kaang BK. 2008. Synaptic protein degradation underlies destabilization of retrieved fear memory. *Science* **319:** 1253–1256.

Lee SH, Kwak C, Shim J, Kim JE, Choi SL, Kim HF, Jang DJ, Lee JA, Lee K, Lee CH, et al. 2012. A cellular model of memory reconsolidation involves reactivation-induced destabilization and restabilization at the sensorimotor synapse in *Aplysia*. *Proc Natl Acad Sci* **109:** 14200–14205.

Lewis DJ. 1979. Psychobiology of active and inactive memory. *Psychol Bull* **86:** 1054–1083.

Li Y, Meloni EG, Carlezon WA Jr, Milad MR, Pitman RK, Nader K, Bolshakov VY. 2013. Learning and reconsolidation implicate different synaptic mechanisms. *Proc Natl Acad Sci* **110:** 4798–4803.

Malenka RC, Nicoll RA. 1999. Long-term potentiation—A decade of progress? *Science* **285:** 1870–1874.

Maren S. 2001. Neurobiology of Pavlovian fear conditioning. *Annu Rev Neurosci* **24:** 897–931.

Martin SJ, Grimwood PD, Morris RG. 2000. Synaptic plasticity and memory: An evaluation of the hypothesis. *Annu Rev Neurosci* **23:** 649–711.

McGaugh JL. 1966. Time-dependent processes in memory storage. *Science* **153:** 1351–1358.

McGaugh JL. 2004. The amygdala modulates the consolidation of memories of emotionally arousing experiences. *Annu Rev Neurosci* **27:** 1–28.

McGaugh JL, Krivanek JA. 1970. Strychnine effects on discrimination learning in mice: Effects of dose and time of administration. *Physiol Behav* **5:** 1437–1442.

Menzies RP. 2012. Propranolol, traumatic memories, and amnesia: A study of 36 cases. *J Clin Psychiatry* **73:** 129–130.

Milekic MH, Alberini CM. 2002. Temporally graded requirement for protein synthesis following memory reactivation. *Neuron* **36:** 521–525.

Miller CA, Marshall JF. 2005. Molecular substrates for retrieval and reconsolidation of cocaine-associated contextual memory. *Neuron* **47:** 873–884.

Miller RR, Springer AD. 1973. Amnesia, consolidation, and retrieval. *Psychol Rev* **80:** 69–79.

Milton AL, Lee JL, Butler VJ, Gardner R, Everitt BJ. 2008. Intra-amygdala and systemic antagonism of NMDA receptors prevents the reconsolidation of drug-associated memory and impairs subsequently both novel and previously acquired drug-seeking behaviors. *J Neurosci* **28:** 8230–8237.

Misanin JR, Miller RR, Lewis DJ. 1968. Retrograde amnesia produced by electroconvulsive shock after reactivation of a consolidated memory trace. *Science* **160:** 203–204.

Morris RG, Inglis J, Ainge JA, Olverman HJ, Tulloch J, Dudai Y, Kelly PA. 2006. Memory reconsolidation: Sensitivity of spatial memory to inhibition of protein synthesis in dorsal hippocampus during encoding and retrieval. *Neuron* **50:** 479–489.

Müller GE, Pilzecker A. 1900. Experimentelle beitrage zur lehre vom gedachtnis. *Z Psychol* **1:** 1–30.

Nader K, Hardt O. 2009. A single standard for memory: The case for reconsolidation. *Nat Rev Neurosci* **10:** 224–234.

Nader K, Schafe GE, Le Doux JE. 2000. Fear memories require protein synthesis in the amygdala for reconsolidation after retrieval. *Nature* **406:** 722–726.

Nader K, Hardt O, Wang SH. 2005. Response to Alberini: Right answer, wrong question. *Trends Neurosci* **28:** 346–347.

Nader K, Hardt O, Lanius R. 2013. Memory as a new therapeutic target. *Dialogues Clin Neurosci* **15:** 475–486.

Pedreira ME, Maldonado H. 2003. Protein synthesis subserves reconsolidation or extinction depending on reminder duration. *Neuron* **38:** 863–869.

Pedreira ME, Perez-Cuesta LM, Maldonado H. 2002. Reactivation and reconsolidation of long-term memory in the crab *Chasmagnathus*: Protein synthesis requirement and mediation by NMDA-type glutamatergic receptors. *J Neurosci* **22:** 8305–8311.

Pedreira ME, Perez-Cuesta LM, Maldonado H. 2004. Mismatch between what is expected and what actually occurs triggers memory reconsolidation or extinction. *Learn Mem* **11:** 579–585.

Przybyslawski J, Sara SJ. 1997. Reconsolidation of memory after its reactivation. *Behav Brain Res* **84:** 241–246.

Przybyslawski J, Roullet P, Sara SJ. 1999. Attenuation of emotional and nonemotional memories after their reactivation: Role of β adrenergic receptors. *J Neurosci* **19:** 6623–6628.

Rodrigues SM, LeDoux JE, Sapolsky RM. 2009. The influence of stress hormones on fear circuitry. *Annu Rev Neurosci* **32:** 289–313.

Rodriguez WA, Rodriguez SB, Phillips MY, Martinez JL Jr. 1993. Post-reactivation cocaine administration facilitates later acquisition of an avoidance response in rats. *Behav Brain Res* **59:** 125–129.

Rodriguez WA, Horne CA, Padilla JL. 1999. Effects of glucose and fructose on recently reactivated and recently acquired memories. *Prog Neuropsychopharmacol Biol Psychiatry* **23:** 1285–1317.

Rose JK, Rankin CH. 2006. Blocking memory reconsolidation reverses memory-associated changes in glutamate receptor expression. *J Neurosci* **26:** 11582–11587.

Rubin RD. 1976. Clinical use of retrograde amnesia produced by electroconvulsive shock: A conditioning hypothesis. *Can Psychiatry Assoc J* **21:** 87–90.

Rubin RD, Fried R, Franks CM. 1969. New application of ECT. In *Advances in behavior therapy* (ed. Rubin RD, Franks C), pp. 37–44. Academic, New York.

Rumpel S, LeDoux J, Zador A, Malinow R. 2005. Postsynaptic receptor trafficking underlying a form of associative learning. *Science* **308:** 83–88.

Saladin ME, Gray KM, McRae-Clark AL, Larowe SD, Yeatts SD, Baker NL, Hartwell KJ, Brady KT. 2013. A double blind, placebo-controlled study of the effects of post-retrieval propranolol on reconsolidation of memory for craving and cue reactivity in cocaine dependent humans. *Psychopharmacology (Berl)* **226:** 721–737.

Sangha S, Scheibenstock A, Lukowiak K. 2003. Reconsolidation of a long-term memory in *Lymnaea* requires new protein and RNA synthesis and the soma of right pedal dorsal 1. *J Neurosci* **23:** 8034–8040.

Sara S. 2000. Retrieval and reconsolidation: Toward a neurobiology of remembering. *Learn Mem* **7:** 73–84.

Schafe GE, LeDoux JE. 2000. Memory consolidation of auditory Pavlovian fear conditioning requires protein synthesis and protein kinase A in the amygdala. *J Neurosci* **20:** RC96.

Schafe GE, Nader K, Blair HT, LeDoux JE. 2001. Memory consolidation of Pavlovian fear conditioning: A cellular and molecular perspective. *Trends Neurosci* **24:** 540–546.

Schiller D, Phelps EA. 2011. Does reconsolidation occur in humans? *Front Behav Neurosci* **5:** 24.

Schiller D, Monfils MH, Raio CM, Johnson DC, Ledoux JE, Phelps EA. 2010. Preventing the return of fear in humans using reconsolidation update mechanisms. *Nature* **463:** 49–53.

Spear N. 1973. Retrieval of memory in animals. *Psychol Rev* **80:** 163–194.

Spring JD, Wood NE, Mueller-Pfeifer C, Milad MR, Pitman RK, Orr SP. 2015. Prereactivation propranolol fails to reduce skin conductance reactivity to prepared fear-conditioned stimuli. *Psychophysiology* **52:** 407–415.

Stollhoff N, Menzel R, Eisenhardt D. 2005. Spontaneous recovery from extinction depends on the reconsolidation of the acquisition memory in an appetitive learning paradigm in the honeybee (*Apis mellifera*). *J Neurosci* **25:** 4485–4492.

Suzuki A, Josselyn SA, Frankland PW, Masushige S, Silva AJ, Kida S. 2004. Memory reconsolidation and extinction have distinct temporal and biochemical signatures. *J Neurosci* **24:** 4787–4795.

Suzuki A, Mukawa T, Tsukagoshi A, Frankland PW, Kida S. 2008. Activation of LVGCCs and CB1 receptors required for destabilization of reactivated contextual fear memories. *Learn Mem* **15:** 426–433.

Taubenfeld SM, Milekic MH, Monti B, Alberini CM. 2001. The consolidation of new but not reactivated memory requires hippocampal C/EBPβ. *Nat Neurosci* **4:** 813–818.

Tronson NC, Wiseman SL, Olausson P, Taylor JR. 2006. Bidirectional behavioral plasticity of memory reconsolidation depends on amygdalar protein kinase A. *Nat Neurosci* **9:** 167–169.

Tsvetkov E, Carlezon WA, Benes FM, Kandel ER, Bolshakov VY. 2002. Fear conditioning occludes LTP-induced presynaptic enhancement of synaptic transmission in the cortical pathway to the lateral amygdala. *Neuron* **34:** 289–300.

Valjent E, Aubier B, Corbille AG, Brami-Cherrier K, Caboche J, Topilko P, Girault JA, Herve D. 2006. Plasticity-associated gene *Krox24/Zif268* is required for long-lasting behavioral effects of cocaine. *J Neurosci* **26:** 4956–4960.

von Hertzen LS, Giese KP. 2005. Memory reconsolidation engages only a subset of immediate-early genes induced during consolidation. *J Neurosci* **25:** 1935–1942.

Walker MP, Brakefield T, Hobson JA, Stickgold R. 2003. Dissociable stages of human memory consolidation and reconsolidation. *Nature* **425:** 616–620.

Wang SH, Ostlund SB, Nader K, Balleine BW. 2005. Consolidation and reconsolidation of incentive learning in the amygdala. *J Neurosci* **25:** 830–835.

Wang SH, de Oliveira Alvares L, Nader K. 2009. Cellular and systems mechanisms of memory strength as a constraint on auditory fear reconsolidation. *Nat Neurosci* **12:** 905–912.

Wood NE, Rosasco ML, Suris AM, Spring JD, Marin MF, Lasko NB, Goetz JM, Fischer AM, Orr SP, Pitman RK. 2015. Pharmacological blockade of memory reconsolidation in posttraumatic stress disorder: Three negative psychophysiological studies. *Psychiatry Res* **225:** 31–39.

Xue YX, Luo YX, Wu P, Shi HS, Xue LF, Chen C, Zhu WL, Ding ZB, Bao YP, Shi J, et al. 2012. A memory retrieval-extinction procedure to prevent drug craving and relapse. *Science* **336:** 241–245.

# Memory Retrieval in Mice and Men

Aya Ben-Yakov[1], Yadin Dudai[1,2], and Mark R. Mayford[3]

[1]Department of Neurobiology, Weizmann Institute of Science, Rehovot 76100, Israel

[2]Center for Neural Science, New York University, New York, New York 10003

[3]Department of Molecular and Cellular Neuroscience, The Scripps Research Institute, La Jolla, California 92037

*Correspondence:* yadin.dudai@weizmann.ac.il; mmayford@scripps.edu

Retrieval, the use of learned information, was until recently mostly terra incognita in the neurobiology of memory, owing to shortage of research methods with the spatiotemporal resolution required to identify and dissect fast reactivation or reconstruction of complex memories in the mammalian brain. The development of novel paradigms, model systems, and new tools in molecular genetics, electrophysiology, optogenetics, in situ microscopy, and functional imaging, have contributed markedly in recent years to our ability to investigate brain mechanisms of retrieval. We review selected developments in the study of explicit retrieval in the rodent and human brain. The picture that emerges is that retrieval involves coordinated fast interplay of sparse and distributed corticohippocampal and neocortical networks that may permit permutational binding of representational elements to yield specific representations. These representations are driven largely by the activity patterns shaped during encoding, but are malleable, subject to the influence of time and interaction of the existing memory with novel information.

Retrieval is the use of learned information, induced by sensory or internal cues. In simple modified reflex behavior, it refers to the postexperience readout of the experience-induced change in behavior and in its underlying synaptic efficacy (Kandel and Schwartz 1982; Byren and Hawkins 2015). In memories encoded and stored in more complex circuits, such as distributed memories in the mammalian brain, retrieval is posited to involve distinct processes, including selection, reactivation or reconstruction of the target representation, and assessment of the outcome (Tulving 1983; Dudai 2002). These sequential and parallel processes can be completed within a fraction of a second (e.g., Thorpe et al. 1996). Retrieval is critical to understanding memory. In fact, once encoding is over, memory unretrieved, whether naturally or by experimental manipulations, is undetected, hence retrieval of the engram or part of it is an essential part of the proof that the specific engram exists.

Despite its central importance in the study of memory and the abundance of data and models of retrieval in human experimental psychology, until fairly recently, retrieval in complex neural circuitry remained mostly an uncharted terrain in the neuroscience of memory. This was owing to a multitude of hindrances, including difficulties in teasing apart retrieval from encoding, limited knowledge on localization of specific candidate memory circuits in humans and animals, and lack of neurobiological methods with the proper spatiotemporal resolution that permits monitoring and manip- ulation of these circuits to observe, block, trigger or enhance retrieval. The development of novel paradigms, model systems, and new tools in molecular genetics, electrophysiology, optoge- netics, in situ microscopy and functional imag- ing, have contributed markedly in recent years to our ability to investigate retrieval and under- stand part of its processes and mechanisms from the cellular to the behavioral level. In this work, we will review some of these devel- opments. We will begin with selected studies of memory retrieval in the rodent brain and pro- ceed to discuss aspects of retrieval of episodic memory in the human brain.

## MEMORY RETRIEVAL IN THE RODENT BRAIN

The study of memory retrieval in the mam- malian brain assumes that the process involves reactivation of patterns of neural activity asso- ciated with the original experience, although not necessarily identical with the activity pat- terns that represented the original experience. Retrieval is hence considered as a reconstructive rather than a replicative process. This activity is likely to be sparse and anatomically distributed, with different brain regions contributing to the quality and strength of the recall. Although the human work focuses on a richer psychology and more complex neuroanatomy than accessi- ble in the rodent, a common theme in both strands of research is a search for coherent pat- terns of activation correlated with retrieval and for correlations between retrieval and initial learning. The rodent work has been spurred re- cently by techniques that allow direct activation of distributed neural ensembles to test their functional involvement in memory. We will fo- cus first on the role of hippocampal and cortical circuits in the retrieval of explicit memories in the mouse and rat.

### Patterned Activity during Retrieval

As discussed in the literature, the hippocampus plays a critical role in explicit forms of memory that in rodents has been investigated extensively in relation to spatial learning (Morris et al.

1982). A striking feature of the rodent hippo- campus is the identification of place cells (O'Keefe and Dostrovsky 1971), neurons that fire when the animals enter specific locations in their environment (Moser et al. 2015). This has led to the view that the hippocampus forms a cognitive map; it encodes a map of space that can be used to allow recognition of specific environ- ments, guide movement through the environ- ment, and identify specific goal areas within an environment (O'Keefe 1990). At a circuit level, the hippocampus (and surrounding structures such as entorhinal and perirhinal cortex) could serve this function by integrating multimodal sensory information to form a unique map of each particular environment, object, or event. During retrieval, when a sufficient partial set of cues is provided, the entire map (memory) is recruited in a manner that likely involves the activation of multiple cortical regions coordi- nated by the hippocampus. A first question we can ask is how similar is the pattern of neural activity during two retrieval trials, or during re- trieval and initial learning?

One means of assessing neural activity is through examination of a group of genes, the immediate early genes (IEGs), whose expres- sion in neurons is responsive to activity (Farivar et al. 2004). The most commonly used IEGs for neural activity mapping are *cfos*, *arc*, and *zif268*. The expression of these genes is modulated by a variety of second messenger signaling pathways, but in excitatory neurons these all seem to be linked to neural activity (Morgan et al. 1987; Sagar et al. 1988). The *cfos* gene has been used most extensively in the analysis of rodent behav- ior and is responsive to burst activity of 30 or more action potentials at frequencies of 10 Hz or above (Schoenenberger et al. 2009). It has been used in many behavioral studies examin- ing different brain areas and generally indicates activity in areas consistent with the known elec- trophysiological responses to the behavior. Al- though the use of IEG expression offers a simple assay for neural activity with cellular resolution, it lacks temporal resolution, the ability to assess low levels of activity—for example, single action potentials, and the ability to determine activity patterns at more than one time point, which is a requirement for determining the stability of neural ensemble activity during learning and retrieval or across multiple retrieval trials.

The problem of determining activity patterns at two different time points with IEGs was addressed with a method called compartmental fluorescent in situ hybridization (catFISH) (Guzowski et al. 1999). This approach takes advantage of the fact that genes that are being actively transcribed will have unspliced mRNA in the nucleus but relatively little processed transcript in the cytoplasm at short time points after induction. Cells that were active in the past, but are currently silent, will have mRNA in the cytoplasm but no nuclear transcript. The approach was used with the *arc* gene to test ensemble activity in the hippocampus in animals that explore an identical environment twice (A-A) compared with animals exploring two distinct environments (A-B). The study found greater ensemble reactivation in CA1 neurons (40%) when the two environments explored were identical (A-A group) versus when they were distinct (A-B group, 15% reactivation). This result is consistent with the idea that the hippocampus is encoding a representation of place that is reactivated when the information is retrieved. However, there are a number of caveats to this interpretation. First, the experiment is really just comparing the activity pattern of two sensory experiences rather than retrieval of a memory. Second, the technique only allows the comparison of ensemble activity at two closely spaced time points (30 min or less) and so does not indicate the stability of these ensembles over long time frames or in multiple retrieval trials. Finally, the link of IEGs to neural activity is crude and does not allow precise assessment of activity patterns or low levels of activity as can be achieved with electrophysiological recording.

Classical hippocampal recording techniques cannot readily detect large ensembles of neurons stably over long periods of time, making it difficult to use for the type of experiment discussed above. An alternative approach is the use of genetically encoded reporters of calcium levels combined with optical imaging of the transient fluorescent signals produced on calci-

um binding (Akerboom et al. 2012). This approach offers a level of temporal resolution and action potential sensitivity that is intermediate between physiological recording and IEG expression, with the advantage that the neurons are precisely identified anatomically and can be stably imaged over long time periods. This approach was used to simultaneously record the activity of between 500 and 1000 CA1 hippocampal neurons over 45 days while the animals ran on a linear track in a constant spatial environment (Ziv et al. 2013). As seen in previous electrophysiological studies, on any given day, the neurons showed clear spatial firing fields with ~20% of neurons meeting the criteria for place cells. However, between any 2 days, the precise ensemble of neurons recruited in each session showed a lower level of stability. At 5 days separation between recording sessions, there was a 25% overlap in the identity of place cells recorded in the two sessions. The overlap dropped to 15% at 30 days separation between recording sessions with only ~3% of cells active in all 10 recording sessions. So while on each day 20% of the CA1 neurons were active as place cells on the track, the majority of those cells were different from day to day in identical spatial environments. If the hippocampus is representing the environment through the activity of place cells, then why is there not greater stability of the ensemble of neurons activated when the animal is reexposed to that environment?

There are several possible explanations for the apparent instability in hippocampal ensembles in the preceding experiment. One possibility is that there are subtle differences in the environment from day to day to which the animal is responding. Another possibility is that the task (running for a water reward) was not sufficiently salient to produce a stable representation. A previous study in mice using tetrode recordings showed that the stability of place cells over several days is modulated by the salience of the task the animals were required to perform in that environment (Kentros et al. 2004). Having non-food-deprived animals collect food pellets during the recording caused the place cells to be unstable from day to day while requiring the animals to attend by demanding that they navigate to a specific location to avoid an aversive light/noise cue produced the greatest temporal stability in place cell firing. Another possibility is that the CA1 neurons are encoding a component of time such that each successive day the ensemble varies to indicate that, although the environmental cues are identical, today's exploration is a different event in time from the previous days. This was suggested by electrophysiological recording of place cells over hours to days where it was found that many CA1 neurons changed their firing rate over time, consistent with the calcium imaging studies above, although the CA3 neurons showed greater session-independent stability in firing pattern (Mankin et al. 2012). Thus, the CA3 region of the hippocampus may hold an environment-specific map with the CA1 region adding information related to the specific time the environment is explored. Finally, it is possible that the critical spatial signal is contained in the small percentage of neurons that do show consistent firing between multiple recording sessions or in subtleties in the activity patterns that are below the threshold of this technique to resolve.

Whatever the explanation, this experiment raises an important question in understanding how memories are represented and retrieved. How consistent is the pattern of brain activity in response to two identical sensory inputs or two memory recall events? What is noise and what is signal in the pattern of neural activity that is observed? Certainly the brain's ability to consistently recognize and learn about elements in the environment implies some coherent signal in the neural activity patterns induced by the same sensory stimulation, but the models and approaches to understanding this information will differ depending on whether the signal is a dominant or a minor component of the sensory-evoked activity.

The examination of activity patterns during learning and retrieval or during two bouts of memory retrieval provide one means of identifying a neural signal associated with memory recall. As we have seen above, these experiments provide somewhat conflicting data regarding the degree of ensemble reactivation fidelity during different retrieval events. These experiments

are also conceptually problematic in that they do not truly differentiate memory retrieval from sensory processing or encoding. When animals are placed in the same environment twice, they experience the same sensory cues and any consistency in the neural ensemble activity could reflect the processing of these cues rather than memory retrieval. Finally, even if the neural activity represents memory retrieval, the data are correlative and would require direct manipulation to test for functional relevance.

## Optogenetic Manipulation during Retrieval

One approach to directly test the function of distributed ensembles of active neurons in the mouse has recently been developed (Reijmers et al. 2007). The approach uses a transgenic mouse that allows the genetic modification of neurons based on their natural, environmentally evoked, activity patterns within an experimentally controlled time window. The IEGs, discussed above, are genes that are expressed in response to neural activity and previous studies have shown that their DNA promoter elements could confer this activity-dependent expression on linked reporters in transgenic mice (Smeyne et al. 1992). This was exploited to develop a binary genetic system in which neural activity at a given point in time could drive expression of any gene of interest (GOI) in the active neurons (Fig. 1). In this system, the cfos promoter is used to drive expression of the tetracycline transactivator (tTA), a transcription factor that can be modulated by the antibiotic doxycycline (Dox), a derivative of tetracycline. The tTA can activate the expression of a second gene when it is linked to a tetracycline-responsive promoter element (TRE). In animals carrying two transgenes, cfos-tTA and TRE-GOI, the expression of tTA is directly linked to natural neural activity by the cfos promoter rising and falling as neurons become active. In the presence of Dox the transcriptional activity of the tTA is blocked, preventing downstream activation from the GOI. When Dox is removed, the GOI will now be expressed (via tTA driven transcription) in all neurons that are sufficiently active to drive the cfos-linked tTA. This allows the genetic modi-

 Cite this article as *Cold Spring Harb Perspect Biol* doi: 10.1101/cshperspect.a021790

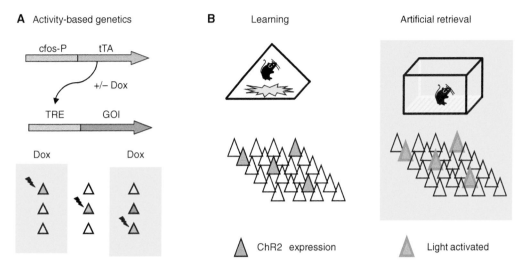

**A** Activity-based genetics

**B** Learning    Artificial retrieval

ChR2 expression    Light activated

Figure 1. Memory retrieval induced by direct circuit reactivation. (*A*) System for introducing genes into neurons based on their natural, sensory-evoked activity patterns (Reijmers et al. 2007). The cfos-promoter drives tetracycline transactivator (tTA) in response to neural activity and the tTA then activates and gene of interest (GOI) that is linked to a TRE-promoter. Doxycycline (Dox-yellow shading) blocks the transcriptional activity of tTA providing temporal control over the time frame in which neural activity drives the GOI. (*B*) Experimental design for demonstrating the functional relevance of distributed neural ensembles. Neurons activated during learning in a fear-conditioning paradigm are genetically tagged with channelrhodopsin (ChR2). The subsequent light-induced firing of these ensembles is able to produce a fear response in a second, emotionally neutral, context.

fication of neurons active at specific points in time in response to specific environmental stimuli, for example, learning or retrieval cues.

The approach was used in several recent publications to test the psychological relevance of distributed neural ensembles activated during learning in a contextual fear-conditioning task (see Fanselow and Wassum 2015). In contextual fear conditioning, animals receive footshocks in a particular context (the conditioning box) that contains multimodal sensory cues (generally distinct visual, tactile, and odor cues). Memory retrieval is assessed by the fear response the animals display when returned to the conditioning chamber. This is a model of explicit memory that is sensitive to hippocampal lesions and, reported by a number of groups, to show the temporal gradient in amnesia seen in many studies of human hippocampal patients with older (consolidated) context memories insensitive to hippocampal lesion (Kim and Fanselow 1992; Anagnostaras et al. 1999; Squire and Bayley 2007). In Liu et al. (2012), *cfos*-based genetic

tagging was used to introduce channelrhodopsin (ChR2) into dentate gyrus (DG) neurons that were activated during contextual fear learning. Mice were placed in one of two contexts A or B (conditioning boxes with different visual, tactile, and odor cues) and the neurons that were naturally activated by this environmental exploration were tagged with ChR2 to allow their subsequent direct activation with light. Both groups of mice were then fear conditioned to context A while on Dox, to prevent any further labeling of active ensembles. When mice in which DG neurons were tagged in context A were also fear conditioned in context A, the subsequent stimulation of the ChR2 expressing neurons produced a fear response in a neutral environment. Stimulation of those neurons that were genetically tagged with ChR2 while animals explored context B failed to produce fear in animals conditioned to fear context A. This suggests that the ensemble of neurons activated during the exploration of context A is capable of producing memory retrieval when directly stimulated. That is, the brain represents the conditioning box through the activity of the ensemble of neurons active in the DG during initial exploration (learning) and presumably through downstream neurons in other brain regions that are activated when this ensemble is directly stimulated via ChR2.

It is quite remarkable that the highly nonphysiological stimulation that is produced with ChR2, which fires all neurons simultaneously and thus eliminates any temporal sequence effects or coordination with endogenous rhythms, can apparently produce a coherent internal representation of a complex environment. To further explore this idea, a complementary experiment was performed to determine whether the ensemble activity in the DG could serve as a conditioned stimulus (CS) in fear conditioning. In this case, *cfos*-activated neurons were genetically tagged with ChR2 while the animals explored context A, as in the previous study. However, now the ChR2 labeled ensemble was activated in a neutral context and paired with footshocks. Thus, the artificial stimulation served as a CS that was paired with a footshock unconditioned stimulus (US). In this case, the animals developed a fear response to context A even though they never actually received the aversive shock US in that environment (Ramirez et al. 2013). Taken together, these results provide support for the notion that the activity of DG neural ensembles is sufficient to represent the context and serve as a cue for memory encoding and retrieval.

The artificial stimulation of small ensembles of neurons in the DG region of the hippocampus allows the retrieval of contextual memories presumably by activating more distributed ensembles of neurons in downstream hippocampal and cortical regions. Given the highly parallel nature of neural connectivity and processing, it is possible that although these ensembles may be sufficient for retrieval, they are not necessary as other pathways could compensate. The question of necessity was addressed in two recent studies using a light-gated proton pump from archaebacteria (ArchT) to hyperpolarize and silence cfos tagged neural ensembles in experiments analogous to those described above with

ChR2 (Denny et al. 2014; Tanaka et al. 2014). Both teams of investigators also used context fear conditioning and examined the requirement for ensembles in three different hippocampal regions, the DG, CA3, and CA1. Expression of ArchT was driven into neurons that were active during learning in context fear conditioning, and these ensembles were subsequently silenced in either the DG, CA3, or CA1 during memory retrieval. In each case, the silencing impaired the retrieval of the contextual fear memory. These results show that in each of the three major hippocampal subregions, the reactivation of the ensemble of neurons active during initial learning is necessary for the subsequent retrieval of the memory. This is consistent with the view that retrieval involves the reconstruction of patterns of brain activity produced during initial learning.

## Retrieval of Consolidated Memories

Another notable aspect of memory and its retrieval is the change in circuit structure of hippocampal-dependent memories over time, originally exemplified in amnesic patients such as H.M. (Squire and Dede 2015). The observation that newly formed memories require the hippocampus for retrieval, but following weeks (in rodents) or months to years (in humans), the hippocampus seems to become dispensable for at least some forms of explicit memory, suggests that there is an anatomical alteration in the memory over time. What is the nature of this circuit-based consolidation of memory ("systems consolidation") (Squire et al. 2015)? What is the anatomical locus of hippocampal-dependent new memories and hippocampal-independent older, consolidated, memories? What is the role of the hippocampus and cortex in the encoding and retrieval of new and old memories? These questions will also resurface below in our discussion of episodic retrieval in humans. The current view of this process posits that during initial encoding and retrieval of an explicit memory, the hippocampus plays a critical role, for example, by encoding an index or map that helps recruit the appropriate cortical regions and specific neural ensembles for recall.

Cite this article as *Cold Spring Harb Perspect Biol* doi: 10.1101/cshperspect.a021790

Over time, these cortical ensembles may become independent of the hippocampus, possibly by strengthening their connectivity through postlearning replay of relevant activity patterns (Wilson and McNaughton 1994), such that with the newly strengthened cortical connections the memory, or at least some processed versions of it, can be retrieved independently of the hippocampus.

The evidence in rodents for this view of consolidated memories comes from lesion and inactivation of specific brain regions and from studies of the neural activity recruited during retrieval of recent and remote memories. One prediction of the classic memory consolidation model is that, although the hippocampus is important in recent memory retrieval, cortical areas should instead be required for remote memory retrieval, or possibly recent and remote retrieval. This has been tested in the consolidation of three different explicit memory paradigms in rodents: spatial memory, contextual memory, and an olfactory-based social memory (Bontempi et al. 1999; Frankland et al. 2004; Frankland and Bontempi 2005). In all three studies, hippocampal lesions impaired recent memory retrieval (1 day posttraining) but spared remote retrieval (30 days posttraining) as expected for hippocampus-dependent tasks involving consolidation. For contextual fear and spatial memories, inactivation of the anterior cingulate cortex produced an impairment in remote memory retrieval but did not affect recent memory, the opposite profile to what is observed with hippocampal inactivation or lesion (Kim and Fanselow 1992; Anagnostaras et al. 1999). A similar result was seen with prefrontal inactivation in the spatial task, while in the olfactory-based social task, inactivation of the orbital frontal cortex impaired both recent and remote memory retrieval.

In addition to inactivation and lesions to probe the anatomical structure of recent and remote memories in rodents, these studies also examined the neural activation patterns during retrieval using IEG expression. In each behavioral paradigm, the activation of the hippocampus was greater with retrieval of recent as opposed to remote memories. Conversely, a wide variety of cortical areas showed increased activity during the retrieval of remote relative to recent memories. Although these studies support the view that remote memories require an increased cortical role in processing, they also raise some questions regarding the initial model of hippocampal coordination of cortical activity during retrieval of recent memory. If, during recent memory retrieval, the hippocampus is coordinating and recruiting cortical activity, then why are these regions often not required for retrieval at this time point and why is there altered cortical activity over time. This suggests that there may be a more fundamental alteration in the circuitry and nature of these memories over time. This is evident at the behavioral level as studies in mice (and humans) have shown that remote memories lose some specificity (Wiltgen and Silva 2007; Winocur et al. 2010; Furman et al. 2012).

One recent study in the mouse has examined the cortical representation of explicit memory using tagging of behaviorally active neural ensembles with ChR2 (Cowansage et al. 2014). In experiments similar to those described above for the DG, ChR2 expression was driven into neurons that were activated during context fear leaning. The investigators examined the retrosplenial cortex, a cortical output area of the hippocampus that is required for encoding and retrieval of both recent and remote contextual memory (Keene and Bucci 2008a,b; see also human studies below). They found that artificial stimulation of the learning-activated ensemble (via ChR2) produced a freezing response suggesting that it instantiated a recall event similar to what was seen in the hippocampal studies. This shows a contextual representation capable of producing retrieval forms in the cortex, in addition to the hippocampus, at the time of learning. More surprisingly, they found that the artificial stimulation of these retrosplenial cortex ensembles produced fear recall even when the hippocampus was pharmacologically silenced 1 day after training, a time point when the hippocampus is still required for natural retrieval of contextual memories. These results show not only that a cortical representation for context memory forms at the time of learning but the normal requirement for the hippocampus in recall can be bypassed by direct activation of this representation. This finding is consistent with the view of the hippocampus as a map or index that recruits the appropriate cortical circuits during memory retrieval but does not directly store a necessary component of the consolidated representation itself.

The experiments using ChR2 to reactivate neural ensembles that were naturally activated during learning provide important information on how the brain stores and represents complex information about the external world. However, it may be incorrect to characterize these neurons as the "engram" or engram-containing neurons as has sometimes been suggested. The term engram refers to the physical changes in the brain that underlie memory (Semon 1904; Lashley 1950). If we take context conditioning as an example, the conditioning box is initially neutral and exposure produces exploratory behavior. When paired with footshock, the same chamber now causes animals to express a fear response. The engram for this memory would be the sites within the brain that lead the same sensory information (the cues in the box) to be processed in a way that produces fear rather than exploratory behavior, for example, by routing the information to activate the amygdala. The stimulation of ChR2 ensembles at any point in the pathway for processing the sensory information of the context might be expected to represent, and therefore substitute for, the context, without actually being the critical site of plasticity required to produce the processing to fear circuits (the engram). For example, if the retina could be artificially stimulated in precisely the same manner as when an animal explored the context then it might be expected to produce a fear response in conditioned animals in the same way as the experiments discussed above but without carrying the "engram" for that memory.

The study of memory in invertebrates sets up a solid conceptual framework for understanding information processing by nervous systems. Sensory input is processed to different motor/behavioral output based on experience and via plastic changes at specific nodes in the

processing network. The mammalian brain introduces a 4 (mouse)- to 7 (human)-order of magnitude increase in the number of neurons between the input and output nodes, but a deep understanding of mammalian memory and retrieval will still require identifying how this information flows through and is represented in the nervous system and the specific sites that are altered with experience to produce the retrievable memory. We have introduced some of the techniques used in mice and rat models and results that suggest a framework for probing this processing at a fine level focused on specific neural ensembles. The added complexity of the mammalian brain has allowed greater specialization of regions for different processing tasks, and memory/retrieval will likely involve a coordinated interplay of more sparse and distributed networks than in simpler systems. The work in humans involving functional brain imaging and, more recently, electrophysiology, is also beginning to identify some of the principles of these processing networks. These techniques allow the assessment of activity across the entire brain during encoding and retrieval and provide a view of the interactions and functional distinctions of different areas during processing.

## MEMORY RETRIEVAL IN THE HUMAN BRAIN

Our discussion of human memory retrieval will focus on declarative and particularly episodic memory (Squire and Dede 2015). We will start by briefly reviewing aspects of the rich phenomenological analyses of human memory retrieval, which has laid the foundations for much of contemporary research on brain substrates and processes of retrieval. We will then survey current knowledge on human brain circuits that subserve retrieval, and conclude with a brief description of functional models that inform many brain studies of long-term memory retrieval.

### Phenomenological Analyses of Human Retrieval Processes

Contemporary research in the neurobiology of human memory retrieval relies heavily on a rich body of research in experimental and cognitive psychology that flourished already half a century ago. This research yielded classifications, models, and questions that are at the forefront of investigation of brain and behavioral mechanisms of retrieval. We will mention only a few examples and then proceed to describe their reflection in brain research.

A major distinction in human declarative retrieval is between recall and recognition. Whereas recall is the reactivation or reconstruction of the internal representation of a target item in the absence of that item, prompted by implicit or explicit cues, recognition is the judgment of previous occurrence in the presence of at least part of the target item. This distinction refers hence both to the test used to probe the memory (i.e., whether in the presence or in the absence of the target item) and to the postulated cognitive underpinning of the memory performance gauged by the test. Recall and recognition were each proposed to involve multiple processes, and these processes themselves were further dissociated. In brief, recall was initially posited by some to consist of two major phases, a "generation phase" followed by a recognition phase (Bahrick 1970; Kintsch 1970; Anderson and Bower 1972). Similarly, "dual-process" models of recognition maintain that recognition judgments can be based on two distinct types of memory, familiarity and recollection (James 1890; Mandler 1980; Jacoby and Dallas 1981; Yonelinas 1994; Diana et al. 2006; Voss and Paller 2010). Multiple experimental paradigms have been used to dissociate the two processes (Yonelinas 2002), probing either a subjective sense of recollection ( participants indicate whether they recollect an item, or find it familiar in the absence of specific recollection [Tulving 1985]), or the objective ability to recollect additional aspects of the study event, such as the context in which it was learned ("source memory") or an item associated with the probe during learning.

Although many investigators agree on the existence of two distinct recognition processes (but see Shimamura 2010, for a single-process model), there remain several contentious debates. The first of these pertains to the nature

of the recollection signal and how recollection and familiarity ultimately contribute to the recognition judgment. The two leading classes of models, each based on both behavioral and functional neuroimaging evidence, differ primarily with respect to their view of recollection as either a threshold process or as a continuous variable. According to the dual-process, signal-detection model (Yonelinas 1994, 2002; Parks and Yonelinas 2007), familiarity and recollection are two independent processes initiated in parallel. Familiarity is considered to reflect a continuous measure of memory strength, best modeled as a signal-detection process. Conversely, recollection is considered to be a threshold process, whereby only items falling above a certain threshold will be recollected, resulting in relatively high-confidence responses (Yonelinas et al. 2010).

According to an alternate approach, recollection is also a continuous measure that is best modeled by a signal-detection process, and recognition judgments are based on an aggregated memory-strength variable (Rotello et al. 2004; Wixted 2007; Mickes et al. 2009; Wixted and Mickes 2010). The continuous dual-process model (Wixted and Mickes 2010) maintains that, during the process of recognition, recollection and familiarity each elicit a separate internal measure along a memory strength axis. The decision criterion for identifying an item as recognized may be based either on one of these axes, or on an aggregated memory-strength axis, which takes into account both the sense of recollection and the sense of familiarity, depending on the task at hand. The different approaches start to converge, with suggestions that recollection may be graded based on the amount of recollected information (Rugg et al. 2012), but still subject to a threshold-like process (Yonelinas and Jacoby 2012).

Another point of dispute is whether the tests typically used to dissociate recollection and familiarity (e.g., remember/know test [Tulving 1985]) indeed dissociate these two processes (e.g., Perfect 1996; Yonelinas 2002; Parks and Yonelinas 2007; Yonelinas et al. 2010), or whether they in fact only separate strong memories from weaker ones (Donaldson 1996; Dunn

2004; Wais et al. 2008; Wixted et al. 2010). Last, but clearly not least, in the context of the present discussion, the debates involve the role of distinct brain regions, primarily in the medial temporal lobe (MTL) in recollection/familiarity processes (see below).

We selected as an example the recollection/familiarity debate as it is currently one of the more heated topics in research on episodic memory retrieval (Voss and Paller 2010). However, it is noteworthy that this far from covers the research into the phenomenology, and ultimately candidate brain mechanisms, of human memory retrieval. An additional distinction is drawn between the content of retrieval and the entering of a state that enables retrieval ("retrieval mode" [Tulving 1983], see Rugg and Wilding 2000, for a more fine-grained fractionation), or more generally between different types of item-specific and item-invariant processes in retrieval (e.g., Nyberg et al. 1995; Buckner et al. 1998; Köhler et al. 1998; Dobbins and Wagner 2005; Duarte et al. 2011; Bergström et al. 2013). Further, different types of memories have been shown to involve different retrieval processes, for example, when comparing autobiographical and laboratory-based memories (Cabeza et al. 2004; Svoboda et al. 2006; McDermott et al. 2009), field versus observer perspective (Eich et al. 2009), objective versus subjective measures of recollection (see Spaniol et al. 2009 for a meta-analysis) and emotional versus neutral memories (Maratos and Rugg 2001; LaBar and Cabeza 2006; Buchanan 2007). Although this limited discussion does not allow for a comprehensive review of the fine-grained analysis of the phenomenology of human retrieval, it is important to bear in mind that the rich differentiation of retrieval process has been shown to manifest in differential brain activity during retrieval, emphasizing the notion that retrieval cannot be investigated as a unitary construct.

## Substrates of Retrieval in the Human Brain

As found in rodents, brain circuits of human explicit retrieval are highly distributed (Maguire et al. 2000; Svoboda et al. 2006; Spaniol et al. 2009; Kim et al. 2010; Mendelsohn et al. 2010).

We will briefly review the role of only a few of the major brain areas involved and their interactions in human episodic retrieval, while relating them to some of the postulated subprocesses of retrieval mentioned earlier.

## Hippocampus and MTL

Much of the emphasis in the study of episodic retrieval has been placed on the roles of the MTL (Squire and Dede 2015), specifically differentiating between the hippocampus and surrounding cortices. As focal damage to the fornix and mammillary bodies results in impaired performance that resembles that of hippocampal damage (Dusoir et al. 1990; Tsivilis et al. 2008; Rudebeck et al. 2009; Vann et al. 2009b), it has been suggested the study of hippocampus-based memory should be extended to include the fornix and mammillary bodies as part of the "extended hippocampal system" (Aggleton and Brown 1999).

One approach to differentiating hippocampal versus MTL cortical contributions to retrieval is based on the aforementioned recollection/familiarity distinction. Several studies have addressed the question of whether the hippocampus is uniquely involved in recollection, or whether it supports both familiarity and recollection (Brown and Aggleton 2001; Eichenbaum et al. 2007; Skinner and Fernandes 2007; Squire et al. 2007; Wixted and Squire 2011; Rugg et al. 2012; Rugg and Vilberg 2013). Both viewpoints are based on findings from studies in patients with MTL damage combined with functional neuroimaging results in healthy subjects.

Whereas many studies report a disproportionate effect of hippocampal damage on recollection and associative memory relative to familiarity (Huppert and Piercy 1978; Vargha-Khadem et al. 1997; Holdstock et al. 2002; Yonelinas et al. 2002; Giovanello et al. 2003; Mayes et al. 2004; Aggleton et al. 2005), other reports find that hippocampal damage impacts familiarity and recollection to a similar extent (Manns and Squire 1999; Stark et al. 2002; Manns et al. 2003; Cipolotti et al. 2006; Wais et al. 2006; Jeneson et al. 2010; Kirwan et al. 2010; Song et al. 2011). Interestingly, a patient with significant perirhinal damage that spared the hippocampus showed impaired familiarity and preserved recollection (Bowles et al. 2007). Such studies indicate a causal role of MTL regions in the different processes, yet do not allow for dissociation between encoding, storage and retrieval, nor do they necessarily reflect memory processes in the healthy brain.

Conversely, functional magnetic resonance imaging (fMRI) studies do not allow for demonstration of causality, but they enable targeted investigation of correlation with different stages of memory. However, even in fMRI studies focusing on retrieval processes, incidental encoding during retrieval tasks may hinder the ability to tease apart encoding and retrieval processes (Buckner et al. 2001; Stark and Okado 2003; Kim 2013; but, see Staresina et al. 2012b; Ben-Yakov et al. 2014 for temporal dissociations between encoding and retrieval). As with the patient studies, fMRI studies have led to divergent results regarding hippocampal involvement in recollection. One set of studies finds that the hippocampus subserves recollection, but not familiarity, of memoranda such as words or pictures (Eldridge et al. 2000, 2005; Weis et al. 2004; Wheeler and Buckner 2004; Yonelinas et al. 2005; Montaldi et al. 2006; Diana et al. 2007, 2010; Cohn et al. 2009; Ford et al. 2010; Rugg et al. 2012; Yu et al. 2012). More specifically, the hippocampus has been found to respond more strongly to words or pictures reported as "remembered" (Eldridge et al. 2000, 2005; Wheeler and Buckner 2004; Yonelinas et al. 2005; Montaldi et al. 2006), in correct versus incorrect retrieval of the encoding context (Weis et al. 2004; Tendolkar et al. 2008; Duarte et al. 2011; see Staresina et al. 2012b for a related intracranial electroencephalography [iEEG] study), and in associative relative to nonassociative recognition (Kirwan and Stark 2004) or recognition of compound words (Ford et al. 2010). Recent studies propose that hippocampal activity is not related to the subjective sense of recollection, but modulated by the amount of contextual information actually retrieved (Rugg et al. 2012; Yu et al. 2012).

According to an alternate view, the hippocampus is not preferentially involved in recol-

lection versus familiarity when controlling for memory strength at the time of the retrieval test (Wais 2008; Wais et al. 2010; Smith et al. 2011; see Montaldi et al. 2006 for opposite results). Wixted and colleagues suggest that the hippocampus and perirhinal cortex are involved in both familiarity and recollection (Squire et al. 2007), but that the hippocampus supports strong memories, whereas the perirhinal supports weak memories (see Wais 2008 for a meta-analysis). However, they propose that memory strength is itself not the parameter that differentiates the hippocampus from the surrounding structures, but rather that different MTL structures process attributes of the memory that are differentially expressed in strong versus weak memories. According to this view, the hippocampus supports both recollection-based and familiarity-based recognition of multiattribute stimuli, and its involvement in retrieval is most evident for strong memories (Wixted and Squire 2011). Despite the differing interpretations with respect to recollection/familiarity, overall, the hippocampus appears to be primarily involved in the retrieval of strong, rich, multiattribute memories, whereas the surrounding cortices can support retrieval of more simple memories without hippocampal involvement.

Recollection/familiarity are often used in the human literature to describe both the behavioral phenomena and the patterns of brain activity correlated with these phenomena, although there may not necessarily exist a one-to-one mapping between the two types of measures (Voss and Paller 2010). A key example is the finding that the perirhinal cortex is involved in associative memory recognition under conditions of unitization, in which the associated elements comprise a unitized item (Diana et al. 2010; Ford et al. 2010), whereas the hippocampus is preferentially involved in recognition of nonunitized versus unitized pairs (Quamme et al. 2007; Ford et al. 2010). This has been interpreted either as evidence that familiarity can support associative recognition when the paired associates can be bound into a compound unit (Ford et al. 2010; Ranganath 2010) or as evidence that the perirhinal cortex is involved in recollection of unitized pairs (Diana et al. 2010; Wixted and Squire 2011). Similarly, findings of hippocampal neurons sensitive to picture novelty/oldness (Fried et al. 1997; Rutishauser et al. 2006, 2008; Viskontas et al. 2006) have been subject to differing interpretations, as it is not clear how neuronal sensitivity to familiarity relates to familiarity at the behavioral level (Parks and Yonelinas 2007; Wixted 2007).

The attempts to delineate the role of hippocampus in recollection versus familiarity rest on the assumption that the hippocampus honors the well-established behavioral dissociation between these two manifestations of retrieval. However, there is an increasing view that the behavioral distinction between recollection and familiarity does not reflect the underlying basic computational role(s) of the hippocampus and its surrounding cortices (Diana et al. 2010; Ranganath 2010; Voss and Paller 2010; Wixted and Squire 2011; Rugg et al. 2012). An alternate approach to interpreting the findings reviewed above is that the hippocampal formation plays a role in the binding items in their context (Diana et al. 2007; Eichenbaum et al. 2007; Ranganath 2010). This postulated role also introduces a functional distinction between perirhinal and parahippocampal cortices, according to which the perirhinal cortex encodes item information, the parahippocampal cortex encodes contextual information, and the hippocampus encodes item–context associations as well as item–item associations. A related view (Montaldi and Mayes 2010) assigns to the perirhinal cortex a role in item–memory and within-domain inter-item associations. All of these models emphasize the role of the hippocampus in retrieval of bound associations relative to single item/context recognition, which are supported by the surrounding cortices. They can also be considered in line with the view, mentioned above in discussing the rodent work, that the hippocampus serves as a map or index that recruits the appropriate cortical circuits.

The aforementioned models predict a role for perirhinal cortex in recognition of single items. This is supported by fMRI studies that find decreased perirhinal activity for highly familiar stimuli (Weis et al. 2004; Gonsalves et al. 2005; Daselaar et al. 2006a; Montaldi et al. 2006; although see Kirwan and Stark 2004 for increased perirhinal activity in response to correct associative recognition), as well as an intracranial recording study that found reduced firing in response to familiar images in perirhinal cortex (Viskontas et al. 2006). A study combining iEEG in patients with fMRI in healthy controls (Staresina et al. 2012b) paints a more complex picture. In this study, the perirhinal cortex showed both an early-item novelty effect (differential response to novel vs. familiar words, potentially reflecting a familiarity process) and a sustained source retrieval effect (differential response to correct vs. incorrect retrieval of the context associated with the word). The hippocampus showed an early source retrieval effect, followed by a late-item novelty effect (potentially underlying encoding of the novel item). These studies show a clear role of perirhinal cortex in the recognition of familiar stimuli.

All in all, these different views converge on a predominant role of the hippocampus in associative retrieval. In addition to the functional division between the hippocampus and surrounding cortices, a within-hippocampus functional dissociation has also been suggested, with anterior hippocampus more involved in encoding and posterior hippocampus more involved in retrieval (Lepage et al. 1998; Strange et al. 1999; Spaniol et al. 2009; Poppenk et al. 2013; but see Schacter and Wagner 1999; Ludowig et al. 2008). High-resolution fMRI (Carr et al. 2010) enables an alternate distinction, between the different hippocampal subfields. The few studies available at the time of writing seem to reveal a dissociation between encoding processes observed in DG/CA3 and retrieval (or recollection) in CA1/subiculum (Zeineh et al. 2003; Eldridge et al. 2005; Viskontas et al. 2009).

## "Default Mode" Network

A set of regions consisting of the medial prefrontal cortex (mPFC), posterior cingulate, in-

Cite this article as *Cold Spring Harb Perspect Biol* doi: 10.1101/cshperspect.a021790

ferior parietal lobule, lateral temporal cortex, and MTL, has been identified as preferentially active in situations in which external stimuli are absent. This set of regions has therefore been named the "rest" or "default mode" network (Raichle et al. 2001; Buckner et al. 2008; Raichle 2010). Multiple functions have been attributed to the default network, including stimulus-independent mental processes ("mind wandering"; Mason et al. 2007), mental scene construction (Hassabis and Maguire 2007) and self-projection into the future/past (Addis et al. 2007; Buckner and Carroll 2007).

There is a remarkable overlap between the network activated in retrieval of autobiographical memory (Svoboda et al. 2006; McDermott et al. 2009) and the default network (Buckner et al. 2008). Further, a recent study of patients with focal lesions in regions within the default network showed correlation of damage to this network with disruption of autobiographical memory retrieval (Philippi et al. 2015). Whereas deficits in semantic elements of autobiographical memory were associated with damage to the left mPFC and MTL, deficits to episodic autobiographical memory were associated with damage to the right mPFC and MTL. A meta-analysis study (Spreng et al. 2009) that compared default mode regions with activations of autobiographical memory retrieval, prospection (imagining oneself in the future), navigation, and intentionality ("theory of mind") found significant overlap between most of the different functions. Retrieval, prospection, and theory of mind tasks all activated similar default mode regions, including multiple prefrontal, parietal, and temporal regions. The abundance of evidence linking episodic retrieval to mental construction or simulation of future/fictitious events (Hassabis and Maguire 2007, 2009; Hassabis et al. 2007; Schacter and Addis 2007; Schacter et al. 2008; Buckner 2010; Rabin et al. 2010; Szpunar 2010; Viard et al. 2012) has led to a renewed emphasis on the constructive nature of memory (Bartlett 1932).

Investigation of the commonalities versus the differences between episodic retrieval, which is mental time travel to the past, and mental time travel to the future (imagination), have

enabled dissociation between purely mnemonic components and processes relating more generally to construction of episodes (Addis et al. 2007, 2009; Weiler et al. 2010; Addis and Schacter 2011; Schacter et al. 2012). Notably, the hippocampus, which is coupled with the default network at retrieval (Huijbers et al. 2011), is one of the main regions implicated in tasks that require the participant to imagine future scenarios (Addis and Schacter 2011; Schacter et al. 2012; Maguire and Mullally 2013).

## Prefrontal Cortex

The prefrontal cortex (PFC), subregions of which were mentioned above in the context of the "default network," was one of the first regions linked to episodic memory retrieval in functional neuroimaging studies (Fig. 2A) (Nyberg et al. 1995; Buckner et al. 1998; Rugg et al. 1998; Wagner et al. 1998; Düzel et al. 1999; Lepage et al. 2000), in line with studies revealing memory impairment in patients with frontal lobe damage (e.g., Luria 1966; Stuss et al. 1994; Wheeler et al. 1995; Duarte et al. 2005; Aly et al. 2011). Different regions of the PFC have been implicated in mnemonic processes that support retrieval, particularly in domain-general processes, such as setting the retrieval mode or orientation, initiating the retrieval attempt, monitoring, goal-directed manipulation of retrieved information and overcoming of interference (Shimamura 1995; Rugg et al. 1996; Moscovitch and Winocur 2002; Badre and Wagner 2007; Blumenfeld and Ranganath 2007). Several functional divisions of the PFC have been suggested, from the original hemispheric encoding/retrieval asymmetry (HERA) model that proposed preferential right PFC involvement in retrieval (Tulving et al. 1994; Habib et al. 2003; see Cabeza et al. 2003 for an alternate hypothesis regarding the roles of left/ right PFC in retrieval), to within-hemispheric divisions according to medial/lateral, anterior/ posterior and dorsal/ventral axes attributing specific mnemonic processes to each subregion (Fletcher and Henson 2001; Wagner et al. 2001; Petrides 2002; Simons et al. 2005; Badre and Wagner 2007).

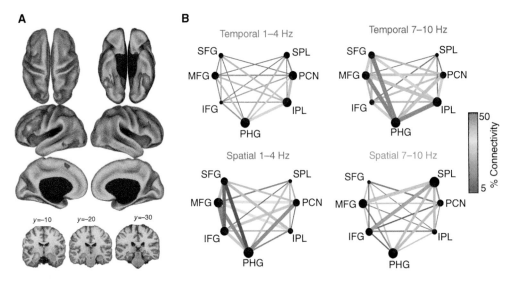

**Figure 2.** Brain correlates of retrieval of human declarative memory. (*A*) Brain network associated with recognition memory. The figure depicts a meta-analysis of areas that show old > new activation in recognition tests in 38 fMRI (functional magnetic resonance imaging) studies. The regions identified include the, angular gyrus, caudate nucleus, dorsolateral prefrontal cortex (DLPFC), dorsomedial prefrontal cortex, dorsal posterior parietal cortex, posterior cingulate cortex, and precuneus. (Based on data in Kim 2013; with permission from the author.) (*B*) Diagrams depicting the dynamics of brain network fast functional connectivity in memory retrieval revealed by electrocorticographical (ECoG) recording in patients undergoing seizure monitoring. The patients were engaged in retrieving spatial and temporal episodic contexts. Phase synchronization between brain areas was used as a measure of connectivity. The panels display the connectivity correlated with correct spatial and temporal retrieval in the 1–4 Hz and 7–10 Hz bands. PHG, Parahippocampal gyrus; MFG, middle frontal gyrus; SFG, superior frontal gyrus; IFG, inferior frontal gyrus; IPL, inferior parietal lobule; PCN, precuneus; SPL, superior parietal lobule. Successful retrieval was associated with greater global connectivity among the sites with the medial temporal lobe (MTL) acting as a hub for the interactions, but whereas correct spatial context retrieval was characterized by lower frequency interactions across the network, temporal context retrieval was characterized by faster frequency interactions. These results provide insight into how multiple contexts associated with a single event can be retrieved in the same network. (Based on data in Watrous et al. 2013; with permission from the authors.)

A meta-analysis comparing encoding and recollection found that both encoding and retrieval of memoranda such as pictures and words primarily activated the left PFC, in contradiction to the HERA model (Spaniol et al. 2009). A comparison between objective/subjective recollection revealed stronger activation for objective recollection in ventrolateral PFC (VLPFC) and dorsolateral PFC (DLPFC) and lateral anterior PFC and stronger activation for subjective recollection in medial anterior PFC, in line with suggested roles for lateral PFC in cognitive control of memory (Simons and Spiers 2003; Badre and Wagner 2007) and medial PFC in self-referential processing (Northoff et al. 2006).

## Posterior Parietal Cortex

In contrast to the functional neuroimaging findings concerning MTL and PFC, there is a striking inconsistency between findings of posterior parietal cortex involvement in retrieval and studies of patients with parietal lesions. Functional neuroimaging studies consistently observed parietal activations in episodic retrieval tasks (Shannon and Buckner 2004; Wagner et al. 2005; Vilberg and Rugg 2008; Hutchinson et al. 2009; Spaniol et al. 2009), with the ventral posterior parietal cortex involved primarily in recollection, and the dorsal posterior parietal involved more generally in recognition (Shannon and Buckner 2004; Wagner et al. 2005; Vilberg and Rugg 2007, 2008, 2012). Conversely, most lesion studies find limited, if any, impairment in episodic retrieval (Berryhill et al. 2007, 2010; Haramati et al. 2008), supported by an repetitive transcranial magnetic stimulation (rTMS) study that found disruption of the posterior parietal cortex does not significantly impair memory (Rossi et al. 2006). A study directly comparing patients lesioned in this cortex with fMRI of healthy individuals found an anatomical overlap between the parietal recollection-related activations and the locus of the damage in patients that showed normal performance on the same task (Simons et al. 2008). The robust parietal involvement in intact recollection taken together with the relatively subtle impairments yielded by damage to this cortex have led to several models regarding its role in memory, particularly in attention to memory (Cabeza et al. 2008, 2012; Ciaramelli et al. 2008), in serving as a short-lived episodic buffer (Vilberg and Rugg 2008; Wagner et al. 2005), and in binding relational activity (Shimamura 2011).

## Additional Contributions to the Retrieval Network

We focused so far on the region's most commonly identified with human episodic memory retrieval, yet additional brain regions have also been implicated in episodic retrieval. The thalamus, in particular, the anterior thalamic nuclei, has received particular attention because of their connections with the hippocampus. Aggleton and Brown (1999) suggested that the anterior thalamic nuclei and mammillary bodies are necessary for recollection, whereas the medial dorsal nucleus supports familiarity. A recent review (Aggleton et al. 2011) supports this proposal only in part. Damage to anterior thalamic nuclei, which are connected to the hippocampus and mammillary bodies, has been shown to impair recognition memory, although the impairment is significantly less severe than damage to the hippocampus or mammillary bodies. An fMRI study identified a linear increase of activity in the dorsomedial thalamus in response to pictures eliciting a

stronger sense of familiarity (Montaldi et al. 2006), yet it is unclear whether damage to this nucleus directly impairs recognition. An additional region that has received renewed interest is the retrosplenial cortex (see also rodent studies above). Although the retrosplenial cortex is consistently activated in studies of autobiographical retrieval, and damage to the region impairs retrieval of recent episodic memories, its role in memory was originally thought to be indirect (Vann et al. 2009a). Specifically, this cortex is preferentially involved in recollection of recent (Gilboa et al. 2004) and self-relevant memories (Summerfield et al. 2009), and it has been suggested to support translation from egocentric to allocentric views (Byrne et al. 2007). Taken together, these findings suggest a role for the retrosplenial cortex in semantization of episodic memories, which may entail a shift to a more allocentric representation of the memory. The striatum, traditionally linked to nondeclarative memory, has also become a focus of studies in the context of declarative memory retrieval, and it has been suggested to support cognitive control of episodic retrieval in concert with the PFC (Scimeca and Badre 2012).

Cabeza and Moscovitch (2013) propose that each memory task recruits a different combination of processing components, some of which are likely to be shared by nonmemory tasks as well. The set of processing components that interact to support a given task, for example, retrieval of an episodic memory, is denoted a "process-specific alliance," underlining the effect of transient interregional interactions on the contribution of individual brain regions to the task at hand. In the context of our discussion, this emphasizes the importance of understanding not only the role of each brain region in retrieval, but also how different regions interact to support distinct mnemonic functions, including retrieval (Simons and Spiers 2003; Mendelsohn et al. 2010; Cabeza et al. 2011; Huijbers et al. 2011; St Jacques et al. 2011; Fornito et al. 2012; Watrous et al. 2013). Further, the notion that the same region could also be involved in nonmnemonic tasks, may at times blur the distinction between brain signatures of perception and retrieval, a possibility hinted above in the discussion of recent data on artificial activation of candidate engrams in the rodent brain.

## Distributed Interactions in Retrieval

fMRI studies have shown that, during retrieval, the interactions between distal brain regions is modulated by the type of memory retrieved (Maguire et al. 2000; Greenberg et al. 2005), the age of the memory (Viard et al. 2010; Söderlund et al. 2012), and its veridicality (Mendelsohn et al. 2010). However, fMRI provides us with only snapshots of brain states, averaged over a period of time that is much longer than that in which the machinery of retrieval functions (see above). Electrophysiological studies in human patients undergoing evaluation before neurosurgery has led to finer spatiotemporal dissection and novel insights into the detailed mechanisms of retrieval in the human brain (Suthana and Fried 2012). A prime example is interregional phase locking. Activated ensembles of neurons show oscillatory patterns of activity, and it has been suggested that synchronization in the oscillation phase between different regions may enable communication and mutual interaction between them (Fries 2005). Further, synchronization at different frequency ranges corresponds to different spatial scales, with synchronization at lower frequencies associated with interactions between more distal regions (Canolty et al. 2010). Thus, phase coupling between low and high frequencies may enable synchronization of local activity in distal brain regions (Canolty et al. 2006; Jacobs et al. 2007; Canolty and Knight 2010; Rutishauser et al. 2010; Maris et al. 2011; van der Meij et al. 2012). In regard to memory, there has been particular interest in the theta band (3–8 Hz including "slow theta"), and in theta/gamma (30–150 Hz including "high gamma") coupling, which have been linked to various mnemonic processes of encoding, working memory and long-term memory (Nyhus and Curran 2010; Fell and Axmacher 2011). Specifically, successful memory retrieval has been linked to intraregional increases in theta and gamma os-

cillations (Osipova et al. 2006; Gruber et al. 2008) unveiled in EEG/MEG as well as phase synchronization between prefrontal, parietal, and medial temporal regions (Anderson et al. 2009; Polanía et al. 2012; Watrous et al. 2013) and phase-amplitude coupling between prefrontal theta and parietal gamma oscillations (Köster et al. 2014).

Beyond the role of these oscillations in enabling interregional synchronization, several models suggest that the theta oscillations could serve to encode specific temporal and spatial contexts (Hasselmo 2012; Watrous et al. 2013, 2015; Hasselmo and Stern 2014). In one recent study, Watrous and colleagues (2013) examined low-frequency (1–10 Hz) phase synchronization in electrocorticographical recordings from the parahippocampal gyrus, the parietal cortex and lateral prefrontal cortex during retrieval of spatial and temporal information regarding locations previously encountered during virtual navigation. They found an overall increase in presumed network connectivity (measured by phase synchronization) during successful retrieval, primarily centered on the parahippocampal gyrus. When comparing successful retrieval of spatial and temporal context, both were characterized by an increase in network connectivity, yet correct retrieval of spatial context involved lower frequency interactions relative to retrieval of temporal context (Fig. 2B). This suggests a mechanism that could enable simultaneous representation of spatial and temporal contextual information in the same circuit through phase/frequency multiplexing (Watrous et al. 2013, 2015). In addition, theta oscillations may support rapid switching between encoding and retrieval modes in the MTL (Hasselmo et al. 2002; Hasselmo and Stern 2014). This model is also supported by evidence of differing oscillatory phases following study items versus test probes, albeit in a working memory task (Rizzuto et al. 2006).

## Reinstatement of Encoding Processes during Retrieval

A widely presented view of episodic memory retrieval, mentioned above, also in referring to

Cite this article as *Cold Spring Harb Perspect Biol* doi: 10.1101/cshperspect.a021790

rodent studies, is that retrieval entails reinstatement of brain activity that was elicited during encoding (Damasio 1989; McClelland et al. 1995; reviewed in Buckner and Wheeler 2001; Rugg et al. 2008; Danker and Anderson 2010; Rissman and Wagner 2012; Levy and Wagner 2013). Influential data-driven cognitive theories posited that the match between encoding and retrieval cues and between level of processing (e.g., in verbal memoranda) is pertinent for successful retrieval (Tulving and Thomson 1973; Morris et al. 1977). This gives rise to the question of how the representation of a stored memory may relate to the representation of the experience. With the advent of fMRI it became possible to probe this question in the human brain more directly, testing whether retrieval entails reinstatement of specific blood oxygenation level–dependent (BOLD) activity patterns elicited during encoding in distinct brain areas.

Initial studies showed that sensory cortices active at encoding were also active at retrieval (Nyberg et al. 2000, 2001; Wheeler et al. 2000, 2006; Vaidya et al. 2002; Gottfried et al. 2004; Kahn et al. 2004; Wheeler and Buckner 2004; Woodruff et al. 2005). This reactivation was suggested to be related to the subjective experience of recollection (Wheeler and Buckner 2004), for both veridical recollection and false alarms (Kahn et al. 2004). Johnson and Rugg (2007) directly compared the task-specific recollection activity to the brain activity at time of encoding. They found an overlap between brain regions underlying task-specific successful encoding and regions underlying successful recollection of words encoded under the same task (but no cross-task overlap). This set of studies provides compelling evidence in favor of cortical reinstatement, that is, reinstatement of cortical activity patterns elicited at encoding during retrieval of the same event. However, these studies reveal reinstatement only at a rather coarse level, indicating a general overlap between brain regions involved in content-specific encoding and those involved retrieval of the corresponding stimuli. Multivariate approaches for analysis of fMRI data enable more fine-grained tests of reinstatement of specific spatial activity patterns

within regions. Traditional fMRI analyses focus on identifying voxels (volumetric pixels) that change their activity in response to certain tasks or cognitive states, testing each voxel in isolation. More recent multivariate approaches, such as multivoxel pattern analysis (Haxby et al. 2001; Norman et al. 2006) and representational similarity analysis (Kriegeskorte et al. 2008), shift the focus to multivoxel patterns of activity, allowing for the identification of reinstatement at a higher spatial resolution.

In the first multivoxel pattern analysis of episodic memory, Polyn et al. (2005) trained a classifier (i.e., an algorithm that enables classification of brain states based on distinct activation patterns) on activity patterns during encoding of pictures in three categories, and were able to predict the category of pictures recalled in a free recall test. Notably, reinstatement of activity patterns during retrieval has been found to occur in the absence of subjective recollection (Johnson et al. 2009) as well as in situations of competing memories (Kuhl et al. 2011, 2012), suggesting that reinstatement may occur even in implicit retrieval. The increased spatial resolution afforded by such multivariate approaches enables investigation of event-specific cortical reinstatement. Two recent studies (Staresina et al. 2012a; Ritchey et al. 2013) were able to identify such item-specific reinstatement. Staresina et al. (2012a) examined whether item-specific neural representations elicited during episodic encoding are later reinstated during successful recollection. They presented participants with word–scene pairs and later tested their memory using the words as cues, measuring brain activity (using fMRI) during both encoding and retrieval. The parahippocampal cortex showed increased similarity between encoding and retrieval activation patterns (encoding–retrieval similarity) during successful recollection. This reactivation was observed only during successful recollection, not during recognition, and was event-specific (i.e., the encoding–retrieval similarity was driven by reinstatement of specific word–scene combinations rather than by the scene itself ). Notably, the hippocampal encoding–retrieval similarity was not predictive of memory, but the amplitude of the hippocampal response was correlated with the parahippocampal encoding–retrieval similarity during retrieval, suggesting it may mediate the reinstatement observed in the parahippocampus.

EEG studies have also shown encoding–retrieval overlap (Rösler et al. 1995; Gratton et al. 1997; Khader et al. 2005). Given the low temporal resolution of fMRI, which does not enable comparing the temporal dynamics of the activity at encoding versus retrieval at a resolution finer than several seconds, a promising approach may be to test encoding–retrieval overlap using MEG (Jafarpour et al. 2014) or by combining EEG and fMRI (Khader et al. 2007; see Johnson and Rugg 2007 and Johnson et al. 2008 for two studies using similar protocols with fMRI and EEG, respectively). Eletrophysiological measures obtained from depth electrodes in neurosurgical patients are also used to obtain high temporal resolution data in the investigation of human retrieval. For example, using this approach, Gelbard-Sagiv et al. (2008) showed that episodic recall, using movie clips as memoranda, entails the activation of episode-specific activity patterns in the hippocampus.

Although there is compelling evidence that encoding-related brain activity is partially reinstated during retrieval, the overlapping activity is still a small fraction of the overall brain activity elicited during encoding and retrieval (Johnson and Rugg 2007; Rugg et al. 2008; Daselaar et al. 2009; Kim et al. 2010). This may be caused by several contributing factors (Rugg et al. 2008), such as features of the original experience that may not be encoded, or encoded but not retrieved. In addition, given the reconstructive nature of memory retrieval (Loftus and Palmer 1974; Schacter et al. 1998), additional information, unrelated to the original event, may be incorporated into the representation.

Indeed, false memories have been shown to elicit false reinstatement effects that mimic those of veridical reinstatement (Kahn et al. 2004), in line with findings that the majority of brain regions involved in retrieval are sensitive to the subjective feeling of retrieval rather than the veridicality of retrieval (Mendelsohn et al. 2010; Rissman et al. 2010; Dennis et al.

2012). Interestingly, the main regions found to differentiate true from false memories are in sensory cortices and MTL (Cabeza et al. 2001; Slotnick and Schacter 2004; Daselaar et al. 2006b; Kirwan et al. 2009; Dennis et al. 2012), the same regions that may be expected to show reinstatement.

More generally, dominant models of systems consolidation (see above, also Squire et al. 2015), suggest that memory representations change over time. For example, systems consolidation in humans has been reported to involve semantization of episodic experience (Cermak 1984; Conway et al. 1997; Nadel et al. 2007; Conway 2009; Piolino et al. 2009; Furman et al. 2012). One of the major debates in this regard (see also discussion of rodent studies above) pertains to whether only retrieval of recent memories is dependent on the hippocampus (e.g., Squire et al. 2001; Squire and Bayley 2007; Bartsch et al. 2011), or whether it is required for retrieval of vivid memories, regardless of memory age (Moscovitch et al. 2005, 2006; Hassabis and Maguire 2007; Rosenbaum et al. 2008; Dudai 2012; Bonnici et al. 2013). Evidence from human neuroimaging studies suggests that both vividness and memory age affect different aspects of the hippocampal activity and its connectivity with the cortical retrieval network (e.g., Gilboa et al. 2004; Piolino et al. 2009; Viard et al. 2010; Furman et al. 2012; Harand et al. 2012; Söderlund et al. 2012). Two recent multivoxel pattern analysis studies directly probed how passage of time affects the representation in both hippocampal and cortical regions by decoding specific recent/remote autobiographical memories (Bonnici et al. 2012, 2013). They found that both vmPFC and hippocampus contained information about recent and remote memories, with more remote information in vmPFC and posterior hippocampus.

Memory representations may also be subject to change when encountering in retrieval novel events that are related to the existing memories (e.g., in reconsolidation; Nader 2015). Such conditions may result in generalization/inference (Moses et al. 2006; Shohamy and Wagner 2008; Kumaran and McClelland 2012), interference (Müller and Pilzecker 1900; Anderson 2003; Wixted 2004), or creation of false memories (Schacter and Slotnick 2004; Loftus 2005). Increasingly sensitive analyses methods of human brain imaging provide improved abilities to probe the interactions between encoding of new information and reactivation of previous representations (Staresina et al. 2012b; Ben-Yakov et al. 2014; Brown et al. 2015).

All in all, the studies described in this section show that episodic memory retrieval in the human brain entails activation of cortico-hippocampal or cortical representations. Time-locking of transient spatiotemporal patterns of activation in distinct distributed cortical or corticohippocampal ensembles may permit permutational binding of representational elements to yield specific representations, and may also allow different representations to be stored in the same network. These representations are driven largely by the activity during encoding, but they are malleable, subject to the influence of time, semantization, and interaction of the existing memory with novel information.

## CONCLUSIONS

Our brief review of the investigation of memory retrieval in the mammalian brain shows how the combination of multiple levels of analysis, methods, and experimental systems and the introduction of novel techniques have markedly advanced the study of retrieval in recent years. But it also shows how much has yet to be learned. Research on declarative retrieval builds on a rich body of studies in experimental psychology, which has also generated influential models of retrieval and of the relevance of the information retrieved to the information encoded. Studies of brain damage in amnesia have identified brain areas, particularly in the hippocampus and surrounding neocortex, that play a role in both encoding and retrieval, and indicated that the role of these areas changes as the memory consolidates and with the passage of time. However, until recently, the methods available did not allow the spatiotemporal resolution required to dissect the processes involved in retrieval of complex memory, which

is accomplished within a fraction of a second. This is now possible, with the introduction of powerful methods in molecular biology, neurogenetics, and in vivo molecular and cellular imaging in rodents. The development of improved noninvasive functional brain imaging in humans, electrophysiology in patients, and more sophisticated data analyses, now also enable teasing apart subprocesses and brain substrates of retrieval of high-level memory unique to humans, such as autonoetic (i.e., involving self-awareness) episodic memory, verbal recollections, and autobiographical memory. The picture that emerges from both rodent and human studies is that retrieval of complex memory involves coordinated fast interplay of sparse and distributed networks, and yet leaves open the question whether some principles and elements of the codes that embody the expressed representations are still unknown.

## REFERENCES

*Reference is also in this collection.

Addis DR, Schacter DL. 2011. The hippocampus and imagining the future: Where do we stand? *Front Hum Neurosci* **5:** 173.

Addis DR, Wong AT, Schacter DL. 2007. Remembering the past and imagining the future: Common and distinct neural substrates during event construction and elaboration. *Neuropsychologia* **45:** 1363–1377.

Addis DR, Pan L, Vu MA, Laiser N, Schacter DL. 2009. Constructive episodic simulation of the future and the past: Distinct subsystems of a core brain network mediate imagining and remembering. *Neuropsychologia* **47:** 2222–2238.

Aggleton JP, Brown MW. 1999. Episodic memory, amnesia, and the hippocampal–anterior thalamic axis. *Behav Brain Sci* **22:** 425–444.

Aggleton JP, Vann SD, Denby C, Dix S, Mayes AR, Roberts N, Yonelinas AP. 2005. Sparing of the familiarity component of recognition memory in a patient with hippocampal pathology. *Neuropsychologia* **43:** 1810–1823.

Aggleton JP, Dumont JR, Warburton EC. 2011. Unraveling the contributions of the diencephalon to recognition memory: A review. *Learn Mem* **18:** 384–400.

Akerboom J, Chen TW, Wardill TJ, Tian L, Marvin JS, Mutlu S, Calderon NC, Esposti F, Borghuis BG, Sun XR, et al. 2012. Optimization of a GCaMP calcium indicator for neural activity imaging. *J Neurosci* **32:** 13819–13840.

Aly M, Yonelinas AP, Kishiyama MM, Knight RT. 2011. Damage to the lateral prefrontal cortex impairs familiarity but not recollection. *Behav Brain Res* **225:** 297–304.

Anagnostaras SG, Maren S, Fanselow MS. 1999. Temporally graded retrograde amnesia of contextual fear after hippocampal damage in rats: Within-subjects examination. *J Neurosci* **19**: 1106–1114.

Anderson MC. 2003. Rethinking interference theory: Executive control and the mechanisms of forgetting. *J Mem Lang* **49**: 415–445.

Anderson JR, Bower GH. 1972. Recognition and retrieval processes in free recall. *Psychol Rev* **79**: 97.

Anderson KL, Rajagovindan R, Ghacibeh GA, Meador KJ, Ding M. 2009. Theta oscillations mediate interaction between prefrontal cortex and medial temporal lobe in human memory. *Cereb Cortex* **20**: 1604–1612.

Baddeley AD. 1982. Domains of recollection. *Psychol Rev* **89**: 708.

Baddeley A. 2000. The episodic buffer: A new component of working memory? *Trends Cogn Sci* **4**: 417–423.

Badre D, Wagner AD. 2007. Left ventrolateral prefrontal cortex and the cognitive control of memory. *Neuropsychologia* **45**: 2883–2901.

Bahrick HP. 1970. Two-phase model for prompted recall. *Psychol Rev* **77**: 215.

Bakker A, Kirwan CB, Miller M, Stark CE. 2008. Pattern separation in the human hippocampal CA3 and dentate gyrus. *Science* **319**: 1640–1642.

Bartlett FC. 1932. *Remembering: A study in experimental and social psychology.* Cambridge University Press, Cambridge.

Bartsch T, Döhring J, Rohr A, Jansen O, Deuschl G. 2011. CA1 neurons in the human hippocampus are critical for autobiographical memory, mental time travel, and autonoetic consciousness. *Proc Natl Acad Sci* **108**: 17562–17567.

Ben-Yakov A, Rubinson M, Dudai Y. 2014. Shifting gears in hippocampus: Temporal dissociation between familiarity and novelty signatures in a single event. *J Neurosci* **34**: 12973–12981.

Bergström ZM, Henson RN, Taylor JR, Simons JS. 2013. Multimodal imaging reveals the spatiotemporal dynamics of recollection. *Neuroimage* **68**: 141–153.

Berryhill ME, Phuong L, Picasso L, Cabeza R, Olson IR. 2007. Parietal lobe and episodic memory: Bilateral damage causes impaired free recall of autobiographical memory. *J Neurosci* **27**: 14415–14423.

Berryhill ME, Picasso L, Arnold R, Drowos D, Olson IR. 2010. Similarities and differences between parietal and frontal patients in autobiographical and constructed experience tasks. *Neuropsychologia* **48**: 1385–1393.

Blumenfeld RS, Ranganath C. 2007. Prefrontal cortex and long-term memory encoding: An integrative review of findings from neuropsychology and neuroimaging. *Neuroscientist* **13**: 280–291.

Bonnici HM, Chadwick MJ, Lutti A, Hassabis D, Weiskopf N, Maguire EA. 2012. Detecting representations of recent and remote autobiographical memories in vmPFC and hippocampus. *J Neurosci* **32**: 16982–16991.

Bonnici HM, Chadwick MJ, Maguire EA. 2013. Representations of recent and remote autobiographical memories in hippocampal subfields. *Hippocampus* **23**: 849–854.

Bontempi B, Laurent-Demir C, Destrade C, Jaffard R. 1999. Time-dependent reorganization of brain circuitry underlying long-term memory storage. *Nature* **400**: 671–675.

Bowles B, Crupi C, Mirsattari SM, Pigott SE, Parrent AG, Pruessner JC, Yonelinas AP, Köhler S. 2007. Impaired familiarity with preserved recollection after anterior temporal-lobe resection that spares the hippocampus. *Proc Natl Acad Sci* **104**: 16382–16387.

Brown MW, Aggleton JP. 2001. Recognition memory: What are the roles of the perirhinal cortex and hippocampus? *Nat Rev Neurosci* **2**: 51–61.

* Brown TI, Staresina BP, Wagner AD. 2015. Noninvasive functional and anatomical imaging of the human medial temporal lobe. *Cold Spring Harb Perspect Biol* **7**: 10.1101/cshperspect.a021840.

Buchanan TW. 2007. Retrieval of emotional memories. *Psychol Bull* **133**: 761.

Buckner RL. 2010. The role of the hippocampus in prediction and imagination. *Annu Rev Psychol* **61**: 27–48.

Buckner RL, Carroll DC. 2007. Self-projection and the brain. *Trends Cogn Sci* **11**: 49–57.

Buckner RL, Wheeler ME. 2001. The cognitive neuroscience of remembering. *Nat Rev Neurosci* **2**: 624–634.

Buckner RL, Koutstaal W, Schacter DL, Dale AM, Rotte M, Rosen BR. 1998. Functional–anatomic study of episodic retrieval. II: Selective averaging of event-related fMRI trials to test the retrieval success hypothesis. *Neuroimage* **7**: 163–175.

Buckner RL, Wheeler ME, Sheridan MA. 2001. Encoding processes during retrieval tasks. *J Cogn Neurosci* **13**: 406–415.

Buckner RL, Andrews-Hanna JR, Schacter DL. 2008. The brain's default network. *Ann NY Acad Sci* **1124**: 1–38.

* Byrne JH, Hawkins RD. 2015. Nonassociative learning in invertebrates. *Cold Spring Harb Perspect Biol* **7**: a021675.

Byrne P, Becker S, Burgess N. 2007. Remembering the past and imagining the future: A neural model of spatial memory and imagery. *Psychol Rev* **114**: 340.

Cabeza R. 2008. Role of parietal regions in episodic memory retrieval: The dual attentional processes hypothesis. *Neuropsychologia* **46**: 1813–1827.

Cabeza R, Moscovitch M. 2013. Memory systems, processing modes, and components functional neuroimaging evidence. *Perspect Psychol Sci* **8**: 49–55.

Cabeza R, Rao SM, Wagner AD, Mayer AR, Schacter DL. 2001. Can medial temporal lobe regions distinguish true from false? An event-related functional MRI study of veridical and illusory recognition memory. *Proc Natl Acad Sci* **98**: 4805–4810.

Cabeza R, Locantore J, Anderson N. 2003. Lateralization of prefrontal activity during episodic memory retrieval: Evidence for the production-monitoring hypothesis. *J Cogn Neurosci* **15**: 249–259.

Cabeza R, Prince SE, Daselaar SM, Greenberg DL, Budde M, Dolcos F, LaBar KS, Rubin DC. 2004. Brain activity during episodic retrieval of autobiographical and laboratory events: An fMRI study using a novel photo paradigm. *J Cogn Neurosci* **16**: 1583–1594.

Cite this article as *Cold Spring Harb Perspect Biol* doi: 10.1101/cshperspect.a021790

Cabeza R, Ciaramelli E, Olson IR, Moscovitch M. 2008. The parietal cortex and episodic memory: An attentional account. *Nat Rev Neurosci* **9**: 613–625.

Cabeza R, Mazuz YS, Stokes J, Kragel JE, Woldorff MG, Ciaramelli E, Olson IR, Moscovitch M. 2011. Overlapping parietal activity in memory and perception: Evidence for the attention to memory model. *J Cogn Neurosci* **23**: 3209–3217.

Cabeza R, Ciaramelli E, Moscovitch M. 2012. Cognitive contributions of the ventral parietal cortex: An integrative theoretical account. *Trends Cogn Sci* **16**: 338–352.

Canolty RT, Knight RT. 2010. The functional role of cross-frequency coupling. *Trends Cogn Sci* **14**: 506–515.

Canolty RT, Edwards E, Dalal SS, Soltani M, Nagarajan SS, Kirsch HE, Berger MS, Barbaro NM, Knight RT. 2006. High gamma power is phase-locked to theta oscillations in human neocortex. *Science* **313**: 1626–1628.

Canolty RT, Ganguly K, Kennerley SW, Cadieu CF, Koepsell K, Wallis JD, Carmena JM. 2010. Oscillatory phase coupling coordinates anatomically dispersed functional cell assemblies. *Proc Natl Acad Sci* **107**: 17356–17361.

Carr VA, Rissman J, Wagner AD. 2010. Imaging the human medial temporal lobe with high-resolution fMRI. *Neuron* **65**: 298–308.

Cermak LS. 1984. The episodic-semantic distinction in amnesia. In *Neuropsychology of memory* (ed. Squire LR, Butter N), pp. 55–62. Guilford, New York.

Chadwick MJ, Hassabis D, Weiskopf N, Maguire EA. 2010. Decoding individual episodic memory traces in the human hippocampus. *Curr Biol* **20**: 544–547.

Ciaramelli E, Grady CL, Moscovitch M. 2008. Top-down and bottom-up attention to memory: A hypothesis (AtoM) on the role of the posterior parietal cortex in memory retrieval. *Neuropsychologia* **46**: 1828–1851.

Cipolotti L, Bird C, Good T, Macmanus D, Rudge P, Shallice T. 2006. Recollection and familiarity in dense hippocampal amnesia: A case study. *Neuropsychologia* **44**: 489–506.

Cohn M, Moscovitch M, Lahat A, McAndrews MP. 2009. Recollection versus strength as the primary determinant of hippocampal engagement at retrieval. *Proc Natl Acad Sci* **106**: 22451–22455.

Conway MA. 2009. Episodic memories. *Neuropsychologia* **47**: 2305–2313.

Conway MA, Gardiner JM, Perfect TJ, Anderson SJ, Cohen GM. 1997. Changes in memory awareness during learning: The acquisition of knowledge by psychology undergraduates. *J Exp Psychol Gen* **126**: 393.

Cowansage KK, Shuman T, Dillingham BC, Chang A, Golshani P, Mayford M. 2014. Direct reactivation of a coherent neocortical memory of context. *Neuron* **84**: 432–441.

Damasio AR. 1989. Time-locked multiregional retroactivation: A systems-level proposal for the neural substrates of recall and recognition. *Cognition* **33**: 25–62.

Danker JF, Anderson JR. 2010. The ghosts of brain states past: Remembering reactivates the brain regions engaged during encoding. *Psychol Bull* **136**: 87.

Daselaar SM, Fleck MS, Cabeza R. 2006a. Triple dissociation in the medial temporal lobes: Recollection, familiarity, and novelty. *J Neurophysiol* **96**: 1902–1911.

Daselaar SM, Fleck MS, Prince SE, Cabeza R. 2006b. The medial temporal lobe distinguishes old from new independently of consciousness. *J Neurosci* **26**: 5835–5839.

Daselaar SM, Prince SE, Dennis NA, Hayes SM, Kim H, Cabeza R. 2009. Posterior midline and ventral parietal activity is associated with retrieval success and encoding failure. *Front Hum Neurosci* **3**: 13.

Dennis NA, Bowman CR, Vandekar SN. 2012. True and phantom recollection: An fMRI investigation of similar and distinct neural correlates and connectivity. *Neuroimage* **59**: 2982–2993.

Denny CA, Kheirbek MA, Alba EL, Tanaka KF, Brachman RA, Laughman KB, Tomm NK, Turi GF, Losonczy A, Hen R. 2014. Hippocampal memory traces are differentially modulated by experience, time, and adult neurogenesis. *Neuron* **83**: 189–201.

Diana RA, Reder LM, Arndt J, Park H. 2006. Models of recognition: A review of arguments in favor of a dual-process account. *Psychon Bull Rev* **13**: 1–21.

Diana RA, Yonelinas AP, Ranganath C. 2007. Imaging recollection and familiarity in the medial temporal lobe: A three-component model. *Trends Cogn Sci* **11**: 379–386.

Diana RA, Yonelinas AP, Ranganath C. 2010. Medial temporal lobe activity during source retrieval reflects information type, not memory strength. *J Cogn Neurosci* **22**: 1808–1818.

Dobbins IG, Wagner AD. 2005. Domain-general and domain-sensitive prefrontal mechanisms for recollecting events and detecting novelty. *Cereb Cortex* **15**: 1768–1778.

Donaldson W. 1996. The role of decision processes in remembering and knowing. *Mem Cognit* **24**: 523–533.

Duarte A, Ranganath C, Knight RT. 2005. Effects of unilateral prefrontal lesions on familiarity, recollection, and source memory. *J Neurosci* **25**: 8333–8337.

Duarte A, Henson RN, Graham KS. 2011. Stimulus content and the neural correlates of source memory. *Brain Res* **1373**: 110–123.

Dudai Y. 2002. *Memory from A to Z keywords, concepts and beyond*. Oxford University Press, Oxford.

Dudai Y. 2012. The endless engram: Consolidations never end. *Annu Rev Neurosci* **35**: 227–247.

Duncan K, Sadanand A, Davachi L. 2012. Memory's penumbra: Episodic memory decisions induce lingering mnemonic biases. *Science* **337**: 485–487.

Dunn JC. 2004. Remember-know: A matter of confidence. *Psychol Rev* **111**: 524.

Dusoir H, Kapur N, Byrnes DP, McKinstry S, Hoare RD. 1990. The role of diencephalic pathology in human memory disorder. Evidence from a penetrating paranasal brain injury. *Brain* **113**: 1695.

Düzel E, Cabeza R, Picton TW, Yonelinas AP, Scheich H, Heinze HJ, Tulving E. 1999. Task-related and item-related brain processes of memory retrieval. *Proc Natl Acad Sci* **96**: 1794–1799.

Eich E, Nelson AL, Leghari MA, Handy TC. 2009. Neural systems mediating field and observer memories. *Neuropsychologia* **47**: 2239–2251.

Eichenbaum H, Yonelinas AR, Ranganath C. 2007. The medial temporal lobe and recognition memory. *Annu Rev Neurosci* **30**: 123.

Eldridge LL, Knowlton BJ, Furmanski CS, Bookheimer SY, Engel SA. 2000. Remembering episodes: A selective role for the hippocampus during retrieval. *Nat Neurosci* **3**: 1149–1152.

Eldridge LL, Engel SA, Zeineh MM, Bookheimer SY, Knowlton BJ. 2005. A dissociation of encoding and retrieval processes in the human hippocampus. *J Neurosci* **25**: 3280–3286.

* Fanselow MS, Wassum KM. 2015. The origin and organization of vertebrate Pavlovian conditioning. *Cold Spring Harb Perspect Biol* doi: 10.1101/cshperspect.a021717.

Farivar R, Zangenehpour S, Chaudhuri A. 2004. Cellular-resolution activity mapping of the brain using immediate-early gene expression. *Front Biosci* **9**: 104–109.

Fell J, Axmacher N. 2011. The role of phase synchronization in memory processes. *Nat Rev Neurosci* **12**: 105–118.

Fletcher PC, Henson RNA. 2001. Frontal lobes and human memory insights from functional neuroimaging. *Brain* **124**: 849–881.

Ford JH, Verfaellie M, Giovanello KS. 2010. Neural correlates of familiarity-based associative retrieval. *Neuropsychologia* **48**: 3019–3025.

Fornito A, Harrison BJ, Zalesky A, Simons JS. 2012. Competitive and cooperative dynamics of large-scale brain functional networks supporting recollection. *Proc Natl Acad Sci* **109**: 12788–12793.

Frankland PW, Bontempi B. 2005. The organization of recent and remote memories. *Nat Rev Neurosci* **6**: 119–130.

Frankland PW, Bontempi B, Talton LE, Kaczmarek L, Silva AJ. 2004. The involvement of the anterior cingulate cortex in remote contextual fear memory. *Science* **304**: 881–883.

Fried I, MacDonald KA, Wilson CL. 1997. Single neuron activity in human hippocampus and amygdala during recognition of faces and objects. *Neuron* **18**: 753–765.

Fries P. 2005. A mechanism for cognitive dynamics: Neuronal communication through neuronal coherence. *Trends Cogn Sci* **9**: 474–480.

Furman O, Mendelsohn A, Dudai Y. 2012. The episodic engram transformed: Time reduces retrieval-related brain activity but correlates it with memory accuracy. *Learn Mem* **19**: 575–587.

Gelbard-Sagiv H, Mukamel R, Harel M, Malach R, Fried I. 2008. Internally generated reactivation of single neurons in human hippocampus during free recall. *Science* **322**: 96–101.

Gilboa A, Winocur G, Grady CL, Hevenor SJ, Moscovitch M. 2004. Remembering our past: Functional neuroanatomy of recollection of recent and very remote personal events. *Cereb Cortex*, **14**: 1214–1225.

Giovanello KS, Verfaellie M, Keane MM. 2003. Disproportionate deficit in associative recognition relative to item recognition in global amnesia. *Cogn Affect Behav Neurosci* **3**: 186–194.

Godden DR, Baddeley AD. 1975. Context-dependent memory in two natural environments: On land and underwater. *Br J Psychol* **66**: 325–331.

Gonsalves BD, Kahn I, Curran T, Norman KA, Wagner AD. 2005. Memory strength and repetition suppression: Multimodal imaging of medial temporal cortical contributions to recognition. *Neuron* **47**: 751–761.

Gottfried JA, Smith AP, Rugg MD, Dolan RJ. 2004. Remembrance of odors past: Human olfactory cortex in cross-modal recognition memory. *Neuron* **42**: 687–695.

Gratton G, Corballis PM, Jain S. 1997. Hemispheric organization of visual memories. *J Cogn Neurosci* **9**: 92–104.

Greenberg DL, Rice HJ, Cooper JJ, Cabeza R, Rubin DC, LaBar KS. 2005. Co-activation of the amygdala, hippocampus and inferior frontal gyrus during autobiographical memory retrieval. *Neuropsychologia* **43**: 659–674.

Gruber T, Tsivilis D, Giabbiconi CM, Müller MM. 2008. Induced electroencephalogram oscillations during source memory: Familiarity is reflected in the gamma band, recollection in the theta band. *J Cogn Neurosci* **20**: 1043–1053.

Guzowski JF, McNaughton BL, Barnes CA, Worley PF. 1999. Environment-specific expression of the immediate-early gene *Arc* in hippocampal neuronal ensembles. *Nat Neurosci* **2**: 1120–1124.

Habib R, Nyberg L, Tulving E. 2003. Hemispheric asymmetries of memory: The HERA model revisited. *Trends Cogn Sci* **7**: 241–245.

Haramati S, Soroker N, Dudai Y, Levy DA. 2008. The posterior parietal cortex in recognition memory: A neuropsychological study. *Neuropsychologia* **46**: 1756–1766.

Harand C, Bertran F, La Joie R, Landeau B, Mézenge F, Desgranges B, Peigneux P, Eustache F, Rauchs G. 2012. The hippocampus remains activated over the long term for the retrieval of truly episodic memories. *PLoS ONE* **7**: e43495.

Hassabis D, Maguire EA. 2007. Deconstructing episodic memory with construction. *Trends Cogn Sci* **11**: 299–306.

Hassabis D, Maguire EA. 2009. The construction system of the brain. *Philos Trans R Soc B Lond B Biol Sci* **364**: 1263–1271.

Hassabis D, Kumaran D, Maguire EA. 2007. Using imagination to understand the neural basis of episodic memory. *J Neurosci* **27**: 14365–14374.

Hasselmo ME. 2012. *How we remember: Brain mechanisms of episodic memory.* MIT Press, Cambridge, MA.

Hasselmo ME, Stern CE. 2014. Theta rhythm and the encoding and retrieval of space and time. *Neuroimage* **85**: 656–666.

Hasselmo M, Bodelón C, Wyble B. 2002. A proposed function for hippocampal theta rhythm: Separate phases of encoding and retrieval enhance reversal of prior learning. *Neural Comput* **14**: 793–817.

Haxby JV, Gobbini MI, Furey ML, Ishai A, Schouten JL, Pietrini P. 2001. Distributed and overlapping representations of faces and objects in ventral temporal cortex. *Science* **293**: 2425–2430.

Holdstock JS, Mayes AR, Roberts N, Cezayirli E, Isaac CL, O'Reilly RC, Norman KA. 2002. Under what conditions is recognition spared relative to recall after selective hippocampal damage in humans? *Hippocampus* **12**: 341–351.

Huijbers W, Pennartz CM, Cabeza R, Daselaar SM. 2011. The hippocampus is coupled with the default network during memory retrieval but not during memory encoding. *PLoS ONE* **6**: e17463.

Huppert FA, Piercy M. 1978. The role of trace strength in recency and frequency judgements by amnesic and control subjects. *Q J Exp Psychol* **30:** 347–354.

Hutchinson JB, Uncapher MR, Wagner AD. 2009. Posterior parietal cortex and episodic retrieval: Convergent and divergent effects of attention and memory. *Learn Mem* **16:** 343–356.

Jacobs J, Kahana MJ, Ekstrom AD, Fried I. 2007. Brain oscillations control timing of single-neuron activity in humans. *J Neurosci* **27:** 3839–3844.

Jacoby LL, Dallas M. 1981. On the relationship between autobiographical memory and perceptual learning. *J Exp Psychol Gen* **110:** 306.

Jafarpour A, Fuentemilla L, Horner AJ, Penny W, Düzel E. 2014. Replay of very early encoding representations during recollection. *J Neurosci* **34:** 242–248.

James W. 1890. *The principles of psychology.* Dover, New York.

Jeneson A, Kirwan CB, Hopkins RO, Wixted JT, Squire LR. 2010. Recognition memory and the hippocampus: A test of the hippocampal contribution to recollection and familiarity. *Learn Mem* **17:** 63–70.

Johnson JD, Rugg MD. 2007. Recollection and the reinstatement of encoding-related cortical activity. *Cereb Cortex* **17:** 2507–2515.

Johnson JD, Minton BR, Rugg MD. 2008. Content dependence of the electrophysiological correlates of recollection. *Neuroimage* **39:** 406–416.

Johnson JD, McDuff SG, Rugg MD, Norman KA. 2009. Recollection, familiarity, and cortical reinstatement: A multivoxel pattern analysis. *Neuron* **63:** 697–708.

Kahn I, Davachi L, Wagner AD. 2004. Functional-neuroanatomic correlates of recollection: Implications for models of recognition memory. *J Neurosci* **24:** 4172–4180.

Kandel ER, Schwartz JH. 1982. Molecular biology of learning: Modulation of transmitter release. *Science* **218:** 433–443.

Keene CS, Bucci DJ. 2008a. Contributions of the retrosplenial and posterior parietal cortices to cue-specific and contextual fear conditioning. *Behav Neuurosci* **122:** 89–97.

Keene CS, Bucci DJ. 2008b. Neurotoxic lesions of retrosplenial cortex disrupt signaled and unsignaled contextual fear conditioning. *Behav Neurosci* **122:** 1070–1077.

Kentros CG, Agnihotri NT, Streater S, Hawkins RD, Kandel ER. 2004. Increased attention to spatial context increases both place field stability and spatial memory. *Neuron* **42:** 283–295.

Khader P, Heil M, Rösler F. 2005. Material-specific long-term memory representations of faces and spatial positions: Evidence from slow event-related brain potentials. *Neuropsychologia* **43:** 2109–2124.

Khader P, Knoth K, Burke M, Ranganath C, Bien S, Rösler F. 2007. Topography and dynamics of associative long-term memory retrieval in humans. *J Cogn Neurosci* **19:** 493–512.

Kim H. 2010. Dissociating the roles of the default-mode, dorsal, and ventral networks in episodic memory retrieval. *Neuroimage* **50:** 1648–1657.

Kim H. 2013. Differential neural activity in the recognition of old versus new events: An activation likelihood estimation meta-analysis. *Hum Brain Mapp* **34:** 814–836.

Kim JJ, Fanselow MS. 1992. Modality-specific retrograde amnesia of fear. *Science* **256:** 675–677.

Kim H, Daselaar SM, Cabeza R. 2010. Overlapping brain activity between episodic memory encoding and retrieval: Roles of the task-positive and task-negative networks. *Neuroimage* **49:** 1045–1054.

Kintsch W. 1970. Models for free recall and recognition. In *Models of human memory*, Vol. 124. Academic, New York.

Kirwan CB, Stark CE. 2004. Medial temporal lobe activation during encoding and retrieval of novel face-name pairs. *Hippocampus* **14:** 919–930.

Kirwan CB, Shrager Y, Squire LR. 2009. Medial temporal lobe activity can distinguish between old and new stimuli independently of overt behavioral choice. *Proc Natl Acad Sci* **106:** 14617–14621.

Kirwan CB, Wixted JT, Squire LR. 2010. A demonstration that the hippocampus supports both recollection and familiarity. *Proc Natl Acad Sci* **107:** 344–348.

Köhler S, Moscovitch M, Winocur G, Houle S, McIntosh AR. 1998. Networks of domain-specific and general regions involved in episodic memory for spatial location and object identity. *Neuropsychologia* **36:** 129–142.

Köster M, Friese U, Schöne B, Trujillo-Barreto N, Gruber T. 2014. Theta-gamma coupling during episodic retrieval in the human EEG. *Brain Res* **1577:** 57–68.

Kriegeskorte N, Mur M, Bandettini P. 2008. Representational similarity analysis-connecting the branches of systems neuroscience. *Front Systems Neurosci* **2:** 4. *10.3389/neuro.06.004.2008*

Kuhl BA, Shah AT, DuBrow S, Wagner AD. 2010. Resistance to forgetting associated with hippocampus-mediated reactivation during new learning. *Nat Neurosci* **13:** 501–506.

Kuhl BA, Rissman J, Chun MM, Wagner AD. 2011. Fidelity of neural reactivation reveals competition between memories. *Proc Natl Acad Sci* **108:** 5903–5908.

Kuhl BA, Bainbridge WA, Chun MM. 2012. Neural reactivation reveals mechanisms for updating memory. *J Neurosci* **32:** 3453–3461.

Kumaran D, McClelland JL. 2012. Generalization through the recurrent interaction of episodic memories: A model of the hippocampal system. *Psychol Rev* **119:** 573.

LaBar KS, Cabeza R. 2006. Cognitive neuroscience of emotional memory. *Nat Rev Neurosci* **7:** 54–64.

Lashley KS. 1950. In search of the engram. *Symp Soc Exp Biol* **4:** 454–482.

Lepage M, Habib R, Tulving E. 1998. Hippocampal PET activations of memory encoding and retrieval: The HIPER model. *Hippocampus* **8:** 313–322.

Lepage M, Ghaffar O, Nyberg L, Tulving E. 2000. Prefrontal cortex and episodic memory retrieval mode. *Proc Natl Acad Sci* **97:** 506–511.

Levy BJ, Wagner AD. 2013. Measuring memory reactivation with functional MRI implications for psychological theory. *Perspect Psychol Sci* **8:** 72–78.

Liu X, Ramirez S, Pang PT, Puryear CB, Govindarajan A, Deisseroth K, Tonegawa S. 2012. Optogenetic stimulation of a hippocampal engram activates fear memory recall. *Nature* **484:** 381–385.

Loftus EF. 2005. Planting misinformation in the human mind: A 30-year investigation of the malleability of memory. *Learn Mem* **12:** 361–366.

Loftus EF, Palmer JC. 1974. Reconstruction of automobile destruction: An example of the interaction between language and memory. *J Verbal Learning Verbal Behav* **13:** 585–589.

Ludowig E, Trautner P, Kurthen M, Schaller C, Bien CG, Elger CE, Rosburg T. 2008. Intracranially recorded memory-related potentials reveal higher posterior than anterior hippocampal involvement in verbal encoding and retrieval. *J Cogn Neurosci* **20:** 841–851.

Luria AR. 1966. *Higher cortical functions in man.* Basic Books, New York.

Maguire EA, Mullally SL. 2013. The hippocampus: A manifesto for change. *J Exp Psychol Gen* **142:** 1180.

Maguire EA, Mummery CJ, Büchel C. 2000. Patterns of hippocampal-cortical interaction dissociate temporal lobe memory subsystems. *Hippocampus* **10:** 475–482.

Mandler G. 1980. Recognizing: The judgment of previous occurrence. *Psychol Rev* **87:** 252.

Mankin EA, Sparks FT, Slayyeh B, Sutherland RJ, Leutgeb S, Leutgeb JK. 2012. Neuronal code for extended time in the hippocampus. *Proc Natl Acad Sci* **109:** 19462–19467.

Manns JR, Squire LR. 1999. Impaired recognition memory on the doors and people test after damage limited to the hippocampal region. *Hippocampus* **9:** 495–499.

Manns JR, Hopkins RO, Reed JM, Kitchener EG, Squire LR. 2003. Recognition memory and the human hippocampus. *Neuron* **37:** 171–180.

Maratos EJ, Rugg MD. 2001. Electrophysiological correlates of the retrieval of emotional and non-emotional context. *J Cogn Neurosci* **13:** 877–891.

Maris E, van Vugt M, Kahana M. 2011. Spatially distributed patterns of oscillatory coupling between high-frequency amplitudes and low-frequency phases in human iEEG. *Neuroimage* **54:** 836–850.

Mason MF, Norton MI, Van Horn JD, Wegner DM, Grafton ST, Macrae CN. 2007. Wandering minds: The default network and stimulus-independent thought. *Science* **315:** 393–395.

Mayes AR, Holdstock JS, Isaac CL, Montaldi D, Grigor J, Gummer A, Cariga P, Downes JJ, Tsivilis D, Gaffan D, et al. 2004. Associative recognition in a patient with selective hippocampal lesions and relatively normal item recognition. *Hippocampus* **14:** 763–784.

McClelland JL, McNaughton BL, O'Reilly RC. 1995. Why there are complementary learning systems in the hippocampus and neocortex: Insights from the successes and failures of connectionist models of learning and memory. *Psychol Rev* **102:** 419–457.

McDermott KB, Szpunar KK, Christ SE. 2009. Laboratory-based and autobiographical retrieval tasks differ substantially in their neural substrates. *Neuropsychologia* **47:** 2290–2298.

Mendelsohn A, Furman O, Dudai Y. 2010. Signatures of memory: Brain coactivations during retrieval distinguish correct from incorrect recollection. *Front Behav Neurosci* **4:** 18.

Mickes L, Wais PE, Wixted JT. 2009. Recollection is a continuous process implications for dual-process theories of recognition memory. *Psychol Sci* **20:** 509–515.

Montaldi D, Mayes AR. 2010. The role of recollection and familiarity in the functional differentiation of the medial temporal lobes. *Hippocampus* **20:** 1291–1314.

Montaldi D, Spencer TJ, Roberts N, Mayes AR. 2006. The neural system that mediates familiarity memory. *Hippocampus* **16:** 504–520.

Morgan JI, Cohen DR, Hempstead JL, Curran T. 1987. Mapping patterns of *c-fos* expression in the central nervous system after seizure. *Science* **237:** 92–197.

Morris CD, Bransford JD, Franks JJ. 1977. Levels of processing versus transfer appropriate processing. *J Verbal Learning Verbal Behav* **16:** 519–533.

Morris RG, Garrud P, Rawlins JN, O'Keefe J. 1982. Place navigation impaired in rats with hippocampal lesions. *Nature* **297:** 681–683.

Moscovitch M, Winocur G. 2002. The frontal cortex and working with memory. In *Principles of frontal lobe function* (ed. Stuss DT, Knight RT), pp. 188–209. Oxford University Press, Oxford.

Moscovitch M, Rosenbaum RS, Gilboa A, Addis DR, Westmacott R, Grady C, McAndrews MP, Levine B, Black S, Winocur G, et al. 2005. Functional neuroanatomy of remote episodic, semantic and spatial memory: A unified account based on multiple trace theory. *J Anat* **207:** 35–66.

Moscovitch M, Nadel L, Winocur G, Gilboa A, Rosenbaum RS. 2006. The cognitive neuroscience of remote episodic, semantic and spatial memory. *Curr Opin Neurobiol* **16:** 179–190.

* Moser M-B, Rowland DC, Moser EI. 2015. Place cells, grid cells, and memory. *Cold Spring Harb Perspect Biol* **7:** a021808.

Moses SN, Villate C, Ryan JD. 2006. An investigation of learning strategy supporting transitive inference performance in humans compared to other species. *Neuropsychologia* **44:** 1370–1387.

Müller GE, Pilzecker A. 1900. Experimental contributions to the theory of memory. *Z Psychol Z Angew Psychol* **1:** 1–288.

Nadel L, Campbell J, Ryan L. 2007. Autobiographical memory retrieval and hippocampal activation as a function of repetition and the passage of time. *Neural Plast* **2007:** 90472.

* Nader K. 2015. Reconsolidation and the dynamic nature of memory. *Cold Spring Harb Perspect Biol* **7:** a021782.

Norman KA, Polyn SM, Detre GJ, Haxby JV. 2006. Beyond mind-reading: Multi-voxel pattern analysis of fMRI data. *Trends Cogn Sci* **10:** 424–430.

Northoff G, Heinzel A, de Greck M, Bermpohl F, Dobrowolny H, Panksepp J. 2006. Self-referential processing in our brain—A meta-analysis of imaging studies on the self. *Neuroimage* **31:** 440–457.

Nyberg L, Tulving E, Habib R, Nilsson LG, Kapur S, Houle S, Cabeza R, McIntosh AR. 1995. Functional brain maps of retrieval mode and recovery of episodic information. *Neuroreport* **7:** 249–252.

Cite this article as *Cold Spring Harb Perspect Biol* doi: 10.1101/cshperspect.a021790

Nyberg L, Habib R, McIntosh AR, Tulving E. 2000. Reactivation of encoding-related brain activity during memory retrieval. *Proc Natl Acad Sci* 97: 11120–11124.

Nyberg L, Petersson KM, Nilsson LG, Sandblom J, Åberg C, Ingvar M. 2001. Reactivation of motor brain areas during explicit memory for actions. *Neuroimage* 14: 521–528.

Nyhus E, Curran T. 2010. Functional role of gamma and theta oscillations in episodic memory. *Neurosci Biobehav Rev* 34: 1023–1035.

O'Keefe J. 1990. A computational theory of the hippocampal cognitive map. *Prog Brain Res* 83: 301–312.

O'Keefe J, Dostrovsky J. 1971. The hippocampus as a spatial map. Preliminary evidence from unit activity in the freely-moving rat. *Brain Res* 34: 171–175.

Osipova D, Takashima A, Oostenveld R, Fernández G, Maris E, Jensen O. 2006. Theta and gamma oscillations predict encoding and retrieval of declarative memory. *J Neurosci* 26: 7523–7531.

Parks CM, Yonelinas AP. 2007. Moving beyond pure signal-detection models: Comment on Wixted (2007). *Psychol Rev* 114: 188–202; discussion 203–209.

Perfect TJ. 1996. Does context discriminate recollection from familiarity in recognition memory? *Q J Exp Psychol A* 49: 797–813.

Petrides M. 2002. The mid-ventrolateral prefrontal cortex and active mnemonic retrieval. *Neurobiol Learn Mem* 78: 528–538.

Philippi CL, Tranel D, Duff M, Rudrauf D. 2015. Damage to the default mode network disrupts autobiographical memory retrieval. *Soc Cogn Affect Neurosci* 10: 318–326.

Piolino P, Desgranges B, Eustache F. 2009. Episodic autobiographical memories over the course of time: Cognitive, neuropsychological and neuroimaging findings. *Neuropsychologia* 47: 2314–2329.

Polanía R, Nitsche MA, Korman C, Batsikadze G, Paulus W. 2012. The importance of timing in segregated theta phase-coupling for cognitive performance. *Curr Biol* 22: 1314–1318.

Polyn SM, Natu VS, Cohen JD, Norman KA. 2005. Category-specific cortical activity precedes retrieval during memory search. *Science* 310: 1963–1966.

Poppenk J, Evensmoen HR, Moscovitch M, Nadel L. 2013. Long-axis specialization of the human hippocampus. *Trends Cogn Sci* 17: 230–240.

Ritchey M, Wing EA, LaBar KS, Cabeza R. 2013. Neural similarity between encoding and retrieval is related to memory via hippocampal interactions. *Cereb Cortex* 23: 2818–2828.

Quamme JR, Yonelinas AP, Norman KA. 2007. Effect of unitization on associative recognition in amnesia. *Hippocampus* 17: 192–200.

Rabin JS, Gilboa A, Stuss DT, Mar RA, Rosenbaum RS. 2010. Common and unique neural correlates of autobiographical memory and theory of mind. *J Cogn Neurosci* 22: 1095–1111.

Raichle ME. 2010. Two views of brain function. *Trends Cogn Sci* 14: 180–190.

Raichle ME, MacLeod AM, Snyder AZ, Powers WJ, Gusnard DA, Shulman GL. 2001. A default mode of brain function. *Proc Natl Acad Sci* 98: 676–682.

Ramirez S, Liu X, Lin PA, Suh J, Pignatelli M, Redondo RL, Ryan TJ, Tonegawa S. 2013. Creating a false memory in the hippocampus. *Science* 341: 387–391.

Ranganath C. 2010. A unified framework for the functional organization of the medial temporal lobes and the phenomenology of episodic memory. *Hippocampus* 20: 1263–1290.

Reijmers LG, Perkins BL, Matsuo N, Mayford M. 2007. Localization of a stable neural correlate of associative memory. *Science* 317: 1230–1233.

Rissman J, Wagner AD. 2012. Distributed representations in memory: Insights from functional brain imaging. *Annu Rev Psychol* 63: 101–128.

Rissman J, Greely HT, Wagner AD. 2010. Detecting individual memories through the neural decoding of memory states and past experience. *Proc Natl Acad Sci* 107: 9849–9854.

Rizzuto DS, Madsen JR, Bromfield EB, Schulze-Bonhage A, Kahana MJ. 2006. Human neocortical oscillations exhibit theta phase differences between encoding and retrieval. *Neuroimage* 31: 1352–1358.

Rosenbaum RS, Moscovitch M, Foster JK, Schnyer DM, Gao F, Kovacevic N, Verfaellie M, Black SE, Levine B. 2008. Patterns of autobiographical memory loss in medial-temporal lobe amnesic patients. *J Cogn Neurosci* 20: 1490–1506.

Rösler F, Heil M, Hennighausen E. 1995. Exploring memory functions by means of brain electrical topography: A review. *Brain Topogr* 7: 301–313.

Rossi S, Pasqualetti P, Zito G, Vecchio F, Cappa SF, Miniussi C, Babiloni C, Rossini PM. 2006. Prefrontal and parietal cortex in human episodic memory: An interference study by repetitive transcranial magnetic stimulation. *Eur J Neurosci* 23: 793–800.

Rotello CM, Macmillan NA, Reeder JA. 2004. Sum-difference theory of remembering and knowing: A two-dimensional signal-detection model. *Psychol Rev* 111: 588.

Rudebeck SR, Scholz J, Millington R, Rohenkohl G, Johansen-Berg H, Lee AC. 2009. Fornix microstructure correlates with recollection but not familiarity memory. *J Neurosci* 29: 14987–14992.

Rugg MD, Vilberg KL. 2013. Brain networks underlying episodic memory retrieval. *Curr Opin Neurobiol* 23: 255–260.

Rugg MD, Wilding EL. 2000. Retrieval processing and episodic memory. *Trends Cogn Sci* 4: 108–115.

Rugg MD, Fletcher PC, Frith CD, Frackowiak RSJ, Dolan RJ. 1996. Differential activation of the prefrontal cortex in successful and unsuccessful memory retrieval. *Brain* 119: 2073–2083.

Rugg MD, Fletcher PC, Allan K, Frith CD, Frackowiak RSJ, Dolan RJ. 1998. Neural correlates of memory retrieval during recognition memory and cued recall. *Neuroimage* 8: 262–273.

Rugg MD, Johnson JD, Park H, Uncapher MR. 2008. Encoding-retrieval overlap in human episodic memory: A functional neuroimaging perspective. *Prog Brain Res* 169: 339–352.

Rugg MD, Vilberg KL, Mattson JT, Yu SS, Johnson JD, Suzuki M. 2012. Item memory, context memory and the hippocampus: fMRI evidence. *Neuropsychologia* **50:** 3070–3079.

Rutishauser U, Mamelak AN, Schuman EM. 2006. Single-trial learning of novel stimuli by individual neurons of the human hippocampus–amygdala complex. *Neuron* **49:** 805–813.

Rutishauser U, Schuman EM, Mamelak AN. 2008. Activity of human hippocampal and amygdala neurons during retrieval of declarative memories. *Proc Natl Acad Sci* **105:** 329–334.

Rutishauser U, Ross IB, Mamelak AN, Schuman EM. 2010. Human memory strength is predicted by theta-frequency phase-locking of single neurons. *Nature* **464:** 903–907.

Sagar SM, Sharp FR, Curran T. 1988. Expression of c-fos protein in brain: Metabolic mapping at the cellular level. *Science* **240:** 1328–1331.

Schacter DL, Addis DR. 2007. The cognitive neuroscience of constructive memory: Remembering the past and imagining the future. *Philos Trans R Soc Lond B Biol Sci* **362:** 773–786.

Schacter DL, Slotnick SD. 2004. The cognitive neuroscience of memory distortion. *Neuron* **44:** 149–160.

Schacter DL, Wagner AD. 1999. Medial temporal lobe activations in fMRI and PET studies of episodic encoding and retrieval. *Hippocampus* **9:** 7–24.

Schacter DL, Norman KA, Koutstaal W. 1998. The cognitive neuroscience of constructive memory. *Annu Rev Psychol* **49:** 289–318.

Schacter DL, Addis DR, Buckner RL. 2008. Episodic simulation of future events. *Ann NY Acad Sci* **1124:** 39–60.

Schacter DL, Addis DR, Hassabis D, Martin VC, Spreng RN, Szpunar KK. 2012. The future of memory: Remembering, imagining, and the brain. *Neuron* **76:** 677–694.

Schoenenberger P, Gerosa D, Oertner TG. 2009. Temporal control of immediate early gene induction by light. *PLoS ONE* **4:** e8185.

Scimeca JM, Badre D. 2012. Striatal contributions to declarative memory retrieval. *Neuron* **75:** 380–392.

Semon R. 1904/1921. *The mneme.* G. Allen & Unwin, London.

Shannon BJ, Buckner RL. 2004. Functional-anatomic correlates of memory retrieval that suggest nontraditional processing roles for multiple distinct regions within posterior parietal cortex. *J Neurosci* **24:** 10084–10092.

Shimamura AP. 1995. Memory and frontal lobe function. *Ann NY Acad Sci* **15:** 769.

Shimamura AP. 2010. Hierarchical relational binding in the medial temporal lobe: The strong get stronger. *Hippocampus* **20:** 1206–1216.

Shimamura AP. 2011. Episodic retrieval and the cortical binding of relational activity. *Cogn Affect Behav Neurosci* **11:** 277–291.

Shohamy D, Wagner AD. 2008. Integrating memories in the human brain: Hippocampal-midbrain encoding of overlapping events. *Neuron* **60:** 378–389.

Simons JS, Spiers HJ. 2003. Prefrontal and medial temporal lobe interactions in long-term memory. *Nat Rev Neurosci* **4:** 637–648.

Simons JS, Owen AM, Fletcher PC, Burgess PW. 2005. Anterior prefrontal cortex and the recollection of contextual information. *Neuropsychologia* **43:** 1774–1783.

Simons JS, Peers PV, Hwang DY, Ally BA, Fletcher PC, Budson AE. 2008. Is the parietal lobe necessary for recollection in humans? *Neuropsychologia* **46:** 1185–1191.

Skinner EI, Fernandes MA. 2007. Neural correlates of recollection and familiarity: A review of neuroimaging and patient data. *Neuropsychologia* **45:** 2163–2179.

Slotnick SD, Schacter DL. 2004. A sensory signature that distinguishes true from false memories. *Nat Neurosci* **7:** 664–672.

Smeyne RJ, Schilling K, Robertson L, Luk D, Oberdick J, Curran T, Morgan JI. 1992. fos-lacZ transgenic mice: Mapping sites of gene induction in the central n nervous system. *Neuron* **8:** 13–23.

Smith CN, Wixted JT, Squire LR. 2011. The hippocampus supports both recollection and familiarity when memories are strong. *J Neurosci* **31:** 15693–15702.

Söderlund H, Moscovitch M, Kumar N, Mandic M, Levine B. 2012. As time goes by: Hippocampal connectivity changes with remoteness of autobiographical memory retrieval. *Hippocampus* **22:** 670–679.

Song Z, Wixted JT, Hopkins RO, Squire LR. 2011. Impaired capacity for familiarity after hippocampal damage. *Proc Natl Acad Sci* **108:** 9655–9660.

Spaniol J, Davidson PS, Kim AS, Han H, Moscovitch M, Grady CL. 2009. Event-related fMRI studies of episodic encoding and retrieval: Meta-analyses using activation likelihood estimation. *Neuropsychologia* **47:** 1765–1779.

Spreng RN, Mar RA, Kim AS. 2009. The common neural basis of autobiographical memory, prospection, navigation, theory of mind, and the default mode: A quantitative meta-analysis. *J Cogn Neurosci* **21:** 489–510.

Squire LR, Bayley PJ. 2007. The neuroscience of remote memory. *Curr Opin Neurobiol* **17:** 185–196.

* Squire LR, Dede AJO. 2015. Conscious and unconscious memory systems. *Cold Spring Harb Perspect Biol* doi: 10.1101/cshperspect.a021667.

Squire LR, Clark RE, Knowlton BJ. 2001. Retrograde amnesia. *Hippocampus* **11:** 50–55.

Squire LR, Wixted JT, Clark RE. 2007. Recognition memory and the medial temporal lobe: A new perspective. *Nat Rev Neurosci* **8:** 872–883.

* Squire LR, Morris RG, Genzel L, Wixted JT. 2015. Memory consolidation. *Cold Spring Harb Perspect Biol* doi: 10.1101/cshperspect.a021766.

Staresina BP, Henson RN, Kriegeskorte N, Alink A. 2012a. Episodic reinstatement in the medial temporal lobe. *J Neurosci* **32:** 18150–18156.

Staresina BP, Fell J, Do Lam AT, Axmacher N, Henson RN. 2012b. Memory signals are temporally dissociated in and across human hippocampus and perirhinal cortex. *Nat Neurosci* **15:** 1167–1173.

Stark CE, Okado Y. 2003. Making memories without trying: Medial temporal lobe activity associated with incidental memory formation during recognition. *J Neurosci* **23:** 6748–6753.

Stark CE, Bayley PJ, Squire LR. 2002. Recognition memory for single items and for associations is similarly impaired following damage to the hippocampal region. *Learn Mem* **9:** 238–242.

St Jacques PL, Kragel PA, Rubin DC. 2011. Dynamic neural networks supporting memory retrieval. *Neuroimage* **57:** 608–616.

Strange BA, Fletcher PC, Henson RNA, Friston KJ, Dolan RJ. 1999. Segregating the functions of human hippocampus. *Proc Natl Acad Sci* **96:** 4034–4039.

Stuss DT, Alexander MP, Palumbo CL, Buckle L, Sayer L, Pogue J. 1994. Organizational strategies with unilateral or bilateral frontal lobe injury in word learning tasks. *Neuropsychology* **8:** 355.

Summerfield JJ, Hassabis D, Maguire EA. 2009. Cortical midline involvement in autobiographical memory. *Neuroimage* **44:** 1188–1200.

Suthana N, Fried I. 2012. Percepts to recollections: Insights from single neuron recordings in the human brain. *Trends Cogn Sci* **16:** 427–436.

Svoboda E, McKinnon MC, Levine B. 2006. The functional neuroanatomy of autobiographical memory: A meta-analysis. *Neuropsychologia* **44:** 2189–2208.

Szpunar KK. 2010. Episodic future thought an emerging concept. *Perspect Psychol Sci* **5:** 142–162.

Tanaka KZ, Pevzner A, Hamidi AB, Nakazawa Y, Graham J, Wiltgen BJ. 2014. Cortical representations are reinstated by the hippocampus during memory retrieval. *Neuron* **84:** 347–354.

Tendolkar I, Arnold J, Petersson KM, Weis S, Brockhaus-Dumke A, van Eijndhoven P, Buitelaar J, Fernández G. 2008. Contributions of the medial temporal lobe to declarative memory retrieval: Manipulating the amount of contextual retrieval. *Learn Mem* **15:** 611–617.

Thorpe S, Fine D, Marlot C. 1996. Speed of processing in the human visual system. *Nature* **381:** 520–522.

Tsivilis D, Vann SD, Denby C, Roberts N, Mayes AR, Montaldi D, Aggleton JP. 2008. A disproportionate role for the fornix and mammillary bodies in recall versus recognition memory. *Nat Neurosci* **11:** 834–842.

Tulving E. 1983. *Elements of episodic memory.* Oxford University Press, Oxford.

Tulving E. 1985. Memory and consciousness. *Can Psychol* **26:** 1.

Tulving E, Thomson DM. 1973. Encoding specificity and retrieval processes in episodic memory. *Psychol Rev* **80:** 352.

Tulving E, Kapur S, Craik FI, Moscovitch M, Houle S. 1994. Hemispheric encoding/retrieval asymmetry in episodic memory: Positron emission tomography findings. *Proc Natl Acad Sci* **91:** 2016–2020.

Vaidya CJ, Zhao M, Desmond JE, Gabrieli JD. 2002. Evidence for cortical encoding specificity in episodic memory: Memory-induced re-activation of picture processing areas. *Neuropsychologia* **40:** 2136–2143.

van der Meij R, Kahana M, Maris E. 2012. Phase–amplitude coupling in human electrocorticography is spatially distributed and phase diverse. *J Neurosci* **32:** 111–123.

Vann SD, Aggleton JP, Maguire EA. 2009a. What does the retrosplenial cortex do? *Nat Rev Neurosci* **10:** 792–802.

Vann SD, Tsivilis D, Denby CE, Quamme JR, Yonelinas AP, Aggleton JP, Montaldi D, Mayes AR. 2009b. Impaired recollection but spared familiarity in patients with extended hippocampal system damage revealed by 3 convergent methods. *Proc Natl Acad Sci* **106:** 5442–5447.

Vargha-Khadem F, Gadian DG, Watkins KE, Connelly A, Van Paesschen W, Mishkin M. 1997. Differential effects of early hippocampal pathology on episodic and semantic memory. *Science* **277:** 376–380.

Viard A, Lebreton K, Chételat G, Desgranges B, Landeau B, Young A, De La Sayette V, Eustache F, Piolino P. 2010. Patterns of hippocampal–neocortical interactions in the retrieval of episodic autobiographical memories across the entire life-span of aged adults. *Hippocampus* **20:** 153–165.

Viard A, Desgranges B, Eustache F, Piolino P. 2012. Factors affecting medial temporal lobe engagement for past and future episodic events: An ALE meta-analysis of neuroimaging studies. *Brain Cogn* **80:** 111–125.

Vilberg KL, Rugg MD. 2007. Dissociation of the neural correlates of recognition memory according to familiarity, recollection, and amount of recollected information. *Neuropsychologia* **45:** 2216–2225.

Vilberg KL, Rugg MD. 2008. Memory retrieval and the parietal cortex: A review of evidence from a dual-process perspective. *Neuropsychologia* **46:** 1787–1799.

Vilberg KL, Rugg MD. 2012. The neural correlates of recollection: Transient versus sustained fMRI effects. *J Neurosci* **32:** 15679–15687.

Viskontas IV, Knowlton BJ, Steinmetz PN, Fried I. 2006. Differences in mnemonic processing by neurons in the human hippocampus and parahippocampal regions. *J Cogn Neurosci* **18:** 1654–1662.

Viskontas IV, Carr VA, Engel SA, Knowlton BJ. 2009. The neural correlates of recollection: Hippocampal activation declines as episodic memory fades. *Hippocampus* **19:** 265–272.

Voss JL, Paller KA. 2010. Bridging divergent neural models of recognition memory: Introduction to the special issue and commentary on key issues. *Hippocampus* **20:** 1171–1177.

Wagner AD, Desmond JE, Glover GH, Gabrieli JD. 1998. Prefrontal cortex and recognition memory. Functional-MRI evidence for context-dependent retrieval processes. *Brain* **121:** 1985–2002.

Wagner AD, Maril A, Bjork RA, Schacter DL. 2001. Prefrontal contributions to executive control: fMRI evidence for functional distinctions within lateral prefrontal cortex. *Neuroimage* **14:** 1337–1347.

Wagner AD, Shannon BJ, Kahn I, Buckner RL. 2005. Parietal lobe contributions to episodic memory retrieval. *Trends Cogn Sci* **9:** 445–453.

Wais PE. 2008. fMRI signals associated with memory strength in the medial temporal lobes: A meta-analysis. *Neuropsychologia* **46:** 3185–3196.

Wais P, Wixted J, Hopkins R, Squire L. 2006. The hippocampus supports both the recollection and the familiarity components of recognition memory. *Neuron* **49:** 459–466.

Wais PE, Mickes L, Wixted JT. 2008. Remember/know judgments probe degrees of recollection. *J Cogn Neurosci* **20:** 400–405.

Wais PE, Squire LR, Wixted JT. 2010. In search of recollection and familiarity signals in the hippocampus. *J Cogn Neurosci* **22:** 109–123.

Watrous AJ, Tandon N, Conner CR, Pieters T, Ekstrom AD. 2013. Frequency-specific network connectivity increases underlie accurate spatiotemporal memory retrieval. *Nat Neurosci* **16:** 349–356.

Watrous AJ, Fell J, Ekstrom AD, Axmacher N. 2015. More than spikes: Common oscillatory mechanisms for content specific neural representations during perception and memory. *Curr Opin Neurobiol* **31:** 33–39.

Weiler JA, Suchan B, Daum I. 2010. When the future becomes the past: Differences in brain activation patterns for episodic memory and episodic future thinking. *Behav Brain Res* **212:** 196–203.

Weis S, Specht K, Klaver P, Tendolkar I, Willmes K, Ruhlmann J, Elger CE, Fernández G. 2004. Process dissociation between contextual retrieval and item recognition. *Neuroreport* **15:** 2729–2733.

Wheeler ME, Buckner RL. 2004. Functional-anatomic correlates of remembering and knowing. *Neuroimage* **21:** 1337–1349.

Wheeler MA, Stuss DT, Tulving E. 1995. Frontal lobe damage produces episodic memory impairment. *J Int Neuropsychol Soc* **1:** 525–536.

Wheeler ME, Petersen SE, Buckner RL. 2000. Memory's echo: Vivid remembering reactivates sensory-specific cortex. *Proc Natl Acad Sci* **97:** 11125–11129.

Wheeler ME, Shulman GL, Buckner RL, Miezin FM, Velanova K, Petersen SE. 2006. Evidence for separate perceptual reactivation and search processes during remembering. *Cereb Cortex* **16:** 949–959.

Wilson MA, McNaughton BL. 1994. Reactivation of hippocampal ensemble memories during sleep. *Science* **265:** 676–679.

Wiltgen BJ, Silva AJ. 2007. Memory for context becomes less specific with time. *Learn Mem* **14:** 313–317.

Winocur G, Moscovitch M, Bontempi B. 2010. Memory formation and long-term retention in humans and in animals: Convergence towards a transformation account of hippocampal-neocortical interactions. *Neuropsychologia* **48:** 2339–2356.

Wixted JT. 2004. The psychology and neuroscience of forgetting. *Annu Rev Psychol* **55:** 235–269.

Wixted JT. 2007. Dual-process theory and signal-detection theory of recognition memory. *Psychol Rev* **114:** 152.

Wixted JT, Mickes L. 2010. A continuous dual-process model of remember/know judgments. *Psychol Rev* **117:** 1025.

Wixted JT, Squire LR. 2011. The medial temporal lobe and the attributes of memory. *Trends Cogn Sci* **15:** 210–217.

Wixted JT, Mickes L, Squire LR. 2010. Measuring recollection and familiarity in the medial temporal lobe. *Hippocampus* **20:** 1195–1205.

Woodruff CC, Johnson JD, Uncapher MR, Rugg MD. 2005. Content-specificity of the neural correlates of recollection. *Neuropsychologia* **43:** 1022–1032.

Yonelinas AP. 1994. Receiver-operating characteristics in recognition memory: Evidence for a dual-process model. *J Exp Psychol Learn Mem Cogn* **20:** 1341.

Yonelinas AP. 2002. The nature of recollection and familiarity: A review of 30 years of research. *J Mem Lang* **46:** 441–517.

Yonelinas AP, Jacoby LL. 2012. The process-dissociation approach two decades later: Convergence, boundary conditions, and new directions. *Mem Cognit* **40:** 663–680.

Yonelinas AP, Kroll NE, Quamme JR, Lazzara MM, Sauvé MJ, Widaman KF, Knight RT. 2002. Effects of extensive temporal lobe damage or mild hypoxia on recollection and familiarity. *Nat Neurosci* **5:** 1236–1241.

Yonelinas AP, Otten LJ, Shaw KN, Rugg MD. 2005. Separating the brain regions involved in recollection and familiarity in recognition memory. *J Neurosci* **25:** 3002–3008.

Yonelinas AP, Aly M, Wang WC, Koen JD. 2010. Recollection and familiarity: Examining controversial assumptions and new directions. *Hippocampus* **20:** 1178–1194.

Yu SS, Johnson JD, Rugg MD. 2012. Hippocampal activity during recognition memory co-varies with the accuracy and confidence of source memory judgments. *Hippocampus* **22:** 1429–1437.

Zeineh MM, Engel SA, Thompson PM, Bookheimer SY. 2003. Dynamics of the hippocampus during encoding and retrieval of face-name pairs. *Science* **299:** 577–580.

Ziv Y, Burns LD, Cocker ED, Hamel EO, Ghosh KK, Kitch LJ, El Gamal A, Schnitzer MJ. 2013. Long-term dynamics of CA1 hippocampal place codes. *Nat Neurosci* **16:** 264–266.

Cite this article as *Cold Spring Harb Perspect Biol* doi: 10.1101/cshperspect.a021790

# Place Cells, Grid Cells, and Memory

May-Britt Moser, David C. Rowland, and Edvard I. Moser

Centre for Neural Computation, Kavli Institute for Systems Neuroscience, Norwegian University of Science and Technology, 7489 Trondheim, Norway

*Correspondence:* edvard.moser@ntnu.no

The hippocampal system is critical for storage and retrieval of declarative memories, including memories for locations and events that take place at those locations. Spatial memories place high demands on capacity. Memories must be distinct to be recalled without interference and encoding must be fast. Recent studies have indicated that hippocampal networks allow for fast storage of large quantities of uncorrelated spatial information. The aim of this article is to review and discuss some of this work, taking as a starting point the discovery of multiple functionally specialized cell types of the hippocampal–entorhinal circuit, such as place, grid, and border cells. We will show that grid cells provide the hippocampus with a metric, as well as a putative mechanism for decorrelation of representations, that the formation of environment-specific place maps depends on mechanisms for long-term plasticity in the hippocampus, and that long-term spatiotemporal memory storage may depend on offline consolidation processes related to sharp-wave ripple activity in the hippocampus. The multitude of representations generated through interactions between a variety of functionally specialized cell types in the entorhinal–hippocampal circuit may be at the heart of the mechanism for declarative memory formation.

The scientific study of human memory started with Herman Ebbinghaus, who initiated the quantitative investigation of associative memory processes as they take place (Ebbinghaus 1885). Ebbinghaus described the conditions that influence memory formation and he determined several basic principles of encoding and recall, such as the law of frequency and the effect of time on forgetting. With Ebbinghaus, higher mental functions were brought to the laboratory. In parallel with the human learning tradition that Ebbinghaus started, a new generation of experimental psychologists described the laws of associative learning in animals. With behaviorists like Pavlov, Watson, Hull, Skinner, and

Tolman, a rigorous program for identifying the laws of animal learning was initiated. By the middle of the 20th century, a language for associative learning processes had been developed, and many of the fundamental relationships between environment and behavior had been described. What was completely missing, though, was an understanding of the neural activity underlying the formation of the memory. The behaviorists had deliberately shied away from physiological explanations because of the intangible nature of neural activity at that time.

Then the climate began to change. Karl Lashley had shown that lesions in the cerebral cortex had predictable effects on behavior in animals (Lashley 1929, 1950), and Donald Hebb introduced concepts and ideas to account for complex brain functions at the neural circuit level, many of which have retained a place in modern neuroscience (Hebb 1949). Both Lashley and Hebb searched for the engram, but they found no specific locus for it. A significant turning point was reached when Scoville and Milner (1957) reported severe loss of memory in an epileptic patient, patient H.M., after bilateral surgical removal of the hippocampal formation and the surrounding medial temporal lobe areas. "After operation this young man could no longer recognize the hospital staff nor find his way to the bathroom, and he seemed to recall nothing of the day-to-day events of his hospital life." This tragic misfortune inspired decades of research on the function of the hippocampus in memory. H.M.'s memory impairment could be reproduced in memory tasks in animals and studies of H.M., as well as laboratory animals, pointed to a critical role for the hippocampus in declarative memory—memory, which, in humans, can be consciously recalled and declared, such as memories of experiences and facts (Milner et al. 1968; Mishkin 1978; Cohen and Squire 1980; Squire 1992; Corkin 2002). What was missing from these early studies, however, was a way to address the neuronal mechanisms that led information to be stored as memory.

The aim of this article is to show how studies of hippocampal neuronal activity during the past few decades have brought us to a point at which a mechanistic basis of memory formation is beginning to surface. An early landmark in this series of investigations was the discovery of place cells, cells that fire selectively at one or few locations in the environment. At first, these cells seemed to be part of the animal's instantaneous representation of location, independent of memory, but gradually, over the course of several decades, it has become clear that place cells express current as well as past and future locations. In many ways, place cells can be used as readouts of the memories that are stored in the hippocampus. More recent work has also shown that place cells are part of a wider network of spatially modulated neurons, including grid, border, and head direction cells, each with distinct roles in the representation of space and spatial memory. In this article, we shall discuss potential mechanisms by which these cell types, particularly place and grid cells, in conjunction with synaptic plasticity, may form the basis of a mammalian system for fast high-capacity declarative memory.

## PLACE CELLS, SYNAPTIC PLASTICITY, AND MEMORY

The growing interest in hippocampal function and memory led John O'Keefe and John Dostrovsky (O'Keefe and Dostrovsky 1971) and Jim Ranck (Ranck 1973) to introduce methods for recording activity from hippocampal neurons in awake and freely moving animals. Using miniaturized electrodes for extracellular single-cell recording, they were able to show reliable links between neural activity and behavior. The most striking relationship was noted by O'Keefe and Dostrovsky, who found that hippocampal cells responded specifically to the current location of the animal. They called these cells "place cells" (Fig. 1). Different place cells were found to have different firing locations, or place fields (O'Keefe 1976). Place was mapped nontopographically in the sense that place fields of neighboring cells were no more similar than those of cells that were far apart (O'Keefe 1976; Wilson and McNaughton 1993), although the size of the firing fields increased from dorsal to ventral hippocampus (Jung et al. 1994; Kjelstrup et al. 2008). The combination of cells that were active at each location in the environment was unique, despite the lack of location topography, leading O'Keefe and Nadel (1978) to suggest that the hippocampus is the locus of the brain's internal map of the spatial environment, a manifestation of the cognitive map proposed from purely behavioral experiments by Edward Tolman several decades earlier (Tolman 1948).

The discovery of place cells changed the way many experimental neuroscientists thought about hippocampal functions. Clinical studies starting with patient H.M. pointed to a role for the hippocampus in declarative memory (Squire 1992), but the fact that hippocampal

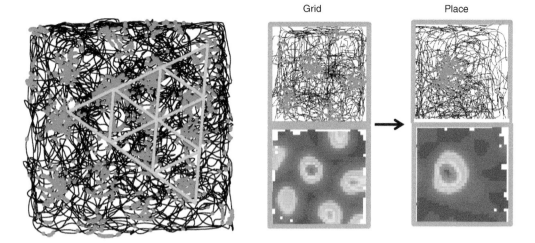

Figure 1. Grid cells and place cells. (*Left*) A grid cell from the entorhinal cortex of the rat brain. The black trace shows the trajectory of a foraging rat in part of a 1.5-m-diameter-wide square enclosure. Spike locations of the grid cell are superimposed in red on the trajectory. Each red dot corresponds to one spike. Blue equilateral triangles have been drawn on top of the spike distribution to illustrate the regular hexagonal structure of the grid pattern. (*Right*) Grid cell and place cell. (*Top*) Trajectory with spike locations, as in the *left* part. (*Bottom*) Color-coded rate map with red showing high activity and blue showing low activity. Grid cells are thought to provide much, but not all, of the entorhinal spatial input to place cells.

neurons were so strongly modulated by location suggested that space was primary. Moreover, for the most part, place cells represented current space, not as expected if the function of the hippocampus was purely mnemonic. Reconciling space and memory functions remained a challenge for several decades after the discovery of place cells.

A framework that accounts for both lines of observation has now emerged. Converging evidence has suggested that hippocampal neurons respond also to nonspatial features of the environment, such as odors (Eichenbaum et al. 1987; Wood et al. 1999; Igarashi et al. 2014), tactile inputs (Young et al. 1994), and timing (Hampson et al. 1993). The same cells that respond to nonspatial stimuli fire like place cells when animals move around in space, suggesting that place cells express the location of the animal in combination with information about events that take place or took place there (Leutgeb et al. 2005b; Moser et al. 2008). The representation of space does not exclude a central role of the hippocampus in declarative memory, as space is a central element of all episodic and

many semantic memories (Buzsáki and Moser 2013).

A role for place cells in hippocampal memory was apparent already in the earliest studies of place cells. It was shown in these studies that ensembles of place cells represent not only the animal's current location but also locations that the animal had visited earlier. In maze tasks, place cells fired when the animal made errors, as if the animal was in the location where the cell fired normally (O'Keefe and Speakman 1987). In spatial alternation tasks, firing patterns reflected locations that the animal came from, as well as upcoming locations (Frank et al. 2000; Woods et al. 2000; Ferbinteanu and Shapiro 2003), and during sequential testing in multiple environments, place-cell activity was found to carry over from one environment to the next (Leutgeb et al. 2004, 2005a). Moreover, sequences of spatial firing during exploration were shown to be replayed during rest or sleep subsequent to the behavioral experience, as if those patterns were stored in the hippocampal network during exploration and retrieved later in offline mode, when the animal was not acquiring new information (Pavlides and Winson 1989; Wilson and McNaughton 1994; Foster and Wilson 2006; O'Neill et al. 2006).

The fact that place cells express past experience raises the question whether ensembles of place cells are completely formed by experience or if there is an underlying component that is hardwired in the circuit. Hill (1978) sought to address this issue by recording place fields as rats entered a novel environment. Of the 12 cells that he recorded, 10 appeared to have spatial firing fields immediately, supporting the idea that the place-cell map was largely predetermined. Subsequently, studies with larger ensembles of cells found that place fields often took several minutes of exploration before settling into a stable firing field (McNaughton and Wilson 1993; Frank et al. 2004) and the formation of new and stable place fields was dependent on the animal's behavior and attention to the spatial features of the environment (Kentros et al. 2004; Monaco et al. 2014). These results point to a critical role for experience in forming the hippocampal map of space. However, the plasticity can occur extremely rapidly (Leutgeb et al. 2006) and, just as Hill observed, some place cells show stable firing fields immediately (Frank et al. 2004). Thus, place maps are expressed, in some form, from the very moment when animals are put into an environment for the first time, although the map may evolve further with experience. The findings raise the possibility that a skeletal map of a novel environment is drawn from a set of preexisting maps, and then gets modified to fit the specifics of the environment through experience-dependent plasticity (Samsonovich and McNaughton 1997; Dragoi and Tonegawa 2011, 2013).

The role of synaptic plasticity in the formation of place maps has been tested experimentally. In agreement with the proposed existence of prewired maps, neither systemic pharmacological blockade of $N$-methyl-$D$-aspartate (NMDA) receptors, nor subfield-specific targeted knockouts of such receptors, have a large effect on the basic firing patterns of place cells in familiar or novel environments (McHugh et al. 1996; Kentros et al. 1998), suggesting that place-field expression is quite independent of at least one major form of long-term synaptic plasticity. However, cellular mechanisms involved in long-term plasticity are clearly required for the long-term stability of newly formed maps (Rotenberg et al. 1996; Kentros et al. 1998). These studies suggest that the place-cell map of the environment is stored and stabilized through changes in synaptic weights, similar to other memory systems (Kandel and Schwartz 1982).

NMDA receptors also play a role in more subtle forms of experience-dependent modifications of place fields. One example is the experience-dependent asymmetric expansion of place fields observed following repeated traversals of place fields on a linear track (Mehta et al. 1997, 2000). It was suggested in theoretical studies in the 1990s that as a rat moves through locations A, B, and C along a linear track, the cells coding for location A will repeatedly activate the cells coding for location B and the cells coding for location B will, in turn, activate cells coding for location C. By the logic of Hebbian plasticity, the connections from A to B and B to C should become strengthened, with the result that place fields of cells A, B, and C are shifted forward on the track, against the direction of motion (Abbott and Blum 1996; Blum and Abbott 1996). Experimental evidence for such experience-dependent asymmetric expansion was obtained by Mehta and colleagues (1997, 2000). Subsequently, studies found that the asymmetric shift depends on NMDA receptor activation (Ekstrom et al. 2001), consistent with the suggestion that place maps are refined by experience-dependent long-term synaptic plasticity.

## MEMORY ENCODING

What are the factors that determine whether new place maps are stabilized? One of the hallmarks of episodic memories is that attended information is more likely to be encoded and stored long term (Chun and Turk-Browne 2007). It is simply impossible to remember everything, and as Ebbinghaus's curve of memory shows, most memories will fade over time. However, some particularly meaningful mem-

Cite this article as *Cold Spring Harb Perspect Biol* doi: 10.1101/cshperspect.a021808

ories become permanent. On this background, Kentros et al. (2004) considered whether attention to spatial cues could improve the long-term stability of place fields. They trained mice to find an unmarked goal location in a cylinder (similar to the Morris water maze) while recording hippocampal place cells. The mice that learned the task had more stable place fields than mice that were simply running in the same cylinder with no task requirements. To test whether the driving force was true selective attention, as opposed to general arousal, Muzzio et al. (2009) trained mice to attend to odor cues and ignore spatial cues or vice versa. When the odors were the relevant cues, the hippocampal neurons acquired stable odor representations, but had less stable spatial representations. The reverse was true when space was relevant. Taken together with recent evidence suggesting that place fields can be induced by attentive scanning (Monaco et al. 2014), the findings point to selective attention, and not merely general arousal, as a major determinant of experience-dependent stabilization of hippocampal place maps.

What could be the mechanisms for selective attention in the hippocampus? Recently, Igarashi et al. (2014) recorded simultaneously from the lateral entorhinal cortex and CA1 region of the rat hippocampus as the animals learned an odor–place association. As the animals learned the association, the two structures showed an increasing degree of synchronous oscillatory activity in the 20- to 40-Hz range and a corresponding increase in spiking activity to the rewarded odors. The development of temporal coherence between activity in the hippocampus and entorhinal cortex may allow CA1 cells to respond to particular entorhinal inputs at the same time as the cells are closest to firing threshold (Singer 1993). The 20- to 40-Hz oscillation is substantially lower than the fast (60-100 Hz) gamma oscillation found in the medial entorhinal cortex (Colgin et al. 2009). The two subdivisions of the entorhinal cortex may, therefore, convey relevant information to the hippocampus via distinct frequency channels, each leading to a different firing pattern in the hippocampus.

## MEMORY CONSOLIDATION AND RETRIEVAL

Once encoded, the memories must be consolidated. In an early theoretical paper, Buzsáki (1989) proposed that hippocampal memory formation occurs in two stages. First, there is a stage in which memory is encoded via weak synaptic potentiation in the CA3 network when the network is in theta-oscillation mode during exploratory behavior. Then, there is a memory consolidation stage, which can take place hours later during sharp-wave activity, associated with sleep and resting. In this stage, synapses that were weakly potentiated during the preceding exploration participate in sharp-wave activity that, in turn, evokes ripple activity in the CA1 area of the hippocampus. Ripples occur at a frequency that is optimal for induction of long-term potentiation (LTP) in efferent synapses of CA1 cells, possibly including long-distance targets in the cortex. By this mechanism, memory was thought to be slowly induced in the neocortex, consistent with a large body of evidence pointing to gradual recruitment of neocortical memory circuits in long-term storage of hippocampal memories (McClelland et al. 1995; Squire and Alvarez 1995; Frankland et al. 2001). Over the years, considerable evidence has accumulated to point to a role for sharp waves and ripples in the formation of hippocampus-dependent long-term memories. Selectively disrupting sharp-wave ripple activity during posttraining rest periods impairs learning, providing a causal link between sharp-wave ripples and consolidation (Girardeau et al. 2009; Ego-Stengel and Wilson 2010). Moreover, it is now clear that sequences of firing among place cells are replayed during subsequent sharp-wave ripples in the same or reverse order that the cells were active during experience (Wilson and McNaughton 1994; Foster and Wilson 2006; Diba and Buzsáki 2007). Structured replay is seen across many brain regions (Hoffman and McNaughton 2002), indicating that the sequence information from the hippocampus may be conferred on downstream cortical targets.

Recent work points to a wider role for replay in which replay may contribute not only to consolidation and recall of memory, but also to planning of future behavior. Studies in human subjects show that overlapping hippocampal networks are activated during episodic recall and imagination of fictitious experiences (Hassabis et al. 2007). In animals, sharp-wave ripples can activate cells along both past and future trajectories (Karlsson and Frank 2009; Gupta et al. 2010; Pfeiffer and Foster 2013). Pfeiffer and Foster (2013), for example, trained rats to find a rewarded well within a large environment while sharp-wave ripple-associated replay events were recorded in the hippocampus. In many of the events, the sequence of active cells began at the current location and ended at the goal location, followed by the animal taking the path defined by the place-cell activity. Although the sequence of activated cells clearly preceded behavior, the phenomenon also depended on previous experience with the environment and the rules of the task. Thus, the replay can either lead or follow the behavior once the map of space is established. In that sense, the replay phenomenon may support "mental time travel" (Suddendorf and Corballis 2007) through the spatial map, both forward and backward in time. Whether the sharp-wave ripple-mediated replay in rats represents conscious recall is impossible to know, but observations in humans during free recall provide a clue (Gelbard-Sagiv et al. 2008; Miller et al. 2013). Miller et al. (2013), for example, recorded from the medial temporal lobe of human subjects as they navigated a virtual town (the subjects were awaiting surgery for epilepsy and had electrodes placed in their medial temporal lobe to localize the origin of the seizures, affording Miller et al. the rare opportunity to record place cells in humans). After an initial familiarization period, subjects were asked to deliver items to one of the stores in the town and when all the deliveries were complete, the subjects were asked to recall only the items they delivered. Remarkably, the place cells responsive to the area where the item was delivered became active during recall of the item, closely mirroring the reactivation of place cells during replay events in rodents. Although free recall in humans is not likely to correspond to sharp-wave ripple events (Watrous et al. 2013),

the time course of reactivation was similar to a typical sharp-wave ripple event in rodents, and may therefore reflect a qualitatively similar phenomenon. The place cell activity during recall of events or items likely brings to mind the spatial context in which the events and items were experienced, creating a fully reconstructed memory for what was experienced, along with where it was experienced.

## UPSTREAM OF PLACE CELLS: GRID CELLS AND OTHER CELL TYPES

To get a better insight into the mechanisms of memory formation in hippocampal place-cell circuits, it may pay off to consider how place cells interact with cells in adjacent brain systems. The origin of the place-cell signal was long thought to be intrahippocampal, considering that early recordings upstream in the entorhinal cortex showed only weak spatial modulation (Barnes et al. 1990; Quirk et al. 1992; Frank et al. 2000). At the turn of the millennium, we started a series of experiments aimed at localizing the sources of the place signal. First, we isolated the CA1 region of the hippocampus from the earlier parts of the hippocampal excitatory circuit, that is, the dentate gyrus and the CA3 (Brun et al. 2002). Activity was then recorded from the remaining CA1. Place cells were still present, suggesting that intrahippocampal circuits are not necessary for spatial signals to develop. The findings pointed to direct inputs from the entorhinal cortex as an alternative source of incoming spatial information to the hippocampus. Thus, in a subsequent study, we recorded directly from the entorhinal cortex, not in the deep ventral areas where cells had been recorded in previous studies, but in the dorsal parts that projected directly to the hippocampal recording locations used by O'Keefe and others (Fyhn et al. 2004). Electrodes were placed in the medial part of the entorhinal cortex. We found that many neurons in this area were as sharply modulated by position as place cells in the hippocampus. Entorhinal neurons had multiple firing fields with clear regions of silence between the fields. In a third study, we expanded the size of the recording environment to determine the spatial structure of the many firing fields (Hafting et al. 2005). The multiple firing fields of individual entorhinal neurons formed a regularly spaced triangular or hexagonal grid pattern, which repeated itself across the entire available space. We named these cells "grid cells." Grid cells were organized in a non-topographic manner, much like place cells. The firing fields of neighboring grid cells were no more similar than those of grid cells recorded at different brain locations. However, the scale of the grid increased from dorsal to ventral medial entorhinal cortex (Fyhn et al. 2004; Hafting et al. 2005), suggesting that the earliest recordings in the entorhinal cortex had missed the grid pattern because the period of the firing pattern was too large for repeated fields to be observed in conventionally sized recording boxes. The discovery of grid cells was followed by studies showing that these cells were part of a wider spatial network comprising other cell types as well, such as head direction–modulated cells (Sargolini et al. 2006) and cells that fire specifically along one or several borders of the local environment (border cells) (Savelli et al. 2008; Solstad et al. 2008). Head direction cells had previously been observed in a number of brain systems, from the dorsal tegmental nucleus in the brain stem to the pre- and parasubiculum in the parahippocampal cortex (Ranck 1985; Taube et al. 1990; Taube 2007). Border cells were described at the same time in the subiculum (Barry et al. 2006; Lever et al. 2009). Thus, by the end of the first decade of the new millennium, it was clear that place and grid cells were part of a diverse and entangled network of cell types with distinct functions in spatial representation.

How place cells are formed from the diversity of cell types remains to be determined. An obvious possibility is that place cells are generated by transformation of spatial input from grid cells. The presence of grid cells in the superficial layers of the entorhinal cortex, the main cortical input to the hippocampus, led investigators to propose that place fields form by linear combination of periodic firing fields from grid cells with a common central peak, but different grid spacing and orientation (O'Keefe and Burgess 2005; Fuhs and Touretzky 2006; McNaughton et al. 2006; Solstad et al. 2006). The suggestion was that, because the wavelength of the individual grid patterns is different, the patterns cancel each other except at the central peak, which becomes the place field of the receiving cell (Fig. 2).

Experimental observations have suggested that the mechanisms are more complex, however. If place cells were generated exclusively from grid cells, grid and place cells would be expected to appear simultaneously in developing animals or with a faster time course for grid cells than place cells. Recordings from rat pups suggest that this is not the case (Langston et al. 2010; Wills et al. 2010). When pups leave the nest for the first time at 2–2.5 weeks of age, sharp and confined firing fields are present in a large proportion of the hippocampal pyramidal-cell population. In contrast, grid cells show only weakly periodic fields at that age. Strong periodicity is not expressed until 3–4 weeks of age. The delayed maturation of the grid cells offers at least two interpretations. First, weak spatial inputs may be sufficient for place-cell formation. Sharply confined firing fields may be generated by local mechanisms in the hippocampal network, such as recurrent inhibition (de Almeida et al. 2009; Monaco and Abbott 2011), Hebbian plasticity (Rolls et al. 2006; Savelli and Knierim 2010), or active dendritic properties (Smith et al. 2013). Alternatively, place cells may be generated from other classes of spatially modulated cells, such as border cells, which have adult-like properties from the very first day of exploration outside the nest (Bjerknes et al. 2014). Retrograde labeling studies suggest that border cells have projections to the hippocampus that may be equally dense as those from grid cells, although the latter are more abundant (Zhang et al. 2013). A potential role for border cells in place-cell formation would be consistent with early models, suggesting that place cells arise by linear combination of inputs from cells with firing fields defined by their proximity to geometric boundaries (O'Keefe and Burgess 1996; Hartley et al. 2000). Recordings in the medial entorhinal cortex have, so far, identified such cells only near the boundaries of the environment (Solstad et al. 2008; Zhang et al. 2013; Bjerknes et al. 2014), suggesting that a contri-

Map #1                                          Map #2

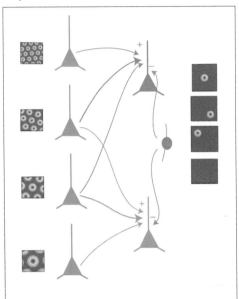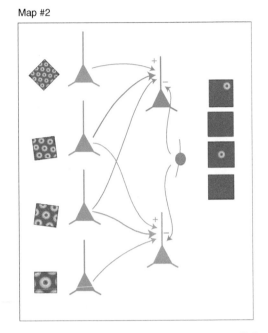

**Figure 2.** Schematic illustration of how periodic grid cells could be transformed to nonperiodic place cells by linear summation of output from grid cells with overlapping firing fields, but different spacing and orientation, and how differential responses among modules of grid cells might give rise to remapping in the hippocampus. (*Left*) Map 1, grid cells with different spacing converge to generate place cells in a subset of the hippocampal place-cell population. Each grid cell belongs to a different grid module. (*Right*) Map 2, differential realignment of each of the grid maps induces recruitment of a new subset of place cells. (From images in Solstad et al. 2006 and Fyhn et al. 2007; modified, with permission, from the authors and Nature Publishing Group © 2006 and 2007, respectively.)

bution by these cells may be limited to place cells with peripheral firing fields.

The exact function of different entorhinal cell types in place-cell formation remains to be determined, but it is not unlikely that individual place cells receive inputs from both grid and border cells, possibly with grid cells providing self-motion-based distance information and border cells providing position in relation to geometric boundaries (Bush et al. 2014; Zhang et al. 2014). The strongest input may originate from grid cells, which, in the superficial layers of the medial entorhinal cortex, are several times more abundant than border cells (Sargolini et al. 2006; Solstad et al. 2008; Boccara et al. 2010). Under most circumstances, the two classes of input are likely to be coherent and redundant. If one is absent, the other may often be sufficient to generate localized firing in the hippocampus.

## REMAPPING AND MEMORY

One of the events that pointed to place cells as an expression of declarative memory was the discovery of remapping, or the fact that any place cell is part of not one, but many independent representations. In 1987, Bob Muller and John Kubie found that place cells can alter their firing patterns in response to minor changes in the experimental task, such as alterations in the shape of the recording enclosure (Fig. 3) (Muller and Kubie 1987; Bostock et al. 1991). Place cells may begin firing, stop firing, or change their firing location. The changes are expressed widely across the place-cell population, such that a new map is installed for each occasion. Remapping could also be induced by changes in motivational state or behavioral context (Markus et al. 1995; Frank et al. 2000; Wood et al. 2000; Moita et al. 2004).

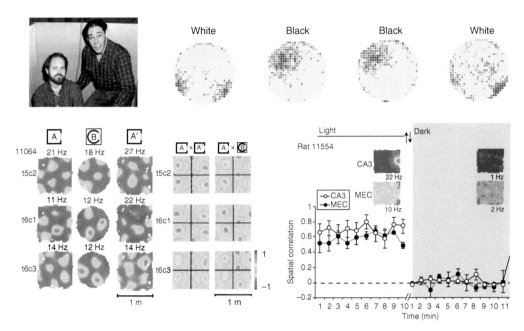

**Figure 3.** Remapping in place cells and grid cells. (*Top left*) John Kubie and Bob Muller in 1983. (*Top right*) Color-coded firing rate map for a hippocampal place cell from an early remapping experiment ( purple, high rate; yellow, low rate). The cell fired at different locations in different versions of the recording cylinder, one with a black cue card and one with a white cue card. (*Bottom left*) Realignment of entorhinal grid cells under conditions that generate global remapping in the hippocampus. The rat was tested in boxes with square or circular surfaces. The *left* panel shows color-coded rate maps for three grid cells (t5c2, t6c1, and t6c3) (color coded as in Fig. 1). The *right* panel shows cross-correlation maps for pairs of rate maps (same grid cells as in the *left* panel; repeated trials in A or one trial in A and one trial in B). The cross-correlation maps are color-coded, with red corresponding to high correlation and blue to low (negative) correlation. Note that the center of the cross-correlation map is shifted in the same direction and at a similar distance from the origin in all three grid cells, suggesting that all grid cells in an ensemble respond coherently to changes in the environment very much unlike the remapping that is observed in the hippocampus. (*Bottom right*) Response to a change in the environment (darkness) in a simultaneously recorded pair of grid and place cells. (*Top left* photo courteously provided by John Kubie; *top right* image is modified from data in Bostock et al. 1991; *bottom* image from Fyhn et al. 2007; reprinted, with permission, from the authors and Nature Publishing Group © 2007.)

The remapping experiments showed that place cells participate in multiple spatial maps. Different maps could be recruited not only in different environments, but also when animals are tested under different conditions in the same location (Markus et al. 1995; Leutgeb et al. 2005b). Maps for different conditions or places were often completely uncorrelated (global remapping) (Leutgeb et al. 2004; Fyhn et al. 2007), as if a pattern-separation process takes place when information enters the hippocampus from the surrounding cortex (Marr 1971; Mc-Naughton and Morris 1987; Leutgeb et al. 2004,

2007). The discovery of remapping and the uncorrelated nature of place maps was important because it showed that place cells participate in multiple orthogonal representations, as expected if the hippocampus plays a role in accurate storage and retrieval of high-capacity declarative memory. The number of place maps stored in the hippocampus is not known, but if place maps are expressions of individual memories, that number should be very large. Remapping is, thus, a necessity if place cells express memories.

Do spatial inputs from medial entorhinal cortex contribute to remapping in the hippocampus? The first clue to the underlying mechanism is that remapping is unique to the hippocampus. The orthogonal nature of place-cell maps is not shared by any of the known spatial cell types upstream of the hippocampus. In the hippocampus, and particularly in the CA3 subfield, different subsets of the place-cell population are active in different environments. The overlap between active subsets in two environments is not larger than expected by chance (Leutgeb et al. 2004). The apparent independence of the place-cell maps contrasts with the functional rigidity of the grid-cell population (Fig. 2). Changes in the environment, which lead to global remapping in the hippocampus, induce changes in the firing locations of simultaneously recorded grid cells, but these changes are always coherent among numbers of grid cells (Fyhn et al. 2007). Among grid cells with similar grid spacing, the firing locations of the grid cells shift in the *xy* plane from one environment to the other, but the distance and direction of grid displacements are similar across the cell population. Similarly, internal coherence is observed in head direction and border cells. When animals are moved from one task to another, head direction cells in the presubiculum and anterior nuclei of the thalamus rotate coherently such that the magnitude of the difference in directional preference among any pair of head direction cells is retained from one condition to the next (Taube et al. 1990; Taube and Burton 1995; Yoganarasimha et al. 2006). A similar spatial coherence is seen among border cells (Solstad et al. 2008). Pairs of cells that fire along the same wall in one environment also fire along the same wall in another environment; cells that fire along opposite walls in one box fire along opposite walls in another box. Changes in orientation are coherent also across entorhinal cell types; if border fields switch to the opposite wall, this is accompanied by a 180-degree change in the orientation of head direction cells, as well as grid cells (Solstad et al. 2008). Taken together, these observations suggest that remapping is generated not in the entorhinal cortex, but in the hippocampus itself.

The findings do not rule out, however, that inputs from realigned or reoriented entorhinal cells give rise to remapping in the hippocampus. Two classes of explanations were put forward when we observed that remapping in the hippocampus is accompanied by coherent realignment in the grid-cell population (Fyhn et al. 2007). The first class assumed a continuous map of space in the medial entorhinal cortex. In this scenario, different portions of a universal entorhinal map would be activated in different environments. Different subsets of hippocampal cells would be activated from independent portions of the entorhinal map and global remapping would be seen in the hippocampus. The second class of explanation assumes that grid cells have a modular organization and that different modules of grid cells respond independently to changes in the environment. Place cells were thought to receive input from several modules. Differential realignment across modules would lead to different overlap of incoming grid signals in hippocampal target cells; the subset of hippocampal cells activated by entorhinal grid-cell inputs would be entirely dependent on the difference in realignment between different modules.

Subsequently, experimental studies have provided evidence for a modular organization of grid cells, consistent with the second explanation (Stensola et al. 2012). For many years, the low number of simultaneously recorded grid cells prevented a clear answer to the question of whether grid cells were modular or not, although early studies pointed in that direction (Barry et al. 2007). With a more than 10-fold increase in the number of grid cells from the same animal, it was possible to show that grid cells cluster into modules with distinct grid scale and grid orientation (Stensola et al. 2012). Four modules could be detected in most animals, but the number may be larger, considering that only a part of the medial entorhinal cortex was sampled. It was not only the properties of the grid pattern that differed between modules, however; they also responded independently to changes in the environment (Stensola et al. 2012). When the recording environment was compressed, changing it from a square to a rectangle, grid cells in the module with the smallest grid spacing maintained their firing locations,

Cite this article as *Cold Spring Harb Perspect Biol* doi: 10.1101/cshperspect.a021808

whereas cells in the larger modules rescaled completely and consistently, firing at shorter spatial wavelengths in the compressed direction, but maintaining wavelengths in the orthogonal unaltered direction. The apparent independence between grid modules contrasts with the strong coherence observed in earlier recordings from grid cells (Fyhn et al. 2007). The difference is likely to reflect the fact that the earlier recordings were all made from the same location and, probably, mostly from a single module.

The new data suggest that modules respond with different degrees of displacement and reorientation when animals move from one environment to another. Computational simulations have shown that independent realignment in four or fewer modules is sufficient to generate complete or global remapping in the hippocampus (Monaco and Abbott 2011). Independent responses among only a handful of grid modules may be sufficient to create an enormous diversity of firing patterns in the hippocampus because the number of displacements or phases that each module may take is large. The mechanism would be similar to that of a combination lock in which 10,000 combinations may be generated with only four modules of 10 possible values each (Rowland and Moser 2014), or that of an alphabet in which all words of a language can be generated by combining only 30 letters or less. The proposed mechanism is only a hypothesis, however. Whether hippocampal remapping actually requires independent realignment among grid modules remains to be determined. It should also be noted that a possible connection between grid modules and remapping does not rule out roles for other cell types, such as border cells, in inducing hippocampal remapping, although modular organization has not yet been observed in any of the other functional cell populations (Giocomo et al. 2014).

Finally, we would like to emphasize that, up to this point, we have mostly discussed the entorhinal–hippocampal space circuit as if interactions between cell types were constant over time. However, the connectivity of this network is dynamic (Buzsáki and Moser 2013). Whether entorhinal and hippocampal neurons influence each other depends strongly on the state of theta and gamma oscillations, which, during active awake behavior, predominates frequency spectra in both regions (Buzsáki et al. 1983; Bragin et al. 1995; Chrobak and Buzsáki 1998; Csicsvari et al. 2003; Colgin et al. 2009). Theta oscillations are generally coherent across most of the entorhinal–hippocampal network, but the coherence of beta and gamma oscillations is more local and fluctuates at subsecond timescales (Colgin et al. 2009; Igarashi et al. 2014). Such fluctuations may enable place cells to interact with different entorhinal subpopulations at different times. Coincidence of pre- and postsynaptic activity may be a prerequisite not only for synaptic strengthening of connections between entorhinal and hippocampal cell pairs (Singer 1993; Bi and Poo 1998), but also for pattern-completion processes during retrieval of already-stored information. Whether a place cell responds to inputs from grid or border cells may change with time, as may the influence of different modules of grid cells. Recordings from CA1 and lateral entorhinal cortex suggest that place cells also respond dynamically to nonspatial inputs, such as odors, with learned relationships to locations in the environment (Igarashi et al. 2014). Beta and gamma oscillations may enable place cells to respond temporarily to information about the content of locations in the spatial environment.

## CONCLUSION

We have known for almost six decades that certain types of memory depend on the hippocampus and surrounding areas. The discovery of place cells showed that space is a critical element of the information that is stored and expressed by neurons in the hippocampus; however, it is, perhaps, with studies of place cells at the ensemble or population level and interventions that selectively change synaptic plasticity in specific brain circuits, that the mechanisms of memory processing have become accessible. Today, we know that hippocampal networks can rapidly store a multitude of uncorrelated representations, a property that any high-capacity episodic memory network must have. We know that place cells are only one element of a wider network for spatial mapping. Place cells coexist with grid, head direction, and border cells, all likely to interact with each other to yield a global representation of the animal's changing position, which may be used to guide the animal to particular locations in the environment. With a modular organization of grid cells, the network may be able to generate not only one map of the external environment, but thousands or millions. Whether and how these maps contribute to declarative memory remains to be determined, but the investigation of the hippocampal–entorhinal circuit is now at a stage in which the computational mechanisms underlying specific memory processes are fully addressable.

## REFERENCES

Abbott LF, Blum KI. 1996. Functional significance of long-term potentiation for sequence learning and prediction. *Cereb Cortex* **6:** 406–416.

Barnes CA, McNaughton BL, Mizumori SJ, Leonard BW, Lin LH. 1990. Comparison of spatial and temporal characteristics of neuronal activity in sequential stages of hippocampal processing. *Prog Brain Res* **83:** 287–300.

Barry C, Lever C, Hayman R, Hartley T, Burton S, O'Keefe J, Jeffery K, Burgess N. 2006. The boundary vector corpcell model of place cell firing and spatial memory. *Rev Neurosci* **17:** 71–97.

Barry C, Hayman R, Burgess N, Jeffery KJ. 2007. Experience-dependent rescaling of entorhinal grids. *Nat Neurosci* **10:** 682–684.

Bi GQ, Poo MM. 1998. Synaptic modifications in cultured hippocampal neurons: Dependence on spike timing, synaptic strength, and postsynaptic cell type. *J Neurosci* **18:** 10464–10472.

Bjerknes TL, Moser EI, Moser MB. 2014. Representation of geometric borders in the developing rat. *Neuron* **82:** 71–78.

Blum KI, Abbott LF. 1996. A model of spatial map formation in the hippocampus of the rat. *Neural Comput* **8:** 85–93.

Boccara CN, Sargolini F, Thoresen VH, Solstad T, Witter MP, Moser EI, Moser M-B. 2010. Grid cells in pre- and parasubiculum. *Nat Neurosci* **13:** 987–994.

Bostock E, Muller RU, Kubie JL. 1991. Experience-dependent modifications of hippocampal place cell firing. *Hippocampus* **1:** 193–205.

Bragin A, Jandó G, Nádasdy Z, Hetke J, Wise K, Buzsáki G. 1995. Gamma (40–100 Hz) oscillation in the hippocampus of the behaving rat. *J Neurosci* **15:** 47–60.

Brun VH, Otnass MK, Molden S, Steffenach HA, Witter MP, Moser MB, Moser EI. 2002. Place cells and place recognition maintained by direct entorhinal–hippocampal circuitry. *Science* **296:** 2243–2246.

Bush D, Barry C, Burgess N. 2014. What do grid cells contribute to place cell firing? *Trends Neurosci* **37:** 136–145.

Buzsáki G. 1989. Two-stage model of memory trace formation: A role for "noisy" brain states. *Neuroscience* **31:** 551–570.

Buzsáki G, Moser EI. 2013. Memory, navigation and theta rhythm in the hippocampal-entorhinal system. *Nat Neurosci* **16:** 130–138.

Buzsáki G, Leung LW, Vanderwolf CH. 1983. Cellular bases of hippocampal EEG in the behaving rat. *Brain Res* **287:** 139–171.

Chrobak JJ, Buzsáki G. 1998. Gamma oscillations in the entorhinal cortex of the freely behaving rat. *J Neurosci* **18:** 388–398.

Chun MM, Turk-Browne NB. 2007. Interactions between attention and memory. *Curr Opin Neurobiol* **17:** 177–184.

Cohen NJ, Squire LR. 1980. Preserved learning and retention of pattern analyzing skill in amnesia: Dissociation of knowing how and knowing that. *Science* **210:** 207–209.

Colgin LL, Denninger T, Fyhn M, Hafting T, Bonnevie T, Jensen O, Moser MB, Moser EI. 2009. Frequency of gamma oscillations routes flow of information in the hippocampus. *Nature* **462:** 353–357.

Corkin S. 2002. What's new with the amnesic patient H.M.? *Nature Rev Neurosci* **3:** 153–160.

Csicsvari J, Jamieson B, Wise KD, Buzsáki G. 2003. Mechanisms of gamma oscillations in the hippocampus of the behaving rat. *Neuron* **37:** 311–322.

de Almeida L, Idiart M, Lisman JE. 2009. The input–output transformation of the hippocampal granule cells: From grid cells to place fields. *J Neurosci* **29:** 7504–7512.

Diba K, Buzsaki G. 2007. Forward and reverse hippocampal place-cell sequences during ripples. *Nat Neurosci* **10:** 1241–1242.

Dragoi G, Tonegawa S. 2011. Preplay of future place cell sequences by hippocampal cellular assemblies. *Nature* **469:** 397–401.

Dragoi G, Tonegawa S. 2013. Distinct preplay of multiple novel spatial experiences in the rat. *Proc Natl Acad Sci* **110:** 9100–9105.

Ebbinghaus H. 1885. *Über das Gedächtnis Untersuchungen zur Experimentellen Psychologie* [*Memory: A contribution to experimental psychology*]. von Duncker and Humber, Leipzig, Germany.

Ego-Stengel V, Wilson MA. 2010. Disruption of ripple-associated hippocampal activity during rest impairs spatial learning in the rat. *Hippocampus* **20:** 1–10.

Eichenbaum H, Kuperstein M, Fagan A, Nagode J. 1987. Cue-sampling and goal-approach correlates of hippocampal unit activity in rats performing an odor-discrimination task. *J Neurosci* **7:** 716–732.

Ekstrom AD, Meltzer J, McNaughton BL, Barnes CA. 2001. NMDA receptor antagonism blocks experience-dependent expansion of hippocampal "place fields." *Neuron* **31:** 631–638.

Ferbinteanu J, Shapiro ML. 2003. Prospective and retrospective memory coding in the hippocampus. *Neuron* **40:** 1227–1239.

Foster DJ, Wilson MA. 2006. Reverse replay of behavioural sequences in hippocampal place cells during the awake state. *Nature* **440:** 680–683.

Frank LM, Brown EN, Wilson M. 2000. Trajectory encoding in the hippocampus and entorhinal cortex. *Neuron* **27:** 169–178.

Frank LM, Stanley GB, Brown EN. 2004. Hippocampal plasticity across multiple days of exposure to novel environments. *J Neurosci* **24:** 7681–7689.

Frankland PW, O'Brien C, Ohno M, Kirkwood A, Silva AJ. 2001. α-CaMKII-dependent plasticity in the cortex is required for permanent memory. *Nature* **411:** 309–313.

Fuhs MC, Touretzky DS. 2006. A spin glass model of path integration in rat medial entorhinal cortex. *J Neurosci* **26:** 4266–4276.

Fyhn M, Molden S, Witter MP, Moser EI, Moser M-B. 2004. Spatial representation in the entorhinal cortex. *Science* **305:** 1258–1264.

Fyhn M, Hafting T, Treves A, Moser M-B, Moser EI. 2007. Hippocampal remapping and grid realignment in entorhinal cortex. *Nature* **446:** 190–194.

Gelbard-Sagiv H, Mukamel R, Harel M, Malach R, Fried I. 2008. Internally generated reactivation of single neurons in human hippocampus during free recall. *Science* **322:** 96–101.

Giocomo LM, Stensola T, Bonnevie T, Van Cauter T, Moser M-B, Moser EI. 2014. Topography of head direction cells in medial entorhinal cortex. *Curr Biol* **24:** 252–262.

Girardeau G, Benchenane K, Wiener SI, Buzsaki G, Zugaro MB. 2009. Selective suppression of hippocampal ripples impairs spatial memory. *Nat Neurosci* **12:** 1222–1223.

Gupta AS, van der Meer MA, Touretzky DS, Redish AD. 2010. Hippocampal replay is not a simple function of experience. *Neuron* **65:** 695–705.

Hafting T, Fyhn M, Molden S, Moser M-B, Moser EI. 2005. Microstructure of a spatial map in the entorhinal cortex. *Nature* **436:** 801–806.

Hampson RE, Heyser CJ, Deadwyler SA. 1993. Hippocampal cell firing correlates of delayed-match-to-sample performance in the rat. *Behav Neurosci* **107:** 715–739.

Hartley T, Burgess N, Lever C, Cacucci F, O'Keefe J. 2000. Modeling place fields in terms of the cortical inputs to the hippocampus. *Hippocampus* **10:** 369–379.

Hassabis D, Kumaran D, Maguire EA. 2007. Using imagination to understand the neural basis of episodic memory. *J Neurosci* **27:** 14365–14374.

Hebb DO. 1949. *The organization of behavior.* Wiley, New York.

Hill AJ. 1978. First occurrence of hippocampal spatial firing in a new environment. *Exp Neurol* **62:** 282–297.

Hoffman KL, McNaughton BL. 2002. Coordinated reactivation of distributed memory traces in primate neocortex. *Science* **297:** 2070–2073.

Igarashi KM, Lu L, Colgin LL, Moser M-B, Moser EI. 2014. Coordination of entorhinal–hippocampal ensemble activity during associative learning. *Nature* **510:** 143–147.

Jung MW, Wiener SI, McNaughton BL. 1994. Comparison of spatial firing characteristics of units in dorsal and ventral hippocampus of the rat. *J Neurosci* **14:** 7347–7356.

Kandel ER, Schwartz JH. 1982. Molecular biology of learning: Modulation of transmitter release. *Science* **218:** 433–443.

Karlsson MP, Frank LM. 2009. Awake replay of remote experiences in the hippocampus. *Nat Neurosci* **12:** 913–918.

Kentros C, Hargreaves E, Hawkins RD, Kandel ER, Shapiro M, Muller RV. 1998. Abolition of long-term stability of new hippocampal place cell maps by NMDA receptor blockade. *Science* **280:** 2121–2126.

Kentros CG, Agnihotri NT, Streater S, Hawkins RD, Kandel ER. 2004. Increased attention to spatial context increases both place field stability and spatial memory. *Neuron* **42:** 283–295.

Kjelstrup KB, Solstad T, Brun VH, Hafting T, Leutgeb S, Witter MP, Moser EI, Moser M-B. 2008. Finite scales of spatial representation in the hippocampus. *Science* **321:** 140–143.

Langston RF, Ainge JA, Couey JJ, Canto CB, Bjerknes TL, Witter MP, Moser EI, Moser M-B. 2010. Development of the spatial representation system in the rat. *Science* **328:** 1576–1580.

Lashley KS. 1929. *Brain mechanisms and intelligence: A qualitative study of injuries to the brain.* University of Chicago Press, Chicago.

Lashley KS. 1950. In search of the engram. In *Symposium of the society for experimental biology,* Vol. 4. Cambridge University Press, New York.

Leutgeb S, Leutgeb JK, Treves A, Moser M-B, Moser EI. 2004. Distinct ensemble codes in hippocampal areas CA3 and CA1. *Science* **305:** 1295–1298.

Leutgeb JK, Leutgeb S, Treves A, Meyer R, Barnes CA, McNaughton BL, Moser M-B, Moser EI. 2005a. Progressive transformation of hippocampal neuronal representations in "morphed" environments. *Neuron* **48:** 345–358.

Leutgeb S, Leutgeb JK, Barnes CA, Moser EI, McNaughton BL, Moser M-B. 2005b. Independent codes for spatial and episodic memory in hippocampal neuronal ensembles. *Science* **309:** 619–623.

Leutgeb S, Leutgeb JK, Moser EI, Moser MB. 2006. Fast rate coding in hippocampal CA3 cell ensembles. *Hippocampus* **16:** 765–774.

Leutgeb JK, Leutgeb S, Moser MB, Moser EI. 2007. Pattern separation in the dentate gyrus and CA3 of the hippocampus. *Science* **315:** 961–966.

Lever C, Burton S, Jeewajee A, O'Keefe J, Burgess N. 2009. Boundary vector cells in the subiculum of the hippocampal formation. *J Neurosci* **29:** 9771–9777.

Markus EJ, Qin YL, Leonard B, Skaggs WE, McNaughton BL, Barnes CA. 1995. Interactions between location and task affect the spatial and directional firing of hippocampal neurons. *J Neurosci* **15:** 7079–7094.

Marr D. 1971. Simple memory: A theory for archicortex. *Philos Trans R Soc Lond B Biol Sci* **262:** 23–81.

McClelland JL, McNaughton BL, O'Reilly RC. 1995. Why there are complementary learning systems in the hippocampus and neocortex: Insights from the successes and failures of connectionist models of learning and memory. *Psychol Rev* **102:** 419–457.

McHugh TJ, Blum KI, Tsien JZ, Tonegawa S, Wilson MA. 1996. Impaired hippocampal representation of space in

CA1-specific NMDAR1 knockout mice. *Cell* **87:** 1339–1349.

McNaughton BL, Battaglia FP, Jensen O, Moser EI, Moser M-B. 2006. Path integration and the neural basis of the "cognitive map." *Nature Rev Neurosci* **7:** 663–678.

Mehta MR, Barnes CA, McNaughton BL. 1997. Experience-dependent, asymmetric expansion of hippocampal place fields. *Proc Natl Acad Sci* **94:** 8918–8921.

Mehta MR, Quirk MC, Wilson MA. 2000. Experience-dependent asymmetric shape of hippocampal receptive fields. *Neuron* **25:** 707–715.

Miller JF, Neufang M, Solway A, Brandt A, Trippel M, Mader I, Hefft S, Merkow M, Polyn SM, Jacobs J, et al. 2013. Neural activity in human hippocampal formation reveals the spatial context of retrieved memories. *Science* **342:** 1111–1114.

Milner B, Corkin S, Teuber HL. 1968. Further analysis of the hippocampal amnesic syndrome: 14-year follow-up study of H.M. *Neuropsychologia* **6:** 215–234.

Mishkin M. 1978. Memory in monkeys severely impaired by combined but not by separate removal of amygdala and hippocampus. *Nature* **273:** 297–298.

Moita MA, Rosis S, Zhou Y, LeDoux JE, Blair HT. 2004. Putting fear in its place: Remapping of hippocampal place cells during fear conditioning. *J Neurosci* **24:** 7015–7023.

Monaco JD, Abbott LF. 2011. Modular realignment of entorhinal grid cell activity as a basis for hippocampal remapping. *J Neurosci* **31:** 9414–9425.

Monaco JD, Rao G, Roth ED, Knierim JJ. 2014. Attentive scanning behavior drives one-trial potentiation of hippocampal place fields. *Nat Neurosci* **17:** 725–731.

Moser EI, Kropff E, Moser M-B. 2008. Place cells, grid cells, and the brain's spatial representation system. *Annu Rev Neurosci* **31:** 69–89.

Muller RU, Kubie JL. 1987. The effects of changes in the environment on the spatial firing of hippocampal complex-spike cells. *J Neurosci* **7:** 1951–1968.

Muzzio IA, Levita L, Kulkarni J, Monaco J, Kentros C, Stead M, Abbott LF, Kandel ER. 2009. Attention enhances the retrieval and stability of visuospatial and olfactory representations in the dorsal hippocampus. *PLoS Biol* **7:** e1000140.

O'Keefe J. 1976. Place units in the hippocampus of the freely moving rat. *Exp Neurol* **51:** 78–109.

O'Keefe J, Burgess N. 1996. Geometric determinants of the place fields of hippocampal neurons. *Nature* **381:** 425–428.

O'Keefe J, Burgess N. 2005. Dual phase and rate coding in hippocampal place cells: Theoretical significance and relationship to entorhinal grid cells. *Hippocampus* **15:** 853–866.

O'Keefe J, Dostrovsky J. 1971. The hippocampus as a spatial map. Preliminary evidence from unit activity in the freely-moving rat. *Brain Res* **34:** 171–175.

O'Keefe J, Nadel L. 1978. *The hippocampus as a cognitive map.* Clarendon, Oxford.

O'Keefe J, Speakman A. 1987. Single unit activity in the rat hippocampus during a spatial memory task. *Exp Brain Res* **68:** 1–27.

O'Neill J, Senior T, Csicsvari J. 2006. Place-selective firing of CA1 pyramidal cells during sharp wave/ripple network patterns in exploratory behavior. *Neuron* **49:** 143–155.

Pavlides C, Winson J. 1989. Influences of hippocampal place cell firing in the awake state on the activity of these cells during subsequent sleep episodes. *J Neurosci* **9:** 2907–2918.

Pfeiffer BE, Foster DJ. 2013. Hippocampal place-cell sequences depict future paths to remembered goals. *Nature* **497:** 74–79.

Quirk GJ, Muller RU, Kubie JL, Ranck JB Jr. 1992. The positional firing properties of medial entorhinal neurons: Description and comparison with hippocampal place cells. *J Neurosci* **12:** 1945–1963.

Ranck JB Jr. 1973. Studies on single neurons in dorsal hippocampal formation and septum in unrestrained rats: I. Behavioral correlates and firing repertoires. *Exp Neurol* **41:** 461–531.

Ranck JB. 1985. Head direction cells in the deep cell layer of dorsal presubiculum in freely moving rats. In *Electrical activity of the archicortex* (ed. Buzsáki G, Vanderwolf CH), pp. 217–220. Akademiai Kiado, Budapest.

Rolls ET, Stringer SM, Elliot T. 2006. Entorhinal cortex grid cells can map to hippocampal place cells by competitive learning. *Network* **17:** 447–465.

Rotenberg A, Mayford M, Hawkins RD, Kandel ER, Muller RU. 1996. Mice expressing activated CaMKII lack low frequency LTP and do not form stable place cells in the CA1 region of the hippocampus. *Cell* **87:** 1351–1361.

Rowland DC, Moser M-B. 2014. From cortical modules to memories. *Curr Opin Neurobiol* **24C:** 22–27.

Samsonovich A, McNaughton BL. 1997. Path integration and cognitive mapping in a continuous attractor neural network model. *J Neurosci* **17:** 5900–5920.

Sargolini F, Fyhn M, Hafting T, McNaughton BL, Witter MP, Moser M-B, Moser EI. 2006. Conjunctive representation of position, direction and velocity in entorhinal cortex. *Science* **312:** 754–758.

Savelli F, Knierim JJ. 2010. Hebbian analysis of the transformation of medial entorhinal grid-cell inputs to hippocampal place fields. *J Neurophysiol* **103:** 3167–3183.

Savelli F, Yoganarasimha D, Knierim JJ. 2008. Influence of boundary removal on the spatial representations of the medial entorhinal cortex. *Hippocampus* **18:** 1270–1282.

Scoville WB, Milner B. 1957. Loss of recent memory after bilateral hippocampal lesions. *J Neurol Neurosurg Psychiatry* **20:** 11–21.

Singer W. 1993. Synchronization of cortical activity and its putative role in information processing and learning. *Annu Rev Physiol* **55:** 349–374.

Smith SL, Smith IT, Branco T, Häusser M. 2013. Dendritic spikes enhance stimulus selectivity in cortical neurons in vivo. *Nature* **503:** 115–120.

Solstad T, Moser EI, Einevoll GT. 2006. From grid cells to place cells: A mathematical model. *Hippocampus* **16:** 1026–1031.

Solstad T, Boccara CN, Kropff E, Moser M-B, Moser EI. 2008. Representation of geometric borders in the entorhinal cortex. *Science* **322:** 1865–1868.

Squire LR. 1992. Memory and the hippocampus: A synthesis from findings with rats, monkeys, and humans. *Psychol Rev* **99:** 195–231.

Squire LR, Alvarez P. 1995. Retrograde amnesia and memory consolidation: A neurobiological perspective. *Curr Opin Neurobiol* **5:** 169–177.

Stensola H, Stensola T, Solstad T, Frøland K, Moser M-B, Moser EI. 2012. The entorhinal map is discretized. *Nature* **492:** 72–78.

Suddendorf T, Corballis MC. 2007. The evolution of foresight: What is mental time travel, and is it unique to humans? *Behav Brain Sci* **30:** 299–313.

Taube JS. 2007. The head direction signal: Origins and sensory-motor integration. *Annu Rev Neurosci* **30:** 181–207.

Taube JS, Burton HL. 1995. Head direction cell activity monitored in a novel environment and during a cue conflict situation. *J Neurophysiol* **74:** 1953–1971.

Taube JS, Muller RU, Ranck JB Jr. 1990. Head-direction cells recorded from the postsubiculum in freely moving rats: I. Description and quantitative analysis. *J Neurosci* **10:** 420–435.

Tolman EC. 1948. Cognitive maps in rats and men. *Psychol Rev* **55:** 189–208.

Watrous AJ, Tandon N, Conner CR, Pieters T, Ekstrom AD. 2013. Frequency-specific network connectivity increases underlie accurate spatiotemporal memory retrieval. *Nat Neurosci* **16:** 349–356.

Wills TJ, Cacucci F, Burgess N, O'Keefe J. 2010. Development of the hippocampal cognitive map in preweanling rats. *Science* **328:** 1573–1576.

Wilson MA, McNaughton BL. 1993. Dynamics of the hippocampal ensemble code for space. *Science* **261:** 1055–1058.

Wilson MA, McNaughton BL. 1994. Reactivation of hippocampal ensemble memories during sleep. *Science* **265:** 676–679.

Wood ER, Dudchenko PA, Eichenbaum H. 1999. The global record of memory in hippocampal neuronal activity. *Nature* **397:** 613–616.

Wood ER, Dudchenko PA, Robitsek RJ, Eichenbaum H. 2000. Hippocampal neurons encode information about different types of memory episodes occurring in the same location. *Neuron* **27:** 623–633.

Yoganarasimha D, Yu X, Knierim JJ. 2006. Head direction cell representations maintain internal coherence during conflicting proximal and distal cue rotations: Comparison with hippocampal place cells. *J Neurosci* **26:** 622–631.

Young BJ, Fox GD, Eichenbaum H. 1994. Correlates of hippocampal complex-spike cell activity in rats performing a nonspatial radial maze task. *J Neurosci* **14:** 6553–6563.

Zhang SJ, Ye J, Miao CL, Tsao A, Cerniauskas I, Ledergerber D, Moser M-B, Moser EI. 2013. Optogenetic dissection of entorhinal-hippocampal functional connectivity. *Science* **340:** 1232627.

Zhang S-J, Ye J, Couey JJ, Witter MP, Moser EI, Moser M-B. 2014. Functional connectivity of the entorhinal-hippocampal space circuit. *Philos Trans R Soc Lond B Biol Sci* **369:** 20120516.

Zola-Morgan S, Squire LR, Amaral DG. 1986. Human amnesia and the medial temporal region: Enduring memory impairment following a bilateral lesion limited to field CA1 of the hippocampus. *J Neurosci* **6:** 2950–2967.

# Working Memory: Maintenance, Updating, and the Realization of Intentions

Lars Nyberg and Johan Eriksson

Umeå Center for Functional Brain Imaging (UFBI), Umeå University, 901 87 Umeå, Sweden

*Correspondence:* lars.nyberg@umu.se

"Working memory" refers to a vast set of mnemonic processes and associated brain networks, relates to basic intellectual abilities, and underlies many real-world functions. Working-memory maintenance involves frontoparietal regions and distributed representational areas, and can be based on persistent activity in reentrant loops, synchronous oscillations, or changes in synaptic strength. Manipulation of content of working memory depends on the dorsofrontal cortex, and updating is realized by a frontostriatal '"gating" function. Goals and intentions are represented as cognitive and motivational contexts in the rostrofrontal cortex. Different working-memory networks are linked via associative reinforcement-learning mechanisms into a self-organizing system. Normal capacity variation, as well as working-memory deficits, can largely be accounted for by the effectiveness and integrity of the basal ganglia and dopaminergic neurotransmission.

Imagine the following scenario:

You enjoy a sabbatical semester and visit a close colleague to work on a joint review paper. You have generously been offered a room on the floor where your colleague sits. One floor below, a nice library holds many texts (not accessible on the Internet) that you may want to consult when writing your review. One morning you realize that a volume in the library would be relevant for the section of the review you currently work on, and walk down only to find out that the door is locked. You head upstairs to your colleague's office to borrow her key card, and are told that the code is "1, 9, 6, 9, 3." While rushing back down, you repeat the code silently to yourself, noticing that subtracting the last digit from the first four will give your birth year (1966). However, after you punched the code, the door will not open. Puzzled, you think you may have entered the wrong code and try again, but the door remains locked. So you head back to your colleague and tell her that "1, 9, 6, 9, 3" did not work. She responds, "I'm sorry, that's the code for the parking garage; the correct code should be 3, 7, 4, 9, 8." You repeat the new code to yourself while heading down, and this time it works. You enter into the library and quickly forget all about any door problems when you start to think about your section of the review and try to locate the relevant volume. You find it and bring it back to your desk and continue writing. Later on, in the afternoon of the same day, work on a new section of the review prompts you to return to the library to pick up another volume. You still have your colleague's key card and head downstairs. Once there you realize that you need the code and to save yourself from yet another stair climb you try to retrieve the code from memory. "1, 9, 6, 9, 3" pops up and you try it with no success. You think, "maybe that was the first code I tried (the one for the garage), but what was the correct one"?

This little scenario, which most readers may have experienced in real life in some form or another (while cooking/following a recipe, doing carpentry/construction, or while doing errands in a shopping mall), highlights several key defining features of working memory:

1. Working memory can guide behavior by means of "mnemonic representations of stimuli" in the absence of the stimuli themselves, as above when the code had to be retained in memory until the door was reached. Such active "online maintenance" of information is at the heart of the working-memory concept. The influential multicomponent model of Baddeley and Hitch (1974) postulated distinct "buffers" for maintaining verbal or visuospatial information. That model further suggested that an attentional control system, the "central executive," controls information maintenance, for example, via active rehearsal processes. The ability to maintain information in working memory can be tested in many different ways, including with delayed-match-to-sample and simple span tasks.

2. Working memory "interacts with long-term memory" in many ways. State-based working-memory models assume that maintaining information in working memory critically depends on allocating attentional resources to internal long-term memory representations (see D'Esposito and Postle 2015). Long-term memory can support the clustering or "chunking" of information in working memory (Miller 1956), as in the situation when a birth year was derived from the five digits, which can greatly reduce working-memory demands. Working memory may also rely, at least in part, on some of the same principles for information storage as long-term memory, and working-memory processes might critically contribute to the encoding and retrieval of long-term memory, as in the instance when the first code was remembered hours after it had been used.

3. Information that is maintained in working memory can be replaced with other information by means of an "updating" process, as in the case when the incorrect code (to the garage) was replaced with the correct one. Thus, successful and adaptive working memory requires both stability (when information is actively and robustly maintained) and flexibility (when information needs to be replaced and updated). Biologically based computational models have been proposed to capture these complex dynamics, and they can be taxed with tests such as running span tasks (e.g., keep track).

4. Information that is maintained in working memory can be "manipulated" or operated on by additional processes, as in the case when "3" was subtracted from "1 9 6 9" to give a birth year (1966). The possibility to actively process and manipulate information that is maintained in working memory is a salient defining feature of working memory and likely a foundation for a wide range of complex abilities. The famous reading span task (Daneman and Carpenter 1980) and related complex span tasks, such as the operation span task, all combine maintenance with the requirement to perform additional manipulation processes.

5. Working memory is essential for the "execution of a plan," as in the example we considered above when the idea emerged to go to the library to collect a volume. In their excellent review of the cognitive neuroscience of working memory, D'Esposito and Postle (2015) recently noted that the functional characterization of working memory as underlying the ability to execute complex plans dates back >50 years. More recently, Patricia Goldman-Rakic (1987) noted, in the same spirit, that plans can govern behavior. A particularly critical role for working memory should be in situations when the execution of a plan is interrupted, as in the scenario when the core plan of picking up a volume only could be realized after several "door-related" interruptions, and also when multiple goals are concurrently active. Inherent in

the notion of coding for the execution of a plan is motivation and incentives, and recent models try to combine cognitive and motivational control of behavior (e.g., Fuster 2013; Watanabe 2013).

Thus, working memory can narrowly be defined as temporary online maintenance of information for the performance of a task in the (near) future, but more broadly to also include manipulation and updating of the aforementioned information, as well as coordinating behavior when multiple goals are active. In the present work, we will discuss these key features of working memory from a cognitive neuroscience perspective, attempting to synthesize behavioral and neurobiological data. A variety of sources of neurobiological data will be considered, including functional magnetic resonance imaging (fMRI), molecular imaging with positron emission tomography (PET), electrophysiological registrations with electroencephalography (EEG), and magnetoencephalography (MEG), transcranial magnetic stimulation (TMS), lesion studies, cell recordings from primate neurons, and also computational modeling.

It should be stated up front that the treatment of past studies must, out of necessity, be highly selective. D'Esposito and Postle (2015) reported the results of a PubMed search on "working memory" performed late in 2014 that returned more than 18,000 results! Here, the discussion of empirical findings will be guided by a "processing-component" theoretical framework (Fig. 1; cf. Fuster 2009, 2013; see also Moscovitch and Winocur 2002; Eriksson et al. 2015). According to this framework, there is no dedicated "working-memory system" in the brain in the sense of corresponding systems for visual perception. Rather, working memory is seen as a computational and cognitive faculty emerging from the interaction among various basic processes, some of which are used in various combinations in the service of other forms of memory, such as declarative (episodic and semantic), long-term memory (cf. Nyberg and Cabeza 2001; D'Esposito and Postle 2015). In line with this framework, functional brain-im-

aging studies have found overlapping activity patterns, notably in the prefrontal cortex, for working-memory challenges, as well as for several other cognitive demands (Cabeza and Nyberg 2000; Duncan and Owen 2000; Cabeza et al. 2002; Nyberg et al. 2002, 2003; Naghavi and Nyberg 2005). The specific components of the framework outlined in Figure 1 will be presented in more detail in the next sections.

We start by discussing working-memory maintenance, including interactions between working memory and long-term memory. Next, we will discuss manipulations and updating of working memory, followed by a section on the complex topic of how intentions can be maintained and guide behavior. In the concluding section, Working Memory in Action, we will relate findings from studies of working memory at the brain and behavioral levels to variation in working-memory functioning. This will include variation among healthy younger adults, in aging and Parkinson's disease, and in psychiatric disorder.

## MAINTENANCE IN WORKING MEMORY AND ITS RELATION TO LONG-TERM MEMORY

### Where Is Working Memory Maintained?

In the spirit of the fundamental perception–action cycle for the temporal organization of behavior (e.g., Fuster 2013; see arrow 1 that connects the light blue [perception] and orange [action] ellipses in Fig. 1), we begin the discussion of working memory at the perception stage in which different kinds of sensory information are processed by dedicated brain systems. A long-standing basic hypothesis concerning information storage in the brain is that of "distributed storage," according to which the specific sites in the brain where information is stored are determined by how the brain was engaged during initial perception/learning (schematically shown by the overlap between the perception and representation ellipses in Fig. 1). This hypothesis is supported by studies of declarative (episodic and sematic) long-term memory (Nyberg et al. 2000; Martin and Chao 2001; Danker and Anderson 2010). Thus, when

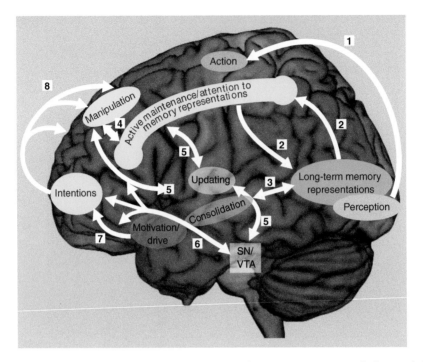

**Figure 1.** Working-memory processes and interactions within the perception-action cycle (arrow 1). The color-coding of processes has been freely adapted after Fuster (2013), with perception in light blue, action in orange, motivation/drive in red, long-term memory representation and consolidation in blue, and, in yellow, the working-memory processes that will be discussed. Arrow 2 represents reverberating activity in frontal, parietal, and representational areas during maintenance. Arrow 3 represents consolidation of working-memory information into long-term memory via interactions with the medial-temporal lobe system. Arrow 4 represents associations between manipulation networks, mainly in dorsofrontal cortex, and frontoparietal maintenance/ attention processes. Arrow 5 represents nigrostriatal dopaminergic neurotransmission and striatocortical interactions during working-memory updating. Arrow 6 represents diffuse dopamine gating signals from the ventral tegmental area (VTA) to the frontal cortex. Arrow 7 represents emotional input to rostrofrontal cortical regions, and arrow 8 represents how neurons in the rostrofrontal cortex, coding for cognitive and motivational context, influence other working-memory networks to support goal-directed behavior. Transparency of ellipses indicates major subcortical nodes for updating (striatum), consolidation (hippocampus), and motivation (amygdala, brain stem, hypothalamus). SN, Substantia nigra.

a long-term memory is retrieved, some of the perceptual regions that were recruited during learning become reactivated in a material-specific sense.

A similar principle seems to hold true for maintenance of working memory. That is, in the context of working-memory maintenance, attention to semantic representations (e.g., letters and digits), as well as sensorimotor representations (e.g., colors, line orientations), have been found to engage material-specific brain areas. In particular, analyses of fMRI data by means of multivariate "pattern analysis" techniques have revealed that modality-specific regions retain sensory-specific working-memory representations during the delay period (see Sreenivasan et al. 2014). Also, if TMS is applied to visual cortex during maintenance of visual information, there is a reduction in the performance of working memory (van de Ven et al. 2012). Collectively, these and related studies converge to suggest that the maintenance of working memory does not rely on any specialized storage "buffers," but instead shares the same representational zones as retrieval from long-term memory.

## How Is Information in Working Memory Maintained?

How, then, is activity in neuronal representational populations maintained during a working-memory delay? In the scenario above, this concerns how the five-digit code was upheld in working memory from the time when it was heard (perception) to when the code was used (action). Similarly, the majority of working-memory studies involve online information maintenance over a few seconds. As discussed in Where Is Working Memory Maintained?, we would predict that posterior cortical neuronal populations engaged in the representation of digits will be involved, but also the prefrontal cortex. Recordings by Joaquín Fuster of cells in the primate cortex (Fuster and Alexander 1971; see also Fuster 2013) revealed that cells in the frontal cortex showed sustained activity during a working-memory delay. Subsequently, such cell populations have been discovered outside the frontal cortex as well, and persistent neural activity in frontal and select posterior neuronal populations jointly define working memory for a specific type of stimulus. These kinds of memory networks, or cognits (see Fuster 2013), are established by means of associative principles when the participating cell populations jointly become active during the performance of a task (later, we will return to the issue of memory networks being related to other kinds of networks, for example, those representing contextual information).

Cells in one and the same zone of the frontal cortex may contribute to working memory of different kinds of information by interacting with select posterior cell populations. Thus, the frontal cortex is likely not a storage buffer per se but exerts top-down control of other neuronal populations in the network that actually represent information (for discussions about material specificity in frontal maintenance activity through posterior feedback signals, see, e.g., Sala and Courtney 2007). Typically, along with the frontal cortex, superior parietal regions also show elevated sustained activity during the delay period, suggesting that parts of the parietal cortex may serve general top-down

functions (e.g., sustain attention to internal representation; see Fig. 1). Likely, there are multiple sources (frontal, parietal, other) of top-down signals to lower-order areas (see arrow 2 from the yellow maintenance ellipse in Fig. 1 to the blue representation ellipse), which jointly contribute to maintenance of specific representations via persistent activity (cf. D'Esposito and Postle 2015).

If the top-down signal is interfered with, for example, by local cooling of the lateral frontal prefrontal cortex, this has a negative impact on posterior activity, as well as on performance, and comparable effects are seen after cooling of posterior cortical sites (see Fuster 2013). One likely interpretation of the findings of such cooling experiments, in which nodes of a distributed memory network are inhibited, is that it interrupts reentrant loops between frontal and posterior cortical zones that maintain information in working memory by reverberation (see double arrows 2 that form a circuit in Fig. 1).

Synchronous oscillations also offer a basis for how distributed brain regions in frontal, parietal, and posterior cortices might interact in the service of working-memory maintenance. In an MEG study, Jensen and Tesche (2002) examined maintenance of visually presented digits in an experimental protocol in which the memory load was parametrically varied between one and seven items. During a 3-sec maintenance period, they found ongoing $\theta$ activity over frontal sensors, and this $\theta$ activity changed in relation to the number of items maintained in working memory. The investigators suggested that the elevated $\theta$ activity resulted from sustained neuronal activity related to active maintenance of memory representations. This interpretation is supported by the findings from a study that examined interactions between frontal cortex and a posterior cortical region, V4 (Liebe et al. 2012). During the maintenance period, these two areas showed synchronized local field potentials in the $\theta$ range. Oscillations in other frequencies may also be relevant, such as in the $\gamma$ band (Roux and Uhlhaas 2014).

A complementary mechanism by which working memories may be maintained is via rapid changes in synaptic weights (Barak and

Tsodyks 2014). Neuronal (Rainer and Miller 2002) and fMRI (Lewis-Peacock et al. 2012) recordings have shown that elevated persistent activity may not always characterize the working-memory delay phase—and still behavior can be successful. Moreover, the metabolic demands of persistent-activity coding are high (Mongillo et al. 2008). Synaptic plasticity as one basis for working memory has been incorporated in various forms in computational models (e.g., O'Reilly et al. 1999) based both on non-Hebbian (Mongillo et al. 2008) and Hebbian (Lansner et al. 2013) synaptic plasticity.

## Does Working-Memory Maintenance Promote Long-Term Memory Formation?

The formation of new declarative long-term memories is critically dependent on the mediotemporal/hippocampal brain system and a cascade of cellular events, including long-term potentiation (LTP) (see Squire and Kandel 2000). Relatedly, a computational model has been suggested in which synaptic weight changes in the hippocampus underlie working-memory encoding and subsequent maintenance (O'Reilly et al. 1999; see also Hasselmo and Stern 2006; Fiebig and Lansner 2014). There is imaging evidence that the hippocampus is engaged during working-memory maintenance under certain circumstances, such as during maintenance of novel information (e.g., Ranganath et al. 2001; Cabeza 2004), and a recent high-resolution fMRI study of mediotemporal lobe (MTL) subregions provided evidence that neurons in these regions may act as a working-memory buffer for novel information (Schon et al. 2015). In the introductory hypothetical scenario, the first code was retained hours later, and some studies have linked working-memory-related brain activity to subsequent long-term memory and found that parahippocampal-sustained fMRI activity during active maintenance was correlated with later memory performance (Schon et al. 2004; Axmacher et al. 2008; see also Rudner et al. 2007; for related intracranial EEG evidence, see Axmacher et al. 2007). Relatedly, findings by Ben-Yakov and Dudai (2011) show that poststimulus hippocampus activity contributes to the registration into long-term memory of complex real-life information.

Thus, maintenance of novel information in working memory can lead to the formation of new long-term memories by engaging the MTL "consolidation" system, which is consistent with the current processing-component framework postulating shared stores for working and long-term memory. The possibility that working-memory processing leads to long-term memory formation is shown by arrow 3 between the "representation" and "consolidation" ellipses in Figure 1. At the same time, it must be noted that patients with MTL lesions can perform well on tests of working-memory maintenance, even when the task situation requires binding of items with locations (Allen et al. 2014), but they show impairment if the material to be learned exceeds working-memory capacity (Jeneson and Squire 2012). These data highlight the dynamic relation between working memory and long-term memory. At subspan challenges, an intact MTL system is not a prerequisite for working-memory maintenance, which, however, does not preclude the possibility that the MTL system becomes engaged and information encoded into long-term memory in individuals with no MTL lesions. At supraspan challenges, the core working-memory maintenance system is insufficient for supporting performance, so long-term memory processes and an intact MTL system become vital. Further work will be crucial for elucidating the factors that influence whether information maintained in working memory becomes consolidated into long-term memory (see, e.g., Wagner 1999; Jensen and Lisman 2005; Sneve et al. 2015).

## HOW CAN INFORMATION IN WORKING MEMORY BE MANIPULATED AND UPDATED?

### Frontal Cortex and Working-Memory Manipulation

Until now, we have discussed the role of frontal cortex in the maintenance of working memory in terms of providing a source of top-down signals to posterior cortical areas involved in long-term memory storage. Some models hold that ventrocaudal frontal zones (BA 44, 45, 47) constitute the main loci where maintenance signals are generated (Smith and Jonides 1999; see also Pudas et al. 2009), whereas other findings indicate that maintenance can be supported by both ventro- and dorsolateral frontal regions (e.g., Postle et al. 1999). To the degree that vast portions of the frontal cortex can contribute top-down maintenance signals, along with other "sources," such as the parietal cortex and hippocampus (Fuster 2013), one might predict that restricted frontal lesions will not profoundly impair the performance on simple maintenance tasks, such as digit-span forward. This prediction is supported by the results from human lesion studies showing that frontal lesions have weak or no negative effects on simple working-memory maintenance (e.g., Volle et al. 2008; Barbey et al. 2013). In contrast, frontal lesions, notably in the dorsolateral cortex, have marked impact on working-memory manipulation operations, and functional imaging studies have shown elevated dorsolateral prefrontal activity as a function of the complexity of working-memory operations (e.g., Nagel et al. 2009; Nyberg et al. 2009b). It should be stressed that increased complexity can be instantiated by using tasks that require that some operation is performed on the information that is maintained in working memory (e.g., letter–number sequencing or digit-span backward) or by increasing the number of items that have to be maintained (as in Sternberg type of tasks). Both of these procedures have been associated with elevated dorsolateral prefrontal activity (Veltman et al. 2003). Increased dorsolateral activity when maintenance gets harder likely reflects the use of manipulation processes (cf. Rypma et al. 1999), such as attempts at chunking (cf. discussion above about working-memory and long-term memory interactions).

Thus, in the literal sense of the working-memory concept (cf. Moscovitch and Winocur 2002), the role of the prefrontal cortex is particularly salient when the situation actually requires one to work with the content of memory—not only passively hold it (the latter is sometimes referred to as "short-term memo-

ry"). In the spirit of the processing component framework (Fig. 1), one may ask whether distinct frontal regions become engaged depending on the specific "kind" of manipulation process that is engaged by a given task. There is some meta-analytic evidence for regional specificity, including interactions between manipulation demand and material type (Wager and Smith 2003). However, a more recent meta-analysis of 36 event-related fMRI studies found only limited support for specificity (Nee et al. 2013; see further the updating section below). Instead, it found that four manipulation processes (protect from external distraction, prevent intrusion of irrelevant memories, shifting of attention, and updating the contents of working memory) engaged several overlapping frontal regions. Sites of strong convergence were observed in widespread medial and lateral frontal cortex, including the middle frontal gyrus and the caudal superior frontal sulcus (illustrated by the yellow "manipulation ellipse" in Fig. 1), and also regions of parietal cortex that form anatomical circuits with prefrontal cortical regions. The superior frontal activation was assigned an attention function, possibly of a spatial nature, that should characterize all four forms of manipulation processes examined.

Lesion studies provide converging support for a general role of the mid-dorsolateral prefrontal cortex (Petrides 2000), medial prefrontal cortex/anterior cingulate (Mesulam 1981), and parietal cortex (Koenigs et al. 2009) in working-memory manipulation, including conflict monitoring (see Botvinick et al. 2001). Prefrontal and parietal cortex regions form circuits with cerebellar regions, and both imaging studies (Stoodley and Schmahmann 2009; Marvel and Desmond 2010) and studies of patients with brain damage (Malm et al. 1998) implicate the cerebellum in working-memory manipulation. Moreover, a frontostriatal–cerebellar circuit is thought to have a general control function in resource-demanding long-term and working-memory tasks (Marklund et al. 2007a).

The notion of a "general control system" that becomes more engaged under challenging working-memory manipulation conditions is in good agreement with the proposition by Baddeley and Hitch (1974) that an attentional control system (the "central executive") guides various kinds of working-memory storage (cf. Awh et al. 2006). Importantly, however, the instantiation of manipulation networks, general as well as more specific ones, is likely not via some form of superordinate executive or regulator. Instead, as long as the task poses only limited challenges (e.g., a forward digit-span task with maintenance of three to four items), it can be performed without much executive control or manipulation operations, but if the situation changes into a more demanding one (e.g., change to "backward" digit-span or supraspan levels), relevant manipulations will be triggered by means of association, which translates into the orderly activation of networks and their subnetworks within and outside of the frontal cortex (Fuster 2013). This principle is shown by the reciprocal arrow 4 in Figure 1. The much more prolonged developmental trajectory for manipulation versus maintenance processes (Crone et al. 2006), in addition to structural maturation of relevant brain areas and connections, could reflect the gradual establishment of associative networks that can support complex goal-directed behavior.

## A Subcortical Dopaminergic Updating System

Stable maintenance of information in working memory is imperative for goal-directed behavior, but so is flexible and rapid updating of the contents of working memory. The necessity of a dynamic relation between maintenance and updating has been captured in terms of a "gating mechanism"; a closed gate promotes maintenance, an open gate allows updating. There is evidence from functional imaging that parts of the frontal cortex are more active during working-memory updating relative to other forms of working-memory manipulation, such as inhibition (Dahlin et al. 2008) and shifting of attention (Nee et al. 2013). However, it has been argued that it is only through interactions with the basal ganglia that the prefrontal cortex realizes this gating function (O'Reilly 2006), such that the striatum provides gating signals to the frontal cortex (Go/update or NoGo/

maintain) via interactions with the thalamus and the substantia nigra. In support of this model, fMRI studies have observed striatal activity during working-memory tasks that involve updating (Lewis et al. 2004; Dahlin et al. 2008), and high-resolution fMRI of the midbrain revealed activation in or near the substantia nigra (D'Ardenne et al. 2012). Relatedly, frontostriatal interactions have also been implicated in the control of access to working-memory storage (McNab and Klingberg 2008), which may contribute to interindividual differences in working memory. We will return to the latter issue in the concluding section of this review. Working-memory updating is illustrated by arrow 5 in Figure 1, connecting the brain stem (substantia nigra), the basal ganglia (striatum), and frontoparietal cortices.

A related model for updating includes a dopamine-based gating mechanism (Braver and Cohen 1999; Durstewitz et al. 2000; Cools and Robbins 2004; see also Gabrieli et al. 1996; O'Reilly 2006). The neurotransmitter dopamine has long been implicated in higher-order cognitive functions, such as working memory (e.g., Williams and Goldman-Rakic 1993). The two major dopamine receptors, D1 and D2, have been associated with distinct working-memory functions (Grace 2000; Cohen et al. 2002). The extrasynaptic D1 receptor has been linked to tonic (sustained) dopamine action and maintenance in working memory, whereas the synaptic D2 system is implicated in phasic (transient) dopamine functions that are relevant for flexible updating of working memory. Updating gating signals take the form of phasic bursts of dopamine that activate D2 receptors and destabilize the maintenance state that is upheld by lower concentrations of tonic dopamine D1 firing. This can happen through gating signals from the ventral tegmental area to the prefrontal cortex (mesocortical pathway; arrow 6 in Fig. 1), but also through the nigrostriatal pathway (Fig. 1, bottom arrow 5). Indeed, the relative density of D2 receptors is much higher in the striatum than in the frontal cortex, making dopamine D2 action in the striatum well suited to serve a gating function (Hazy et al. 2006; Cools and D'Esposito 2011). Correspondingly, PET imaging during a letter-memory updating task revealed that working-memory updating affects striatal D2 binding (Bäckman et al. 2011).

Via the mesolimbic dopaminergic pathway, dopamine can also influence long-term memory formation (e.g., Lisman et al. 2011), and partly overlapping frontostriatal circuits have been shown to be implicated in updating of long-term memory as they have for working memory (Nyberg et al. 2009a). There is evidence that updating of long-term memory involves adding extra information to already existing memory networks rather than overwriting the older preexisting information (Eriksson et al. 2014). Similarly, by using a three-back working-memory task it was shown that no-longer-relevant items (presented more than three items back) still interfered with ongoing processing (Gray et al. 2003). Thus, although the working-memory content had been updated so that these items no longer were actively maintained, they still resided in memory to the degree that they could influence performance. These and related observations indicate that working-memory updating makes old information less accessible but not "erased," possibly because "familiarity" effects from previously but not currently maintained information can be supported by long-term memory.

## MAINTAINING AND REALIZING INTENTIONS

In the preceding section, we discussed a gating mechanism for maintaining and updating the contents of working memory, which begs the question of how the decision is made as to when to open or close the gate? In turn, this question relates to the more general topic of how goals and intentions are formed, maintained, and updated in working memory, and how they can influence and guide more basic working-memory operations. These complex questions map on to several aspects of our introductory scenario and, in particular, to the ability to maintain the goal from when the plan was formed to when it finally was realized—despite the many interfering and distracting events that happened in between. Indeed, as

has been emphasized by Fuster (2013), the concept of working memory goes well beyond the act of maintaining discrete information (as in a laboratory delayed matching to sample (DMS) task or maintaining the code in our introductory example) to networks that represent the task, the objective, and the current specific context. All of these networks will be linked by association (Asaad et al. 1998; see Fuster 2013) and recruited to various degrees depending on the complexity and familiarity of a given task and context. If, as in the introductory example, the situation is unfamiliar or there are interruptions to the execution of a plan, the highest integrative networks involving rostro-prefrontal cortex (the "intention" ellipse in Fig. 1) become engaged (Fuster 2013; for similar hierarchical views on frontal cortex functional organization, see, e.g., Christoff and Gabrieli 2000; Lepage et al. 2000; Braver and Bongiolatti 2002; Koechlin et al. 2003; Ramnani and Owen 2004; Koechlin and Summerfield 2007; Badre and D'Esposito 2009).

One source of evidence that task context is coded by the prefrontal cortex comes from findings that prefrontal neurons show differential pre-cue baseline activities depending on the task situation (e.g., Wallis et al. 2001; see Watanabe 2013). Such findings are consistent with a view that frontal neurons monitor and maintain the currently relevant cognitive context (cf. Miller and Cohen 2001). Functional imaging studies contribute converging evidence by showing that rostro-prefrontal regions code for the nature of future processing (Sakai and Passingham 2003) and that frontopolar regions show a sustained activation profile throughout a task—even when no stimuli are present (Marklund et al. 2007b; see also Velanova et al. 2003; Dosenbach et al. 2007). Also, frontopolar regions have been implicated in prospective memory (Burgess et al. 2003) and in forming new intentions (Kalpouzos et al. 2010). Collectively, these kinds of findings link the rostrofrontal cortex to prospective coding of the cognitive context in a given situation.

Relatedly, prefrontal neurons are involved in coding the motivational context of a task (see Watanabe 2013), for example, as induced by different kinds of rewards (Wallis and Kennerley 2013). Reward signals from the orbitofrontal cortex can influence the current cognitive context via lateral prefrontal neurons. More generally, the orbitofrontal cortex is a site where emotional signals from the brain stem, amygdala, and the hypothalamus can interact with the cognitive processing of the perception–action cycle (arrow 7 in Fig 1; cf. Fuster 2013). By affecting updating of task-context representations (D'Ardenne et al. 2012) and by driving associative reinforcement-learning mechanisms (Schultz 1998), dopamine also plays a key role in this context. A dual role for dopamine in gating and learning opens up for a self-organizing system that learns when to update goals and contexts to maximize rewards and minimize punishments (cf. Miller and Cohen 2001; McClure et al. 2003; see also Jonasson et al. 2014). The self-organizing nature of executive control processes is also emphasized in Fuster's (e.g., 2013) associative perception–action cycle. As such, there is no need for "a controller of the control processes," and avoids the concept of a "homunculus."

Thus, the intention to go and get a reference text (or a cup of coffee) while writing a paper can be elicited via bottom-up input from the internal milieu or by associative mechanisms, and its subsequent successful realization is dependent on neurons in the rostrofrontal cortex that code the current cognitive and motivational contexts. Prefrontal context representations can be updated by dopamine signals and strengthened by reinforcement if they lead to successful outcomes. The rostrofrontal cortex (area 10) has diffuse and extensive nonreciprocal projections with more caudal frontal regions (see Badre and D'Esposito 2009) and forms functional networks with dorsofrontal and parietal regions (Vincent et al. 2008). These patterns of connectivity allow rostrofrontal cortex to affect processing in other networks that maintain, update, and manipulate information (represented by arrow 8 in Fig. 1). Anterior goal networks may influence other networks by means of synchronous oscillations (Voytek and Knight 2015; Voytek et al. 2015; see also Engel et al. 2001; Cavanagh and Frank 2014),

possibly in conjunction with the thalamus (Saalmann et al. 2012). "Stepwise" feedback signals in the hierarchical frontal pathways to the rostrofrontal cortex (cf. Badre and D'Esposito 2009) could contribute to maintaining the current goal. When the realization of an intention is much delayed and/or interfered with by various distracting events, long-term memory processes will be critical for reactivating the relevant frontal goal representations based on sensorimotor cues (Rainer et al. 1999; cf. Miller and Cohen 2001; Kalpouzos et al. 2010).

## WORKING MEMORY IN ACTION

In this final section, we briefly consider how the present view of working memory, schematically outlined in Figure 1, can be related to normal variation in working-memory functioning, as well as working-memory deficits induced by aging and disease. Consider first normal variation in the capacity of working memory. A characteristic feature of working memory is limitations on how much information can be retained at any given time. Average capacity has been estimated to be 3–4 items in young, healthy individuals (Cowan 2001), but exactly "how" much information depends on a number of factors, including how well the information can be related to preexisting representations that can serve as "schemas." For instance, in the opening example, the first access code could be linked to a birth year. Such mnemonics can greatly increase how much information can be retained. In a sense, then, any claim for a fixed capacity will be somewhat artificial in that it depends on what processing components are allowed to be at play (Saults and Cowan 2007). Individual differences in working-memory capacity on simple, as well as more complex tasks, have received much attention as they have been related to basic intellectual abilities, such as general intelligence (Kane and Engle 2002; Gray et al. 2003; Fukuda et al. 2010), and found to predict real-world achievements, such as learning math (De Smedt et al. 2009; see Raghubar et al. 2010) and school grades (Cowan et al. 2005).

Working memory is tightly linked to a frontoparietal cortical network (see Fig. 1) and parietal activity correlates strongly with load and plateaus when capacity limits are reached (Todd and Marois 2004; Vogel and Machizawa 2004). Also, parietal activity increases when distracting information is unintentionally encoded into working memory, which is more likely to happen to individuals with lower capacity (Vogel et al. 2005). McNab and Klingberg (2008) found that parietal load effects, reflecting unnecessary storage of distractors, were negatively correlated with basal ganglia activity. They further showed that prefrontal and basal-ganglia activity was positively associated with capacity. Collectively, these findings suggest that one determinant of working-memory capacity is how efficiently an individual can exert control over the encoding of working memory, and how frontostriatal regions serve a gatekeeping function in this regard. Clearly, this function resembles the dopaminergic gating function discussed above in the context of updating, and McNab and Klingberg (2008) also note the potential role of dopamine in gating access to working memory. Evidence for a link between working-memory capacity and dopamine also comes from human PET studies (Cools et al. 2008) and genetic (e.g., Bilder et al. 2004) and pharmacological (e.g., Garrett et al. 2015) studies (for a comprehensive review, see Cools and D'Esposito 2011).

Dysfunctional dopamine neurotransmission may, at least in part, account for working-memory deficits in aging (Karlsson et al. 2009; Fischer et al. 2010; Nyberg et al. 2014; see Bäckman et al. 2010) and in Parkinson's disease (Marklund et al. 2009; Ekman et al. 2012). Dopamine has also been implicated as a source of working-memory difficulties in schizophrenia (Cohen and Servan-Schreiber 1992; Goldman-Rakic 1999; Castner et al. 2000; Abi-Dargham et al. 2002), attention-deficit/hyperactivity disorder (Castellanos and Tannock 2002; Martinussen et al. 2005; Sagvolden et al. 2005), and other psychiatric and neurological conditions (see Maia and Frank 2011).

Thus, the networks we have outlined herein (see Fig. 1) may be of fundamental importance for adequate functioning in a variety of situations. The disturbance of dopaminergic neurotransmission may be a common basis for deficits in higher-order cognition in many conditions. Cognitive interventions have shown some promise in modulating dopamine D1 and D2 systems (McNab et al. 2009; Bäckman et al. 2011; for reviews, see Klingberg 2010; Bäckman and Nyberg 2013). We are currently exploring the potential role of long-term physical interventions in strengthening dopamine and related cognitive functions, and psychopharmacological approaches hold promise in this regard (Wang et al. 2011). An important task for future research is to examine further ways of supporting deficient working-memory networks, as this in turn may influence significant aspects of everyday functioning.

## CONCLUSIONS

In this review, we have argued that working-memory maintenance is the result of directing attention to semantic or sensorimotor representations. This process can be realized as persistent top-down signals from the frontal cortex and other sources to lower-order areas, and synchronous network oscillations and rapid changes in synaptic weights may also contribute to maintenance. Although attention is central to working memory, specific working-memory functions may be performed with little or no attentional processing. For example, integrating pieces of information to be maintained in working memory (i.e., "chunking") can be relatively unaffected by attention-demanding concurrent tasks (Baddeley et al. 2009), and there are demonstrations of short-term maintenance of information made nonconscious by diverting attention from the target (Bergström and Eriksson 2014). The details of when and how attention is critical for working-memory processes should be further specified in future research.

The concept of working memory goes well beyond the act of maintaining discrete information to networks that underlie manipulation and updating of the contents of working memory, as well as to networks that represent the task, the objective, and the specific context. All of these networks are linked by association and are differentially recruited depending on the current demands.

## ACKNOWLEDGMENTS

The writing of this review was supported by Torsten and Ragnar Söderberg's Foundation (L.N.), the Swedish Science Council (J.E., L.N.), and the European Union Seventh Framework Program (FP7/2007–2013) under Grant Agreement No. 604102 (Human Brain Project) to L.N. We thank our colleagues and collaborators for important (direct or indirect) contributions to the content of this work.

## REFERENCES

Abi-Dargham A, Mawlawi O, Lombardo I, Gil R, Martinez D, Huang Y, Hwang D-R, Keilp J, Kochan L, Van Heertum R, et al. 2002. Prefrontal dopamine D1 receptors and working memory in schizophrenia. *J Neurosci* **22:** 3708–3719.

Allen RJ, Vargha-Khadem F, Baddeley AD. 2014. Item-location binding in working memory: Is it hippocampus-dependent? *Neuropsychologia* **59:** 74–84.

Asaad WF, Rainer G, Miller EK. 1998. Neural activity in the primate prefrontal cortex during associative learning. *Neuron* **21:** 1399–1407.

Awh E, Vogel E, Oh S. 2006. Interactions between attention and working memory. *Neuroscience* **139:** 201–208.

Axmacher N, Mormann F, Fernández G, Cohen MX, Elger CE, Fell J. 2007. Sustained neural activity patterns during working memory in the human medial temporal lobe. *J Neurosci* **27:** 7807–7816.

Axmacher N, Schmitz DP, Weinreich I, Elger CE, Fell J. 2008. Interaction of working memory and long-term memory in the medial temporal lobe. *Cereb Cortex* **18:** 2868–2878.

Bäckman L, Nyberg L. 2013. Dopamine and training-related working-memory improvement. *Neurosci Biobehav Rev* **37:** 2209–2219.

Bäckman L, Lindenberger U, Li S-C, Nyberg L. 2010. Linking cognitive aging to alterations in dopamine neurotransmitter functioning: Recent data and future avenues. *Neurosci Biobehav Rev* **34:** 670–677.

Bäckman L, Nyberg L, Soveri A, Johansson J, Andersson M, Dahlin E, Neely AS, Virta J, Laine M, Rinne JO. 2011. Effects of working-memory training on striatal dopamine release. *Science* **333:** 718.

Baddeley AD, Hitch GJ. 1974. Working memory. In *Recent advances in learning and motivation* (ed. Bower GA), pp. 47–89. Academic, New York.

Baddeley AD, Hitch GJ, Allen RJ. 2009. Working memory and binding in sentence recall. *J Mem Lang* **61:** 438–456.

Badre D, D'Esposito M. 2009. Is the rostro-caudal axis of the frontal lobe hierarchical? *Nat Rev Neurosci* **10:** 659–69.

Barak O, Tsodyks M. 2014. Working models of working memory. *Curr Opin Neurobiol* **25:** 20–24.

Barbey AK, Koenigs M, Grafman J. 2013. Dorsolateral prefrontal contributions to human working memory. *Cortex* **49:** 1195–1205.

Ben-Yakov A, Dudai Y. 2011. Constructing realistic engrams: Poststimulus activity of hippocampus and dorsal striatum predicts subsequent episodic memory. *J Neurosci* **31:** 9032–9042.

Bergström F, Eriksson J. 2014. Maintenance of non-consciously presented information engages the prefrontal cortex. *Front Hum Neurosci* **8:** 1–10.

Bilder RM, Volavka J, Lachman HM, Grace AA. 2004. The catechol-*O*-methyltransferase polymorphism: Relations to the tonic-phasic dopamine hypothesis and neuropsychiatric phenotypes. *Neuropsychopharmacology* **29:** 1943–1961.

Botvinick MM, Braver TS, Barch DM, Carter CS, Cohen JD. 2001. Conflict monitoring and cognitive control. *Psychol Rev* **108:** 624–652.

Braver TS, Bongiolatti SR. 2002. The role of frontopolar cortex in subgoal processing during working memory. *NeuroImage* **15:** 523–536.

Braver TS, Cohen JD. 1999. Dopamine, cognitive control, and schizophrenia: The gating model. *Prog Brain Res* **121:** 327–349.

Burgess PW, Scott SK, Frith CD. 2003. The role of the rostral frontal cortex (area 10) in prospective memory: A lateral versus medial dissociation. *Neuropsychologia* **41:** 906–918.

Cabeza R. 2004. Task-independent and task-specific age effects on brain activity during working memory, visual attention and episodic retrieval. *Cereb Cortex* **14:** 364–375.

Cabeza R, Nyberg L. 2000. Imaging cognition. II: An empirical review of 275 PET and fMRI studies. *J Cogn Neurosci* **12:** 1–47.

Cabeza R, Dolcos F, Graham R, Nyberg L. 2002. Similarities and differences in the neural correlates of episodic memory retrieval and working memory. *NeuroImage* **16:** 317–330.

Castellanos FX, Tannock R. 2002. Neuroscience of attention-deficit/hyperactivity disorder: The search for endophenotypes. *Nat Rev Neurosci* **3:** 617–628.

Castner SA, Williams GV, Goldman-Rakic PS. 2000. Reversal of antipsychotic-induced working memory deficits by short-term dopamine D1 receptor stimulation. *Science* **287:** 2020–2022.

Cavanagh JF, Frank MJ. 2014. Frontal θ as a mechanism for cognitive control. *Trends Cogn Sci* **18:** 414–421.

Christoff K, Gabrieli JDE. 2000. The frontopolar cortex and human cognition: Evidence for a rostrocaudal hierarchical organization within the human prefrontal cortex. *Psychobiology* **28:** 168–186.

Cohen JD, Servan-Schreiber D. 1992. Context, cortex, and dopamine: A connectionist approach to behavior and biology in schizophrenia. *Psychol Rev* **99:** 45–77.

Cohen JD, Braver TS, Brown JW. 2002. Computational perspectives on dopamine function in prefrontal cortex. *Curr Opin Neurobiol* **12:** 223–229.

Cools R, D'Esposito M. 2011. Inverted-U-shaped dopamine actions on human working memory and cognitive control. *Biol Psychiatry* **69:** e113–e125.

Cools R, Robbins TW. 2004. Chemistry of the adaptive mind. *Philos Trans A* **362:** 2871–2888.

Cools R, Gibbs SE, Miyakawa A, Jagust W, D'Esposito M. 2008. Working memory capacity predicts dopamine synthesis capacity in the human striatum. *J Neurosci* **28:** 1208–1212.

Cowan N. 2001. The magical number 4 in short-term memory: A reconsideration of mental storage capacity. *Behav Brain Sci* **24:** 87–114; discussion 114–185.

Cowan N, Elliott EM, Saults SJ, Morey CC, Mattox S, Hismjatullina A, Conway ARA. 2005. On the capacity of attention: Its estimation and its role in working memory and cognitive aptitudes. *Cogn Psychol* **51:** 42–100.

Crone EA, Wendelken C, Donohue S, van Leijenhorst L, Bunge SA. 2006. Neurocognitive development of the ability to manipulate information in working memory. *Proc Natl Acad Sci* **103:** 9315–9320.

Dahlin E, Neely AS, Larsson A, Bäckman L, Nyberg L. 2008. Transfer of learning after updating training mediated by the striatum. *Science* **320:** 1510–1512.

Daneman M, Carpenter PA. 1980. Individual differences in working memory and reading. *J Verbal Learning Verbal Behav* **19:** 450–466.

Danker JF, Anderson JR. 2010. The ghosts of brain states past: Remembering reactivates the brain regions engaged during encoding. *Psychol Bull* **136:** 87–102.

D'Ardenne K, Eshel N, Luka J, Lenartowicz A, Nystrom LE, Cohen JD. 2012. Role of prefrontal cortex and the midbrain dopamine system in working memory updating. *Proc Natl Acad Sci* **109:** 19900–19909.

De Smedt B, Janssen R, Bouwens K, Verschaffel L, Boets B, Ghesquière P. 2009. Working memory and individual differences in mathematics achievement: A longitudinal study from first grade to second grade. *J Exp Child Psychol* **103:** 186–201.

D'Esposito M, Postle BR. 2015. The cognitive neuroscience of working memory. *Annu Rev Psychol* **66:** 115–142.

Dosenbach NUF, Fair DA, Miezin FM, Cohen AL, Wenger KK, Dosenbach RT, Fox MD, Snyder AZ, Vincent JL, Raichle ME, et al. 2007. Distinct brain networks for adaptive and stable task control in humans. *Proc Natl Acad Sci* **104:** 11073–11078.

Duncan J, Owen AM. 2000. Common regions of the human frontal lobe recruited by diverse cognitive demands. *Trends Neurosci* **23:** 475–483.

Durstewitz D, Seamans JK, Sejnowski TJ. 2000. Neurocomputational models of working memory. *Nat Neurosci* 1184–1191.

Ekman U, Eriksson J, Forsgren L, Mo SJ, Riklund K, Nyberg L. 2012. Functional brain activity and presynaptic dopamine uptake in patients with Parkinson's disease and mild cognitive impairment: A cross-sectional study. *Lancet Neurol* **11:** 679–687.

Engel AK, Fries P, Singer W. 2001. Dynamic predictions: Oscillations and synchrony in top-down processing. *Nat Rev Neurosci* **2:** 704–716.

Eriksson J, Stiernstedt M, Öhlund M, Nyberg L. 2014. Changing Zaire to Congo: The fate of no-longer relevant mnemonic information. *NeuroImage* **101:** 1–7.

Eriksson J, Vogel EK, Lansner A, Bergström F, Nyberg L. 2015. The cognitive architecture of working memory. *Neuron* (in press).

Fiebig F, Lansner A. 2014. Memory consolidation from seconds to weeks: A three-stage neural network model with autonomous reinstatement dynamics. *Front Comput Neurosci* **8:** 64.

Fischer H, Nyberg L, Karlsson S, Karlsson P, Brehmer Y, Rieckmann A, MacDonald SWS, Farde L, Bäckman L. 2010. Simulating neurocognitive aging: Effects of a dopaminergic antagonist on brain activity during working memory. *Biol Psychiatry* **67:** 575–580.

Fukuda K, Vogel E, Mayr U, Awh E. 2010. Quantity, not quality: The relationship between fluid intelligence and working memory capacity. *Psychon Bull Rev* **17:** 673–679.

Fuster JM. 2009. Cortex and memory: Emergence of a new paradigm. *J Cogn Neurosci* **21:** 2047–2072.

Fuster JM. 2013. Cognitive functions of the prefrontal cortex. In *Principles of frontal lobe function* (ed. Stuss D, Knight R), pp. 11–22. Oxford University Press, Oxford.

Fuster J, Alexander G. 1971. Neuron activity related to short-term memory. *Science* **173:** 652–654.

Gabrieli JDE, Singh J, Stebbins GT, Goetz CG. 1996. Reduced working memory span in Parkinson's disease: Evidence for the role of frontostriatal system in working and strategic memory. *Neuropsychology* **10:** 322–332.

Garrett DD, Nagel IE, Preuschhof C, Burzynska AZ, Marchner J, Wiegert S, Jungehülsing GJ, Nyberg L, Villringer A, Li S-C, et al. 2015. Amphetamine modulates brain signal variability and working memory in younger and older adults. *Proc Natl Acad Sci* **112:** 7593–7598.

Goldman-Rakic PS. 1987. Circuitry of primate prefrontal cortex and regulation of behaviour by representational memory. In *Handbook of physiology: The nervous system*, pp. 373–417. Wiley, New York.

Goldman-Rakic PS. 1999. The physiological approach: Functional architecture of working memory and disordered cognition in schizophrenia. *Biol Psychiatry* **46:** 650–661.

Grace AA. 2000. The tonic/phasic model of dopamine system regulation and its implications for understanding alcohol and psychostimulant craving. *Addiction* **95:** S119–S128.

Gray JR, Chabris CF, Braver TS. 2003. Neural mechanisms of general fluid intelligence. *Nat Neurosci* **6:** 316–322.

Hasselmo ME, Stern CE. 2006. Mechanisms underlying working memory for novel information. *Trends Cogn Sci* **10:** 487–493.

Hazy TE, Frank MJ, O'Reilly RC. 2006. Banishing the homunculus: Making working memory work. *Neuroscience* **139:** 105–118.

Jeneson A, Squire LR. 2012. Working memory, long-term memory, and medial temporal lobe function. *Learn Mem* **19:** 15–25.

Jensen O, Lisman JE. 2005. Hippocampal sequence-encoding driven by a cortical multi-item working memory buffer. *Trends Neurosci* **28:** 67–72.

Jensen O, Tesche CD. 2002. Frontal θ activity in humans increases with memory load in a working memory task. *Eur J Neurosci* **15:** 1395–1399.

Jonasson LS, Axelsson J, Riklund K, Braver TS, Ögren M, Bäckman L, Nyberg L. 2014. Dopamine release in nucleus accumbens during rewarded task switching measured by [$^{11}$C]raclopride. *NeuroImage* **99:** 357–364.

Kalpouzos G, Eriksson J, Sjölie D, Molin J, Nyberg L. 2010. Neurocognitive systems related to real-world prospective memory. *PLoS ONE* **5:** e13304.

Kane MJ, Engle RW. 2002. The role of prefrontal cortex in working-memory capacity, executive attention, and general fluid intelligence: An individual-differences perspective. *Psychon Bull Rev* **9:** 637–671.

Karlsson S, Nyberg L, Karlsson P, Fischer H, Thilers P, MacDonald S, Brehmer Y, Rieckmann A, Halldin C, Farde L, et al. 2009. Modulation of striatal dopamine D1 binding by cognitive processing. *NeuroImage* **48:** 398–404.

Klingberg T. 2010. Training and plasticity of working memory. *Trends Cogn Sci* **14:** 317–324.

Koechlin E, Summerfield C. 2007. An information theoretical approach to prefrontal executive function. *Trends Cogn Sci* **11:** 229–235.

Koechlin E, Ody C, Kouneiher F. 2003. The architecture of cognitive control in the human prefrontal cortex. *Science* **302:** 1181–1185.

Koenigs M, Barbey AK, Postle BR, Grafman J. 2009. Superior parietal cortex is critical for the manipulation of information in working memory. *J Neurosci* **29:** 14980–14986.

Lansner A, Marklund P, Sikström S, Nilsson L-G. 2013. Reactivation in working memory: An attractor network model of free recall. *PLoS ONE* **8:** e73776.

Lepage M, Ghaffar O, Nyberg L, Tulving E. 2000. Prefrontal cortex and episodic memory retrieval mode. *Proc Natl Acad Sci* **97:** 506–511.

Lewis SJG, Dove A, Robbins TW, Barker RA, Owen AM. 2004. Striatal contributions to working memory: A functional magnetic resonance imaging study in humans. *Eur J Neurosci* **19:** 755–760.

Lewis-Peacock JA, Drysdale AT, Oberauer K, Postle BR. 2012. Neural evidence for a distinction between short-term memory and the focus of attention. *J Cogn Neurosci* **24:** 61–79.

Liebe S, Hoerzer GM, Logothetis NK, Rainer G. 2012. θ Coupling between V4 and prefrontal cortex predicts visual short-term memory performance. *Nat Neurosci* **15:** 456–462.

Lisman J, Grace AA, Duzel E. 2011. A neoHebbian framework for episodic memory; role of dopamine-dependent late LTP. *Trends Neurosci* **34:** 536–547.

Maia TV, Frank MJ. 2011. From reinforcement learning models to psychiatric and neurological disorders. *Nat Neurosci* **14:** 154–162.

Malm J, Kristensen B, Karlsson T, Carlberg B, Fagerlund M, Olsson T. 1998. Cognitive impairment in young adults with infratentorial infarcts. *Neurology* **51:** 433–440.

Marklund P, Fransson P, Cabeza R, Larsson A, Ingvar M, Nyberg L. 2007a. Unity and diversity of tonic and phasic executive control components in episodic and working memory. *NeuroImage* **36:** 1361–1373.

Marklund P, Fransson P, Cabeza R, Petersson KM, Ingvar M, Nyberg L. 2007b. Sustained and transient neural modulations in prefrontal cortex related to declarative long-term memory, working memory, and attention. *Cortex* **43:** 22–37.

Marklund P, Larsson A, Elgh E, Linder J, Riklund KA, Forsgren L, Nyberg L. 2009. Temporal dynamics of basal ganglia under-recruitment in Parkinson's disease: Transient caudate abnormalities during updating of working memory. *Brain* **132:** 336–346.

Martin A, Chao LL. 2001. Semantic memory and the brain: Structure and processes. *Curr Opin Neurobiol* **11:** 194–201.

Martinussen R, Hayden J, Hogg-Johnson S, Tannock R. 2005. A meta-analysis of working memory impairments in children with attention-deficit/hyperactivity disorder. *J Am Acad Child Adolesc Psychiatry* **44:** 377–384.

Marvel CL, Desmond JE. 2010. The contributions of cerebro-cerebellar circuitry to executive verbal working memory. *Cortex* **46:** 880–895.

McClure SM, Berns GS, Montague PR. 2003. Temporal prediction errors in a passive learning task activate human striatum. *Neuron* **38:** 339–346.

McNab F, Klingberg T. 2008. Prefrontal cortex and basal ganglia control access to working memory. *Nat Neurosci* **11:** 103–107.

McNab F, Varrone A, Farde L, Jucaite A, Bystritsky P, Forssberg H, Klingberg T. 2009. Changes in cortical dopamine D1 receptor binding associated with cognitive training. *Science* **323:** 800–802.

Mesulam MM. 1981. A cortical network for directed attention and unilateral neglect. *Ann Neurol* **10:** 309–325.

Miller GA. 1956. The magical number seven plus or minus two: Some limits on our capacity for processing information. *Psychol Rev* **63:** 81–97.

Miller EK, Cohen JD. 2001. An integrative theory of prefrontal cortex function. *Annu Rev Neurosci* **24:** 167–202.

Mongillo G, Barak O, Tsodyks M. 2008. Synaptic theory of working memory. *Science* **319:** 1543–1546.

Moscovitch M, Winocur G. 2002. The frontal cortex and working with memory. In *Principles of frontal lobe function* (ed. Stuss D, Knight R), pp. 188–209. Oxford University Press, Oxford.

Nagel IE, Preuschhof C, Li S-C, Nyberg L, Bäckman L, Lindenberger U, Heekeren HR. 2009. Performance level modulates adult age differences in brain activation during spatial working memory. *Proc Natl Acad Sci* **106:** 22552–22557.

Naghavi HR, Nyberg L. 2005. Common fronto-parietal activity in attention, memory, and consciousness: Shared demands on integration? *Conscious Cogn* **14:** 390–425.

Nee DE, Brown JW, Askren MK, Berman MG, Demiralp E, Krawitz A, Jonides J. 2013. A meta-analysis of executive components of working memory. *Cereb Cortex* **23:** 264–282.

Nyberg L, Cabeza R. 2001. The versatile frontal lobes: A meta-analysis of 1000 PET and fMRI activations. *Brain Cogn* **47:** 106–110.

Nyberg L, Habib R, McIntosh R, Tulving E. 2000. Reactivation of encoding-related brain activity during memory retrieval. *Proc Natl Acad Sci* **97:** 11120–11124.

Nyberg L, Forkstam C, Petersson KM, Cabeza R, Ingvar M. 2002. Brain Imaging of human memory systems: Between-systems similarities and within-system differences. *Brain Res Cogn Brain Res* **13:** 281–292.

Nyberg L, Marklund P, Persson J, Cabeza R, Forkstam C, Petersson KM, Ingvar M. 2003. Common prefrontal activations during working memory, episodic memory, and semantic memory. *Neuropsychologia* **41**: 371–377.

Nyberg L, Andersson M, Forsgren L, Jakobsson-Mo S, Larsson A, Marklund P, Nilsson L-G, Riklund K, Bäckman L. 2009a. Striatal dopamine D2 binding is related to frontal BOLD response during updating of long-term memory representations. *NeuroImage* **46**: 1194–1199.

Nyberg L, Dahlin E, Stigsdotter Neely A, Bäckman L. 2009b. Neural correlates of variable working memory load across adult age and skill: Dissociative patterns within the fronto-parietal network. *Scand J Psychol* **50**: 41–46.

Nyberg L, Andersson M, Kauppi K, Lundquist A, Persson J, Pudas S, Nilsson L-G. 2014. Age-related and genetic modulation of frontal cortex efficiency. *J Cogn Neurosci* **26**: 746–754.

O'Reilly RC. 2006. Biologically based computational models of high-level cognition. *Science* **314**: 91–94.

O'Reilly RC, Braver TS, Cohen JD. 1999. A biologically-based computational model of working memory. In *Models of working memory mechanisms of active maintenance and executive control* (ed. Miyake A, Shah P). Cambridge University Press, Cambridge.

Petrides M. 2000. The role of the mid-dorsolateral prefrontal cortex in working memory. *Exp Brain Res* **133**: 44–54.

Postle BR, Berger JS, D'Esposito M. 1999. Functional neuroanatomical double dissociation of mnemonic and executive control processes contributing to working memory performance. *Proc Natl Acad Sci* **96**: 12959–12964.

Pudas S, Persson J, Nilsson L-G, Nyberg L. 2009. Maintenance and manipulation in working memory: Differential ventral and dorsal frontal cortex fMRI activity. *Acta Psychol Sin* **41**: 1054–1062.

Raghubar KP, Barnes MA, Hecht SA. 2010. Working memory and mathematics: A review of developmental, individual difference, and cognitive approaches. *Learn Individ Differ* **20**: 110–122.

Rainer G, Miller EK. 2002. Timecourse of object-related neural activity in the primate prefrontal cortex during a short-term memory task. *Eur J Neurosci* **15**: 1244–1254.

Rainer G, Rao SC, Miller EK. 1999. Prospective coding for objects in primate prefrontal cortex. *J Neurosci* **19**: 5493–5505.

Ramnani N, Owen AM. 2004. Anterior prefrontal cortex: Insights into function from anatomy and neuroimaging. *Nat Rev Neurosci* **5**: 184–194.

Ranganath C, Esposito MD, Wills H. 2001. Medial temporal lobe activity associated with active maintenance of novel information. *Neuron* **31**: 865–873.

Roux F, Uhlhaas PJ. 2014. Working memory and neural oscillations: α-γ versus θ-γ codes for distinct WM information? *Trends Cogn Sci* **18**: 16–25.

Rudner M, Fransson P, Ingvar M, Nyberg L, Rönnberg J. 2007. Neural representation of binding lexical signs and words in the episodic buffer of working memory. *Neuropsychologia* **45**: 2258–2276.

Rypma B, Prabhakaran V, Desmond JE, Glover GH, Gabrieli JD. 1999. Load-dependent roles of frontal brain regions in the maintenance of working memory. *NeuroImage* **9**: 216–226.

Saalmann YB, Pinsk MA, Wang L, Li X, Kastner S. 2012. The pulvinar regulates information transmission between cortical areas based on attention demands. *Science* **337**: 753–756.

Sagvolden T, Aase H, Johansen EB, Russell VA. 2005. A dynamic developmental theory of attention-deficit/hyperactivity disorder (ADHD) predominantly hyperactive/impulsive and combined subtypes. *Behav Brain Res* **28**: 397–468.

Sakai K, Passingham RE. 2003. Prefrontal interactions reflect future task operations. *Nat Neurosci* **6**: 75–81.

Sala JB, Courtney SM. 2007. Binding of what and where during working memory maintenance. *Cortex* **43**: 5–21.

Saults JS, Cowan N. 2007. A central capacity limit to the simultaneous storage of visual and auditory arrays in working memory. *J Exp Psychol Gen* **136**: 663–684.

Schon K, Hasselmo ME, Lopresti ML, Tricarico MD, Stern CE. 2004. Persistence of parahippocampal representation in the absence of stimulus input enhances long-term encoding: A functional magnetic resonance imaging study of subsequent memory after a delayed match-to-sample task. *J Neurosci* **24**: 11088–11097.

Schon K, Newmark RE, Ross RS, Stern CE. 2015. A working memory buffer in parahippocampal regions: Evidence from a load effect during the delay period. *Cereb Cortex* doi: 10.1093/cercor/bhv013.

Schultz W. 1998. Predictive reward signal of dopamine neurons. *J Neurophysiol* **80**: 1–27.

Smith EE, Jonides J. 1999. Storage and executive processes in the frontal lobes. *Science* **283**: 1657–1661.

Sneve MH, Grydeland H, Nyberg L, Bowles B, Amlien IK, Langnes E, Walhovd KB, Fjell AM. 2015. Mechanisms underlying encoding of short-lived versus durable episodic memories. *J Neurosci* **35**: 5202–5212.

Squire LS, Kandel ER. 2000. *Memory: From mind to molecules.* Macmillan, New York.

Sreenivasan KK, Curtis CE, D'Esposito M. 2014. Revisiting the role of persistent neural activity during working memory. *Trends Cogn Sci* 1–8.

Stoodley CJ, Schmahmann JD. 2009. Functional topography in the human cerebellum: A meta-analysis of neuroimaging studies. *NeuroImage* **44**: 489–501.

Todd JJ, Marois R. 2004. Capacity limit of visual short-term memory in human posterior parietal cortex. *Nature* **428**: 751–754.

Van de Ven V, Jacobs C, Sack AT. 2012. Topographic contribution of early visual cortex to short-term memory consolidation: A transcranial magnetic stimulation study. *J Neurosci* **32**: 4–11.

Velanova K, Jacoby LL, Wheeler ME, McAvoy MP, Petersen SE, Buckner RL. 2003. Functional-anatomic correlates of sustained and transient processing components engaged during controlled retrieval. *J Neurosci* **23**: 8460–8470.

Veltman DJ, Rombouts SA, Dolan RJ. 2003. Maintenance versus manipulation in verbal working memory revisited: An fMRI study. *NeuroImage* **18**: 247–256.

Vincent JL, Kahn I, Snyder AZ, Raichle ME, Buckner RL. 2008. Evidence for a frontoparietal control system revealed by intrinsic functional connectivity. *J Neurophysiol* **100**: 3328–3342.

Vogel EK, Machizawa MG. 2004. Neural activity predicts individual differences in visual working memory capacity. *Nature* **428**: 748–751.

Vogel EK, McCollough AW, Machizawa MG. 2005. Neural measures reveal individual differences in controlling access to working memory. *Nature* **438:** 500–503.

Volle E, Kinkingnéhun S, Pochon JB, Mondon K, Thiebaut De Schotten M, Seassau M, Duffau H, Samson Y, Dubois B, Levy R. 2008. The functional architecture of the left posterior and lateral prefrontal cortex in humans. *Cereb Cortex* **18:** 2460–2469.

Voytek B, Knight RT. 2015. Dynamic network communication as a unifying neural basis for cognition, development, aging, and disease. *Biol Psychiatry* **77:** 1089–1097.

Voytek B, Kayser AS, Badre D, Fegen D, Chang EF, Crone NE, Parvizi J, Knight RT, D'Esposito M. 2015. Oscillatory dynamics coordinating human frontal networks in support of goal maintenance. *Nat Neurosci* **18:** 1318–1324.

Wager TD, Smith EE. 2003. Neuroimaging studies of working memory: A meta-analysis. *Cogn Affect Behav Neurosci* **3:** 255–274.

Wagner AD. 1999. Working memory contributions to human learning and remembering. *Neuron* **22:** 19–22.

Wallis JD, Kennerley SW. 2013. The functional role of reward signals on different prefrontal areas. In *Principles of frontal lobe function* (ed. Stuss DT, Knight RT), pp. 69–78. Oxford University Press, Oxford.

Wallis JD, Anderson KC, Miller EK. 2001. Single neurons in prefrontal cortex encode abstract rules. *Nature* **411:** 953–956.

Wang M, Gamo NJ, Yang Y, Jin LE, Wang X-J, Laubach M, Mazer JA, Lee D, Arnsten AFT. 2011. Neuronal basis of age-related working memory decline. *Nature* **476:** 210–213.

Watanabe M. 2013. How context impacts cognitive control and motivational control of behavior in the primate prefrontal cortex. In *Principles of frontal lobe function* (ed. Stuss D, Knight R), pp. 211–225. Oxford University Press, Oxford.

Williams SM, Goldman-Rakic PS. 1993. Characterization of the dopaminergic innervation of the primate frontal cortex using a dopamine-specific antibody. *Cereb Cortex* **3:** 199–222.

Cite this article as *Cold Spring Harb Perspect Biol* doi: 10.1101/cshperspect.a021816

# Large-Scale Fluorescence Calcium-Imaging Methods for Studies of Long-Term Memory in Behaving Mammals

Pablo Jercog[1], Thomas Rogerson[1], and Mark J. Schnitzer[1,2,3]

[1]CNC Program, Stanford University, Stanford, California 94305

[2]Howard Hughes Medical Institute, Stanford University, Stanford, California 94305

[3]James H. Clark Center for Biomedical Engineering & Sciences, Stanford University, Stanford, California 94305

*Correspondence:* mschnitz@stanford.edu

During long-term memory formation, cellular and molecular processes reshape how individual neurons respond to specific patterns of synaptic input. It remains poorly understood how such changes impact information processing across networks of mammalian neurons. To observe how networks encode, store, and retrieve information, neuroscientists must track the dynamics of large ensembles of individual cells in behaving animals, over timescales commensurate with long-term memory. Fluorescence $Ca^{2+}$-imaging techniques can monitor hundreds of neurons in behaving mice, opening exciting avenues for studies of learning and memory at the network level. Genetically encoded $Ca^{2+}$ indicators allow neurons to be targeted by genetic type or connectivity. Chronic animal preparations permit repeated imaging of neural $Ca^{2+}$ dynamics over multiple weeks. Together, these capabilities should enable unprecedented analyses of how ensemble neural codes evolve throughout memory processing and provide new insights into how memories are organized in the brain.

Recent years have brought major new capabilities for manipulating mammalian neural circuits involved in memory processing. Advances in genetics, viral delivery methods, and optogenetic and pharmacogenetic techniques have allowed researchers to explore how specific cell types and neural projection pathways contribute to memory processing in behaving rodents (Ciocchi et al. 2010; Goshen et al. 2011; Letzkus et al. 2011; Garner et al. 2012; Liu et al. 2012; Nguyen-Vu et al. 2013; Cowansage et al. 2014; Redondo et al. 2014; Senn et al. 2014; Wolff et al. 2014). By using optogenetic or pharmacogenetic modes of neural excitation in associative learning paradigms, one can substitute for conditioned (Choi et al. 2011; Kwon et al. 2014) or unconditioned (Johansen et al. 2010; Kimpo et al. 2014) stimuli, create synthetic memory traces (Garner et al. 2012), artificially evoke previously learned behaviors (Liu et al. 2012; Cowansage et al. 2014; Kim et al. 2014), and even condition mice to associate pairs of stimuli that were never actually presented together (Ramirez et al. 2013; Redondo et al.

2014). One optogenetic study comparing the behavioral effects of fast neural inhibition against those of inactivation over longer timescales has also demonstrated that the set of brain circuits required for memory function depends strongly on the timescale at which one probes this necessity (Goshen et al. 2011), likely because slower methods of inactivation allow the brain to attain partial or total recoveries of function. Overall, recent advances in the causal manipulation of mammalian neural circuits have yielded important gains in the delineation of necessary and sufficient neural events for memory performance.

To complement the new approaches for perturbing circuits, neuroscientists also need improved observational methods for probing the neural representations the mammalian brain uses normally for memory processing and storage. In many cases, these representations appear to be distributed over large networks of cells and multiple brain areas (Tse et al. 2007; Goshen et al. 2011; Cowansage et al. 2014; Redondo et al. 2014). As computer scientists well appreciate, the manner in which information is represented in storage can have a major impact on the ease of retrieving information, encoding objects' attributes along with their identities, classifying and grouping items according to their characteristics, and associating items that have defined relationships such as in space or time (Date 2000). To understand how the brain accomplishes these feats, we need in vivo recording methods capable of revealing large-scale neural representations and tracking their evolution over timescales pertinent to long-term memory. Such methods would open the door to sophisticated studies of how large-scale neural dynamics and codes may facilitate or hinder different forms of information management in the mammalian brain (Sadtler et al. 2014).

Large-scale $Ca^{2+}$-imaging techniques offer considerable promise toward achieving studies of this kind (Dombeck et al. 2010; Komiyama et al. 2010; Harvey et al. 2012; Huber et al. 2012; Ziv et al. 2013; Heys et al. 2014; Low et al. 2014; Modi et al. 2014; Peters et al. 2014; Rickgauer et al. 2014; Hamel et al. 2015). Key technical advantages include the capacity to monitor

the dynamics of hundreds of cells concurrently, target specific neuron types for study based on their genetic identities or connectivity patterns, reliably track individual cells for many weeks in behaving animals, extract the signals of individual neurons nearly regardless of their activity rates, and visualize the anatomical organization of memory storage at the cellular scale. Optical methods for $Ca^{2+}$ imaging also lend themselves naturally to combined usages with other optical methods (Prakash et al. 2012; Deisseroth and Schnitzer 2013; Jorgenson et al. 2015), such as optogenetics (Rickgauer et al. 2014; Szabo et al. 2014; Grosenick et al. 2015; Packer et al. 2015), fluorescent tagging of activated neurons (Reijmers et al. 2007; Liu et al. 2012; Guenthner et al. 2013; Kawashima et al. 2013; Ramirez et al. 2013; Redondo et al. 2014), and postmortem imaging of optically cleared tissues (Hama et al. 2011; Chung et al. 2013; Ke et al. 2013; Susaki et al. 2014; Yang et al. 2014a). In the future, integrated optical studies will likely combine observations of neurons' normal activity patterns, precise optogenetic manipulations of these dynamics, and detailed postmortem examinations of the dendritic morphologies, axonal projections, and macromolecular architectures of the very same cells (Deisseroth and Schnitzer 2013; Jorgenson et al. 2015).

This review aims to introduce in vivo $Ca^{2+}$-imaging methods to researchers studying mammalian learning and memory. The next sections provide technical information for practitioners of in vivo $Ca^{2+}$ imaging, emphasizing methodological strengths and limitations. Later sections discuss the types of experiments on learning and memory that $Ca^{2+}$-imaging techniques are likely to enable in the near future.

## $Ca^{2+}$ IMAGING AS A MEANS OF INFERRING NEURONAL SPIKING DYNAMICS

Nearly all neuron types express voltage-sensitive $Ca^{2+}$ channels (Trimmer and Rhodes 2004). Fluorescence $Ca^{2+}$ imaging as a means of detecting somatic spiking dynamics relies on increases in intracellular $Ca^{2+}$ ion concentration, $[Ca^{2+}]$, which occur in response to neural membrane depolarization (Helmchen et al.

1996; Grienberger and Konnerth 2012; Hamel et al. 2015). An essential part of all $Ca^{2+}$-imaging methodologies is the $Ca^{2+}$ indicator, a fluorescent reporter molecule that changes its photophysical properties in response to variations in intracellular $[Ca^{2+}]$. The detailed biophysical processes underlying $Ca^{2+}$ imaging have been reviewed at length elsewhere (Grienberger and Konnerth 2012). Thus, this section focuses more on the enabling features and pitfalls that memory researchers should consider when contemplating the use of a $Ca^{2+}$ indicator as a means of visualizing spiking activity.

A fluorescent $Ca^{2+}$ indicator molecule can bind one or more $Ca^{2+}$ ions, which thereby alters the indicator's optical properties (Tian et al. 2009; Grienberger and Konnerth 2012). In general, there are distinct classes of indicators that undergo different types of photophysical changes, such as in fluorescence intensity, absorption or emission spectra, or emission lifetime (Grienberger and Konnerth 2012). For studies of long-term memory, researchers will generally want to choose a $Ca^{2+}$ indicator that is genetically encoded (Mank et al. 2008; Grienberger and Konnerth 2012; Looger and Griesbeck 2012; Tian et al. 2012) owing to the capacity for stably expressing these protein sensors over time courses commensurate with those of long-term memory storage, such as for the types of experiments considered here in later sections.

For $Ca^{2+}$-imaging studies in behaving mammals, the genetically encoded indicator that is most widely used at present (e.g., based on gene requests to Addgene) is GCaMP6, which can bind up to four $Ca^{2+}$ ions and reports $[Ca^{2+}]$ rises via increases in the intensity of its green fluorescence emissions (Chen et al. 2013). GCaMP6 has a greater dynamic range of fluorescence signaling than prior genetically encoded $Ca^{2+}$ indicators. It also has $Ca^{2+}$ binding kinetics rivaling those of the best synthetic $Ca^{2+}$ indicators used for acute in vivo imaging studies (Garaschuk et al. 2006, Komiyama et al. 2010, Modi et al. 2014). Emerging types of genetically encoded red fluorescent $Ca^{2+}$ indicators with spectrally distinct emissions from GCaMP6 open up interesting possibilities for

monitoring the dynamics of two distinct cell types simultaneously (Inoue et al. 2015).

Despite the impressive capabilities of GCaMP6, no $Ca^{2+}$ indicator provides an exact readout of membrane voltage dynamics (Fig. 1A–H). The biophysical processes that govern the $[Ca^{2+}]$ rise in response to an action potential include $Ca^{2+}$ channel activation and intracellular $Ca^{2+}$ buffering and are distinct from those setting the action potential's electrical waveform (Helmchen et al. 1996). The time-dependence of the signals from a $Ca^{2+}$ indicator also depends strongly on its $Ca^{2+}$ handling properties, particularly the kinetic rates, equilibrium constant and Hill coefficient for $Ca^{2+}$ binding and unbinding (Fig. 1A–D,G,H) (Sun et al. 2013). Hence, beyond the intrinsic differences between the dynamics of transmembrane voltage and intracellular $Ca^{2+}$ excitation, the signals from a fluorescent $Ca^{2+}$ indicator represent a temporally filtered version of the intracellular $[Ca^{2+}]$ time course. Because of the experimental difficulty of directly relating a cell's time-varying fluorescence emissions to its intracellular $[Ca^{2+}]$ dynamics in a live brain, virtually all researchers express fluorescence time traces from in vivo $Ca^{2+}$-imaging studies in units of a percentage change from each cell's baseline fluorescence level, typically denoted $\Delta F(t)/F_0$.

There are three main variants of GCaMP6 (slow, medium, and fast) that have different $Ca^{2+}$-binding kinetics and, hence, temporally filter intracellular $[Ca^{2+}]$ dynamics to different degrees (Fig. 1) (Chen et al. 2013). The slow variant, GCaMP6s, binds $Ca^{2+}$ more tightly than the fast variant, GCaMP6f. This difference confers greater $Ca^{2+}$ sensitivity to GCaMP6s and leads to fluorescence transients of longer duration (Fig. 1C), but sacrifices signaling speed and the ability to distinguish action potentials occurring in quick succession (Fig. 1A,G,H). For GCaMP6s, the fluorescence signals in response to an action potential have a rise time-constant of $\sim$180 msec and a decay time constant of $\sim$580 msec, whereas, for GCaMP6f, these values are $\sim$30 msec and $\sim$150 msec, respectively (Fig. 1G,H) (Chen et al. 2013). Cooperation and saturation effects in the binding of multiple $Ca^{2+}$ ions to a GCaMP6 mole-

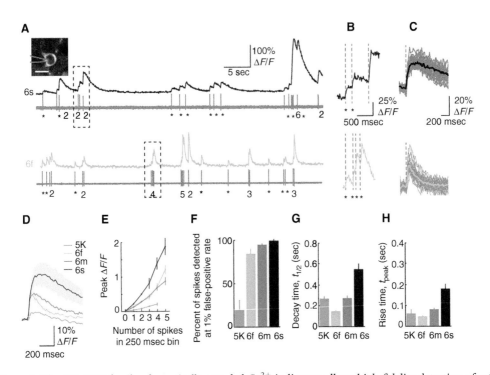

**Figure 1.** The GCaMP6 family of genetically encoded $Ca^{2+}$ indicators allows high-fidelity detection of action potentials from neocortical pyramidal cells in live mice and $\sim$20–200 msec spike timing estimation accuracy. (*A*) In pyramidal neurons expressing GCaMP6s (black traces) or GCaMP6f (cyan traces) in neocortical visual area V1 of live mice, simultaneous loose-seal cell-attached electrical and two-photon fluorescence $Ca^{2+}$-imaging recordings show the reliability of the optical response to each action potential. Asterisks below the electrical traces (*lower* traces in each pair) mark occurrences of individual action potentials; numerals mark action potential bursts and report the number of spikes in each burst. (*Inset*) Two-photon image of a neuron expressing GCaMP6s; the recording pipette is indicated by the red lines. Scale bar, 10 μm. (*B*) Expanded view of the fluorescence traces in *A* over the time periods delineated by the dashed boxes. Dashed vertical lines mark action potential occurrences. (*C*) Fluorescence changes in response to one action potential for GCaMP6s (*upper*) and GCaMP6f (*lower*). Gray traces show the optical responses to individual spikes. Black and cyan traces show the average response. (*D*) Traces of the median fluorescence change in response to one action potential for the genetically encoded $Ca^{2+}$ indicators GCaMP5K (green trace), GCaMP6f (cyan), GCaMP6m (magenta), and GCaMP6s (black). Shading indicates S.E.M. (*E*) Peak fluorescence changes (mean $\pm$ S.E.M.) observed by $Ca^{2+}$ imaging as a function of the number of spikes detected electrically in a 250 msec time bin during simultaneously acquired optical and electrical recordings in V1 neurons in live mice. Color key is the same as in *D*. (*F–H*) Comparative performance metrics of the different indicators. (F) Percentage of action potentials correctly detected when the detection threshold is set to yield only 1% false positives. (G) Time constant for the decay of fluorescence following an action potential occurrence. (H) Time constant for the fluorescence rise in response to an action potential. All error bars are S.E.M. All panels derived and modified from Chen et al. (2013).

cule also cause nonlinearities in the temporal filtering.

In addition to these aspects of indicator dynamics, stochastic fluctuations in the emission and detection of fluorescence photons constrain the fidelity of action potential detection and the accuracy of spike-timing estimation (Wilt et al. 2013; Hamel et al. 2015). Photon "shot noise" has its origins in quantum mechanics and sets physical limits on spike detection fidelity and timing accuracy—even when additional noise sources, such as instrumentation noise, are minuscule. Signal detection and estimation theories provide theoretical lower bounds on the spike detection and timing errors incurred because of fluctuations

 Cite this article as *Cold Spring Harb Perspect Biol* doi: 10.1101/cshperspect.a021824

in photon counts (Wilt et al. 2013; Hamel et al. 2015).

Because of these biophysical and optical facets of $Ca^{2+}$ imaging, there is substantial variability in the relationship between the amplitude and waveform of a somatic $Ca^{2+}$ transient, as seen by fluorescence imaging, and the number of action potentials underlying the transient (Fig. 1C). Moreover, the relationship between membrane voltage and somatic $[Ca^{2+}]$ dynamics differs between different types of neurons. Cell types vary regarding the expression and properties of their $Ca^{2+}$ channels, levels of intracellular $[Ca^{2+}]$ buffering, and other aspects of $Ca^{2+}$ signaling such as $Ca^{2+}$-induced $Ca^{2+}$ release from intracellular stores. Whenever feasible, it is best to empirically determine the relationship between $Ca^{2+}$-related fluorescence signals and the underlying spike trains in the specific cell type of interest. A common approach is to perform simultaneous electrical recordings and fluorescence measurements in the exact same cells, such as in a brain slice (Tian et al. 2009), or by using in vivo $Ca^{2+}$ imaging to guide the targeting of an electrode to a selected cell in a live animal (Fig. 1) (Chen et al. 2013).

Simultaneous optical and electrical recordings from visual cortical pyramidal neurons in live mice have revealed approximately linear relationships between the peak amplitude of a somatic $Ca^{2+}$ transient and the underlying number of action potentials (Fig. 1E) (Chen et al. 2013). Other types of pyramidal cells likely also exhibit similar linear relationships between these parameters, but not necessarily with the same linear function. More generally, in neurons with temporally sparse spiking patterns, in vivo $Ca^{2+}$ imaging usually reveals discrete $Ca^{2+}$ transients representing one or more spikes.

Inhibitory interneurons have distinct $Ca^{2+}$ signaling and buffering attributes from those of excitatory neurons, and they often fire spikes at fast rates or in bursts. In such fast-spiking neurons, it is typically infeasible to assign precise spike numbers to individual $Ca^{2+}$ transients. Still, $Ca^{2+}$ imaging generally reveals the overall patterns of how interneurons modulate their time-dependent activity rates; for instance, recent $Ca^{2+}$-imaging studies of visual cortical interneurons have monitored visually evoked increases in somatic fluorescence intensity, rather than incidences of individual $Ca^{2+}$ transients (Kerlin et al. 2010; Runyan et al. 2010; El-Boustani and Sur 2014).

Regardless of cell type, in live or behaving animals, there are yet additional noise sources that can make it challenging to infer the exact number of spikes underlying any $Ca^{2+}$ transient. These noise sources are generally nonstationary over time and can originate from hemodynamic events, brain motion artifacts induced by physiological rhythms or voluntary movements, or fluorescence $Ca^{2+}$ signals from neuropil activation. Notwithstanding, there is almost always a useful, monotonic relationship between the mean intensity of somatic fluorescence emissions from a $Ca^{2+}$ indicator and the rate of somatic spiking.

In vivo $Ca^{2+}$ imaging generally does not reliably reveal membrane hyperpolarization or other subthreshold aspects of somatic voltage dynamics. However, outside the soma, by sparsely expressing a $Ca^{2+}$ indicator in isolated cells, one can detect in vivo the subcellular $Ca^{2+}$ activation patterns of individual dendrites, dendritic spines, and even axons (Bock et al. 2011; Petreanu et al. 2012; Xu et al. 2012; Chen et al. 2013; Glickfeld et al. 2013; Kaifosh et al. 2013; Lovett-Barron et al. 2014; Sheffield and Dombeck 2015).

Overall, although in vivo $Ca^{2+}$ imaging may not be able to adjudicate issues that hinge on exact spike counts and millisecond-scale timing, there are many questions about neural coding and memory processing that $Ca^{2+}$ imaging can answer. The latter questions might focus on the identities of cells that encode specific types of information, the acquisition and long-term stability of coding properties under different behavioral conditions, the relationships between coding properties and other cellular attributes, the extent to which ensembles of neurons encode information in a cooperative manner, how neural codes change across different brain states, or the anatomical relationships between cells involved in storing specific memories.

## TARGETING SPECIFIC SUBSETS OF CELLS FOR EXPRESSION OF A GENETICALLY ENCODED $Ca^{2+}$ INDICATOR

To study large-scale neural coding by $Ca^{2+}$ imaging, it is important to consider the various ways in which one can express a genetically encoded indicator in the cells of interest. To date, most $Ca^{2+}$-imaging studies in behaving mammals based on a genetically encoded $Ca^{2+}$ indicator have used an adeno-associated viral (AAV) vector to target indicator expression to specific neuron types at the virus injection site (Dombeck et al. 2010; Komiyama et al. 2010; Harvey et al. 2012; Huber et al. 2012; Ziv et al. 2013). A key consideration with the use of an AAV concerns the viral serotype, because distinct serotypes of AAV enter different neuron types to varying degrees (Aschauer et al. 2013). Another key factor is the genetic promoter used to express the $Ca^{2+}$ indicator. For instance, an AAV whose genome incorporates a *CaMK2a* promoter can target indicator expression to the specific subset of virally infected neurons in which this promoter is active, such as pyramidal neurons (Ziv et al. 2013). A pan-neuronal promoter, such as the *Synapsin* promoter, drives expression in nearly all neuron types (Dombeck et al. 2010; Huber et al. 2012). However, many promoters of interest are too large to be packaged within an AAV. There are other viruses capable of delivering larger genetic payloads, but they tend to be either toxic to neurons or less effective at expressing a fluorescent indicator at the levels needed for adequate brightness.

Thus, a widely used alternate means of targeting specific cell types is to create an AAV that selectively drives indicator expression over the long-term in cells expressing the enzyme Cre-recombinase (Sauer 1998). Many different Cre-driver mouse lines, transgenic animals that express this enzyme in particular classes of cells, are widely available (Madisen et al. 2010, 2015). Hence, a single viral construct that expresses the $Ca^{2+}$ indicator in a Cre-dependent manner can be fruitfully combined with a wide variety of Cre-driver lines. This modular approach is generally more efficient and feasible than creating a separate virus for each cell type of interest.

In addition to facilitating studies of somatic $Ca^{2+}$ activity, AAV-based strategies for GCaMP expression have enabled observations of $Ca^{2+}$ activity in specific neural pathways, by imaging the $Ca^{2+}$ dynamics of axons of neurons whose cell bodies lie elsewhere (e.g., in a brain area distal to the imaging field of view). This approach selectively confines the $Ca^{2+}$ signals to neurons that have their cell bodies at the viral injection site and project axons to the imaging site (Petreanu et al. 2012; Xu et al. 2012; Glickfeld et al. 2013; Kaifosh et al. 2013; Lovett-Barron et al. 2014). For example, in higher visual areas of the mouse neocortex, this approach allowed selective imaging of axons from neurons with cell bodies in visual area V1 (Glickfeld et al. 2013). A promising alternative to this approach for imaging pathway specific $Ca^{2+}$ activity is to use canine adenovirus-2 (CAV-2), which efficiently infects neuronal axons (Hnasko et al. 2006; Bru et al. 2010; Boender et al. 2014).

Other emerging strategies for expressing a genetically encoded $Ca^{2+}$ indicator include the use of transgenic mice (Chen et al. 2012b; Zariwala et al. 2012; Dana et al. 2014; Madisen et al. 2015) or transsynaptic viruses (Osakada et al. 2011). Until recently, transgenic mice created to express genetically encoded $Ca^{2+}$ indicators usually suffered from dimmer fluorescence levels than those attainable by viral delivery methods. This is because transgenic mice typically had one, or only a few, copies of the $Ca^{2+}$ indicator gene. By comparison, many AAV particles can enter an individual infected cell, which boosts each cell's expression of the protein fluorophore. This issue is now widely appreciated, and a promising transgenic mouse strategy that uses a triple transgene approach to increase $Ca^{2+}$ indicator expression levels has recently emerged (Madisen et al. 2015).

Transsynaptic neurotropic viruses that traverse neuronal synapses also offer interesting prospects for targeting $Ca^{2+}$-imaging studies to cells receiving or providing specific synaptic inputs. Retrograde transsynaptic viruses include rabies (Kelly and Strick 2000; Taber et al. 2005; Wickersham et al. 2007a,b; Miyamichi et al. 2011; Osakada et al. 2011; Osakada and Callaway 2013) and pseudorabies (Enquist et al. 2002;

Card et al. 2011; Card and Enquist 2014; Oyibo et al. 2014). The use of these viruses for anatomical mapping studies is established but applications to $Ca^{2+}$ imaging remain somewhat experimental (Osakada et al. 2011). A key advantage of rabies has been that genetically modified versions of rabies exist that can pass retrograde across a single synapse, but not across disynaptic or polysynaptic projections (Etessami et al. 2000; Wickersham et al. 2007a,b). This restriction confines the set of cells under study to those with well-defined monosynaptic connections. Analogous versions of pseudorabies are emerging (Oyibo et al. 2014). Transsynaptic labeling approaches based on anterograde viruses remain less well developed. Anterograde labeling methods exist based on the herpes simplex virus but are highly toxic for the infected neurons (Hoover and Strick 1999; Lo and Anderson 2011).

Another strategy for studying specific cells involved in memory processing is to target cells that undergo immediate early gene (IEG) activation at specific phases of a behavioral protocol. Memory researchers often use IEG expression as a neuronal marker of engagement in memory formation, such as via spiking or synaptic plasticity, depending on the particular IEG (Guzowski and Worley 2001; Czajkowski et al. 2014). There are several different types of transgenic mice that permit fluorescence tagging of specific cells according to their patterns of *Fos* or *Arc* gene activation (Reijmers et al. 2007; Garner et al. 2012; Liu et al. 2012; Guenthner et al. 2013; Ramirez et al. 2013; Czajkowski et al. 2014; Redondo et al. 2014). Beyond readout of cell activation, IEG-based tagging has chiefly been used in combination with pharmacogenetic or optogenetic means of selectively reactivating or silencing tagged neurons to probe their causal roles in memory processing (Garner et al. 2012; Liu et al. 2012; Ramirez et al. 2013; Cowansage et al. 2014; Redondo et al. 2014; Yiu et al. 2014).

Similar strategies involving IEG-based tagging appear promising for $Ca^{2+}$-imaging studies. For instance, one might selectively image $Ca^{2+}$ activity in neurons that underwent IEG activation at an earlier timepoint in a behavioral protocol, to probe how the activity of neurons

involved in memory encoding (based on IEG expression) relates to their spiking dynamics at memory recall. More broadly, joint monitoring of IEG expression and $Ca^{2+}$ imaging might allow comparisons of how information is represented across different sets of neurons, such as those in which *Fos*, *Arc*, or other IEGs are activated on different days in a long-term experiment. This type of study might help adjudicate recent hypotheses proposing that the activation of certain IEGs may prime neurons for upcoming bouts of memory storage—before the remembered events have even occurred (Han et al. 2007; Rogerson et al. 2014). Such issues might be fruitfully studied through the use of genetic constructs that express a pair of differently colored fluorescence indicators, one reporting the level of IEG activation and the other signaling $Ca^{2+}$ activity (Kawashima et al. 2013).

## OPTICAL INSTRUMENTATION FOR IN VIVO $Ca^{2+}$ IMAGING IN AWAKE BEHAVING ANIMALS

$Ca^{2+}$-imaging studies in behaving rodents generally use one of two main types of optical instrumentation. In some studies, the animal is head-fixed and behaves in place under the objective lens of a conventional upright two-photon fluorescence microscope (Dombeck et al. 2007; Nimmerjahn et al. 2009). In other studies, the animal carries a miniature fluorescence microscope on its head, thereby allowing $Ca^{2+}$-imaging studies of neuronal activity during unconstrained animal behavior (Helmchen et al. 2001; Flusberg et al. 2008). Both approaches allow long-term imaging over weeks and have enabled $Ca^{2+}$-imaging studies of the neurobiology of memory (Komiyama et al. 2010; Harvey et al. 2012; Huber et al. 2012; Ziv et al. 2013; Peters et al. 2014). For a comparison of the optical issues arising in the two imaging formats and discussion of the image analysis algorithms used to extract individual cells' time-varying $Ca^{2+}$ signals from the fluorescence videos, we refer readers to a recent review of these engineering matters (Hamel et al. 2015). Here we focus on considerations important to the design of scientific studies.

A key consideration is whether limiting an animal's range of behavior via head fixation will be a benefit or a drawback. To address many scientific questions, it is crucial to gather data from a large set of stereotyped trials, and the behavioral constraints imposed by head fixation can facilitate both the controlled delivery of sensory stimuli (Verhagen et al. 2007; Carey et al. 2009; Andermann et al. 2011; Blauvelt et al. 2013; Patterson et al. 2013; Miller et al. 2014) and behavioral stereotypy in an animal's responses (Huber et al. 2012; Masamizu et al. 2014; Peters et al. 2014). Together, these capabilities have allowed $Ca^{2+}$-imaging studies in head-fixed animals trained to perform perceptual discrimination (Andermann et al. 2010; Komiyama et al. 2010; O'Connor et al. 2010) and motor execution tasks (Huber et al. 2012; Masamizu et al. 2014; Peters et al. 2014). A recent $Ca^{2+}$-imaging study even examined a version of associative fear conditioning adapted for head-restrained animals (Lovett-Barron et al. 2014).

In addition to its compatibility with conventional two-photon microscopy, the head-restrained imaging format is also well suited for use with custom-designed fluorescence microscopy setups that provide novel imaging capabilities (Horton et al. 2013; Heys et al. 2014; Lecoq et al. 2014; Low et al. 2014; Quirin et al. 2014; Stirman et al. 2014; Bouchard et al. 2015) or combine two-photon $Ca^{2+}$ imaging and two-photon optogenetic capabilities in behaving mammals (Packer et al. 2012; Prakash et al. 2012; Rickgauer et al. 2014; Packer et al. 2015). As new optical brain-imaging modalities emerge, the steadily increasing number of behavioral assays for head-fixed rodents will provide valuable test beds for validating novel optical approaches and then capitalizing on the resulting opportunities for biological experimentation.

Head-fixation can also be combined with approaches in which an animal navigates a virtual reality while walking or running in place (e.g., on a spherical treadmill) (Fig. 2) (Holscher et al. 2005; Harvey et al. 2009; Dombeck et al. 2010; Keller et al. 2012; Ravassard et al. 2013; Sofroniew et al. 2014; Aghajan et al. 2015). Virtual reality methods permit sensory manipulations that would be difficult or impossible to achieve in a freely behaving animal (Keller et al. 2012). To date, most virtual reality systems for rodents have involved projection of a visual scene around the head-fixed animal, such as by using a toroidal screen (Fig. 2A,B) (Holscher et al. 2005; Harvey et al. 2009; Dombeck et al. 2010), or arrays of two or more flat video monitors (Fig. 2C) (Keller et al. 2012). However, virtual reality methods that provide tactile feedback are now also emerging (Sofroniew et al. 2014).

A noteworthy caveat is that in an animal exploring a virtual environment, neural dynamics may deviate from their normal forms during unrestrained animal behavior (Ravassard et al. 2013), perhaps because of altered vestibular, self-motion, or visual or tactile sensory inputs. The extent of these effects remains unclear, especially for long-term experiments in which neural plastic effects might accrue over time, and the degree to which different forms of virtual reality impact neural coding remains under active investigation (Ravassard et al. 2013; Aronov and Tank 2014; Aghajan et al. 2015). As computer graphics and reality simulation methods advance, neuroscientists will gain increasingly sophisticated capabilities for performing sensory manipulations during large-scale $Ca^{2+}$ imaging. Notably, virtual reality methods have been used to study spatial memory in humans in subjects undergoing functional magnetic resonance brain imaging (Pine et al. 2002) or intracranial electrical recordings (Suthana et al. 2012). This commonality might enable parallel studies in which animal and human subjects perform similar tasks, but during different forms of brain imaging suitable for each species.

Complementary to $Ca^{2+}$-imaging techniques that require head fixation of an alert animal, miniature head-mounted fluorescence microscopes that have flexible cables (optical fibers or floppy electrical lines) allow $Ca^{2+}$-imaging studies in freely behaving mammals (Helmchen et al. 2001; Flusberg et al. 2008; Sawinski et al. 2009; Ghosh et al. 2011; Ziv et al. 2013; Berdyyeva et al. 2014; Betley et al. 2015; Jennings et al. 2015). The ability to study unrestrained forms of animal behavior is an impor-

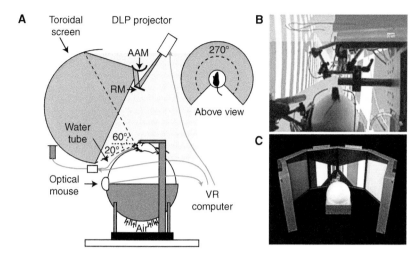

**Figure 2.** Apparatus for projecting a virtual reality environment. (*A*) Schematic of a virtual reality apparatus. An image of a virtual space is projected by a digital light-processing (DLP) projector, deflects off a reflecting mirror (RM), and is magnified by a curved, angular amplification mirror (AAM) onto a toroidal screen. A head-fixed rodent is at liberty to walk or run in place on an air-suspended spherical treadmill. An optical computer mouse monitors the treadmill's rotations in both angular dimensions (modified from Harvey et al. 2009; see also Figures 1 and 2 in Holscher et al. 2005). (*B,C*) Examples of commercially available apparatus, using a toroidal screen, *B*, and a set of six computer screens, *C*, that create virtual environments in which head-fixed rodents can navigate. (Images obtained from and used with permission of PhenoSys GmbH.)

tant capacity, because many animal behaviors are incompatible with or poorly suited to study under conditions of head restraint. Examples include the social behaviors such as fighting, mating, care giving, and other forms of interaction; behaviors probing fear, stress, or anxiety in which head restraint may cause confounding behavioral or physiological effects; and many motor and vestibular-dependent behaviors that involve self-motion cues. All of these behaviors can have learning and memory components. Miniaturized microscopes are also compatible with most of the behavioral assays that are already widely deployed and validated across neuroscience research institutions and in the pharmaceutical industry.

At present, the most widely used miniature fluorescence microscope in the neuroscience research field is a two-gram integrated microscope for $Ca^{2+}$ imaging in freely behaving mice (Fig. 3A–C) (Ghosh et al. 2011; Ziv et al. 2013). This device is optically integrated in the sense that each two-gram unit contains all required optical components, including a light-emitting diode

(LED) that provides illumination, miniature lenses, a tiny fluorescence filter cube, and a cell phone camera chip that captures the fluorescence images (Fig. 3A,B). Fine electrical wires carry power and control signals to the microscope and convey digital images from the camera chip to an external data-acquisition box. The integrated microscope is commercially available and generally compatible with the spatial mazes, operant chambers, and fear-conditioning boxes commonly used for behavioral studies in mice. With multiple integrated microscopes, one can perform fluorescence $Ca^{2+}$ imaging in several freely behaving mice in parallel.

When used in conjunction with microendoscope probes (350–1000 μm diameters) that can be permanently implanted deep in brain tissue (Jung and Schnitzer 2003; Jung et al. 2004; Levene et al. 2004; Barretto et al. 2011), the integrated microscope allows brain imaging in a wide variety of different brain areas. These include neocortex, hippocampus, striatum, amygdala, hypothalamus, nucleus accumbens, substantia nigra, and cerebellar cortex

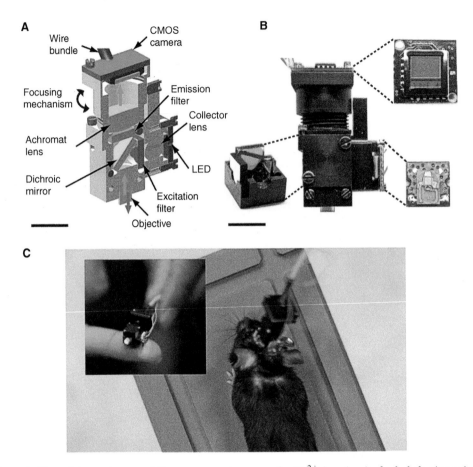

**Figure 3.** The miniature integrated fluorescence microscope for $Ca^{2+}$ imaging in freely behaving mice. (*A*) Computer-assisted design of an integrated fluorescence microscope, showing the optical and mechanical components in cross section. (*B*) Photograph of an assembled microscope. *Insets* show the printed circuit boards holding the cell phone camera chip (*upper right*), the blue light-emitting diode (*lower right*), and the fluorescence filter cube (*lower left*). (*C*) Integrated microscope mounted on a freely behaving mouse. (*Inset*) An integrated microscope on the tip of a finger. Scale bars, 5 mm (*A* and *B*) and apply also to the *insets*. Panels *A* and *B* are from Ghosh et al. (2011). (Images in panel *C* are courtesy of Inscopix.)

(Ghosh et al. 2011; Ziv et al. 2013; Grewe et al. 2014; Lefort et al. 2014; Parker et al. 2014; Betley et al. 2015; Hamel et al. 2015; Jennings et al. 2015). To monitor the $Ca^{2+}$ dynamics of large sets of individual cells in these brain areas across multiple weeks, researchers can repeatedly image the same field of view simply by reattaching the integrated microscope to a permanently affixed cranial base plate at the start of each imaging session (Ziv et al. 2013). Electroencephalography or electromyography recordings can also be performed simultaneously with $Ca^{2+}$ imaging, such as for monitoring brain or loco-

motor states (Berdyyeva et al. 2014). A single field of view ($\sim$0.5–0.6 mm$^2$) from the integrated microscope can routinely reveal many hundreds of neurons over the course of an extended $Ca^{2+}$-imaging study (Ziv et al. 2013), and sometimes >1000 cells (Fig. 4) (Alivisatos et al. 2013).

We expect that the optical methods for $Ca^{2+}$ imaging in head-restrained and freely behaving animals will both continue to advance in their capabilities. Likely improvements include optical access to greater numbers of cells per animal, fast volumetric imaging, multicolored $Ca^{2+}$

**A**

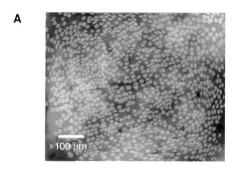

**B**

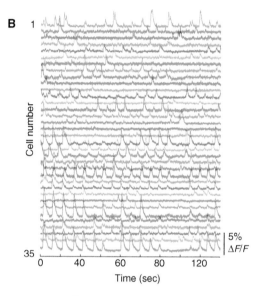

**Figure 4.** A Ca$^{2+}$-imaging dataset from a single freely behaving mouse can contain ∼1000 neurons. (*A*) 1202 GCaMP-expressing CA1 hippocampal pyramidal neurons (red somata), identified in Ca$^{2+}$-imaging data taken in a freely moving mouse using the integrated microscope, shown atop a mean fluorescence image (green) of CA1. Vessels appear as dark shadows (from Alivisatos et al. 2013). (*B*) 35 example traces of Ca$^{2+}$ activity from CA1 hippocampal pyramidal neurons expressing GCaMP3 under the control of the *Camk2a* promoter (modified from supplementary Fig. 1b in Ziv et al. 2013).

imaging for monitoring two targeted subpopulations of cells simultaneously, imaging in multiple brain areas concurrently, and superior capabilities for simultaneous Ca$^{2+}$ imaging and optogenetic manipulation. These capacities will further expand the pregnant set of possibilities for optical studies of memory in behaving animals.

## ADVANTAGES OF Ca$^{2+}$ IMAGING FOR ANALYSES OF LONG-TERM MEMORY

Beyond its capacities for recording from many individual cells and targeting cells of specific types, both of which are generically useful for systems neuroscience, Ca$^{2+}$ imaging offers additional capabilities that are especially valuable for studies of long-term memory. The latter include the abilities to track individual neurons over many weeks, isolate the signals of neurons with very low activity rates, and visualize the anatomical organization of information processing at cellular resolution. These three facilities have specific import for the learning and memory field.

Long-term memories can persist from timescales of hours to years. To understand how neural ensembles maintain stored information, it is important to have recording techniques that can monitor individual cells' dynamics over time spans commensurate with the memory durations of interest. Hence, chronic animal preparations for time-lapse in vivo imaging offer key benefits for studies of long-term memory. Time-lapse fluorescence imaging in live adult mice has provided a potent means for repeatedly inspecting the same neurons and even the same dendritic spines, over periods ranging from weeks to over a year (Grutzendler et al. 2002; Trachtenberg et al. 2002; Zuo et al. 2005; Chen et al. 2008, 2012a; Holtmaat et al. 2009; Yang et al. 2009, 2014b; Cruz-Martin et al. 2010; Fu et al. 2012; Lai et al. 2012; Mostany et al. 2013; Attardo et al. 2015). When this capability is combined with Ca$^{2+}$ imaging, one can track individual neurons' coding properties over weeks with an ease and efficiency that has been lacking from electrophysiological techniques (Huber et al. 2012; Ziv et al. 2013; Peters et al. 2014).

Traditional electrophysiological recording methods have allowed researchers to track the spiking dynamics of modest numbers of individual cells over days to weeks (Thompson and Best 1990). A majority of such long-term electrophysiological studies has focused on the rodent hippocampus, a crucial brain area for the formation of spatial and episodic memories

(Eichenbaum 2000; Buzsaki and Moser 2013). Early reports (Muller et al. 1987; Thompson and Best 1990) of long-term electrical recordings in the hippocampus described longer recording durations than those typical of more recent studies (Lever et al. 2002; Kentros et al. 2004; Muzzio et al. 2009; Mankin et al. 2012; McKenzie et al. 2014). For instance, in rats that repeatedly explored a familiar spatial environment, Thompson and Best (1990) reported that 10 hippocampal neurons maintained stable place fields for 6–153 days. By comparison, recent studies of hippocampal coding have generally tracked greater numbers of cells over briefer durations. For instance, Mankin et al. (2012) reported that they maintained electrical recordings from 30 neurons in three rats over 3 days. Muzzio et al. (2009) followed 65 neurons in 14 mice for 4 days. McKenzie et al. (2014) tracked 38 neurons in four rats over 2 days.

This trend toward more cells and shorter recording durations is probably a reflection of increasingly sophisticated analyses of neural coding and the resulting need for datasets with greater numbers of cells, rising appreciation for the importance of repeatability within individual studies and reproducibility by other labs, and progressive improvements in the statistical rigor of the spike sorting procedures used to identify and track individual cells' spike waveforms. Crucially, studies based on chronic electrical recordings should include statistical tests confirming that the variations between individual cells' spike waveforms across consecutive recording days are smaller than the variations between the waveforms of neighboring cells recorded on the same day. Without this, the argument that electrical recordings suffer minimal drift over time is compromised. Overall, the technical challenges inherent to maintaining stable electrical recordings in behaving animals have limited the recording durations and numbers of neurons used for analyses of memory codes. This in turn has constrained the complexity of the questions about long-term memory that researchers have been able to address empirically. Hence, a major advantage of $Ca^{2+}$ imaging is the ability to follow hundreds of individual neurons in a single an-

imal over weeks (Ziv et al. 2013; Peters et al. 2014) yielding datasets that electrical recordings cannot match in statistical power for analyses of neural coding.

In addition to increasing the number of cells whose activity can be tracked in vivo, $Ca^{2+}$ imaging allows researchers to reliably track neurons with very low rates of activity. The ability to monitor such cells is important, because even neurons with very low activity rates can make substantial contributions to information coding if their spikes convey reliable signals. With extracellular electrical recording methods, a certain minimum number of spikes is required to isolate and extract the action potentials from an individual cell by spike sorting, because of the statistical variability of action potential waveforms. This makes it nearly impossible for researchers to track a neuron across 1 or more days of electrical recording in which the neuron fired no spikes. By comparison, in long-term $Ca^{2+}$-imaging studies one tracks individual cells across days by image registration (i.e., by aligning the fluorescence images across imaging sessions). Such image alignment can often be done to <1 μm scale accuracy and does not depend critically on cells' activity rates (Ziv et al. 2013). Hence, with $Ca^{2+}$ imaging one can track a neuron's dynamics even across long recording epochs in which it is silent. An electrophysiologist experienced with extracellular recordings might argue that this advantage of $Ca^{2+}$ imaging is mainly a theoretical one, because in practical experience neurons are rarely perfectly silent for extended durations. It is important to remember that any such appeal to experience is based on the specific subset of neurons that can be followed electrically and neglects the cells that are too silent for spike sorting.

Indeed, neuroscientists are increasingly realizing that extracellular electrical recordings may often provide a sampling of cells that is biased toward active cells (Lutcke et al. 2013). Studies using intracellular whole-cell patch electrical recordings in both head fixed (Margrie et al. 2002) and freely behaving (Lee et al. 2006) animals support this observation. To achieve whole cell patch electrode recordings the physiologist identifies a neuron from its membrane impedance, not its spiking activity (Margrie et al. 2002). Hence, like $Ca^{2+}$ imaging, these recordings can sample very quiet cells, and the mean neuronal activity rates obtained with the two techniques are broadly consistent. For example, whole cell recording and $Ca^{2+}$-imaging studies of the rodent motor cortex reported spiking and $Ca^{2+}$ transient rates of $\sim 0.01-1.8$ Hz (Lee et al. 2006) and $\sim 0.001-0.15$ Hz (Komiyama et al. 2010; Huber et al. 2012), respectively, one or more orders of magnitude lower than those observed by extracellular recording ($\sim 1-2$ Hz) (Laubach et al. 2000). In one $Ca^{2+}$-imaging study of motor learning the rates of motor cortical neuron activation were as low as $\sim 3.6-0.7$ spikes per hour (Huber et al. 2012). Neurons with activity rates this low would be highly challenging to isolate by extracellular electrical recording methods. The degree to which the different methods will yield activity rates out of accord with one another will clearly vary by brain area, cell type, and behavioral state. Regardless of the exact extent of such differences, the ability to follow quiet neurons is especially important for studies of long-term memory, because different subsets of cells may be active on different days (Ziv et al. 2013; Peters et al. 2014).

Recent time-lapse imaging studies of neural ensembles in hippocampus and motor cortex well illustrate $Ca^{2+}$ imaging's methodological advantages. To study the long-term dynamics of CA1 ensemble place codes, Ziv et al. (2013) performed $Ca^{2+}$ imaging in mice exploring a familiar environment and tracked 515–1040 individual CA1 hippocampal pyramidal cells across 45 days in each of four mice (Fig. 5). It was important to verify that when monitored by $Ca^{2+}$ imaging, CA1 neurons displayed coding attributes normally expected of place cells. As anticipated, many neurons exhibited $Ca^{2+}$ activity when the mouse explored a specific portion of its arena. Many of the optically determined place fields remapped when the mouse was transferred to a different arena placed at the same location in the laboratory room, as reported previously (Leutgeb et al. 2005). When the mouse explored a linear track, the set of place fields fully covered the track, and many neurons

had statistically significant place fields when the mouse was heading in only one of the two possible running directions, matching past observations (McHugh et al. 1996). About 20% of the cells seen in individual imaging sessions had significant place fields for one or both running directions, and the sizes of the place fields determined by $Ca^{2+}$ imaging were consistent with prior reports (Dombeck et al. 2010; McHugh et al. 1996; Rotenberg et al. 1996; Nakazawa et al. 2003). These observations lent confidence that the pyramidal cells' $Ca^{2+}$ activation patterns accurately conveyed their place coding properties. By tracking these coding properties as the mice explored the linear track during sessions spaced at 5-day intervals, Ziv et al. (2013) attained some unexpected findings regarding the long-term dynamics of the CA1 representation of a familiar environment.

Notably, there was substantial turnover from session to session in the set of neurons that displayed $Ca^{2+}$ activity (Fig. 5A,B). Only $31 \pm 1\%$ (mean $\pm$ s.d.) of all neurons seen across the entire experiment were active in any one session, and this percentage was constant over the entire study (Fig. 5B, inset). Across sessions, individual cells came in and out of this active subset in an apparently random manner (Fig. 5A,B), and the overlap in the active subsets from different days declined moderately from $\sim 60\%$ for sessions 5 days apart to $\sim 40\%$ for 30 days apart (Fig. 5C). Among the cells that had statistically significant place fields in one or more sessions, the overlap in these coding subsets from different days was $\sim 25\%$ for sessions 5 days apart and $\sim 15\%$ for sessions 30 days apart (Fig. 5C). Thus, even in a familiar environment there is substantial dynamism in the CA1 ensemble representation of space. Strikingly, the odds of a cell's recurrence in either the active or coding ensembles were uncorrelated with the rate and amplitude of its $Ca^{2+}$ activity and with the stability of its place field within individual sessions. However, when individual cells did show place fields in more than one session, the place fields' locations on the linear track were nearly always unchanged (Fig. 5D). Thus, individual cells' place fields were spatially invariant but temporally stochastic in their day-to-day appearances.

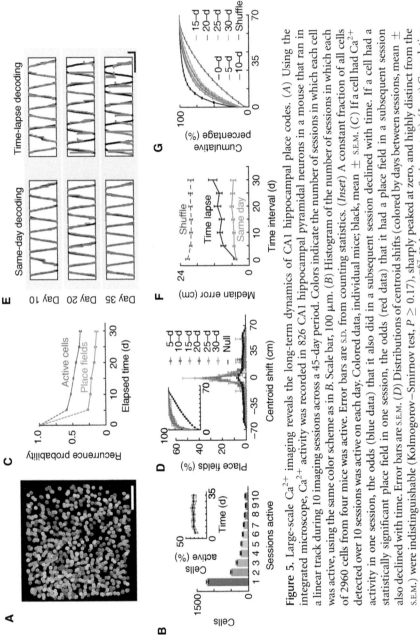

**Figure 5.** Large-scale Ca$^{2+}$ imaging reveals the long-term dynamics of CA1 hippocampal place codes. (*A*) Using the integrated microscope, Ca$^{2+}$ activity was recorded in 826 CA1 hippocampal pyramidal neurons in a mouse that ran in a linear track during 10 imaging sessions across a 45-day period. Colors indicate the number of sessions in which each cell was active, using the same color scheme as in *B*. Scale bar, 100 μm. (*B*) Histogram of the number of sessions in which each of 2960 cells from four mice was active. Error bars are S.D. from counting statistics. (*Inset*) A constant fraction of all cells detected over 10 sessions was active on each day. Colored data, individual mice; black, mean ± S.E.M. (*C*) If a cell had Ca$^{2+}$ activity in one session, the odds (blue data) that it also did in a subsequent session declined with time. If a cell had a statistically significant place field in one session, the odds (red data) that it had a place field in a subsequent session also declined with time. Error bars are S.E.M. (*D*) Distributions of centroid shifts (colored by days between sessions, mean ± S.E.M.) were indistinguishable (Kolmogorov–Smirnov test, $P \geq 0.17$), sharply peaked at zero, and highly distinct from the null hypothesis that place fields would randomly relocate ($P = 4 \times 10^{-67}$, Kolmogorov–Smirnov test). (*Inset*) Cumulative histograms of shift magnitudes; 74%–83% were ≤ 7 cm. Median shift (3.5 cm) was much less than the median place field width (24 cm). (*E*) A Bayesian decoding algorithm accurately estimated the mouse's trajectory in the linear track based on the Ca$^{2+}$-imaging data from area CA1 and retained its fidelity across 30 days of experimentation. Reconstructions of the mouse's trajectory (colored curves) and its actual position (black curves); three paired reconstructions comparing time-lapse decoders trained on data from day 5 (*right column*), using all cells with place fields on both days of each pair, and decoders trained on data from the same day as the test trial (*left column*). (*Legend continues on following page.*)

Is ~15%–25% overlap in the set of cells with significant place fields sufficient to maintain a stable representation of space over weeks? To address this question, Ziv et al. (2013) used computational decoding methods to analyze the spatial information content of the ensemble activity patterns. With datasets containing 515–1040 cells per mouse, it was possible to reconstruct each animal's locomotor trajectory using the representations of space encoded within the ensemble neural dynamics. Despite individual cells' fluctuating contributions to place coding across different imaging sessions, the ensemble representations of space maintained their coding fidelity for at least a month (Fig. 5D–G) (Ziv et al. 2013). In essence, this fidelity at the ensemble level arose from the spatial invariance of the individual cells' place fields when they recurred.

Overall, the results of Ziv et al. (2013) suggest quite a different picture of hippocampal spatial coding than had emerged from electrophysiological studies. Certainly, the $Ca^{2+}$ imaging and electrophysiological data are mutually consistent, in that both reveal individual CA1 neurons that stably express place fields over the long term. However, through its capabilities for tracking neurons across sessions in which the neurons are silent, $Ca^{2+}$ imaging reveals that such long-term stability is not the norm. For each location in a familiar environment, it seems there are more than enough cells with corresponding place fields to encode the mouse's position; only a portion of these cells are active in any one session, but this portion is sufficient for stable spatial coding. What is the functional role of the ~75%–85% of coding cells that do not overlap between any two sessions?

One possibility is that this ~75%–85% of cells endows each episode with a unique neural

signature, potentially allowing the hippocampus to retain distinct memories of the different episodes while preserving information about the shared environment in the other ~15%–25%. Although speculative, this proposition about episodic memory storage should be experimentally testable, by using long-term $Ca^{2+}$ imaging, decoding analyses, and well-controlled behavioral assays to dissect different cells' contributions to memories' spatial and episodic components. By combining $Ca^{2+}$ imaging with molecular manipulations, or with other optical approaches such as fluorescence tagging (Reijmers et al. 2007; Liu et al. 2012; Guenthner et al. 2013; Kawashima et al. 2013; Ramirez et al. 2013; Redondo et al. 2014) and long-term imaging of CA1 dendritic spines (Attardo et al. 2015), it should also be possible to study the mechanisms underlying the spatial stability of the place fields and the temporal variability of their expression.

Another illustration of the substantial statistical analyses that are feasible using large-scale $Ca^{2+}$-imaging data comes from a study of the mouse motor cortex that tracked ~200 neurons in each of 10 mice as the animals learned a lever-press task over 14 days (Fig. 6A–G) (Peters et al. 2014). The data revealed substantial reorganization and refinement of the ensemble neural dynamics over the course of motor learning. During the first 3 days, the number of neurons with movement-related activity patterns approximately doubled, from ~10% to ~20%; across subsequent days, the number of such cells gradually returned to near its initial value (Fig. 6A). Throughout training, the timing of neural responses relative to movement onset gradually became less variable (Fig. 6B) and more correlated across movement trials (Fig. 6C–E).

Moreover, as the mice became expert and learned a stereotyped movement, the neural en-

---

**Figure 5.** (*Continued*) Each pair used an equal number of cells, optimally chosen at *left* to minimize errors. Scale bars, 2 sec (horizontal) and 10 cm (vertical). (*F*) Median errors in estimating the mouse's position were ~7–13 cm, even for decoders trained on data from 30 days prior (black, mean ± s.e.m.). (Red) Decoders trained on data from the same day as test data, using equal numbers of cells as black points and optimally chosen to minimize errors. (Gray) Errors using shuffled traces of $Ca^{2+}$ activity from the same day as training data (averaged over 10,000 shuffles). (*G*) Cumulative distributions of decoding error magnitudes (mean ± s.e.m.) for test and training data separated by the indicated times or (gray) for decoders tested on shuffled data (all panels are from Ziv et al. 2013).

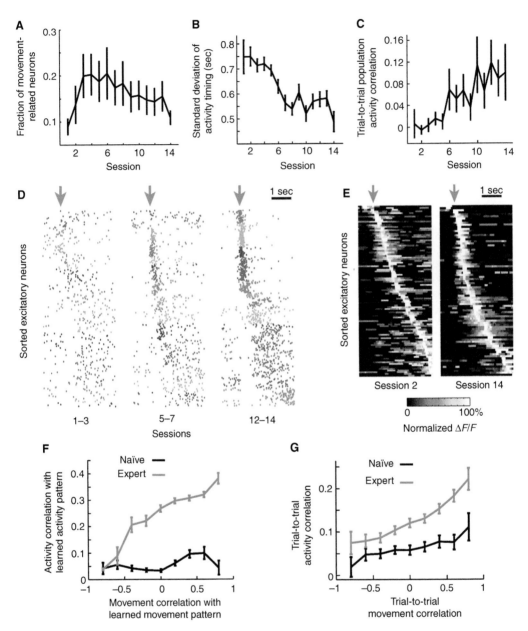

**Figure 6.** Large-scale Ca$^{2+}$ imaging reveals the reorganization of ensemble neural dynamics in mouse layer 2/3 motor cortical neurons across a 2-week regimen of daily motor learning. (*A*) Fraction of excitatory neurons classified as movement related in each of the 14 training sessions, during which mice learned a lever-press task. (*B*) Standard deviation of the timing of activity onsets for movement-related excitatory neurons for neurons that were active in five or more sessions. The trial-by-trial variability in the onset of single cell activity declined as learning progressed. (*C*) Ensemble neural activity became more stereotyped as learning advanced, as shown by the increase in the pairwise trial-to-trial correlation of temporal population activity vectors. The temporal population activity vector was a concatenation of the activity traces of all movement-related neurons and thus maintained temporal information within each movement. (*D*) Neuronal activity increased in temporal structure over the course of motor learning. Raster plots of neuronal Ca$^{2+}$ activity for movement-related neurons that were active in at least 10% of trials on the sessions indicated, shown for an example animal. Red arrows indicate movement onset. (*Legend continues on following page.*)

sembles gradually acquired stereotyped activity patterns, and the degree to which the ensemble dynamics on individual trials resembled these stereotyped patterns was predictive of the degree to which the animal's movement resembled its expert, stereotyped form (Fig. 6F,G). However, even in mice that were expert at performing stereotyped movements, there remained substantial trial-to-trial variability in neurons' responses (Fig. 6F,G). Like the variability in CA1 place cells' dynamics that left ensemble spatial coding fidelity largely intact, this finding suggests that stable retention of information at the level of neural populations does not require purely stable neuronal firing properties. On the contrary, neural ensembles may be far more robust in their information-processing capabilities than can be inferred from the activity patterns of individual cells.

Notably, though $Ca^{2+}$ imaging affords access to the topographic organization of memory at a cellular scale, the recent imaging studies of both hippocampus and motor cortex did not find any clear anatomical patterns by which cells with similar coding properties are organized (Dombeck et al. 2010; Ziv et al. 2013; Peters et al. 2014). Instead, the arrangements appeared to be random. Similarly, a recent study of piriform cortex yielded evidence that cells involved in storing learned odor memories are also randomly organized (Choi et al. 2011). By comparison, a recent $Ca^{2+}$-imaging study of medial entorhinal cortex found a clear micro-organization in the anatomical arrangements of grid

cells (Heys et al. 2014). In sensory and motor brain areas there are well-known topographic maps that relate neurons' anatomic locations and coding properties to specific features of the sensory world or the animal's body plan (Albright et al. 2000). Particularly in cognitive but also in motor areas, the manner in which information processing is organized at the cellular scale remains far less explored. Plainly, large-scale in vivo $Ca^{2+}$ imaging provides a potent new tool to do so, for multiple forms of information processing (Dombeck et al. 2010; Ziv et al. 2013; Heys et al. 2014; Peters et al. 2014).

Overall, $Ca^{2+}$ imaging permits memory researchers to probe aspects of neural coding that have been inaccessible from electrical recordings. The sheer number of neurons that neuroscientists can now reliably image over weeks enables wholly new analyses of how ensemble codes form, evolve, and are refined over the course of memory encoding, consolidation, maintenance, and recall. As individual neurons enter, exit, or return to the ensemble codes involved in memory processing, neuroscientists can track these cellular transitions, even over extended time periods during which some of the individual cells are silent. By using $Ca^{2+}$ imaging, memory researchers can also examine how cells' anatomical arrangements relate to their memory processing roles. Each of these new capacities will likely provide important insights regarding the brain's rules and mechanisms for handling memory data.

---

**Figure 6.** (*Continued*) Colors distinguish individual neurons that have been sorted according to their preferred timing. Note that the same color across different sessions does not necessarily mark the same neurons. (*E*) Maximum-normalized average activity from all movement-related neurons from all animals in session 2 (106 neurons) and session 14 (84 neurons) aligned to movement onset (red arrow). The timing of neuronal activity was more refined in session 14, as evidenced by the narrower peaks and lower levels of background activity, and the timing shifts toward movement onset. (*F*) In expert but not naïve mice, the extent to which ensemble neural activity in individual trials correlated with the average learned neural activity pattern from expert sessions was predictive of the extent to which the animal's movement pattern correlated with the learned movement pattern. In sessions during which the mice were naïve, when the animals made movements that resembled the learned movement patterns acquired later in training, the underlying neural activity patterns were nevertheless very different from the learned neural activity pattern late in training. (*G*) Pairwise trial-to-trial correlation of temporal population neural activity vectors (as in *C*) plotted as a function of movement correlation on those trials. A stronger relationship between ensemble neural activity and movement emerges during learning. All error bars are s.e.m. (all panels are modified from Peters et al. 2014).

## PROSPECTS FOR STUDYING THE ORGANIZATION OF MEMORY USING Ca$^{2+}$ IMAGING

Memories are often linked to one another, such as through associations and shared attributes, and organized into structures such as categories, sequences, and hierarchies. By providing the means to examine the ensemble neural codes supporting two or more different memories, in vivo Ca$^{2+}$ imaging offers intriguing prospects for directly visualizing the biological instantiation of memory organization. The large numbers of neurons that can be monitored currently by Ca$^{2+}$ imaging will likely be crucial for this pursuit, because substantial statistical power will be necessary to determine the extent to which related memories are supported by overlapping neural representations, and whether these representations use local or distributed coding schemes. Moreover, the brain's memory representations are generally dynamic and involve multiple brain regions to varying extents at different stages of memory processing. As a memory becomes more remote in time, the reshaping of its representation across different brain areas is known as systems memory consolidation (Dudai 2004; Frankland and Bontempi 2005). Time-lapse Ca$^{2+}$-imaging studies may provide a window onto these events at the cellular scale in live animals.

A large body of research in psychology and cognitive neuroscience has focused on the extent to which memory representations are local or distributed, and how such representations encode the relationships between different memories. There is a spectrum of theoretical possibilities (Quian Quiroga and Kreiman 2010). In theories at the "localist" end of the spectrum (Page 2000; Bowers 2009), individual memories reside within individual cells; cell identities provide individual memories their unique codes or signatures and synaptic connections link different memories together. By comparison, in theories involving fully distributed representations, individual cells are generally involved in the encoding of multiple memories (Rumelhart et al. 1986; Masson 1987; Plaut and McClelland 2010); in these theories,

it is the distributed patterns of neural activity that provide unique signatures to individual memories, hence, memories can only be deciphered from the population-level activity patterns and not the dynamics of individual neurons. Clearly, there are many intermediate possibilities as well, melding these aspects of representation for different facets of memory and across different brain structures. Notably, the vast majority of such ideas from cognitive science have not yet been tested empirically through biological studies of neural dynamics. In vivo Ca$^{2+}$-imaging studies might change this situation, by enabling direct observations of the extent to which cell identities or activity patterns provide memories their signatures—at least in ensembles of cells that can be imaged simultaneously.

In addition to the newfound experimental possibilities for examining whether memory representations are local or distributed, Ca$^{2+}$ imaging also offers new means to examine the manner in which the brain encodes associations, which can link different parts of an individual memory and relate different memories to each other. Consider an example Ca$^{2+}$-imaging study of second-order associative conditioning. Presently, there are ongoing Ca$^{2+}$-imaging studies of first-order associative learning in brain structures such as the amygdala and nucleus accumbens (Grewe et al. 2014; Parker et al. 2014), and it is reasonable to envision these experiments could be suitably extended. In second-order conditioning (Gewirtz and Davis 2000; Debiec et al. 2006), an animal first experiences paired presentations of an initially neutral conditioned stimulus (CS$_1$) and an unconditioned stimulus (US) that evokes an unconditioned behavioral response. With learning, the animal gains a conditioned response to CS$_1$. A second conditioned stimulus (CS$_2$) is then paired with CS$_1$; afterward, a presentation of CS$_2$ evokes the conditioned response despite that CS$_2$ was never directly paired with the US. How is the relationship between the US and CS$_2$ encoded?

One proposal is that the individual representations of CS$_1$ and CS$_2$ involve overlapping populations of neurons (Debiec et al. 2006). However, any such overlap has not been directly

Cite this article as *Cold Spring Harb Perspect Biol* doi: 10.1101/cshperspect.a021824

observed, in part because of the long-standing inability to visualize ensemble neural activity throughout learning and memory formation. By using in vivo $Ca^{2+}$ imaging in a brain area that has been implicated in forming the associations for a particular form of learning, it may be feasible to directly examine the neural codes involved. Specifically, it should be possible to visualize the extent to which the cellular representations of $CS_2$, $CS_1$, and the US overlap, as determined from the subsets of neurons activated by each stimulus. Of key interest is the manner in which the representation of the association between $CS_2$ and the US depends on the prior association between $CS_1$ and the US. Will the associative relationships be encoded by overlapping cellular memberships in the representations (Debiec et al. 2006), distinctive signatures in the neural ensemble activity (Buzsaki 2005; Buzsaki and Moser 2013), combinations thereof, or means that are not observable by $Ca^{2+}$ imaging? More generally, a second-order association is only one type of relationship between memories, and $Ca^{2+}$ imaging might help elucidate the organization of neural codes and their interrelationships for multiple other forms of learning. For example, categorical memories may arise from more elementary ones, and future $Ca^{2+}$-imaging studies might allow sophisticated analyses of this process.

A related set of issues concerns the temporal relationships between different memories and the manner in which new information is encoded in relation to prior memories. The biological mechanisms that govern the encoding of temporal relationships are only starting to be uncovered. Notably, neurons expressing virally induced, elevated levels of the cAMP-response element-binding protein (CREB) appear to have increased electrical excitability and may be primed for upcoming bouts of memory storage (Han et al. 2007, 2009; Zhou et al. 2009; Yiu et al. 2014). Further, neurons that have recently encoded a memory, as assessed by CREB phosphorylation, maintain their elevated levels of phosphorylated CREB and presumed heightened excitability for several hours following learning; this persistence of CREB activation may bias the storage of subsequent memories

toward the same set of excitable neurons encoding the first memory (Silva et al. 2009). Thus, two learning episodes that are temporally separated, but within the time window of CREB activation, may be encoded in overlapping populations of neurons, thereby forming an association between the two memories in the brain (Rogerson et al. 2014). Similarly, behavioral (Ballarini et al. 2009) and cell biological (Frey and Morris 1997) phenomena in which one set of events impacts the encoding of other temporally separated events indicate that memories are not encoded in isolation, but rather in relation to prior and subsequent events (Moncada and Viola 2007). To investigate issues of this kind, multiple studies to date have examined memory representations via patterns of IEG activation (Guzowski et al. 1999; Frankland and Bontempi 2005; Reijmers et al. 2007; Goshen et al. 2011; Tse et al. 2011; Czajkowski et al. 2014), but $Ca^{2+}$-imaging datasets might offer richer possibilities for analyses of the spatiotemporal features of ensemble neural coding.

Another important aspect of memory organization that may be ripe for study by $Ca^{2+}$ imaging is systems memory consolidation (Dudai 2004; Frankland and Bontempi 2005). Substantial literature suggests that as a stored memory becomes more remote, neocortical memory systems may gradually extract and reorganize memory information from an initially acting memory system such as hippocampus (McClelland et al. 1995), eventually obviating the necessity of hippocampus for memory recall (Scoville and Milner 1957; Kim and Fanselow 1992; Squire and Alvarez 1995; Wiltgen et al. 2004; Bontempi and Durkin 2007; Tse et al. 2007, 2011). These rearrangements of memories across brain areas might allow the neocortex to organize information in a more abstract form (Fuster 2009), such as by using schemas that establish higher-order associations and categories of information or knowledge (Piaget et al. 1929; Bartlett 1932; van Kesteren et al. 2012; McKenzie et al. 2013; Preston and Eichenbaum 2013).

However, because of the extended time course of systems consolidation, from weeks in rodents (Kim and Fanselow 1992) to years in humans (Scoville and Milner 1957; Squire and Alvarez 1995), the relevant interactions between neural ensembles in hippocampus and neocortex remain unknown. By using time-lapse $Ca^{2+}$-imaging ability to track neuronal ensembles over extended durations, researchers might conceivably be able to watch how large sets of individual neurons in different brain regions exchange information. Optical methods for imaging multiple brain areas concurrently (Lecoq et al. 2014), including hippocampus and neocortex, may be especially useful.

Recent studies in rodents have also challenged the long-held notions that the engagement of neocortex in systems consolidation is always slow (Tse et al. 2007; Tse et al. 2011), and that the necessity of hippocampus for memory recall always diminishes over the course of long-term memory storage (Wiltgen et al. 2010; Goshen et al. 2011). By using IEG expression as a readout of cellular involvement in memory formation, studies in rats have shown that once a schema is established, the neocortex can be directly activated by new information that fits in the schema, in a necessary manner for learning (Tse et al. 2007; Tse et al. 2011). Recent studies in mice indicate a sustained role for hippocampus in the activation of neocortical circuits (Goshen et al. 2011) and in the recall of detailed memories (Wiltgen et al. 2010) at 2–4 weeks after memory storage. Overall, the anatomical and time-dependent reorganization of memory representations following their initial formation deserves intensive further study; $Ca^{2+}$ imaging may allow a much richer set of observations regarding how schemas are established and impact the subsequent encoding of new information and the engagement of multiple brain areas. Other brain structures beyond hippocampus and neocortex also seem to undergo processes similar to systems consolidation (Repa et al. 2001; Do-Monte et al. 2015), indicating the general need for methods capable of revealing how the brain reorganizes memories over time.

## OUTLOOK

The advent of powerful technologies for large-scale $Ca^{2+}$ imaging in behaving mammals has provided major new opportunities to study the ensemble neural dynamics that underlie learning and memory. We expect active advancements on multiple research fronts that will continue to improve the potency of $Ca^{2+}$ imaging for such studies. Areas of technical progress will almost certainly include advances in new $Ca^{2+}$ indicators (Fosque et al. 2015; Inoue et al. 2015), viral and genetic labeling methods (Oyibo et al. 2014; Madisen et al. 2015), optical instrumentation for imaging in one or more brain areas (Lecoq et al. 2014; Stirman et al. 2014), the combination of $Ca^{2+}$ imaging and optogenetics (Grosenick et al. 2015), behavioral assays expressly designed or optimized for $Ca^{2+}$-imaging studies (Lovett-Barron et al. 2014; Sofroniew et al. 2014), and computational analyses of large-scale neural-imaging data (Ziv et al. 2013; Peters et al. 2014). By combining such technologies, memory researchers will have unprecedented capabilities to decipher the ensemble neural codes that support long-term memory across a range of animal behaviors and brain areas, and to dissect the rules that govern memory storage. To bring the suite of available techniques to full fruition, collaborations between scientists with diverse expertise will become ever more crucial, as cutting-edge experiments increasingly require multidisciplinary knowledge in fields such as genetics, virology, cognitive and systems neuroscience, optics, and computer science. However, if the research community can effectively harness its collective know-how, we stand to gain important insights into long-standing questions about the organization of memory in the mammalian brain.

## ACKNOWLEDGMENTS

We thank Alessio Attardo, Benjamin Grewe, Cristina Irimia, Margaret C. Larkin, Lacey J. Kitch, and Y. Ziv for helpful suggestions on the manuscript. M.J.S. acknowledges research support from NIBIB, NIDA, NINDS, NIMH, NSF, DARPA, and the Ellison Foundation. M.J.S. is a co-founder and consults scientifically for Inscopix, which has commercialized the miniature integrated microscope technology of Figure 3. P.J. is the recipient of a Marie Curie Fellowship.

## REFERENCES

Aghajan ZM, Acharya L, Moore JJ, Cushman JD, Vuong C, Mehta MR. 2015. "Impaired spatial selectivity and intact phase precession in two-dimensional virtual reality." *Nat Neurosci* **18**: 121–128.

Albright TD, Jessell TM, Kandel ER, Posner MI. 2000. "Neural science: A century of progress and the mysteries that remain. *Neuron* **25**: S1–S55.

Alivisatos AP, Andrews AM, Boyden ES, Chun M, Church GM, Deisseroth K, Donoghue JP, Fraser SE, Lippincott-Schwartz J, Looger LL, et al. 2013. Nanotools for neuroscience and brain activity mapping. *ACS Nano* **7**: 1850–1866.

Andermann ML, Kerlin AM, Reid RC. 2010. Chronic cellular imaging of mouse visual cortex during operant behavior and passive viewing. *Front Cell Neurosci* **4**: 3.

Andermann ML, Kerlin AM, Roumis DK, Glickfeld LL, Reid RC. 2011. Functional specialization of mouse higher visual cortical areas. *Neuron* **72**: 1025–1039.

Aronov D, Tank DW. 2014. Engagement of neural circuits underlying 2D spatial navigation in a rodent virtual reality system. *Neuron* **84**: 442–456.

Aschauer DF, Kreuz S, Rumpel S. 2013. Analysis of transduction efficiency, tropism and axonal transport of AAV serotypes 1, 2, 5, 6, 8 and 9 in the mouse brain. *PLoS ONE* **8**: e76310.

Attardo A, Fitzgerald JE, Schnitzer MJ. 2015. Impermanence of dendritic spines in live adult CA1 hippocampus. *Nature* **523**: 592–596.

Ballarini F, Moncada D, Martinez MC, Alen N, Viola H. 2009. Behavioral tagging is a general mechanism of long-term memory formation. *Proc Natl Acad Sci USA* **106**: 14599–14604.

Barretto RP, Ko TH, Jung JC, Wang TJ, Capps G, Waters AC, Ziv Y, Attardo A, Recht L, Schnitzer MJ. 2011. Time-lapse imaging of disease progression in deep brain areas using fluorescence microendoscopy. *Nat Med* **17**: 223–228.

Bartlett FC. 1932. *Remembering: A study in experimental and social psychology.* Cambridge University Press, Cambridge.

Berdyyeva T, Otte S, Aluisio L, Ziv Y, Burns LD, Dugovic C, Yun S, Ghosh KK, Schnitzer MJ, Lovenberg T, et al. 2014. Zolpidem reduces hippocampal neuronal activity in freely behaving mice: A large scale calcium imaging study with miniaturized fluorescence microscope. *PLoS ONE* **9**: e112068.

Betley JN, Xu S, Cao ZF, Gong R, Magnus CJ, Yu Y, Sternson SM. 2015. Neurons for hunger and thirst transmit a negative-valence teaching signal. *Nature* **521**: 180–185.

Blauvelt DG, Sato TF, Wienisch M, Knopfel T, Murthy VN. 2013. Distinct spatiotemporal activity in principal neurons of the mouse olfactory bulb in anesthetized and awake states. *Front Neural Circuits* **7**: 46.

Bock DD, Lee WC, Kerlin AM, Andermann ML, Hood G, Wetzel AW, Yurgenson S, Soucy ER, Kim HS, Reid RC.

2011. Network anatomy and in vivo physiology of visual cortical neurons. *Nature* **471**: 177–182.

Boender AJ, de Jong JW, Boekhoudt L, Luijendijk MC, Plasse Gvander, Adan RA. 2014. Combined use of the canine adenovirus-2 and DREADD-technology to activate specific neural pathways in vivo. *PLoS ONE* **9**: e95392.

Bontempi B, Durkin TP. 2007. Memories: Molecules and circuits. In *Dynamics of hippocampal–cortical interactions during memory consolidation: Insights from functional brain imaging*, pp. 19–39. Springer, New York.

Bouchard MB, Voleti V, Mendes CS, Lacefield C, Grueber WB, Mann RS, Bruno RM, Hillman EM. 2015. Swept confocally-aligned planar excitation (SCAPE) microscopy for high speed volumetric imaging of behaving organisms. *Nat Photonics* **9**: 113–119.

Bowers JS. 2009. On the biological plausibility of grandmother cells: Implications for neural network theories in psychology and neuroscience. *Psychol Rev* **116**: 220–251.

Bru T, Salinas S, Kremer EJ. 2010. An update on canine adenovirus type 2 and its vectors. *Viruses* **2**: 2134–2153.

Buzsaki G. 2005. Theta rhythm of navigation: Link between path integration and landmark navigation, episodic and semantic memory. *Hippocampus* **15**: 827–840.

Buzsaki G, Moser EI. 2013. Memory, navigation and θ rhythm in the hippocampal–entorhinal system. *Nat Neurosci* **16**: 130–138.

Card JP, Enquist LW. 2014. Transneuronal circuit analysis with pseudorabies viruses. *Curr Protoc Neurosci* doi: 10.1002/0471142301.ns0105s68.

Card JP, Kobiler O, McCambridge J, Ebdlahad S, Shan Z, Raizada MK, Sved AF, Enquist LW. 2011. Microdissection of neural networks by conditional reporter expression from a Brainbow herpesvirus. *Proc Natl Acad Sci USA* **108**: 3377–3382.

Carey RM, Verhagen JV, Wesson DW, Pírez N, Wachowiak M. 2009. Temporal structure of receptor neuron input to the olfactory bulb imaged in behaving rats. *J Neurophysiol* **101**: 1073–1088.

Chen BE, Trachtenberg JT, Holtmaat AJ, Svoboda K. 2008. Long-term, high-resolution imaging in the neocortex in vivo. *CSH Protoc* **2008**: pdb prot4902.

Chen JL, Villa KL, Cha JW, So PT, Kubota Y, Nedivi E. 2012a. Clustered dynamics of inhibitory synapses and dendritic spines in the adult neocortex. *Neuron* **74**: 361–373.

Chen Q, Cichon J, Wang W, Qiu L, Lee SJ, Campbell NR, Destefino N, Goard MJ, Fu Z, Yasuda R, et al. 2012b. Imaging neural activity using Thy1-GCaMP transgenic mice. *Neuron* **76**: 297–308.

Chen TW, Wardill TJ, Sun Y, Pulver SR, Renninger SL, Baohan A, Schreiter ER, Kerr RA, Orger MB, Jayaraman V, et al. 2013. Ultrasensitive fluorescent proteins for imaging neuronal activity. *Nature* **499**: 295–300.

Choi GB, Stettler DD, Kallman BR, Bhaskar ST, Fleischmann A, Axel R. 2011. Driving opposing behaviors with ensembles of piriform neurons. *Cell* **146**: 1004–1015.

Chung K, Wallace J, Kim SY, Kalyanasundaram S, Andalman AS, Davidson TJ, Mirzabekov JJ, Zalocusky KA, Mattis J, Denisin AK, et al. 2013. Structural and molecular interrogation of intact biological systems. *Nature* **497:** 332–337.

Ciocchi S, Herry C, Grenier F, Wolff SB, Letzkus JJ, Vlachos I, Ehrlich I, Sprengel R, Deisseroth K, Stadler MB, et al. 2010. Encoding of conditioned fear in central amygdala inhibitory circuits. *Nature* **468:** 277–282.

Cowansage KK, Shuman T, Dillingham BC, Chang A, Golshani P, Mayford M. 2014. Direct reactivation of a coherent neocortical memory of context. *Neuron* **84:** 432–441.

Cruz-Martin A, Crespo M, Portera-Cailliau C. 2010. Delayed stabilization of dendritic spines in fragile X mice. *J Neurosci* **30:** 7793–7803.

Czajkowski R, Jayaprakash B, Wiltgen B, Rogerson T, Guzman-Karlsson MC, Barth AL, Trachtenberg JT, Silva AJ. 2014. Encoding and storage of spatial information in the retrosplenial cortex. *Proc Natl Acad Sci USA* **111:** 8661–8666.

Dana H, Chen TW, Hu A, Shields BC, Guo C, Looger LL, Kim DS, Svoboda K. 2014. Thy1-GCaMP6 transgenic mice for neuronal population imaging in vivo. *PLoS ONE* **9:** e108697.

Date CJ. 2000. *The database relational model: A retrospective review and analysis.* Pearson, London.

Debiec J, Doyere V, Nader K, Ledoux JE. 2006. Directly reactivated, but not indirectly reactivated, memories undergo reconsolidation in the amygdala. *Proc Natl Acad Sci USA* **103:** 3428–3433.

Deisseroth K, Schnitzer MJ. 2013. Engineering approaches to illuminating brain structure and dynamics. *Neuron* **80:** 568–577.

Dombeck DA, Harvey CD, Tian L, Looger LL, Tank DW. 2010. Functional imaging of hippocampal place cells at cellular resolution during virtual navigation. *Nat Neurosci* **13:** 1433–1440.

Dombeck DA, Khabbaz AN, Collman F, Adelman TL, Tank DW. 2007. Imaging large-scale neural activity with cellular resolution in awake, mobile mice. *Neuron* **56:** 43–57.

Do-Monte FH, Quinones-Laracuente K, Quirk GJ. 2015. A temporal shift in the circuits mediating retrieval of fear memory. *Nature* **519:** 460–463.

Dudai Y. 2004. The neurobiology of consolidations, or, how stable is the engram? *Annu Rev Psychol* **55:** 51–86.

Eichenbaum H. 2000. A cortical-hippocampal system for declarative memory. *Nat Rev Neurosci* **1:** 41–50.

El-Boustani S, Sur M. 2014. Response-dependent dynamics of cell-specific inhibition in cortical networks in vivo. *Nat Commun* **5:** 5689.

Enquist LW, Tomishima MJ, Gross S, Smith GA. 2002. Directional spread of an α-herpesvirus in the nervous system. *Vet Microbiol* **86:** 5–16.

Etessami R, Conzelmann KK, Fadai-Ghotbi B, Natelson B, Tsiang H, Ceccaldi PE. 2000. Spread and pathogenic characteristics of a G-deficient rabies virus recombinant: An in vitro and in vivo study. *J Gen Virol* **81:** 2147–2153.

Flusberg BA, Nimmerjahn A, Cocker ED, Mukamel EA, Barretto RP, Ko TH, Burns LD, Jung JC, Schnitzer MJ. 2008. High-speed, miniaturized fluorescence microscopy in freely moving mice. *Nat Methods* **5:** 935–938.

Fosque BF, Sun Y, Dana H, Yang CT, Ohyama T, Tadross MR, Patel R, Zlatic M, Kim DS, Ahrens MB, et al. 2015. Neural

circuits. Labeling of active neural circuits in vivo with designed calcium integrators. *Science* **347:** 755–760.

Frankland PW, Bontempi B. 2005. The organization of recent and remote memories. *Nat Rev Neurosci* **6:** 119–130.

Frey U, Morris RG. 1997. Synaptic tagging and long-term potentiation. *Nature* **385:** 533–536.

Fu M, Yu X, Lu J, Zuo Y. 2012. Repetitive motor learning induces coordinated formation of clustered dendritic spines in vivo. *Nature* **483:** 92–95.

Fuster JM. 2009. Cortex and memory: Emergence of a new paradigm. *J Cogn Neurosci* **21:** 2047–2072.

Garaschuk O, Milos RI, Konnerth A. 2006. Targeted bulk-loading of fluorescent indicators for two-photon brain imaging in vivo. *Nat Protoc* **1:** 380–386.

Garner AR, Rowland DC, Hwang SY, Baumgaertel K, Roth BL, Kentros C, Mayford M. 2012. Generation of a synthetic memory trace. *Science* **335:** 1513–1516.

Gewirtz JC, Davis M. 2000. Using pavlovian higher-order conditioning paradigms to investigate the neural substrates of emotional learning and memory. *Learn Mem* **7:** 257–266.

Ghosh KK, Burns LD, Cocker ED, Nimmerjahn A, Ziv Y, Gamal AE, Schnitzer MJ. 2011. Miniaturized integration of a fluorescence microscope. *Nat Methods* **8:** 871–878.

Glickfeld LL, Andermann ML, Bonin V, Reid RC. 2013. Cortico–cortical projections in mouse visual cortex are functionally target specific. *Nat Neurosci* **16:** 219–226.

Goshen I, Brodsky M, Prakash R, Wallace J, Gradinaru V, Ramakrishnan C, Deisseroth K. 2011. Dynamics of retrieval strategies for remote memories. *Cell* **147:** 678–689.

Grewe BF, Lecoq J, Kitch L, Li J, Marshall J, Venkataraman G, Grundemann J, Luthi A, Schnitzer MJ. 2014. *Calcium imaging of stimulus specific neuronal responses in the lateral amygdala during fear learning.* Society for Neuroscience, Washington, DC.

Grienberger C, Konnerth A. 2012. Imaging calcium in neurons. *Neuron* **73:** 862–885.

Grosenick L, Marshel JH, Deisseroth K. 2015. Closed-loop and activity-guided optogenetic control. *Neuron* **86:** 106–139.

Grutzendler J, Kasthuri N, Gan WB. 2002. Long-term dendritic spine stability in the adult cortex. *Nature* **420:** 812–816.

Guenthner CJ, Miyamichi K, Yang HH, Heller HC, Luo L. 2013. Permanent genetic access to transiently active neurons via TRAP: Targeted recombination in active populations. *Neuron* **78:** 773–784.

Guzowski JF, McNaughton BL, Barnes CA, Worley PF. 1999. Environment-specific expression of the immediate-early gene Arc in hippocampal neuronal ensembles. *Nat Neurosci* **2:** 1120–1124.

Guzowski JF, Worley PF. 2001. Cellular compartment analysis of temporal activity by fluorescence in situ hybridization (catFISH). *Curr Protoc Neurosci* doi: 10.1002/0471142301.ns0108s15.

Hama H, Kurokawa H, Kawano H, Ando R, Shimogori T, Noda H, Fukami K, Sakaue-Sawano A, Miyawaki A. 2011. Scale: A chemical approach for fluorescence imaging and reconstruction of transparent mouse brain. *Nat Neurosci* **14:** 1481–1488.

Hamel EJ, Grewe BF, Parker JG, Schnitzer MJ. 2015. Cellular level brain imaging in behaving mammals: An engineering approach. *Neuron* **86:** 140–159.

Han JH, Kushner SA, Yiu AP, Cole CJ, Matynia A, Brown RA, Neve RL, Guzowski JF, Silva AJ, Josselyn SA. 2007. Neuronal competition and selection during memory formation. *Science* **316:** 457–460.

Han JH, Kushner SA, Yiu AP, Hsiang HL, Buch T, Waisman A, Bontempi B, Neve RL, Frankland PW, Josselyn SA. 2009. Selective erasure of a fear memory. *Science* **323:** 1492–1496.

Harvey CD, Collman F, Dombeck DA, Tank DW. 2009. Intracellular dynamics of hippocampal place cells during virtual navigation. *Nature* **461:** 941–946.

Harvey CD, Coen P, Tank DW. 2012. Choice-specific sequences in parietal cortex during a virtual-navigation decision task. *Nature* **484:** 62–68.

Helmchen F, Imoto K, Sakmann B. 1996. $Ca^{2+}$ buffering and action potential-evoked $Ca^{2+}$ signaling in dendrites of pyramidal neurons. *Biophys J* **70:** 1069–1081.

Helmchen F, Fee MS, Tank DW, Denk W. 2001. A miniature head-mounted two-photon microscope. High-resolution brain imaging in freely moving animals. *Neuron* **31:** 903–912.

Heys JG, Rangarajan KV, Dombeck DA. 2014. The functional micro-organization of grid cells revealed by cellular-resolution imaging. *Neuron* **84:** 1079–1090.

Hnasko TS, Perez FA, Scouras AD, Stoll EA, Gale SD, Luquet S, Phillips PE, Kremer EJ, Palmiter RD. 2006. Cre recombinase-mediated restoration of nigrostriatal dopamine in dopamine-deficient mice reverses hypophagia and bradykinesia. *Proc Natl Acad Sci USA* **103:** 8858–8863.

Holscher C, Schnee A, Dahmen H, Setia L, Mallot HA. 2005. Rats are able to navigate in virtual environments. *J Exp Biol* **208:** 561–569.

Holtmaat A, Bonhoeffer T, Chow DK, Chuckowree J, De Paola V, Hofer SB, Hubener M, Keck T, Knott G, Lee WC, et al. 2009. Long-term, high-resolution imaging in the mouse neocortex through a chronic cranial window. *Nat Protoc* **4:** 1128–1144.

Hoover JE, Strick PL. 1999. The organization of cerebellar and basal ganglia outputs to primary motor cortex as revealed by retrograde transneuronal transport of herpes simplex virus type 1. *J Neurosci* **19:** 1446–1463.

Horton NG, Wang K, Kobat D, Clark CG, Wise FW, Schaffer CB, Xu C. 2013. In vivo three-photon microscopy of subcortical structures within an intact mouse brain. *Nat Photonics* **7:** 205–209.

Huber D, Gutnisky DA, Peron S, O'Connor DH, Wiegert JS, Tian L, Oertner TG, Looger LL, Svoboda K. 2012. Multiple dynamic representations in the motor cortex during sensorimotor learning. *Nature* **484:** 473–478.

Inoue M, Takeuchi A, Horigane S, Ohkura M, Gengyo-Ando K, Fujii H, Kamijo S, Takemoto-Kimura S, Kano M, Nakai J, et al. 2015. Rational design of a high-affinity, fast, red calcium indicator R-CaMP2. *Nat Methods* **12:** 64–70.

Jennings JH, Ung RL, Resendez SL, Stamatakis AM, Taylor JG, Huang J, Veleta K, Kantak PA, Aita M, Shilling-Scrivo K, et al. 2015. Visualizing hypothalamic network dynam-

ics for appetitive and consummatory behaviors. *Cell* **160:** 516–527.

Johansen JP, Hamanaka H, Monfils MH, Behnia R, Deisseroth K, Blair HT, LeDoux JE. 2010. Optical activation of lateral amygdala pyramidal cells instructs associative fear learning. *Proc Natl Acad Sci USA* **107:** 12692–12697.

Jorgenson LA, Newsome WT, Anderson DJ, Bargmann CI, Brown EN, Deisseroth K, Donoghue JP, Hudson KL, Ling GS, MacLeish PR, et al. 2015. The BRAIN Initiative: Developing technology to catalyse neuroscience discovery. *Philos Trans R Soc Lond B Biol Sci* **370:** 20140164.

Jung JC, Schnitzer MJ. 2003. Multiphoton endoscopy. *Opt Lett* **28:** 902–904.

Jung JC, Mehta AD, Aksay E, Stepnoski R, Schnitzer MJ. 2004. In vivo mammalian brain imaging using one- and two-photon fluorescence microendoscopy. *J Neurophysiol* **92:** 3121–3133.

Kaifosh P, Lovett-Barron M, Turi GF, Reardon TR, Losonczy A. 2013. Septo-hippocampal GABAergic signaling across multiple modalities in awake mice. *Nat Neurosci* **16:** 1182–1184.

Kawashima T, Kitamura K, Suzuki K, Nonaka M, Kamijo S, Takemoto-Kimura S, Kano M, Okuno H, Ohki K, Bito H. 2013. Functional labeling of neurons and their projections using the synthetic activity-dependent promoter E-SARE. *Nat Methods* **10:** 889–895.

Ke MT, Fujimoto S, Imai T. 2013. SeeDB: A simple and morphology-preserving optical clearing agent for neuronal circuit reconstruction. *Nat Neurosci* **16:** 1154–1161.

Keller GB, Bonhoeffer T, Hubener M. 2012. Sensorimotor mismatch signals in primary visual cortex of the behaving mouse. *Neuron* **74:** 809–815.

Kelly RM, Strick PL. 2000. Rabies as a transneuronal tracer of circuits in the central nervous system. *J Neurosci Methods* **103:** 63–71.

Kentros CG, Agnihotri NT, Streater S, Hawkins RD, Kandel ER. 2004. Increased attention to spatial context increases both place field stability and spatial memory. *Neuron* **42:** 283–295.

Kerlin AM, Andermann ML, Berezovskii VK, Reid RC. 2010. Broadly tuned response properties of diverse inhibitory neuron subtypes in mouse visual cortex. *Neuron* **67:** 858–871.

Kim JJ, Fanselow MS. 1992. Modality-specific retrograde amnesia of fear. *Science* **256:** 675–677.

Kim J, Kwon JT, Kim HS, Josselyn SA, Han JH. 2014. Memory recall and modifications by activating neurons with elevated CREB. *Nat Neurosci* **17:** 65–72.

Kimpo RR, Rinaldi JM, Kim CK, Payne HL, Raymond JL. 2014. Gating of neural error signals during motor learning. *eLife* **3:** e02076.

Komiyama T, Sato TR, O'Connor DH, Zhang Y-X, Huber D, Hooks BM, Gabitto M, Svoboda K. 2010. Learning-related fine-scale specificity imaged in motor cortex circuits of behaving mice. *Nature* **464:** 1182–1186.

Komiyama T, Sato TR, O'Connor DH, Zhang YX, Huber D, Hooks BM, Gabitto M, Svoboda K. 2010. Learning-related fine-scale specificity imaged in motor cortex circuits of behaving mice. *Nature* **464:** 1182–1186.

Kwon JT, Nakajima R, Kim HS, Jeong Y, Augustine GJ, Han JH. 2014. Optogenetic activation of presynaptic inputs in lateral amygdala forms associative fear memory. *Learn Mem* **21:** 627–633.

Lai CS, Franke TF, Gan WB. 2012. Opposite effects of fear conditioning and extinction on dendritic spine remodelling. *Nature* **483:** 87–91.

Laubach M, Wessberg J, Nicolelis MAL. 2000. Cortical ensemble activity increasingly predicts behaviour outcomes during learning of a motor task. *Nature* **405:** 567–571.

Lecoq J, Savall J, Vucinic D, Grewe BF, Kim H, Li JZ, Kitch LJ, Schnitzer MJ. 2014. Visualizing mammalian brain area interactions by dual-axis two-photon calcium imaging. *Nat Neurosci* **17:** 1825–1829.

Lee AK, Manns ID, Sakmann B, Brecht M. 2006. Whole-cell recordings in freely moving rats. *Neuron* **51:** 399–407.

Lefort S, O'Connor EC, Luscher C. 2014. *In vivo calcium imaging of nucleus accumbens neurons in freely feeding mice.* Society for Neuroscience, Washington, DC.

Letzkus JJ, Wolff SB, Meyer EM, Tovote P, Courtin J, Herry C, Luthi A. 2011. A disinhibitory microcircuit for associative fear learning in the auditory cortex. *Nature* **480:** 331–335.

Leutgeb S, Leutgeb JK, Barnes CA, Moser EI, McNaughton BL, Moser MB. 2005. Independent codes for spatial and episodic memory in hippocampal neuronal ensembles. *Science* **309:** 619–623.

Levene MJ, Dombeck DA, Kasischke KA, Molloy RP, Webb WW. 2004. In vivo multiphoton microscopy of deep brain tissue. *J Neurophysiol* **91:** 1908–1912.

Lever C, Wills T, Cacucci F, Burgess N, O'Keefe J. 2002. Long-term plasticity in hippocampal place-cell representation of environmental geometry. *Nature* **416:** 90–94.

Liu X, Ramirez S, Pang PT, Puryear CB, Govindarajan A, Deisseroth K, Tonegawa S. 2012. Optogenetic stimulation of a hippocampal engram activates fear memory recall. *Nature* **484:** 381–385.

Lo L, Anderson DJ. 2011. A Cre-dependent, anterograde transsynaptic viral tracer for mapping output pathways of genetically marked neurons. *Neuron* **72:** 938–950.

Looger LL, Griesbeck O. 2012. Genetically encoded neural activity indicators. *Curr Opin Neurobiol* **22:** 18–23.

Lovett-Barron M, Kaifosh P, Kheirbek MA, Danielson N, Zaremba JD, Reardon TR, Turi GF, Hen R, Zemelman BV, Losonczy A. 2014. Dendritic inhibition in the hippocampus supports fear learning. *Science* **343:** 857–863.

Low RJ, Gu Y, Tank DW. 2014. Cellular resolution optical access to brain regions in fissures: Imaging medial prefrontal cortex and grid cells in entorhinal cortex. *Proc Natl Acad Sci USA* **111:** 18739–18744.

Lutcke H, Margolis DJ, Helmchen F. 2013. Steady or changing? Long-term monitoring of neuronal population activity. *Trends Neurosci* **36:** 375–384.

Madisen L, Zwingman TA, Sunkin SM, Oh SW, Zariwala HA, Gu H, Ng LL, Palmiter RD, Hawrylycz MJ, Jones AR, et al. 2010. "A robust and high-throughput Cre reporting and characterization system for the whole mouse brain." *Nat Neurosci* **13:** 133–140.

Madisen L, Garner AR, Shimaoka D, Chuong AS, Klapoetke NC, Li L, Bourg Avander, Niino Y, Egolf L, Monetti C, et al. 2015. Transgenic mice for intersectional targeting of neural sensors and effectors with high specificity and performance. *Neuron* **85:** 942–958.

Mank M, Santos AF, Direnberger S, Mrsic-Flogel TD, Hofer SB, Stein V, Hendel T, Reiff DF, Levelt C, Borst A, et al. 2008. A genetically encoded calcium indicator for chronic in vivo two-photon imaging. *Nat Methods* **5:** 805–811.

Mankin EA, Sparks FT, Slayyeh B, Sutherland RJ, Leutgeb S, Leutgeb JK. 2012. Neuronal code for extended time in the hippocampus. *Proc Natl Acad Sci USA* **109:** 19462–19467.

Margrie TW, Brecht M, Sakmann B. 2002. In vivo, low-resistance, whole-cell recordings from neurons in the anaesthetized and awake mammalian brain. *Pflugers Arch-Eur J Physiol* **444:** 491–498.

Masamizu Y, Tanaka YR, Tanaka YH, Hira R, Ohkubo F, Kitamura K, Isomura Y, Okada T, Matsuzaki M. 2014. Two distinct layer-specific dynamics of cortical ensembles during learning of a motor task. *Nat Neurosci* **17:** 987–994.

Masson JME. 1987. A distributed memory model of context effects in word identification. *Bull Psychon Soc* **25:** 335–335.

McClelland JL, McNaughton BL, O'Reilly RC. 1995. Why there are complementary learning systems in the hippocampus and neocortex: Insights from the successes and failures of connectionist models of learning and memory. *Psychol Rev* **102:** 419–457.

McHugh TJ, Blum KI, Tsien JZ, Tonegawa S, Wilson MA. 1996. Impaired hippocampal representation of space in CA1-specific NMDAR1 knockout mice. *Cell* **87:** 1339–1349.

McKenzie S, Robinson NT, Herrera L, Churchill JC, Eichenbaum H. 2013. "Learning causes reorganization of neuronal firing patterns to represent related experiences within a hippocampal schema." *J Neurosci* **33:** 10243–10256.

McKenzie S, Frank AJ, Kinsky NR, Porter B, Riviere PD, Eichenbaum H. 2014. Hippocampal representation of related and opposing memories develop within distinct, hierarchically organized neural schemas. *Neuron* **83:** 202–215.

Miller JE, Ayzenshtat I, Carrillo-Reid L, Yuste R. 2014. Visual stimuli recruit intrinsically generated cortical ensembles. *Proc Natl Acad Sci USA* **111:** E4053–E4061.

Miyamichi K, Amat F, Moussavi F, Wang C, Wickersham I, Wall NR, Taniguchi H, Tasic B, Huang ZJ, He Z, et al. 2011. Cortical representations of olfactory input by trans-synaptic tracing. *Nature* **472:** 191–196.

Modi MN, Dhawale AK, Bhalla US. 2014. CA1 cell activity sequences emerge after reorganization of network correlation structure during associative learning. *eLife* **3:** e01982.

Moncada D, Viola H. 2007. Induction of long-term memory by exposure to novelty requires protein synthesis: Evidence for a behavioral tagging. *J Neurosci* **27:** 7476–7481.

Mostany R, Anstey JE, Crump KL, Maco B, Knott G, Portera-Cailliau C. 2013. Altered synaptic dynamics during normal brain aging. *J Neurosci* **33:** 4094–4104.

Muller RU, Kubie JL, Ranck JB Jr, 1987. Spatial firing patterns of hippocampal complex-spike cells in a fixed environment. *J Neurosci* **7:** 1935–1950.

Muzzio IA, Levita L, Kulkarni J, Monaco J, Kentros C, Stead M, Abbott LF, Kandel ER. 2009. Attention enhances the retrieval and stability of visuospatial and olfactory representations in the dorsal hippocampus. *PLoS Biol* **7**: e1000140.

Nakazawa K, Sun LD, Quirk MC, Rondi-Reig L, Wilson MA, Tonegawa S. 2003. Hippocampal CA3 NMDA receptors are crucial for memory acquisition of one-time experience. *Neuron* **38**: 305–315.

Nguyen-Vu TD, Kimpo RR, Rinaldi JM, Kohli A, Zeng H, Deisseroth K, Raymond JL. 2013. Cerebellar Purkinje cell activity drives motor learning. *Nat Neurosci* **16**: 1734–1736.

Nimmerjahn A, Mukamel EA, Schnitzer MJ. 2009. Motor behavior activates Bergmann glial networks. *Neuron* **62**: 400–412.

O'Connor DH, Peron SP, Huber D, Svoboda K. 2010. Neural activity in barrel cortex underlying vibrissa-based object localization in mice. *Neuron* **67**: 1048–1061.

Osakada F, Callaway EM. 2013. Design and generation of recombinant rabies virus vectors. *Nat Protoc* **8**: 1583–1601.

Osakada F, Mori T, Cetin AH, Marshel JH, Virgen B, Callaway EM. 2011. New rabies virus variants for monitoring and manipulating activity and gene expression in defined neural circuits. *Neuron* **71**: 617–631.

Oyibo HK, Znamenskiy P, Oviedo HV, Enquist LW, Zador AM. 2014. Long-term Cre-mediated retrograde tagging of neurons using a novel recombinant pseudorabies virus. *Front Neuroanat* **8**: 86.

Packer AM, Peterka DS, Hirtz JJ, Prakash R, Deisseroth K, Yuste R. 2012. Two-photon optogenetics of dendritic spines and neural circuits. *Nat Methods* **9**: 1202–1205.

Packer AM, Russell LE, Dalgleish HW, Hausser M. 2015. Simultaneous all-optical manipulation and recording of neural circuit activity with cellular resolution in vivo. *Nat Methods* **12**: 140–146.

Page M. 2000. "Connectionist modelling in psychology: A localist manifesto. *Behav Brain Sci* **23**: 443–467; discussion 467–512.

Parker JG, Marshall JD, Ahanonu B, Grewe B, J Zhong LI, Ehlers MD, J MJSM. 2014. *Parallel remodeling of direct and indirect pathway neuronal activity in the nucleus accumbens during Pavlovian reward conditioning.* Society for Neuroscience, Washington, DC.

Patterson MA, Lagier S, Carleton A. 2013. Odor representations in the olfactory bulb evolve after the first breath and persist as an odor afterimage. *Proc Natl Acad Sci USA* **110**: E3340–E3349.

Peters AJ, Chen SX, Komiyama T. 2014. Emergence of reproducible spatiotemporal activity during motor learning. *Nature* **510**: 263–267.

Petreanu L, Gutnisky DA, Huber D, Xu N-l, O'Connor DH, Tian L, Looger L, Svoboda K. 2012. Activity in motorsensory projections reveals distributed coding in somatosensation. *Nature* **489**: 299–303.

Piaget J, Tomlinson J, Tomlinson A. 1929. *The child's conception of the world.* Rowman & Littlefield, Lanham, MD.

Pine DS, Grun J, Maguire EA, Burgess N, Zarahn E, Koda V, Fyer A, Szeszko PR, Bilder RM. 2002. Neurodevelopmental aspects of spatial navigation: A virtual reality fMRI study. *Neuroimage* **15**: 396–406.

Plaut DC, McClelland JL. 2010. Locating object knowledge in the brain: Comment on Bowers's (2009) attempt to revive the grandmother cell hypothesis. *Psychol Rev* **117**: 284–288.

Prakash R, Yizhar O, Grewe B, Ramakrishnan C, Wang N, Goshen I, Packer AM, Peterka DS, Yuste R, Schnitzer MJ, et al. 2012. Two-photon optogenetic toolbox for fast inhibition, excitation and bistable modulation. *Nat Methods* **9**: 1171–1179.

Preston AR, Eichenbaum H. 2013. Interplay of hippocampus and prefrontal cortex in memory. *Curr Biol* **23**: R764–R773.

Quian Quiroga R, Kreiman G. 2010. Measuring sparseness in the brain: Comment on Bowers (2009). *Psychol Rev* **117**: 291–297.

Quirin S, Jackson J, Peterka DS, Yuste R. 2014. Simultaneous imaging of neural activity in three dimensions. *Front Neural Circuits* **8**: 29.

Ramirez S, Liu X, Lin PA, Suh J, Pignatelli M, Redondo RL, Ryan TJ, Tonegawa S. 2013. Creating a false memory in the hippocampus. *Science* **341**: 387–391.

Ravassard P, Kees A, Willers B, Ho D, Aharoni D, Cushman J, Aghajan ZM, Mehta MR. 2013. Multisensory control of hippocampal spatiotemporal selectivity. *Science* **340**: 1342–1346.

Redondo RL, Kim J, Arons AL, Ramirez S, Liu X, Tonegawa S. 2014. Bidirectional switch of the valence associated with a hippocampal contextual memory engram. *Nature* **513**: 426–430.

Reijmers LG, Perkins BL, Matsuo N, Mayford M. 2007. Localization of a stable neural correlate of associative memory. *Science* **317**: 1230–1233.

Repa JC, Muller J, Apergis J, Desrochers TM, Zhou Y, LeDoux JE. 2001. Two different lateral amygdala cell populations contribute to the initiation and storage of memory. *Nat Neurosci* **4**: 724–731.

Rickgauer JP, Deisseroth K, Tank DW. 2014. Simultaneous cellular-resolution optical perturbation and imaging of place cell firing fields. *Nat Neurosci* **17**: 1816–1824.

Rogerson T, Cai DJ, Frank A, Sano Y, Shobe J, Lopez-Aranda MF, Silva AJ. 2014. Synaptic tagging during memory allocation. *Nat Rev Neurosci* **15**: 157–169.

Rotenberg A, Mayford M, Hawkins RD, Kandel ER, Muller RU. 1996. Mice expressing activated CaMKII lack low frequency LTP and do not form stable place cells in the CA1 region of the hippocampus. *Cell* **87**: 1351–1361.

Rumelhart DE, McClelland JL; Group University of California San Diego PDP Research. 1986. Parallel distributed processing explorations in the microstructure of cognition, Vol. 1, Foundations. In *Computational models of cognition and perception*. MIT Press, Cambridge, MA.

Runyan CA, Schummers J, Van Wart A, Kuhlman SJ, Wilson NR, Huang ZJ, Sur M. 2010. Response features of parvalbumin-expressing interneurons suggest precise roles for subtypes of inhibition in visual cortex. *Neuron* **67**: 847–857.

Sadtler PT, Quick KM, Golub MD, Chase SM, Ryu SI, Tyler-Kabara EC, Yu BM, Batista AP. 2014. Neural constraints on learning. *Nature* **512**: 423–426.

Sauer B. 1998. Inducible gene targeting in mice using the Cre/lox system. *Methods* **14**: 381–392.

Sawinski J, Wallace DJ, Greenberg DS, Grossmann S, Denk W, Kerr JN. 2009. Visually evoked activity in cortical cells imaged in freely moving animals. *Proc Natl Acad Sci USA* **106:** 19557–19562.

Scoville WB, Milner B. 1957. Loss of recent memory after bilateral hippocampal lesions. *J Neurol Neurosurg Psychiatry* **20:** 11–21.

Senn V, Wolff SB, Herry C, Grenier F, Ehrlich I, Grundemann J, Fadok JP, Muller C, Letzkus JJ, Luthi A. 2014. Long-range connectivity defines behavioral specificity of amygdala neurons. *Neuron* **81:** 428–437.

Sheffield ME, Dombeck DA. 2015. Calcium transient prevalence across the dendritic arbour predicts place field properties. *Nature* **517:** 200–204.

Silva AJ, Zhou Y, Rogerson T, Shobe J, Balaji J. 2009. Molecular and cellular approaches to memory allocation in neural circuits. *Science* **326:** 391–395.

Sofroniew NJ, Cohen JD, Lee AK, Svoboda K. 2014. Natural whisker-guided behavior by head-fixed mice in tactile virtual reality. *J Neurosci* **34:** 9537–9550.

Squire LR, Alvarez P. 1995. Retrograde amnesia and memory consolidation: A neurobiological perspective. *Curr Opin Neurobiol* **5:** 169–177.

Stirman JN, Smith IT, Kudenov MW, Smith SL. 2014. Wide field-of-view, twin-region two-photon imaging across extended cortical networks. *bioRxiv* doi: 10.1101/011320.

Sun XR, Badura A, Pacheco DA, Lynch LA, Schneider ER, Taylor MP, Hogue IB, Enquist LW, Murthy M, Wang SS. 2013. Fast GCaMPs for improved tracking of neuronal activity. *Nat Commun* **4:** 2170.

Susaki EA, Tainaka K, Perrin D, Kishino F, Tawara T, Watanabe TM, Yokoyama C, Onoe H, Eguchi M, Yamaguchi S, et al. 2014. Whole-brain imaging with single-cell resolution using chemical cocktails and computational analysis. *Cell* **157:** 726–739.

Suthana N, Haneef Z, Stern J, Mukamel R, Behnke E, Knowlton B, Fried I. 2012. Memory enhancement and deep-brain stimulation of the entorhinal area. *N Engl J Med* **366:** 502–510.

Szabo V, Ventalon C, De Sars V, Bradley J, Emiliani V. 2014. Spatially selective holographic photoactivation and functional fluorescence imaging in freely behaving mice with a fiberscope. *Neuron* **84:** 1157–1169.

Taber KH, Strick PL, Hurley RA. 2005. Rabies and the cerebellum: New methods for tracing circuits in the brain. *J Neuropsychiatry Clin Neurosci* **17:** 133–139.

Thompson LT, Best PJ. 1990. Long-term stability of the place-field activity of single units recorded from the dorsal hippocampus of freely behaving rats. *Brain Res* **509:** 299–308.

Tian L, Hires SA, Mao T, Huber D, Chiappe ME, Chalasani SH, Petreanu L, Akerboom J, McKinney SA, Schreiter ER, et al. 2009. Imaging neural activity in worms, flies and mice with improved GCaMP calcium indicators. *Nat Methods* **6:** 875–881.

Tian L, Akerboom J, Schreiter ER, Looger LL. 2012. Neural activity imaging with genetically encoded calcium indicators. *Prog Brain Res* **196:** 79–94.

Trachtenberg JT, Chen BE, Knott GW, Feng G, Sanes JR, Welker E, Svoboda K. 2002. Long-term in vivo imaging of experience-dependent synaptic plasticity in adult cortex. *Nature* **420:** 788–794.

Trimmer JS, Rhodes KJ. 2004. Localization of voltage-gated ion channels in mammalian brain. *Annu Rev Physiol* **66:** 477–519.

Tse D, Langston RF, Kakeyama M, Bethus I, Spooner PA, Wood ER, Witter MP, Morris RG. 2007. Schemas and memory consolidation. *Science* **316:** 76–82.

Tse D, Takeuchi T, Kakeyama M, Kajii Y, Okuno H, Tohyama C, Bito H, Morris RG. 2011. Schema-dependent gene activation and memory encoding in neocortex. *Science* **333:** 891–895.

van Kesteren MT, Ruiter DJ, Fernandez G, Henson RN. 2012. How schema and novelty augment memory formation. *Trends Neurosci* **35:** 211–219.

Verhagen JV, Wesson DW, Netoff TI, White JA, Wachowiak M. 2007. Sniffing controls an adaptive filter of sensory input to the olfactory bulb. *Nat Neurosci* **10:** 631–639.

Wickersham IR, Finke S, Conzelmann KK, Callaway EM. 2007a. Retrograde neuronal tracing with a deletion-mutant rabies virus. *Nat Methods* **4:** 47–49.

Wickersham IR, Lyon DC, Barnard RJ, Mori T, Finke S, Conzelmann KK, Young JA, Callaway EM. 2007b. Monosynaptic restriction of transsynaptic tracing from single, genetically targeted neurons. *Neuron* **53:** 639–647.

Wilt BA, Fitzgerald JE, Schnitzer MJ. 2013. Photon shot noise limits on optical detection of neuronal spikes and estimation of spike timing. *Biophys J* **104:** 51–62.

Wiltgen BJ, Brown RA, Talton LE, Silva AJ. 2004. New circuits for old memories: the role of the neocortex in consolidation. *Neuron* **44:** 101–108.

Wiltgen BJ, Zhou M, Cai Y, Balaji J, Karlsson MG, Parivash SN, Li W, Silva AJ. 2010. The hippocampus plays a selective role in the retrieval of detailed contextual memories. *Curr Biol* **20:** 1336–1344.

Wolff SB, Grundemann J, Tovote P, Krabbe S, Jacobson GA, Muller C, Herry C, Ehrlich I, Friedrich RW, Letzkus JJ, et al. 2014. Amygdala interneuron subtypes control fear learning through disinhibition. *Nature* **509:** 453–458.

Xu N-l, Harnett MT, Williams SR, Huber D, O'Connor DH, Svoboda K, Magee JC. 2012. Nonlinear dendritic integration of sensory and motor input during an active sensing task. *Nature* **492:** 247–251.

Yang G, Pan F, Gan WB. 2009. Stably maintained dendritic spines are associated with lifelong memories. *Nature* **462:** 920–924.

Yang B, Treweek JB, Kulkarni RP, Deverman BE, Chen CK, Lubeck E, Shah S, Cai L, Gradinaru V. 2014a. Single-cell phenotyping within transparent intact tissue through whole-body clearing. *Cell* **158:** 945–958.

Yang G, Lai CS, Cichon J, Ma L, Li W, Gan WB. 2014b. Sleep promotes branch-specific formation of dendritic spines after learning. *Science* **344:** 1173–1178.

Yiu AP, Mercaldo V, Yan C, Richards B, Rashid AJ, Hsiang HL, Pressey J, Mahadevan V, Tran MM, Kushner SA, et al. 2014. Neurons are recruited to a memory trace based on relative neuronal excitability immediately before training. *Neuron* **83:** 722–735.

Zariwala HA, Borghuis BG, Hoogland TM, Madisen L, Tian L, De Zeeuw CI, Zeng H, Looger LL, Svoboda K, Chen

TW. 2012. A Cre-dependent GCaMP3 reporter mouse for neuronal imaging in vivo. *J Neurosci* **32:** 3131–3141.

Zhou Y, Won J, Karlsson MG, Zhou M, Rogerson T, Balaji J, Neve R, Poirazi P, Silva AJ. 2009. CREB regulates excitability and the allocation of memory to subsets of neurons in the amygdala. *Nat Neurosci* **12:** 1438–1443.

Ziv Y, Burns LD, Cocker ED, Hamel EO, Ghosh KK, Kitch LJ, Gamal AEl, Schnitzer MJ. 2013. Long-term dynamics of CA1 hippocampal place codes. *Nat Neurosci* **16:** 264–266.

Zuo Y, Yang G, Kwon E, Gan WB. 2005. Long-term sensory deprivation prevents dendritic spine loss in primary somatosensory cortex. *Nature* **436:** 261–265.

# Exploring Memory Representations with Activity-Based Genetics

Mark Mayford[1] and Leon Reijmers[2]

[1]Molecular and Cellular Neurosciences Department, The Scripps Research Institute, La Jolla, California 92037

[2]Department of Neuroscience, School of Medicine, Tufts University, Boston, Massachusetts 02111

Correspondence: mmayford@scripps.edu; leon.reijmers@tufts.edu

The brain is thought to represent specific memories through the activity of sparse and distributed neural ensembles. In this review, we examine the use of immediate early genes (IEGs), genes that are induced by neural activity, to specifically identify and genetically modify neurons activated naturally by environmental experience. Recent studies using this approach have identified cellular and molecular changes specific to neurons activated during learning relative to their inactive neighbors. By using opto- and chemogenetic regulators of neural activity, the neurons naturally recruited during learning can be artificially reactivated to directly test their role in coding external information. In contextual fear conditioning, artificial reactivation of learning-induced neural ensembles in the hippocampus or neocortex can substitute for the context itself. That is, artificial stimulation of these neurons can apparently cause the animals to "think" they are in the context. This represents a powerful approach to testing the principles by which the brain codes for the external world and how these circuits are modified with learning.

A central feature of nervous systems is that, to function properly, specific neurons must become active in response to specific stimuli. The nature of this selective activation and its modification with experience is the focus of much neuroscience research, ranging from studies of sensory processing in experimental animals to disorders of thought such as schizophrenia in humans. The central dogma of neuroscience is that perceptions, memories, thoughts, and higher mental functions arise from the pattern and timing of the activity in neural ensembles in specific parts of the brain at specific points in time. Until quite recently, the investigation of these "circuit"-based questions has primarily been limited to observational techniques, such as single unit recording, functional magnetic resonance imagery (fMRI), and calcium imaging, to document the patterns of neural activity evoked by sensory experience or even complex psychological contingencies in human fMRI studies. These techniques have been enormously successful and created a framework for understanding information processing in the brain. For example, recordings in the visual system have indicated that, in the primary visual cortex, neurons are tuned to the orientation of linear stimuli (Hubel and Wiesel 1962). In contrast, neurons in higher brain areas can respond to discrete items. The most striking example of this specificity comes from in vivo recording in the human medial temporal lobe in which single units have been identified that respond to photos of the actress Halle Berry as well as her written name (Quiroga et al. 2005). This highly selective tuning of neural activity is suggestive of function, but how can this be directly tested? What would be the effect of stimulating just this rare population of neurons, a memory of the actress, a sensory illusion of her image? How does this type of specific firing arise? Do these neurons differ from their nonresponsive neighbors in terms of biochemistry, cell biology, or connectivity? Do they undergo molecular alterations when new information is learned about this individual and are these changes required for the learning? These types of questions have recently become accessible to study in mice through the use of activity-based genetic manipulation, in which neurons that are activated by a specific sensory stimulus can be altered to express any gene of experimental interest. These studies and approaches will be the focus of this work.

## IMMEDIATE EARLY GENES FOR ACTIVITY MAPPING

The first observation that the expression of certain genes was responsive to neural activity was made almost 30 years ago with the identification of increased expression of *cfos* in the brain following seizure (Morgan et al. 1987). *cfos* is one of a class of genes known as immediate early genes (IEGs) that are defined by their rapid induction by pre-existing transcription factors without the need for de novo protein synthesis (Greenberg et al. 1986). This allows rapid transcription initiation, within 1 min of electrical stimulation, supported by RNA polymerase II that is bound to the promoter region under resting conditions (Saha et al. 2011). On neuronal activation, the stalled RNA polymerase is released, thereby enabling the extremely rapid induction of IEG transcription. In addition to rapid induction, the cfos protein and messenger RNA (mRNA) have a relatively short half-life, such that following the end of active transcription the levels of expressed protein rapidly re-

turn to baseline (Rahmsdorf et al. 1987; Wellington et al. 1993). In the induction of *cfos* with seizure, the expression peaks at 1 h and returns to baseline by 3 h, allowing the expression of this gene to provide a snapshot view of brain activity with this 3-h time window (Morgan et al. 1987). Since the initial identification of *cfos*, genetic screens have identified a wide array of immediate early genes that show similar responses to neural activity (Nedivi et al. 1993; Qian et al. 1993; Yamagata et al. 1993). The most notable are *arc* and *zif268/erg1*, which, along with *cfos*, have been used extensively as a surrogate measure for neural ensemble activity in experimental animals (Worley et al. 1991; Lyford et al. 1995).

There is an extensive literature on the regulation of IEGs in cultured neurons and nonneuronal cells in which transcription can be triggered in response to multiple classes of stimuli, including growth factors and cAMP as well as strong depolarization (Ghosh et al. 1994). The degree of synaptic or intrinsic neural activity required to induce IEG expression is not straightforward and varies between genes and brain regions. For example, in vivo stimulation of dentate gyrus (DG) granule cells with high-frequency bursts that are sufficient to produce long-term potentiation (LTP) causes strong induction of *zif268*, but *cfos* is only induced in these neurons with more prolonged stimulation protocols. In both cases, the induction requires *N*-methyl-D-aspartate (NMDA) receptor function (French et al. 2001). In hippocampal CA1 neurons, similar stimulation protocols failed to induce any of the IEGs. In hippocampal slice, the cfos gene is induced with high fidelity by 30 or more action potentials delivered at 10 Hz or greater frequency (Schoenenberger et al. 2009). Although the precise nature of activity required to induce IEG expression is not clear, the expression of IEGs has been used extensively as a measure of neural activation in response to environmental stimulation and has given results that often track with activity determined by other techniques, such as electrophysiological recording or with known anatomical function and connectivity. For example, *cfos* is activated in a tonotopic pattern in the auditory brainstem with auditory stimulation (Saint Marie et al. 1999) and in a somatotopic pattern in somatosentory cortex following whisker stimulation (Wagener et al. 2010). In learning and memory paradigms, such as fear conditioning, there is *cfos* activation in an array of areas, such as hippocampus, amygdala, prefrontal cortex, and anterior cingulate cortex, that are known to be active and functionally necessary for the learning or retrieval of contextual fear memories (Milanovic et al. 1998; Radulovic et al. 1998; Bontempi et al. 1999; Frankland et al. 2004; Knapska and Maren 2009).

Until the development of $Ca^{2+}$ imaging approaches (Jercog et al. 2015), detection of IEG expression was one of the only ways of assessing large ensembles of neural activity with cellular resolution. A critical question in learning and memory (and neural processing more generally) is how similar is neural circuit activation with learning and with retrieval or during multiple retrieval trials. One approach to this question that takes advantage of the rapid and transient nature of IEG expression is known as catFISH (for cellular compartment analysis of temporal activity by fluorescence in situ hybridization) (Guzowski et al. 1999). Because mRNA is initially produced in the nucleus in an unspliced form and is later transported to the cytoplasm following splicing, the use of probes targeting unspliced nuclear RNA (currently active neurons) and processed cytoplasmic RNA (previously active neurons) can be used to determine the activity of neurons at two different but closely spaced time points. This approach was used to examine the activity of CA1 hippocampal neurons following the exploration of two identical environments relative to the exploration of two distinct environments. The results showed that there was a greater overlap in *arc* expression in CA1 neurons following the two epochs exploring identical environments. This is consistent with the role of the hippocampus in place recognition and with in vivo electrophysiological studies of place cell activity and its role in encoding location. It also supports the idea that IEG expression reliably reports behaviorally relevant neural activity.

## IEG-Based Transgenics

Although the IEGs offer a temporally crude measure of neural activation, they do parallel other methods of assessing activity in many brain areas. In addition, genetic studies have suggested a role for *cfos*, *zif268*, and *arc* in synaptic plasticity and memory, suggesting that their activation may be particularly important in identifying circuits undergoing plastic changes (Fleischmann et al. 2003; Plath et al. 2006; Baumgartel et al. 2008). One advantage of IEGs over more direct and temporally precise measures of neural activity is that by providing a transcriptional response they offer a molecular genetic conduit into environmentally activated neural ensembles. This was first shown in a simple transgenic mouse in which a promoter element from the *cfos* gene was used to drive expression of a lacZ marker protein (Smeyne et al. 1992). The expression of the marker was strongly induced throughout the brain following seizure. In addition, the lacZ expression was responsive to environmental stimuli showing strong induction in the superchiasmatic nucleus, an area involved in circadian rhythm regulation, following a light pulse during the animal's dark cycle. The transgene incorporated just 600 bp of promoter region containing four different transcriptional response elements, all of which were required for normal stimulus-induced gene expression (Robertson et al. 1995).

More recently, a number of transgenic mouse lines have been developed that drive green fluorescent protein (GFP) in an activity-dependent manner (see Table 1 for a list of published IEG-based transgenic lines) (Barth et al. 2004; Wang et al. 2006; Reijmers et al. 2007). The use of a fluorescent marker compatible with live tissue imaging allows the recording of neurons in brain slices or in anesthetized animals specifically from recently active *cfos*-positive cells. This approach was used to examine spontaneously active neurons in the somatosentory cortex, where approximately 15% of neurons in layer 2/3 are cfos-GFP positive in the absence of any specific environmental stimuli (Yassin et al. 2010). These cfos-positive neurons showed an increase in excitatory drive, a reduc-

**Table 1.** Currently available immediate early gene (IEG)-based transgenic mouse lines

| Gene | Type | References | JAX stock No. |
|------|------|-----------|---------------|
| Fos-lacZ | Transgenic | Smeyne et al. 1992 | – |
| Fos-taulacZ | Transgenic | Mehta et al. 2002 | – |
| Fos-GFP | Transgenic | Barth et al. 2004 | 014135 |
| Fos-GFP/tTA | Transgenic | Reijmers et al. 2007 | 018306 |
| Fos-CreER$^{T2}$ | Knockin | Guenthner et al. 2013 | 021882 |
| Arc-CreER$^{T2}$ | Knockin | Guenthner et al. 2013 | 021881 |
| Arc-CreER$^{T2}$ | BAC transgenic | Denny et al. 2014 | 022357 |
| Arc-GFP | Knockin | Wang et al. 2006 | 007662 |
| ERG1-GFP | BAC transgenic | Xie et al. 2014 | 014709-UCD |

List of IEG-based transgenic and knockin mouse lines. Most are available from the Jackson Laboratory repository (www.jax .org).

tion in inhibitory drive, and a greater degree of interconnectivity than their cfos-negative neighbors. These results indicate that ensembles of neurons with increased activity are maintained for several hours (the timing between cfos expression and electrophysiological recording) and that there may be a stable structure and connectivity in these spontaneously active ensembles. The presence of such spontaneously active ensembles is consistent with electrophysiological recordings that have suggested the importance of posttraining replay of neural activity patterns evoked during learning in the consolidation of memory (Wilson and Mc-Naughton 1994; Foster and Wilson 2006; Karlsson and Frank 2009).

IEG-linked GFP reporter mice have also been used as a convenient means of repeatedly measuring neural activation patterns in intact animals by two-photon imaging through cranial windows. To examine the role of *arc* in the orientation selectivity responses of neurons in the primary visual cortex, Wang et al. (2006) generated a mouse in which GFP replaced the functional arc protein to produce a knockout in which the activation of neural ensembles could be imaged through measurement of GFP expression levels. Heterozygous mice (carrying one functional *arc* allele and the activity-dependent *arc*-GFP) showed normal orientation tuning, whereas the homozygous, *arc* knockout mice showed impaired tuning of individual neurons measured electrophysiologically. Wang et al. then examined the activity of all neurons in the field using the GFP reporter

and determined that the overall loss of tuning specificity was because of an increase in the number of weakly tuned cells rather than a loss of specificity in the highly tuned neurons. In another study, a transgenic mouse in which GFP was driven by the *zif268* promoter was used to examine the formation of contextual representations in the neocortex following fear conditioning (Xie et al. 2014). These investigators found slowly developing and very sparse contextual representations that were stable for up to 2 months. Similar results were obtained using a *cfos*-GFP reporter to examine the formation of spatial representations in the retrosplenial cortex (Czajkowski et al. 2014).

Finally, IEG-linked GFP reporters have been used to detect neural activity on a brain-wide scale (Kim et al. 2015; Vousden et al. 2015). Serial two-photon tomography is a method for obtaining fluorescent images in an automated manner with the capacity to capture 260 50 μm serial sections throughout an entire mouse brain at 10× resolution within several hours (Ragan et al. 2012). This imaging approached was applied to *cfos*-GFP transgenic mice and combined with an analysis pipeline for automated cell counting and anatomical partitioning to obtain whole brain activity maps at cellular resolution in social behavior (Kim et al. 2015). The results were generally consistent with previous studies showing activation in appropriate areas of the hypothalamus and amygdala with social interaction and served as a proof of principle for automated whole brain imaging of activity using an IEG-based reporter.

Cite this article as *Cold Spring Harb Perspect Biol* doi: 10.1101/cshperspect.a021832

The use of activity-dependent promoters to drive expression of fluorescent markers allows for targeted recording, repeated imaging for prolonged time periods, and whole brain activity mapping at a single time point. To probe the *function* of active neural ensembles, several groups have developed IEG-based transgenic lines that allow the expression of effector molecules in neurons that are active at a specific time point. By using the *cfos* promoter in conjunction with the tetracycline system for control of gene expression, Reijmers et al. 2007 produced a transgenic mouse in which neurons that are naturally active in a given time window could be genetically modified to express essentially any gene, as shown in Figure 1A. The approach requires two transgenes introduced into the same animal. The first transgene consists of the *cfos* promoter driving expression of the tetracycline transactivator (tTA), which is a transcription factor that can be regulated with the antibiotic

doxycycline (Dox) (Gossen and Bujard 1992). The second transgene carries a tetracycline-responsive element promoter (TRE) to drive expression of any gene of interest. In the presence of Dox delivered in the animal's diet, the *cfos* promoter will drive expression of tTA in active neurons; however, transcription of the second TRE-linked gene will be blocked. When Dox is removed, a time window is open during which neurons that are sufficiently active to induce the *cfos*-tTA will express the TRE-linked transgene via tTA-driven transcription at the TRE promoter. The investigators used these mice to examine the neural ensembles in the amygdala that are activated with learning and retrieval in contextual fear conditioning. They used *fos*-tTA to drive expression of a long-lasting lacZ reporter protein in neurons activated during learning and compared the degree to which these neurons were reactivated following memory retrieval. They found that the amygdala

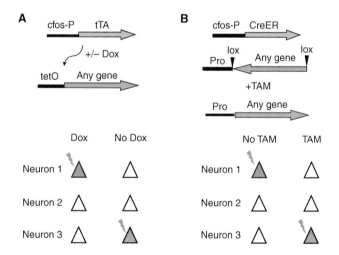

**Figure 1.** Two systems for the genetic manipulation of active neural ensembles. (*A*) In this tetracycline (TET)-based system, two transgenes are required, a cfos promoter-driven tetracycline transactivator (tTA) and a tetracycline-responsive element (TRE) promoter-driven gene of interest. In the presence of doxycycline (Dox) the tTA is expressed in electrically active (cfos⁺) neurons but is prevented from activating expression of the gene of interest by the presence of Dox. In the absence of Dox, a window is opened during which active neurons that express tTA drive expression of the gene of interest from the TRE promoter. (*B*) This Cre-based system also uses two transgenes, a cfos promoter-driven CreER[t2] and a gene of interest that is flanked by loxP sites and positioned in an inverted orientation to any neuronal promoter (Pr). The loxP sites are arranged such that Cre activity will lead to a single inversion event of the flanked DNA. In the absence of tamoxifin (TAM) the Cre recombinase is inactive so that no recombination takes place even in active neurons. On administration of TAM, any active (cfos⁺) neurons will express the Cre, which is now active, and inverts the orientation of the gene of interest. This gene is then constitutively and permanently expressed from the neurons specific promoter.

neurons activated during learning are responsive primarily to shock, the unconditioned stimulus (US), and that after training these shock-US-responsive neurons were activated by the context, the conditioned stimulus (CS), alone and that the degree of reactivation was correlated with the strength of the expressed fear memory. These results suggest a model whereby pairing of CS and US during training produces plasticity within the circuit that allows the CS to now recruit US neurons, linking the two pathways during recall. These results are consistent with the literature on amygdala circuitry in fear conditioning (and also with fear conditioning in *Aplysia*; see Byrne and Hawkins 2015; Hawkins and Byrne 2015) and provide a validation of this genetic model, which has been used extensively in studies of the circuitry underlying memory as discussed in later sections.

More recently, several groups have developed mouse lines in which IEG-promoters are used to drive expression of a tamoxifen-regulated form of Cre recombinase (CreER$^{T2}$) (Guenthner et al. 2013; Denny et al. 2014). As shown in Figure 1B, these mice can be used to activate or delete genes from subsets of neurons that are active at a particular point in time. In the absence of tamoxifen, the *cfos*-linked CreER$^{T2}$ will be expressed in active neurons but fail to induce loxP-mediated recombination because the protein is restricted to the cytoplasm by the estrogen receptor (ER$^{T2}$) component of the molecule. Administration of tamoxifen opens a time window during which CreER$^{T2}$ expressed in any *cfos*-positive neurons will be translocated to the nucleus to induce loxP-based recombination to activate or delete target genes. The use of Cre-based systems offers the advantage of permanent genetic tagging of active neural ensembles, the ability to delete endogenous genes, and the availability of a greater number of Cre-responsive genetic tools than are available for the tTA system.

## IEG-Based Manipulation of Circuit Function

As discussed in the introductory paragraphs, there is an extensive body of literature characterizing the patterns of neuronal firing in re-

sponse to sensory stimuli. For example, neurons in the medial temporal lobe of human patients have been identified that seem to respond only to a specific individual (Quiroga et al. 2005). The IEG-based genetic tools provide a mechanism for directly probing the function of these sparse and dispersed ensembles in the representation of sensory information and memories. By driving the expression of genetic regulators of neural activity into the neural ensembles activated naturally by a specific environmental stimulus, the activity of these distributed ensembles can be experimentally controlled to test their behavioral relevance. In this section, we will discuss several recent studies using this approach to the study of the representation of context in fear conditioning.

Contextual fear conditioning is a hippocampal-dependent task in which animals learn to fear a place (the context) where they receive aversive foot shocks. How is the context (characterized by specific visual, tactile, and olfactory cues) represented by dispersed neural activity? Would artificially stimulating neurons naturally activated by the context be sufficient to substitute for the actual sensory experience of the context during fear conditioning? One attempt to address this question used the *cfos*-tTA system to drive expression of the excitatory hM3Dq receptor into neural ensembles that were active during contextual learning (Fig. 2A) (Garner et al. 2012). hM3Dq is a human Gq coupled muscarinic receptor that has been mutated so that it no longer responds to acetylcholine but responds instead to the synthetic ligand clozapine-*N*-oxide (CNO) (Alexander et al. 2009). In neurons expressing hM3Dq, application of CNO causes an 8–10-mV depolarization of the membrane potential and a subsequent increase in action potential firing. In this study, mice carrying both the *cfos*-tTA and TRE-hM3Dq transgenes were allowed to explore a distinct context (ctxA) causing the cfos ensembles activated by this environmental stimulation to express the hM3Dq transgene. The animals were then fear conditioned in a separate context (ctxB), and CNO was delivered to depolarize the hM3Dq-expressing ctxA ensemble of neurons. Under these conditions, the animals

Cite this article as *Cold Spring Harb Perspect Biol* doi: 10.1101/cshperspect.a021832

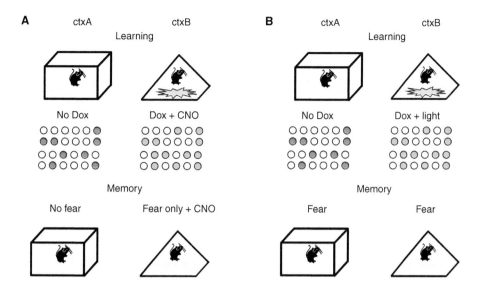

**Figure 2.** Manipulating contextual memory representations. This figure shows the arrangement of two sets of experiments to test the role of distributed neural ensembles in the coding of contextual memory. Both experiments use the cfos-tTA system discussed in Figure 1A to label neurons that are activated by exploration of a context (ctxA). In *A*, the active neurons express hM3Dq and in *B*, the neurons express ChR2. The animals are then fear conditioned in a distinct context (ctxB), whereas the ctxA neurons are activated with either clozapine *N*-oxide (CNO) or light, panels *A* and *B*, respectively. Memory retrieval is then tested in ctxA and ctxB. In panel *A*, the mice only show a fear response when the ctxA neurons are artificially activated while the animal is in ctxB, suggesting formation of a hybrid representation. In panel *B*, the animals show fear in ctxA even though they never received a shock in that context, suggesting that the artificial stimulation of the ChR2 positive neurons tagged in ctxA was able to substitute for (represent) that context. (Red circles) Neural ensembles expressing the genetic effector hM3Dq in *A* and ChR2 in *B*. (Blue circles) Neural ensembles naturally active by sensory input during training in ctxB.

fail to develop a fear response to either ctxA or ctxB, suggesting that the artificial activity produced in the ctxA neurons was not sufficient to substitute as a neural representation of ctxA but did interfere with the ability of the animals to represent the context in which they actually received foot shocks (ctxB). However, when the animals were placed in ctxB and received CNO to concurrently activate the ctxA neural ensemble, they showed a fear response that approached that of control animals. This suggests that the experimental animals formed an artificial contextual representation that incorporated both ctxA (artificial-internally generated) and ctxB (natural-sensory driven) neural ensembles.

Although these results failed to show that artificial activation of a distributed pattern of sensory evoked neurons could function to rep-

resent that sensory experience, it provided a number of important insights. First, the neurons that were activated artificially in this study were widely distributed throughout the neocortex and hippocampus. Previous studies had shown that local stimulation of a particular brain region or fiber bundle could serve as a CS in a conditioning paradigm (e.g., Huber et al. 2008), but the current results suggest that widely distributed, even global, patterns of activity can be meaningfully incorporated into neural representations. In fact, the brain is not silent in the absence of experimenter-provided stimuli yet this ongoing and dispersed "spontaneous" activity is rarely taken into account. Studies in the hippocampus, but also in the neocortex, indicate that spontaneous activity may not be random but instead reflect the replay of patterned activity associated with recent experiences (Kenet et al. 2003; Foster and Wilson 2006; Ji and Wilson 2007). The study by Garner et al. 2012 suggests that the integration of internally activated ensembles of neurons (ctxA in these experiments) at the time of new sensory-evoked learning (ctxB in this case) may represent a mechanism for integrating new information with previous experience, an important component of all higher forms of learning.

A more direct test of the functional relevance of distributed neural ensembles used the *cfos*-based genetic tagging approach in conjunction with channelrhodopsin (ChR2) (Boyden et al. 2005) and local stimulation in the hippocampus (Liu et al. 2012). In this study, the *cfos*-tTA transgenic mouse was used to express a virally delivered TRE-ChR2 transgene in the DG region of the hippocampus (Fig. 2B). Animals received contextual fear conditioning to allow the expression of ChR2 in the learning-activated population of neurons. The animals were then moved to a neutral context that did not induce fear. Light-induced firing of the ChR2-labeled DG neurons in this neutral context was able to produce a fear response and this response was tightly linked to the light activation of the DG. Would stimulating any set of DG neurons produce a fear response, or do the neurons carry specific information about the context in which they were labeled? To address this question, active DG neurons were labeled with ChR2, while the animal explored one context (ctxA) but did not receive foot shocks. The animals were then fear conditioned in a distinct context (ctxB) while receiving Dox to prevent further labeling of active neurons. Stimulation of the ChR2-containing ctxA neurons did not produce a response in animals trained to fear only ctxB, indicating that the neurons carry context specific information. Taken together, the results suggest that the activation of a sparse ($<5\%$ of total neurons in the DG), distributed, population of neurons, originally activated during learning, was sufficient to produce an apparent context-specific memory retrieval event.

In a follow-up study using the same techniques (Ramirez et al. 2013), the neurons activated by exploration of a specific context (ctxA), in the absence of shock, were labeled

with ChR2. The artificial stimulation of these ChR2-expressing neurons with light was then paired with foot shock in a different context (ctxB). When animals were placed back into ctxA, where they had never received the foot-shock US, they showed a fear response. This result suggests that the artificial stimulation of the distributed cfos-positive population of neurons labeled during the natural exploration of ctxA was sufficient to produce a neural representation of that complex environment. In essence, the artificial stimulation and the natural sensory experience were interchangeable, at least for the purposes of fear conditioning. The arrangement of this experiment is similar to that discussed previously in Garner et al. (2012) with the training taking place in ctxB, while a competing neural ensemble representing ctxA was stimulated. The results were similar; the animal's fear memory in the natural conditioning chamber (ctxB) was strongly reduced by the stimulation of ctxA neurons during training but recovered to normal levels during concurrent activation of the ctxA cells. Whether this reflects the integration of the A and B ensembles into a single complex representation, as suggested by Garner et al. (2012), or simply reflects the effects of the compound CS-A and CS-B each acquiring a lesser amount of associative strength is unclear.

Taken together these results indicate that the artificial activation of a sparse and distributed sensory-evoked population of neurons is sufficient to instantiate a neural representation of that sensory experience. The observation that artificial stimulation with ChR2, which will tend to activate all neurons simultaneously and was not synchronized to endogenous rhythms, is sufficient to produce this effect suggests that the temporal patterning seen in natural activity may not be critical, at least for this particular brain region and behavior. There are many remaining questions relating to this effect, such as in what other brain regions can it be produced, how many neurons need to be stimulated, what frequency parameters are required to produce these effects, what downstream circuits are required to produce the behavior, and how are they recruited? Perhaps the most interesting but difficult question to address is what is the animal experiencing when the neurons for ctxA are activated. Is it experiencing a memory of that context, a hallucinatory sensory experience, or a poorly defined sense of fear or anxiety independent of coherent contextual features?

Some of these questions are examined in another recent paper studying contextual representations in the retrosplenial cortex (Cowansage et al. 2014). The retrosplenial cortex is one of the output areas of the hippocampus and projects to a wide variety of other cortical areas. It is required for the encoding and retrieval of both recent and remote contextual fear memories (Keene and Bucci 2008a,b). Similar to what was seen in the DG studies, Cowansage et al. (2014) showed that stimulation of the ChR2 labeled neural ensembles naturally activated in the retrosplenial cortex at the time of contextual fear learning was sufficient to produce a fear response. The hippocampus has been the focus of intense study regarding its role in spatial and contextual learning; however, the cortical contribution has been less well characterized. This study shows that at the time of learning, a neural ensemble representing the context forms in the neocortex as well as the hippocampus. To investigate the circuit relationship and relative contribution of these two neural representations of context, the investigators inactivated the hippocampus pharmacologically and found that, although this produced the well-characterized amnesia for natural context recall (Kim and Fanselow 1992), this effect could be overcome by direct stimulation of the retrosplenial neural ensembles. This shows that the retrosplenial contextual representation is downstream from and functions independent of the hippocampal representation.

Early studies by Penfield (1968) showed that, in a small number of cases, direct brain stimulation resulted in apparent perceptions or memories. Although it is not clear whether these results represent normal or pathological responses or actual past experience, they do suggest that direct stimulation of the brain can, in some cases, produce a coherent psychological representation. Because mice are unable to report subjective experience, it is difficult to

know the psychological effect of artificial stimulation. To provide some insight into this question, Cowansage et al. (2014) compared the activation of downstream circuit elements with both natural recall (exposure to the context in which animals were shocked) and artificial ChR2 stimulation-induced recall. Using the catFISH approach they found that stimulation of retrosplenial ctxA ensembles and natural memory retrieval in ctxA-activated overlapping subsets of neurons in the entorhinal cortex, as well as basolateral and central amygdala, suggesting that the two methods of producing memory retrieval are processed in a similar manner. This is perhaps the best insight that can be achieved in a nonverbal species; whatever is experienced with artificial stimulation in a mouse, we can say that it produces a state of brain activity that is similar to posttraining context exposure, which we define as natural memory retrieval.

The artificial stimulation of small ensembles of neurons in the DG region of the hippocampus allows the retrieval of contextual memories, presumably by activating more distributed ensembles of neurons in downstream hippocampal and cortical regions. Given the highly parallel nature of neural connectivity and processing, it is possible that although these ensembles may be sufficient for retrieval, they may not be necessary because other pathways could compensate. The question of necessity was addressed in two recent studies using a light-gated proton pump from archaebacteria (ArchT) (Chow et al. 2010) to hyperpolarize and silence cfos-tagged neural ensembles in experiments analogous to those described above with ChR2 (Denny et al. 2014; Tanaka et al. 2014). Both papers also used contextual fear conditioning and examined the requirement for ensembles in three different hippocampal regions, the DG, CA3, and CA1. Expression of ArchT was driven into neurons that were active during learning in context fear conditioning and these ensembles were subsequently silenced in either the DG, CA3, or CA1 during memory retrieval. In each case, the silencing impaired the retrieval of the contextual fear memory. These results show that in each of the three major hippocampal subregions, the reactivation of the ensemble of neurons active during initial learning is necessary for the subsequent retrieval of the memory. This is consistent with the view that retrieval involves the recruitment of patterns of brain activity produced during initial learning.

In associative conditioning tasks, the unconditioned stimulus (US) generally has an emotional valence; for example, food is generally positive or rewarding, whereas shock, the US in fear conditioning, is aversive. Two recent studies examined the circuits that encode emotional valence in the hippocampus and amygdala using the cfos-tagging approach. The first study extended on the finding of a neural representation for context fear in the DG (Redondo et al. 2014). The investigators genetically tagged DG neurons active during fear conditioning with channelrhodopsin (ChR2), and when stimulated with light these neurons produced a fear response. They next asked whether the emotional valence of this fear memory could be reversed by pairing light stimulation of the DG ensembles with a positive valence US. Following the pairing, animals that previously avoided the context in which they received a shock now sought out the context, suggesting that there had been a switch in emotional valence associated with the context. This result is consistent with the view that the DG neural ensemble encodes a contextual representation (i.e., substitutes for the conditioning box), so that when the artificial activation of these ensembles is paired with a positive or negative stimulus, the box itself acquires this emotional valence as occurs in natural associative conditioning. Interestingly, when the same experiment was conducted using stimulation of cfos-labeled amygdala neural ensembles, the switch in valence was not observed. This suggests that the amygdala ensembles activated during learning are hard wired for a particular emotional valence.

The amygdala circuitry for emotional valence was explored in another recent paper (Gore et al. 2015). The study genetically introduced ChR2 into neurons activated by either a positive valence US (nicotine), or a negative valence US (shock). The investigators found two sparse (approximately 3%) and distinct populations of basolateral amygdala neurons were labeled by the stimuli, with little overlap in the cells activated by positive and negative valence stimuli. They went on to show that the artificial stimulation of these neurons could serve as a US (positive or negative valence depending on the cell population stimulated) in an array of associative and instrumental conditioning tasks. To examine the effects of conditioning on the US neurons, animals were conditioned by pairing a CS with artificial stimulation of the US neurons to produce an associative memory. After this training, the CS alone was now able to activate the US neurons of the paired emotional valence, similar to what was seen in previous studies with fear conditioning (Reijmers et al. 2007). Taken together, this study presents a view in which the amygdala contains distinct populations of neurons that confer positive or negative valence to a stimulus, presumably because of their output targets. During associative conditioning, the CS develops the ability to activate the subpopulation of neurons that were activated by the US to which it was paired and those neurons output the appropriate emotional valence to guide behavior.

## Biochemical and Cellular Studies in Active Circuits

The formation of memories presumably involves cellular and molecular changes at specific points within the processing stream of incoming sensory information such that it is altered to produce new behavioral outputs following learning. For example, in our fear-conditioning model, when exposed to a new environment (context), mice show exploratory behavior. However, if the animal received an aversive foot shock while in this context, their subsequent behavior to the same sensory cues of the context will be immobility or freezing. The physical instantiation of this context memory, the engram, is presumably reflected in cellular and molecular changes at specific points in processing of the contextual cues, which diverts or enhances activity to the fear-output circuits in the amygdala. Where these changes occur, whether they are distributed or localized, their cellular and molecular nature and how they alter the processing of information from sensory input to motor output are unknown but required for a deep understanding of the engram in this behavioral model. The IEG-based genetic technology enables the targeting of cellular and biochemical studies specifically to active neurons as opposed to their inactive neighbors and has been used in several recent studies to look for specific changes that may be relevant to the underlying cellular mechanisms of memory.

Forms of activity-dependent synaptic plasticity, such as LTP and long-term depression (LTD), represent attractive candidates for a cellular mechanism of learning. The idea remains controversial and a detailed discussion of the issue can be found in Angelakos and Abel (2015) and Basu and Siegelbaum (2015) and in several recent reviews (Morris 2013; Kandel et al. 2014; Mayford 2014). Perhaps the best support for LTP and LTD as a synaptic mechanism of memory comes from a recent study in which pairing optogenetic stimulation of inputs to the amygdala with foot shock produced a fear memory in which the optogenetic stimulation served as the CS. When LTD and LTP were subsequently produced by the appropriate stimulation of the ChR2 positive fibers carrying the CS, the behavioral fear response showed a corresponding decrease and increase, respectively, providing a direct link between these forms of plasticity and behavior, albeit in a nonnatural memory. The molecular signaling involved in the production of LTP is complex but may converge on at least one common mechanism, the insertion of new 2-amino-3-5-methyl-3-oxo-1,2-oxazol-4-yl propanoic acid (AMPA)-type glutamate receptors into dendritic spines that have undergone plasticity leading to an increased postsynaptic response (Shi et al. 1999; Rumpel et al. 2005; Kessels and Malinow 2009).

To examine glutamate receptor trafficking specifically in circuits activated during learning, Matsuo et al. (2008) used the *cfos*-tTA-based genetic tagging approach to drive expression of a GFP-linked GluR1 receptor subunit. Animals received either standard contextual fear conditioning or an unpaired protocol in which they were exposed to the conditioning chamber and later received immediate foot shocks in a separate context. Novel context exposure induced GFP-GluR1 expression in approximately 25% of hippocampal CA1 neurons and this was not altered by the shock US, consistent with a role for the hippocampus in encoding contextual information. Twenty-four hours after conditioning, the investigators examined the distribution of GFP-tagged glutamate receptors on dendritic spines and found that the paired training caused an increase in receptor trafficking specifically to mushroom-type spines. Because the expression of the GFP-GluR1 transgene was induced by the training itself and the protein did not reach the dendrites for several hours, the results were consistent with the idea that the training produced a molecular change or tag at some synapses allowing them to capture the newly synthesized receptors when they arrived. The notion of synaptic tagging was developed in studies of LTP to explain how newly expressed genes, required for long-term maintenance of the plasticity and of memories, could influence only those synapses that underwent the initial short-term plasticity (Frey and Morris 1997, 1998; see also Martin et al. 1997 in *Aplysia* and Si and Kandel 2015). Although these results are purely correlative, they are consistent with a subtle associative change in hippocampal synaptic function following fear conditioning. They also serve to indicate some of the difficulties likely to be encountered in trying to identify and functionally test the role of molecular changes with learning in mammalian systems that are likely to occur in a limited group of neurons and on a limited number or type of synapse on those neurons.

One of the weakest links in the LTP-memory connection is the paucity of studies that directly observe synaptic potentiation with learning. This may be attributable to the presumed paucity of neurons and synapses that contribute to any single learning event. To overcome this problem, Whitlock et al. (2006) used an array of recording electrodes implanted in the apical dendrites of the hippocampal CA1 region to measure synaptic responses before and after learning in an inhibitory avoidance task. They found a modest increase in synaptic response on 14 of 44 electrodes with the conditioning, indicating a sparse but localized potentiation. Several studies have reported LTP-like synaptic potentiation in the lateral amygdala with auditory fear conditioning (McKernan and Shinnick-Gallagher 1997; Rogan et al. 1997; Rumpel et al. 2005). The ability to detect plasticity in this region may reflect the recruitment of a large number of neurons in this particular behavioral task. In one study, it was estimated that one-third of the lateral amygdala neurons showed plasticity following conditioning (Rumpel et al. 2005). The identification of plastic changes in more sparsely recruited networks may be more difficult.

This issue was recently addressed in a study using the cfos-tagging approach to look for physiological changes in DG neurons activated with contextual fear conditioning, the same neurons that can produce an apparent contextual representation when artificially stimulated (Ryan et al. 2015). Whole-cell recording of the cfos-positive DG neurons from fear-conditioned animals showed an increase in excitatory postsynaptic current (EPSC) amplitude and in the ratio of AMPA/NMDA current relative to their negative neighbors. The cfos-positive neurons also showed an increase in dendritic spine density. Both of these changes were blocked by immediate posttraining administration of the protein synthesis inhibitor anisomycin, which also blocked the contextual fear memory. Taken together, these results suggest that the active ensemble of neurons undergo a learning-related increase in synaptic strength and connectivity. However, a similar change was also found in neurons from the animals that received only context exposure. This could reflect plasticity resulting from the encoding of a novel context, which might be the expected function of hippocampal ensembles. Alternately, it could reflect differences in a constitutively active subpopulation of neurons as the results are quite similar to what has been described previously in cfos-positive cortical neurons from animals not receiving any environmental stimuli (Yassin et al. 2010).

## Memory Allocation

Are the neurons that are recruited to encode a specific representation or memory determined exclusively by their preexisting connectivity and synaptic responses within the circuit processing the relevant sensory cues or is there a greater flexibility in the precise set of neurons recruited in any particular circumstance? A number of recent studies have been put forward to support a memory allocation model, which posits that in a given brain area many neurons have the potential to be recruited for the encoding of a particular stimulus and that the specific neurons are chosen for this task based on their preexisting state of excitability.

The initial idea for this model came from a study attempting to rescue memory deficits in a cyclic AMP response element-binding protein (CREB)-deficient mouse line (Han et al. 2007). Viral vectors carrying a wild-type copy of CREB were injected into the lateral amygdala of CREB-deficient animals and resulted in a complete rescue of the impairments in auditory fear conditioning, even though the virus infected only about 20% of the neurons. How could such a small number of restored cells support normal memory encoding? The viral construct carried a GFP marker to allow identification of the infected cells. When the investigators examined the neurons activated during a memory retrieval trial, using staining for the IEG *arc*, they found that almost all of the arc was expressed in GFP-positive cells, indicating that the infected neurons were selectively recruited to the fear memory representation. This was not an effect that was specific to the CREB mutant mouse line as viral elevation CREB levels in wild-type animals produce the same effect. In subsequent studies, it was shown that the specific silencing or lesioning of the CREB-elevated neurons impaired the fear memory showing that these specific neural ensembles contributed to expression of the memory (Han et al. 2009; Zhou et al. 2009).

How could CREB, a transcription factor implicated in the consolidation of memory, alter the recruitment of neurons during learning? The first clue came from the observation that CREB-transfected neurons showed a significant increase in excitability (Zhou et al. 2009). This suggests that it was not necessarily CREB itself that was responsible but simply that neurons with elevated excitability at the time of learning are those more likely to be recruited to the representation of that event. This idea was tested by increasing neuronal excitability directly using expression of a dominant negative mutant of the potassium channel KCNQ2 or the hM3Dq receptor in viral transfection experiments analogous to those described above for CREB (Yiu et al. 2014). In both cases, increased excitability was sufficient to recruit the transfected neurons to the fear memory representation. Using the regulated hM3Dq system, it was further shown that the increase in excitability was effective only immediately before training but not after learning had occurred. Although these studies were restricted to the lateral amygdala and fear conditioning, similar effects have recently been described in the insular cortex with conditioned taste aversion, suggesting that memory allocation may be a more general phenomena (Sano et al. 2014). The memory allocation model is attractive as a mechanism for linking memories for events that are closely spaced in time (Silva et al. 2009). According to this view, neural activity and the formation of long-term memory evokes CREB-dependent transcription, which in turn opens a time window of increased excitability in those active neurons. When new information is acquired during this time window, it will be more likely to recruit neurons that participated in the previous learning event creating a link at the circuit level between events that are closely spaced in time.

## CONCLUSIONS

One of the critical difficulties in studying information processing and its modification with learning in the mammalian brain is identifying and manipulating the neurons involved in representing any specific environmental stimulus or memory. A particular complex item, a place, or an individual, is likely represented in higher brain areas in the activity pattern of a relatively small and anatomically dispersed group of neurons. In this review, we have discussed the use of

genetic tools in the mouse that allow the molecular identification and genetic modification of neurons that are electrically active (because of natural environmental stimuli) at a specific point in time. This allows experiments to focus on just those neurons that were active in response to a given stimulus. This has been used to identify cellular and molecular changes that are specific to active neurons relative to their inactive neighbors. Active neural ensembles in the DG and somatosensory cortex have been shown to have increased excitability, interconnectivity, and dendritic spine density. In the hippocampus, learning induced an increase in glutamate receptor trafficking specifically in neurons activated during context learning. Although the functional significance of these cellular changes is still unclear, the approach is revealing that active neural ensembles show underlying molecular differences compared to apparently identical neurons in the same brain region that are not activated by behaviorally relevant stimuli.

In a second line of research, effectors of neural activity were introduced into naturally active ensembles to explore the nature of neural representations. Previous experiments were limited to recording the sensory- or learning-evoked activity of neurons and inferring a role in producing the psychological or perceptual manifestation of that sensory experience or memory. By introducing ChR2 into cfos-active neural ensembles during learning of a specific location in context fear conditioning, the activity of naturally activated neurons could actually be replayed artificially to directly test the relevance of that specific pattern of neural activity in encoding that specific context. The results suggest that even with the relatively crude IEG-based labeling of active ensembles, their reactivation via CHR2 produced a coherent representation/perception of the context. This result has been found in both the hippocampus and retrosplenial cortex with complementary studies in the hippocampus showing that suppression of these ensembles is sufficient to impair context recall. This is the first direct link between dispersed patterns of sensory-evoked neural activity and a coherent perceptual representation. In relation to learning and memory, these results should be viewed as an important first step in identifying the relevant principles of information processing rather than identification of the elusive "engram." The studies show that the stimulation of the correct pattern of neurons in one brain area is able to substitute for, or represent, a complex environmental stimulus (the context box), not that these neurons undergo the physical changes necessary to link that context to the fear-evoking shock. For example, if one could stimulate retinal neurons in precisely the same pattern as when an animal explored a specific context, then the animal would presumably perceive that they were in the box, and freeze if fear conditioned, but we would not suggest that the engram for that memory lies in the retina.

The mammalian brain is an immensely complex information-processing system that transforms an ongoing stream of sensory information into evolutionarily advantageous behavioral outputs. Learning is the process by which specific environmental contingencies (e.g., paring a specific location with an aversive stimulus) alter this processing to produce altered behavioral outputs. The IEG-based genetic tagging approach provides the ability to focus studies specifically on those neurons that are activated by any given set of environmental contingencies. It should prove useful in future studies to identify and genetically test specific models for molecular changes that may underlie learning, such as Hebbian forms of plasticity. A deep understanding of learning and memory will require not just identifying these molecular mechanisms and their location(s) in the brain, but also identifying how they alter the processing of information to evoke the learned responses. The new approaches for recording and manipulating active neural ensembles described in this and previous work offers an experimental framework in which this type of question can begin to be addressed.

## REFERENCES

*Reference is also in this collection.

Alexander GM, Rogan SC, Abbas AI, Armbruster BN, Pei Y, Allen JA, Nonneman RJ, Hartmann J, Moy SS, Nicolelis MA, et al. 2009. Remote control of neuronal activity in transgenic mice expressing evolved G protein-coupled receptors. Neuron 63: 27–39.

* Angelakos CC, Abel T. 2015. Molecular genetic strategies in the study of corticohippocampal circuits. Cold Spring Harb Perspect Biol doi: 10.1101/cshperspect.a021725.

Barth AL, Gerkin RC, Dean KL. 2004. Alteration of neuronal firing properties after in vivo experience in a FosGFP transgenic mouse. J Neurosci 24: 6466–6475.

* Basu J, Siegelbaum SA. 2015. The corticohippocampal circuit, synaptic plasticity, and memory. Cold Spring Harb Perspect Biol doi: 10.1101/cshperspect.a021733.

Baumgartel K, Genoux D, Welzl H, Tweedie-Cullen RY, Koshibu K, Livingstone-Zatchej M, Mamie C, Mansuy IM. 2008. Control of the establishment of aversive memory by calcineurin and Zif268. Nat Neurosci 11: 572–578.

Bontempi B, Laurent-Demir C, Destrade C, Jaffard R. 1999. Time-dependent reorganization of brain circuitry underlying long-term memory storage. Nature 400: 671–675.

Boyden ES, Zhang F, Bamberg E, Nagel G, Deisseroth K. 2005. Millisecond-timescale, genetically targeted optical control of neural activity. Nat Neurosci 8: 1263–1268.

* Byrne JH, Hawkins RD. 2015. Nonassociative learning in invertebrates. Cold Spring Harb Perspect Biol 7: a021675.

Chow BY, Han X, Dobry AS, Qian X, Chuong AS, Li M, Henninger MA, Belfort GM, Lin Y, Monahan PE, et al. 2010. High-performance genetically targetable optical neural silencing by light-driven proton pumps. Nature 463: 98–102.

Cowansage KK, Shuman T, Dillingham BC, Chang A, Golshani P, Mayford M. 2014. Direct reactivation of a coherent neocortical memory of context. Neuron 84: 432–441.

Czajkowski R, Jayaprakash B, Wiltgen B, Rogerson T, Guzman-Karlsson MC, Barth AL, Trachtenberg JT, Silva AJ. 2014. Encoding and storage of spatial information in the retrosplenial cortex. Proc Natl Acad Sci 111: 8661–8666.

Denny CA, Kheirbek MA, Alba EL, Tanaka KF, Brachman RA, Laughman KB, Tomm NK, Turi GF, Losonczy A, Hen R. 2014. Hippocampal memory traces are differentially modulated by experience, time, and adult neurogenesis. Neuron 83: 189–201.

Fleischmann A, Hvalby O, Jensen V, Strekalova T, Zacher C, Layer LE, Kvello A, Reschke M, Spanagel R, Sprengel R, et al. 2003. Impaired long-term memory and NR2A-type NMDA receptor-dependent synaptic plasticity in mice lacking c-Fos in the CNS. J Neurosci 23: 9116–9122.

Foster DJ, Wilson MA. 2006. Reverse replay of behavioural sequences in hippocampal place cells during the awake state. Nature 440: 680–683.

Frankland PW, Bontempi B, Talton LE, Kaczmarek L, Silva AJ. 2004. The involvement of the anterior cingulate cortex in remote contextual fear memory. Science 304: 881–883.

French PJ, O'Connor V, Jones MW, Davis S, Errington ML, Voss K, Truchet B, Wotjak C, Stean T, Doyere V, et al. 2001. Subfield-specific immediate early gene expression associated with hippocampal long-term potentiation in vivo. Eur J Neurosci 13: 968–976.

Frey U, Morris RG. 1997. Synaptic tagging and long-term potentiation. Nature 385: 533–536.

Frey U, Morris RG. 1998. Synaptic tagging: Implications for late maintenance of hippocampal long-term potentiation. *Trends Neurosci* **21**: 181–188.

Garner AR, Rowland DC, Hwang SY, Baumgaertel K, Roth BL, Kentros C, Mayford M. 2012. Generation of a synthetic memory trace. *Science* **335**: 1513–1516.

Ghosh A, Ginty DD, Bading H, Greenberg ME. 1994. Calcium regulation of gene expression in neuronal cells. *J Neurobiol* **25**: 294–303.

Gore F, Schwartz EC, Brangers BC, Aladi S, Stujenske JM, Likhtik E, Russo MJ, Gordon JA, Salzman CD, Axel R. 2015. Neural representations of unconditioned stimuli in basolateral amygdala mediate innate and learned responses. *Cell* **162**: 134–145.

Gossen M, Bujard H. 1992. Tight control of gene expression in mammalian cells by tetracycline-responsive promoters. *Proc Natl Acad Sci* **89**: 5547–5551.

Greenberg ME, Hermanowski AL, Ziff EB. 1986. Effect of protein synthesis inhibitors on growth factor activation of c-fos, c-myc, and actin gene transcription. *Mol Cell Biol* **6**: 1050–1057.

Guenthner CJ, Miyamichi K, Yang HH, Heller HC, Luo L. 2013. Permanent genetic access to transiently active neurons via TRAP: Targeted recombination in active populations. *Neuron* **78**: 773–784.

Guzowski JF, McNaughton BL, Barnes CA, Worley PF. 1999. Environment-specific expression of the immediate-early gene Arc in hippocampal neuronal ensembles. *Nat Neurosci* **2**: 1120–1124.

Han JH, Kushner SA, Yiu AP, Cole CJ, Matynia A, Brown RA, Neve RL, Guzowski JF, Silva AJ, Josselyn SA. 2007. Neuronal competition and selection during memory formation. *Science* **316**: 457–460.

Han JH, Kushner SA, Yiu AP, Hsiang HL, Buch T, Waisman A, Bontempi B, Neve RL, Frankland PW, Josselyn SA. 2009. Selective erasure of a fear memory. *Science* **323**: 1492–1496.

* Hawkins RD, Byrne JH. 2015. Associative learning in invertebrates. *Cold Spring Harb Perspect Biol* **7**: a021709.

Hubel DH, Wiesel TN. 1962. Receptive fields, binocular interaction and functional architecture in the cat's visual cortex. *J Physiol* **160**: 106–154.

Huber D, Petreanu L, Ghitani N, Ranade S, Hromadka T, Mainen Z, Svoboda K. 2008. Sparse optical microstimulation in barrel cortex drives learned behaviour in freely moving mice. *Nature* **451**: 61–64.

Ji D, Wilson MA. 2007. Coordinated memory replay in the visual cortex and hippocampus during sleep. *Nat Neurosci* **10**: 100–107.

* Jercog P, Rogerson T, Schnitzer MJ. 2015. Large-scale fluorescence calcium-imaging methods for studies of long-term memory in behaving mammals. *Cold Spring Harb Perspect Biol* doi: 10.1101/cshperspect.a021824.

Kandel ER, Dudai Y, Mayford MR. 2014. The molecular and systems biology of memory. *Cell* **157**: 163–186.

Karlsson MP, Frank LM. 2009. Awake replay of remote experiences in the hippocampus. *Nat Neurosci* **12**: 913–918.

Keene CS, Bucci DJ. 2008a. Contributions of the retrosplenial and posterior parietal cortices to cue-specific and contextual fear conditioning. *Behav Neurosci* **122**: 89–97.

Keene CS, Bucci DJ. 2008b. Neurotoxic lesions of retrosplenial cortex disrupt signaled and unsignaled contextual fear conditioning. *Behav Neurosci* **122**: 1070–1077.

Kenet T, Bibitchkov D, Tsodyks M, Grinvald A, Arieli A. 2003. Spontaneously emerging cortical representations of visual attributes. *Nature* **425**: 954–956.

Kessels HW, Malinow R. 2009. Synaptic AMPA receptor plasticity and behavior. *Neuron* **61**: 340–350.

Kim JJ, Fanselow MS. 1992. Modality-specific retrograde amnesia of fear. *Science* **256**: 675–677.

Kim Y, Venkataraju KU, Pradhan K, Mende C, Taranda J, Turaga SC, Arganda-Carreras I, Ng L, Hawrylycz MJ, Rockland KS, et al. 2015. Mapping social behavior-induced brain activation at cellular resolution in the mouse. *Cell Rep* **10**: 292–305.

Knapska E, Maren S. 2009. Reciprocal patterns of c-Fos expression in the medial prefrontal cortex and amygdala after extinction and renewal of conditioned fear. *Learn Mem* **16**: 486–493.

Liu X, Ramirez S, Pang PT, Puryear CB, Govindarajan A, Deisseroth K, Tonegawa S. 2012. Optogenetic stimulation of a hippocampal engram activates fear memory recall. *Nature* **484**: 381–385.

Lyford GL, Yamagata K, Kaufmann WE, Barnes CA, Sanders LK, Copeland NG, Gilbert DJ, Jenkins NA, Lanahan AA, Worley PF. 1995. Arc, a growth factor and activity-regulated gene, encodes a novel cytoskeleton-associated protein that is enriched in neuronal dendrites. *Neuron* **14**: 433–445.

Martin KC, Casadio A, Zhu H, Yaping E, Rose JC, Chen M, Bailey CH, Kandel ER. 1997. Synapse-specific, long-term facilitation of aplysia sensory to motor synapses: A function for local protein synthesis in memory storage. *Cell* **91**: 927–938.

Matsuo N, Reijmers L, Mayford M. 2008. Spine-type-specific recruitment of newly synthesized AMPA receptors with learning. *Science* **319**: 1104–1107.

Mayford M. 2014. The search for a hippocampal engram. *Philos Trans R Soc Lond B Biol Sci* **369**: 20130161.

McKernan MG, Shinnick-Gallagher P. 1997. Fear conditioning induces a lasting potentiation of synaptic currents in vitro. *Nature* **390**: 607–611.

Mehta MR, Lee AK, Wilson MA. 2002. Role of experience and oscillations in transforming a rate code into a temporal code. *Nature* **417**: 741–746.

Milanovic S, Radulovic J, Laban O, Stiedl O, Henn F, Spiess J. 1998. Production of the Fos protein after contextual fear conditioning of C57BL/6N mice. *Brain Res* **784**: 37–47.

Morgan JI, Cohen DR, Hempstead JL, Curran T. 1987. Mapping patterns of c-fos expression in the central nervous system after seizure. *Science* **237**: 192–197.

Morris RG. 2013. NMDA receptors and memory encoding. *Neuropharmacology* **74**: 32–40.

Nedivi E, Hevroni D, Naot D, Israeli D, Citri Y. 1993. Numerous candidate plasticity-related genes revealed by differential cDNA cloning. *Nature* **363**: 718–722.

Penfield W. 1968. Engrams in the human brain. Mechanisms of memory. *Proc R Soc Med* **61:** 831–840.

Plath N, Ohana O, Dammermann B, Errington ML, Schmitz D, Gross C, Mao X, Engelsberg A, Mahlke C, Welzl H, et al. 2006. Arc/Arg3.1 is essential for the consolidation of synaptic plasticity and memories. *Neuron* **52:** 437–444.

Qian Z, Gilbert ME, Colicos MA, Kandel ER, Kuhl D. 1993. Tissue-plasminogen activator is induced as an immediate-early gene during seizure, kindling and long-term potentiation. *Nature* **361:** 453–457.

Quiroga RQ, Reddy L, Kreiman G, Koch C, Fried I. 2005. Invariant visual representation by single neurons in the human brain. *Nature* **435:** 1102–1107.

Radulovic J, Kammermeier J, Spiess J. 1998. Relationship between fos production and classical fear conditioning: Effects of novelty, latent inhibition, and unconditioned stimulus preexposure. *J Neurosci* **18:** 7452–7461.

Ragan T, Kadiri LR, Venkataraju KU, Bahlmann K, Sutin J, Taranda J, Arganda-Carreras I, Kim Y, Seung HS, Osten P. 2012. Serial two-photon tomography for automated ex vivo mouse brain imaging. *Nat Methods* **9:** 255–258.

Rahmsdorf HJ, Schonthal A, Angel P, Litfin M, Ruther U, Herrlich P. 1987. Posttranscriptional regulation of c-fos mRNA expression. *Nucleic Acids Res* **15:** 1643–1659.

Ramirez S, Liu X, Lin PA, Suh J, Pignatelli M, Redondo RL, Ryan TJ, Tonegawa S. 2013. Creating a false memory in the hippocampus. *Science* **341:** 387–391.

Redondo RL, Kim J, Arons AL, Ramirez S, Liu X, Tonegawa S. 2014. Bidirectional switch of the valence associated with a hippocampal contextual memory engram. *Nature* **513:** 426–430.

Reijmers LG, Perkins BL, Matsuo N, Mayford M. 2007. Localization of a stable neural correlate of associative memory. *Science* **317:** 1230–1233.

Robertson LM, Kerppola TK, Vendrell M, Luk D, Smeyne RJ, Bocchiaro C, Morgan JI, Curran T. 1995. Regulation of c-fos expression in transgenic mice requires multiple interdependent transcription control elements. *Neuron* **14:** 241–252.

Rogan MT, Staubli UV, LeDoux JE. 1997. Fear conditioning induces associative long-term potentiation in the amygdala. *Nature* **390:** 604–607.

Rumpel S, LeDoux J, Zador A, Malinow R. 2005. Postsynaptic receptor trafficking underlying a form of associative learning. *Science* **308:** 83–88.

Ryan TJ, Roy DS, Pignatelli M, Arons A, Tonegawa S. 2015. Engram cells retain memory under retrograde amnesia. *Science* **348:** 1007–1013.

Saha RN, Wissink EM, Bailey ER, Zhao M, Fargo DC, Hwang JY, Daigle KR, Fenn JD, Adelman K, Dudek SM. 2011. Rapid activity-induced transcription of Arc and other IEGs relies on poised RNA polymerase II. *Nat Neurosci* **14:** 848–856.

Saint Marie RL, Luo L, Ryan AF. 1999. Spatial representation of frequency in the rat dorsal nucleus of the lateral lemniscus as revealed by acoustically induced c-fos mRNA expression. *Hearing Res* **128:** 70–74.

Sano Y, Shobe JL, Zhou M, Huang S, Shuman T, Cai DJ, Golshani P, Kamata M, Silva AJ. 2014. CREB regulates memory allocation in the insular cortex. *Curr Biol* **24:** 2833–2837.

Schoenenberger P, Gerosa D, Oertner TG. 2009. Temporal control of immediate early gene induction by light. *PLoS ONE* **4:** e8185.

Shi SH, Hayashi Y, Petralia RS, Zaman SH, Wenthold RJ, Svoboda K, Malinow R. 1999. Rapid spine delivery and redistribution of AMPA receptors after synaptic NMDA receptor activation. *Science* **284:** 1811–1816.

* Si K, Kandel ER. 2015. The role of functional prion-like proteins in the persistence of memory. *Cold Spring Harb Perspect Biol* doi: 10.1101/cshperspect.a021774.

Silva AJ, Zhou Y, Rogerson T, Shobe J, Balaji J. 2009. Molecular and cellular approaches to memory allocation in neural circuits. *Science* **326:** 391–395.

Smeyne RJ, Schilling K, Robertson L, Luk D, Oberdick J, Curran T, Morgan JI. 1992. fos-lacZ transgenic mice: Mapping sites of gene induction in the central nervous system. *Neuron* **8:** 13–23.

Tanaka KZ, Pevzner A, Hamidi AB, Nakazawa Y, Graham J, Wiltgen BJ. 2014. Cortical representations are reinstated by the hippocampus during memory retrieval. *Neuron* **84:** 347–354.

Vousden DA, Epp J, Okuno H, Nieman BJ, van Eede M, Dazai J, Ragan T, Bito H, Frankland PW, Lerch JP, et al. 2015. Whole-brain mapping of behaviourally induced neural activation in mice. *Brain Struct Function* **220:** 2043–2057.

Wagener RJ, David C, Zhao S, Haas CA, Staiger JF. 2010. The somatosensory cortex of *reeler* mutant mice shows absent layering but intact formation and behavioral activation of columnar somatotopic maps. *J Neurosci* **30:** 15700–15709.

Wang KH, Majewska A, Schummers J, Farley B, Hu C, Sur M, Tonegawa S. 2006. In vivo two-photon imaging reveals a role of arc in enhancing orientation specificity in visual cortex. *Cell* **126:** 389–402.

Wellington CL, Greenberg ME, Belasco JG. 1993. The destabilizing elements in the coding region of c-fos mRNA are recognized as RNA. *Mol Cell Biol* **13:** 5034–5042.

Whitlock JR, Heynen AJ, Shuler MG, Bear MF. 2006. Learning induces long-term potentiation in the hippocampus. *Science* **313:** 1093–1097.

Wilson MA, McNaughton BL. 1994. Reactivation of hippocampal ensemble memories during sleep. *Science* **265:** 676–679.

Worley PF, Christy BA, Nakabeppu Y, Bhat RV, Cole AJ, Baraban JM. 1991. Constitutive expression of zif268 in neocortex is regulated by synaptic activity. *Proc Natl Acad Sci* **88:** 5106–5110.

Xie H, Liu Y, Zhu Y, Ding X, Yang Y, Guan JS. 2014. In vivo imaging of immediate early gene expression reveals layer-specific memory traces in the mammalian brain. *Proc Natl Acad Sci* **111:** 2788–2793.

Yamagata K, Andreasson KI, Kaufmann WE, Barnes CA, Worley PF. 1993. Expression of a mitogen-inducible cyclooxygenase in brain neurons: Regulation by synaptic activity and glucocorticoids. *Neuron* **11:** 371–386.

Yassin L, Benedetti BL, Jouhanneau JS, Wen JA, Poulet JF, Barth AL. 2010. An embedded subnetwork of highly active neurons in the neocortex. *Neuron* **68:** 1043–1050.

Yiu AP, Mercaldo V, Yan C, Richards B, Rashid AJ, Hsiang HL, Pressey J, Mahadevan V, Tran MM, Kushner SA, et al. 2014. Neurons are recruited to a memory trace based on relative neuronal excitability immediately before training. *Neuron* **83:** 722–735.

Zhou Y, Won J, Karlsson MG, Zhou M, Rogerson T, Balaji J, Neve R, Poirazi P, Silva AJ. 2009. CREB regulates excitability and the allocation of memory to subsets of neurons in the amygdala. *Nat Neurosci* **12:** 1438–1443.

# Noninvasive Functional and Anatomical Imaging of the Human Medial Temporal Lobe

Thackery I. Brown[1], Bernhard P. Staresina[1], and Anthony D. Wagner[1,2]

[1]Department of Psychology, Stanford University, Stanford, California 94305-2130
[2]Neurosciences Program, Stanford University, Stanford, California 94305-2130

*Correspondence:* thackery@stanford.edu

The ability to remember life's events, and to leverage memory to guide behavior, defines who we are and is critical for everyday functioning. The neural mechanisms supporting such mnemonic experiences are multiprocess and multinetwork in nature, which creates challenges for studying them in humans and animals. Advances in noninvasive neuroimaging techniques have enabled the investigation of how specific neural structures and networks contribute to human memory at its many cognitive and mechanistic levels. In this review, we discuss how functional and anatomical imaging has provided novel insights into the types of information represented in, and the computations performed by, specific medial temporal lobe (MTL) regions, and we consider how interactions between the MTL and other cortical and subcortical structures influence what we learn and remember. By leveraging imaging, researchers have markedly advanced understanding of how the MTL subserves declarative memory and enables navigation of our physical and mental worlds.

One of the central aims of cognitive neuroscience research is to understand how human brain function relates to the mnemonic experiences that define much of who we are as individuals. Recent advances in noninvasive human neuroimaging have given rise to novel insights about the neural foundations of human memory, and have allowed neuroscientists to draw important connections between human and nonhuman animal research. Although noninvasive imaging techniques remain limited in their spatial resolution (we cannot yet describe the behavior of individual neurons in the human brain without implanting electrodes in patients undergoing brain surgery), they also have

important strengths that have allowed researchers to significantly advance mechanistic accounts of learning and memory. In this review, we begin with a historical perspective on noninvasive neuroimaging techniques and their application to the study of memory encoding. We then introduce more recent cutting-edge methodological approaches, as we discuss specific domains of memory theory that they have helped advance. Drawing on research examining the medial temporal lobe (MTL), we emphasize the power of such imaging techniques to allow scientists to make inferences about the types of mnemonic information represented by distinct brain areas, and to understand how the functions of different regions and neural networks underlie our ability to learn and remember the details of our lives.

Although it can be challenging to assay what nonverbal animals are thinking and why, working directly with humans allows researchers to directly probe the subjective experience of remembering in people, and to relate their cognition to underlying neural processes. For example, it is easy to ask a human participant about the contents of their memories, or to assay how confident they are about having previously encountered a stimulus. In this review, we document how the ability to relate behavioral indices of memory to measures of neural activity allows research to advance our thinking about the biological underpinnings of human cognition. We focus primarily on recent functional magnetic resonance imaging (fMRI), and anatomical magnetic resonance imaging (MRI) research addressing the role of MTL subregions in human declarative memory (principally, episodic memory for individual life events), and also highlight findings addressing how declarative memory mechanisms interact with those of procedural memory in service of efficient memory-guided behavior.

## EPISODIC MEMORY ENCODING

In the course of a day, humans encounter a steady stream of sensations, emotions, thoughts, and actions elicited by the external world and internal states. Encoding mechanisms transform this stream of information into long-term neural representations of co-occurring event features —that is, into episodic memory traces—and, thus, establish lasting footprints in our minds of life's events.

### Imaging and the Subsequent Memory Paradigm

How to noninvasively study memory-encoding processes in humans is not a trivial question. In healthy individuals, we cannot directly measure the synaptic and cellular dynamics that allow an experience to be encoded such that it can be later remembered. Instead, cognitive neuroscientists

have leveraged noninvasive imaging methods to identify neural correlates of memory formation in humans, evidenced as neural predictors of subsequent memory expression (Paller et al. 1987; Wagner et al. 1999; Paller and Wagner 2002; Spaniol et al. 2009; Uncapher and Wagner 2009; Kim 2011). Subsequent memory paradigms involve recording neural activity, using either electromagnetic or hemodynamic measures, while participants encounter and process stimuli (e.g., a series of words) during an "encoding period." Subsequently, memory for each stimulus is tested after a delay, and encoding period activity is examined conditioned on the behavioral expressions of memory, such as whether the stimulus was subsequently remembered or forgotten, or remembered with high or low confidence. Differences in encoding period activity as a function of later memory outcomes, such as subsequently remembered and forgotten stimuli (termed "subsequent memory effects" or difference as a result of memory—"Dm"—effects), can then be interpreted in terms of their potential contributions to the formation of memories.

Early evidence for subsequent memory effects came from electroencephalography (EEG) research in humans (e.g., Sanquist et al. 1980; reviewed in Wagner et al. 1999). EEG is a noninvasive technique in which electrodes, placed on the scalp of a participant, passively record voltages at the surface of the scalp induced by ion currents in the underlying cortex (Niedermeyer and Lopes da Silva 2005). Because EEG signals reflect electrical current from neural activity, the temporal resolution of EEG is on the order of milliseconds, allowing researchers to address hypotheses about the time course of neural events. One of the most common ways to analyze EEG data is to study the profile of event-related potentials (ERPs), which are electrical responses time-locked to or evoked by the onset of a stimulus or response. By analyzing the averaged ERP time-course associated with stimulus encoding, Sanquist and colleagues (1980) provided initial evidence of a subsequent memory effect in humans, finding that items that were subsequently remembered had a more positive deflection in the ERP time-course

Cite this article as *Cold Spring Harb Perspect Biol* doi: 10.1101/cshperspect.a021840

~450 to 750 msec after stimulus presentation (relative to items later forgotten).

Following this early finding, extensive EEG research corroborated the presence of multiple subsequent memory effects underlying successful encoding (e.g., Long et al. 2014; reviewed in Rugg 1995; Wagner et al. 1999; Friedman and Johnson 2000; Paller and Wagner 2002; Nyhus and Curran 2010). Among their discoveries, EEG studies showed that the encoding of memories is a multifaceted process characterized by no single neural response. Instead, this rich literature revealed that there are distinct electrophysiological signatures associated with encoding, which manifest both in different poststimulus time periods and different spatial locations on the scalp. Furthermore, distinct ERPs are associated with different types of subsequent memory, such as memory for specific stimuli versus the contextual details surrounding an encoding experience (Johnson et al. 1997; Bridger and Wilding 2010; Angel et al. 2013). Such data suggest that multiple neural structures and mechanisms underlie successful episodic encoding.

Although providing some leverage on the temporal dynamics of encoding activity, EEG data present an important challenge for understanding the neural bases of memory formation in humans. As a measure of electrical signal at the scalp, EEG is most sensitive to postsynaptic potentials generated in superficial layers of the cortex. Furthermore, the signals from millions of neurons in adjacent brain areas converge and are diffuse at the level of the scalp. As a result, although EEG provides excellent resolution about "when" a neural response happens, it is challenging to discern "where" this processing occurs. This limitation is less pronounced in fMRI, which suffers from low temporal resolution but affords spatial resolution at the millimeter scale. fMRI inherits these strengths and limitations from its use of oxygenated blood flow as an indirect measure of neural activity. The mechanistic link between the blood oxygen level–dependent (BOLD) fMRI signal and neural activity remains an active area of research (Logothetis and Wandell 2004). Most fMRI studies obtain data on the order of 3 to 4 mm$^3$ resolution, and the technology is continually improv-

ing, with recent advances in MRI hardware and fMRI pulse sequences allowing for whole-brain imaging at substantially higher spatial and temporal resolution (Feinberg et al. 2010; Moeller et al. 2010). For example, acquisition and analysis techniques now allow for individual subject and group-level analyses of human brain activity at spatial resolutions approaching 1 mm (Zeineh et al. 2000; Grill-Spector et al. 2006; Ekstrom et al. 2009; Yassa and Stark 2009; Yassa et al. 2010), and in some cases even higher (Yacoub et al. 2008; Heidemann et al. 2012), positioning fMRI as a powerful tool for researchers to noninvasively pinpoint where activity related to memory, perception, and cognition occurs in the brain.

For memory researchers, this increased spatial resolution is particularly beneficial as it allows fine-grained investigation of functional distinctions within the human MTL (e.g., Kirwan et al. 2007; Bakker et al. 2008; Ekstrom et al. 2009; reviewed in Carr et al. 2010; Chadwick et al. 2011; Bonnici et al. 2012, 2013; Libby et al. 2012; LaRocque et al. 2013; Brown et al. 2014a). The MTL has long been understood to be critical for episodic memory, based on decades of research that was spurred by the landmark case of Henry Molaison (HM), whose ability to remember new experiences from his daily life was dramatically impaired following bilateral resection of the hippocampus, and portions of the neighboring MTL cortical structures (perirhinal, parahippocampal, and entorhinal cortex) (Scoville and Milner 1957; Corkin 2013; Squire and Dede 2015). Although studies of HM and other patients suffering MTL lesions have yielded a multitude of novel insights about the role of the MTL in declarative memory (Eichenbaum and Cohen 2001; Squire et al. 2004; Moscovitch et al. 2006; Squire and Bayley 2007; Graham et al. 2010; Greenberg and Verfaellie 2010; Montaldi and Mayes 2010; Rosenbaum et al. 2014), a challenge for neuropsychological studies is to identify whether the observed memory deficits reflect impairments at encoding, retrieval, or both. Noninvasive human imaging techniques complement lesion studies, as they provide a critical set of tools for studying healthy brain function by measuring neural responses at the time memories are encoded and, as we will later discuss, at the time that they are retrieved. Indeed, the first fMRI studies using the subsequent memory paradigm (Brewer et al. 1998; Wagner et al. 1998) provided early evidence that event-related levels of encoding period activation in the human MTL, as well as in the lateral prefrontal cortex, predict whether a stimulus will be later remembered or forgotten. We next consider how fMRI subsequent memory data have informed theory about MTL mnemonic function.

## Testing Theories of MTL Functional Differentiation

Episodic memories often contain a remarkable wealth of detail from an object that caught our attention to the environment in which we encountered it. At retrieval, sometimes we only have a vague sense of recognition when viewing a previously encountered stimulus, whereas, at other times, we vividly recollect many details of the prior experience. A number of theoretical frameworks have been formulated in an effort to capture the link between this diversity in mnemonic experience and the functions of different subregions of the MTL, most notably the hippocampus, perirhinal cortex, and parahippocampal cortex (Cohen and Eichenbaum 1993; Murray and Bussey 1999; Brown and Aggleton 2001; Davachi 2006; Diana et al. 2007; Eichenbaum et al. 2007; Mayes et al. 2007; Graham et al. 2010).

Leveraging the subsequent memory paradigm combined with fMRI's ability to distinguish activity from distinct, but spatially proximal MTL cortical regions (Carr et al. 2010), researchers have begun to test competing hypotheses about how specific MTL subregions contribute to episodic memory. For example, fMRI studies have shown content specialization within the MTL cortex that is predictive of subsequent memory. Specifically, consistent with their differing connectivity with other cortical areas (Suzuki 2009), activity in the parahippocampal cortex at encoding has been shown to predict later memory for scenes, while activity in the perirhinal cortex at encoding predicts later memory for faces, objects (Litman et al.

2009; Preston et al. 2010; Staresina and Davachi 2010; Staresina et al. 2011), and their associations (Staresina and Davachi 2008; Watson et al. 2012). Convergent evidence suggests that such functional differentiation along the anterior–posterior axis of the MTL cortex is best understood as a continuous gradient; for example, Liang and colleagues (2013) showed that face/object and scene representations are coded to differing degrees across the MTL cortex, with the greatest specialization for scene memory in the posterior parahippocampal region and greatest specialization for face/object processing in the perirhinal cortex (see also Lee et al. 2008; Barense et al. 2010).

Although specific types of event content appear to be differentially represented along the anterior–posterior axis of the MTL cortex, the convergence of perirhinal and parahippocampal inputs on the hippocampus, via the entorhinal cortex, is thought to enable the binding of the distinct facets (i.e., "items" or "items and context") of an event into a conjunctive memory trace. This theoretical perspective has garnered some support from fMRI subsequent memory studies, which have revealed that MTL cortical activity at encoding (principally in the perirhinal cortex) differentially predicts later item recognition memory, whereas hippocampal activity at encoding differentially predicts later memory for item–context and item–item associations (e.g., Davachi et al. 2003; Kirwan and Stark 2004; Ranganath et al. 2004).

## Multivariate fMRI Analyses and Memory Theory

The high spatial resolution of MRI has recently been combined with multivariate analysis techniques (Norman et al. 2006; Kriegeskorte et al. 2008), providing a powerful new means to address the mechanisms giving rise to, and the representational contents of, episodic memories. One such technique—representational similarity analysis (RSA)—has been particularly informative for testing hypotheses about the neural mechanisms that support successful encoding. Briefly, RSA measures the similarity (correlation) between event- or stimulus-specific patterns of activation, which can be used to query how the similarity (or dissimilarity) of multivoxel patterns measured during encoding relate to subsequent memory. As we next illustrate, such measures provide important tests of theoretical predictions about MTL functional differentiation underlying memory.

One dominant theoretical perspective on the functional contributions of different MTL subregions to memory, the "complementary learning systems" model, holds that a hippocampal "pattern separation" mechanism supports memory for events by orthogonalizing (making more distinct) neural representations during encoding, allowing later retrieval cues to trigger memory for a unique event (O'Reilly and McClelland 1994; McClelland et al. 1995; O'Reilly and Rudy 2001; Norman and O'Reilly 2003; Norman 2010). Conversely, this perspective holds that the neighboring MTL cortex sup-

ports memory by gradually encoding neural representations that capture the commonalities across similar stimuli, permitting later item recognition on the basis of global similarity between the present and past. Using RSA, coupled with high-resolution fMRI of the MTL, La Rocque and colleagues (2013) computed the similarity of the multivoxel fMRI pattern elicited by a stimulus at encoding to the patterns elicited by other encoded stimuli. By computing these across-item pattern similarities for each MTL subregion—perirhinal cortex, parahippocampal cortex, and these researchers obtained strong support for the "complementary learning systems" framework: namely, greater across-item similarity in the MTL cortex, but reduced across-item similarity in the hippocampus, was predictive of later successful memory (Fig. 1).

RSA has also been used to gain leverage on a longstanding debate about whether the stabil-

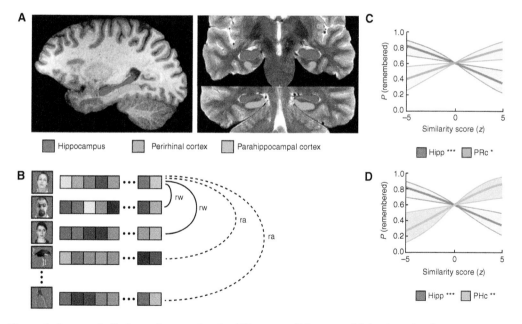

Figure 1. Pattern similarity and separation in different medial temporal lobe (MTL) subregions supports encoding. (A) Anatomically defined hippocampus, perirhinal cortex, and parahippocampal cortex regions of interest, overlaid on a structural magnetic resonance imaging (MRI). (B) Across-stimuli representational similarity (within category, rw; across category, ra) was computed using Pearson correlations between the patterns of neural activity, across voxels, for pairs of stimuli. (C,D) Critically, logistic regression revealed that the similarity of an item's encoding pattern to those of other items in the perirhinal cortex (PRc) and parahippocampal cortex (PHc) positively predicted subsequent memory. Conversely, hippocampal (Hipp) pattern similarity negatively predicted subsequent memory. $^*p < 0.05$; $^{**}p < 0.01$; $^{***}p < 0.005$. (From LaRocque et al. 2013; adapted from several in the source, with permission, from the Society for Neuroscience © 2013.)

ity of a neural representation across encoding experiences is beneficial for subsequent memory, or whether variability in the neural representations of a stimulus across experiences is beneficial (the "encoding-variability" hypothesis) (Martin 1968). In particular, RSA analyses of fMRI encoding data suggest that, in some contexts, representational stability may be beneficial to later remembering (Xue et al. 2010; Ward et al. 2013; but see Wagner et al. 2000), demonstrating that greater pattern similarity of an item's neural representations across multiple encoding trials predicts better subsequent memory for the item. Although some questions remain (Xue et al. 2013; Davis et al. 2014), researchers are now positioned to measure, at the individual trial level and within an individual human brain, the large-scale distributed neural representations that underlie important aspects of memory behavior.

A related multivariate technique—multivoxel pattern analysis (MVPA)—has also been effectively leveraged to advance understanding of episodic memory, including testing the prediction that the "strength" of a neural representation at encoding predicts later memory expression. In MVPA, multivoxel activation patterns for two or more classes of stimuli (e.g., faces vs. scenes) are used to train a "classifier" to identify the characteristic activity patterns that maximally discriminate between the stimulus classes. As such, from training data, a classifier learns to partition neural patterns into class-labeled decision regions (e.g., patterns representative of faces and patterns representative of scenes), and then can be used to estimate where a novel test pattern falls with respect to the boundaries between these decision regions. Thus, when presented with a new test pattern, the classifier is used to predict to which of the learned classes the new event belongs (Norman et al. 2006; Rissman and Wagner 2012). Importantly, pattern classifiers can output probabilistic predictions about the new event's likely class, which provides a trial-specific quantitative measure of the strength of neural evidence. Recently, researchers have shown that the "strength" of content-specific (i.e., face vs. scene) neural evidence in the visual cortex correlates with the magni-

tude of hippocampal univariate fMRI activity at encoding and, critically, predicts whether that information will later be remembered or forgotten (Kuhl et al. 2012; Gordon et al. 2013). These data suggest that the success of hippocampally mediated encoding of event details is influenced by, or at least covaries with, the strength or fidelity of the corresponding cortical representations. We expect that these, and other recent MVPA and RSA observations (Johnson et al. 2009; McDuff et al. 2009; Rissman et al. 2010; Ward et al. 2013), will be the first of many instances, in which the application of multivariate analytic techniques, combined with the spatial resolution of fMRI, allows researchers to make critical progress on open questions about the neurobiological mechanisms governing memory.

## EPISODIC MEMORY RETRIEVAL

At its core, retrieval can be considered as the reinstatement or reconstruction of information that was encoded in memory (Ben-Yakov et al. 2015). With the ability to measure large-scale neural networks involved in retrieval, researchers studying memory in humans are able to consider not only where features of encoded memories are "stored" in the brain, but which regions and processes contribute to what is successfully remembered (and in how much detail). In this section, we discuss noninvasive functional imaging findings that provide critical insights into (1) the role of MTL subregions at retrieval, and (2) how distributed neural networks can interact to guide goal-directed memory retrieval and memory-guided behavior.

### Multivariate fMRI and Connectivity Analyses of Episodic Retrieval

As with encoding, multivariate pattern analyses provide a powerful approach to examine the expression of memory content at retrieval. For instance, early MVPA work from Polyn et al. (2005) showed that a classifier, trained to distinguish between neural patterns elicited by the encoding of face, object, and scene classes of stimuli, can predict which category of stimuli will be imminently freely recalled based on shifts in these categorical neural patterns during retrieval. Other data indicate that the representational strength of face and scene patterns in the visual cortex during encoding is not only predictive of subsequent memory behavior, but also of the strength of pattern reinstatement at retrieval (Gordon et al. 2013), providing an important demonstration of the link between representations of experiences as they unfold and of memory for those experiences as they are retrieved.

Episodic memories are characterized by the representation of unique experiences in our lives. One strength of RSA for assessing episodic reinstatement lies in the ability to examine stimulus- or event-specific representations at retrieval. For instance, in a recent fMRI study, participants were presented with word–scene combinations during encoding and then, at retrieval, they were provided with only the word as a retrieval cue. It was found that stimulus-specific scene patterns were reactivated in the para-

hippocampal cortex when participants indicated they had recollected the target scene, but not when they indicated they had no recollection (Fig. 2) (Staresina et al. 2012; see also, Ritchey et al. 2008; Bosch et al. 2014). This is an important step toward demonstrating that episodic reinstatement in the MTL can be observed at the level of individual memory representations.

Although multivariate pattern analyses facilitate the study of representational content in specific brain regions, functional connectivity measures provide a means of noninvasively studying how different regions and networks interact in support of memory and cognition. Connectivity measures leverage variability in the magnitude and timing of fMRI responses across trials to index functional relationships between distinct regions (e.g., Friston et al. 1993, 2003; McIntosh and Gonzalez-Lima 1994; Cordes et al. 2000; Lowe et al. 2000; Greicius et al. 2003; Rissman et al. 2004; Sun et al. 2004; reviewed in Stephan and Friston 2010; Friston

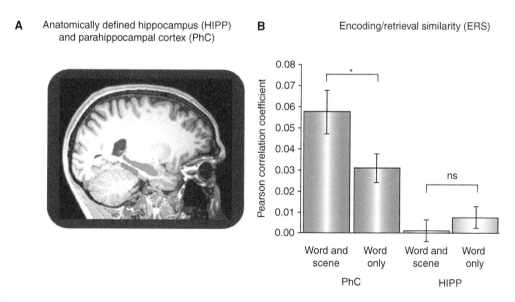

**A** Anatomically defined hippocampus (HIPP) and parahippocampal cortex (PhC)

**B** Encoding/retrieval similarity (ERS)

Figure 2. Encoding period patterns in the medial temporal lobe (MTL) are reinstated during successful retrieval. (A) Anatomically defined hippocampus and parahippocampal cortex regions of interest, overlaid on a structural magnetic resonance imaging (MRI). (B) Representational similarity (Pearson correlation) in the parahippocampal cortex (PhC) showed that when word–scene stimulus pairs are successfully recollected, neural patterns at retrieval are significantly more similar to those elicited for the unique word–scene combinations during encoding than when only the word is remembered. These reinstatement measures were not present in the hippocampus (HIPP), but trial-by-trial hippocampal activity predicted reinstatement, consistent with pattern completion mechanisms supporting retrieval. ns, not significant; $^*p < 0.05$. (From Staresina et al. 2012; reprinted, with permission, from the Society for Neuroscience 2012.)

2011; Friston et al. 2013) and, within the context of memory, their relation to memory behavior. Such functional relationships can be either direct (i.e., activity in one region directly influences activity in the other) or indirect (i.e., mediated by the function of another region). The ability to study large-scale networks across the brain is a particular strength of noninvasive functional imaging measures in humans.

Connectivity measures provide an important means for studying how information is shared within the MTL system, and between the MTL and other networks in the brain. In the preceding section on encoding, we discussed how different features of memories, specifically object/item information and scene/spatial information, are differentially represented by the perirhinal and parahippocampal cortex, respectively. The integration of these signals via the hippocampus is theorized to be a key neurobiological step in episodic memory encoding, and provides a basis for stimuli to cue associated memories at retrieval (Marr 1971; Teyler and DiScenna 1986; Treves and Rolls 1994). For example, how is it that when we look at our desk lamp it can trigger a memory for the location where the lamp was purchased last week? Researchers have posited that this might occur via a systematic flow of information, whereby information about the lamp might elicit a spread of activity through associations in the hippocampus (known as "pattern completion") that leads to reactivation of the other representations associated with the lamp. A recent development in functional connectivity approaches, known as effective connectivity, allows researchers to test evidence for such directional predictions about the flow of information in brain networks (McIntosh and Gonzalez-Lima 1994; Stephan and Friston 2010; Smith et al. 2011; Friston et al. 2013). Effective connectivity measures seek to support inferences about causality in connectivity, by modeling the fit of predicted directional relationships with an fMRI signal from a target network of brain regions. Using one form of effective connectivity, known as dynamic causal modeling (DCM) (Friston et al. 2003), researchers have provided evidence that retrieval of a memory from a cue arises from

information transfer within the MTL. Specifically, activity caused by processing a scene cue in the parahippocampal cortex can drive activity for an associated object in the perirhinal cortex, with this interaction being mediated by the hippocampus (Staresina et al. 2013). These data provide novel support for a fundamental prediction about the role of the hippocampus in mediating the link between segregated item or feature representations (here, scene and object information) in support of rich episodic remembering.

## SPATIAL MEMORY AND THE CONTRIBUTIONS OF MULTIPLE MEMORY SYSTEMS

The functional interaction between separate spatial and nonspatial representations, mediated via the hippocampus, has important implications for the direction of future memory research. Although the early findings in patient HM provided a framework for decades of research on the role of the hippocampus in episodic memory, the existence of "place cells" in the rodent hippocampus (O'Keefe and Dostrovsky 1971; O'Keefe 1976; Moser et al. 2008, 2015) led to a separate hypothesis that the hippocampus creates internal "cognitive maps" of environments (Tolman 1948; O'Keefe and Nadel 1978). Although research examining the roles of the hippocampus in episodic and spatial memories increasingly crosses paths, these have historically remained two distinct areas of study in human cognitive neuroscience. One means of bridging the gap between these areas of research is to consider hippocampal representations of location as one mechanism underlying its broader role in associating stimuli and experiences across space and time (Eichenbaum and Cohen 2014).

In real-world scenarios, episodic memories encompass the "who, what, when, and where" of an experience and, thus, require the ability to embed nonspatial information (e.g., faces and objects) in memory for environments (e.g., Burgess et al. 2001; reviewed in Burgess et al. 2002; Bird and Burgess 2008). The effective connectivity data described in the preceding section

(Staresina et al. 2013) are an important step toward understanding how spatial and nonspatial memory signals combine within the hippocampus to support the expression (in addition to the encoding and construction) of such integrated knowledge. Moreover, recent high-resolution fMRI data show that, when scene information is presented as a cue for memory of a specific navigational episode, trial-by-trial responses in the parahippocampal cortex and the hippocampal CA1 subfield during cue processing correlate with prospective retrieval of the desired navigational event (Brown et al. 2014a). CA1 represents a final stage of processing in the hippocampal circuit, and theories of hippocampal function propose that the convergence in CA1 of representations from the CA3 subfield and the MTL cortex facilitates sequential retrieval, and gates hippocampal output to memories that are congruent with current context (Hasselmo and Wyble 1997; Hasselmo and Eichenbaum 2005; Kesner 2007). Brown and colleagues' high-resolution fMRI data, among other recent navigation work (Brown et al. 2010; Brown and Stern 2013), suggest that parahippocampal scene representations could underlie hippocampal reactivation of navigational episodes, and support key predictions about CA1's role in the flexible retrieval of goal-relevant memories. Moreover, complementary fMRI data (Suthana et al. 2009) have associated CA1 activity with the learning of allocentric (map level) representations of environments; such flexible spatial representations may be critical as a scaffold for remembering the locations of, and relationships between, specific events in our lives.

Building from the hypothesis that the hippocampus represents cognitive maps of environments, spatial navigation research has often contrasted hippocampal-dependent memory for complex spatial relationships between locations with navigation based on striatal-dependent motor associations for specific cues and landmarks (Hartley et al. 2003; Iaria et al. 2003; Doeller et al. 2008). Noninvasive measures of anatomical morphology have recently linked these concepts. Specifically, using one class of anatomical MRI analysis known as voxel-based morphometry (VBM) (Mechelli et al. 2005),

researchers have shown that volume estimates in the hippocampus and caudate nucleus differentially correlate with the predisposition of a person to rely on spatial knowledge or response-based strategies to solve navigational problems (Bohbot et al. 2007; Konishi and Bohbot 2013), as well as with the level of an individual's "expertise" as a spatial navigator (Maguire et al. 2006). Briefly, VBM leverages regional volumetric differences between anatomical MRI images for each participant and a standardized template brain. By examining the degree to which a brain region (e.g., hippocampus) must be enlarged or compressed to fit the template brain, researchers can infer volumetric differences between participants in their dataset. VBM analyses have been combined with automated methods for segmenting MRI images into gray matter structures (Fischl et al. 2002; Patenaude et al. 2011) to show that hippocampal volume predicts an individual's ability to learn and remember map-level information (Hartley and Harlow 2012; Schinazi et al. 2013) (such techniques have also been used with nonspatial memory paradigms to show functional specialization within the hippocampus—linking posterior hippocampal volume, specifically, with contextual memory performance [Poppenk and Moscovitch 2011]). Similarly, researchers have used diffusion tensor imaging (a method for tracking water molecule diffusion along white matter tracts in the brain) to show that greater directionality in water molecule diffusion in the hippocampus, putatively indicative of greater white matter integrity and organization, correlates with improved ability of participants to learn and retrieve cognitive map information (Iaria et al. 2008).

Critically, in the real world, we need a mechanism for flexibly translating both spatial and nonspatial forms of episodic memory into goal-directed actions. The hippocampus is not anatomically positioned to directly control motor behavior, and early evidence from rodent navigation studies (Devan and White 1999) led to the prediction that the hippocampus may direct behavior via engagement with striatal circuitry. Therefore, although hippocampal and striatal forms of memory may differ in fundamental ways (see Graybiel and Grafton 2015), given that navigational memories are often complex and can incorporate both spatial and behavioral information, our ability to navigate in real-world settings may draw on both regions; more generally, integration of MTL and frontostriatal computations may be important for memory and memory-guided behavior in many scenarios as a function of their combined relevance to current task demands. The interplay between these systems can be compensatory; for example, fMRI research in Huntington's disease patients, who suffer from progressive pathology affecting striatal circuitry, has shown that route navigation—potentially supported by response associations in the striatum (Hartley et al. 2003)—can fall back on hippocampal computations as the striatal system fails (Voermans et al. 2004). Importantly, navigation of familiar routes in healthy populations can also rely on the hippocampus when navigational responses depend on explicit knowledge of the current navigational context, that is, when navigation draws more strongly on features of episodic memory.

A real-world example that most of us are familiar with is the experience of traversing an intersection between two familiar navigational routes. In this scenario, we need to choose between two possible directions based on memory for which path is most relevant to the goal of the current navigational episode. Consistent with the episodic memory demands of this scenario, recent VBM data show that hippocampal gray matter volume in young adults correlates with the ability to perform such context-dependent route navigation (Fig. 3A) (Brown et al. 2014b). Moreover, fMRI research has shown that people faced with alternative route memories in a virtual navigation task draw on both hippocampal and striatal processes to identify and select which path to take (Brown et al. 2010). In fact, not only do the hippocampus and the medial caudate (a striatal subregion implicated in behavioral flexibility) support learning of navigational episodes (Brown and Stern 2013), but their recruitment during navigation of familiar routes increases as memories for new alternative

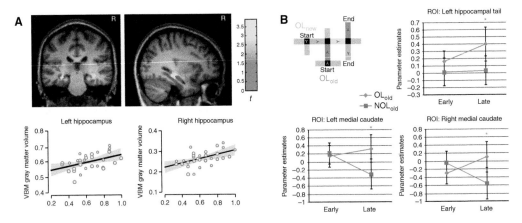

Figure 3. Hippocampus and striatum support flexible navigational learning and memory. (*A*) Statistical map of bilateral hippocampal region, in which volume positively correlated with context-dependent route retrieval ability in healthy young adults (*upper* frames). Volume estimates extracted from the hippocampus tightly predict individual differences in performance (*lower* frames). (*B*) Navigation of familiar routes (overlapping [OL]$_{old}$; blue path on map) becomes increasingly reliant on hippocampal and striatal mechanisms as a novel interfering route memory is introduced (OL$_{new}$; green path on map)—fMRI activity in the left hippocampus and bilateral medial caudate increased from early to late navigation trials (blue line in graphs) as participants became more familiar with the novel competing route memory. In contrast, navigation of familiar nonoverlapping routes (nonoverlapping [NOL]$_{old}$; red line in graphs), for which contextual retrieval demands were limited, relied minimally on these medial temporal lobe (MTL) and striatal subregions from early to late trials (even decreasing for the medial caudate with continued practice). ROI, region of interest; VBM, voxel-based morphometry; R, right hemisphere; *$p < 0.05$. (From Brown et al. 2014b and Brown and Stern 2013; adapted, with permission, from the authors.)

Cite this article as *Cold Spring Harb Perspect Biol* doi: 10.1101/cshperspect.a021840

paths are introduced to the environment, increasing the need to draw on episodic memory to guide selection of behavior (Fig. 3B). Importantly, functional connectivity research suggests that the hippocampus and striatal circuitry interact cooperatively at retrieval when people need to use episodic memory to guide navigation (Brown et al. 2012). (Recent evidence also suggests that corecruitment of, and functional interactions between, the hippocampus and striatum may also be important in non-navigation settings for forming episodic memories [Ben-Yakov and Dudai 2011; Sadeh et al. 2011] and making episodic and relational memory judgments [Moses et al. 2010; Ross et al. 2011]). Furthermore, Brown and colleagues (2012) showed that the hippocampus and striatum functionally interact with regions of the prefrontal cortex during contextual navigation, suggesting that the prefrontal cortex may mediate goal-directed memory and the interaction between these systems. Ultimately, this line of research illustrates how episodic memory supported by the hippocampus plays a critical part in spatial navigation, and demonstrates that the distinct functions of the MTL and striatal systems can combine to support navigation in real-world scenarios in humans; more broadly, the imaging data described in this section have laid important groundwork for understanding (1) how we leverage multiple memory systems to achieve long-term goals, and (2) the importance of network interactions for constructing and navigating mnemonic representations of our lives.

## CONCLUSIONS

The ability to remember an event from last week or to plan which route to take to the grocery store are fundamentally multiprocess and multinetwork acts, integrating declarative memory with systems of attention, and cognitive and behavioral control. Noninvasive functional imaging techniques are essential for advancing memory neuroscience as a field, enabling the study of human memory at its many cognitive and mechanistic levels. These techniques have yielded novel insights into the types of information represented in distinct brain regions, the mnemonic computations that specific regions perform, and how the functions of different regions interact to influence what we learn and remember about our world. Moreover, data addressing functional organization and neural network interactions from human research can serve as a crucial guide for directing neural recordings, invasive high-resolution imaging, and genetic and pharmacological manipulations in nonhuman animals. By leveraging these techniques to study how representational features, functional interactions, and anatomical morphology of brain areas support mnemonic experience, researchers will be able to better understand the neural basis of declarative memory and how changes associated with both development and disease affect this core element of the human condition.

## ACKNOWLEDGMENTS

Supported by grants from the Wallenberg Foundation's Network Initiative on Culture, Brain, and Learning (A.D.W.) and the Wellcome Trust (B.P.S.).

## REFERENCES

*Reference is also in this collection.

Angel L, Isingrini M, Bouazzaoui B, Fay S. 2013. Neural correlates of encoding processes predicting subsequent cued recall and source memory. *Neuroreport* **24:** 176–180.

Bakker A, Kirwan CB, Miller M, Stark CE. 2008. Pattern separation in the human hippocampal CA3 and dentate gyrus. *Science* **319:** 1640–1642.

Barense MD, Henson RNA, Lee ACH, Graham KS. 2010. Medial temporal lobe activity during complex discrimination of faces, objects, and scenes: Effects of viewpoint. *Hippocampus* **20:** 389–401.

Ben-Yakov A, Dudai Y. 2011. Constructing realistic engrams: Poststimulus activity of hippocampus and dorsal striatum predicts subsequent episodic memory. *J Neurosci* **31:** 9032–9042.

Bird CM, Burgess N. 2008. The hippocampus and memory: Insights from spatial processing. *Nat Rev Neurosci* **9:** 182–194.

Bohbot VD, Lerch J, Thorndycraft B, Iaria G, Zijdenbos AP. 2007. Gray matter differences correlate with the spontaneous strategies in a human virtual navigation task. *J Neurosci* **27:** 10078–10083.

Bonnici HM, Kumaran D, Chadwick MJ, Weiskopf N, Hassabis D, Maguire EA. 2012. Decoding representations of scenes in the medial temporal lobes. *Hippocampus* **22:** 1143–1153.

Bonnici HM, Chadwick MJ, Maguire EA. 2013. Representations of recent and remote autobiographical memories in hippocampal subfields. *Hippocampus* **23:** 849–854.

Bosch SE, Jehee JFM, Fernandez G, Doeller CF. 2014. Reinstatement of associative memories in early visual cortex is signaled by the hippocampus. *J Neurosci* **34:** 7493–7500.

Brewer JB, Zhao Z, Desmond JE, Glover GH, Gabrieli JD. 1998. Making memories: Brain activity that predicts how well visual experience will be remembered. *Science* **281:** 1185–1187.

Bridger EK, Wilding EL. 2010. Requirements at retrieval modulate subsequent memory effects: An event-related potential study. *Cogn Neurosci* **1:** 254–260.

Brown MW, Aggleton JP. 2001. Recognition memory: What are the roles of the perirhinal cortex and hippocampus? *Nat Rev Neurosci* **2:** 51–61.

Brown TI, Stern CE. 2013. Contributions of medial temporal lobe and striatal memory systems to learning and retrieving overlapping spatial memories. *Cereb Cortex* **24:** 1906–1922.

Brown TI, Ross RS, Keller JB, Hasselmo ME, Stern CE. 2010. Which way was I going? Contextual retrieval supports the disambiguation of well learned overlapping navigational routes. *J Neurosci* **30:** 7414–7422.

Brown TI, Ross RS, Tobyne SM, Stern CE. 2012. Cooperative interactions between hippocampal and striatal systems support flexible navigation. *Neuroimage* **60:** 1316–1330.

Brown TI, Hasselmo ME, Stern CE. 2014a. A high resolution study of hippocampal and medial temporal lobe correlates of spatial context and prospective overlapping route memory. *Hippocampus* **24:** 819–839.

Brown TI, Whiteman AS, Aselcioglu I, Stern CE. 2014b. Structural differences in hippocampal and prefrontal gray matter volume support flexible context-dependent navigation ability. *J Neurosci* **34:** 2314–2320.

Burgess N, Maguire EA, Spiers HJ, O'Keefe J. 2001. A temporoparietal and prefrontal network for retrieving the spatial context of lifelike events. *Neuroimage* **14:** 439–453.

Burgess N, Maguire EA, O'Keefe J. 2002. The human hippocampus and spatial and episodic memory. *Neuron* **35:** 625–641.

Carr VA, Rissman J, Wagner AD. 2010. Imaging the human medial temporal lobe with high-resolution fMRI. *Neuron* **65:** 298–308.

Chadwick MJ, Hassabis D, Maguire EA. 2011. Decoding overlapping memories in the medial temporal lobes using high-resolution fMRI. *Learn Mem* **18:** 742–746.

Cohen NJ, Eichenbaum HB. 1993. *Memory, amnesia, and the hippocampal system.* MIT Press, Cambridge, MA.

Cordes D, Haughton VM, Arfanakis K, Wendt GJ, Turski PA, Moritz CH, Quigley MA, Meyerand ME. 2000. Mapping functionally related regions of brain with functional connectivity MR imaging. *AJNR Am J Neuroradiol* **21:** 1636–1644.

Corkin S. 2013. *Permanent present tense: The unforgettable life of the amnesic patient.* Basic Books, New York.

Davachi L. 2006. Item, context and relational episodic encoding in humans. *Curr Opin Neurobiol* **16:** 693–700.

Davachi L, Mitchell JP, Wagner AD. 2003. Multiple routes to memory: Distinct medial temporal lobe processes build item and source memories. *Proc Natl Acad Sci* **100:** 2157–2162.

Davis T, LaRocque KF, Mumford J, Norman KA, Wagner AD, Poldrack RA. 2014. What do differences between multi-voxel and univariate analysis mean? How subject-, voxel-, and trial-level variance impact fMRI analysis. *Neuroimage* **97:** 271–283.

Devan BD, White NM. 1999. Parallel information in processing in the dorsal striatum: Relation to hippocampal function. *J Neurosci* **19:** 2789–2798.

Diana RA, Yonelinas AP, Ranganath C. 2007. Imaging recollection and familiarity in the medial temporal lobe: A three-component model. *Trends Cogn Sci* **11:** 379–386.

Doeller CF, King JA, Burgess N. 2008. Parallel striatal and hippocampal systems for landmarks and boundaries in spatial memory. *Proc Natl Acad Sci* **105:** 5915–5920.

* Ben-Yakov A, Dudai Y, Mayford MR. 2015. Memory retrieval in mice and men. *Cold Spring Harb Perspect Biol* doi: 10.1101/cshperspect.a021790.

Eichenbaum HB, Cohen NJ. 2001. *From conditioning to conscious recollection: Memory systems of the brain.* Oxford University Press, New York.

Eichenbaum H, Cohen NJ. 2014. Can we reconcile the declarative memory and spatial navigation views on hippocampal function? *Neuron* **83:** 764–770.

Eichenbaum H, Yonelinas AP, Ranganath C. 2007. The medial temporal lobe and recognition memory. *Annu Rev Neurosci* **30:** 123–152.

Ekstrom AD, Bazih AJ, Suthana NA, Al-Hakim R, Ogura K, Zeineh M, Burggren AC, Bookheimer SY. 2009. Advances in high-resolution imaging and computational unfolding of the human hippocampus. *Neuroimage* **47:** 42–49.

Feinberg DA, Moeller S, Smith SM, Auerbach E, Ramanna S, Gunther M, Glasser MF, Miller KL, Ugurbil K, Yacoub E. 2010. Multiplexed echo planar imaging for sub-second whole brain FMRI and fast diffusion imaging. *PLoS ONE* **5:** e15710.

Fischl B, Salat DH, Busa E, Albert M, Dieterich M, Haselgrove C, van der Kouwe A, Killiany R, Kennedy D, Klaveness S, et al. 2002. Whole brain segmentation: Automated labeling of neuroanatomical structures in the human brain. *Neuron* **33:** 341–355.

Friedman D, Johnson R. 2000. Event-related potential (ERP) studies of memory encoding and retrieval: A selective review. *Microsc Res Tech* **51:** 6–28.

Friston KJ. 2011. Functional and effective connectivity: A review. *Brain Connect* **1:** 13–36.

Friston KJ, Frith CD, Liddle PF, Frackowiak RS. 1993. Functional connectivity: The principal-component analysis of large (PET) data sets. *J Cereb Blood Flow Metab* **13:** 5–14.

Friston KJ, Harrison L, Penny W. 2003. Dynamic causal modelling. *Neuroimage* **19:** 1273–1302.

Friston K, Moran R, Seth AK. 2013. Analysing connectivity with Granger causality and dynamic causal modelling. *Curr Opin Neurobiol* **23:** 172–178.

Gordon AM, Rissman J, Kiani R, Wagner AD. 2013. Cortical reinstatement mediates the relationship between content-specific encoding activity and subsequent recollection decisions. *Cereb Cortex* **24:** 3350–3364.

 Cite this article as *Cold Spring Harb Perspect Biol* doi: 10.1101/cshperspect.a021840

Graham KS, Barense MD, Lee ACH. 2010. Going beyond LTM in the MTL: A synthesis of neuropsychological and neuroimaging findings on the role of the medial temporal lobe in memory and perception. *Neuropsychologia* **48**: 831–853.

* Graybiel AM, Grafton ST. 2015. The striatum: Where skills and habits meet. *Cold Spring Harb Perspect Biol* **7**: a021691.

Greenberg DL, Verfaellie M. 2010. Interdependence of episodic and semantic memory: Evidence from neuropsychology. *J Int Neuropsychol Soc* **16**: 748–753.

Greicius MD, Krasnow B, Reiss AL, Menon V. 2003. Functional connectivity in the resting brain: A network analysis of the default mode hypothesis. *Proc Natl Acad Sci* **100**: 253–258.

Grill-Spector K, Sayres R, Ress D. 2006. High-resolution imaging reveals highly selective nonface clusters in the fusiform face area. *Nat Neurosci* **9**: 1177–1185.

Hartley T, Harlow R. 2012. An association between human hippocampal volume and topographical memory in healthy young adults. *Front Hum Neurosci* **6**: 1–11.

Hartley T, Maguire EA, Spiers HJ, Burgess N. 2003. The well-worn route and the path less traveled: Distinct neural bases of route following and wayfinding in humans. *Neuron* **37**: 877–888.

Hasselmo ME, Eichenbaum H. 2005. Hippocampal mechanisms for the context-dependent retrieval of episodes. *Neural Networks* **18**: 1172–1190.

Hasselmo ME, Wyble BP. 1997. Free recall and recognition in a network model of the hippocampus: Simulating effects of scopolamine on human memory function. *Behav Brain Res* **89**: 1–34.

Heidemann RM, Ivanov D, Trampel R, Fasano F, Meyer H, Pfeuffer J, Turner R. 2012. Isotropic submillimeter fMRI in the human brain at 7 T: Combining reduced field-of-view imaging and partially parallel acquisitions. *Magn Reson Med* **68**: 1506–1516.

Iaria G, Petrides M, Dagher A, Pike B, Bohbot VD. 2003. Cognitive strategies dependent on the hippocampus and caudate nucleus in human navigation: Variability and change with practice. *J Neurosci* **23**: 5945–5952.

Iaria G, Lanyon LJ, Fox CJ, Giaschi D, Barton JJS. 2008. Navigational skills correlate with hippocampal fractional anisotropy in humans. *Hippocampus* **18**: 335–339.

Johnson MK, Kounios J, Nolde SF. 1997. Electrophysiological brain activity and memory source monitoring. *Neuroreport* **8**: 1317–1320.

Johnson JD, McDuff SGR, Rugg MD, Norman KA. 2009. Recollection, familiarity, and cortical reinstatement: A multivoxel pattern analysis. *Neuron* **63**: 697–708.

Kesner RP. 2007. Behavioral functions of the CA3 subregion of the hippocampus. *Learn Mem* **14**: 771–781.

Kim H. 2011. Neural activity that predicts subsequent memory and forgetting: A meta-analysis of 74 fMRI studies. *Neuroimage* **54**: 2446–2461.

Kirwan CB, Stark CEL. 2004. Medial temporal lobe activation during encoding and retrieval of novel face-name pairs. *Hippocampus* **14**: 919–930.

Kirwan CB, Jones CK, Miller MI, Stark CEL. 2007. High-resolution fMRI investigation of the medial temporal lobe. *Hum Brain Mapp* **28**: 959–966.

Konishi K, Bohbot VD. 2013. Spatial navigational strategies correlate with gray matter in the hippocampus of healthy older adults tested in a virtual maze. *Front Aging Neurosci* **5**: 1.

Kriegeskorte N, Mur M, Bandettini P. 2008. Representational similarity analysis—Connecting the branches of systems neuroscience. *Front Syst Neurosci* **2**: 4.

Kuhl BA, Rissman J, Wagner AD. 2012. Multi-voxel patterns of visual category representation during episodic encoding are predictive of subsequent memory. *Neuropsychologia* **50**: 458–469.

LaRocque KF, Smith ME, Carr VA, Witthoft N, Grill-Spector K, Wagner AD. 2013. Global similarity and pattern separation in the human medial temporal lobe predict subsequent memory. *J Neurosci* **33**: 5466–5474.

Lee ACH, Scahill VL, Graham KS. 2008. Activating the medial temporal lobe during oddity judgment for faces and scenes. *Cereb Cortex* **18**: 683–696.

Liang JC, Wagner AD, Preston AR. 2013. Content representation in the human medial temporal lobe. *Cereb Cortex* **23**: 80–96.

Libby LA, Ekstrom AD, Ragland JD, Ranganath C. 2012. Differential connectivity of perirhinal and parahippocampal cortices within human hippocampal subregions revealed by high-resolution functional imaging. *J Neurosci* **32**: 6550–6560.

Litman L, Awipi T, Davachi L. 2009. Category-specificity in the human medial temporal lobe cortex. *Hippocampus* **19**: 308–319.

Logothetis NK, Wandell BA. 2004. Interpreting the BOLD signal. *Annu Rev Physiol* **66**: 735–769.

Long NM, Burke JF, Kahana MJ. 2014. Subsequent memory effect in intracranial and scalp EEG. *Neuroimage* **84**: 488–494.

Lowe MJ, Dzemidzic M, Lurito JT, Mathews VP, Phillips MD. 2000. Correlations in low-frequency BOLD fluctuations reflect cortico–cortical connections. *Neuroimage* **12**: 582–587.

Maguire EA, Woollett K, Spiers HJ. 2006. London taxi drivers and bus drivers: A structural MRI and neuropsychological analysis. *Hippocampus* **16**: 1091–1101.

Marr D. 1971. Simple memory: A theory for archicortex. *Philos Trans R Soc Lond B Biol Sci* **262**: 23–81.

Martin E. 1968. Stimulus meaningfulness and paired-associate transfer: An encoding variability hypothesis. *Psychol Rev* **75**: 421–441.

Mayes A, Montaldi D, Migo E. 2007. Associative memory and the medial temporal lobes. *Trends Cogn Sci* **11**: 126–135.

McClelland JL, McNaughton BL, O'Reilly RC. 1995. Why there are complementary learning systems in the hippocampus and neocortex: Insights from the successes and failures of connectionist models of learning and memory. *Psychol Rev* **102**: 419–457.

McDuff SGR, Frankel HC, Norman KA. 2009. Multivoxel pattern analysis reveals increased memory targeting and reduced use of retrieved details during single-agenda source monitoring. *J Neurosci* **29**: 508–516.

McIntosh AR, Gonzalez-Lima F. 1994. Structural equation modeling and its application to network analysis in functional brain imaging. *Hum Brain Mapp* **2**: 2–22.

Mechelli A, Price CJ, Friston KJ, Ashburner J. 2005. Voxel-based morphometry of the human brain: Methods and applications. *Curr Med Imag Rev* **1**: 105–113.

Moeller S, Yacoub E, Olman CA, Auerbach E, Strupp J, Harel N, Uğurbil K. 2010. Multiband multislice GE-EPI at 7 tesla, with 16-fold acceleration using partial parallel imaging with application to high spatial and temporal whole-brain fMRI. *Magn Reson Med* **63**: 1144–1153.

Montaldi D, Mayes AR. 2010. The role of recollection and familiarity in the functional differentiation of the medial temporal lobes. *Hippocampus* **20**: 1291–1314.

Moscovitch M, Nadel L, Winocur G, Gilboa A, Rosenbaum RS. 2006. The cognitive neuroscience of remote episodic, semantic and spatial memory. *Curr Opin Neurobiol* **16**: 179–190.

Moser EI, Kropff E, Moser M-B. 2008. Place cells, grid cells, and the brain's spatial representation system. *Annu Rev Neurosci* **31**: 69–89.

* Moser M-B, Rowland DC, Moser EI. 2015. Place cells, grid cells, and memory. *Cold Spring Harb Perspect Biol* **7**: a021808.

Moses SN, Brown TM, Ryan JD. 2010. Neural system interactions underlying human transitive inference. *Hippocampus* **20**: 894–901.

Murray E, Bussey T. 1999. Perceptual–mnemonic functions of the perirhinal cortex. *Trends Cogn Sci* **3**: 142–151.

Niedermeyer E, Lopes da Silva F (eds.). 2005. *Electroencephalography: Basic principles, clinical applications, and related fields*, 5th ed. Lippincott Williams and Wilkins, Philadelphia.

Norman KA. 2010. How hippocampus and cortex contribute to recognition memory: Revisiting the complementary learning systems model. *Hippocampus* **20**: 1217–1227.

Norman KA, O'Reilly RC. 2003. Modeling hippocampal and neocortical contributions to recognition memory: A complementary-learning-systems approach. *Psychol Rev* **110**: 611–646.

Norman KA, Polyn SM, Detre GJ, Haxby JV. 2006. Beyond mind-reading: Multi-voxel pattern analysis of fMRI data. *Trends Cogn Sci* **10**: 424–430.

Nyhus E, Curran T. 2010. Functional role of γ and θ oscillations in episodic memory. *Neurosci Biobehav Rev* **34**: 1023–1035.

O'Keefe J. 1976. Place units in the hippocampus of the freely moving rat. *Exp Neurol* **51**: 78–109.

O'Keefe J, Dostrovsky J. 1971. The hippocampus as a spatial map. Preliminary evidence from unit activity in the freely moving rat. *Brain Res* **34**: 171–175.

O'Keefe J, Nadel L. 1978. *The hippocampus as a cognitive map*. Oxford University Press, Oxford.

O'Reilly RC, McClelland JL. 1994. Hippocampal conjunctive encoding, storage, and recall: Avoiding a trade-off. *Hippocampus* **4**: 661–682.

O'Reilly RC, Rudy JW. 2001. Conjunctive representations in learning and memory: Principles of cortical and hippocampal function. *Psychol Rev* **108**: 311–345.

Paller KA, Wagner AD. 2002. Observing the transformation of experience into memory. *Trends Cogn Sci* **6**: 93–102.

Paller KA, Kutas M, Mayes AR. 1987. Neural correlates of encoding in an incidental learning paradigm. *Electroencephalogr Clin Neurophysiol* **67**: 360–371.

Patenaude B, Smith SM, Kennedy DN, Jenkinson M. 2011. A Bayesian model of shape and appearance for subcortical brain segmentation. *Neuroimage* **56**: 907–922.

Polyn SM, Natu VS, Cohen JD, Norman KA. 2005. Category-specific cortical activity precedes retrieval during memory search. *Science* **310**: 1963–1966.

Poppenk J, Moscovitch M. 2011. A hippocampal marker of recollection memory ability among healthy young adults: Contributions of posterior and anterior segments. *Neuron* **72**: 931–937.

Preston AR, Bornstein AM, Hutchinson JB, Gaare ME, Glover GH, Wagner AD. 2010. High-resolution fMRI of content-sensitive subsequent memory responses in human medial temporal lobe. *J Cogn Neurosci* **22**: 156–173.

Ranganath C, Yonelinas AP, Cohen MX, Dy CJ, Tom SM, D'Esposito M. 2004. Dissociable correlates of recollection and familiarity within the medial temporal lobes. *Neuropsychologia* **42**: 2–13.

Rissman J, Wagner AD. 2012. Distributed representations in memory: Insights from functional brain imaging. *Annu Rev Psychol* **63**: 101–128.

Rissman J, Gazzaley A, D'Esposito M. 2004. Measuring functional connectivity during distinct stages of a cognitive task. *Neuroimage* **23**: 752–763.

Rissman J, Greely HT, Wagner AD. 2010. Detecting individual memories through the neural decoding of memory states and past experience. *Proc Natl Acad Sci* **107**: 9849–9854.

Ritchey M, Dolcos F, Cabeza R. 2008. Role of amygdala connectivity in the persistence of emotional memories over time: An event-related FMRI investigation. *Cereb Cortex* **18**: 2494–2504.

Rosenbaum RS, Gilboa A, Moscovitch M. 2014. Case studies continue to illuminate the cognitive neuroscience of memory. *Ann NY Acad Sci* **1316**: 105–133.

Ross RS, Sherrill KR, Stern CE. 2011. The hippocampus is functionally connected to the striatum and orbitofrontal cortex during context dependent decision making. *Brain Res* **1423**: 53–66.

Rugg MD. 1995. Event-related potential studies of human memory. In *The cognitive neurosciences* (ed. Gazzaniga MS), pp. 789–802. MIT Press, Cambridge.

Sadeh T, Shohamy D, Levy DR, Reggev N, Maril A. 2011. Cooperation between the hippocampus and the striatum during episodic encoding. *J Cogn Neurosci* **23**: 1597–1608.

Sanquist TF, Rohrbaugh JW, Syndulko K, Lindsley DB. 1980. Electrocortical signs of levels of processing: Perceptual analysis and recognition memory. *Psychophysiology* **17**: 568–576.

Schinazi VR, Nardi D, Newcombe NS, Shipley TF, Epstein RA. 2013. Hippocampal size predicts rapid learning of a cognitive map in humans. *Hippocampus* **23**: 515–528.

Scoville WB, Milner B. 1957. Loss of recent memory after bilateral hippocampal lesions. *J Neurol Neurosurg Psychiatry* **20**: 11–21.

Smith SM, Miller KL, Salimi-Khorshidi G, Webster M, Beckmann CF, Nichols TE, Ramsey JD, Woolrich MW.

2011. Network modelling methods for FMRI. *Neuroimage* **54**: 875–891.

Spaniol J, Davidson PSR, Kim ASN, Han H, Moscovitch M, Grady CL. 2009. Event-related fMRI studies of episodic encoding and retrieval: Meta-analyses using activation likelihood estimation. *Neuropsychologia* **47**: 1765–1779.

Squire LR, Bayley PJ. 2007. The neuroscience of remote memory. *Curr Opin Neurobiol* **17**: 185–196.

* Squire LR, Dede AJO. 2015. Conscious and unconscious memory systems. *Cold Spring Harb Perspect Biol* **7**: a021667.

Squire LR, Stark CEL, Clark RE. 2004. The medial temporal lobe. *Annu Rev Neurosci* **27**: 279–306.

Staresina BP, Davachi L. 2008. Selective and shared contributions of the hippocampus and perirhinal cortex to episodic item and associative encoding. *J Cogn Neurosci* **20**: 1478–1489.

Staresina BP, Davachi L. 2010. Object unitization and associative memory formation are supported by distinct brain regions. *J Neurosci* **30**: 9890–9897.

Staresina BP, Duncan KD, Davachi L. 2011. Perirhinal and parahippocampal cortices differentially contribute to later recollection of object- and scene-related event details. *J Neurosci* **31**: 8739–8747.

Staresina BP, Henson RNA, Kriegeskorte N, Alink A. 2012. Episodic reinstatement in the medial temporal lobe. *J Neurosci* **32**: 18150–18156.

Staresina BP, Cooper E, Henson RN. 2013. Reversible information flow across the medial temporal lobe: The hippocampus links cortical modules during memory retrieval. *J Neurosci* **33**: 14184–14192.

Stephan KE, Friston KJ. 2010. Analyzing effective connectivity with fMRI. *Wiley Interdiscip Rev Cogn Sci* **1**: 446–459.

Sun FT, Miller LM, D'Esposito M. 2004. Measuring interregional functional connectivity using coherence and partial coherence analyses of fMRI data. *Neuroimage* **21**: 647–658.

Suthana NA, Ekstrom AD, Moshirvaziri S, Knowlton B, Bookheimer SY. 2009. Human hippocampal CA1 involvement during allocentric encoding of spatial information. *J Neurosci* **29**: 10512–10519.

Suzuki WA. 2009. Comparative analysis of the cortical afferents, intrinsic projections and interconnections of the parahippocampal region in monkeys and rats. In *The cognitive neurosciences*, 4th ed. (ed. Gazzaniga SM), pp. 659–674. MIT Press, Cambridge.

Teyler TJ, DiScenna P. 1986. The hippocampal memory indexing theory. *Behav Neurosci* **100**: 147–154.

Tolman EC. 1948. Cognitive maps in rats and men. *Psychol Rev* **55**: 189–208.

Treves A, Rolls ET. 1994. Computational analysis of the role of the hippocampus in memory. *Hippocampus* **4**: 374–391.

Uncapher MR, Wagner AD. 2009. Posterior parietal cortex and episodic encoding: Insights from fMRI subsequent memory effects and dual-attention theory. *Neurobiol Learn Mem* **91**: 139–154.

Voermans NC, Petersson KM, Daudey L, Weber B, van Spaendonck KP, Kremer HPH, Fernandez G. 2004. Interaction between the human hippocampus and the caudate nucleus during route recognition. *Neuron* **43**: 427–435.

Wagner AD, Schacter DL, Rotte M, Koutstaal W, Maril A, Dale AM, Rosen BR, Buckner RL. 1998. Building memories: Remembering and forgetting of verbal experiences as predicted by brain activity. *Science* **281**: 1188–1191.

Wagner AD, Koutstaal W, Schacter DL. 1999. When encoding yields remembering: Insights from event-related neuroimaging. *Philos Trans R Soc Lond B Biol Sci* **354**: 1307–1324.

Wagner AD, Maril A, Schacter DL. 2000. Interactions between forms of memory: When priming hinders new episodic learning. *J Cogn Neurosci* **12**: 52–60.

Ward EJ, Chun MM, Kuhl BA. 2013. Repetition suppression and multi-voxel pattern similarity differentially track implicit and explicit visual memory. *J Neurosci* **33**: 14749–14757.

Watson HC, Wilding EL, Graham KS. 2012. A role for perirhinal cortex in memory for novel object-context associations. *J Neurosci* **32**: 4473–4481.

Xue G, Dong Q, Chen C, Lu Z, Mumford JA, Poldrack RA. 2010. Greater neural pattern similarity across repetitions is associated with better memory. *Science* **330**: 97–101.

Xue G, Dong Q, Chen C, Lu Z-L, Mumford JA, Poldrack RA. 2013. Complementary role of frontoparietal activity and cortical pattern similarity in successful episodic memory encoding. *Cereb Cortex* **23**: 1562–1571.

Yacoub E, Harel N, Ugurbil K. 2008. High-field fMRI unveils orientation columns in humans. *Proc Natl Acad Sci* **105**: 10607–10612.

Yassa MA, Stark CEL. 2009. A quantitative evaluation of cross-participant registration techniques for MRI studies of the medial temporal lobe. *Neuroimage* **44**: 319–327.

Yassa MA, Stark SM, Bakker A, Albert MS, Gallagher M, Stark CEL. 2010. High-resolution structural and functional MRI of hippocampal CA3 and dentate gyrus in patients with amnestic mild cognitive impairment. *Neuroimage* **51**: 1242–1252.

Zeineh MM, Engel SA, Bookheimer SY. 2000. Application of cortical unfolding techniques to functional MRI of the human hippocampal region. *Neuroimage* **11**: 668–683.

# Index